HEALTH SCIENCE STATISTICS USING R AND R COMMANDER

ROBIN BEAUMONT

Honorary Fellow,
The University of Edinburgh

Scion

First published 2015

A CIP catalogue record for this book is available from the British Library.
ISBN 9781907904318

Scion Publishing Limited
The Old Hayloft, Vantage Business Park, Bloxham Road, Banbury OX16 9UX, UK
www.scionpublishing.com

Important Note from the Publisher

Printed in the UK

Contents

Preface

I have written this book as an introductory text focusing on the key statistical techniques likely to be encountered in the health and medical sciences. 'Introductory' means different things to different people so, to clarify, this is a book for non-specialists who need a reasonable grasp of statistics for their study or research.

Any introductory statistics book relies on graphical (i.e. point and click) software to carry out the analyses. I am using the open source software R, along with a user-friendly interface called R Commander. It is no understatement to say that R is revolutionising the way we do statistics: it is incredibly powerful software and is fast gaining recognition across the globe. But it can be complex to learn and you do have to get to grips with some aspects of the R programming language. So, in writing this book, I have combined a step-by-step discussion of the key statistical techniques you are likely to encounter with a hands-on guide to how to perform these techniques in R. Since it is no use to learn techniques without developing the cognitive abilities to interpret the results, this book aims to nurture an attitude of inventive intellectual curiosity towards them.

Although I wrote this book as a step-by-step textbook, it can be used in a number of ways, depending on your needs: As an introductory course; as a reference supported by the glossary and index or as a practical self-help guide, with the strategy explained in the introductory chapter.

The content of this book is the result of over twenty years teaching of applied statistics to health science, medical, health informatics and psychology students in over half a dozen Universities. I helped to set up the first online health informatics MSc at the RCSED in 1999, supported by the extremely enthusiastic charismatic professor Angus Wallace. This course ran until 2013, and some of the material for this book comes from the many series of notes designed to supplement, and often replace, standard texts where earlier versions of this material have been used for both face to face and other online courses. My online experience of teaching statistics has encouraged me to develop YouTube videos of many basic analyses which has resulted in lively online discussions among the ca. 1,000 followers. The realisation that a collaborative approach to learning statistics is highly beneficial has meant that I have designed this book to support this virtual community of informed and engaged statistics users. The books accompanying website www.robin-beaumont.co.uk/rbook provides additional resources for both students and tutors, with multiple choice questions and learning outcomes for course developers.

The collaborative aspect of this book is reflected in the many people involved in its genesis. I first must thank all the students that have used and continue to use my material; their feedback has been invaluable. I must also thank the various institutions (especially the RCSED) and universities that have employed me over the years to develop the earlier versions of this material.

Many teachers and colleagues have attempted to teach me statistics, notably Bill Kirkup, Johnathan Priest and Chris Dracup – who started his sessions with ritualistically taking off his pullover, hinting at the level of work to come. The Open University provided superb tuition demonstrating excellence long before it was the standard in more traditional institutions. In more recent years Claire Nickerson, Johannes Hönekopp and Vincent Deary have provided many interesting discussions and invaluable feedback to early drafts of the book. Simon Watkins and Jonathan Ray at Scion Publishing have spent many hours working with me transforming what was meant originally to be a simple 100 page practical companion to a fully formed standalone tome.

The R community, including the many developers of the packages and datasets, used in this book must be commended for their generosity. Often a problem was solved within hours by contacting the author who was frequently a highly respected professor at a university half way across the globe

Beyond the academic advice, encouragement came from many quarters, among them Henrike Lähnemann delivering valuable practical knowledge, members of Tynemouth Choral Society constantly asking 'is it finished yet' and my partner Andrew Kowel providing the necessary nutritional support. For the continuing support of my hypochondriasis I must thank Johnathon Caudle and for my real medical needs Leonie Coates. Many more should be named if space permitted. Obviously any failings including omissions and inaccuracies are purely mine.

I have always enjoyed teaching statistics, and this book has been a labour of love and a present to myself. I hope that you also gain some pleasure in using it.

Robin Beaumont Newcastle upon Tyne (UK) October 2014

1 How this book works

This book is a practical manual which covers the common analyses and statistical techniques undertaken in the health, medical and behavioural sciences.

There are many books on the R software; most of these have been written for R specialists. This book is different - it has been written for non-specialists who need to get to grips with R for their studies, research, or work. I start by introducing you to a very easy to use graphical front end to R called R Commander. This works in much the same way as Windows does, so I hope it doesn't feel too alien. As your knowledge increases, I gradually introduce the R language, along with various add-ons (R libraries), which you will need for specific tasks. Finally, whilst R is an excellent software environment, it does have some limitations. Therefore, we will also work through some other free applications (*OpenEpi, Gpower* and *Ωnyx*) to supplement R.

To help you get to grips with the subject, the book provides detailed, often visual, explanations of the results and also demonstrates the advantages and limitations of the various statistical techniques discussed.

1.1 Overview of the book

This book consists of 10 sections:

1. About the book – this is the current short section you are reading, providing an outline of the book, including a general introduction to what statistics are and why using them is so important.
2. Statistics – setting the scene – A general introduction.
3. Getting started with R - this section describes how to install and start up R, along with a brief tutorial.
4. Basic general statistical techniques – this section provides explanations, alongside examples demonstrating how to carry out a wide range of common statistical procedures which help you to describe and summarize your data.
5. Samples and populations - this section moves on from describing your data, which is often a small sample, to making inferences about how this compares to a population. **P-values** and **confidence intervals** are introduced to allow the assessment of the similarity/difference between the two.

highlighted words indicate glossary entries

6. Basic medical/health related statistical techniques – this section focuses on those statistical techniques used in medicine and healthcare. It provides explanations, alongside examples of how to carry out such procedures ranging from survival analysis to Bland Altman plots.
7. More data manipulation and R management techniques – this section expands on the techniques described in the previous chapters including; methods to import/export data, manipulating rows and columns and expanding counts into flat files. There are also details of how to manage the R environment. These are the practical but messy aspects often hidden in textbooks which frequently cause problems in the real world.

8. Advanced statistical techniques – Using R code directly allows many advanced statistical techniques to be used which are demonstrated in these chapters.
9. More regression techniques – this section expands the on the previous regression and correlation chapters to allow analysis of different types of output. It ends with a technique, Structural Equation Modelling (SEM) which allows the simultaneous assessment of several multiple regression models.
10. Reference material: appendixes appendixes – the final section of the book includes a detailed glossary, references and index.

1.2 If you are in a hurry

Follow these five steps:

- Download the R software (page 11).
- Install R Commander (page 15).
- Work through the quick tutorial (page 19).
- Look through the *Getting your data into R* chapter.
- Select the chapter that describes the type of analysis you want to carry out. Work through the example in the chapter, checking out the YouTube version before attempting to analyse your own data.

1.3 Use this book as a first course, practical companion or reference

While this book can be used as a standalone text or reference you can also use it as a practical companion to three introductory statistics books:

- Harris M, Taylor G 2014 (3rd ed.) Medical Statistics Made Easy. Scion. ➔MSME3
- Campbell M J, Swinscow T D V, 2009 Statistics at square one. BMJ. ➔ SAS1
- Campbell M J 2006 Statistics at square two. BMJ. ➔ SAS2

I will refer from now on to the Harris and Taylor book as MSME3 and the Campbell and Swinscow books as SAS1 and SAS2.

Typical stages of data analysis

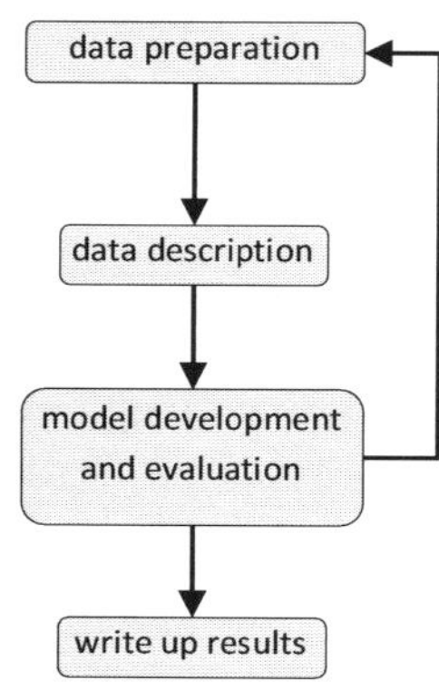

1.4 Typical chapter structure

Most chapters begin with a general introduction including any relevant maths that underlies the topic. The next section then describes how to carry out the analysis following the 4 stages shown opposite. For the less complex analyses this is demonstrated in two ways; first using the point and click menu options (provided by the R Commander add-on), and then by typing directly into R using **R commands**.

Most chapters end by describing some relevant tips and tricks along with helpful links.

1.5 Emoticons

When discussing results I often place a smiley ☺, unhappy ☹ or confused 😐 emoticon beside the result. This is to help you appreciate that most values from a particular analysis are interpreted in a specific way by the analyst.

1.6 Additional help

There is both a glossary and detailed index at the end of this book. Words in the glossary that appear in the text are in **bold** such as **R commands**. Most of the analyses have an accompanying YouTube video offering summary information and additional help with tricky situations.

1.7 Additional online chapters

I have tried to make this book as short as possible but inevitably, it has grown much larger than I ever intended. To help control its growth I have decided to place several preliminary and advanced chapters on the accompanying website. These include chapters on the uses and abuses of basic summary statistics such as the mean, median, and standard deviation. I have combined these to form a section I have called high school revision. There are also chapters concerned with less common statistical techniques and research designs, and a chapter on how to install and use R Commander and R on the Raspberry Pi.

1.8 Topographical conventions (text styles)

This book uses several text styles to illustrate when you need to type in, or select certain menu options. R code is also shown in a different text style, as is reference to it in the text.

Example	Description
Statistics->Summaries Click on the **OK** button	Menu option + Button + Icon + dialog box element title (i.e. list box, textbox titles etc.).

`library(MASS)`
 	
`mydataframe<- data(anorexia, package="MASS")`
 	
`summary(mydataframe)`
	R code example

library(MASS) *mydataframe<- data(anorexia, package="MASS")* To obtain the descriptive statistics for my *dataframe* type *summary(mydataframe)* Type *Ctrl+c*	References to code examples or a snippet of R code in a sentence. Something you need to type in, including key mouse combinations.
Downloading and installing R *R Console*	References to book/chapter titles, objects, packages, applications and window names

2 Statistics and R - Setting the scene

Statistics are used (and misused) in all walks of life. Most people generally have at least a basic grasp of statistics. However, in scientific disciplines statistics form a fundamental role defining the type and amount of data that needs to be collected, dictate how it should be appropriately analysed and form the basis of the decision making process. You can't belong to a scientific discipline without an interest, and skills, in statistics.

This book attempts to provide both the level and breadth of knowledge and skills required by most undergraduates in health sciences, pharmacy and medicine and psychology. It also serves as an introduction to medical statistics for those medical graduates who did not gain much experience (or were asleep!) during their undergraduate years but wish to become more engaged in quantitative research now, hence there are some topic areas which are considered to be at a postgraduate level.

Before one can apply any statistical technique to data it is first necessary to be able to define, collect and store the data.

2.1 Coding frames – helping you define your data

Traditionally each data item (also called a field or variable) was documented in a *coding frame* which served as a reference for the researcher.

Opposite is part of a typical questionnaire (i.e. form) and associated coding frame.

completed by participant

completed by coder/researcher using the coding frame

Section 2 Present Role

For office use only

4. Does your present role involve . . . *(Please tick one box for each question)*

q4a

Being personally involved in the selection of computer systems?

❑ yes ❑ no ❑ no comment

Q4b

The development of computer systems and/or data requirements?

❑ yes ❑ no ❑ no comment

Q5

5. Considering your present role, do you feel your current level of computing skills and knowledge are: *(Please tick one box)*

❑ very inadequate ❑ slightly inadequate ❑ adequate ❑ above that required

Field name: q4a
'Does your present role involve . . . being personally involved in the selection of computer systems?'
Value: integer 1 - n using the following list:

1. YES
2. NO
3. NO COMMENT

single response question = one field = one column of data

Question 4 Field name: q4b
'Does your present role involve . . . the development of computer systems and /or data requirements?'
Value: integer 1 - n using the following list:

1. YES
2. NO
3. NO COMMENT

single response question = one field = one column of data

Question 5 Field name: q5
'Considering your present role, do you feel your current level of computing skills and knowledge are:'
Value: integer 1 - n using the following list:

1. VERY INADEQUATE
2. SLIGHTLY INADEQUATE
3. INADEQUATE
4. ADEQUATE
5. ABOVE THAT REQUIRED

single response question = one field = one column of data

Question types
Question types are discussed more fully, in the accompanying online chapters

When the form is paper based and completed by participants, it is necessary to code each question for incorrect responses such as, *missing* or *spoilt*. Sometimes it is not necessary to differentiate between the two but often at a later date you might want to see if different types of participant failed to complete a question in a specific manner. Often missing and spoilt responses are given impossible values such as 99, and 999.

Each field/variable possesses a **Level of measurement.** It is vital that you are proficient at identifying this, as it partly dictates the appropriate type of statistical analysis. For example in the questionnaire above, questions *4a* and *4b* posses a nominal level of measurement while question *5* has an ordinal one.

2.2 Data collection and storage

When thinking about data collection/storage the three most important questions you need to ask yourself are:

- **Who** will be collecting the data? Is it just yourself or colleagues/researchers or participants/subjects.
- **How** will the data be collected? Directly using sensors or participant completing questionnaires. Alternatively, you might be reanalysing previously collected/published data (i.e. secondary data collection).
- **Where** will the data be stored? The two common options being locally on your laptop/tablet or on a server/cloud.

Online data collection - Today both participants complete questionnaires and researchers fill in forms online. The accompanying online chapter provides details of Surveymonkey, Limesurvey, along with a short Google Drive forms tutorial. Alternative strategies are also discussed.

2.3 Dummy data for practicing

Few people realise that even before you collect and store your 'real' data, so long as you have defined the data you plan to collect you can begin the data analysis process by creating some dummy data from your definitions. Few people do this but is can be very useful in getting to grips with the analysis process and I always recommend students to follow this strategy.

Moving now on to the analysis process.

2.4 The data analysis process

Typical stages of data analysis

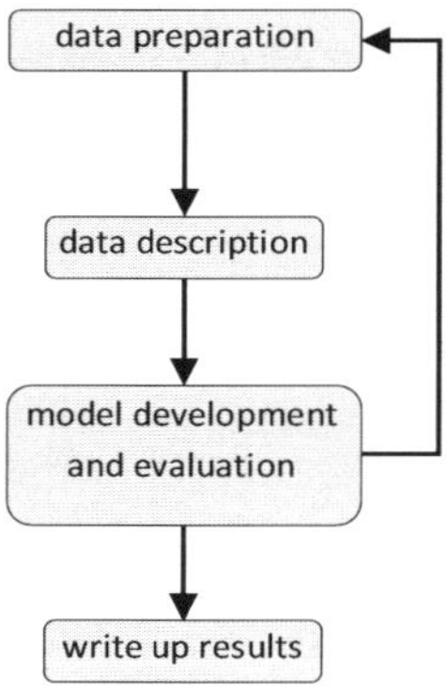

data preparation
data description
model development and evaluation
write up results

Model aspects

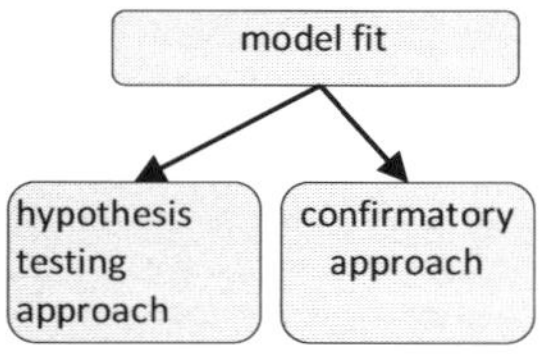

model fit
hypothesis testing approach
confirmatory approach

Statistics come in two main flavours; those that describe your data, and those that compare it to a mathematical model of some kind.

While the 'data description' type of statistics are useful in providing a understanding of the dataset, they do not usually allow us to second guess what a value is for a larger dataset. For example, we might have a small subset of data from a single GP practice but wish to make recommendations for all the GP practices in our area or even across several areas. To achieve this with statistics it is necessary to not only describe our observed data, but also to think how it relates to a larger set of data (called a population) by a process known as **inference**. All this means is that we need to create a mathematical model and then compare how well our observed data fits this model.

Why go to all this extra hassle? Well it means that we can make sensible rational policy decisions from a small dataset. This is very much in contrast to qualitative research techniques which are based firmly at the individual level and do not attempt to jump from the sample to some theoretical population. This is the fundamental reason why the actual research designs of qualitative and quantitative research are so different.

Model types – There are basically two types of model. The null model (also called the null hypothesis) is the least favourable model we want to accept and our aim is to *reject* it. This null model predicts no effect for a particular treatment or no relationship between two variables.

Then if our observed data deviates sufficiently from it we can reject this null model, and by default accept a more pleasant, possibly unspecified alternative model. This is the traditional approach to model evaluation, often called the 'Null hypothesis significance test' (NHST) approach, and uses **p-values** and critical values.

More recent statistical techniques tend instead to use a *confirmatory* approach. Here we develop a model we desire to accept and assess how close it's predicted values are to our observed data, hoping for minimal deviation between the two. This technique makes use of various model fit statistics such as chi-squared, Log likelihood ratios, BIC and AIC. For some of these values we can produce an estimate from which we can calculate a **confidence interval** in which we have a certain level of confidence the actual population value will lie within.

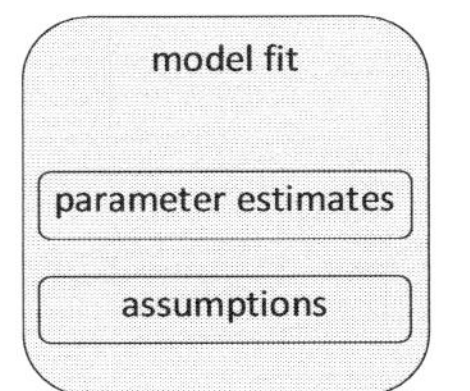

Unfortunately, many introductory statistics books present a rather dictatorial approach to the various types of analysis described above, frequently presenting very much cut and dried binary decisions at the end of an analysis. Such an approach often hides much subtlety and complexity and does not often reflect reality. This dictatorial approach frequently results in the more intelligent students being confused as they feel such a situation hardly reflects even the vaguest approximation to reality. The most common offender propagating this binary decision approach to data analysis is the NHST approach. While many authors recommend that such an approach should be dropped entirely (Cumming, 2012) I take a more measured approach in this book offering both interpretations.

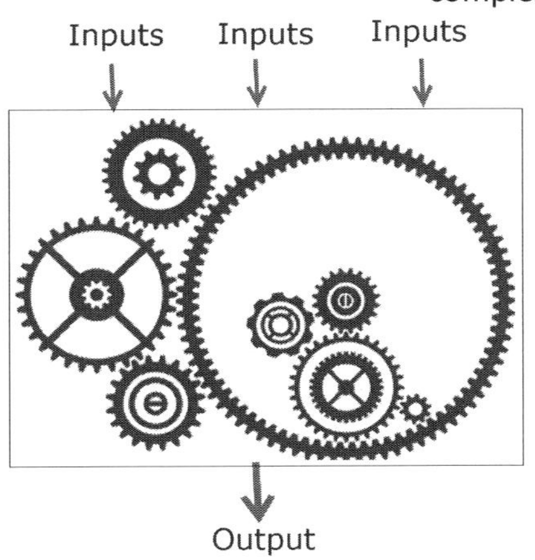

Besides the overall model fit aspect the statistical model is made up of various estimated values. A good analogy is to think of a statistical model as being like a car engine where each component in it is equivalent to what is called a parameter estimate. Also to build a good car engine one needs to make assumptions about the type of fuel it will need and what working conditions it is designed for. Once again, this is analogous to a set of assumptions we make about the data we use to form the statistical model (the inputs). This all sounds very abstract but most chapters in this book provide examples of this approach, and R, as will be demonstrated, provides support for many methods of analysis, which you will remember is all free.

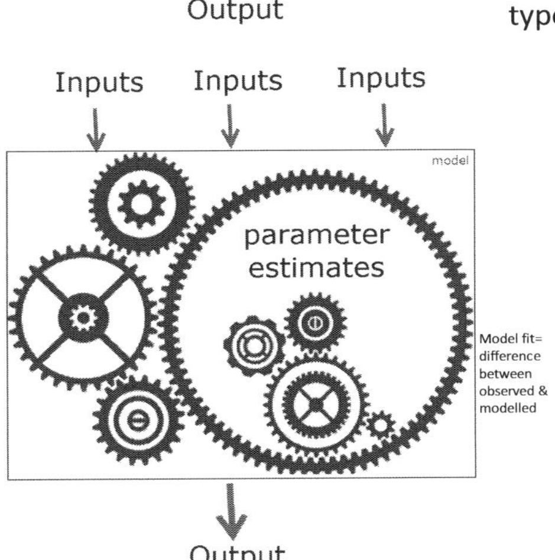

2.5 The new statistics

Cumming dubbed the alternative approach to that of hypothesis testing "*the new statistics*". This involves a more investigative, less objective, approach to the data analysis, stressing the importance of a specific type of statistic called an **Effect size** along with **confidence intervals** and combining results from several experiments using the technique of Meta-analysis (described in a separate chapter).

As you read through the book you will see that I have used these 'new statistics' approaches providing solutions in R. One aspect that is still not regularly reported is that of confidence interval calculation for effect size measures and I have provided a chapter on this topic.

It is now time to start getting your hands dirty....

3 R - What is it? Two ways to use it

R is a very sophisticated piece of statistical software that runs on many different types of computers including mobile phones and the Raspberry Pi. Some important facts:

- It is free
- It is becoming increasingly popular in both the professional and academic settings and is used by a wide range of people from undergraduate students to professors of statistics
- It is very powerful, allowing you to collect, store and analyse data
- It provides many ways of producing sophisticated graphs (flick through this book to see examples)

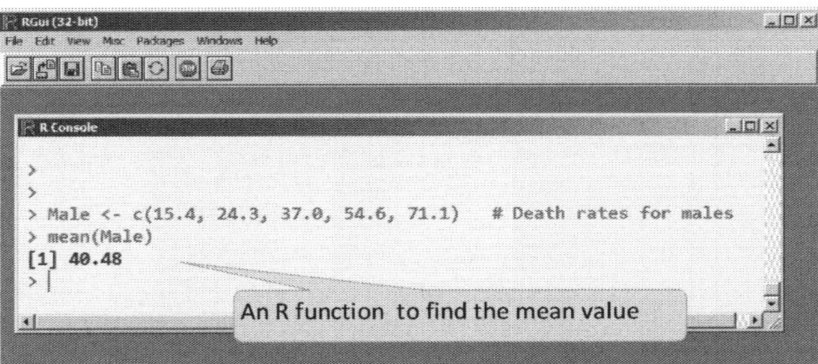

R with no add-ons is of limited use as you can only use it by typing in R commands directly (example opposite).

Luckily, there are many free add-ons, called *packages*, some of which provide a way of using R through drop down menus and dialog boxes (point and click) rather than typing in R commands directly. These are often called *Graphical User Interfaces* (Gui's) or *Graphical front ends* for R. R Commander and Deducer are two such free Gui add-ons for R (shown below).

R Commander

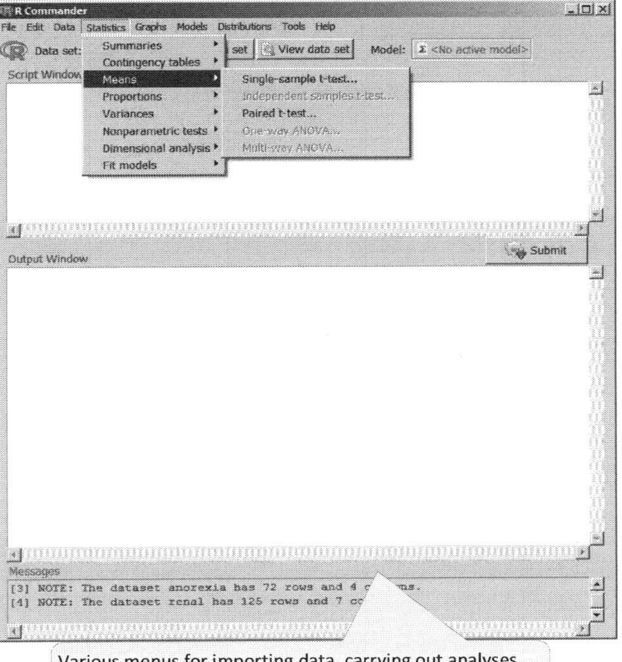

Various menus for importing data, carrying out analyses, plotting and saving results

Deducer

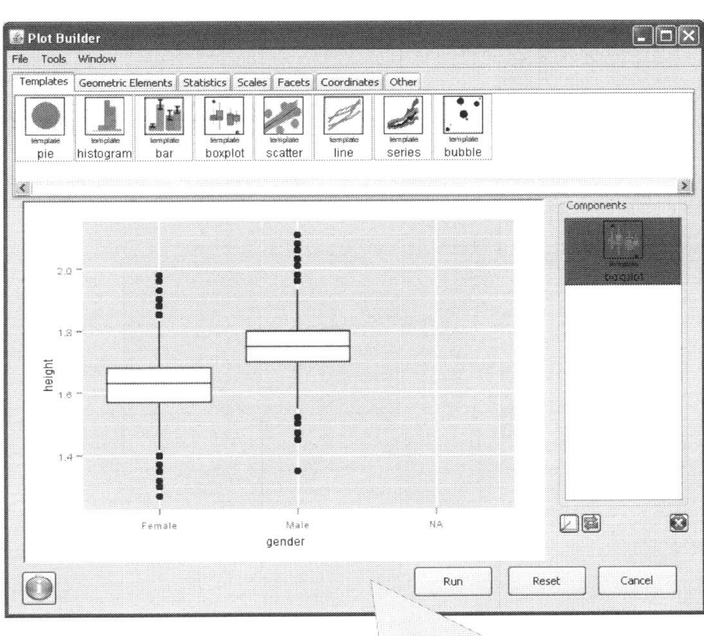

Besides a set of menu options similar to R Commander, Deducer has a plot builder window, details in the *publication graphics* chapter.

4 Downloading and installing the R software - free!

You can obtain the R software, which is free, following the few simple steps described below.

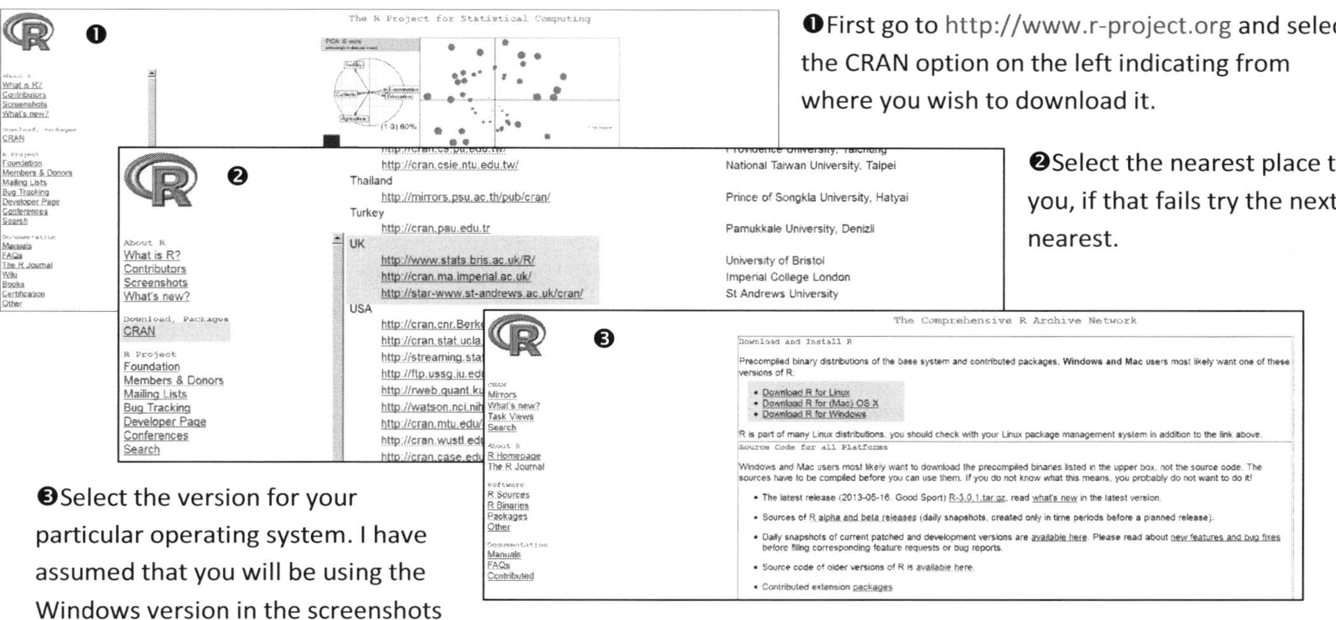

❶First go to http://www.r-project.org and select the CRAN option on the left indicating from where you wish to download it.

❷Select the nearest place to you, if that fails try the next nearest.

❸Select the version for your particular operating system. I have assumed that you will be using the Windows version in the screenshots below, however the installation process for Mac and Linux follows a similar process.

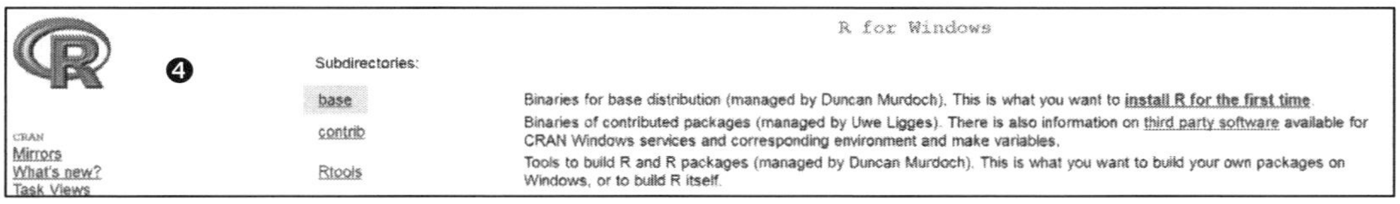

❹Select the *base* subdirectory option.

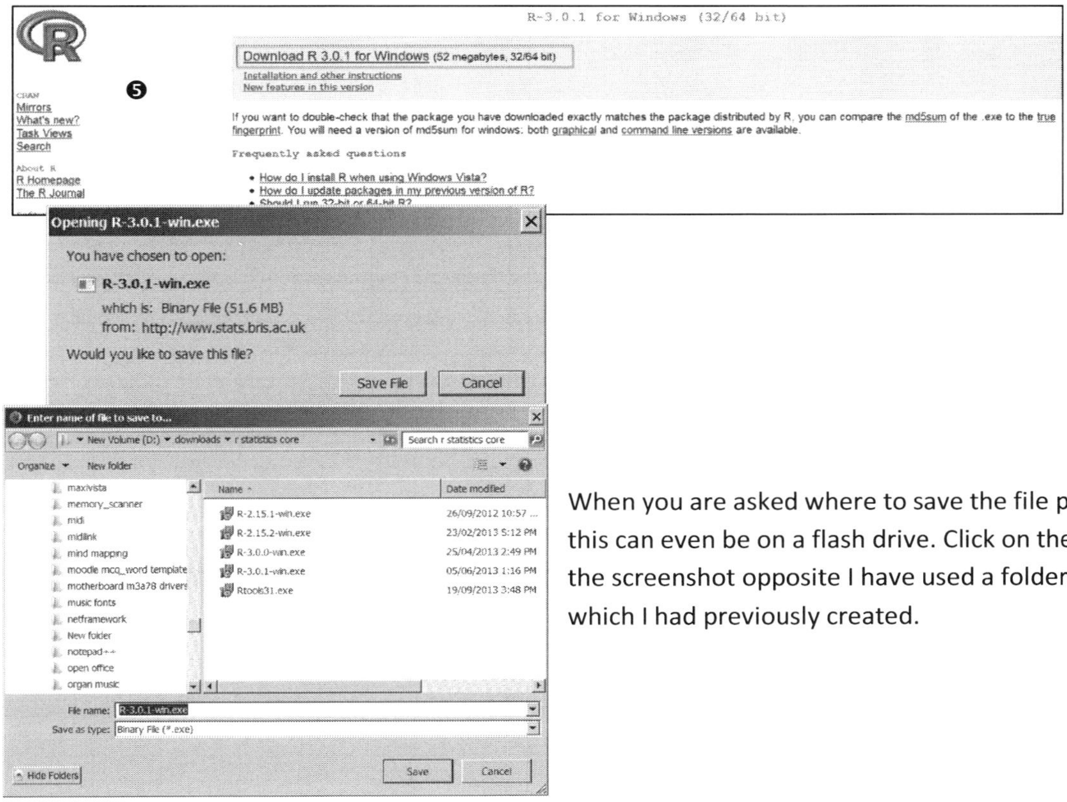

❺Because I have a Windows system I have selected the Download R 3.01 for Windows option.

When you are asked where to save the file put it somewhere sensible, this can even be on a flash drive. Click on the **save** button to save it. In the screenshot opposite I have used a folder called *r statistics core* which I had previously created.

To install the software on your computer just double click on the file in Windows Explorer, if you get an error message try again selecting the ***Run as administrator*** option.

Before the installation you are asked which language you would like, I being English have left it as the default setting of English.

You are then presented with the setup screen (shown on the left). After clicking on the **Next** button several other dialog boxes will appear, where each time you will need to click the **Next** button again before the installation is complete.

If you try to overwrite a previous installation of R, just click the **Yes** button when the warning appears (shown below.)

In the above sequence I have left all the options as defaults. I would recommend this if you are just starting out with R.

4.1 Tips and hints

The above procedure seems like a lengthy process but you only need to do it very occasionally, usually once every few years when you want to completely update R.

YouTube help http://www.youtube.com/watch?v=3zEbzZ9PqBo

5 Starting R

You can start R in one of three ways, all of which are shown below:

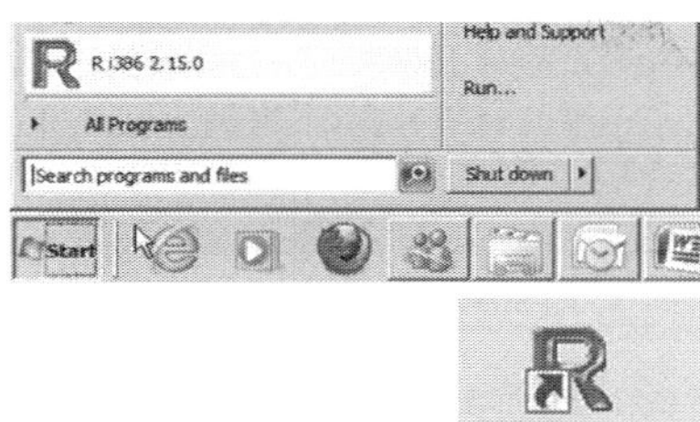

- from the quick menu,

- the desktop icon or

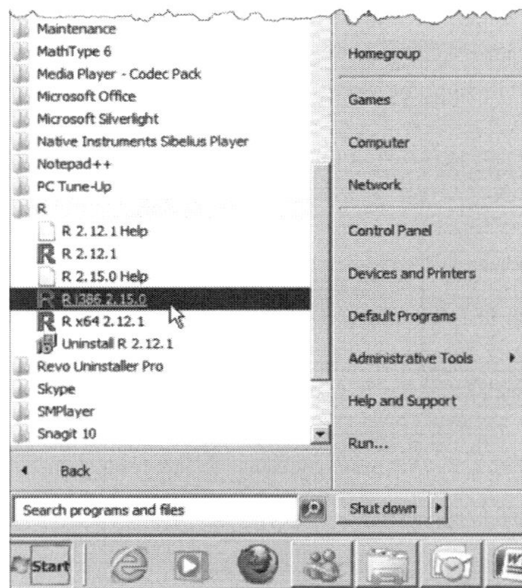

- the full program menu

When you start R you are presented with a window similar to the one below called the *RGui,* in which the console window can be seen (I've shaded it in the screenshot below). This window, the *R Console* window is the one in which you both type commands and see the results. The *RGui* window, contains a drop down menu and eight icons - details of which are provided in the appendix.

When you start to use R, you will notice that the environment is very austere. As you become more proficient with R, you will probably start to use this interface more and more (as it enables you to perform more specific analyses). However, to get you started, we are going to focus on an extension to R called R Commander which will enable you do most of the common things by selecting options with a mouse and filling in various dialog boxes. R Commander is discussed in more detail in the following chapter.

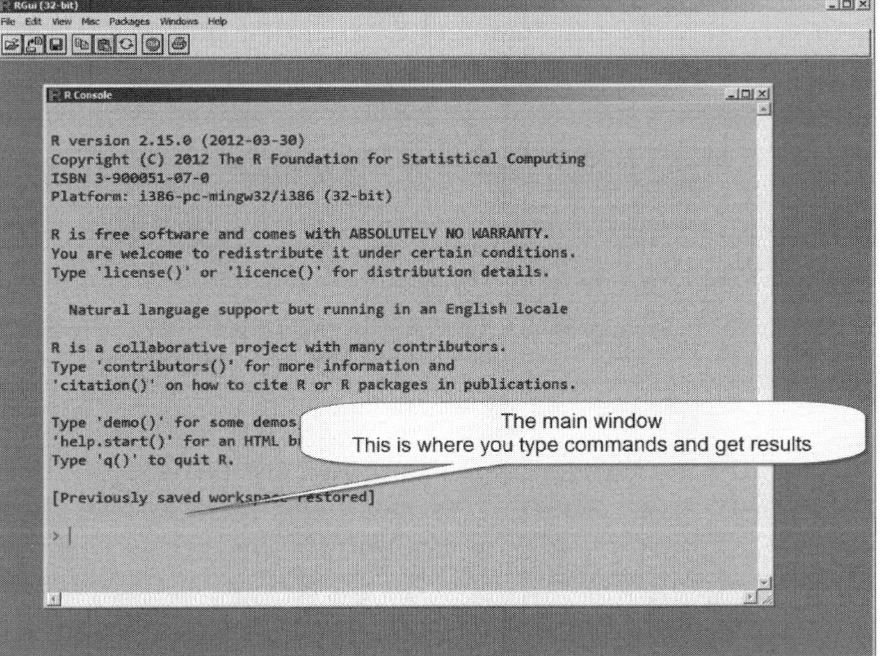

5.1 Other R windows

When you use R other windows can appear which I refer to in subsequent chapters. For example, when you have some data in R selecting the menu option **Edit-> Data editor** brings up the *Data Editor* window. Similarly typing certain commands in the *R Console* window brings up the results in the *R Graphics* window.

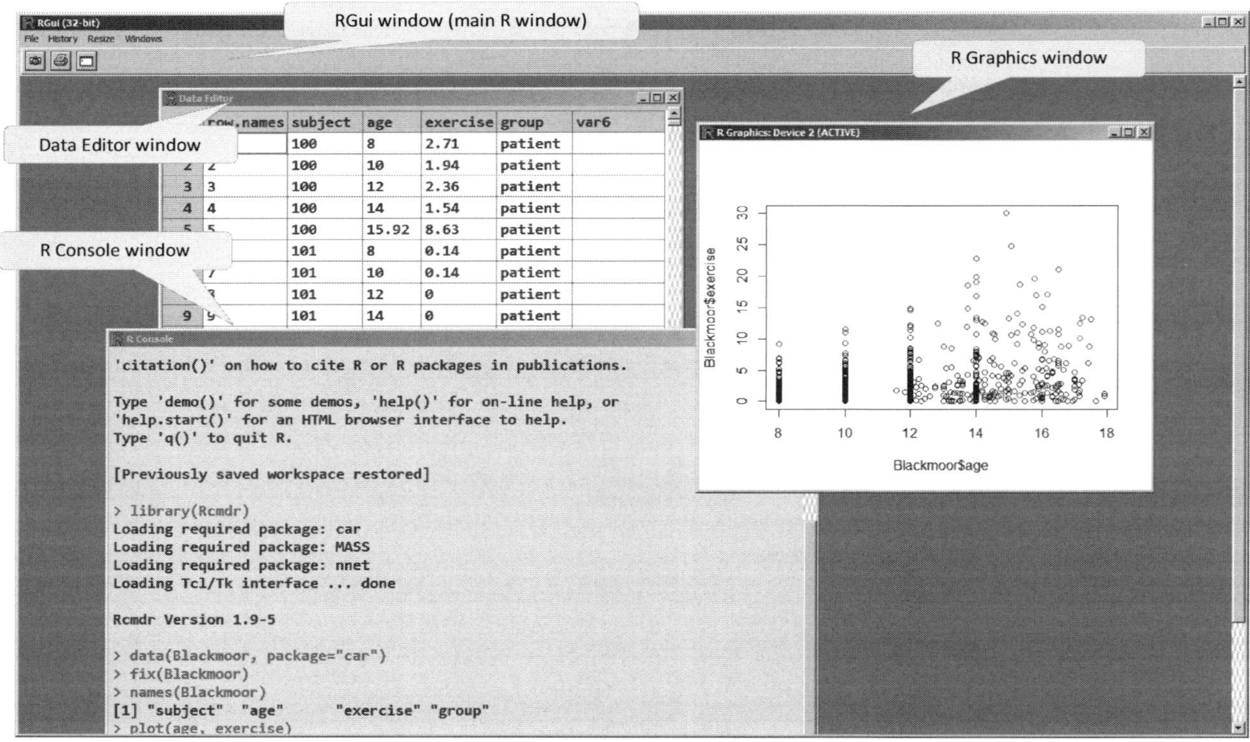

5.2 Tips and hints

You can change how the *R RGui* and *R Console* windows appear by using the menu option

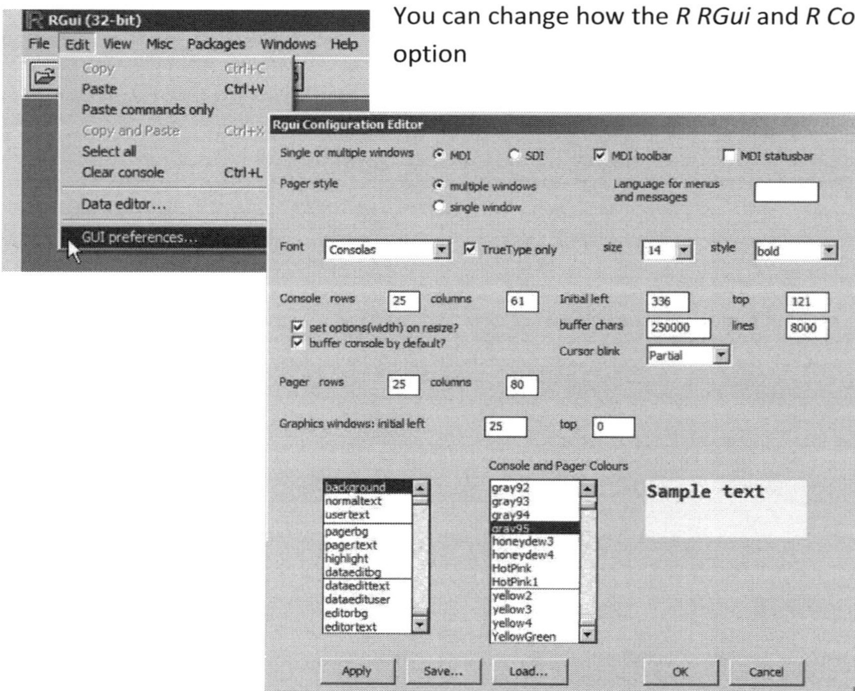

Edit-> GUI preferences

GUI = Graphical user interface i.e. the screen aspect of R. You can also save the changes by clicking on the save button in the GUI preferences window. The GUI configuration file is a text file which you can open up and edit directly if you wish using a free text editor such as *notepad++*.

You can see the variety of fonts R has installed any time by typing demo(Hershey) into the *R Console* window. A series of lists will appear in the *R Graphics* window which you will need to click on to allow you to move onto subsequent screens.

6 R Commander: a graphical front end to R

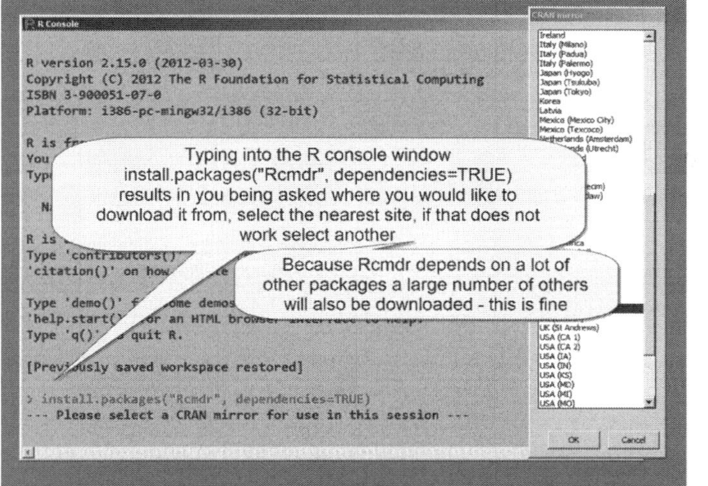

Typing into the R console window install.packages("Rcmdr", dependencies=TRUE) results in you being asked where you would like to download it from, select the nearest site, if that does not work select another
Because Rcmdr depends on a lot of other packages a large number of others will also be downloaded - this is fine

R Commander is a piece of software that delivers a graphical front end to R, that is it provides a complex menu structure and dialog boxes that automatically generate R code. To be able to use R Commander, known as *Rcmdr* for short, you need to both install and load it from within R. To do this you need to have the *R Console* window open (the shaded window opposite) then type into it the following R command:

```
install.packages("Rcmdr", dependencies=TRUE)
```

Notice the capital *R* and *TRUE*. R is case sensitive so *TRUE* means something different to *TruE*. The *CRAN mirror* window will appear asking where you want to download it from, select the nearest site and click on the **OK** button, if that does not work try the next nearest location.

You only need to install R Commander once. But you will need to load the *Rcmdr* package every time you open R. To do this type the following and then press the Enter key (↵):

```
library(Rcmdr)
```

We have now both installed and loaded the R Commander package from within R.

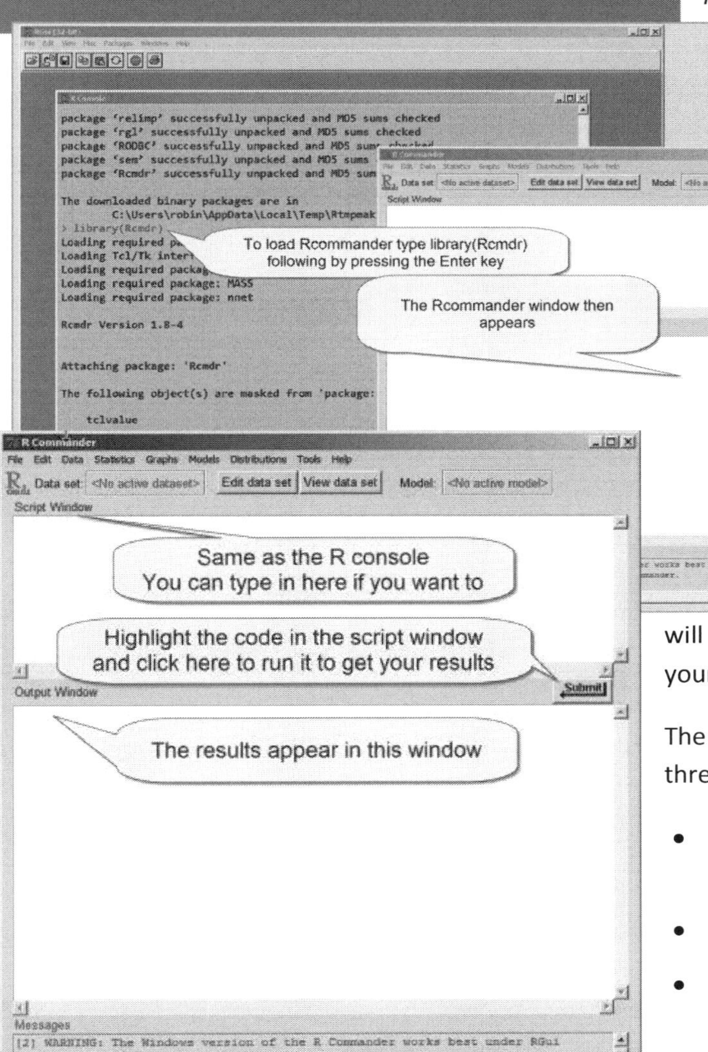

You can type in here if you want to
Highlight the code in the script window and click here to run it to get your results

R Commander is one of many packages we will use in this book, think of them as being like apps for your smartphone or tablet.

The R Commander window shown opposite consists of three panes:

- the top one produces the R commands from the menu options.
- the middle one gives the results.
- the small bottom one provides information (e.g. number of cases in a dataset) and error messages which I often ignore.

R Commander has an extensive menu system of which some of the options are shown below. Many of these will be investigated in depth in the rest of this book.

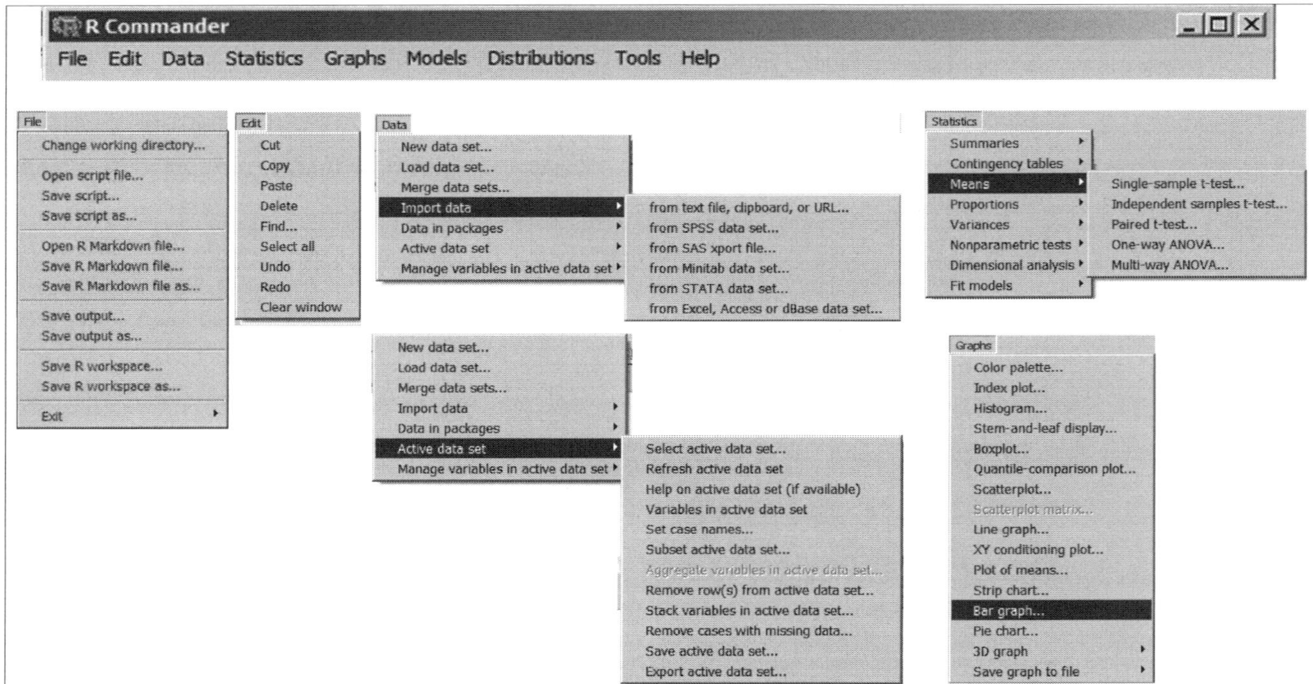

6.1 Tips and hints

Re-opening R Commander

If you close the R Commander window, and want to then subsequently open it again while still in R, you need to type *detach(package:Rcmdr)* before restarting R Commander as just typing *library(Rcmdr)* again does nothing, alternatively typing *Commander()* also works.

The latest version of R Commander has additional file options concerned with creating and managing web reports known as markdown files. The last option in the help menu provides additional information.

Where has R Commander gone?

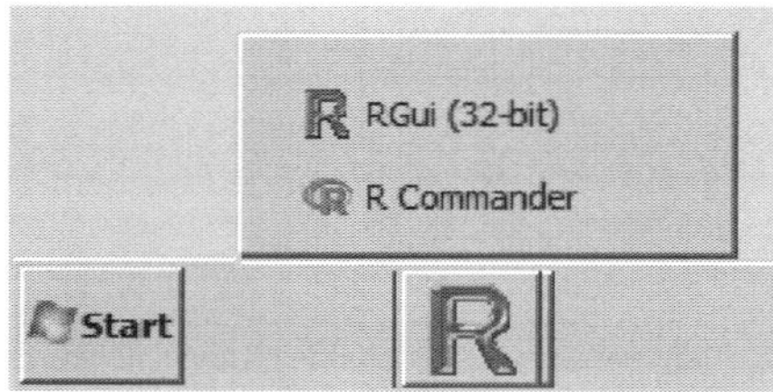

Sometimes the R Commander window seems to disappear but you can always find it by clicking on the R icon at the bottom of the screen and selecting **R Commander**.

YouTube help: http://youtu.be/IEYwLKgKDzA

7 Packages: the apps

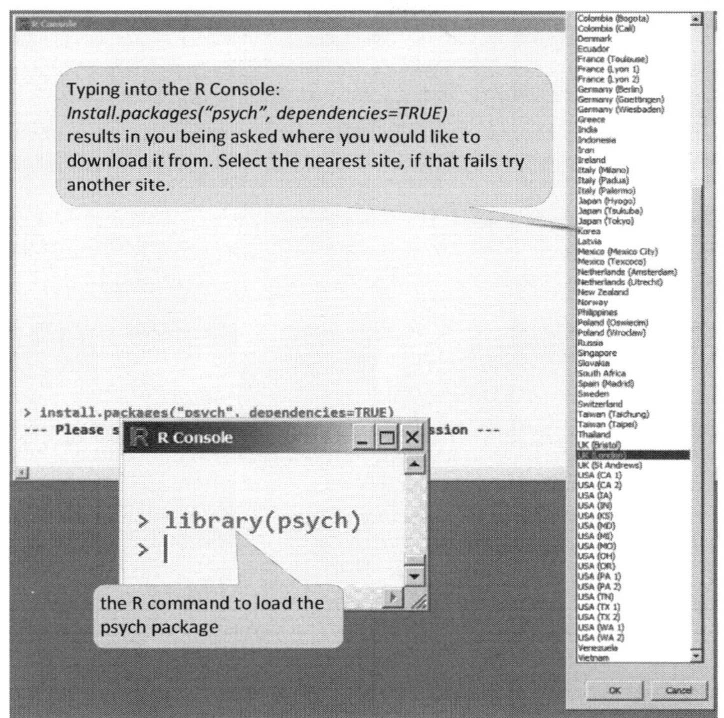

Typing into the R Console:
Install.packages("psych", dependencies=TRUE)
results in you being asked where you would like to download it from. Select the nearest site, if that fails try another site.
R Console
> library(psych)
the R command to load the psych package
OK
Cancel

In the previous chapter we installed and ran the R Commander package. Another package that is very useful is the *psych* package which we will now install.

Type into the *R Console* window:

```
install.packages("psych", dependencies=TRUE)
```

The *CRAN Mirror* window will appear in which you select the nearest location then click on the **OK** button, if that does not work try the next nearest location.

Once it is installed on your computer, to be able to use the package you need to 'load' it. To load the package type the following in the R Console window followed by Enter (↵).

```
library(psych)
```

You have now typed your first R command, more technically called R code.

7.1 Tips and hints

Packages - you install once by typing the following R command replacing *package name* with the specific package name you wish to install:

```
install.packages("package name", dependencies = TRUE)
```

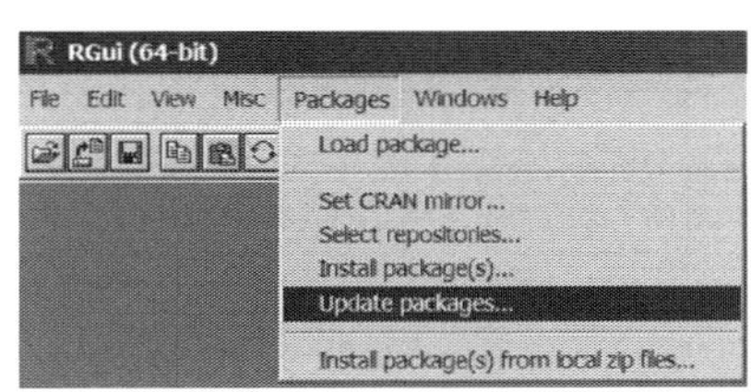

To be able to use the package from within R you need to load it. You need to do this **each** time you start R by typing the following, replacing package name with the specific package name you wish to use:

```
library(package name)
```

To update all the packages you have installed you can select the menu option **Packages->update** packages or type in the *R Console* window:

```
update.packages(ask=FALSE)
```

To find out which packages you have loaded along with their version numbers type:

```
sessionInfo()
```

To find out which packages you have installed but not loaded type *library()*. To unload a package use:

```
detach("package:psych", unload=TRUE)
```

The table opposite lists four of over a thousand packages that are available. There is even a package called, *installr*, designed to help manage them.

Examples of specialist packages

Name	Overview	Details
Epi	Epidemiological statistics	http://cran.r-project.org/web/packages/epiR/epiR.pdf
rpsychi	Psychiatric statistics	http://cran.r-project.org/web/packages/rpsychi/rpsychi.pdf
rhosp	Evaluating risk during hospitalization using simulation	http://cran.r-project.org/web/packages/rhosp/rhosp.pdf
psychometric	Statistics for correlation theory, meta-analysis reliability, item analysis, inter-rater reliability, and classical utility	http://cran.r-project.org/web/packages/psychometric/psychometric.pdf
installr	Supports package management	http://cran.r-project.org/web/packages/installr/installr.pdf

8 A quick tutorial: analysing data shipped with R

The *anorexia* dataset in the MASS package

- *Treat* indicating which group they were in; "Cont" (Control), "CBT" (Cognitive Behavioural Therapy) and "FT" (Family Therapy).
- *Prewt* indicating weight of patient before study period, in lbs.
- *Postwt* indicating weight of patient after study period, in lbs.

In this chapter we will carry out a typical analysis using R Commander.

We will use a dataset which is part of the *MASS* package that was installed when you installed R Commander. It consists of data from 72 anorexic patients, comprising of three columns, detailed opposite.

The aim of the study was to see if either treatment improved weight gain compared to the control group. This will therefore be the focus of our analysis. But first we need to start R Commander and obtain the data.

8.1 Starting R Commander and loading the data

To Run this tutorial you need to have installed R and R Commander, for details see the chapters *Downloading and installing the R software* and *R Commander: a graphical front end to R*. Once you have done this you start R Commander by typing the following in the R Console window followed by Enter (↵).

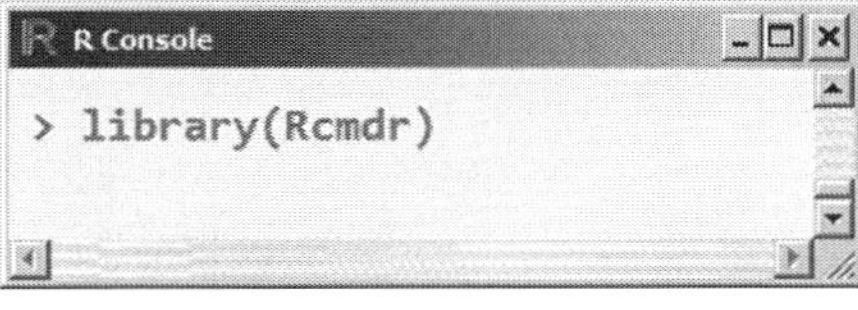

Library(Rcmdr)

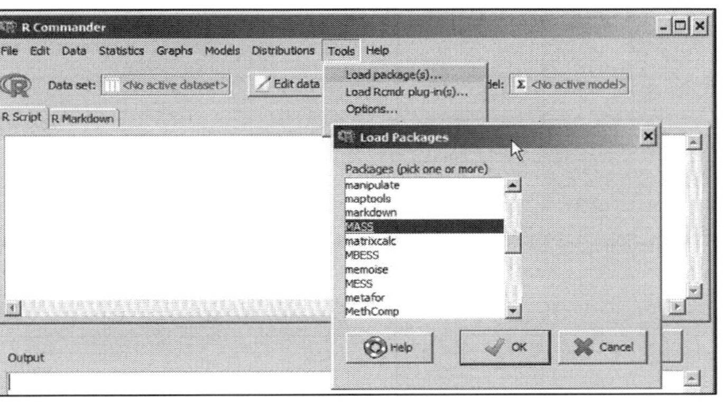

To load the installed package called *MASS* select the R Commander menu option:

Tools->Load packages(s)...

Then:

- Select the *MASS* package
- Click **OK**.

To open the dataset up in R use the menu option:

Data->Data in packages-> Read data set from an attached package

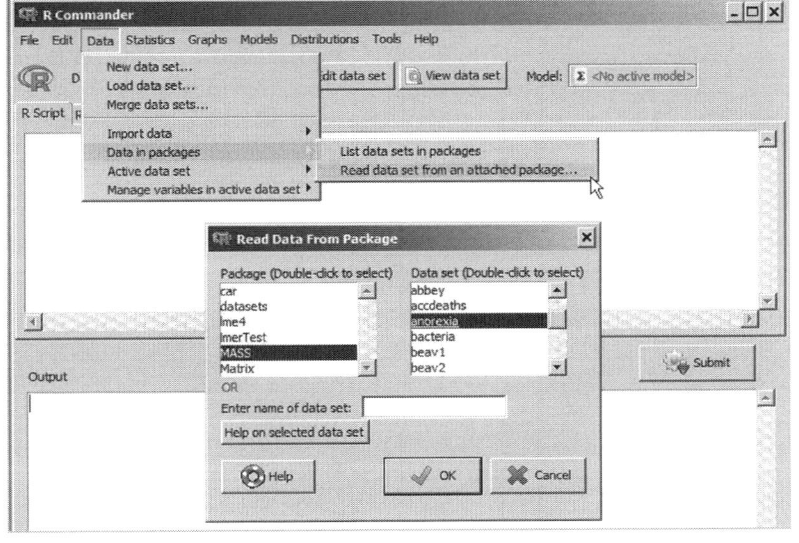

Then from the **Package** listbox:

- Select the *MASS* package

Scroll down the **Data set** listbox to *anorexia* and double left click on it, which moves it into the **Enter name of data set** box. Note that the names of the data sets are ordered alphabetically, first by capital, then by lower case letter.

You can find out more details about a particular data set (i.e. dataset) you have selected by clicking on the **Help** button. Finally click on the **OK** button.

8.2 Checking the dataset

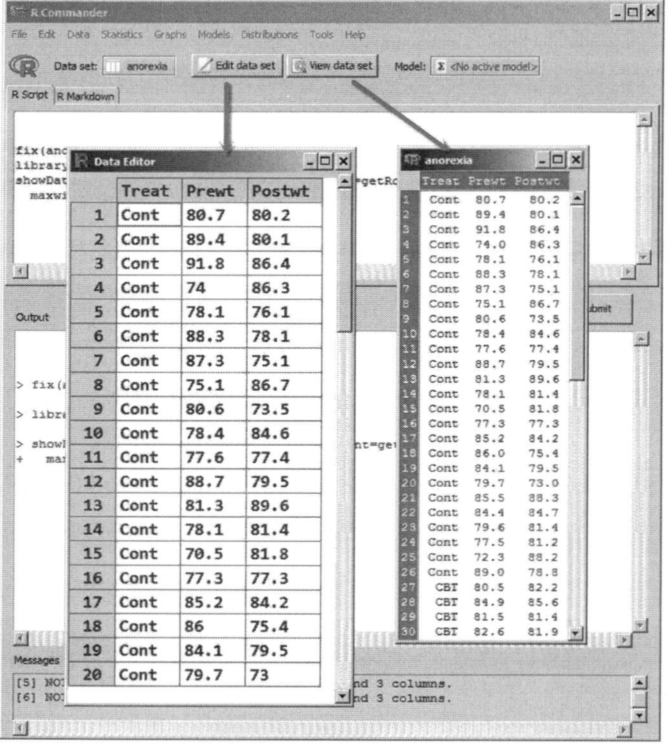

The first thing you should always do is view the data by clicking on the **Edit data set** or **View data set** buttons in the R Commander window.

This basic step of viewing the data is important as it allows you to quickly check to see you have the correct number of columns (variables) and also to assess their quality. A high quality dataset has no or few missing values (shown in R by *NA*) and no inappropriate values such as negative *Prewt* or *Postwt* weights. This dataset is of high quality as it has no missing values and all the values appear sensible.

You should always close the edit/view data window after using it. Failing to do so can prevent other windows, including the R Commander window, from appearing.

If you can't see the edit/view data window after clicking on the button it might be because it is hiding in the *RGui* window. To check this go to the windows toolbar and select the **R** icon and then **RGui**.

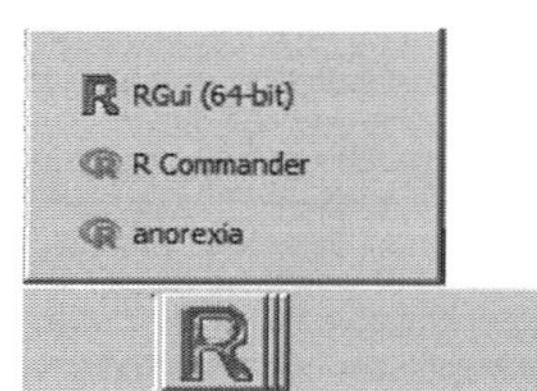

When starting to investigate a dataset there are two complementary strategies we can use to describe it: the production of summary statistics and graphs (often called plots).

8.3 Data description: summary statistics

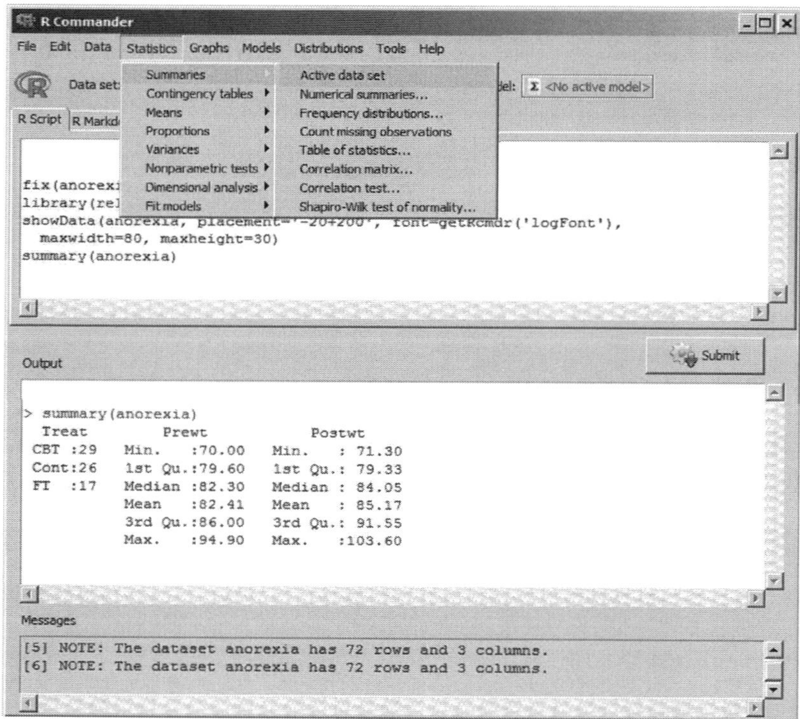

Summary statistics are so named because they **reduce** the entire set of numbers in a variable to either a single value, or possibly two. This idea of reducing a set of numbers down to a smaller set is fundamental to statistics, but is not without its problems. Every summary value is based upon certain assumptions, which if not met, indicates that the summary value might be invalid.

It is remarkably easy to get a set of summary values for each variable. Click on the R Commander menu option:

Statistics-> Summaries-> Active data set

Looking at our dataset we get three columns of results, one for each of the columns (variables) in the dataset.

```
> summary(anorexia)
  Treat         Prewt           Postwt
 CBT :29   Min.    :70.00   Min.   : 71.30
 Cont:26   1st Qu.:79.60   1st Qu.: 79.33
 FT  :17   Median :82.30   Median : 84.05
           Mean   :82.41   Mean   : 85.17
           3rd Qu.:86.00   3rd Qu.: 91.55
           Max.   :94.90   Max.   :103.60
```

The first column provides details of *Treat*; this just gives the number of subjects in each category. We see that the *CBT* group has nearly twice as many participants as the *Family Therapy (FT)* group. We don't know if this was intended or if more people dropped out of the *FT* group or even if people actually chose which group they wished to be in. This latter possibility is highly unlightly as one of the basic rules of research design is to use random group allocation, called simply randomisation, to ensure valid results.

Columns two and three provide more information, including the minimum and maximum values and the mean and median for each variable, ignoring the three different groups. If we look at the mean value, it appears that the group of patients as a whole have gained 2.76 (85.17 – 82.41) pounds. The mean value provides an indication of the typical value for the set of observations but is only really a valid measure when the observations follow a normal distribution, which is checked using either a histogram or a boxplot, discussed below. Even the simplest of the values given above is of use as it provides a method of checking the data. For example, the minimum and maximum values indicate if there are any improper values such as negative weights or excessively heavy people that would not be expected in an anorexia sample.

8.3.1 Why do you get different types of summary values for different columns?

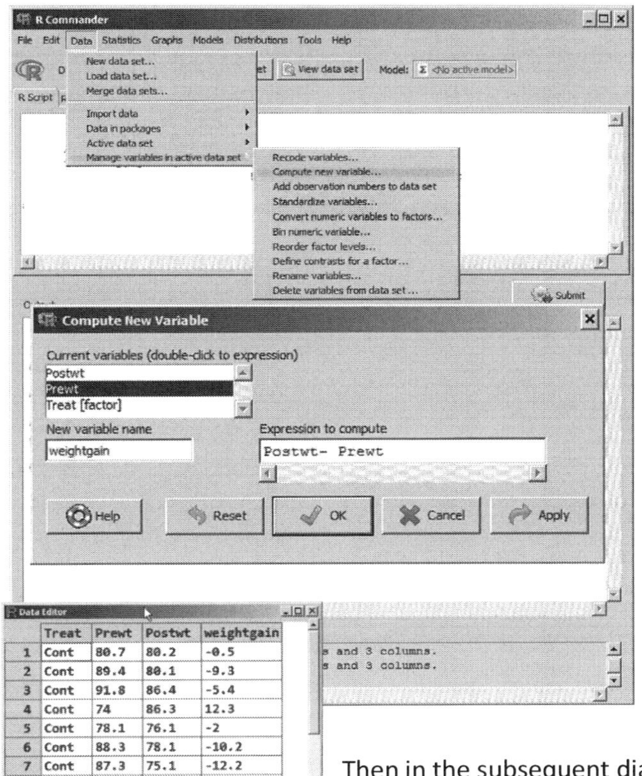

	Treat	Prewt	Postwt	weightgain
1	Cont	80.7	80.2	-0.5
2	Cont	89.4	80.1	-9.3
3	Cont	91.8	86.4	-5.4
4	Cont	74	86.3	12.3
5	Cont	78.1	76.1	-2
6	Cont	88.3	78.1	-10.2
7	Cont	87.3	75.1	-12.2
8	Cont	75.1	86.7	11.6
9	Cont	80.6	73.5	-7.1
10	Cont	78.4	84.6	6.2
11	Cont	77.6	77.4	-0.2
12	Cont	88.7	79.5	-9.2
13	Cont	81.3	89.6	8.3
14	Cont	78.1	81.4	3.3
15	Cont	70.5	81.8	11.3
16	Cont	77.3	77.3	0
17	Cont	85.2	84.2	-1
18	Cont	86	75.4	-10.6
19	Cont	84.1	79.5	-4.6

This is because the data within R each column fall into one of three different types; a character, a factor or a number. Factors are what you might know as nominal or grouping variables (see **Levels of measurement**), all you can do is count them (i.e. tabulate). In this example, the *Treat* column is a factor so all you can do is report the number in each group.

8.3.2 Finding individual weight change: Creating new variables

So we have quickly calculated the overall mean weight gain, but this doesn't tell us anything about the effectiveness of each treatment, we know nothing about individual change. To do this we need to create a new variable indicating the weight change for each individual. We already know the Pre-weight (*Prewt*) and the Post-weight (*Postwt*), so, to find the individual weight gain, we will create a variable that subtracts *Prewt* from *Postwt*. We do this as follows in R Commander by selecting the menu option:

Data -> Manage variables in active data set-> Compute new variable

Then in the subsequent dialog box create the following in the **Expression to compute** box:

```
Postwt - Prewt
```

You do this by double clicking on the variables you wish to analyse in the **Current variables** box. You add the minus sign - by simply typing it in. You will see that I have called the new variable *weightgain*. Note that both of the variables start with a capital letter. Typing *postwt* would mean something else. Finally click on the **OK** button. You can review your changes by clicking on the **Edit** or **view data set** buttons.

8.4 Data description: graphs

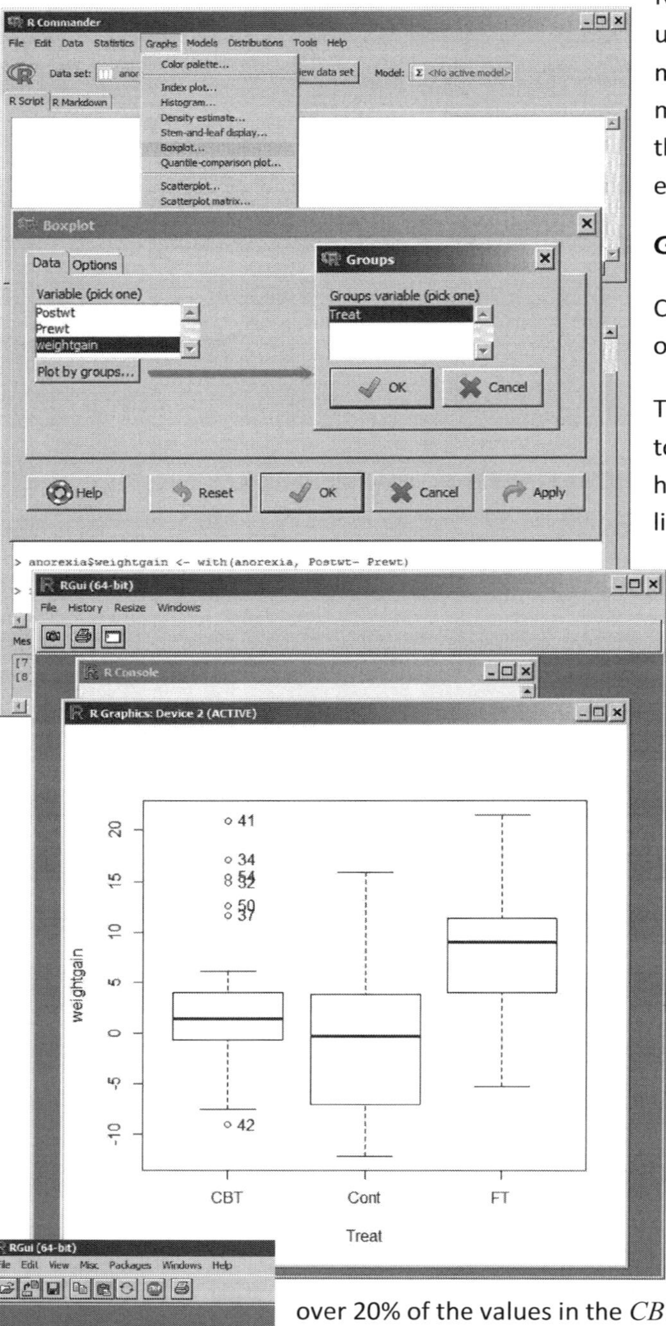

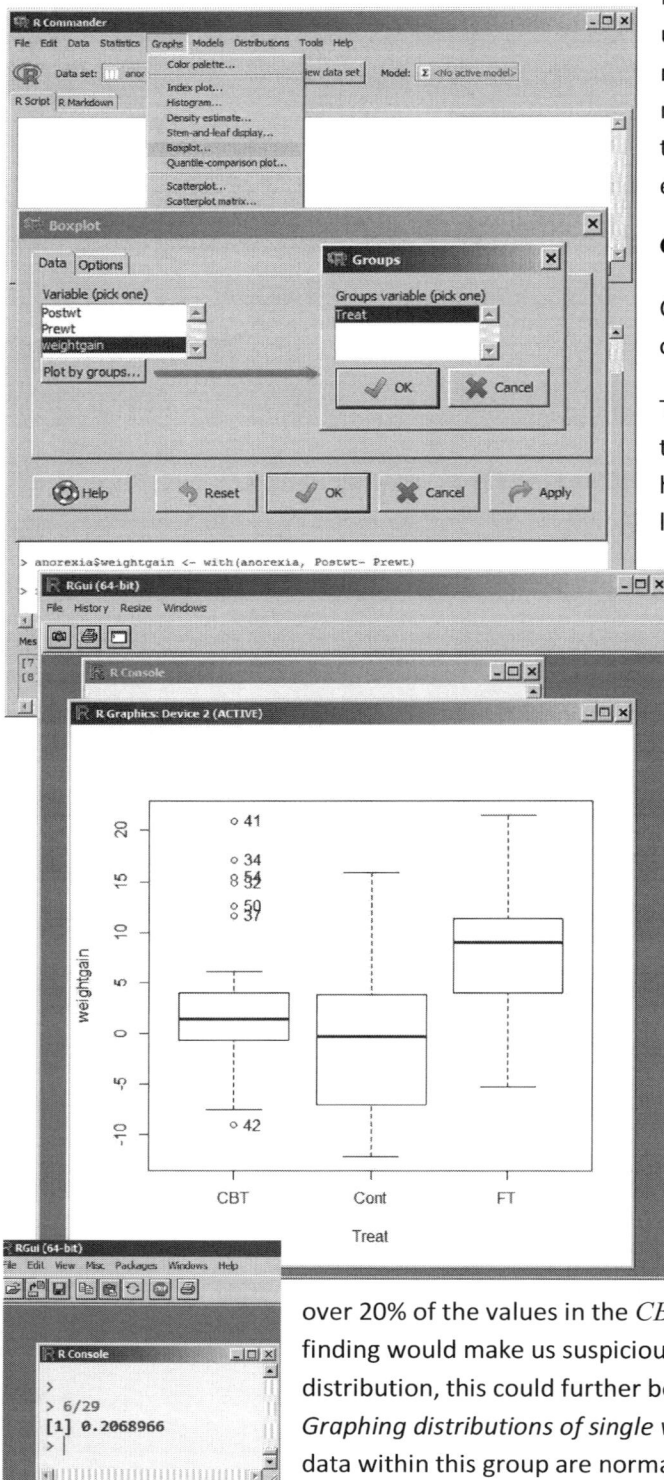

Now that we have created the variable *weightgain* it would be useful to see a plot of the scores for each group along with a mark indicating where 50% of the scores are either less or more than it. This value is called the median. A **Boxplot** (see the glossary entry for further details) fulfils this function and is easily obtained in R Commander selecting the menu option:

Graphs-> Boxplot

Complete the dialog boxes as shown opposite, finally clicking on the **OK** buttons to produce the desired boxplot.

The result appears in the RGui window (which you may need to open, see below). We can see that the Control group (*Cont*) has the smallest median weight gain, represented by the thick line in the box, which is around zero. In contrast the Family Therapy (*FT*) group appears to have the greatest gain.

The boxes represent the middle 50% of the scores, that is 25% each side of the median, a value which is called the interquartile range (IQR). We can see that the control group has the widest box (i.e. IQR) followed by the Family Therapy group and finally the *CBT* group. In fact for the *CBT* group the IQR appears to be less than half that of the Control group. If the IQRs vary widely for the different groups it can affect the validity of some of the statistics we are going to consider later, so it is important to note it at this stage.

In the Boxplot there are a number of '0' marks for the *CBT* group - these are known as extreme values.

Noting there are six such values out of the 29 in the *CBT* group we can quickly use the R Console window as a calculator to work out the percentage of extreme values by typing into it 6/29 and then pressing the return key. The line underneath shows the value .206 indicating that just over 20% of the values in the *CBT* group fall into the high extreme value category. This finding would make us suspicious that the data within the *CBT* group do not follow a normal distribution, this could further be examined by producing a histogram of the data (see chapter *Graphing distributions of single variables*). To keep things simple for now, we will assume the data within this group are normally distributed.

Another important aspect to observe is the degree of overlap between groups, and as the boxes represent the middle 50% of scores we can see that the *FT* group has much less in common with the other two groups than they do between themselves.

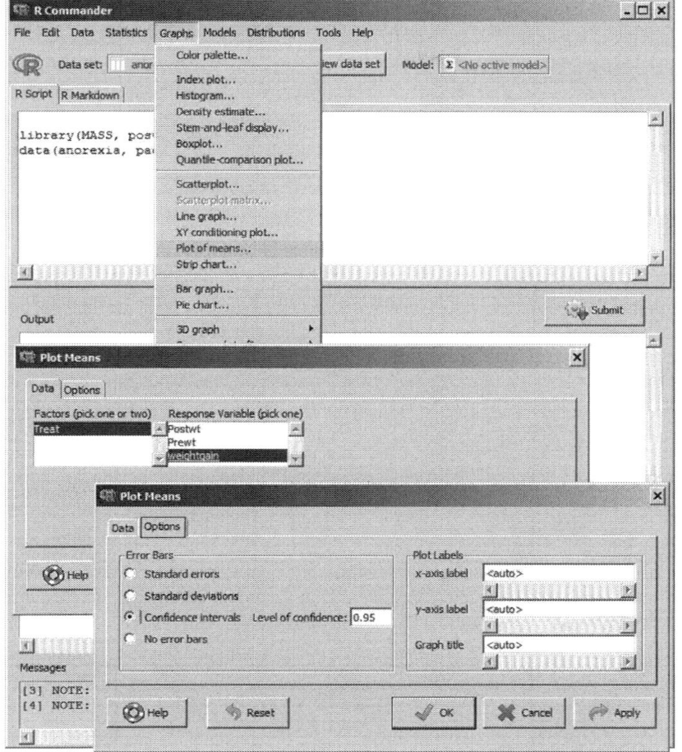

An alternative to the boxplot, if you are satisfied that the data follow a **normal distribution**, is a plot of means with **confidence intervals**. Such a graph is produced in R Commander using the menu option:

words in **bold** indicate a glossary entry providing additional information

Graphs-> Plot of means

Complete the dialog boxes as shown opposite. There is no need to change the default *Level of confidence* setting of 0.95 - meaning 95% confidence intervals. The graph of means and 95% CI of the means reflect the boxplots previously produced but there are differences, now there appears to be a greater difference between the *Cont* and *FT* groups. The *CBT* is midway between the two other groups. This difference is because this time we are assuming that the scores are normally distributed, which we did not with the boxplots.

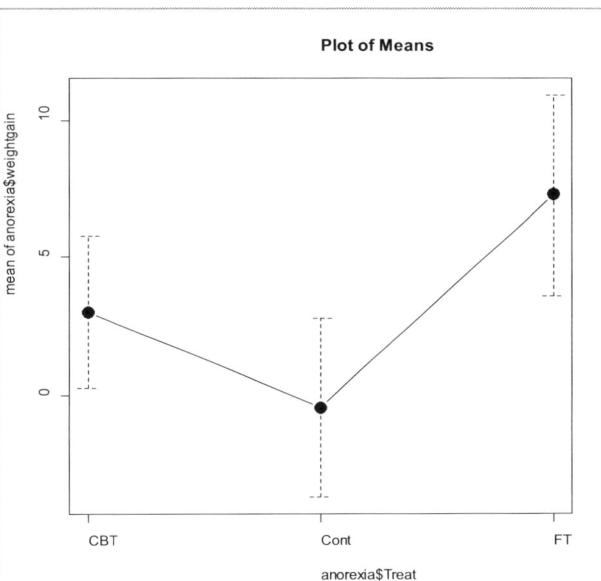

8.5 I can't see the graphics

Sometimes the graphics do not appear which is often because either the *RGui* window is hidden or the actual graphics window within it is minimised. To open the *RGui* window click on the **R** icon at the bottom of the screen along the taskbar and select the **RGu**i menu option.

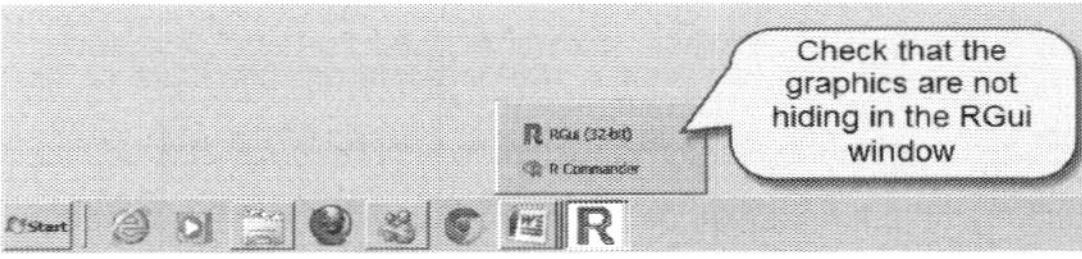

8.6 Inferring from our samples to a population

So far we have been discussing the sample but one of the main goals of statistics is to be able to infer from our often small sample to a much larger population. In other words, we desire to make **inferences**. This is accomplished by creating a statistical model and then evaluating it using **p-values** and **confidence intervals** (see glossary entries for further details).

For this set of data our model will be based upon the assumption that all three groups of data are random samples which came from a single population and the observed differences in the means we have noted are therefore just due to random sampling. In other words, we assume we have randomly collected three samples of our observed sample sizes from a single population. We now want to see if we can reject this 'null' model which would then imply that the samples do come from populations with different means. Therefore the observed means of the groups are truly different and not just due to the vagaries of random sampling. We do this by obtaining the p-value for the data.

8.7 Obtaining a p-value from a statistical test

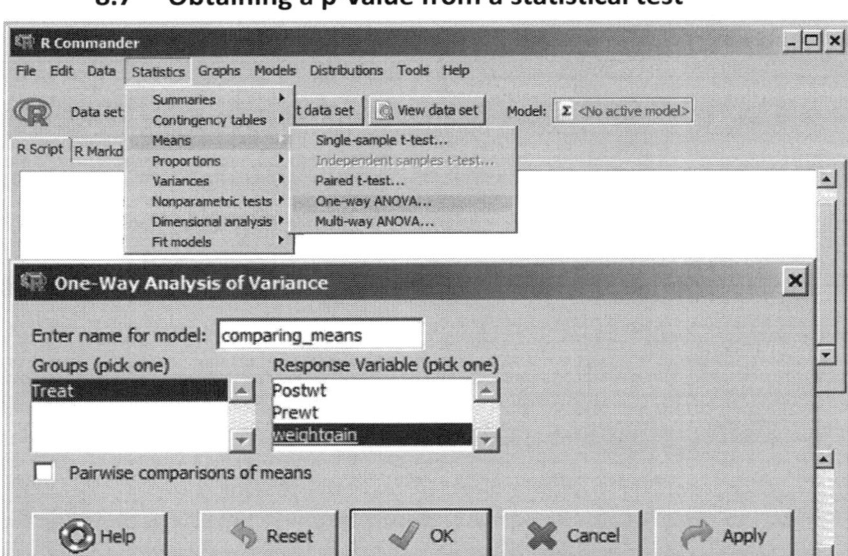

The statistical technique that mimics the above description is called ANOVA (Analysis of Variance).

To carry out a one-way ANOVA select the menu option:

Statistics-> Means-> One-way ANOVA...

You need now to do the following:

- give a name to your model, I have called it *comparing_means*
- select the group/factor variable and outcome variable as shown opposite

Finally click the **OK** button.

Output

```
> comparing_means <- aov(weightgain ~ Treat, data=anorexia)

> summary(comparing_means)
            Df Sum Sq Mean Sq F value  Pr(>F)
Treat        2    615  307.32   5.422 0.0065 **
Residuals   69   3911   56.68
---
Signif. codes:  0 '***' 0.001 '**' 0.01 '*' 0.05 '.' 0.1 ' ' 1

> numSummary(anorexia$weightgain , groups=anorexia$Treat, statistics=c("mean",
+     "sd"))
          mean       sd data:n
CBT   3.006897 7.308504     29
Cont -0.450000 7.988705     26
FT    7.264706 7.157421     17
```

p-value= 0.0065

means, standard deviations and number in each group

The results give us a **p-value** of 0.0065 which we can interpret in the following manner:

We would have obtained a set of means as disparate as those observed or ones even more different 65 times in every ten thousand if our 3 samples had come from a single population.

We usually define a cut off value of: 0.05, 0.01 or 0.001 technically called a critical value (CV) or alpha level, for our (null) model. In health sciences the value is typically set to be 0.05 so with our p-value of 0.0065 we can say our results are statistically significant. This means that we now believe our three samples do not come from the same (i.e. single) population.

Conclusion: There is a statistically significant difference between the means of the three groups.

But exactly where is this statistically significant difference . . .

8.8 Where exactly is the difference? - using pairwise comparisons

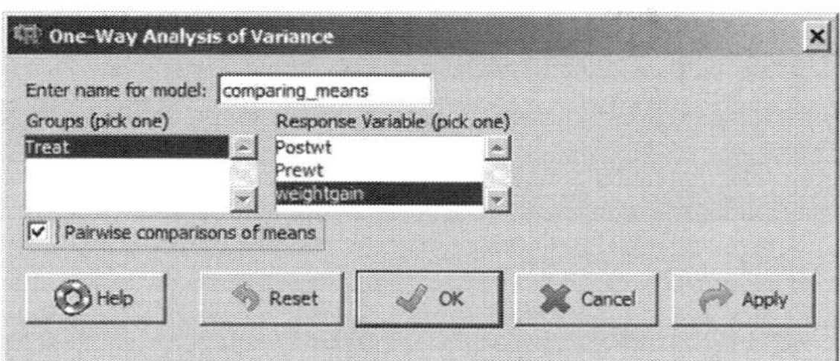

Because we have three groups, the next question is to decide exactly which of the three group means differ in terms of statistical significance. To find out where the 'statistically significant' difference is between the various groups we edit the One-way ANOVA dialog box shown on the previous page by ticking on the option:

Pairwise comparisons of means

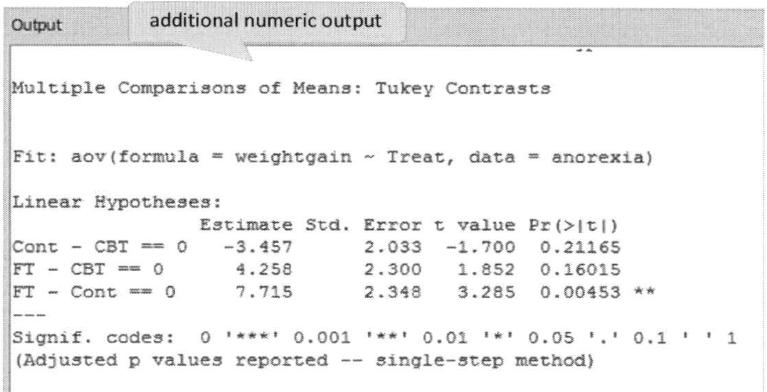


```
Multiple Comparisons of Means: Tukey Contrasts


Fit: aov(formula = weightgain ~ Treat, data = anorexia)

Linear Hypotheses:
                Estimate Std. Error t value Pr(>|t|)
Cont - CBT == 0   -3.457      2.033  -1.700  0.21165
FT - CBT == 0      4.258      2.300   1.852  0.16015
FT - Cont == 0     7.715      2.348   3.285  0.00453 **
---
Signif. codes:  0 '***' 0.001 '**' 0.01 '*' 0.05 '.' 0.1 ' ' 1
(Adjusted p values reported -- single-step method)
```

Selecting this option produces additional output shown opposite. Here we can see that when we consider each pair of groups, only the *FT* against the *Cont* group is statistically significantly different at the 0.05 level, p-value=0.00453.

The above option also produces a 95% family-wise plot using the values shown in the above numeric output and also adding a 95% confidence interval to each of the estimates. Here because the pairwise difference between *FT - Cont* is the only confidence interval that does not extend to the zero x value, once again indicates that it is significantly significant (i.e. p<0.05).

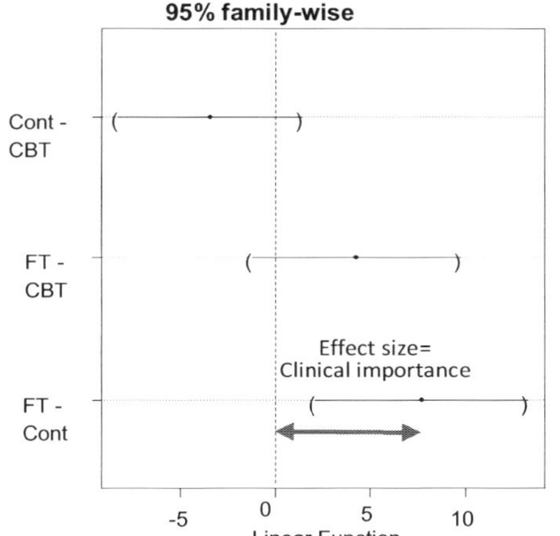

8.9 Examining clinical importance as well as statistical significance

While we have a statistically significant difference between the Family Therapy and Control group the p-value tells us nothing about how large this difference is. To consider this we need to measure the **effect size**. There are several ways of measuring this when data is divided into more than two groups but for now we will just consider differences between pairs of groups. For example the effect size between the Family Therapy and the Control groups is simply the difference between the means; 7.715 lbs. We can see this in both the pairwise comparisons 95% confidence plot and the numerical output. The question is how can we interpret this value? This is where the real problems start as the clinical importance, a way of reinterpreting the effect size, of putting on 7 lbs. on average might be no better than putting nothing on in the long run. Assessing the importance of this value is largely done by both clinical expertise, and further (replication) studies. The glossary entry **Clinical significance** provides further information.

8.10 Closing and saving data and graphs

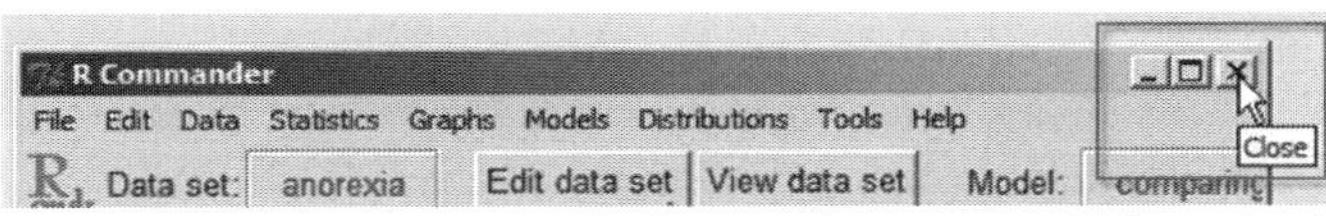

Other chapters in this book describe in detail how to save data and graphs however, for this tutorial, simply click on the **close** button in both the *R Commander* and *RGui* windows, where you will be asked to save various files, each time just click **NO**.

8.11 What now?

You have now completed the R quick tutorial and can now investigate R further. If you wish to start analysing your data now, you will need to import it into R. The chapter *Importing your data into R* explains how to do this.

8.12 Tips and Tricks: investigating other datasets

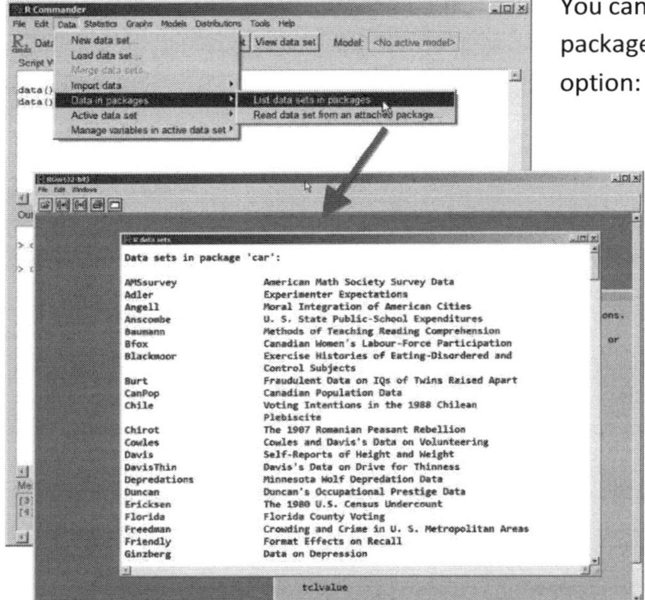

You can see a list of all the datasets provided by R in any of the packages you have installed by selecting the R Commander menu option:

Data->Data in packages-> List data sets in packages

To find out about specific datasets use the R Commander menu option:

Data->Data in packages-> Read data set from an attached package

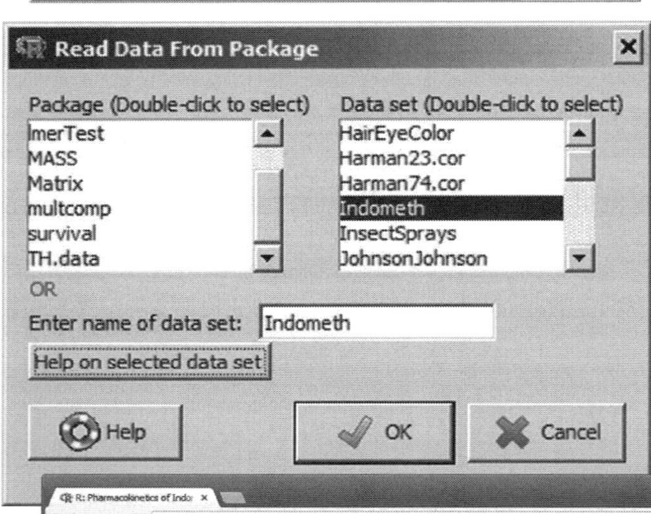

Once you have selected a particular dataset by double clicking on it the name appears in the *Enter name of data set* text box. Then clicking on the **Help on selected data set** opens of a web browser window providing details for the selected dataset. This sometimes also includes a cursory analysis of the data.

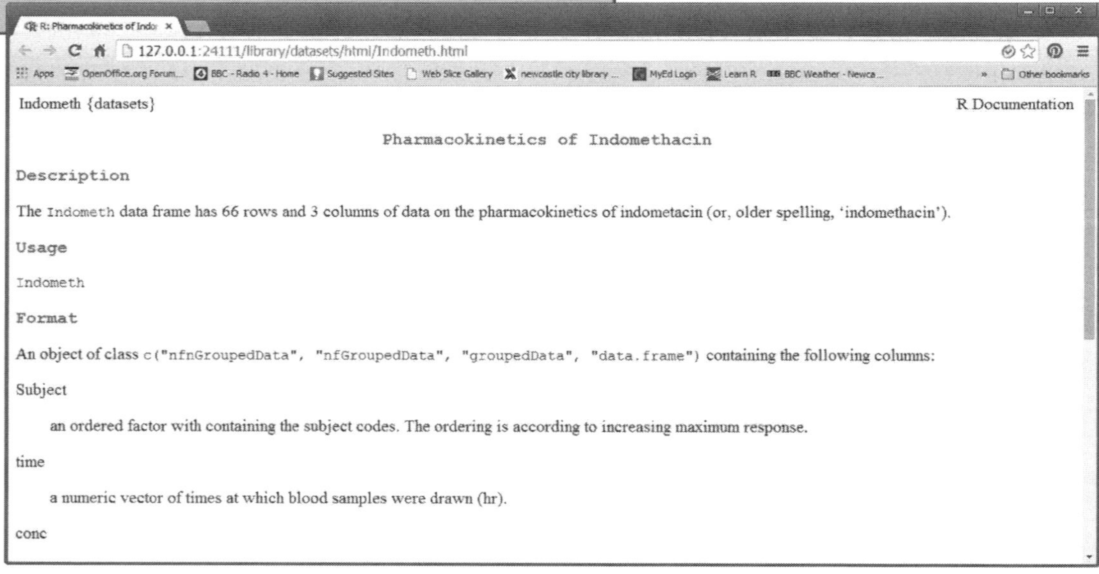

9 A quick introduction to the R language: R

Depending upon your immediate needs, you may wish to skip this chapter for now.

While R Commander allows you to do many things, it only gives you a very constrained view of what is possible in R. To be released from this control you need to learn the R language and, depending upon what you want to do, and how obsessive you are, the required level of R code knowledge might vary from:

- Tweaking the R Commands generated by R Commander - demonstrated in various chapters
- Writing freestanding R code - the focus of this chapter and the R sections in other chapters

Many chapters in this book provide a simple analysis using R Commander and then show how to create a more detailed analysis by writing R code directly.

The *Summary statistics* chapter for example demonstrates using the R Commander menu function for creating tables, followed by a discussion of the R function: *tapply()*.

Using R Commands to produce R code requires you to follow certain syntactical rules which are now described in detail.

9.1 Objects and the assignment/ gets operator <-

bold words in the text indicate entries in the glossary providing more information

R is all about creating and manipulating 'objects'. In R an object can be anything from an individual number, groups of characters, list of numbers, **dataframes**, files containing millions of values, or complex structures such as graphs.

In R an object is created by using the assignment/gets operator which looks like this: **<-**, that is a less than symbol followed by the minus sign. For example, say we want to create an object called *myresult* which consists of the value of 10, we would do this by typing into the R Console or the Script window of R Commander:

the object *myresult* is created and consists of the number 10

```
myresult <- 10
```

R ignores extra spaces unless they are between quotes: myresult <- 10 means the same as myresult <-   10

Similarly we can create an object called *myresult* which is equal to a complex expression such as *myresult <- 2*sqrt(4*pi)*. Here the object, *myresult* is an expression that makes use of multiplication (*), as well as square root *sqrt()* and π *(pi)* functions. We will see latter that an object might also consist of more than one value, where each value is called an *element.*

9.2 Showing the value and structure of an object str()

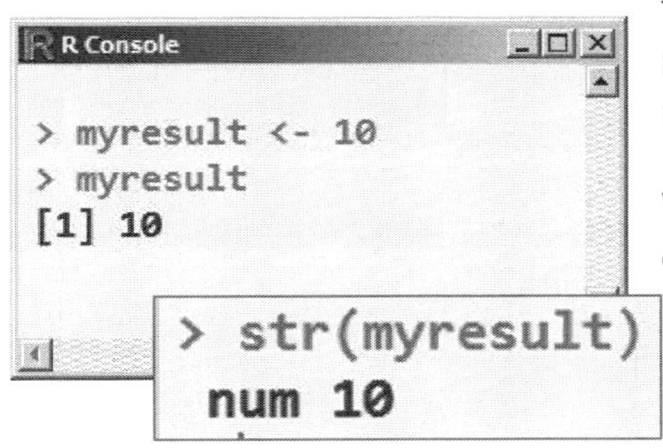
R Console
> myresult <- 10
> myresult
[1] 10
> str(myresult)
num 10

To see the value of an object you simply type its name and press the return key(↵) . Your newly created *myresult* object consists of one element, indicated by the [1], which has the value 10.

You can discover the structure of an object by typing *str(objectname)*. For example, for the *myresult* object you type *str(myresult)*. This shows that *myresult* just consists of the number (num) 10.

9.3 # The comment (#), line extension (+) and multiple command (;) symbols

The # hash character is used to indicate that the rest of the line is to be treated as a comment, R simply ignores it. For example:

R ignores extra spaces unless they are between quotes: myresult <- 10 means the same as myresult <- 10

```
myresult <- 10     # this is all a comment and ignored by R but
                   #  helps me understand it
```

Similarly if you want to create a single command over several lines you just insert a space and plus character + at the end of the previous line of R code. The '+' is added by R at the beginning of the next line. This is shown opposite right.

+ added by R

```
> sum1 <- 2 + 6 + 3
> sum1
[1] 11
>
> sum2 <- 2 + 6 +
+ 3
> sum2
[1] 11
>
> sum3 <- 2 + 6 +
+  + 3
>  sum3
[1] 11
>
> sum4 <- 2 + 6
>  + 3
[1] 3
> sum4
[1] 8
```

You must have a + sign at the end of the line to continue it.

You might also want to put a sequence of small commands on a single line. If so use a semi colon (;) to separate them, for example:

```
myresult2 <- 20;   myresult3 <- 50 ;    myresult4 <- 100 ;   myresult5 <- 700
```

9.4 Functions

A **function** is a special type of R object that includes an instruction. As with all objects it has a name but in addition has a set of parentheses at the end of it in which you place other objects that you want it to either interrogate or change in some way. For example, the function *mean(x)* takes a column from a dataframe (*x*) and returns a single value which is the mean value.

Common R functions
* multiplication
^ exponent i.e. 2^ 3 means 2 cubed
Sqrt(x) square root
max(x), min(x), mean(x), median(x), sum(x), var(x), sd(x), cor(x,y) as named for variable x
summary(data.frame) prints statistics (see screenshot below)
rank(x), sort(x) rank and sort variable x
ave(x,y) averages of x grouped by factor y
sin, cos, tan, asin, acos, atan, atan2, log, log10, exp as named
range(x) range
round(x, n) rounds the elements of x to n decimals
log(x, base) computes the logarithm of x with base = ***base***
mod(x) modulus; it returns the positive value. abs(x) does the same thing

R has a large number of very useful functions, some of which are shown in the reference list opposite. You can see there are functions that provide many different statistical functions such as the mean, standard deviation etc. You can also write your own functions (see the glossary entry for details).

notice that R is case sensitive. Mod does not mean the same as MOD or mOD!!

Before it is possible to demonstrate an R function we need some data which can easily be achieved by using the concatenate function.

9.5 Joining individual values together - the concatenate function c()

The concatenate function *c()*, lower case c, joins a set of values together to form what is called a vector.

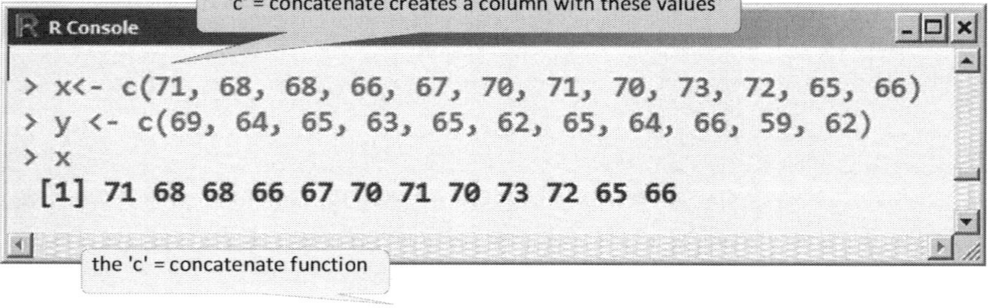

'c' = concatenate creates a column with these values
R Console
> x<- c(71, 68, 68, 66, 67, 70, 71, 70, 73, 72, 65, 66)
> y <- c(69, 64, 65, 63, 65, 62, 65, 64, 66, 59, 62)
> x
[1] 71 68 68 66 67 70 71 70 73 72 65 66
the 'c' = concatenate function

Say we want to create two objects each being a column (vector in technical language) consisting of 12 and 11 values, and we want to call one object *x* and the other *y*. To create the two columns we would type into the R Console the following:

```
x<- c(71, 68, 68, 66, 67, 70, 71, 70, 73, 72, 65, 66)

y <- c(69, 64, 65, 63, 65, 62, 65, 64, 66, 59, 62)

x; y
```

the object I have created, called x, gets these 12 values (elements)

the object I have created, called y, gets these 11 values (elements)

to see the values of x and y you type their names

each element is separated by a comma

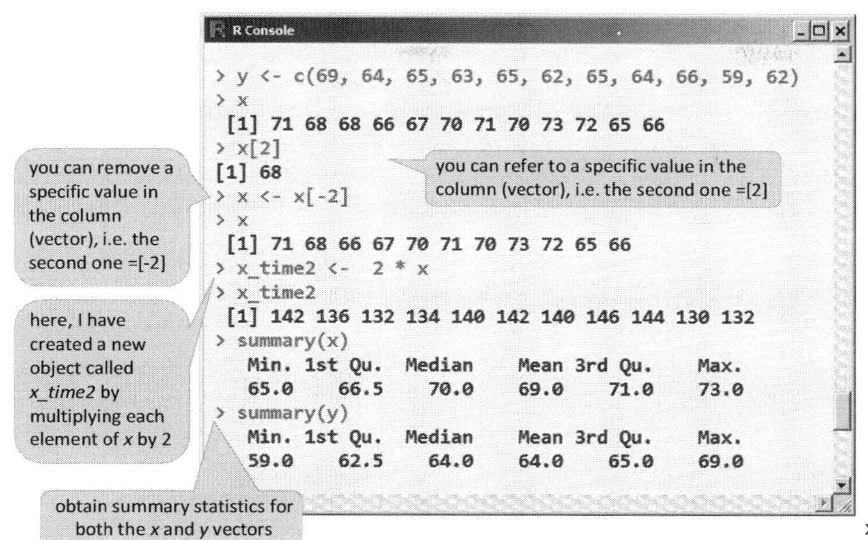

R Console
> y <- c(69, 64, 65, 63, 65, 62, 65, 64, 66, 59, 62)
> x
[1] 71 68 68 66 67 70 71 70 73 72 65 66
> x[2]
[1] 68
> x <- x[-2]
> x
[1] 71 68 66 67 70 71 70 73 72 65 66
> x_time2 <- 2 * x
> x_time2
[1] 142 136 132 134 140 142 140 146 144 130 132
> summary(x)
Min. 1st Qu. Median Mean 3rd Qu. Max.
65.0 66.5 70.0 69.0 71.0 73.0
> summary(y)
Min. 1st Qu. Median Mean 3rd Qu. Max.
59.0 62.5 64.0 64.0 65.0 69.0
you can remove a specific value in the column (vector), i.e. the second one =[-2]
you can refer to a specific value in the column (vector), i.e. the second one =[2]
here, I have created a new object called x_time2 by multiplying each element of x by 2
obtain summary statistics for both the x and y vectors

You can also refer to, and manipulate vectors and their elements. To refer to the second element of x you would type:

x[2]

To remove an element, for example, because we may have accidentally entered *68* twice in the *x* vector you type:

```
x<- x[-2]
```

To create a new vector called *x_time2* that consists of the elements of the *x* vector multiplied by 2 you type:

```
x_time2 <- 2 * x
```

You can also use the statistical functions on the vectors as shown above. To add text together you need to use the *paste()* or *cat()* functions (see index entries under *R code*).

R is a very terse language and you can achieve in a few lines of code what half an hour's pointing and clicking achieves, for example typing into the R Console window *plot(x,y)* will produce you a plot of the two vectors.

To be able to analyse data in various groups we need to structure our vectors by forming **dataframes** described in the next section.

9.6 Lists and dataframes

R prefers it if you create some type of structure for your data in what it calls either a list or a dataframe. A list can have objects with unequal numbers of elements/cases but in a dataframe all the columns (vectors) must be the same length, and you use the *NA* ("not available") value to indicate any missing values.

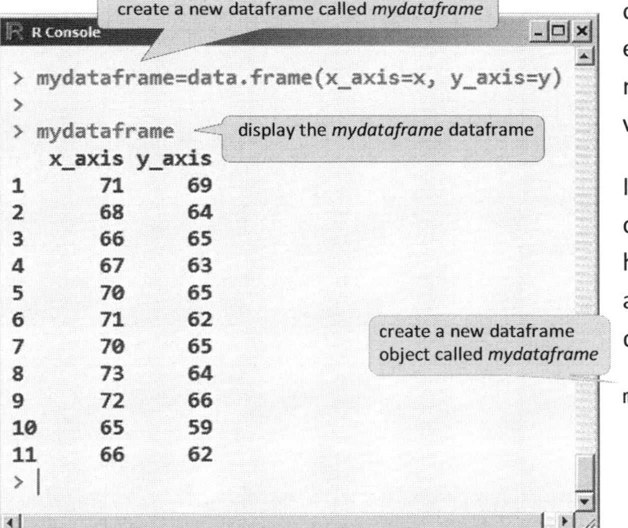

It is always a good idea to give the columns (i.e. vectors) of the dataframe/list names. Considering the dataframe opposite, I have called them, *x_axis* and *y_axis* but they could have been anything including names with spaces if you place them in quotes.

```
mydataframe <- data.frame(x_axis = x, y_axis = y)
```


Many functions work on dataframes, some of which are explained in the following sections.

Notice that when you create a dataframe, each vector (i.e. x, y etc.) becomes a **column**. When you reference a dataframe for editing you refer to it with square brackets [row, column] in the format: [1] or [,1] for column 1 and [1,] for row 1. Note the position of the comma within the brackets.

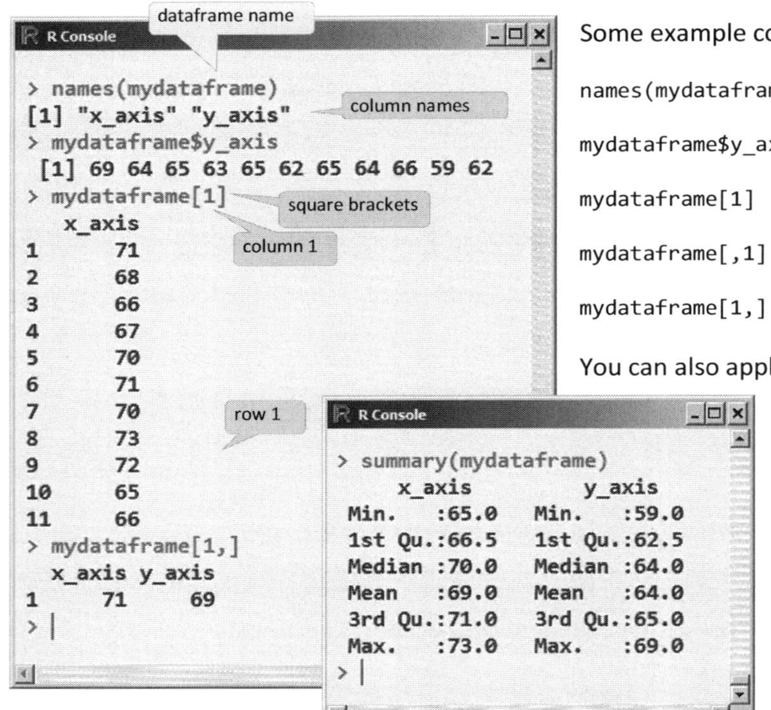
dataframe name
R Console
> names(mydataframe)
[1] "x_axis" "y_axis"
column names
> mydataframe$y_axis
[1] 69 64 65 63 65 62 65 64 66 59 62
> mydataframe[1]
square brackets
x_axis
column 1
1 71
2 68
3 66
4 67
5 70
6 71
7 70
row 1
8 73
9 72
10 65
11 66
> mydataframe[1,]
x_axis y_axis
1 71 69
R Console
> summary(mydataframe)
x_axis y_axis
Min. :65.0 Min. :59.0
1st Qu.:66.5 1st Qu.:62.5
Median :70.0 Median :64.0
Mean :69.0 Mean :64.0
3rd Qu.:71.0 3rd Qu.:65.0
Max. :73.0 Max. :69.0

Some example code is given below demonstrating the above concepts.

```
names(mydataframe)  # returns names of columns

mydataframe$y_axis   # returns the values for the y_axis column

mydataframe[1]       # returns the values for the first column

mydataframe[,1]      # same as above

mydataframe[1,]      # returns the values for the first row
```

You can also apply the statistical functions given earlier to entire dataframes. For example, *summary(mydataframe)* is shown opposite.

9.7 Factors – grouping variables

In R there are two basic types of variables; numbers and **factors** (characters). In the former you can use all the statistical functions like *mean()* etc. whereas with factors you can't. Factors instead allow you to divide data up for analysis. One way to create a factor with 2 possible values is to use the *factor()*function as shown opposite. I have called the factor *grouping_var* here.

```
> grouping_var <- factor(c(1,1,1,1,2,2,2,2))
> grouping_var; str(grouping_var)
[1] 1 1 1 1 2 2 2 2
Levels: 1 2
 Factor w/ 2 levels "1","2": 1 1 1 1 2 2 2 2
```

labels | internal values

the factor I have created is called *grouping_var* and consists of 8 elements with 2 levels

```
grouping_var <- factor(c(1,1,1,1,2,2,2,2))

grouping_var; str(grouping_var)
```

```
> grouping_string <- factor(c("one","one","one","one","two","two","two","two"))
> grouping_string
[1] one one one one two two two two
Levels: one two
> str(grouping_string)
 Factor w/ 2 levels "one","two": 1 1 1 1 2 2 2 2
```

labels | internal values

```
> grouping_gap <- factor(c(5,5,5,5,10,10,10,10))
> grouping_gap
[1] 5  5  5  5  10 10 10 10
Levels: 5 10
> str(grouping_gap)
 Factor w/ 2 levels "5","10": 1 1 1 1 2 2 2 2
```

labels | internal values

You can also create a factor from actual sets of characters, such as 'one and 'two' shown opposite. Notice that regardless of the type of values you supply the *factor()* function, the internal values are always numeric and sequential – you have no control over them. The final example opposite demonstrates the sequential nature of the internal values specifying the values; 5 and 10 still produces internal values of 1's and 2's.

If you are too lazy to type in each value, you can use the **G**enerate **L**evels function: *gl(A,B, labels=c())*. The first value, *A*, is the number of levels, the second, *B*, the number of replications of each and the optional third one the labels you wish to give each level. For example, to create a factor called *grouping_var2* with two levels with labels *control* and *treatment*:

[n] refers to the element number:
[1] = element 1
[10] = element 10
So there are 16 elements in total

```
> grouping_var2 <- gl(2,8, labels= c("control", "treat"))
> grouping_var2; str(grouping_var2)
 [1] control control control control control control control treat
[10] treat   treat   treat   treat   treat   treat   treat
Levels: control treat
 Factor w/ 2 levels "control","treat": 1 1 1 1 1 1 1 1 2 2 ...
```

number of levels=2 | number of replications for each level =8 (16 in total) | labels for the two levels

```
grouping_var2 <- gl(2,8,
labels= c("control", "treat"))
grouping_var2; str(grouping_var2)
```

Note that typing the factor name just produces a list of the labels. Using the str() function produces both the labels and the internal values

Converting a factor to a set of numeric values is a more complex process. This is because you need to decide first if you want to convert the labels or the internal values? If you try to convert a set of labels that do not look like numbers all you get back is a load of Na's. The first example below right demonstrates this.

To extract internal values As they are already numeric we can use the *as.numeric()* function to extract them	**To extract the labels** We need to both extract and convert them
`numeric_vector <- as.numeric(`*factor*`)`	`numeric_vector<- as.numeric(levels(`*factor*`))[as.integer(`*factor*`)]`

`> numeric1<- as.numeric(grouping_var)`
`> numeric1`
`[1] 1 1 1 1 2 2 2 2`
`> numeric12<- as.numeric(grouping_string)`
`> numeric12`
`[1] 1 1 1 1 2 2 2 2`
`> numeric13<- as.numeric(grouping_gap)`
`> numeric13`
`[1] 1 1 1 1 2 2 2 2`
	`> # na example:`
	`> as.numeric(levels(grouping_var2))[as.integer(grouping_var2)]`
	`[1] NA NA NA NA NA NA NA NA NA NA NA NA NA NA NA NA`
	`Warning message:`
	`NAs introduced by coercion`
	`> # working example:`
	`> as.numeric(levels(grouping_gap))[as.integer(grouping_gap)]`
	`[1]  5  5  5  5 10 10 10 10`

9.8 Editing the dataframe

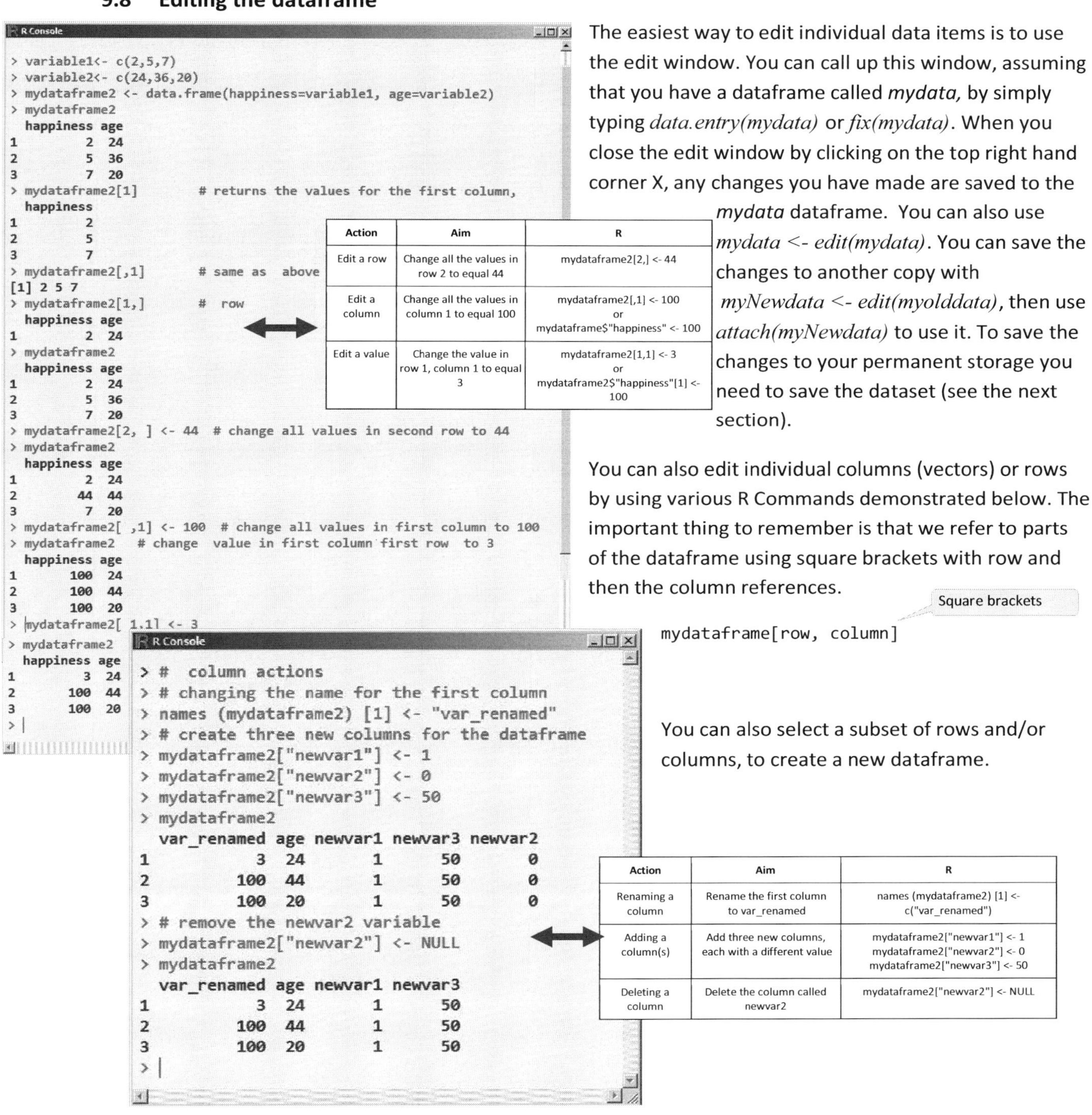

R Console
> variable1<- c(2,5,7)
> variable2<- c(24,36,20)
> mydataframe2 <- data.frame(happiness=variable1, age=variable2)
> mydataframe2
happiness age
1 2 24
2 5 36
3 7 20
> mydataframe2[1]   # returns the values for the first column,
happiness
1 2
2 5
3 7
> mydataframe2[,1]   # same as above
[1] 2 5 7
> mydataframe2[1,]   # row
happiness age
1 2 24
> mydataframe2
happiness age
1 2 24
2 5 36
3 7 20
> mydataframe2[2,] <- 44 # change all values in second row to 44
> mydataframe2
happiness age
1 2 24
2 44 44
3 7 20
> mydataframe2[,1] <- 100 # change all values in first column to 100
> mydataframe2 # change value in first column first row to 3
happiness age
1 100 24
2 100 44
3 100 20
> mydataframe2[1,1] <- 3
> mydataframe2
happiness age
1 3 24
2 100 44
3 100 20
R Console
> # column actions
> # changing the name for the first column
> names (mydataframe2) [1] <- "var_renamed"
> # create three new columns for the dataframe
> mydataframe2["newvar1"] <- 1
> mydataframe2["newvar2"] <- 0
> mydataframe2["newvar3"] <- 50
> mydataframe2
var_renamed age newvar1 newvar3 newvar2
1 3 24 1 50 0
2 100 44 1 50 0
3 100 20 1 50 0
> # remove the newvar2 variable
> mydataframe2["newvar2"] <- NULL
> mydataframe2
var_renamed age newvar1 newvar3
1 3 24 1 50
2 100 44 1 50
3 100 20 1 50
Square brackets

The easiest way to edit individual data items is to use the edit window. You can call up this window, assuming that you have a dataframe called *mydata,* by simply typing *data.entry(mydata)* or *fix(mydata)*. When you close the edit window by clicking on the top right hand corner X, any changes you have made are saved to the *mydata* dataframe. You can also use *mydata <- edit(mydata)*. You can save the changes to another copy with *myNewdata <- edit(myolddata)*, then use *attach(myNewdata)* to use it. To save the changes to your permanent storage you need to save the dataset (see the next section).

Action	Aim	R
Edit a row	Change all the values in row 2 to equal 44	mydataframe2[2,] <- 44
Edit a column	Change all the values in column 1 to equal 100	mydataframe2[,1] <- 100 or mydataframe$"happiness" <- 100
Edit a value	Change the value in row 1, column 1 to equal 3	mydataframe2[1,1] <- 3 or mydataframe2$"happiness"[1] <- 100

You can also edit individual columns (vectors) or rows by using various R Commands demonstrated below. The important thing to remember is that we refer to parts of the dataframe using square brackets with row and then the column references.

```
mydataframe[row, column]
```

You can also select a subset of rows and/or columns, to create a new dataframe.

Action	Aim	R
Renaming a column	Rename the first column to var_renamed	names (mydataframe2) [1] <- c("var_renamed")
Adding a column(s)	Add three new columns, each with a different value	mydataframe2["newvar1"] <- 1 mydataframe2["newvar2"] <- 0 mydataframe2["newvar3"] <- 50
Deleting a column	Delete the column called newvar2	mydataframe2["newvar2"] <- NULL

9.9 Creating a new dataframe from a subset of columns

Suppose you wanted to select only two columns called *age* and *newvar3* to form a new dataframe called *newdataframe_subset,* you could achieve this with the following R code:

```
newdataframe_subset<- mydataframe2[, c("age", "newvar3")]
```

use the concatenate function as we want data from more than one column

9.10 Creating a new dataframe from a subset of rows

You can also select specific rows based upon some criteria. Say we wanted to only select rows where the *var_renamed* vector has a value of 100:

use the double equals sign to mean 'is equal to'

```
newdataframe3 <- mydataframe2[newdataframe2$"var_renamed" ==100, ]
```

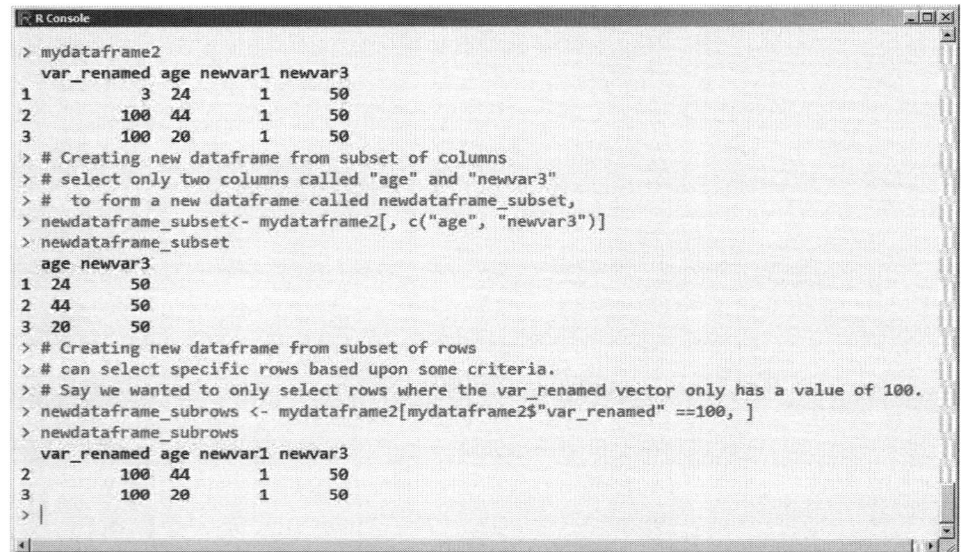
R Console
> mydataframe2
var_renamed age newvar1 newvar3
1 3 24 1 50
2 100 44 1 50
3 100 20 1 50
> # Creating new dataframe from subset of columns
> # select only two columns called "age" and "newvar3"
> # to form a new dataframe called newdataframe_subset,
> newdataframe_subset<- mydataframe2[, c("age", "newvar3")]
> newdataframe_subset
age newvar3
1 24 50
2 44 50
3 20 50
> # Creating new dataframe from subset of rows
> # can select specific rows based upon some criteria.
> # Say we wanted to only select rows where the var_renamed vector only has a value of 100.
> newdataframe_subrows <- mydataframe2[mydataframe2$"var_renamed" ==100,]
> newdataframe_subrows
var_renamed age newvar1 newvar3
2 100 44 1 50
3 100 20 1 50
>

Notice that in the screen shot opposite I have typed the name of the new dataframe at each stage to see what it looks like.

I have also added several comments lines, each beginning with a '#' so that I understand the code.

We can combine the row, column selection approach described above to select a subset of both rows and columns.

9.11 Loading data and giving it focus - *attach()*

R allows you to import many types of data file and R Commander provides comprehensive menu options to do this, see the chapters, *Importing your data into R and Cutting and pasting from Excel/Word.* Often I use R Commander just to load the data and then continue with the analysis using R code.

call the dataframe mydataframe
this function produces a dataframe
read the tab delimited text file
the place where the file is along with its name
each column of data starts with a name of the column

```
mydataframe <- read.delim("http://www.robin-
beaumont.co.uk/virtualclassroom/stats/basics/coursework/data/pain_medication.dat",
header=TRUE)
```

In contrast, if you have a local copy you need to replace the forward slashes '/' with double back slashes '\\' thus

```
mydataframe <- read.delim("D:\\folder\\folder\\pain_medication.dat", header=TRUE)
```

If you can't remember the place or name where you have the file you can annotate the above R code to:

```
mydataframe <-  read.table(file=file.choose())
```

allows you to select the file from the popup dialog box

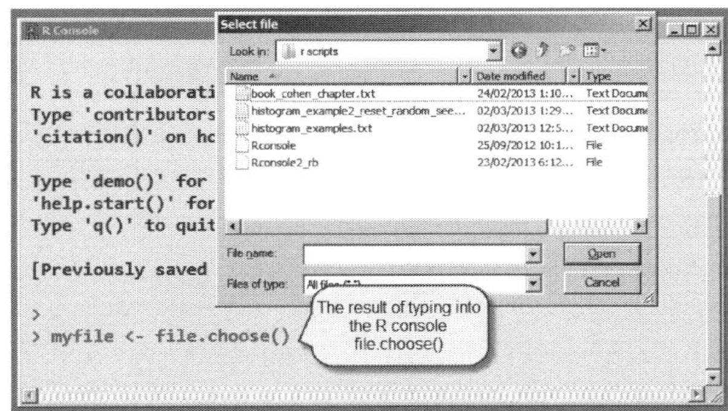

The *file.choose()* function calls up the usual windows *Select file* dialog box shown opposite.

Once you have imported the data and placed it in a dataframe, you can instruct R it has focus by using the *attach()* function. It is important to do this because R can have many dataframes loaded at once. For full details about the *attach()* function see the glossary entry. Finally it is always a good idea for you to get R to display the names of the columns in the dataframe using the *names(mydataframe)* function.

So in general we use a sequence of three commands when starting to work with a dataframe; retrieve data and place it in a dataframe, give it focus, check the names of the columns:

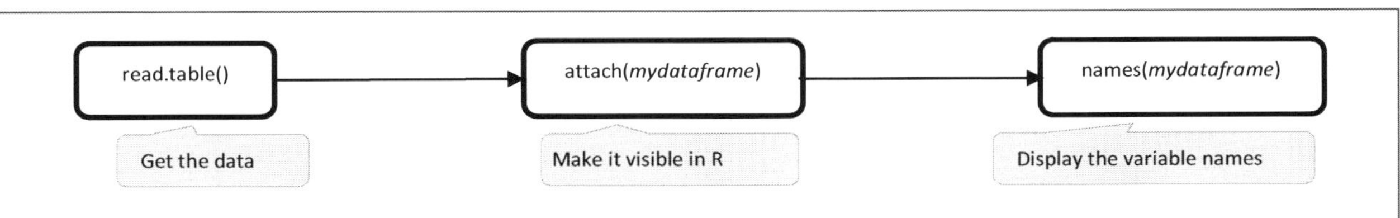

9.12 Seeing and Setting the Working Directory

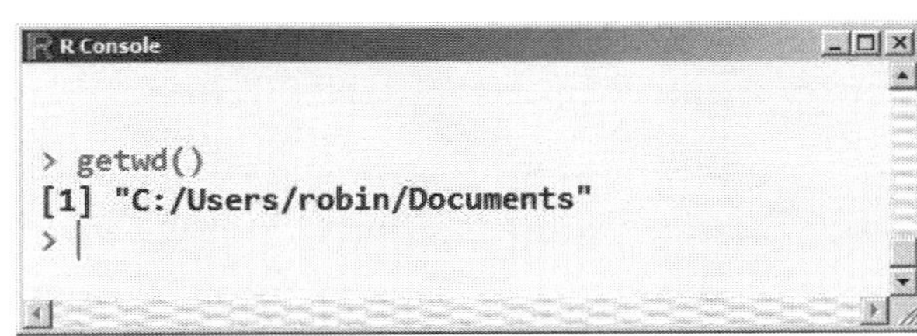

One of the most common questions is, if I save my work where is it being saved to by default? To find this out, type *getwd()* into the R Console window. *Getwd()* stands for get working directory. In the example opposite it can be seen that the default directory is on my C drive in a subdirectory called *Documents.*

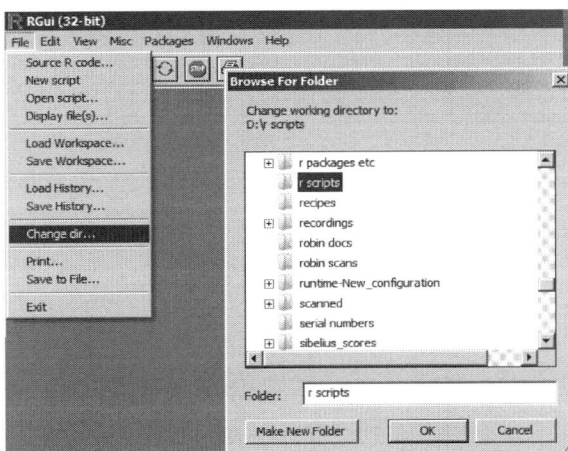

You can change the working directory by using the menu option **File-> Change dir...** in the RGui window. Unfortunately, this change only lasts as long as R is open. When you reopen R, the working directory reverts back to the default one. Instead of using the menu option typing the following into the R Console window, substituting *D:/r scripts* for your desired directory achieves the same result:

```
setwd("D:/r scripts")
```

To permanently change the working directory you would need to edit the *Rprofile.site* text file which is installed with R. The accompanying R code to this chapter, available from the books website explains how to do this.

9.13 Saving and loading your data in R format

You can save the dataframe in R format, as an R object, by using the *save()* function. In the code below you substitute *mydataframe* for the name of the dataframe you wish to save and *mydata.rda* for the path and filename you wish to give it:

```
save(mydataframe,file="mydata.rda")
```

R files need either a *rda* or *rdata* extension

To load the dataframe use the *load()* function:

```
load("mydata.rda")
```

You can also load an R object from the web by wrapping the url in the *url()* function:

```
load(url("http://www.robin-beaumont.co.uk/virtualclassroom/book2data/cummingp245.rdata"))
```

Unfortunately R format files cannot be used by other programs, so it is often necessary to convert your dataframe to some type of text file.

9.14 Saving your data as a text file

To write a tab delimited file, with the names of each column at the top:

```
write.table(mydataframe,"d:/basics/coursework/data/pain_medication.dat", header=TRUE,
sep = "\t")
```

To write a comma separated file with the names of each column at the top:

```
write.csv(mydataframe, "d:/basics/coursework/data/pain_medication.dat", header=TRUE)
```

Both these types of files can be imported into Excel. More details are provided in the *Saving and exporting your work and data* chapter.

9.15 Tips and Tricks

Remember you need to run the *attach(dataframe)* R Command to give the **dataframe** focus.

There are two excellent R Language reference cards that provide summaries of the most common R Commands. One by Jonathan Baron available from http://cran.r-project.org/doc/contrib/refcard.pdf and another by Tom Short available from http://cran.r-project.org/doc/contrib/Short-refcard.pdf I used both of these as a basis for several of the tables in this chapter and the R reference on the inside cover. A list of all the R contributed documentation, including that in at least twenty other languages can be found at: http://cran.r-project.org/other-docs.html.

Two ways of starting to learn the R language:

- When using R Commander look at the R code it generates when you select the various menu options.
- When you produce graphs using R Commander, annotate the R code produced by adding; titles, lines and text. This approach is demonstrated in many of the subsequent chapters.

10 Basic statistical techniques

The next few chapters are going to show you how to use R to perform the following basic statistical techniques.

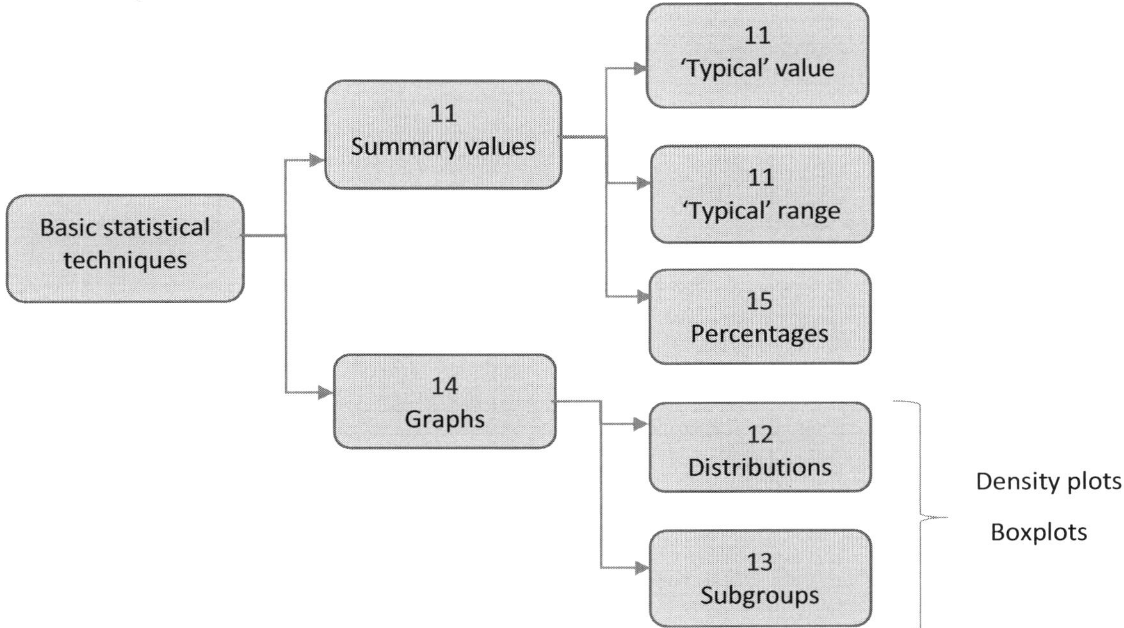

11 Summary statistics

A summary statistic is a value that summarises a dataset in some way either by giving some indication of what a typical value is, or some idea of how broad the range of data is. Essentially we are reducing the individual values down to one or two values technically called *data reduction*. The human brain can only handle a limited number of things at a time and summary statistics provide a way of helping us humans make sense of the world and making informed decisions. We should always be mindful that a particular summary measure may not be appropriate - see the section at the end of this chapter for further details. Many statistics books like MSME3, provide a chapter describing each of the various summary statistics.

The practical question is how do you obtain these values for a dataset yourself? The basis for this example is from MSME3 describing a group of patients who either made emergency appointments with a GP (General Practitioner = Family Doctor) or agreed to see the triage nurse, then at the end of the consultation rated their satisfaction with the consultation ranging from 0 to 5; zero being not satisfied at all to 5 being very satisfied. My sample consists of 65 patients in total.

11.1 Data preparation

The first stage is to download a sample dataset from my website. Go to the following link and save the file locally, making a note of where you saved it
http://www.robin-beaumont.co.uk/virtualclassroom/book2data/mwu_example1.rdata

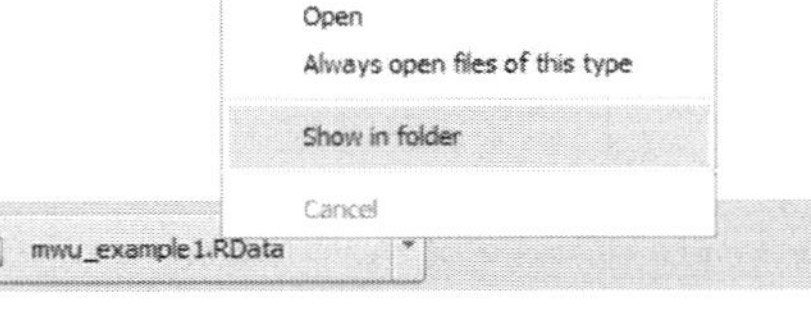

The next stage is to load R Commander and get the dataset into it. To load R Commander from within the *R Console* window type the following:

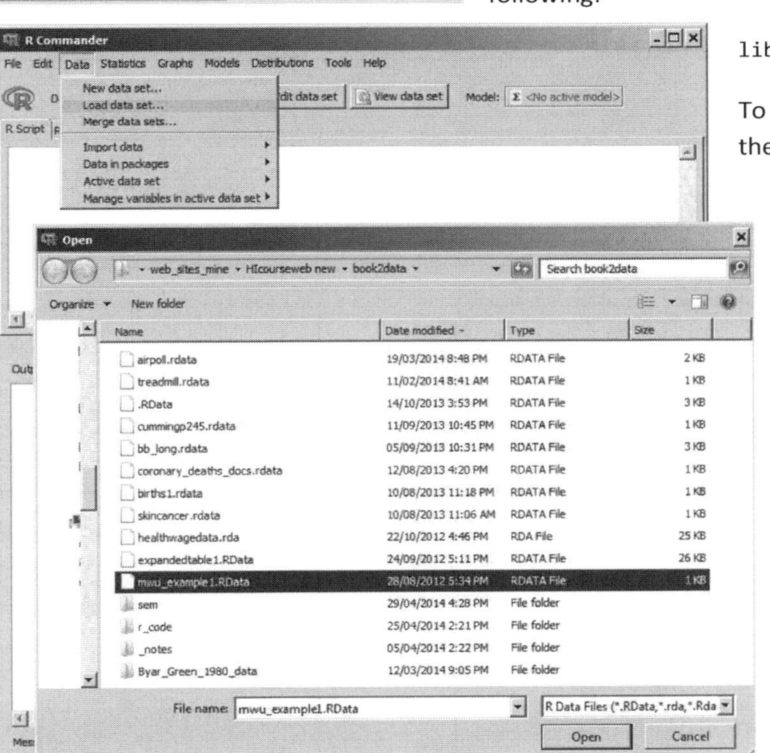


```
library(Rcmdr)
```

To load the dataset into R Commander select the following menu option:

Data->Load data set

Navigate to the folder where you saved the file and then click on the **open** button.

The *mwu_example1.rdata* file contains just a single object, a dataframe called *mydataframe*.

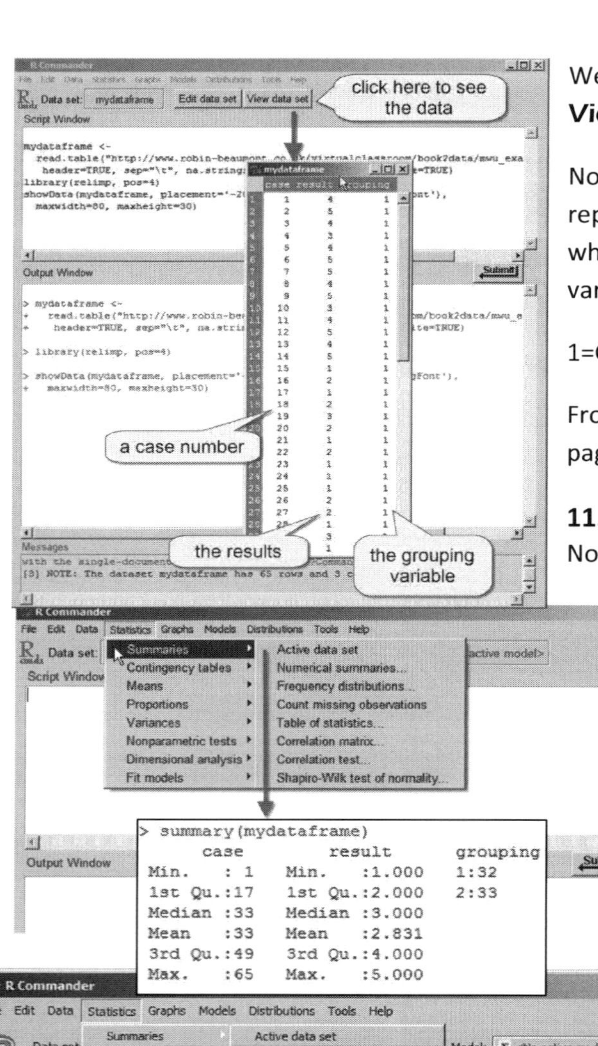

We can see the data within *mydataframe* anytime by clicking on the **View data set** button.

Note that I have all the observations in a single column called result representing patient satisfaction scores, with another one indicating which group they belong to, such a variable is often called a grouping variable where in this instance:

1=GP consultation and 2=Nurse Triage consultation.

From the bottom of the R Commander window (shown bottom of page) you can see that the dataset has 65 rows (i.e. cases).

11.2 R Commander: Obtaining summary statistics

Now we have the dataset within R. One of the first things you should always do is to get a feel for it by considering it on a variable by variable basis. This is easily achieved in R Commander by selecting the R Commander menu option:

Statistics->Summaries->Active data set

Selecting the first menu option, Active dataset produces a set of results for each of the columns in the dataset. Things to note:

Min = Minimum value in the column

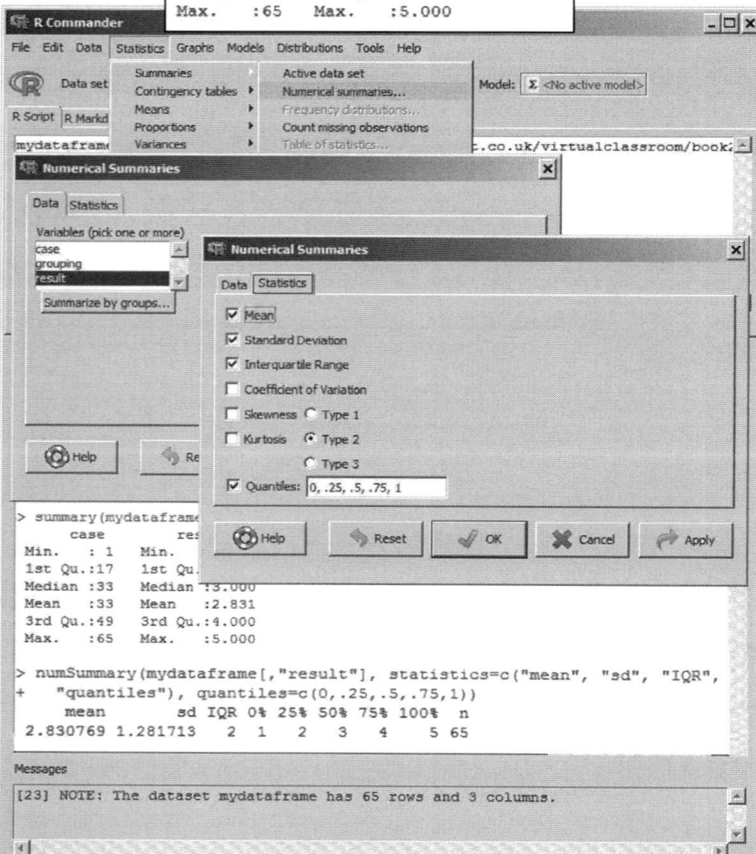

1st Qu = the value of the 1st quartile in the column. This is the value where 25% of the numbers are below it and 75% of the numbers are above it when the numbers are arranged in ascending (increasing) order.

Median = the median value in the column. This is the value where 50% of the numbers are below it and 50% of the numbers are above it when the numbers are arranged in ascending (increasing) order. If there is an even quantity of values in the column the mean value between the two middle ones is taken.

Mean = the mean value in the column. This is all the values in the column added together divided by the number of values in the column.

3rd Qu = the value of the 3rd quartile in the column. This is the value where 75% of the numbers are below it and 25% of the numbers are above it when the numbers are arranged in ascending (increasing) order.

Max = Maximum value in the column

The above set of results also includes the **Standard Deviation** (sd), details of which can be found under the glossary entry.

The *Numerical Summaries* tab also allows us to find the summary values for subgroups of values, for example we might want to see the values for both GP and Nurse Triage consultations.

Note that the results also tell us how many observations there are in each subgroup which is often a very useful and very simple measure. In this example we can see that there are nearly identical numbers in both the GP (n=32) and Nurse Triage consultations (n=33).

11.3 R code: Obtaining summary statistics

Instead of using the R Commander menu option **Statistics -> Summaries->Active data set** we could have obtained the same results by typing the following command into the R Console window directly:

```
summary(mydataframe)
```

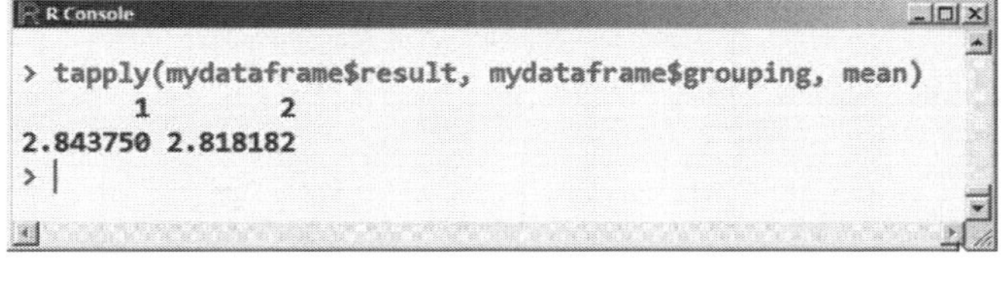
R Console
> tapply(mydataframe$result, mydataframe$grouping, mean)
1 2
2.843750 2.818182
>

To carry out the equivalent analysis described above we use the *tapply()* R function. To do this, you need the names of the two columns for which you want the summary value and the type of summary function you require. Here I'm asking for the mean value:

Common Summary Functions
mean
sum
range
quantile
var
sd
median
max
min
length
(useful if you want frequencies)

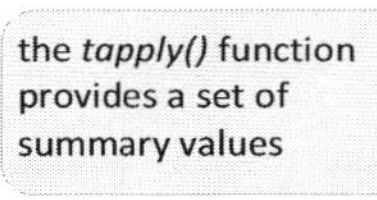
the *tapply()* function provides a set of summary values

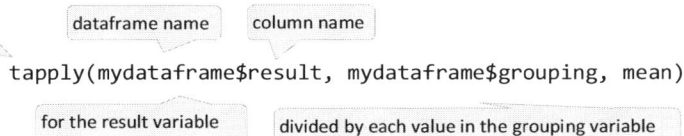

dataframe name

column name

```
tapply(mydataframe$result, mydataframe$grouping, mean)
```

for the result variable

divided by each value in the grouping variable

11.3.1 Getting more

Getting the above is all well and good but often you need more and the *describeBy()* function in the psych package provides that. Assuming you have installed the psych package (see the chapter, *Packages*) typing the following in the *R Console* window provides some additional summary statistics:

it is necessary to load the psych package first by using the *library()* function before you can use any of its functions

I have used the dataframe $ column name format in these commands

```
library(psych)

describeBy(mydataframe$result, mydataframe$grouping)
```

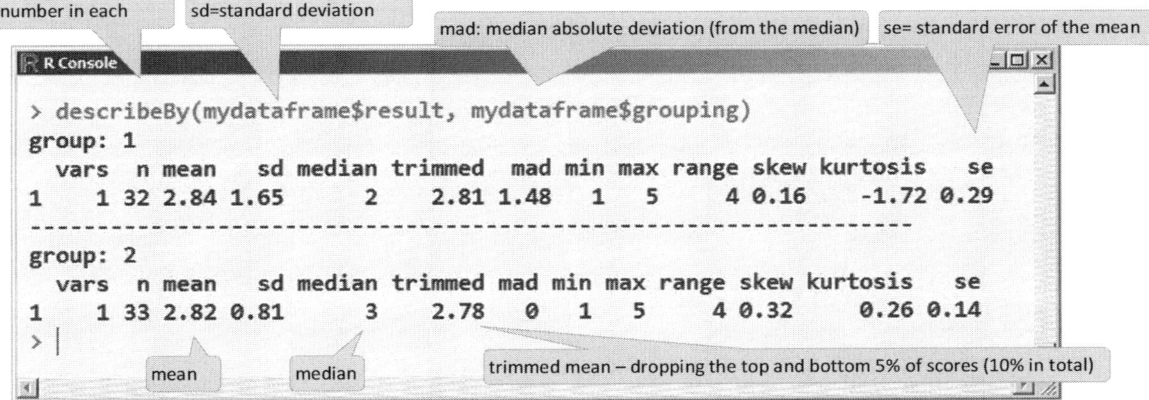
n=number in each
sd=standard deviation
mad: median absolute deviation (from the median)
se= standard error of the mean
R Console
> describeBy(mydataframe$result, mydataframe$grouping)
group: 1
vars n mean sd median trimmed mad min max range skew kurtosis se
1 1 32 2.84 1.65 2 2.81 1.48 1 5 4 0.16 -1.72 0.29
group: 2
vars n mean sd median trimmed mad min max range skew kurtosis se
1 1 33 2.82 0.81 3 2.78 0 1 5 4 0.32 0.26 0.14
>
mean
median
trimmed mean – dropping the top and bottom 5% of scores (10% in total)

11.3.2 Dividing data up by several variables

In contrast to R Commander you can also extend the *describBy()* function, adding more grouping variables. For example, taking the *pain_medication.dat* dataset, which is analysed in more detail in the next chapter, *Graphing distributions of single variables*. This dataset contains among others the following three variables; time which represents the time it took for a drug to take effect, treatment indicating which of two groups each result belongs to, and gender indicating if they were male or female. The dataset is available from:

http://www.robin-beaumont.co.uk/virtualclassroom/book2data/pain_medication.dat

You can load this file into R using the R Commander menu options described at the start of this chapter. Alternatively, you can type the following code into either the R Commander *Script* window (then highlighting and clicking on the **Submit** button), or the *R Console* window and pressing the return key.

```
mydataframe <- read.delim("http://www.robin-
                    beaumont.co.uk/virtualclassroom/book2data/pain_medication.dat", header=TRUE)

names(mydataframe) # displays the names for each column
```

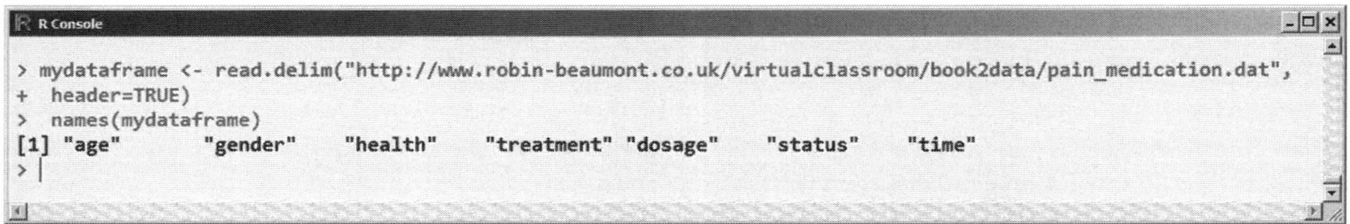
R Console
> mydataframe <- read.delim("http://www.robin-beaumont.co.uk/virtualclassroom/book2data/pain_medication.dat",
+ header=TRUE)
> names(mydataframe)
[1] "age" "gender" "health" "treatment" "dosage" "status" "time"
>

To run the *describeBy()* function requesting a set of summary statistics for time divided by both treatment and gender. Once again you type the following R code into either the R Commander *Script* window (then highlighting and clicking on the **Submit** button), or the *R Console* window and pressing the return key.

it is necessary to load the psych package first by using the *library()* function before you can use any of its functions

```
library(psych)

describeBy(mydataframe$time, list(mydataframe$treatment, mydataframe$gender))
```

it is essential to wrap the variables in the *list()* function for it to work

```
R Console
> describeBy(mydataframe$time, list(mydataframe$treatment, mydataframe$gender))
: Existing drug
: Female
  vars  n mean   sd median trimmed  mad min  max range skew kurtosis   se
1    1 51 4.61 2.75    3.9    4.39 3.11 0.8 11.3  10.5 0.54    -0.49 0.38
---------------------------------------------------------------------------------
: New drug
: Female
  vars  n mean   sd median trimmed  mad min  max range skew kurtosis   se
1    1 50 4.63 2.92   3.75    4.36 3.19 0.6 11.6    11 0.63    -0.69 0.41
---------------------------------------------------------------------------------
: Existing drug
: Male
  vars  n mean   sd median trimmed  mad min max range skew kurtosis   se
1    1 45 4.59 2.78    3.6    4.38 3.11 0.9  11  10.1  0.6    -0.76 0.41
---------------------------------------------------------------------------------
: New drug
: Male
  vars  n mean   sd median trimmed  mad min max range skew kurtosis   se
1    1 54  3.7 2.25    3.4    3.51 2.37 0.6 9.5   8.9 0.72    -0.28 0.31
> 
```

females taking existing drug

females taking new drug

males taking existing drug

males taking new drug

11.4 Tips and hints

Reflecting on the summary values: Is my summary statistic appropriate?

Eyeballing the minimum and maximum values for each variable / group is a good quick way of checking that you have no impossible or unlikely values as you might have been entering the data at 2 am!

Just because you have a summary value for a variable does not imply it is a valid measure. For example the various summary statistics for the case variable, where the numbers simply represent an identifier for each case, are pretty meaningless, case 4 is not four times as important as case 1! Similarly, the average case is not necessarily case number 33, nor do the measures of spread (1st and 3rd quartiles) mean anything here. This is because the case variable has a nominal **level of measurement**. In contrast the maximum number of 65 is useful because, as the cases are numbered sequentially, this value represents the number of cases in our sample. The validity of the Standard Deviation is even more dependent upon the level of measurement and the distribution of the data. For further information, see the glossary entry **Standard Deviation** and the additional online chapters:

Measures of the typical value:
http://www.robin-beaumont.co.uk/virtualclassroom/stats/basics/part2.pdf

Measures of spread:
http://www.robin-beaumont.co.uk/virtualclassroom/stats/basics/part4.pdf

Quick and dirty 95% confidence intervals for the mean

If you want a quick 95% **confidence interval** for the mean value just take the standard error (sem) in the above output from the *psych* package and then multiply it by 1.96 both adding and subtracting this value from your mean and you have the 95% confidence interval.

Take as an example the satisfaction scores for the GP consultations where we have a mean of 2.84 with a sem of .291 so we to calculate; 2.84 - (1.96 x .291) to 2.84 (1.96 x .291).

In R commands you need to use the * to represent the multiplication operator, so we type into the *R Console* window

```
2.84 - ( 1.96  *  .291)

2.84  + ( 1.96 *  .291)
```

Giving the 95% confidence interval of the mean for the GP satisfaction scores to be 2.269 to 3.410.

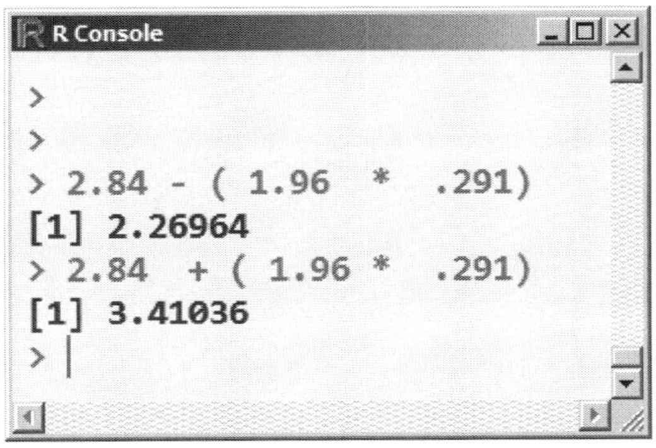

R Console

```
>
>
> 2.84 - ( 1.96   *   .291)
[1] 2.26964
> 2.84  + ( 1.96 *   .291)
[1] 3.41036
> |
```

12 Graphing distributions of single variables: histograms and density plots

A distribution graph uses the information about how often each value occurs in a dataset (its frequency) as a visual summary. Looking at a distribution graph allows you to see immediately the range, most common value in the dataset and the shape the frequencies make when considered over the entire range of the values in the dataset.

Words in **bold** indicate entries in the glossary providing more information.

I'm assuming in this example that your variable possesses ordinal, interval or ratio **level of measurement**. For nominal type variables you would produce bar charts instead.

One of the most important things to do when you have collected your data is to get a feel for the distribution of values for each variable. I find the easiest way to do this is to produce distribution graphs. There are two main types, which we discuss below:

- Histograms - use rectangles showing frequency of values occurring within a particular range.
- Density plots - can be thought of as smoothed histograms. Technically they represent the probability of a particular value occurring because the total area under the graph is equal to one. Most statisticians now consider density plots superior to histograms because they take into account the vagaries of sampling.

12.1 R Commander: Histograms

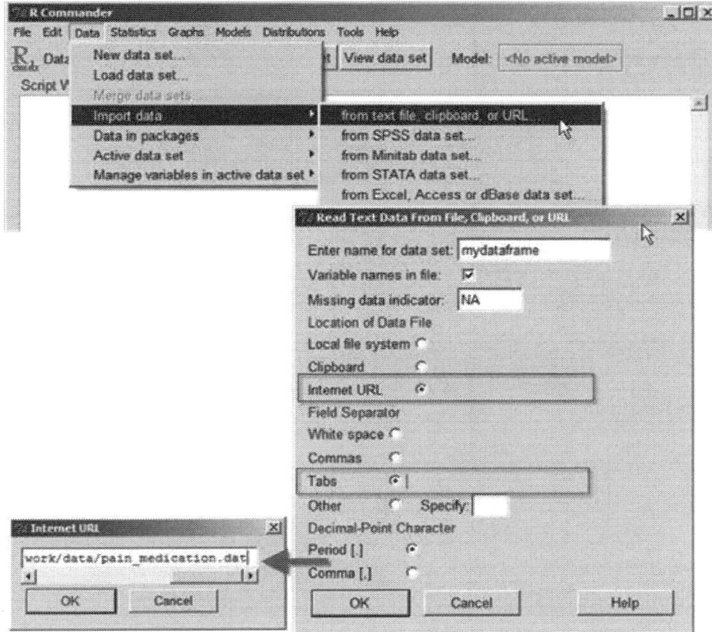

From within R you need to load R Commander by typing in the following command:

```
library(Rcmdr)
```

First of all you need some data and for this example I'll use a sample dataset, and load it directly from my website. You can do this by selecting the R Commander menu option:

Data->Import data->
from text, the clipboard or URL

Then I have given the resultant *dataframe* the name *mydataframe*, also indicating that it is from a URL (i.e. the web) and the columns are separated by *tab* characters.

Clicking on the **OK** button brings up the *Internet URL* box, in which you need to type the following to obtain my sample data:

```
http://www.robin-beaumont.co.uk/virtualclassroom/stats/basics/coursework/data/pain_medication.dat
```

A pharmaceutical company collected this hypothetical dataset. The company were developing an anti-inflammatory medication for treating chronic arthritic pain and comparing the time it took to have an effect compared to an older drug at both low and high dosage levels. The shorter the time it took to take effect the more desirable the drug was considered to be. Some patients failed to respond over the time period of the experiment and this was indicated by the status variable.

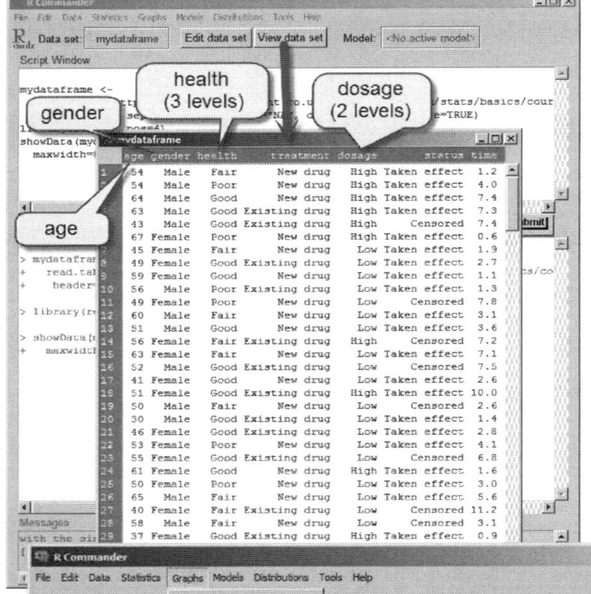

The details of the 7 variables from the 200 patients is given below:

- age — age in years
- gender — 0=male; 1=female
- health — 1=poor; 2=fair; 3=good
- treatment — 0=new drug; 1= existing drug
- dosage — 0=low; 1=high
- status — 0 = censored; 1= drug taken effect
- time — time to take effect

Say we wanted to see the distribution of ages in our dataset by creating a histogram, within R Commander you have three options – *Frequency*, *Percentages* and *Densities*. All three outputs are shown below, but usually you would only show one of these in a report.

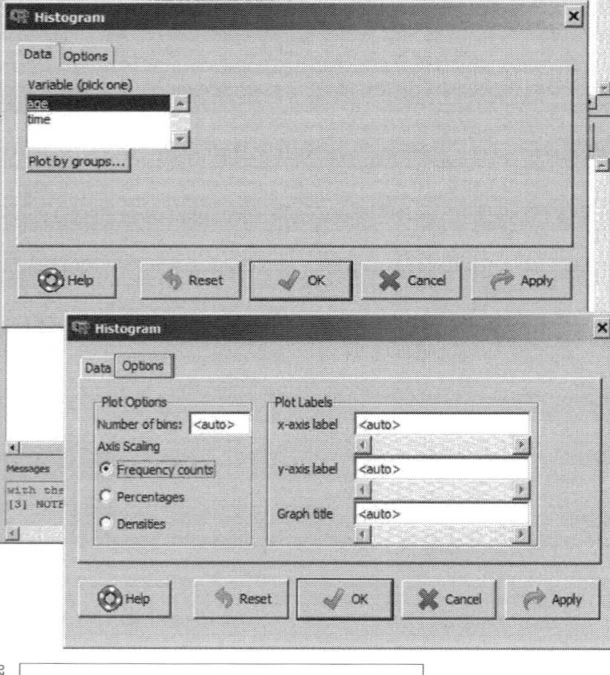

Select the menu option:

Graphs -> Histogram

Complete the dialog box as shown opposite. The **Number of bins** equal the number of columns, usually there is no reason to change this from the *<auto>* option.

The results of selecting the various **Axis scaling** options are shown below.

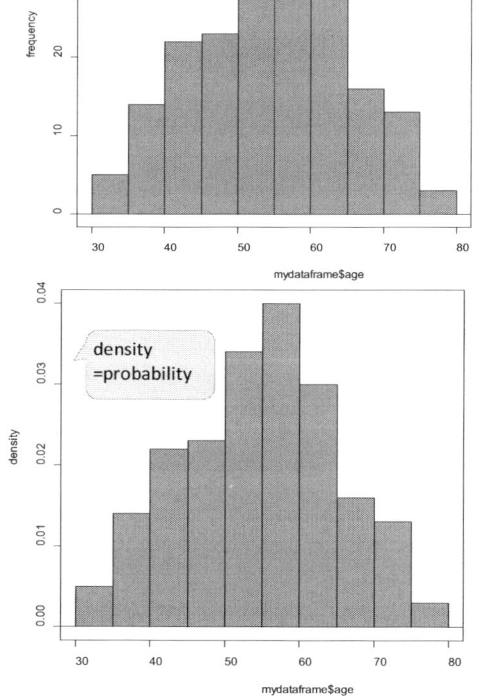

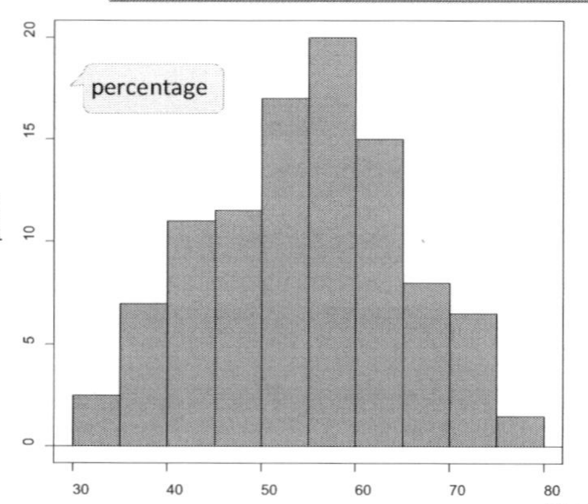

The name on the x axis is in the *dataframe dollar column name format* i.e. *mydataframe$age* the mydataframe dataframe and the age variable.

12.2 R Commander: Density Plots

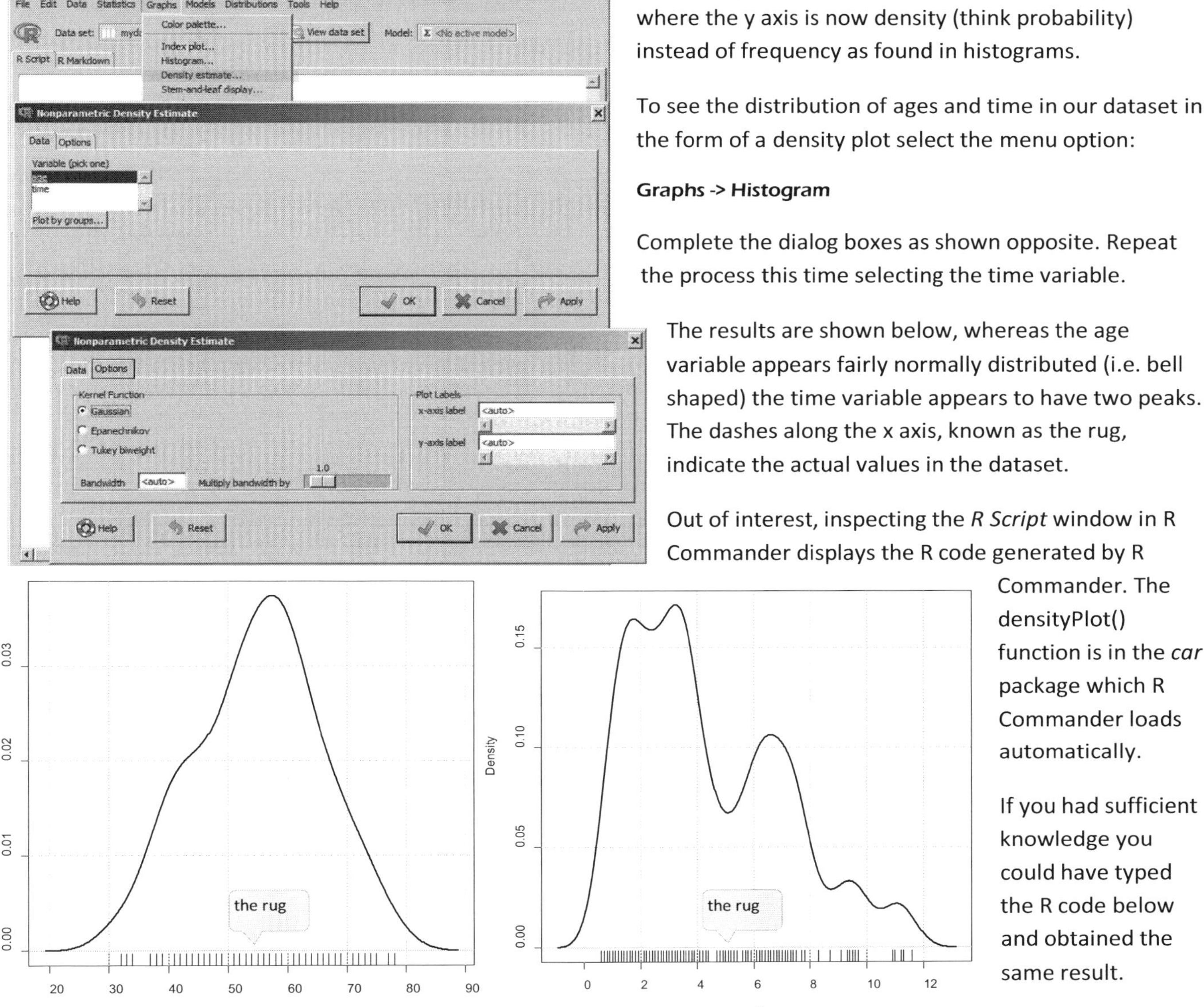

A **density plot** is a smoothed version of a histogram, where the y axis is now density (think probability) instead of frequency as found in histograms.

To see the distribution of ages and time in our dataset in the form of a density plot select the menu option:

Graphs -> Histogram

Complete the dialog boxes as shown opposite. Repeat the process this time selecting the time variable.

The results are shown below, whereas the age variable appears fairly normally distributed (i.e. bell shaped) the time variable appears to have two peaks. The dashes along the x axis, known as the rug, indicate the actual values in the dataset.

Out of interest, inspecting the *R Script* window in R Commander displays the R code generated by R Commander. The densityPlot() function is in the *car* package which R Commander loads automatically.

If you had sufficient knowledge you could have typed the R code below and obtained the same result.

```
densityPlot( ~ time, data=mydataframe, bw="SJ", adjust=1, kernel="gaussian")

densityPlot( ~ age, data=mydataframe, bw="SJ", adjust=1, kernel="gaussian")
```

the technique used to draw the density plot

the variable to plot

the name of the dataframe to provide the data

bandwidth value, "SJ" is the default way of calculating it

bandwidth adjustment value, 1 is the default

N.B. Bandwidth in density plots is analogous to bins in histograms.

12.3 Tips and Tricks

Often researchers want to know if the values for a particular variable are normally distributed (i.e. bell shaped) and the density plot provides a valid way of assessing this. Besides the normal distribution, there are several other shapes which are useful to know about, for details see: http://www.robin-beaumont.co.uk/virtualclassroom/stats/basics/part3.pdf.

13 Histograms and density plots for subgroups defined by factor levels

We will use the dataset described in the previous chapter, *Graphing Distributions of single variables - histograms and density plots,* to demonstrate the use of density plots and histograms for subgroups.

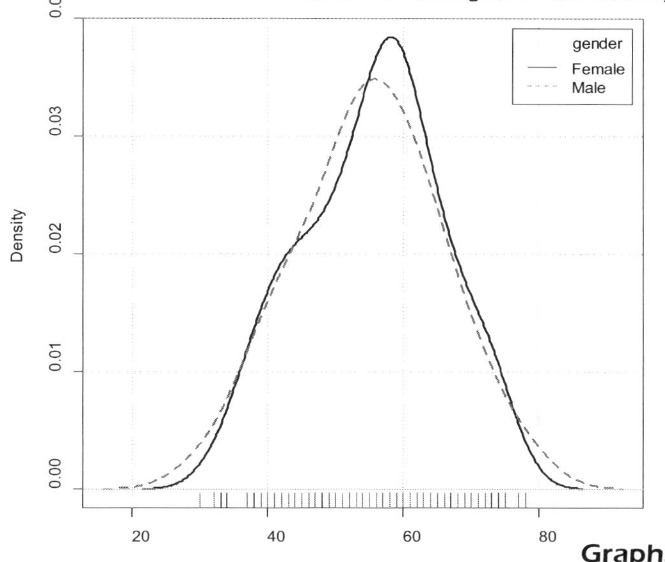

The easiest way to produce these graphs is to use the most recent version of R Commander. The example opposite shows the distribution of ages for the male and female subgroups. Here we have two subgroups and consequently two curves. With more subgroups there would be a matching number of curves. In a dataset a subgroup is modelled as a **Factor** level.

Words in **bold in** the text indicate glossary entries.

13.1 R Commander: Density Plots for subgroups

The Density plot opposite was produced in R Commander using the menu option:

Graphs->Density estimate . . .

Then complete the *Nonparametric Density Estimate* dialog box as described below.

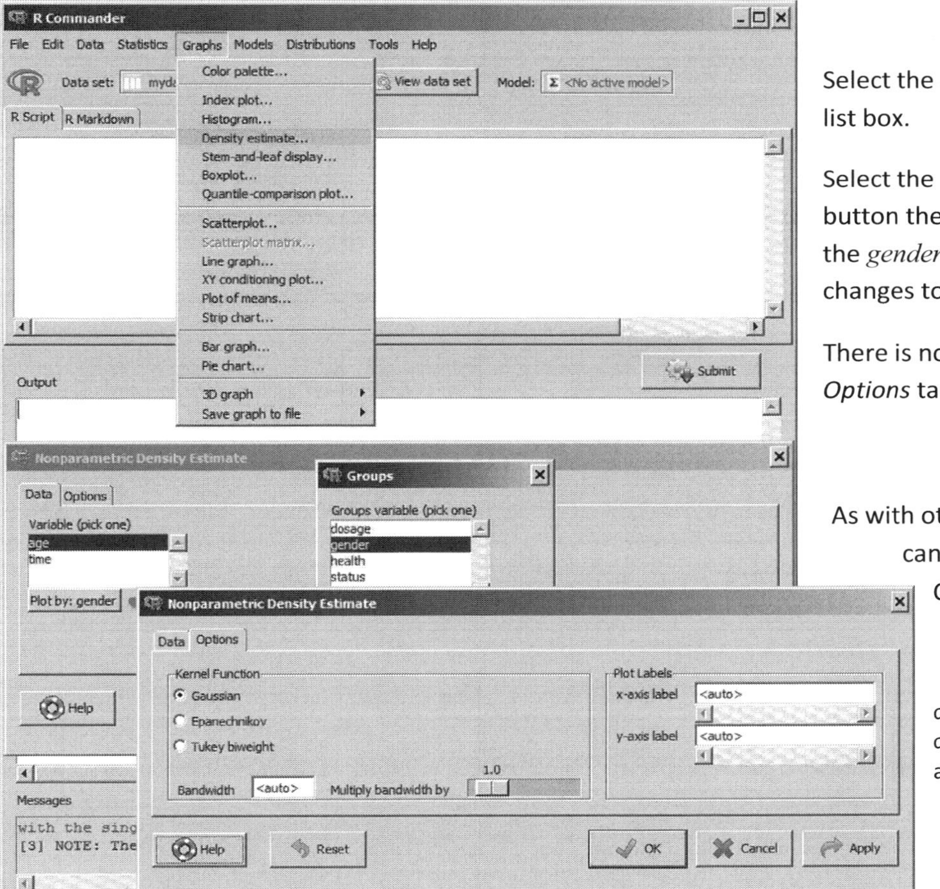

Select the *age* variable from the **Variable** list box.

Select the **Plot by groups . . .** Plot by groups... button then in the *Groups* dialog box select the *gender* variable. The button then changes to *Plot by: gender*.

There is no need to change anything in the *Options* tab.

As with other R Commander options you can see the R code generated by R Commander in the *output* window:

plot age by gender

```
densityPlot(age~gender,
data=mydataframe, bw="SJ",
adjust=1, kernel="gaussian")
```

13.2 R Commander: Histograms for subgroups

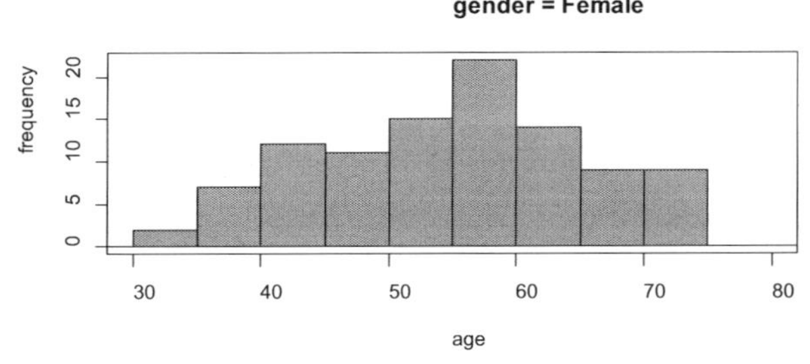

The histograms opposite were produced in R Commander using the menu option:

Graphs->Histogram . . .

Then complete the *Histogram* dialog box as described below.

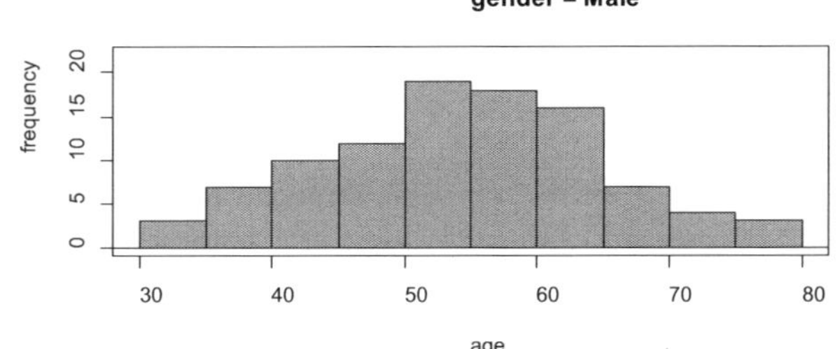

Select the *age* variable from the **Variable** list box.

Select the **Plot by groups . . .** button [Plot by groups...] then in the *Groups* dialog box select the *gender* variable. The button then changes to *Plot by: gender*.

There is no need to change anything in the *Options* tab.

As with other R Commander options you can see the R code generated by R Commander:

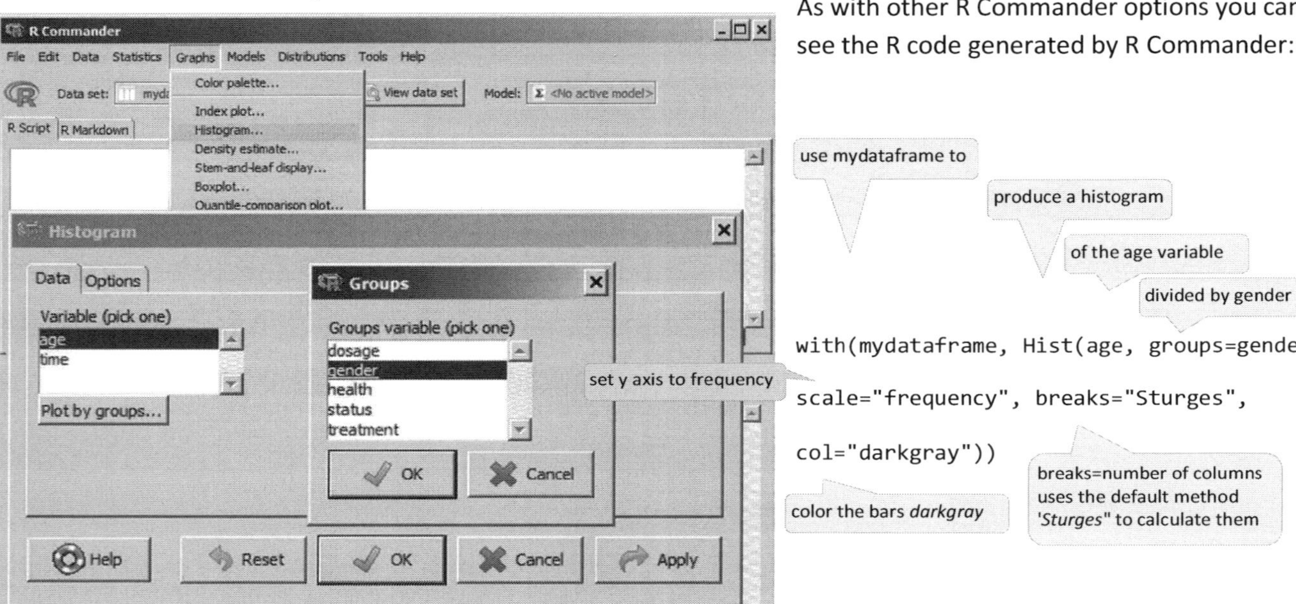


```
with(mydataframe, Hist(age, groups=gender,
scale="frequency", breaks="Sturges",
col="darkgray"))
```

13.3 R Commander: Dividing data up by more than one factor

In both the above examples we are dividing up the data by a single variable, by *gender*. Unfortunately It is not possible to specify more than one in either R Commander or using the *DensityPlot()* or *Hist()* functions. Suppose we want to see the distribution of *ages* for each *treatment group* divided by *gender*. To achieve this we must first create two new dataframes each dividing the original dataset up by *gender* level. To achieve this select the R Commander menu option:

Data->Active data set->Subset active data set. . .

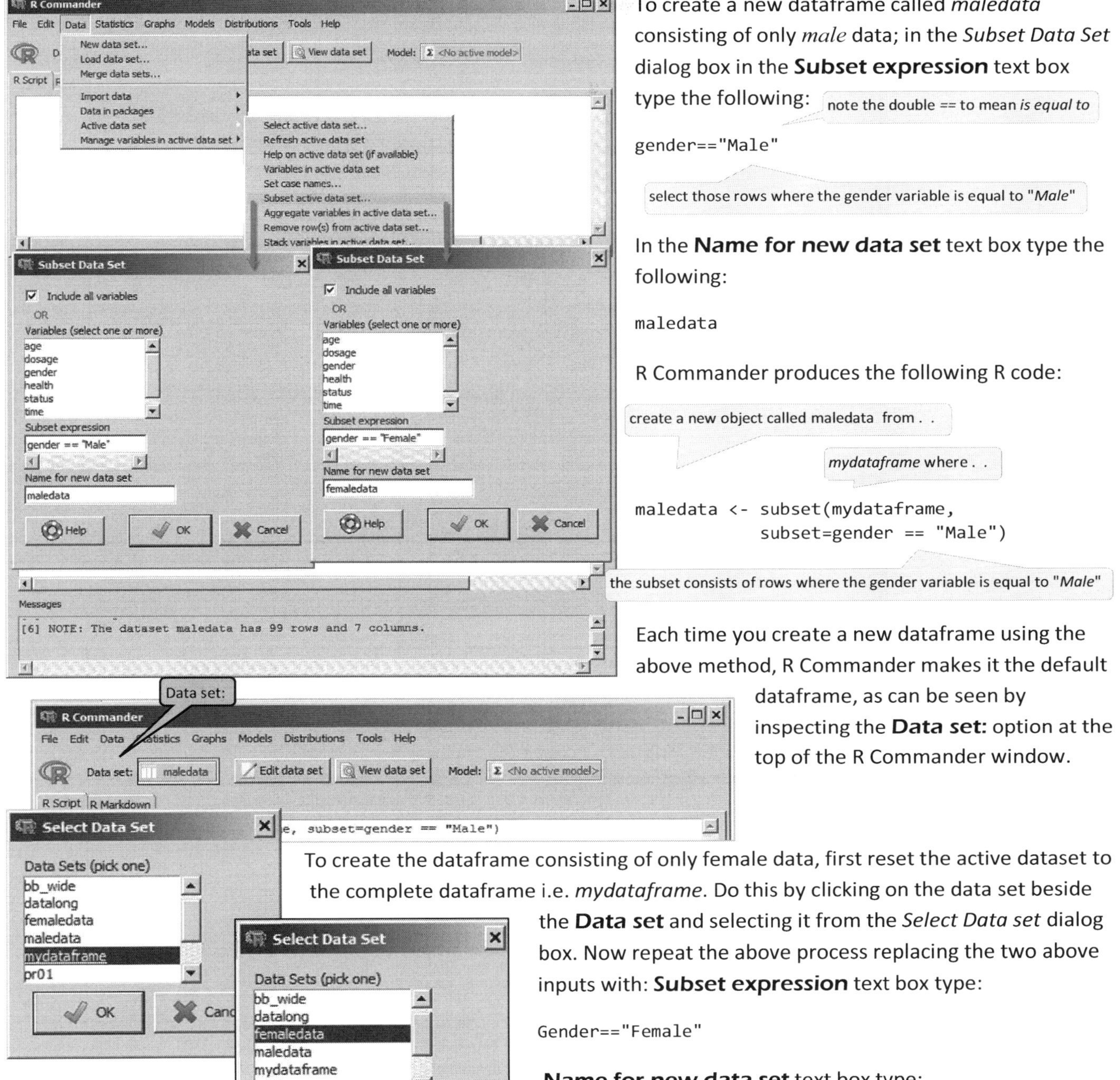

To create a new dataframe called *maledata* consisting of only *male* data; in the *Subset Data Set* dialog box in the **Subset expression** text box type the following: note the double == to mean *is equal to*

```
gender=="Male"
```

select those rows where the gender variable is equal to "*Male*"

In the **Name for new data set** text box type the following:

```
maledata
```

R Commander produces the following R code:

create a new object called maledata from . .

mydataframe where . .

```
maledata <- subset(mydataframe,
          subset=gender == "Male")
```

the subset consists of rows where the gender variable is equal to "*Male*"

Each time you create a new dataframe using the above method, R Commander makes it the default dataframe, as can be seen by inspecting the **Data set:** option at the top of the R Commander window.

To create the dataframe consisting of only female data, first reset the active dataset to the complete dataframe i.e. *mydataframe*. Do this by clicking on the data set beside the **Data set** and selecting it from the *Select Data set* dialog box. Now repeat the above process replacing the two above inputs with: **Subset expression** text box type:

```
Gender=="Female"
```

Name for new data set text box type:

```
femaledata
```

To produce the density plots shown on the following page we select each required dataset (*maledata*, then *femaledata)* in turn and then follow the previous steps in the *Density plots for subgroups* section. The setup for the dialog box is shown opposite.

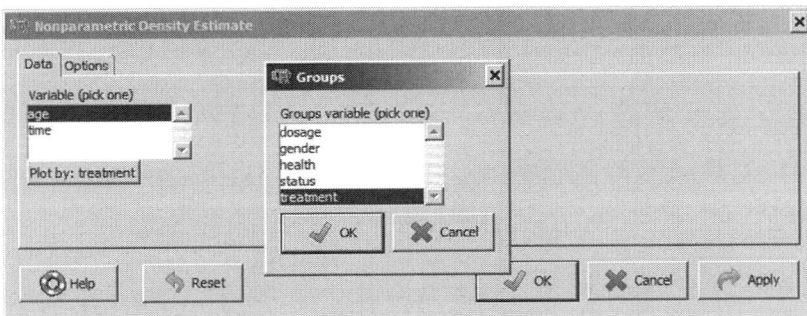

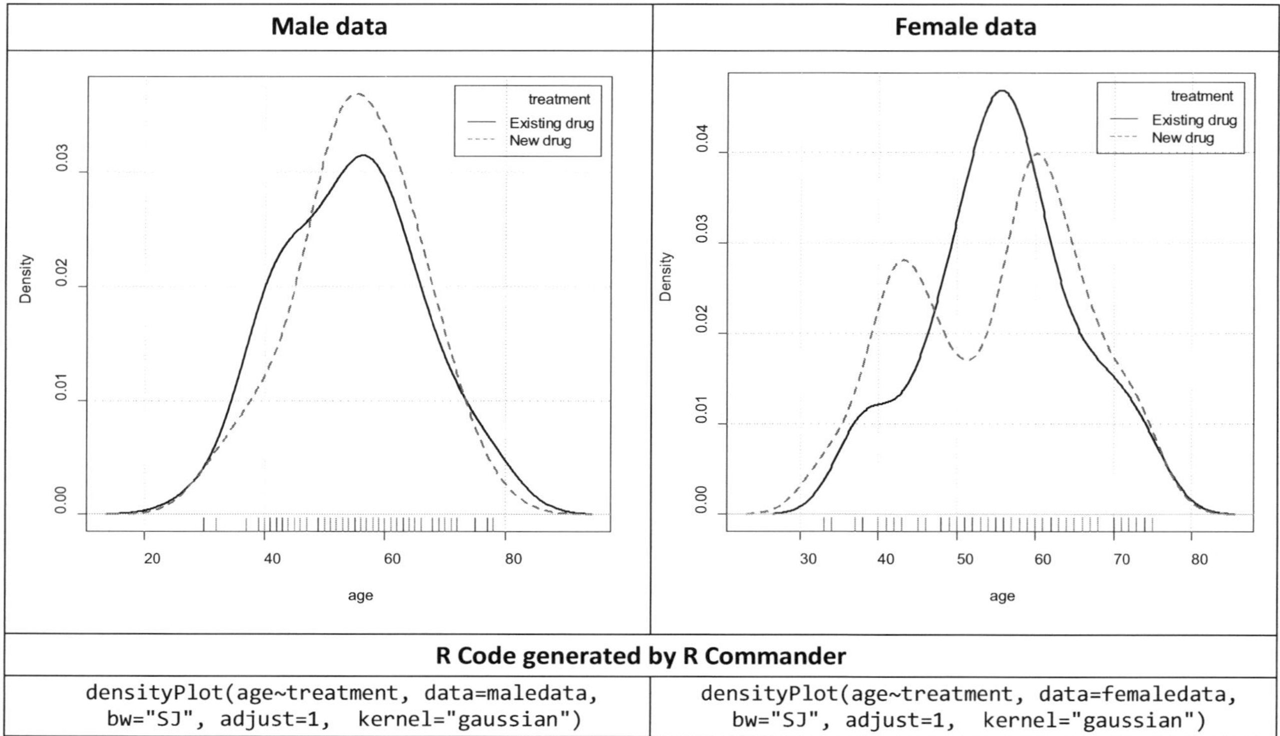
treatment
Existing drug
New drug
Density
0.00
0.01
0.02
0.03
0.04
20
30
40
50
60
70
80
age

Male data	Female data
R Code generated by R Commander	
`densityPlot(age~treatment, data=maledata, bw="SJ", adjust=1, kernel="gaussian")`	`densityPlot(age~treatment, data=femaledata, bw="SJ", adjust=1, kernel="gaussian")`

From the above it appears that the age distributions of those in both treatment groups for males and females appear roughly equivalent.

As an alternative to the above we could have divided the data into those who were allocated to the existing and new drug groups and then compared the distribution of ages for each sex within each. The decision as to which is the best way to divide up the data is the responsibility of the individual researcher dictated by their particular interests.

13.4 R code: Creating subsets

Instead of using the above R Commander dialog boxes to create the subsets of data we could have used R code typing it in the *R Console* or R Commander *R Script* windows. You can either use the *subset()*, function, as used by R Commander above, or select a subset of rows from the dataframe directly.

First copy only the male cases into a dataframe and call it *maledata*:

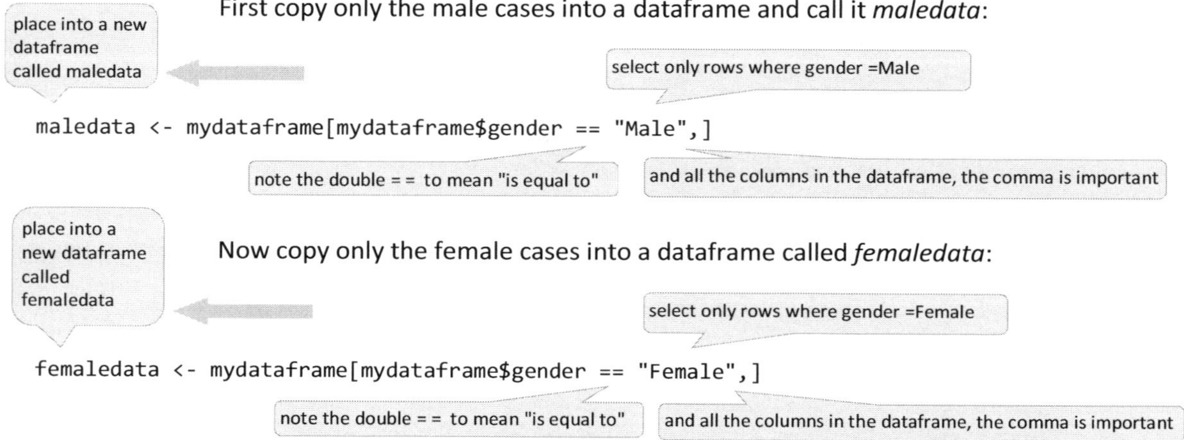
place into a new dataframe called maledata
select only rows where gender =Male
note the double = = to mean "is equal to"
and all the columns in the dataframe, the comma is important
place into a new dataframe called femaledata
select only rows where gender =Female
note the double = = to mean "is equal to"
and all the columns in the dataframe, the comma is important

```
maledata <- mydataframe[mydataframe$gender == "Male",]
```

Now copy only the female cases into a dataframe called *femaledata*:

```
femaledata <- mydataframe[mydataframe$gender == "Female",]
```

13.5 R code: Plotting subsets

Once you have created the above subsets you can then simply use the code that R Commander produced previously to produce the plots:

```
densityPlot(age~ treatment, data=maledata, bw="SJ", adjust=1, kernel="gaussian")

densityPlot(age~ treatment, data=femaledata, bw="SJ", adjust=1, kernel="gaussian")
```

plot() begin by plotting the results for males using existing drug

This produces two separate density plots, equivalent to those on the previous page. If you wish to superimpose all four distributions on a single density plot you need to use the *plot()* and *lines()* functions instead. The *plot()* function sets up the graph specifying its x and y axis limits, the title and the first curve. Then each call to the *lines()* function adds another line to it:

color=red lty=line type =1 (solid) see glossary entry for details

```
plot (density(maledata$age[maledata$treatment == "Existing drug"], data=maledata),
col="red", lty = 1, ylim=c(0,0.05),xlim=c(0,100),
```

set the y and x-axis limits – if you fail to do this you might not see the results. This often requires trial and error

'lty' means line type. lty=1 is a solid line. Other options are described in the glossary entry lty

add a title

```
main="Results for each treatment group according to sex & age")
```

lines() add a line showing results for males using the new drug

```
lines (density(maledata$age[maledata$treatment == "New drug"],
data=maledata), col="green", lty = 2)
```

color=green line type =2 (dashed)

a comment (#) in the code, to help me understand it

```
# now for the female data
```

lines() add a line showing results for females using the existing drug

```
lines (density(femaledata$age[femaledata$treatment == "Existing drug"],
data= femaledata), lty = 1, col="blue")  # females existing drug
```

color=blue line type =1 (solid)

lines() add a line showing results for females using the new drug

```
lines (density(femaledata$age[femaledata$treatment == "New drug"],
data= femaledata), col="black", lty = 2) # females new drug
```

color=black line type =2 (dashed)

You can either type the above code into the *R Console* window or, as shown opposite, type it into the *R Script* window of R Commander, highlight it, then click on the **Submit** button. The result will appear in the *RGui* window.

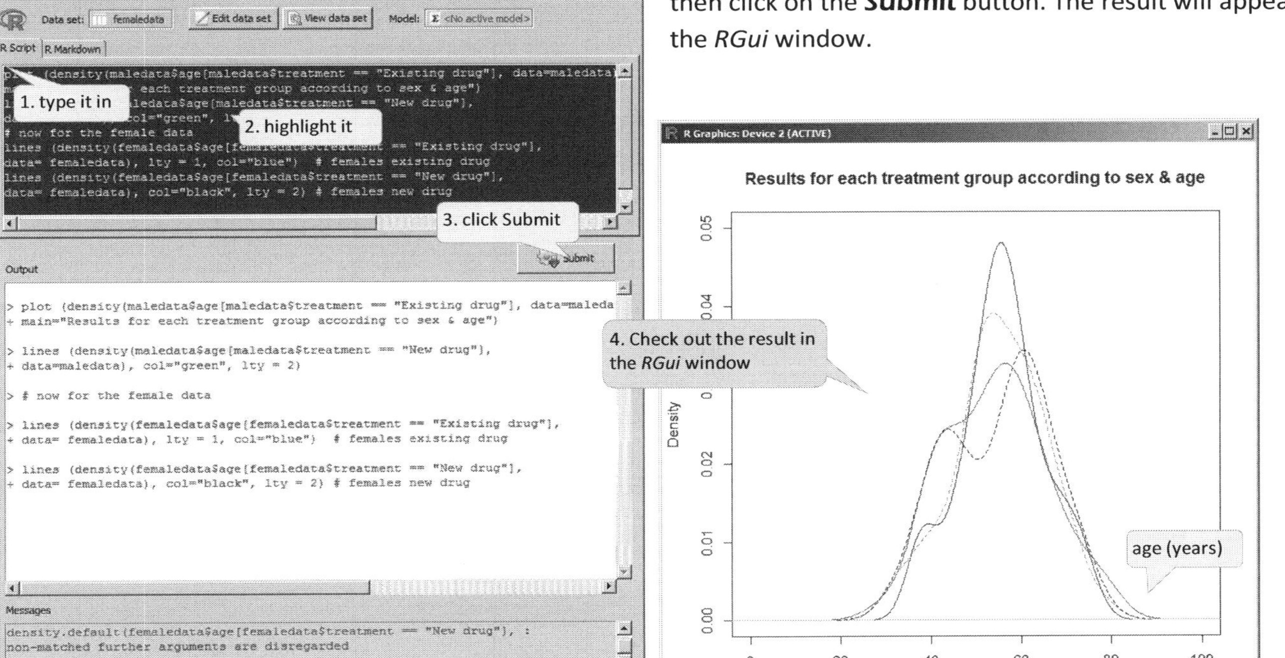

The above graph can be improved by adding a legend using the *legend()* function:

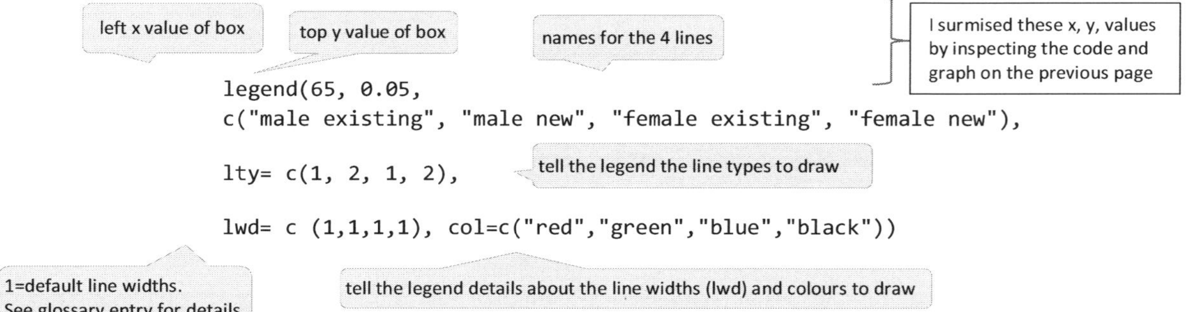


```
legend(65, 0.05,
c("male existing", "male new", "female existing", "female new"),
lty= c(1, 2, 1, 2),
lwd= c (1,1,1,1), col=c("red","green","blue","black"))
```

Results for each treatment group according to sex & age

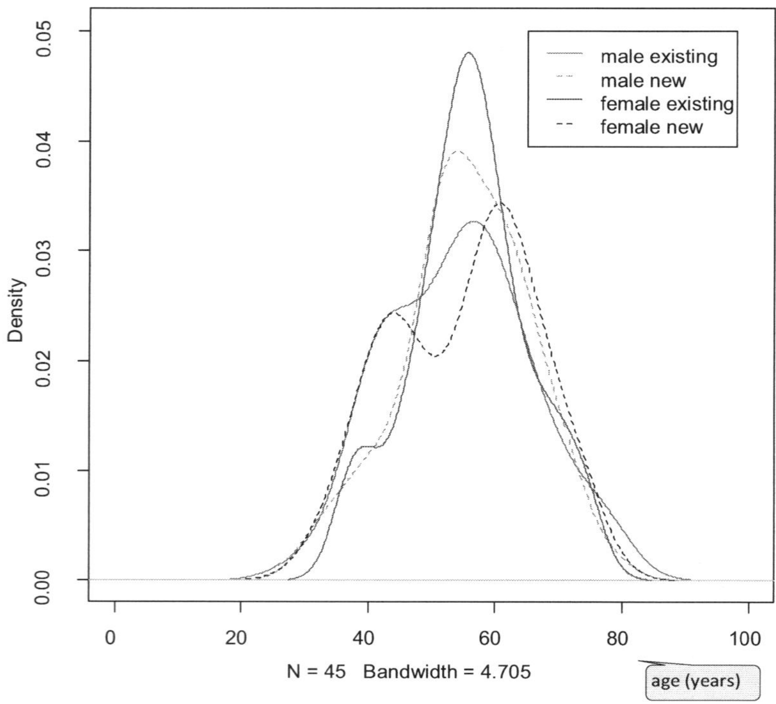

This graph demonstrates nicely how little the age distribution differs when taking into account both gender and treatment allocation. We can say that the groups appear homogenous when taking into account these two factors.

13.6 Tips and Tricks

In this chapter I have concentrated on density plots rather than histograms for subgroups and I would recommend this as a general strategy.

Producing complex density plots requires skill and careful planning and it is often a good idea to sketch out on paper what you want to end up with at the start.

Other chapters in this book offer alternative strategies to producing complex graphs; *Publication quality graphics* discusses the use of the *ggplot2* package along with online alternatives, while the *Repeated measures* chapter introduces the *lattice* package which is another way of obtaining graphs for subsets of data.

14 Boxplots

A highlighted word indicates a glossary entry

Drawing **boxplots** is an essential part of investigating variables that possess either an, ordinal, interval or ratio **level of measurement**.

14.1 R Commander: Creating Boxplots

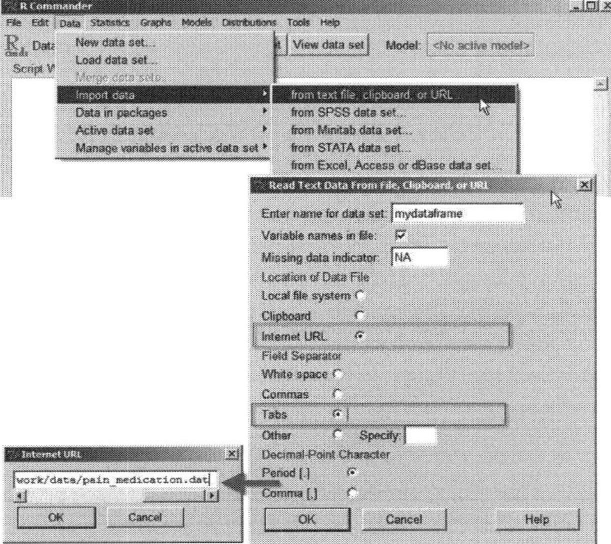

From within R you need to load R Commander by typing in the following command:

```
library(Rcmdr)
```

For this example we will use a sample dataset from my website. You can do this by selecting the R Commander menu option:

Data-> Import Data->from text, the clipboard or URL

I have given the resultant dataframe the name *mydataframe*. I have indicated that it is from a URL (i.e. the web) and the columns are separated by *tab* characters.

Clicking on the **OK** button brings up the *Internet URL* box, type the following to obtain my sample data:

http://www.robin-beaumont.co.uk/virtualclassroom/book2data/pain_medication.dat

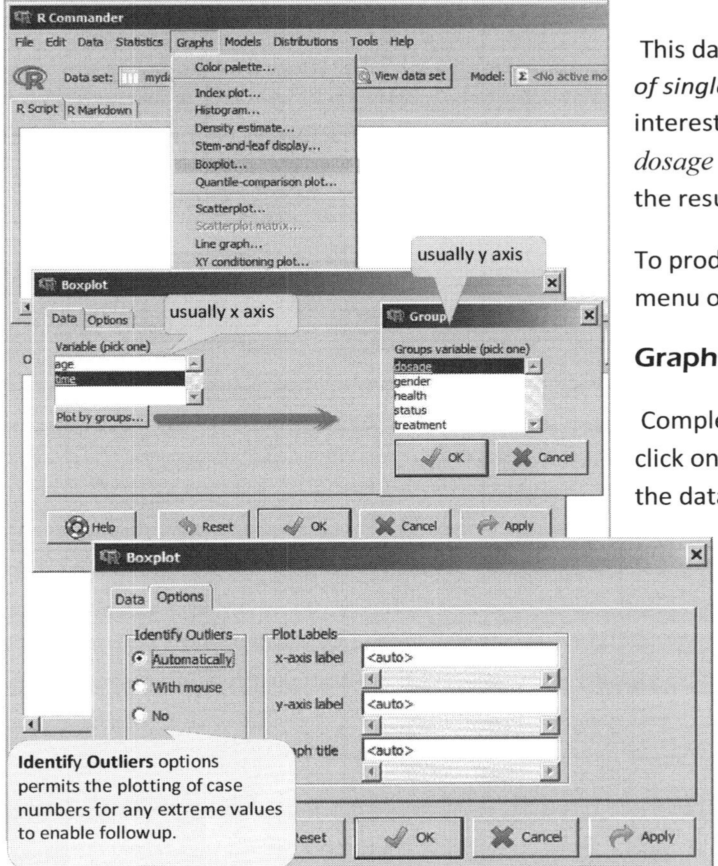

This dataset, described in the chapter *Graphing Distributions of single variables*, has 7 variables of which we are only interested in two here; *time* (the outcome variable) and *dosage* which is a grouping variable indicating which group the result *time* belongs to.

To produce the necessary boxplots select the R Commander menu option:

Graphs->Boxplot

Complete the *Boxplot* dialog box as shown opposite, then click on the **plot by groups** button to specify how you want the data to be divided up. Select the *dosage* variable and then click on the **OK** button to close the window.

Clicking on the **Options** tab allows you to add a title along with x and y axis labels by completing the various text boxes. Click on the **OK** button to close the window.

The **Identify Outliers** options are discussed at the end of this chapter. At present use the default value.

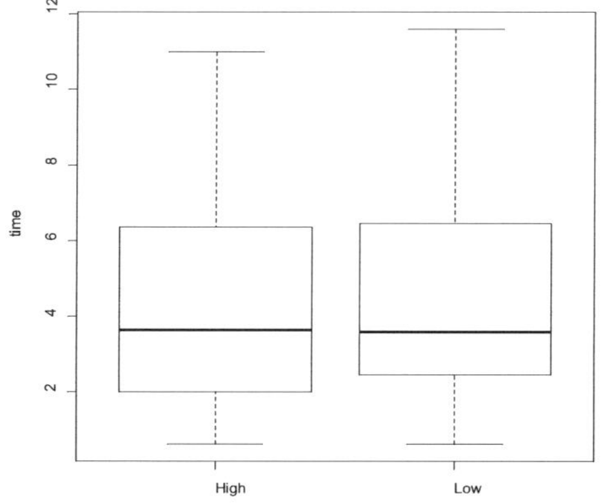

From the boxplots we can immediately see that the two groups are almost identical. See the glossary entry **Boxplot** for further details concerning interpretation.

14.2 Annotating R Commander code to change options

You can inspect the code produced from the R Commander menu options and then annotate it. For example I have added *col="lavender"* to the end of the auto-generated code opposite. Then selecting the line and clicking on the **Submit** button produces lavender filled boxes.

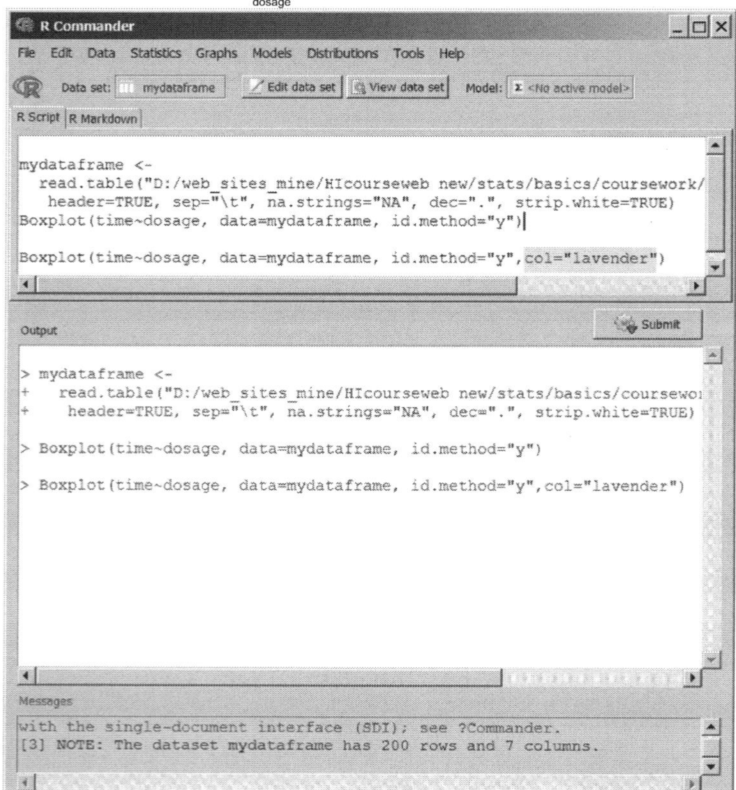

14.3 R Code: Boxplot options

You can also draw much more sophisticated boxplots using R code directly. For example, you could request boxplots by typing the following into the *R Console* window:

split() creates a new dataframe called *newdata* with a separate column for each dosage level

```
newdata <- split(mydataframe$time, mydataframe$dosage)

boxplot(newdata, col = "grey90", notch = TRUE,
         varwidth = TRUE)
```

or

here we use the *boxplot()* function rather than the *Boxplot()* function used by R Commander

```
boxplot(mydataframe$time ~ mydataframe$dosage,
col = "lavender", notch = TRUE, varwidth = TRUE)
```

This code produces a boxplot where the boxes are colored 90% gray (similar to *col=" lavender"*), and where if the notches of the boxes do not overlap there is 'strong evidence' that the two medians differ. Also setting *varwidth* to true means that the boxes are drawn with widths proportional to the square-roots of the number of observations in the groups - in other words, the width reflects the number of observations.

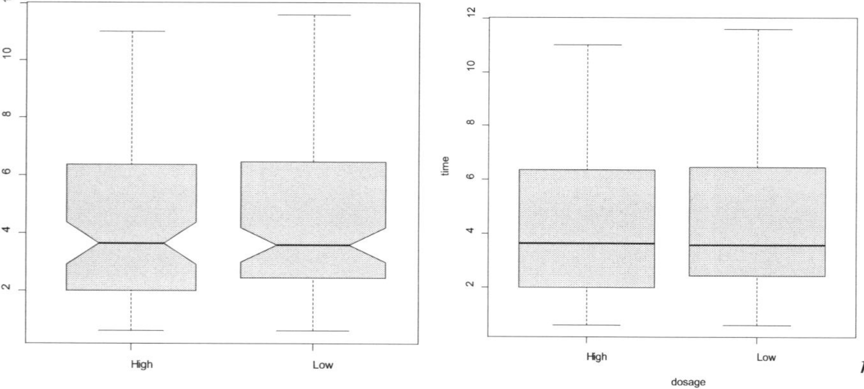

If the data is normally distributed, the *interquartile range* = 1.35 x σ where σ (the Greek letter sigma) is the population standard deviation.

To increase the line thickness of the median line (default=1) add the option *medlwd=5* to the code:

```
boxplot(newdata, col = "grey90", medlwd=5)
```

If you wish to fill the boxes with lines or dots it requires the writing of some complex code discussed at http://stackoverflow.com/questions/17423115/colorfill-boxplot-in-r-cran-with-lines-dots-or-similar. To draw boxplots for several variables see the *Comparing pre-post test means*, and *Factorial Anova* chapters.

14.3.1 Showing outliers

A boxplot is a very useful method for identifying outliers. However, the dataset used in this chapter did not have any to demonstrate this. In contrast, the boxplot presented in the chapter *Comparing two sample means*, does have as shown below. R Commander provides three options for their display; *Automatically* displays them with their case number (default), *No* displays an **O,** and finally *With mouse* allows the user to click on each outlier which then displays the case number. The last option is shown below.

R Commander makes use of the *Boxplot()* function in the *car* package along with the *id.method=* option to specify the outlier identification method.

```
Boxplot(values~ind, data=mydataframe, id.method="y")

Boxplot(values~ind, data=mydataframe, id.method="n")

Boxplot(values~ind, data=mydataframe, id.method="identify")
```

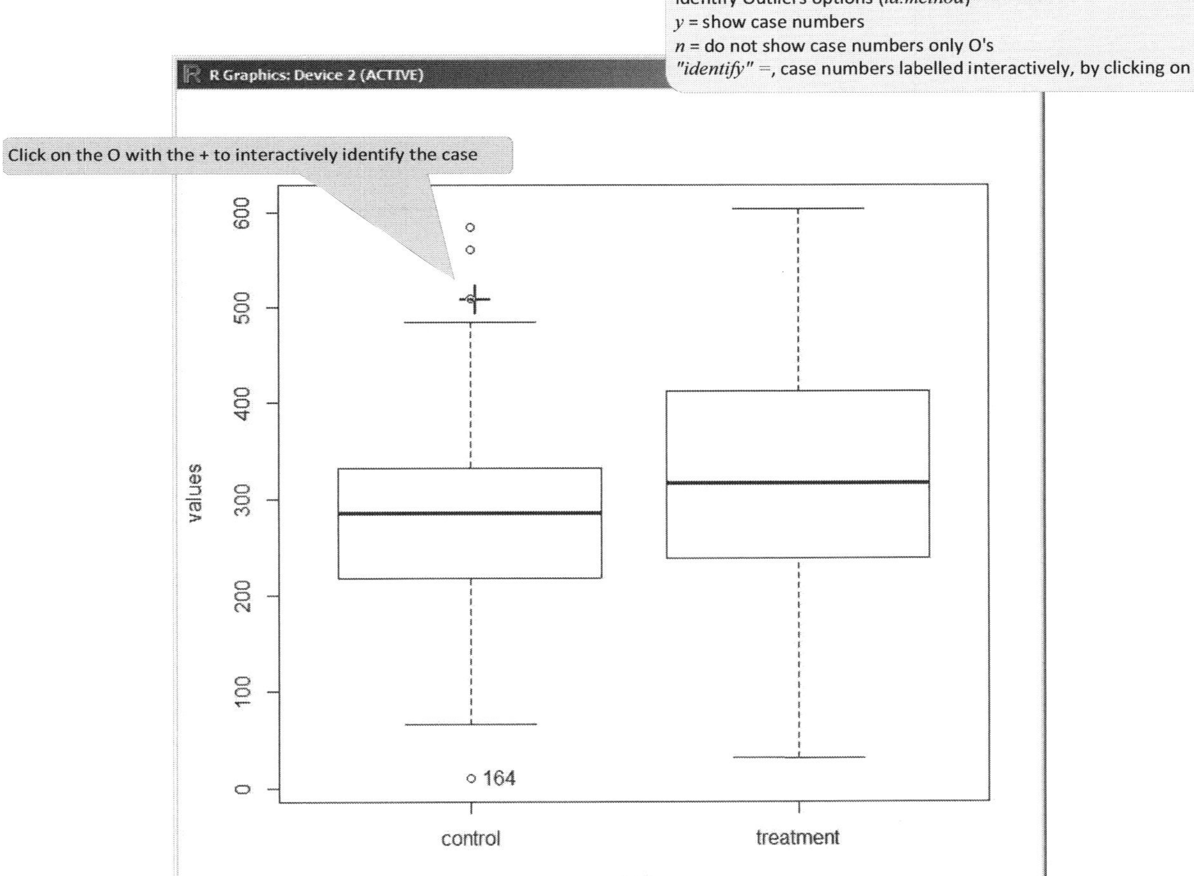

15 Percentages for each category/factor level

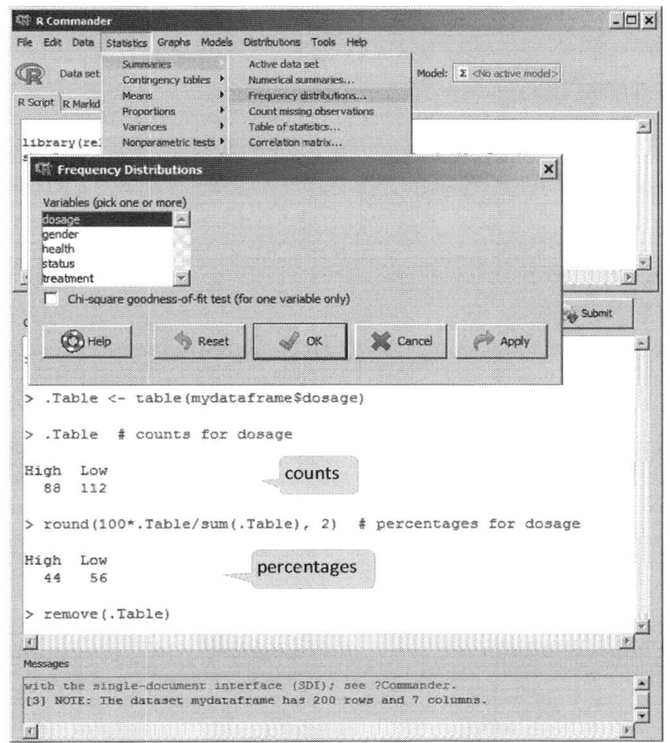
R Commander
Frequency Distributions
Variables (pick one or more)
dosage
gender
health
status
treatment
Chi-square goodness-of-fit test (for one variable only)
Help
Reset
OK
Cancel
Apply
> .Table <- table(mydataframe$dosage)
> .Table # counts for dosage
High Low
88 112
counts
> round(100*.Table/sum(.Table), 2) # percentages for dosage
High Low
44 56
percentages
> remove(.Table)
[3] NOTE: The dataset mydataframe has 200 rows and 7 columns.

Nominal, and to a lesser extent ordinal data, are often presented in the form of tables showing the percentage of subjects which are within each category. If you don't understand the terms nominal or ordinal see the Glossary entry **level of measurement**.

15.1 R Commander: Percentages for each category

Using the *medication.dat* dataset from the *Graphing Distributions of single variables* chapter we can obtain the counts for each category along with their percentages in R Commander using the menu option:

Statistics->Summaries->Frequency distributions...

15.2 R code: Relative and cumulative percentages

Often more detailed values are required such as the relative frequency for each category or the cumulative relative frequency as these values allow you to consider the categories sequentially. To obtain them you can either calculate the values manually from the raw counts/percentages or use the R code given below. This R code can be either typed into the *Script window* of R Commander or into the *R Console* window. The code below explains how you can produce the relative and cumulative relative frequency values for the dosage variable.

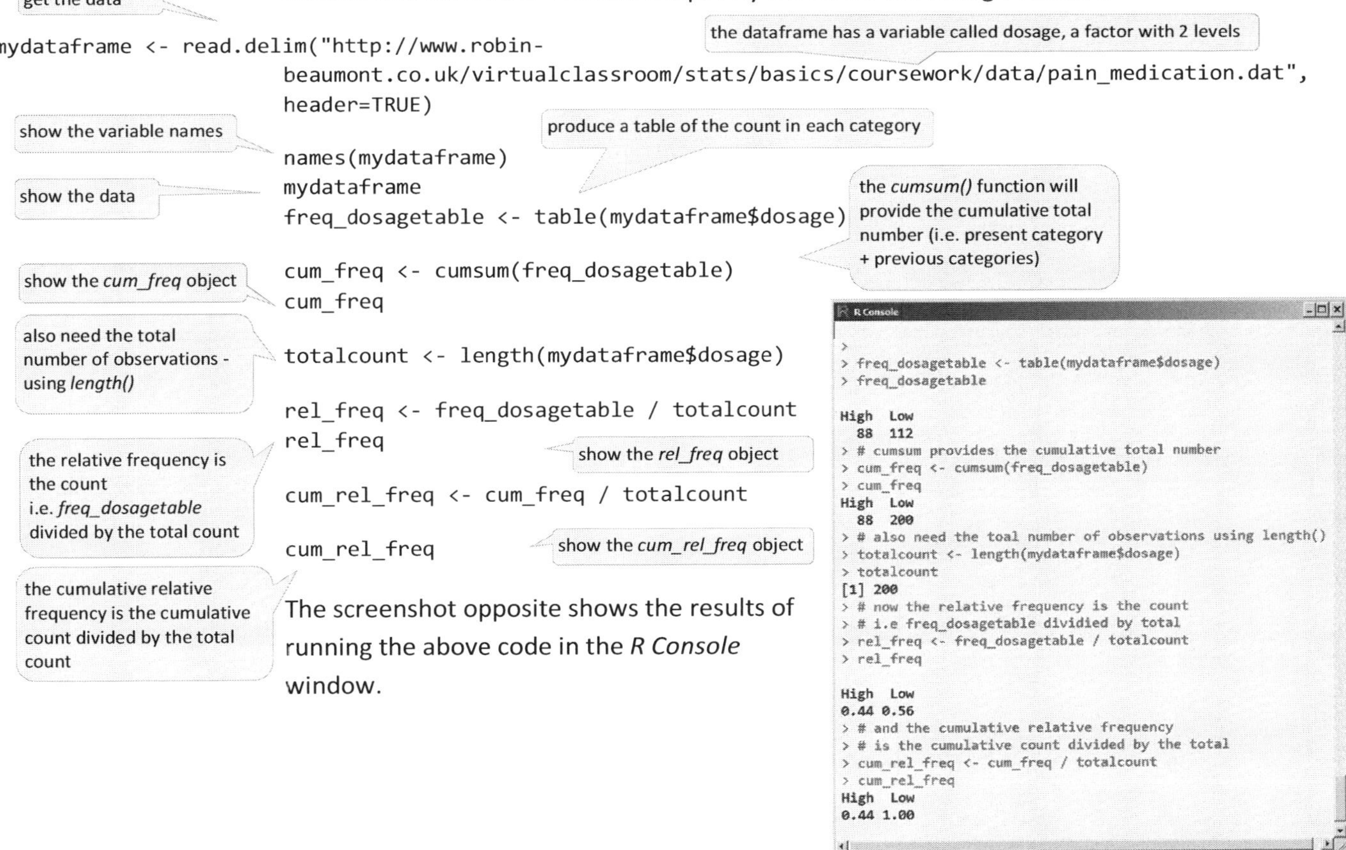
get the data
the dataframe has a variable called dosage, a factor with 2 levels
show the variable names
produce a table of the count in each category
show the data
the *cumsum()* function will provide the cumulative total number (i.e. present category + previous categories)
show the *cum_freq* object
also need the total number of observations - using *length()*
the relative frequency is the count i.e. *freq_dosagetable* divided by the total count
show the *rel_freq* object
show the *cum_rel_freq* object
the cumulative relative frequency is the cumulative count divided by the total count
R Console
> freq_dosagetable <- table(mydataframe$dosage)
> freq_dosagetable
High Low
88 112
> # cumsum provides the cumulative total number
> cum_freq <- cumsum(freq_dosagetable)
> cum_freq
High Low
88 200
> # also need the toal number of observations using length()
> totalcount <- length(mydataframe$dosage)
> totalcount
[1] 200
> # now the relative frequency is the count
> # i.e freq_dosagetable dividied by total
> rel_freq <- freq_dosagetable / totalcount
> rel_freq
High Low
0.44 0.56
> # and the cumulative relative frequency
> # is the cumulative count divided by the total
> cum_rel_freq <- cum_freq / totalcount
> cum_rel_freq
High Low
0.44 1.00

```
mydataframe <- read.delim("http://www.robin-
        beaumont.co.uk/virtualclassroom/stats/basics/coursework/data/pain_medication.dat",
        header=TRUE)

names(mydataframe)
mydataframe
freq_dosagetable <- table(mydataframe$dosage)

cum_freq <- cumsum(freq_dosagetable)
cum_freq

totalcount <- length(mydataframe$dosage)

rel_freq <- freq_dosagetable / totalcount
rel_freq

cum_rel_freq <- cum_freq / totalcount

cum_rel_freq
```

The screenshot opposite shows the results of running the above code in the *R Console* window.

16 Samples and populations

In the previous section, we described our data using summary values including the Mean and Standard Deviation along with various graphs such as the boxplot. Now we up our game, moving the focus from an observed sample to a theoretical population from which our sample or samples have been randomly drawn. This process of inferring from a sample to a population is the linchpin of modern scientific research, distinguishing it from qualitative/interpretive approaches which fail to provide robust methods of inference.

To understand this process you need to be familiar with the concepts of **Inference, sampling distributions** and **standard error (se)**. Please review these glossary entries if you are unfamiliar or unsure of these concepts.

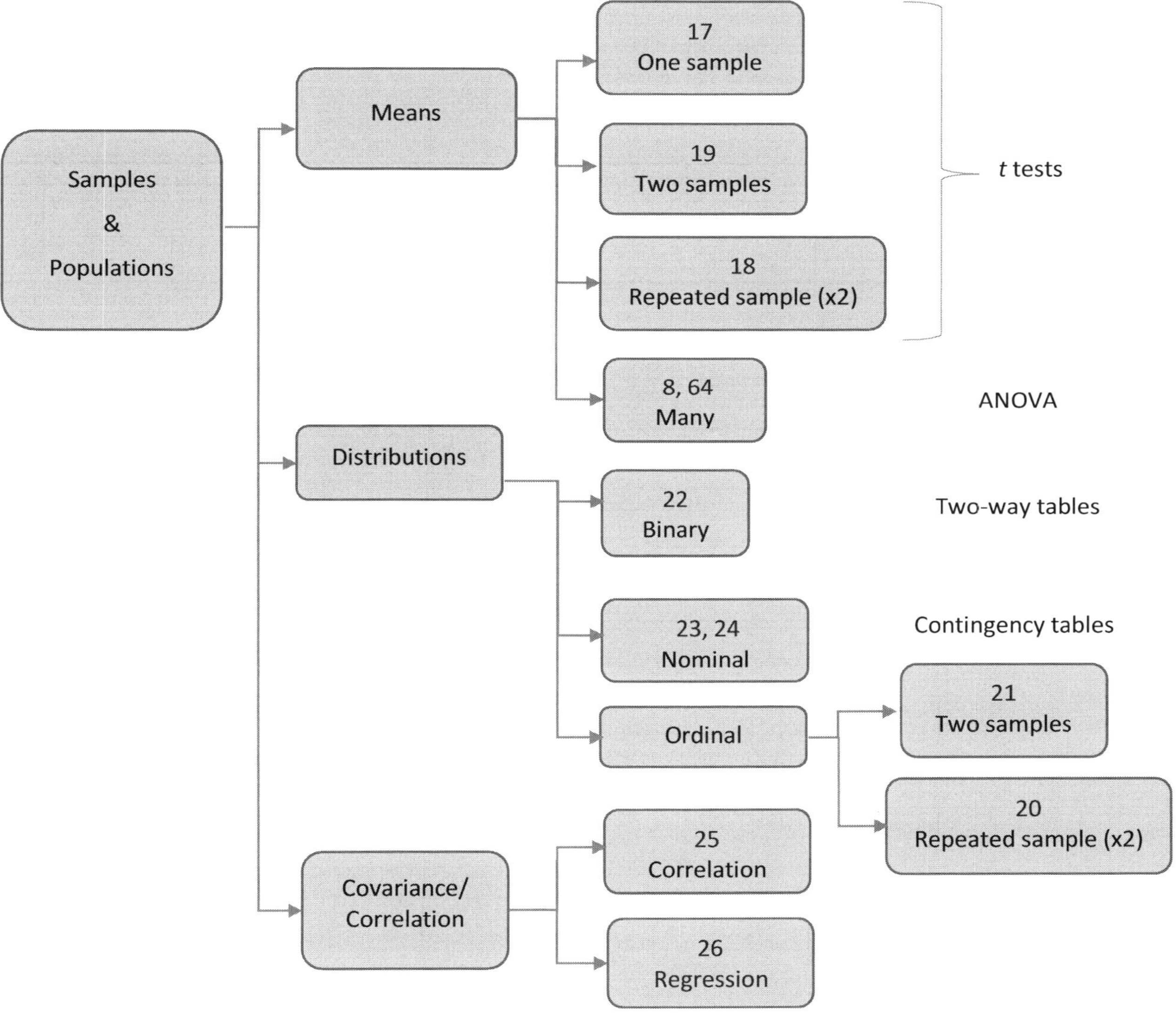

17 Comparing a sample mean to a population mean: single sample *t* test

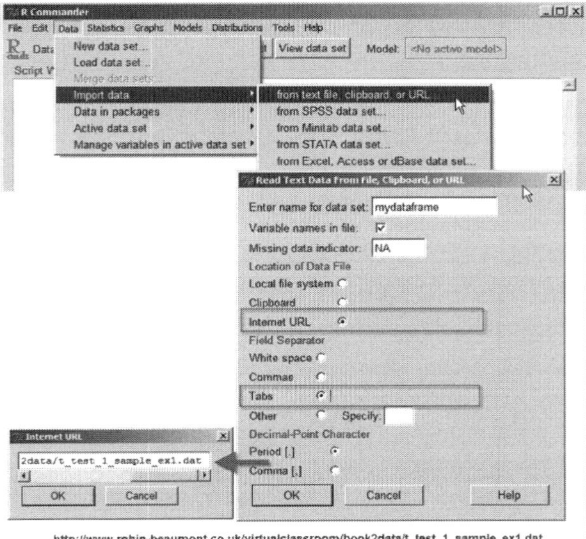

http://www.robin-beaumont.co.uk/virtualclassroom/book2data/t_test_1_sample_ex1.dat

The single sample *t* test allows us to evaluate how different a sample mean is from a population mean. I'll demonstrate this by following the typical stages of a data analysis described in the *How this book works* chapter.

First, we need some data and for this example, I'll use a fictitious sample dataset consisting of 15 observations from Orkney inhabitants with an observed mean of 95.53, which I wish to compare to a healthy population (i.e. rest of Scotland) which has a mean of 100. Imagine these values represent some substance where if the sample (i.e. Orkney) in general (i.e. the mean value) is different from the population value does result in health problems. Although there is an observed difference in means (100-95.5 = 4.5) we want to know if this difference is not simply due to the vagaries of random sampling.

17.1 Data preparation

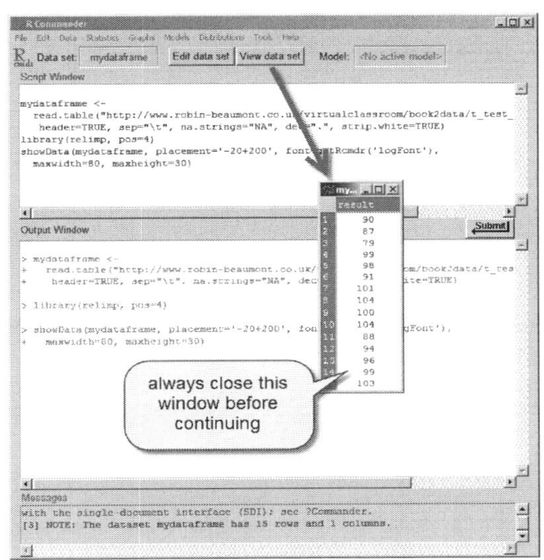

You can download the dataset directly from my website by selecting the R Commander menu option:

Data-> from text, the clipboard or URL

I have given the resultant dataframe the name *mydataframe*, also indicating that it is from a URL (i.e. the web) and the columns are separated by *tab* characters, although this is not necessary as there is only one column.

Clicking on the **OK** button brings up the *Internet URL* box, type the following in it to obtain my sample data:

http://www.robin-beaumont.co.uk/virtualclassroom/book2data/t_test_1_sample_ex1.dat

Click **OK** again.

17.2 Data description

```
> summary(mydataframe)
     result
 Min.   : 79.00
 1st Qu.: 90.50
 Median : 98.00
 Mean   : 95.53
 3rd Qu.:100.50
 Max.   :104.00
```

It is always a good idea to see the data and you can achieve this easily in R Commander by clicking on the **View data set** button.

It is good practice, and may save you a lot of head scratching later, to check there is no missing data, no impossible values (such as - 20) and no extremely unusual values.

We can also gain a feeling for the data by requesting some summary values using the R Commander menu option **Statistics-> Summaries->Active data set**. For an explanation of these values see the glossary entry **Boxplot**.

You can also use the R Commander menu option **Statistics-> Summaries->** to obtain the standard deviation, an alternative to the interquartile range which assumes the data follow a **normal distribution**.

```
> numSummary(mydataframe[,"result"], statistics=c("mean", "sd", "IQR",
+   "quantiles"), quantiles=c(0,.25,.5,.75,1))
     mean       sd IQR 0%  25% 50%   75% 100%  n
 95.53333 7.249302  10 79 90.5  98 100.5  104 15
```

Here we have a Standard Deviation of 7.2 indicating that if the data follows a normal distribution we would expect approximately 68% of the scores to be between 95.5 - 7.2 to 95.5 + 7.2 that is 88.3 to 102.7, slightly wider than the interquartile range and now encompassing a theoretical 68% of the scores.

"or one more extreme"
What I mean when using this expression here is a sample that would produce a mean value deviating more than that observed (4.5 units) from the population value.
Such mean values would include 95, 90, 105, 110

As I explained at the start of the chapter, we now need to compare our sample against a population. What we are trying to find out is whether we would have obtained a mean value of 95.5 or one more extreme from a random sample of 15 observations from within our comparison population (which, if you remember, has a mean value of 100). We can do this by creating a model.

17.3 Model development

In this instance, the model we develop assumes we have collected our random sample of the specified size (15 observations) from a population with a mean of 100. That is there is no real difference between the sample and the population, therefore the model is often known as the *null model* or *null hypothesis*. We then calculate the probability of obtaining a sample of the same size with the observed mean (i.e. 95.5) or one more extreme by assessing a *t* value and its associated **p-value**. Before discussing the values we will obtain them using R Commander.

17.3.1 R Commander: One sample *t* test

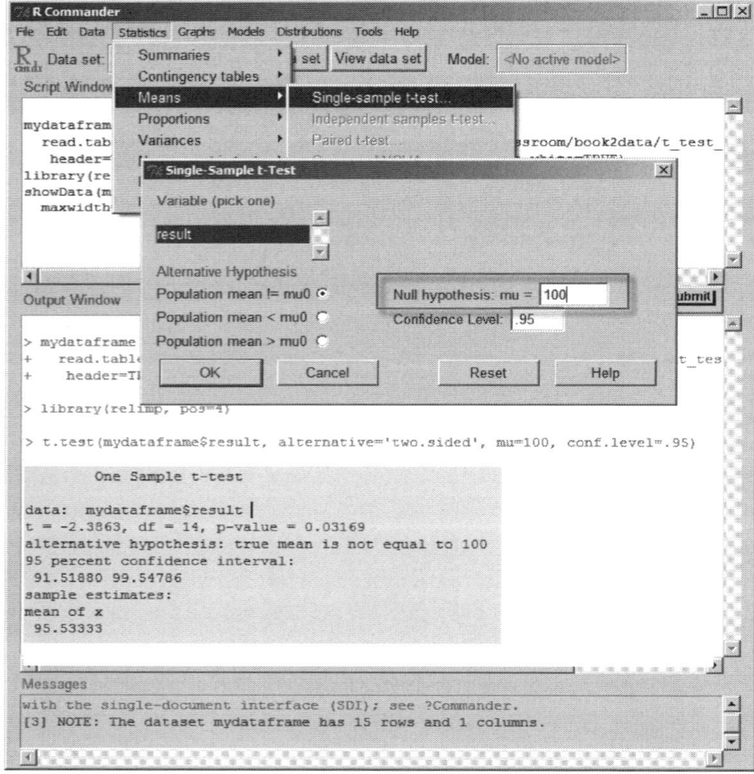

We can carry out the one sample *t* test on our dataset by selecting the R Commander menu option:

Statistics-> Means ->Single sample t- test

The variable is already selected, so all we need to do is specify the hypothesised mean of the population, which we have chosen to be 100, type this in then click on the **OK** button. The results, which I have shaded, are shown opposite. We can interpret this p-value:

We would observe a mean value of 95.5 or one more extreme 31 times in every thousand from a random sample of 15 if our sample had come from a population with a mean of 100.

If we had set a critical value to 0.05 because our obtained p-value is less than this (0.03169) we would reject our original null model and say that we have a statistically significant result and therefore our sample did not come from a population with a mean of 100. The above description contains many hidden steps and assumptions so we will now look in much more depth at this.

17.4 Understanding the *t* and p-value

For a individual observation:

$$\text{Know } z = \frac{X-\mu}{\sigma} = \frac{\text{individual value} - \text{population mean}}{\text{population standard deviation}}$$

When you know the population values:

$$\text{Know } Z_{test} = \frac{\overline{X}-\mu}{\frac{\sigma}{\sqrt{n}}} = \frac{\text{Sample mean} - \text{population mean}}{\text{standard deviation of means}}$$

$$= \frac{\text{Sample mean} - \text{population mean}}{\text{Standard Error of Means (SEM)}}$$

Consider small samples and where population sd not known

Substitute sample estimate for standard deviation into the SEM

$$\text{SEM} = \frac{S}{\sqrt{N}} = \frac{\text{sample standard deviation}}{\sqrt{\text{sample size}}}$$

Assume that the population mean is equal to a value we are interested in

e.g μ_0

Call this a 't' statistic:

$$t_{statistic} = \frac{\overline{X}-\mu_0}{S_{\overline{X}}} = \frac{\overline{X}-\mu_0}{\frac{S}{\sqrt{N}}} = \frac{\text{Sample mean} - \text{comparator mean}}{\text{Sampling standard Error(SEM)}}$$

$$t_{statistic} = \frac{\text{observed difference in means}}{\text{sem}}$$

$$= \frac{\text{observed difference in means}}{\text{Standard Deviation in means due to random samping}}$$

$$= \frac{Signal}{Noise}$$

The description on the previous page of the **p-value** may be rather overwhelming but this is because it is expressing several complex things. To understand this in more detail requires some knowledge of what a *t* value and a p-value are.

The *t* value is simply another summary measure just like the mean or standard deviation and like them the value is a ratio. The top value is the observed difference in means (population verses sample). Whereas the bottom value is the standard error of the mean (*sem*), representing the standard deviation of the means from a infinite number of samples of this size. Basically the *t* value boils down to the observed difference in means divided by the expected standard deviation. Another way of thinking of it is as the signal divided by the expected noise, given our null model is assumed to be true.

To get a feeling for what a specific *t* value means let us consider some examples:

Assuming the null model is true:

t = 0
If we obtained a t statistic value of 0 what would this mean? That is the expression = 0/anything. This would indicate that there was no difference in the value between the observed sample mean and the one we are interested in comparing it against. That is our observed mean is equal to the population mean. **In our null model** this would be the **most common situation**, as we would expect that our sample would equal the comparator mean more often than not.

***t* less than 1**
That is the expression = x/more than x, indicating that the observed difference is less than one standard deviation of the expected variability in means from a infinite number of random samples.

t = 1
That is the expression = x/x. Here the observed difference is equal to the standard deviation (i.e. standard error) of the means due to random sampling.

***t* more than 1**
That is the expression = x/less than x. This occurs when the observed variability in the means is greater than one standard deviation (i.e. standard error) of the expected variability in means from an infinite number of random samples of the same size. We would expect this situation to arise much less frequently and get even less so as the value increased.

We can obtain a graph of all the possible t values we could obtain for a specific sample size by using the R Commander menu option:

Distributions-> continuous distributions-> t distribution->plot t distribution

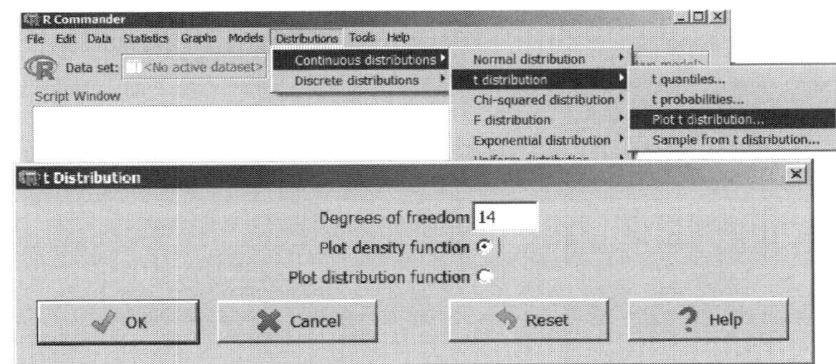

We then specify the sample size by entering the value, called the degrees of freedom, which here is one less than the sample size, into the **Degrees of freedom** box. Select also the **Plot density function** option, and finally click on the **OK** button.

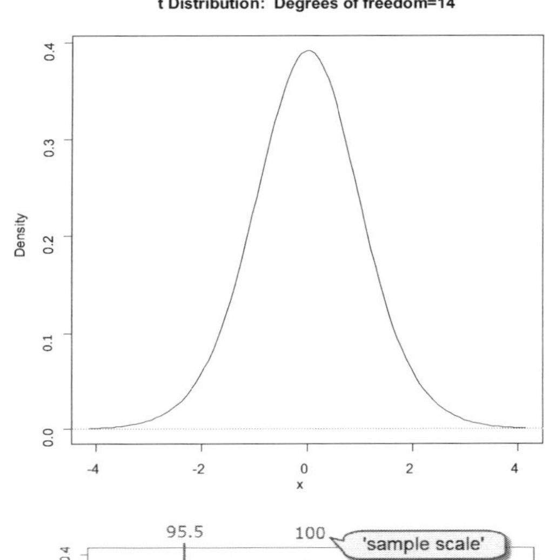

The resultant graph is shown opposite. Notice that the most common values are around the zero value, and are symmetrical. This is because we are assuming that the expected population *t* value is zero. Think of density as being equivalent to frequency here.

The question is how does this graph of expected *t* values relate to our data? The answer is that you can think of the above graph as a rescaled version of the expected variability in the observed mean, when we assume the population mean is equal to 100. I have annotated the graph below to show this. Our hypothesised mean value of 100 is now the zero and the observed mean of 95.5 equates to the t value of -.2.38

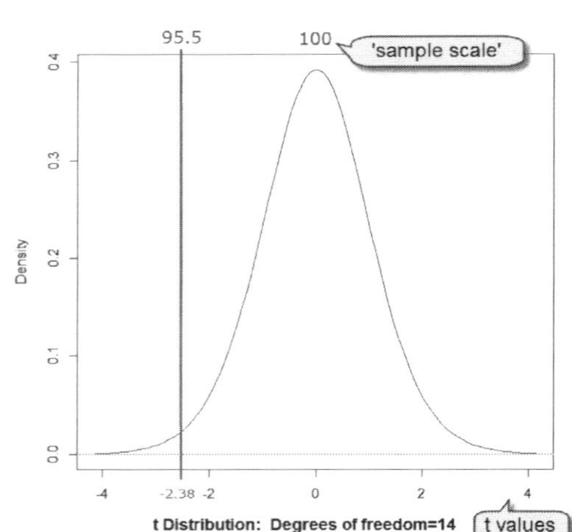

So how does this relate to the **p-value**? This is remarkably simple if you know a few facts:

- The area under the *t* curve is equal to 1
- The area under the *t* curve for more extreme *t* values than -2.38 and +2.38 expresses, in this situation, the chances of obtaining a mean of less than 95.5 or more than 104.5 given that the sample size is 15 and the population mean is 100.

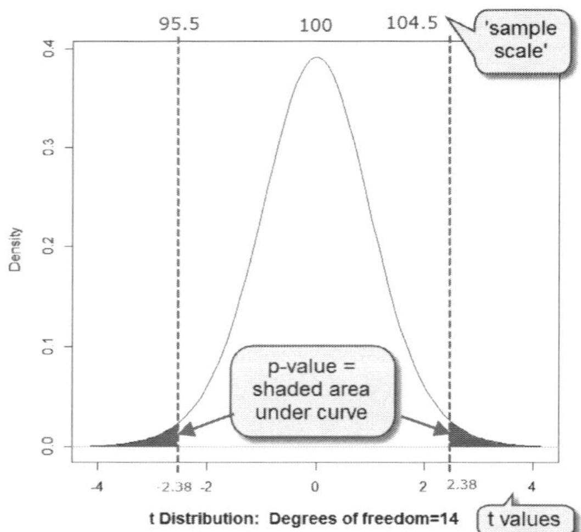

This is equivalent to our p-value which is the probability of obtaining a mean value 'more extreme' than 95.5 from a sample size of 15 given that the population mean is 100.

I hope you can see now how the graph is so much more elegant than the sentence I produced on the previous page explaining the p-value?

17.5 Confidence interval

The R Commander output shown on the previous page includes a confidence interval with the values 91.5 to 99.5. We can interpret this as indicating that we would estimate that 95% of the time the true population mean would lie within the range of 91.5 to 99.5.

17.6 Effect size - clinical significance

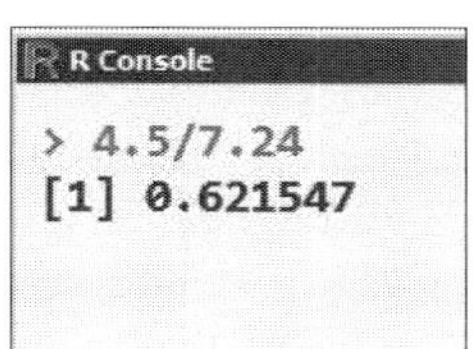

In this situation the effect size is the difference in means between the hypothesised population mean and that observed in the sample 100 - 95.5 = 4.5. This can also be reported as a standardised effect size which is the value in standard deviation units. We obtained the SD of 7.24 from the data description stage of our analysis so the standardised effect size is 4.5/7.24. We can use the R Console window as a calculator to calculate this, giving a value of .62 so the sample mean is just half of a standard deviation from the hypothesised mean.

There are many different effect size measures but probably the best is the simplest here (Baguley, 2009).

17.7 Writing up the result

Key point
Always remember to report the effect size measure and comment upon it in terms of clinical usefulness.
Would an effect size of 4.5 here be of any practical/clinical importance?

Fifteen random subjects from Orkney each provided a single valid value for substance X resulting in a mean of 95.53 (SD 7.2, 95% CI 91.5 to 99.5). A comparison was made between this sample mean and the Scottish population which possesses a mean value of 100 for this substance. A single sample *t* test was performed (*t* = -2.38, df = 14, p-value = 0.031) leading to the conclusion that the sample did not come from a population with a mean of 100.

The effect size was 4.5 units (standardised = .62) indicating the observed difference in just over half a standard deviation between the sample and the population means.

17.8 R Code: One sample t test

R Console

```
> result <- c( 90, 87, 79, 99, 98, 91, 101, 104, 100,  104, 88, 94, 96, 99, 103)
> t.test(result, mu=100)

        One Sample t-test

data:  result
t = -2.3863, df = 14, p-value = 0.03169
alternative hypothesis: true mean is not equal to 100
95 percent confidence interval:
 91.51880 99.54786
sample estimates:
mean of x
 95.53333
```

as would be expected the results are equivalent to those obtained in R Commander.

Instead of using the R Commander menu we could have typed the equivalent R code directly into the *R Console* window. This involves typing a single line of R code using the *t.test(data, mu=)* function where *data* is our data and *mu* the value of the hypothesised population mean, 100 here. In the example opposite I have also first created a dataframe called *result* consisting of the 15 values by using the concatenate **function**, *c()*. The *result* dataframe is then passed to the *t.test()* function:

```
result <- c( 90, 87, 79, 99, 98, 91, 101, 104, 100,  104, 88, 94, 96, 99, 103)

t.test(data, mu=100)
```

17.9 What does df mean and why report it?

df stands for *degrees of freedom*. For practical purposes here it is indicating the size of the sample minus one and subsequently, because the shape of the *t* distribution depends on the sample size, this value affects the shape of the t distribution and therefore the p-value we obtain. More technically it is – the number of observations that cannot be determined after calculating the summary value, in other words these values are free to vary. For example in our sample of 15 values if you know the mean (or *t* value) and pick 14 arbitrary values you then must make the 15th equal a specific value – it is pre-determined, to make the sample equal the specified mean (or *t* value). There are 14 values that are free to vary.

17.10 Tips and Tricks

Remember that it is up to you to specify the population mean in the one sample *t* test, leaving it to the computer is dangerous when the default is often something like zero.

18 Comparing pre-post test means: paired samples *t* test

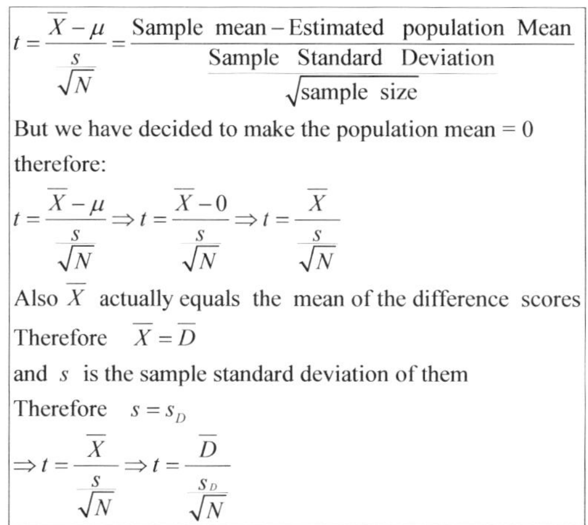

$$t = \frac{\overline{X} - \mu}{\frac{s}{\sqrt{N}}} = \frac{\text{Sample mean} - \text{Estimated population Mean}}{\frac{\text{Sample Standard Deviation}}{\sqrt{\text{sample size}}}}$$

But we have decided to make the population mean = 0 therefore:

$$t = \frac{\overline{X} - \mu}{\frac{s}{\sqrt{N}}} \Rightarrow t = \frac{\overline{X} - 0}{\frac{s}{\sqrt{N}}} \Rightarrow t = \frac{\overline{X}}{\frac{s}{\sqrt{N}}}$$

Also $\overline{X}$ actually equals the mean of the difference scores

Therefore $\overline{X} = \overline{D}$

and s is the sample standard deviation of them

Therefore $s = s_D$

$$\Rightarrow t = \frac{\overline{X}}{\frac{s}{\sqrt{N}}} \Rightarrow t = \frac{\overline{D}}{\frac{s_D}{\sqrt{N}}}$$

The paired samples *t* test allows us to evaluate how different a mean obtained from a sample is to that of a population with a mean of zero. Initially this may seem pretty useless but when the mean of the sample is from a set of difference values it is very useful. Most commonly these difference values, are obtained from paired, matched or two repeated values, such as subjects pre – post tests, or following an exposure study design. The paired samples *t* test therefore only makes use of a single value from each subject - the difference between the two scores.

We can adapt the *t* value described in the previous chapter to the paired sample situation as shown in the box opposite. Once again the top value represents the observed difference in means (this time from zero) divided by the standard deviation of means (*se*) if we have randomly picked an infinite number of samples of the specified size from a population with a mean of zero. As in the previous chapter the *t* value boils down to the observed difference in the mean from zero divided by the expected difference.

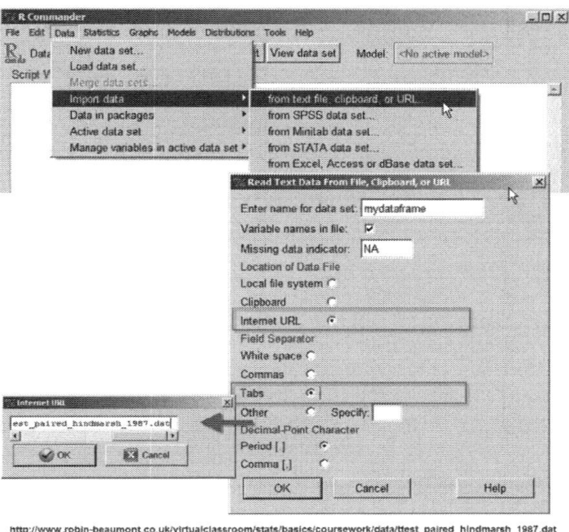

http://www.robin-beaumont.co.uk/virtualclassroom/stats/basics/coursework/data/ttest_paired_hindmarsh_1987.dat

For this example I'll use a sample dataset consisting of 32 observations of standardised heights from 16 subjects, one reading taken before growth hormone therapy and another after one year. The data was reported in Campbell & Machin, 1993 p.76 being taken from Hindmarsh and Brook, 1987.

You can download the dataset directly from my website by selecting the R Commander menu option:

Data-> Import data-> from text, the clipboard or URL

I have given the resultant dataframe the name *mydataframe*, also indicating that it is from a URL (i.e. the web) and the columns are separated by *tab* characters. The details are shown opposite. Clicking on the **OK** button brings up the *Internet URL* box, in which you need to type the following to obtain my sample data:

```
http://www.robin-beaumont.co.uk/virtualclassroom/book2data/ttest_paired_hindmarsh_1987.dat
```

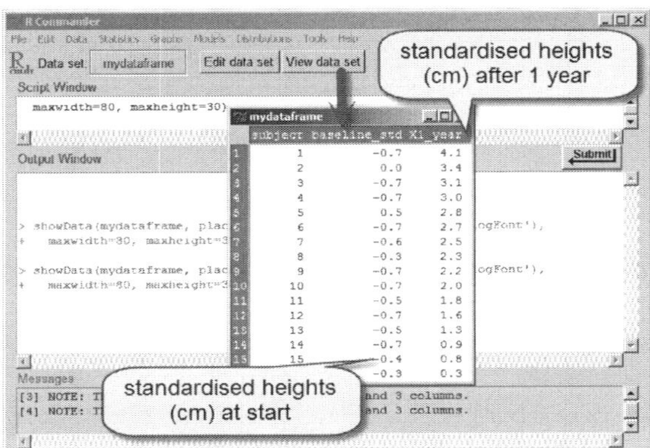

Use the **View data set** button to check your data.

It would appear that all the subjects are below the standardised height at the start of the trial (i.e. all have negative values). However after a year of growth hormone therapy they all have positive values, clearly the therapy has worked well. We will therefore be expecting a statistically significant p-value from our paired *t* test indicating that the sample mean is not likely to have come from the hypothesised population mean of zero.

18.1 Data preparation - Creating a calculated variable in R Commander

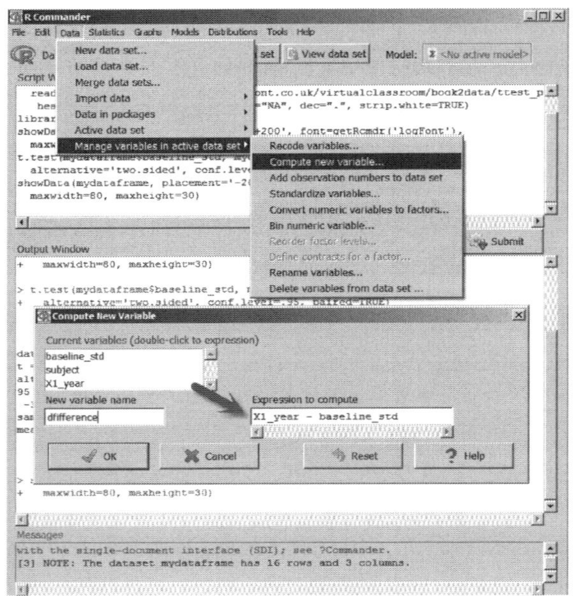

Following the stages of a typical data analysis, described in the introductory chapter, we will prepare the data by creating a calculated variable representing the differences between the pre and post hormone therapy scores for each subject. Creating calculated variables is very easy in R Commander and to obtain the change in each person's height we create the expression:

```
X1_year     -     baseline_std
```

added spaces in R code are ignored by R but often make reading it easier for humans

To create a new variable using this expression select the R Commander menu option:

Data -> Manage variables in active dataset -> Compute new variable

In the screen shot opposite I have given the new variable the name *difference.* After clicking on the **OK** button you can see your changes by clicking on the **View data set** button, remembering to close the window afterwards.You can see the R Code generated from the above dialog box by inspecting the output window in R Commander, and if you felt sufficiently proficient, you could have typed the code directly.

```
mydataframe$difference <- with(mydataframe, X1_year - baseline_std)
```

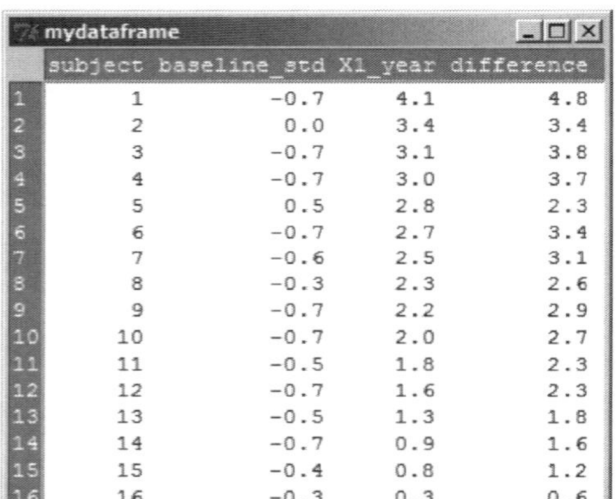

	subject	baseline_std	X1_year	difference
1	1	-0.7	4.1	4.8
2	2	0.0	3.4	3.4
3	3	-0.7	3.1	3.8
4	4	-0.7	3.0	3.7
5	5	0.5	2.8	2.3
6	6	-0.7	2.7	3.4
7	7	-0.6	2.5	3.1
8	8	-0.3	2.3	2.6
9	9	-0.7	2.2	2.9
10	10	-0.7	2.0	2.7
11	11	-0.5	1.8	2.3
12	12	-0.7	1.6	2.3
13	13	-0.5	1.3	1.8
14	14	-0.7	0.9	1.6
15	15	-0.4	0.8	1.2
16	16	-0.3	0.3	0.6

Noting that all the difference scores are positive supporting our feeling that the treatment appears very effective.

18.2 Data description - mean and effect size

As mentioned above the paired *t* test works on the difference scores and this is the focus of our interest. We can obtain summary statistics for the *difference* variable by using the R Commander menu option:

Statistics-> Summaries-> Active data set

We can see the mean is 2.65 representing an average gain of standardised height. We could also produce a boxplot for the variable (see the *Boxplots* chapter). If you cannot see the new variable in the list of available variables see the *tips and tricks* section at the end of this chapter.

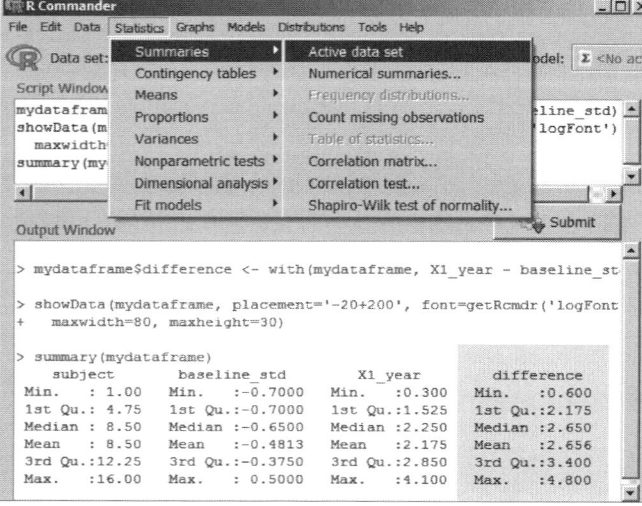

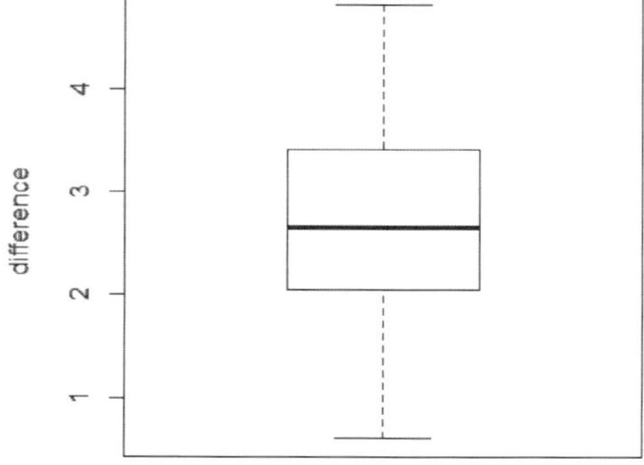

$$Effect\ size = \frac{\text{mean difference}}{SD_{\text{difference}}}$$

The above mean of 2.65 is called the effect size measure. To obtain what is known as the standardised effect size measure (analogous to Cohen's *d* here) we divide it by the Standard Deviation of the difference scores, shown opposite.

We can convert this generic formula into a line of R code, given below. If you type this into the R Commander *Script* window, highlight it and then click on the **Submit button** you will obtain the required result. Alternatively, you can type it into the *R Console* window and then hit the return key (↵).

```
effect_size <- mean(difference)/ sd(difference)
effect_size
```


Produces 1.987866 indicating we have an effect size of 1.98 standard deviations or to put it another way, a standardised effect size of 1.98 for the difference scores. Referring to the glossary entry for **Effect size**, we see that a Cohen's *d* value of 0.8 or more is classed as a large effect size so here we clearly have a large one.

18.3 Model development and evaluation - the paired *t* test in R Commander

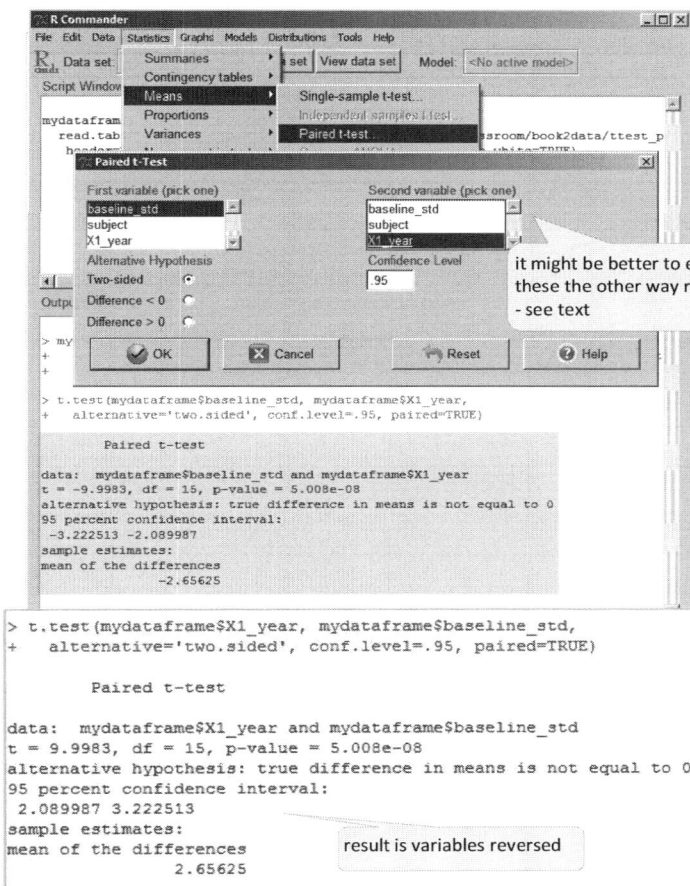

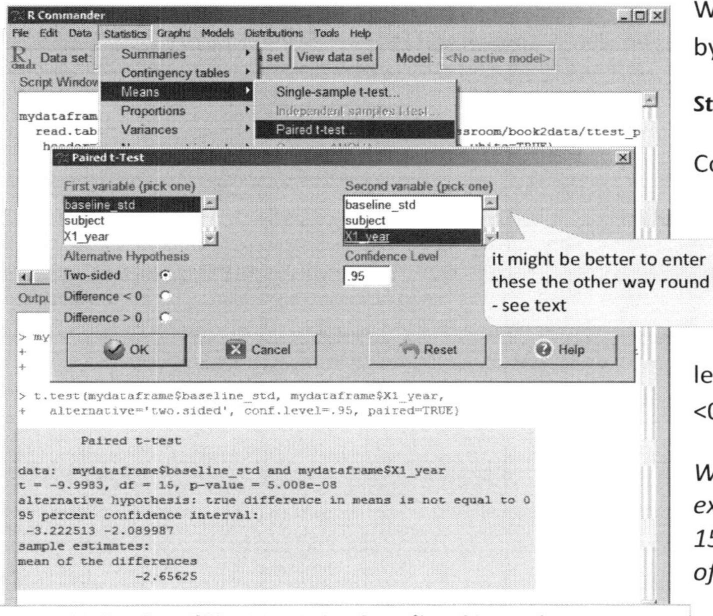


```
> t.test(mydataframe$X1_year, mydataframe$baseline_std,
+   alternative='two.sided', conf.level=.95, paired=TRUE)

        Paired t-test

data:  mydataframe$X1_year and mydataframe$baseline_std
t = 9.9983, df = 15, p-value = 5.008e-08
alternative hypothesis: true difference in means is not equal to 0
95 percent confidence interval:
 2.089987 3.222513
sample estimates:
mean of the differences
               2.65625
```


We can carry out the paired sample *t* test on our dataset by selecting the R Commander menu option:

Statistics-> Means ->Paired sample t- test

Complete the *Paired sample t-Test* dialog box as shown opposite.

We have a *t* value of -9.998 with an associated p-value, given in **E notation**, as 5.008E-08 meaning 0.00000005008 which is a very small value, much less than 0.001 so when reporting it we just say it is <0.001. One interpretation of this p-value:

We would observe a mean value of 2.65 or one even more extreme less than once in a thousand random samples of 15 if our sample had come from a population with a mean of zero.

Assuming a critical value of 0.05, we can see that we have a statistically significant result because our obtained p-value is less than this.

We are also given a 95% confidence interval and the mean value for the differences. Interestingly R Commander has reversed the pre- post values, if you enter the *X1_year* variable in the first variable box and the *baseline_std* in the other box the signs will be reversed, that is positive which makes more sense as we are interested in the change from the start rather than a retrospective perspective. We can interpret this as indicating that we would estimate that 95% of the time the true population mean would lie within the range of 3.22 to 2.08.

The above description is discussed fully in the *Single sample t test* chapter.

18.4 Digging deeper showing the Pre - post changes

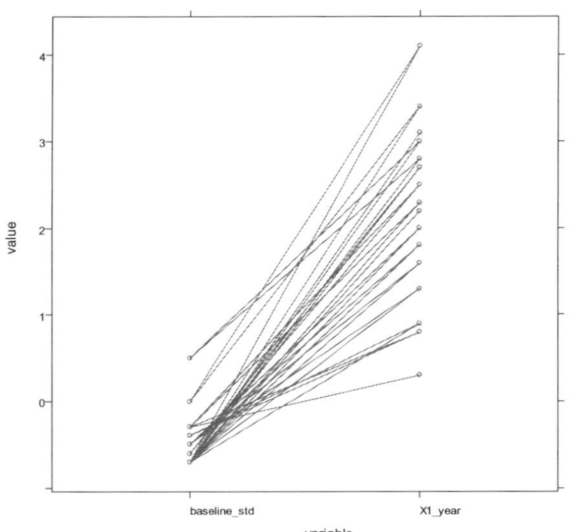

It is often very useful to see the pre-post changes, that is the differences for each subject as shown opposite. To achieve this in R you need to have the dataset in what is known as long format, where unfortunately ours is currently in wide format. Because we only want the see the baseline and subsequent results, not the difference values, we first delete that variable, selecting the menu option:

Data->Manage variables in active data set->
Delete variables from data set

Select the difference variable then click on the **OK** buttons of both the *Delete Variables* and *Confirmation* windows, shown below.

Now to convert the dataset from wide format to long format we need to use R code making use of either the *reshape2* or *lattice* packages. The relevant R code with comments using the *melt()* function from the *reshape2* package is given below.

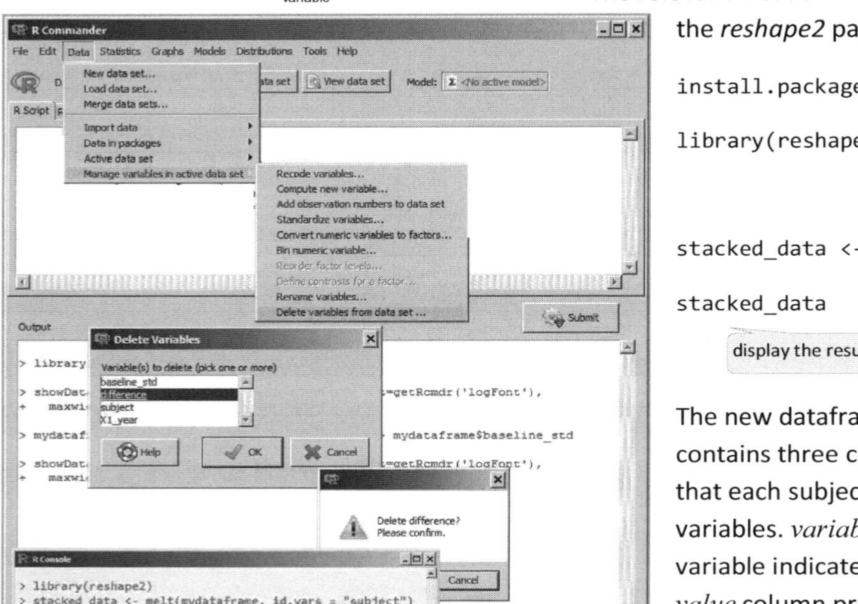

install the package if you don't already have it

```
install.packages("reshape2", dependencies = TRUE)

library(reshape2)
```

load the reshape2 package

the melt function stacks several columns of data

```
stacked_data <- melt(mydataframe, id.vars = "subject")

stacked_data
```

display the result

the melt function defaults to giving the new variables the names '*value*' and '*variable*' in the new dataframe, *stacked_data*

The new dataframe, which I have called *stacked_data* contains three columns. *Subject* is the subject id. You will see that each subject id is repeated twice, once for each of the variables. *variable* indicates which of the values this particular variable indicates (either *baseline_std* or *X1_year*) and the *value* column provides the actual value. Comparing this layout to that presented on the first page of this chapter explains why one is called the long format while the other is the wide.

We now need to order this dataframe on *subject* to produce a line for each subject's data rather than a single line. The following R code achieves this:

```
stacked_data <- stacked_data[order(stacked_data$subject),]

stacked_data
```

order the dataframe on the subject field

display the result

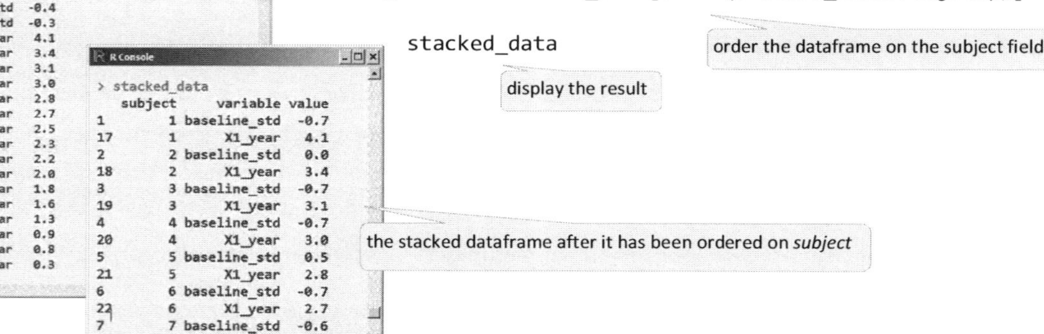

the stacked dataframe after it has been ordered on *subject*

To produce the graph shown on the previous page we then use the *xyplot()* function in the *lattice* package:

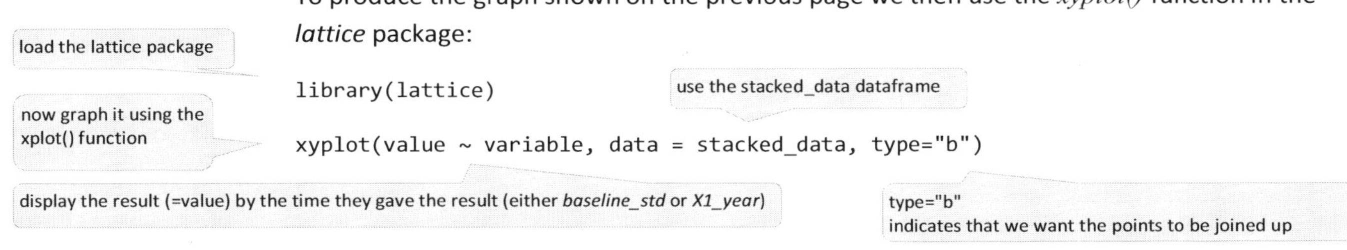
load the lattice package
now graph it using the xplot() function
use the stacked_data dataframe
display the result (=value) by the time they gave the result (either *baseline_std* or *X1_year*)
type="b"
indicates that we want the points to be joined up

```
library(lattice)

xyplot(value ~ variable, data = stacked_data, type="b")
```

18.4.1 Adding enhancements to the graph

We can add a title, axis titles and position the x values using the following R code:

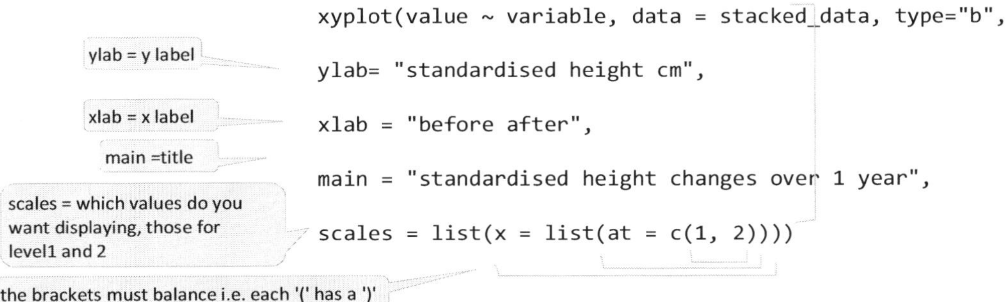
ylab = y label
xlab = x label
main =title
scales = which values do you want displaying, those for level1 and 2
the brackets must balance i.e. each '(' has a ')'

```
xyplot(value ~ variable, data = stacked_data, type="b",
ylab= "standardised height cm",
xlab = "before after",
main = "standardised height changes over 1 year",
scales = list(x = list(at = c(1, 2))))
```

The graph on the previous page as well as the one opposite makes it very easy to see how the individual subjects behave, each demonstrating an improvement, with some appearing quite dramatic.

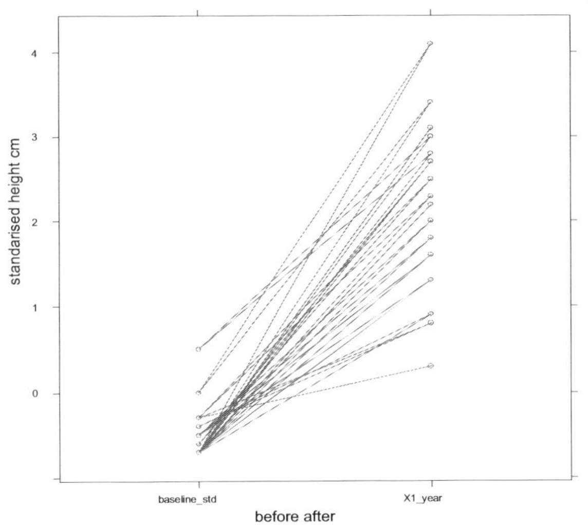
standarised height changes over 1 year
standarised height cm
0
1
2
3
4
baseline_std
X1_year
before after

The above process to create the graphs might seem rather complex but I believe it is worth the effort in helping understand what is going on with the data.

18.5 R code: paired *t* test

The screenshot at the start of the *Model development and evaluation* section to this chapter of the paired shows the R code generated from R Commander, which you could have typed this paired *t* test code in directly:

column of baseline values

column of values at one year

do a two sided test

carry out a t test

```
t.test(mydataframe$baseline_std, mydataframe$X1_year, alternative="two.sided",
conf.level=.95, paired=TRUE)
```

request a 95% CI

we require the results for a paired test

Note that the only difference between this and the code for the 2 independent samples *t* test (see the chapter with the that name) is the *paired=TRUE* option instructing the *t.test()* function that we require a paired *t* test result.

18.6 Reporting the results

A random sample of 16 subjects of below average stature received growth hormone therapy for a year. Standardised height measurements were taken prior to receiving treatment and again after one year of therapy. The sample standardised height mean prior to therapy was -0.48 compared to 2.17 following treatment. Taking into account the paired design the mean difference was 2.65 (SD = 1.06, 95% CI 2.09 to 3.22) producing a statistically significant result (t=9.99, df=15, p-value<0.001) indicating that the pre-post scores come from populations with different means.

It is therefore reasonable to state that the (unstandardized) effect size of 2.65 standardised height units is the average effect of the therapy.

18.7 Tips and Tricks

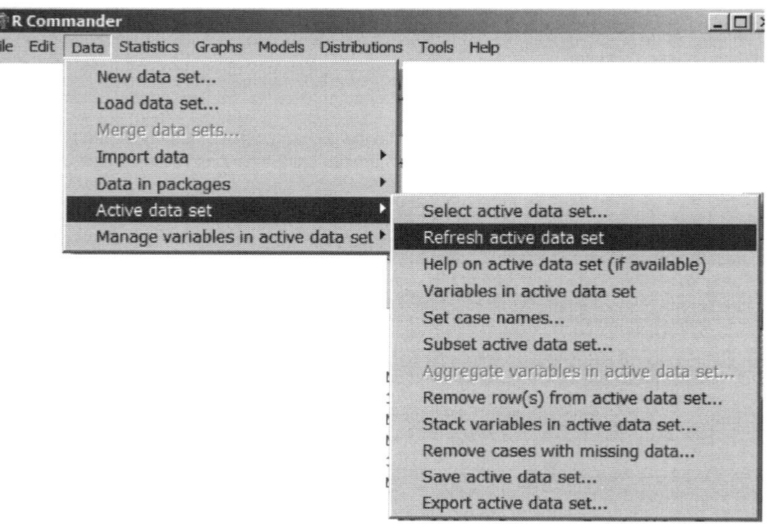

After creating a calculated variable sometimes you can't see the new variable in various selection boxes. This is remedied by selecting the R Commander menu option:

Data->Active data set->Refresh active dataset

The pre-post plot using the *reshape2* package can be adapted for a set of repeated measures when you have more than two measures which is extremely useful. This is developed further in the *Repeated measures* chapter.

Sometimes the pairings are not from the same subject but 'equivalent' subjects that are paired by the researcher. For example a researcher may have given a group of cyclists additional training and wishes to discover if it has any effect on their performance, however he imagines that additional factors such as age and cycling performance level may also affect their results. One way he would **control** for these additional factors (confounding / extraneous variables) is to pair the subjects in the control group with those in the treatment group who have similar characteristics on the confounding variables and whatever he is planning on measuring at the beginning of the research.

A paired design is frequently used because it reduces the size of the sample required. See the chapter *Sample size requirements* for more information.

19 Comparing 2 sample means: independent samples *t* test

The two independent samples *t* test allows the evaluation of means from two sample groups. Each subject can appear in only one of the groups, with subjects being allocated to groups at random. Remember the *t* statistic is basically the observed difference in means divided by the standard deviation of the expected random sampling variability of the means:

t = observed difference in means / SD of random sampling variability of means.

Assume we know each population is normally distributed with the same *Variance* and *Means.* Take two samples of equal size, one from each and repeat this process an infinite number of times. Alternatively, you can imagine a single population where the samples are divided randomly into two subsamples each time. Each time record the difference in the means of the two separate samples. You will discover that, over time, the most common value for this difference between the two means will be zero. As with the other distributions of means they will follow a *t* distribution. Because we also know the distribution of the means from the samples we can calculate its standard deviation which is also called the standard error (*SE*) of the mean.

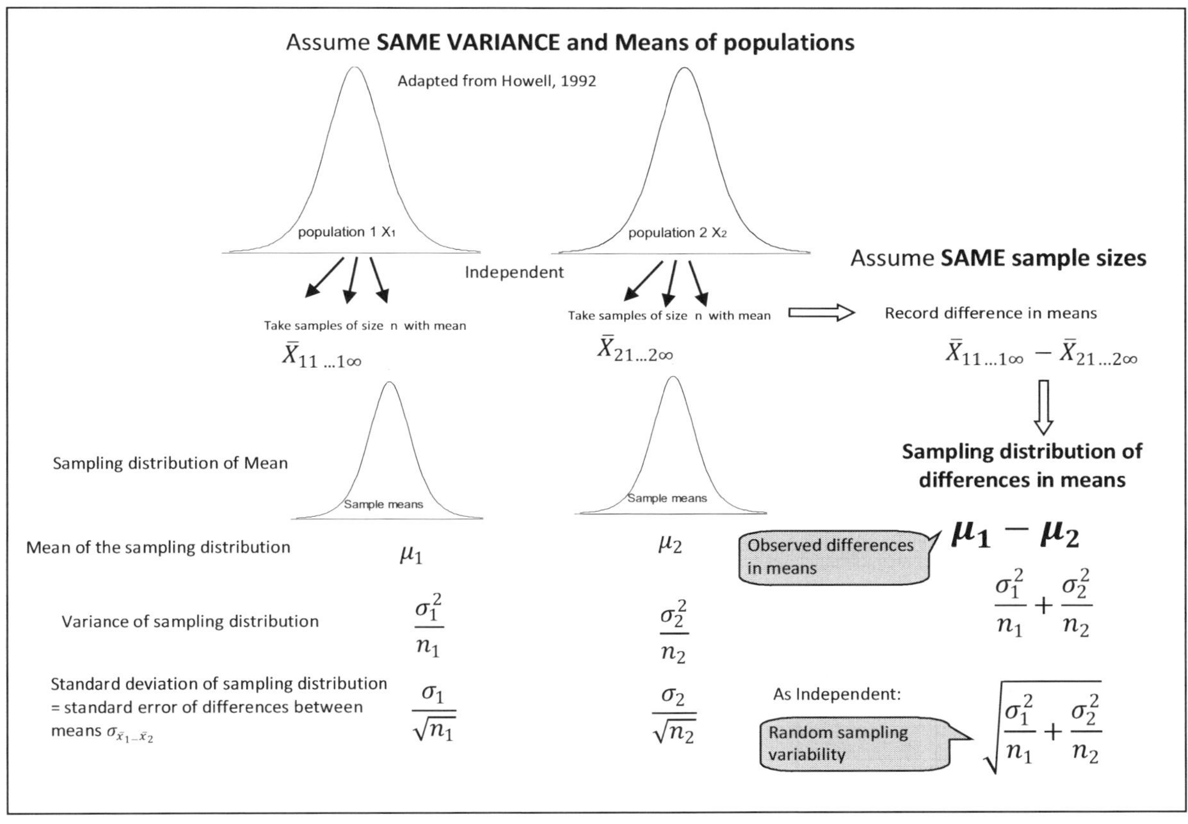

In the above scenario I have made three important assumptions, firstly that the two populations have equal means, secondly the spread (variance) is equal and thirdly that the two samples are of equal size. However, statisticians have discovered ways of adapting the *t* value to cope with violating these assumptions, usually requiring the above standard deviation of the sampling distribution expression to be adapted in some way.

To demonstrate the 2 independent sample t test we will use the example from MSME3 describing 200 subjects where half have been randomly allocated to a particular bronchodilator therapy and the other half given a placebo. I have changed the mean FEV's (Forced Expiratory Volume in mls x10) to equal 322 (bronchodilator therapy) and 289 (placebo).

The aim of bronchodilator therapy is to increase the Forced Expiratory Volume and this will be the focus of our analysis.

I will describe how to carry out the two independent samples *t* test in both R Commander and also directly in the *R Console* window following the typical data analysis stages discussed in the Introductory chapter.

19.1 Preparing the data

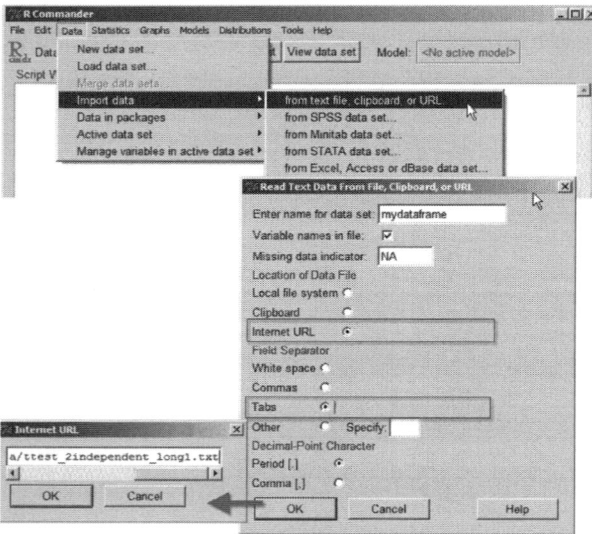

http://www.robin-beaumont.co.uk/virtualclassroom/book2data/ttest_2independent_long1.txt

From within R you need to load R Commander by typing in the following command:

```
library(Rcmdr)
```

To load the sample dataset select the R Commander menu option:

Data-> Import data->from text file, the clipboard, or URL

I have given the resultant dataframe the name *mydataframe*, also indicating that it is from a URL (i.e. the web) and the columns are separated by *tab* characters.

Clicking on the **OK** button brings up the *internet URL* box, type in it the following to obtain my sample data:

```
http://www.robin-beaumont.co.uk/virtualclassroom/book2data/ttest_2independent_long1.txt
```

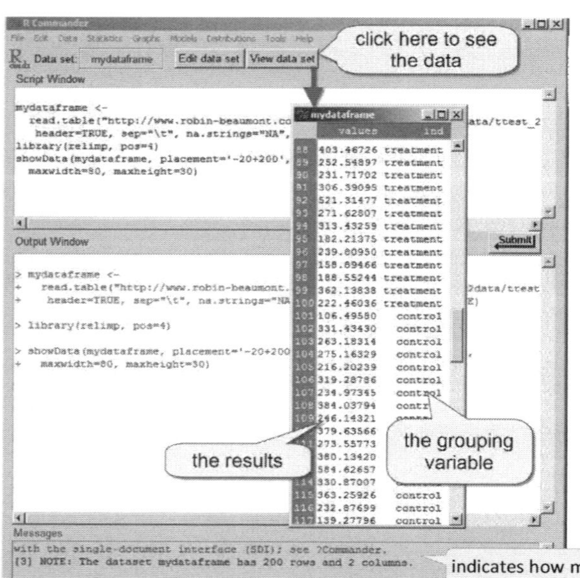

Click **OK** again.

It is always a good idea to see the data and you can achieve this easily in R Commander by clicking on the **View data set** button.

Note that I have all the observations in a single column with another one indicating which group they belong to, such a variable is often called a **grouping variable**.

A highlighted word indicates a glossary entry

19.2 Describing the data

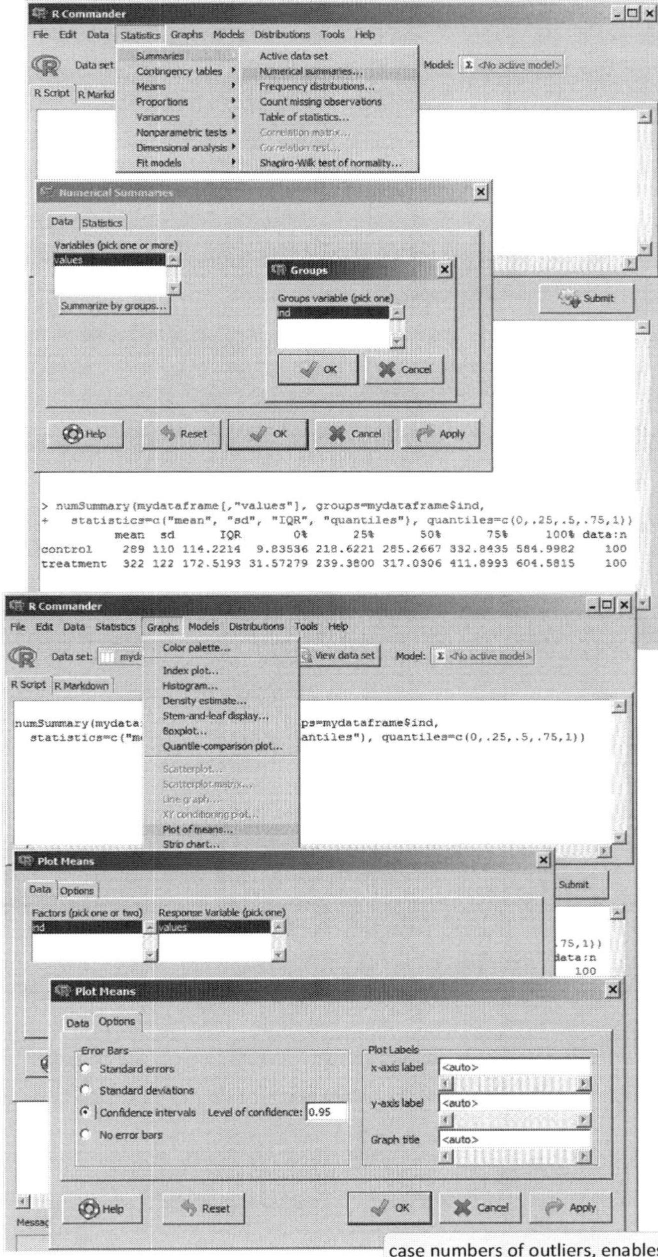

We can obtain a set of descriptive statistics for the two groups by using the R Commander menu option:

Statistics->Summaries-> Numerical summaries

Complete the dialog boxes as shown opposite, making sure you select the **summarize by group** button to get the group values.

You can also obtain the summary values for the entire sample across both groups by ignoring the *summarize by groups...* option.

We can see that the group means (control = 289, treatment = 322) and median (50% quartile) (control = 285.26, treatment = 317.03), both differ by around 30 millilitres x 10.

We can obtain a graph of the confidence intervals of the means for each group by using the R Commander menu option:

Graphs->plots of means

Complete the dialog box as shown opposite then click on both **OK** buttons to produce the results shown below.

We can see that the control group appears to have lower values than the treatment group (☺). However, the plot of means for the confidence intervals assumes normally distributed data. We can then produce boxplots, which do not make this assumption (for a reminder of how to do this, see the *Boxplots* chapter along with the glossary entry). The boxplots suggest that the FEV values from the two groups are more similar (☹) than that suggested by the confidence interval graph with several outliers in the control group. However this is an anomaly due to the different y axis scales of each, as demonstrated opposite.

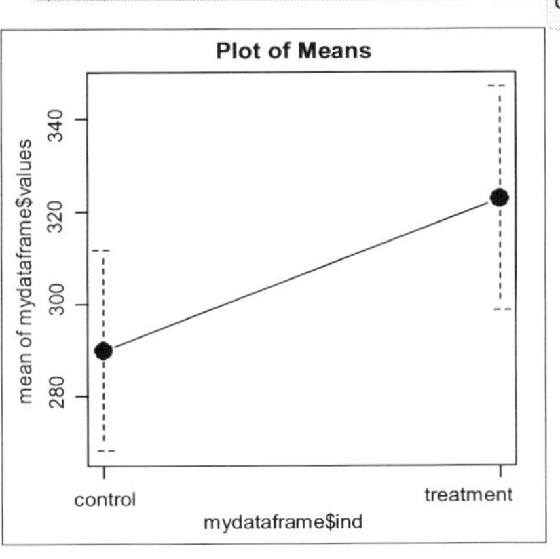

case numbers of outliers, enables follow-up.

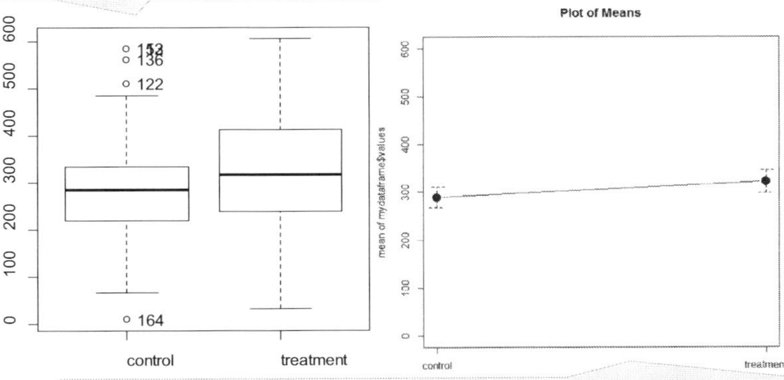

R code generated by R Commander edited to that below to set the y axis limits. *ylim=c(0,600)* added:
plotMeans(mydataframe$values, mydataframe$ind, error.bars="conf.int", level=0.95, ylim=c(0,600))

19.3 R Commander: Model development and evaluation

While the model here assumes that both the control group and treatment group samples come from the same population, in other words the means are equal and consequently the mean difference is zero, there is an additional issue that needs to be considered first.

To obtain a valid *t* value the variability of both samples must be similar, if they are not the *t* value needs to be adjusted to make it produce a valid value. So deciding if we need to adjust the *t* value means that we first must assess the relative variability of the two samples, a process which is called testing the 'equal variance assumption'. We can do this several ways. Informally by inspecting the heights of the boxes in the boxplots (previous page), where similar heights would suggest they are from a single population.

19.3.1 Checking the equal variance assumption

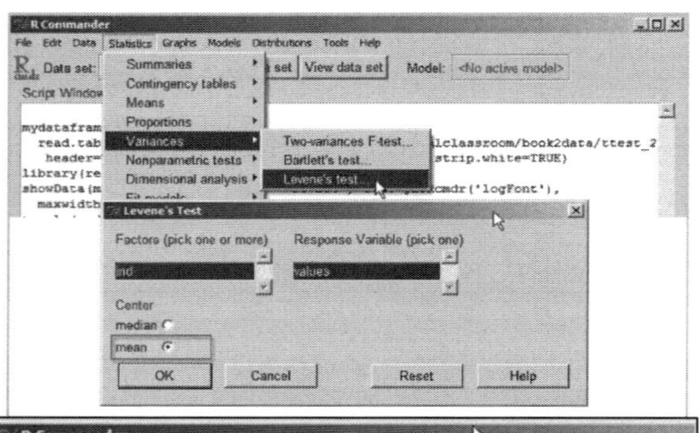

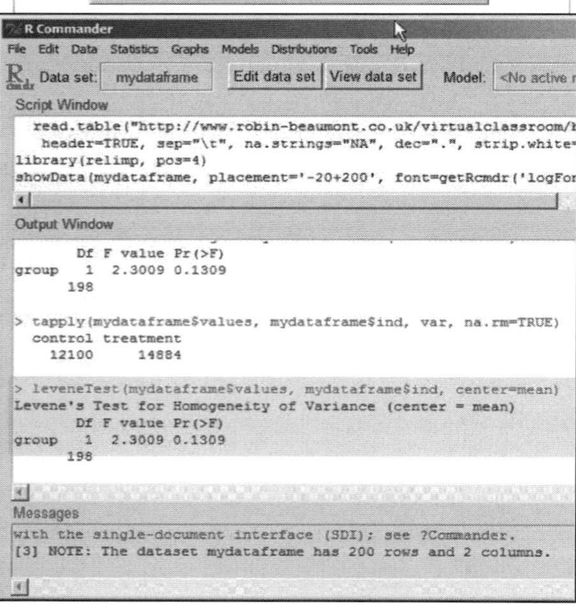

We can formally check the equal variance assumption by a procedure called Levene's test. Here we are looking for a statistically *insignificant* result, which would indicate that we would commonly see variances like those observed (or even more extreme) in our two samples when the samples came from a single population with a single variance.

We can carry out Levene's test on our dataset by selecting the R Commander menu option:

Statistics-> Variances -> Levene's test

R Commander has been clever enough to select the appropriate Factor variable (i.e. grouping variable) and also the Response variable (i.e. the results). The only thing we need to change is setting the centre option to the mean. Clicking on the **OK** button produced the results in the output window.

The associated p-value for Levene's test is 0.1309 and as this is greater than the usual critical value of 0.05, which we set, the result is statistically insignificant. The two degrees of freedom (df) values equal;

v_1 = number of groups minus one =2-1 =1

v_2 = total number of observations minus number of groups
= 200-2 =198.

Levene's test performs an ANOVA (see the *Introductory tutorial* chapter) on a transformed version of the scores (the deviations from the mean or median value for each group).

The statistically insignificant result indicates that we can continue with the 2 sample independent group *t* test without the unequal variance correction.

19.4 Two independent samples *t* test

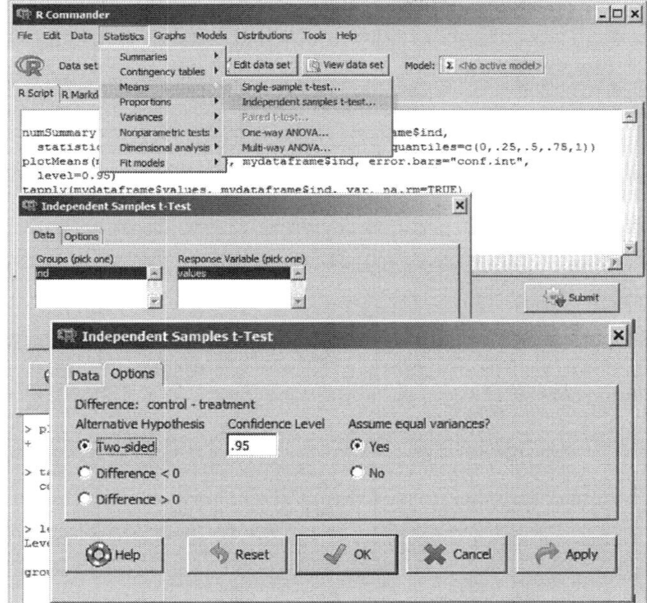

We can carry out the two independent samples *t* test on our dataset by selecting the R Commander menu option:

Statistics-> Means -> Independent samples t- test

Once again R Commander has been clever enough to complete most of the options. We only need to edit the equal variance option, selecting it from the options tab to indicate that we are assuming that the independent samples have equal variance. Technically we are indicating that they come from a population with a single variance and the observed variance differences are purely due to random sampling.

The results are shown below. As discovered previously in the boxplots we can see that the mean FEV value is higher in the treatment group (☺).

```
> t.test(values~ind, alternative='two.sided', conf.level=.95, var.equal=TRUE,
+   data=mydataframe)

        Two Sample t-test

data:  values by ind
t = -2.0089, df = 198, p-value = 0.0459
alternative hypothesis: true difference in means is not equal to 0
95 percent confidence interval:
 -65.3939512  -0.6060488
sample estimates:
  mean in group control mean in group treatment
                    289                     322
```

We can interpret this **p-value**:

We will find means as different as those observed or even more extreme four times in every hundred random samples of size 100 if our samples had come from a single population.

For an explanation of the above paragraph see the *Paired samples t test* chapter. If we had set a critical value to 0.05 because our obtained p-value is less than this we would say that we have, a statistically significant result and therefore our samples do come from populations with different means.

We are also given a 95% **confidence interval** of the mean that is useful when reporting the results and provides a method of making an alternative interpretation. We can interpret this as indicating that we would estimate that 95% of the time the true population mean would lie within the range of -65.39 to -0.60.

19.5 R code: Two independent samples *t* test

The above technique required that we had the data in a particular format, that is all the results in this situation were in a single column with another column acting as a grouping variable. By either using the *R Console* window or the R Commander *Script* window we have many more opportunities some of which are demonstrated below.

19.5.1 Separate groups per column

load the sample data into a dataframe called *mydataframe*, the file contains column names

There is another sample dataset containing the same data as the previous example but with each column representing the data from a single group, we can load it directly from my website into R by typing into the *R Console* or R Commander *Script* window:

```
mydataframe <- read.delim ("http://www.robin-
beaumont.co.uk/virtualclassroom/book2data/ttest_2independent_wide1.txt", header=TRUE)
```

```
names(mydataframe)
```

check the names of the columns by getting R to print them

```
t.test (mydataframe$treatment, mydataframe$control)
```

carry out the 2 independent samples *t* test which automatically adjusts for the possibility that each group may have a different range of scores (variance)

```
t.test  (mydataframe$treatment, mydataframe$control, var.equal=TRUE)
```

dataframe$column name format

stops the automatic variance adjustment

The results:

```
> t.test (mydataframe$treatment, mydataframe$control)

        Welch Two Sample t-test

data:  mydataframe$treatment and mydataframe$control
t = 2.0089, df = 195.915, p-value = 0.04592
alternative hypothesis: true difference in means is not equal to 0
95 percent confidence interval:
  0.6039283 65.3960717
sample estimates:
mean of x mean of y
      322       289
```

```
> t.test  (mydataframe$treatment, mydataframe$control, var.equal=TRUE)

        Two Sample t-test

data:  mydataframe$treatment and mydataframe$control
t = 2.0089, df = 198, p-value = 0.0459
alternative hypothesis: true difference in means is not equal to 0
95 percent confidence interval:
  0.6060488 65.3939512
sample estimates:
mean of x mean of y
      322       289
```

19.5.2 Checking the equal variance assumption

```
> var.test(mydataframe$treatment, mydataframe$control)

        F test to compare two variances

data:  mydataframe$treatment and mydataframe$control
F = 1.2301, num df = 99, denom df = 99, p-value = 0.3046
alternative hypothesis: true ratio of variances is not equal to 1
95 percent confidence interval:
 0.8276509 1.8281904
sample estimates:
ratio of variances
          1.230083
```

statistically insignificant at the 0.05 level ☺

As in R Commander we need to check this assumption, but because the data is in the group per column format we can't do Levene's test directly (see later) but there is an equivalent called the F test which we can do by typing:

```
var.test(mydataframe$treatment, mydataframe$control)
```

While the p-value here is nearly twice that obtained from Levene's test on the previous page it is still statistically insignificant (i.e. above the critical value we have set of 0.05) leading us to the same conclusion as before.

Mathematically the sample variance of one sample divided by the other follows a F distribution with degrees of freedom (v_1 and v_2), where we assume that the variation in the sample variances is purely due to random sampling, in others words they come from the same population.

$\frac{\text{sample 1 variance}}{\text{sample 2 variance}}$ follows a F distribution with degrees of freedom (v_1 and v_2)

$$\frac{S_1^2}{S_2^2} \sim F_{v_2}^{v_1}$$

The two degrees of freedom values this time are equal to;

v_1=number in group 1 minus one = 100-1 = 99

v_2= total number of group minus one = 100-1 = 99.

19.5.2.1 Long format equivalent analysis

You can also carry out the above analysis with the data in long format by selecting the relevant rows. Repeating the f test with the data in long format:

```
var.test(mydataframe$values[ind=="control"], mydataframe$values[ind=="treatment"])
```

or

```
var.test(mydataframe$values[ind=="treatment"], mydataframe$values[ind=="control"])
```

```
> var.test(mydataframe$values[ind=="control"], mydataframe$values[ind=="treatment"])

        F test to compare two variances

data:  mydataframe$values[ind == "control"] and mydataframe$values[ind == "treatment"]
F = 0.813, num df = 99, denom df = 99, p-value = 0.3046
alternative hypothesis: true ratio of variances is not equal to 1
95 percent confidence interval:
 0.546989 1.208239
sample estimates:
ratio of variances
         0.8129535
```

```
> var.test(mydataframe$values[ind=="treatment"], mydataframe$values[ind=="control"])

        F test to compare two variances

data:  mydataframe$values[ind == "treatment"] and mydataframe$values[ind == "control"]
F = 1.2301, num df = 99, denom df = 99, p-value = 0.3046
alternative hypothesis: true ratio of variances is not equal to 1
95 percent confidence interval:
 0.8276509 1.8281904
sample estimates:
ratio of variances
          1.230083
```

The above results also demonstrates the different results you obtain depending upon which group you consider to be the denominator or the numerator, however the important aspect, the p-value is the same for both. It is custom to have the larger variance as the numerator (top value).

19.5.3 Single column with a grouping variable

To do this, we need to return to the original dataset we started the chapter with and let R know which column contains the data (*values*) and which contains the grouping variable (*ind*):

```
mydataframe2 <- read.delim("http://www.robin-beaumont.co.uk/virtualclassroom/book2data/ttest_2independent_long1.txt", header=TRUE)
names(mydataframe2)
t.test (values ~ ind, var.equal=TRUE, data=mydataframe2)
```

load the sample data into a dataframe this time called *mydataframe2*, the file contains column names

check the names of the rows

carry out the 2 independent samples t test assuming there is no need to adjust for different variances for each group

read this as values (i.e. the result) by *ind* (i.e. the group indicator)

this is an alternative way of specifying the data, it means I do not need to use the dataframename$column name format

```
R Console
>
> names(mydataframe2)
[1] "values" "ind"
> t.test (values ~ ind, var.equal=TRUE, data=mydataframe2)

        Two Sample t-test

data:  values by ind
t = -2.0089, df = 198, p-value = 0.0459
alternative hypothesis: true difference in means is not equal to 0
95 percent confidence interval:
 -65.3939512  -0.6060488
sample estimates:
 mean in group control mean in group treatment
                   289                     322
```

The results are the same as presented previously.

When we have a grouping variable we can obtain Levene's test from our data using the following command:

```
leveneTest(mydataframe$values, mydataframe$ind, center=mean)
```

note the capital T

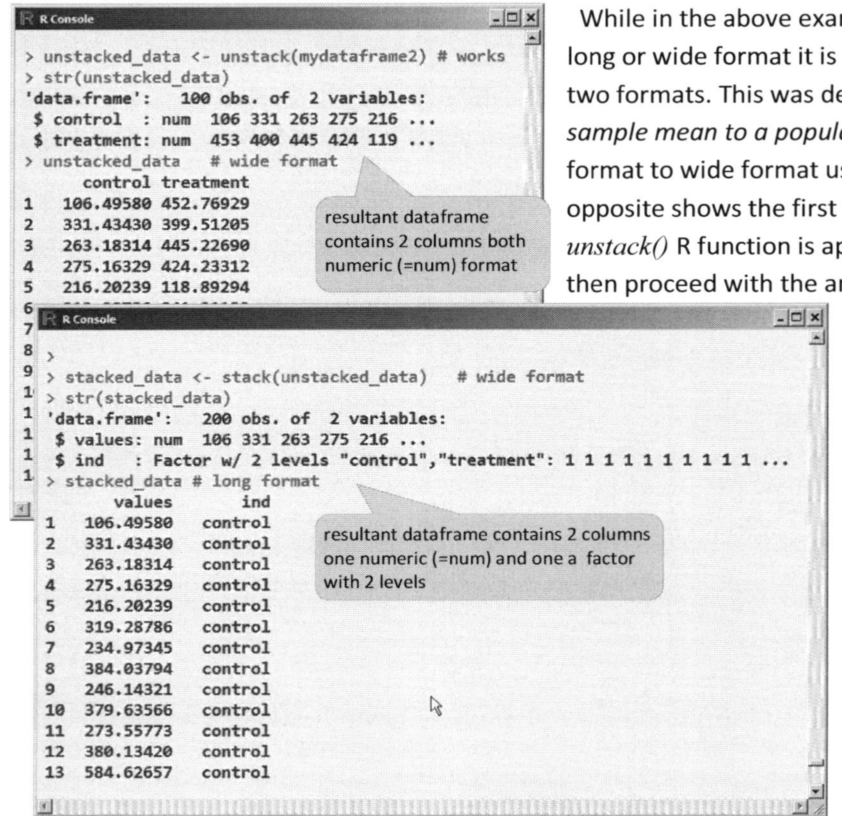

R Console
> unstacked_data <- unstack(mydataframe2) # works
> str(unstacked_data)
'data.frame': 100 obs. of 2 variables:
$ control : num 106 331 263 275 216 ...
$ treatment: num 453 400 445 424 119 ...
> unstacked_data # wide format
control treatment
1 106.49580 452.76929
2 331.43430 399.51205
3 263.18314 445.22690
4 275.16329 424.23312
5 216.20239 118.89294
resultant dataframe contains 2 columns both numeric (=num) format
R Console
> stacked_data <- stack(unstacked_data) # wide format
> str(stacked_data)
'data.frame': 200 obs. of 2 variables:
$ values: num 106 331 263 275 216 ...
$ ind : Factor w/ 2 levels "control","treatment": 1 1 1 1 1 1 1 1 1 1 ...
> stacked_data # long format
values ind
1 106.49580 control
2 331.43430 control
3 263.18314 control
4 275.16329 control
5 216.20239 control
6 319.28786 control
7 234.97345 control
8 384.03794 control
9 246.14321 control
10 379.63566 control
11 273.55773 control
12 380.13420 control
13 584.62657 control
resultant dataframe contains 2 columns one numeric (=num) and one a factor with 2 levels

While in the above examples I have either worked with the data in long or wide format it is possible to convert the dataset between the two formats. This was demonstrated in the chapter *Comparing a sample mean to a population mean*. Here we move from data in long format to wide format using the *unstack()* R function. The screenshot opposite shows the first few rows of the dataframe created when the *unstack()* R function is applied to the *mydataframe2* object. We could then proceed with the analysis described previously for the wide format data.

Working in the opposite direction, from wide to long format, if the dataframe does not contain any factors, as is the case with the *unstacked_data* dataframe, we can use the *stack()* R function to convert it from wide format to long, as shown opposite.

I have also used the *str()* function in the screenshots opposite to check the structure of the dataframes produced.

19.6 Effect size / clinical significance

One should always report the actual measure of the difference between the two sample means, and an appropriate measure is called the (standardised) effect size. In this situation it is the difference in the means divided by the pooled standard deviation.

```
effectsize <- (mean(mydataframe$treatment) -
mean(mydataframe$control))/sapply(stack(mydataframe), sd)
effectsize
```

the difference in the means

divided by

the standard deviation of the entire 200 subjects readings.

```
> # effect size - wide data
> effectsize <- (mean(mydataframe$treatment) - mean(mydataframe$control))/
+ sapply(stack(mydataframe), sd)
> effectsize
    values        ind
 0.2819605 65.8347932
```

The relevant value is the 0.2819 indicating the standardised effect size (also known as Cohen's *d*). Referring to the glossary entry **Effect size**, we see this represents a small effect size. If you have the data in long format you can request the mean for each subset of each subset of rows, representing the control and treatment values.

find the mean of . .

select the *values* column where . .

only those rows where the *ind* variable is equal to *treatment*

minus the mean of . . .

```
effectsize <- (mean(mydataframe2$values[mydataframe2$"ind" =="treatment"])
- mean(mydataframe2$values[mydataframe2$"ind" =="control"]))/sd(mydataframe2$values)
effectsize
```

select the *values* column where . .

only those rows where the *ind* variable is equal to *control*

divide by the Standard Deviation of all the values in the values column.

```
> effectsize <- (mean(mydataframe2$values[mydataframe2$"ind" =="treatment"])
+ - mean(mydataframe2$values[mydataframe2$"ind" =="control"]))/sd(mydataframe2$values)
> effectsize
[1] 0.2819605
```

19.7 Writing up the results

Key point
Always remember to report the effect size measure and comment upon it in terms of clinical usefulness.
Would an effect size of 0.28 here be of any practical/clinical use in terms of symptom control in asthma?

Two hundred subjects were randomly allocated to either receive treatment with a particular bronchodilator (mean FEV 322, SD 122) or receive a placebo (mean FEV 289, SD 110). A F test (F=1.23, p-value=0.304) indicated that both samples did not have statistically significantly different variances. A two sample independent *t* test produced a statistically significant result at the alpha =0.05 level (t = 2.0089, df = 198, p-value = 0.0459) leading to the conclusion that the two samples did not come from the same population with a single mean.

The standardised effect size was small (0.281) indicating an observed difference in means of nearly 3 tenths of a standard deviation between the groups.

19.8 Tips and Tricks

For the two independent sample *t* test, the most helpful data layout is where you have a single column for the results and another for the grouping variable (i.e. factor). Remember this when you start collecting your data.

For the grouping variable values, rather than using names like control and treatment, use numbers such as 1=treatment group and 2=control.

When reporting the *t* test results:

- Remember that a single picture says a thousand words therefore include a boxplot of the two groups. See the chapter *Creating graphical summaries of data* for details of how to do this.
- Include the 95% confidence intervals in with the results beside the p-value.

You can see the R code that R Commander calls up when you select the Levene's test option by typing in *car:::leveneTest.default*. In amongst the code that is displayed you will notice the term *anova* indicating the ANOVA function.

20 Comparing pre-post test median difference: Wilcoxon Matched Pairs Statistic

The Wilcoxon Matched Pairs statistic allows us to evaluate how different a median obtained from a sample is to that of an expected median of zero.

bold words indicate a glossary entry

In the previous chapters a great deal was made about the assumption that samples and the populations from which they come were normally distributed (bell shaped curve that is defined by two parameters; the mean and the standard deviation). But how can we be sure? This question has plagued a large number of researchers and differences of opinion as to how important this assumption is constantly varies. Those statisticians who believe that it is too great a risk to ignore this assumption have developed a set of statistics which do not rely upon the sample being taken from a **normal distribution**. These statistics are called non-parametric or distribution free as the distribution of the parent population is either assumed to be unknown or unable to be described by one or more values (parameters).

Before we can start to discuss what the Wilcoxon Matched Pairs statistic is in detail we need to understand something about ranking data.

20.1 Ranking Data

The process of ordering data and assigning a numerical value is called *Ranking*. Let's take an example by considering the following numbers: 5, 3, 8, 1, 10. Ranking them from smallest to largest and assigning a value to each 'the rank' would produce the result seen here.

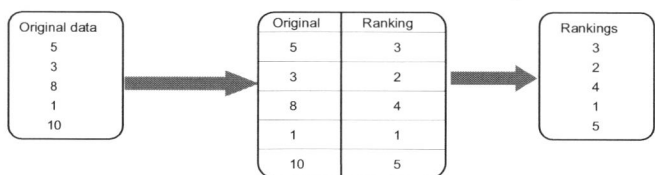

Score (ordered)	Rank
10	1
8	2
5	3
3	5
3	5
3	5
1	7

What do we do if we have the situation of tied scores (**ties**) that is two, or more, with the same value?

Example: Consider the following numbers 5, 3, 8, 3, 1, 3, 10

Placing them in order of magnitude: 10, 8, 5, 3, 3, 3, 1, we note that there are three 3's. These are equivalent to the ranked scores of the 4th, 5th and 6th score. We therefore allocate the average of these ranks (i.e. 4 + 5 + 6 / 3 = 5) to each of them.

20.1.1 Magnitude and Ranking

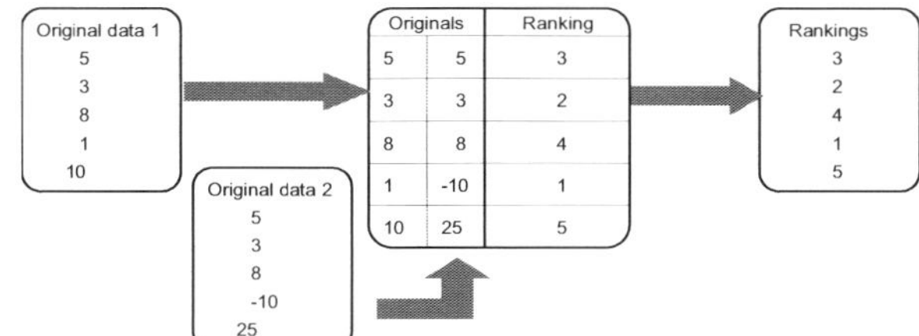

Now consider the following example where instead of one set of data there are two as given below. Notice that increasing the magnitude of the lowest and highest scores has no effect on their rankings. That is by ranking our data we have lost the importance of magnitude in the original dataset.

By ranking our data we lose the magnitude of each observation and we are left with order only.

bold words indicate a glossary entry

Some data is explicitly a set of rankings (i.e. ordinal **level of measurement**) whereas other datasets may contain variables that originally possessed interval/ratio level of measurement but have been converted to rank type data (ordinal). For example weight which may have been originally recorded in Kg. for each subject and subsequently have been converted to an ordinal variable with four values such as (=1), normal weight (=2), obese (=3), and grossly obese (=4). Whenever possible always keep the original values in addition to any ranking you may create, keeping them in a different column.

20.2 The Wilcoxon Matched Pairs statistic

Consider this example. From a group of newly enrolled members to an exercise gym six were randomly selected and asked to indicate on a scale of one to ten how satisfied they were with their personal trainer both after an initial session and again after two months. What we are interested in investigating is how the score has changed over time for each participant rather than the group difference. Because this difference is calculated at the individual level the pair of values are said to be dependent. The values are shown opposite, in the pre and post columns. The Wilcoxon Matched Pairs statistic considers the difference between each pair of scores for each subject which is then ranked and given either a T^+ or T^- value.

Pre	Post	Difference	Rank Difference 1=smallest	pos ranks T+	neg ranks T-
1	7	6 (=7-1)	6+	6	
4	6	2 (=6-4)	2+ (=1+2+3/3)	2	
3	8	5 (=8-3)	5+	5	
5	7	2 (=7-5)	2+ (=1+2+3/3)	2	
4	6	2 (=6-4)	2+ (=1+2+3/3)	2	
5	9	4 (=9-5)	4+	4	
TOTALS				T+=21	T-=0

20.2.1 T^+ and T^- - The positive and negative rankings

The question is what do these T^+ and T^- values indicate? This question is best answered by considering what would their values be if there was no actual difference between the two sets of scores, other than that due to random variability. If this were the case it would be expected that the majority of the difference scores would be zero with some small positive and negative values. This would produce a median (see glossary entry **Boxplot**) of zero. All those with a positive ranking are assigned to the positive rank column (column 5 above). All those with a negative ranking are assigned to the negative rank column (column 6 above) and the zero scores are ignored (equivalent to ties in this context).

In the above example, there are no negative difference scores and therefore no negative rankings. Similarly, there are no pairs that have a difference value of zero where such a pair would not contribute to the analysis and effectively reduce the sample size.

Finally, we sum the ranking values for the negative and positive columns. We consider each of these values as a statistic denoted as T^+ and T^-. Assuming that there is no difference except that of random sampling variation we would expect roughly equal values for T^+ and T^-. So:

T^+ = positive rankings added together = gives us an idea of number of scores that improve
T^- = negative rankings added together = gives us an idea of number of scores that get worse

Let us consider some interesting facts about the above ranks, specifically the positive rank column in the table.

Smallest value = smallest ranking (i.e. 1st ranking); of which we have three pairs with the smallest difference (2) so all three get a rank of '2' – second place. Then considering the other three difference scores in increasing magnitude they achieve rankings of 4, 5, 6 = largest = nth ranking.

The sum of the ranks for both groups always adds up to n(n+1)/2 where n is the total number of paired observations so in this instance it is 6(6+1)/2 =21 as required – this is a good check. This is the maximum possible value of the rankings and we have this value appearing in the positive rankings column!

If the negative and positive differences were roughly equal we would have about half the above total in each column, that is n(n+1)/2÷ 2= n(n+1)/4. For our example 6(6+1)/4=10.5. We will call this the median of T (μ_T), notice that this is the opposite of what we have above. Here n (6) is the number of pairs of observations not the total number (12).

Currently we have a statistic so the next question is can we draw inferences from it (i.e. provide an associated p-value)? To do this we must be able to show that it follows a sampling distribution or somehow we must calculate the associated probabilities (see later).

For small values of n (less than 30 pairs) people used to look up the associated p-value in a table. However with the advent of powerful handheld devices, we can calculate the exact probability value using a method called exact statistics, for details see the *Resampling* chapter.

$$z = \frac{T^{+or-} - \frac{N(N+1)}{4}}{\sqrt{\frac{N(N+1)(2N+1)}{24}}}$$

In contrast, for larger n it has been shown that the sampling distribution of the statistic follows a normal distribution. So we can convert the T^+ or T^- score to a **z score** and then use the standard normal distribution to get the associated p-value. I have provided the equation that converts the T^+ or T^- score to a z score, but usually you let a computer do the work. We will now carry out an appropriate analysis in R Commander and R.

20.3 Data preparation

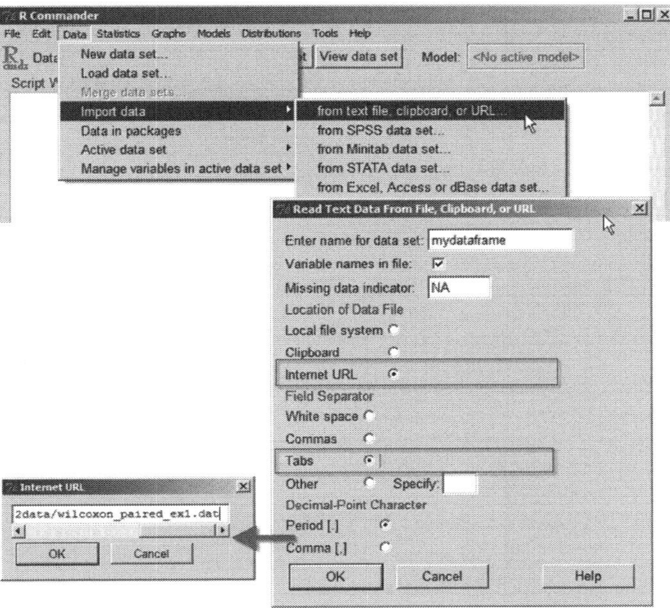

http://www.robin-beaumont.co.uk/virtualclassroom/stats/basics/coursework/data/wilcoxon_paired_ex1.dat

Type the following in the *R Console* window to load R Commander:

```
library(Rcmdr)
```

Load the sample dataset directly from my website by selecting the R Commander menu option:

Data-> from text, the clipboard or URL

I have given the resultant dataframe the name *mydataframe*, also indicating that it is from a URL (i.e. the web) and the columns are separated by *tab* characters.

Clicking on the **OK** button brings up the *Internet URL* box, you need to type in it the following to obtain my sample data:

```
http://www.robin-beaumont.co.uk/virtualclassroom/book2data/wilcoxon_paired_ex1.dat
```

Click **OK**.

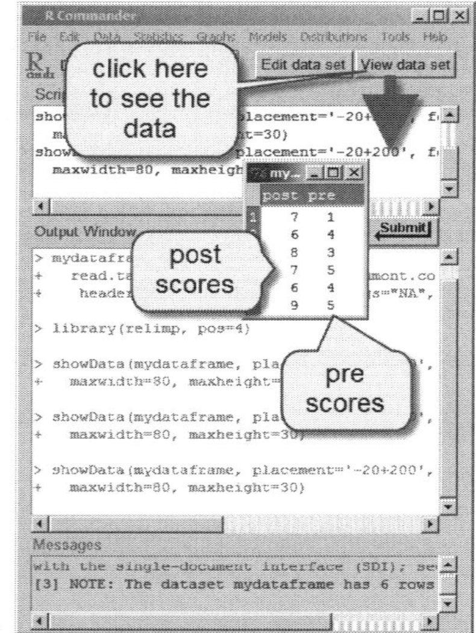

It is always a good idea to see the data and you can achieve this easily in R Commander by clicking on the **View data set** button.

Note that the observations are in two columns.

Close the *view dataset* window.

Because the second score is dependent upon the first for each individual, it is sensible to see the difference each individual achieved by creating a new column (variable) which measures this difference.

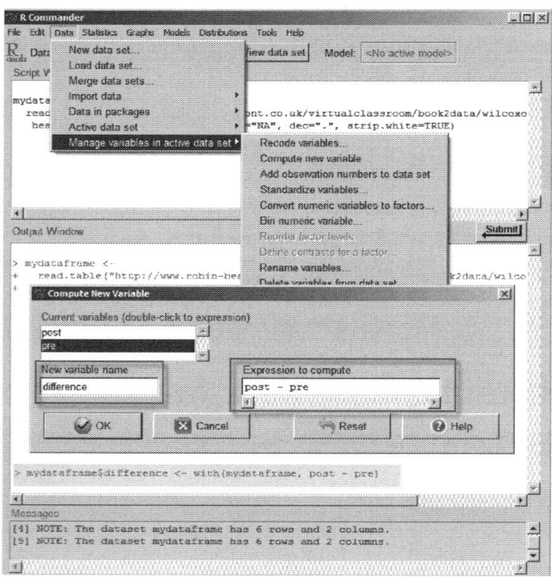

20.3.1 Creating a new column in a dataframe

We can do this either by using R Commander or R code.

20.3.1.1 R Commander

To create the difference column using the R Commander menu option select:

Data-> Manage variables in active data set -> compute new variable

Then complete the dialog box as shown opposite.

20.3.1.2 R code

To create a new column called *difference* in R directly type the following either into the *R Console* or R Commander *Script* window:

```
mydataframe$difference <- mydataframe$post - mydataframe$pre
```

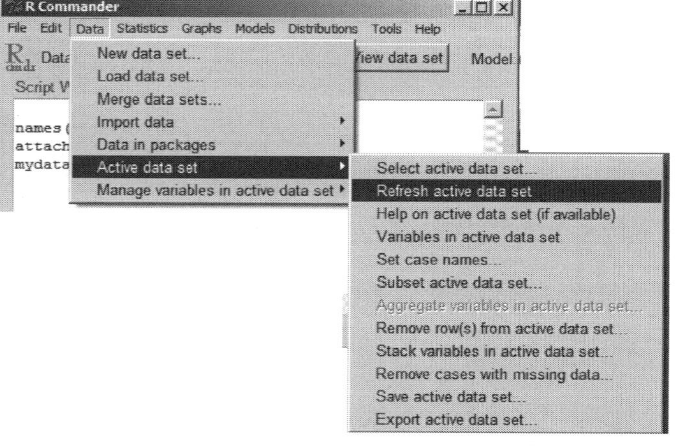

We need then to let R know we have changed the dataframe and this is achieved by selecting the R Commander menu option:

Data-> Active data set -> Refresh active dataset

20.4 Data description: Boxplots for several columns

In R Commander you can produce boxplots for a single column only. What we need here is a boxplot for each of the columns side by side, which means we need to use R code.

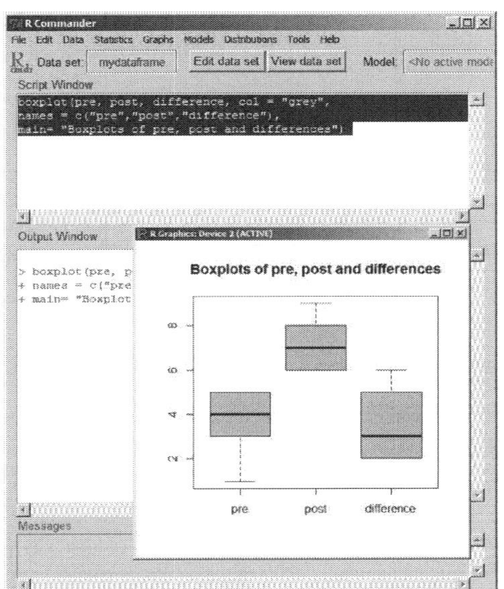
R Commander
Boxplots of pre, post and differences
pre
post
difference

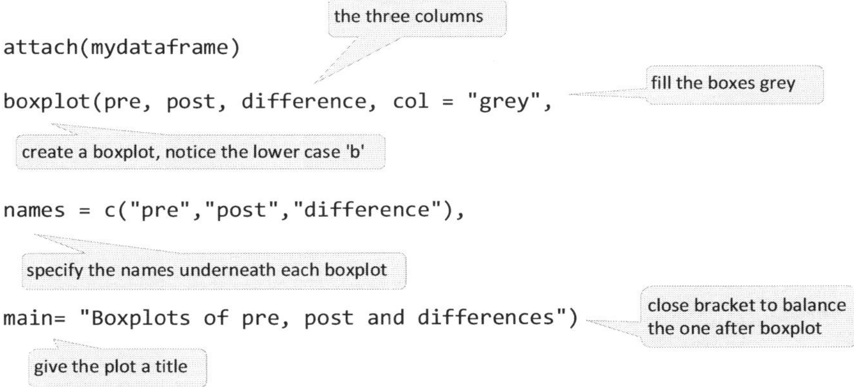
attach(mydataframe)
the three columns
boxplot(pre, post, difference, col = "grey",
fill the boxes grey
create a boxplot, notice the lower case 'b'
names = c("pre","post","difference"),
specify the names underneath each boxplot
main= "Boxplots of pre, post and differences")
close bracket to balance the one after boxplot
give the plot a title

From the difference box plot, that is the important one, we can see that there are no negative scores and the median change is around +3. That is the average person seems to have increased how they rated their personal trainer after two months by 3. ☺

20.5 Model development and evaluation: R Commander

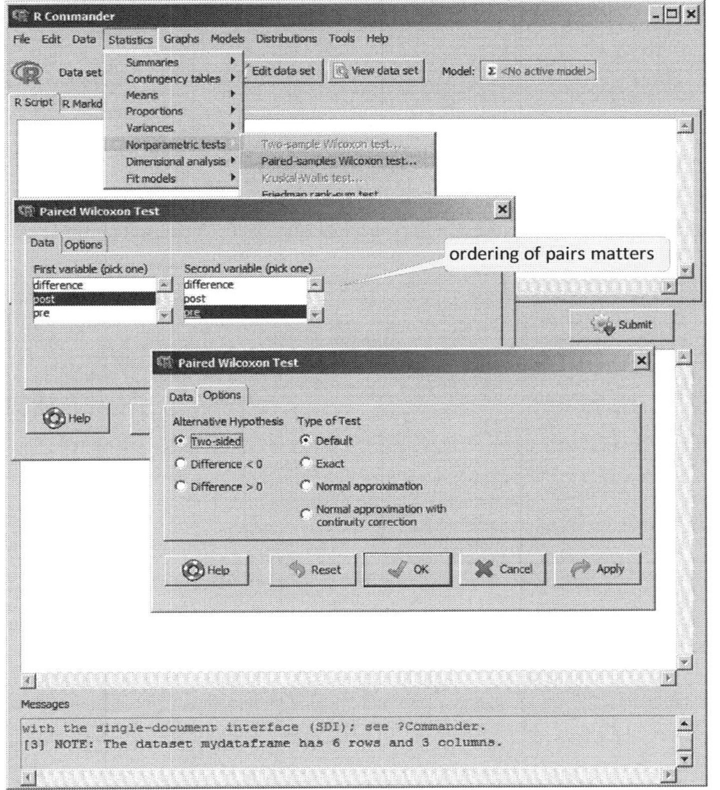
R Commander
Paired Wilcoxon Test
ordering of pairs matters

Because this data is at an ordinal **level of measurement** it is not sensible to examine values such as the mean or Standard Deviation etc. (although there is no harm in noting them). It is even less sensible to consider developing a model where we try to compare the mean of our difference scores to that of a specific population mean, instead we need to develop a different strategy using the idea of rankings.

We can do this by developing a model that assumes a population with an equal number of negative and positive rankings and then compare the observed T^+ or T^- score against this population. So our model is now:

How likely it is to obtain a $T^+ = 21$ or one more extreme from a sample of 5 difference scores randomly selected from a population that has equal numbers of negative and positive rankings?

We can get R Commander to check this model using the menu option:

Statistics-> Nonparametric tests -> Paired samples Wilcoxon test

You then select the type of test as *default*. Setting the **First variable** to *post pre* and the **Second variable** to *pre.* Then click **OK**.

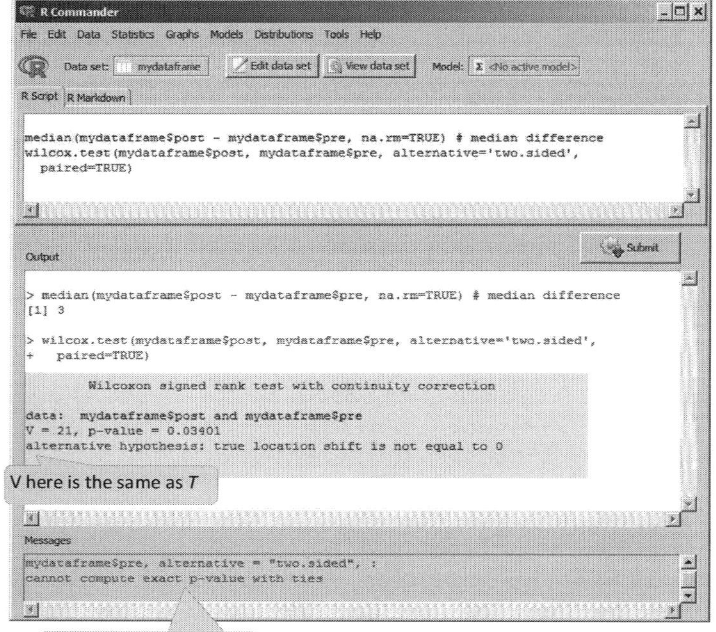
R Commander
File Edit Data Statistics Graphs Models Distributions Tools Help
Data set: mydataframe
Edit data set
View data set
Model: <No active model>
R Script R Markdown
median(mydataframe$post - mydataframe$pre, na.rm=TRUE) # median difference
wilcox.test(mydataframe$post, mydataframe$pre, alternative='two.sided',
paired=TRUE)
Submit
Output
> median(mydataframe$post - mydataframe$pre, na.rm=TRUE) # median difference
[1] 3
> wilcox.test(mydataframe$post, mydataframe$pre, alternative='two.sided',
+ paired=TRUE)
Wilcoxon signed rank test with continuity correction
data: mydataframe$post and mydataframe$pre
V = 21, p-value = 0.03401
alternative hypothesis: true location shift is not equal to 0
V here is the same as T
Messages
mydataframe$pre, alternative = "two.sided", :
cannot compute exact p-value with ties

non-specific warning. See the ties section below.

The type of test options allows you to specify if you want an exact p-value calculated. This can be very slow if you have a dataset with more than 100 pairs but is the better type of p-value for smaller datasets.

The Normal approximation test option sometimes called the *Asymp. Significance,* meaning the asymptotic p-value (2 tailed), is the one that you would traditionally report, however the problem with it is in the name. Asymptotic, basically means "getting ever closer to" and this is what it means here, "getting closer to the correct answer as the sample size increases". This is because, for many statistics, there are no simple formulae which give the perfect (i.e. error free) associated p-value. Instead the formulae give more accurate results as the sample size increases, these are roughly correct with small samples, but get better with larger samples and would be perfect with very large samples.

For small samples (i.e. < n=50) it is best to use the exact probability and R defaults to this method when n<50.

While the ordering of the pair of variables you select matters affecting the median and confidence interval values (below), it has no effect on the p-value.

As usual the result appears in the R Commander *Output Window*.

The p-value of 0.034 is smaller than the usual critical value of 0.05 so we can say that the sample does not come from a population with an equal number of positive and negative rankings (i.e. median of the population is assumed to be zero). Therefore rejecting this model indicates that the sample comes from a population where the median is not equal to zero.

20.5.1 R code: Wilcoxon test

Looking above at the R code generated by R Commander we can see that it uses the *wilcox.test()* command with the option *paired=TRUE.*

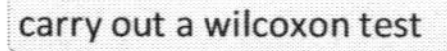
carry out a wilcoxon test

column of values post intervention

column of baseline values

do a two sided test

```
wilcox.test(mydataframe$post, mydataframe$pre, alternative="two.sided", paired=TRUE)
```

we require the results for a paired test

20.6 The two types of ties

ONE important thing to note in the results is the warning concerning ties - if a dataset has a large number of ties it can impinge upon the validity of the results.

Ties are of two types in this situation; zero ties, that is where the pre and post value are the same and are not included in the analysis so effectively reducing the sample size by the number of such ties. The other type of tie is for a specific difference such as we have for the '2's and reduce the effective sample size to a lesser degree.

20.7 Obtaining a confidence interval

You will notice that the previous R Commander output did not produce a confidence interval. This can easily be remedied by editing the code produced from the menu option in the R Commander script window adding the option *conf.int=TRUE*, then selecting it and clicking on the **Submit** button:

```
wilcox.test(mydataframe$post, mydataframe$pre, alternative="two.sided", paired=TRUE, conf.int=TRUE)
```

In the screenshot below I have used the *conf.int=TRUE* option. Two things to note:

```
R Console
> wilcox.test(mydataframe$post, mydataframe$pre, alternative="two.sided", paired=TRUE, conf.int=TRUE)
Warning in wilcox.test.default(mydataframe$post, mydataframe$pre, alternative = "two.sided",  :
  requested conf.level not achievable
Warning in wilcox.test.default(mydataframe$post, mydataframe$pre, alternative = "two.sided",  :
  cannot compute exact p-value with ties
Warning in wilcox.test.default(mydataframe$post, mydataframe$pre, alternative = "two.sided",  :
  cannot compute exact confidence interval with ties

        Wilcoxon signed rank test with continuity correction

data:  mydataframe$post and mydataframe$pre
V = 21, p-value = 0.03401
alternative hypothesis: true location shift is not equal to 0
80 percent confidence interval:
 2.000000 4.999986
sample estimates:
(pseudo)median
      3.500034

> |
```

1. Because of the order I entered the pre, post variables the calculated difference scores is *post-pre*, the difference we are interested in. The *median* and *CI* values are therefore for the improvement from the baseline scores. If you are in doubt about the ordering produce a boxplot of the difference scores to guide your decision.

2. The small sample size and the number of ties meant R was unable to produce an interval with the required (i.e. 95%) confidence. However, in reality we are never going to have such a small sample size! For more information you can check out the R online help for the command at:

http://stat.ethz.ch/R-manual/R-patched/library/stats/html/wilcox.test.html

20.8 Showing the Wilcoxon T+ distribution

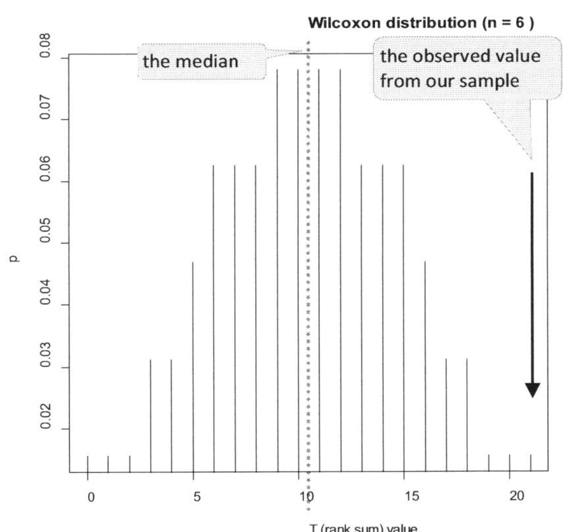

It is possible to show the distribution of all possible values the Wilcoxon $T^{+ \text{ or } -}$ will take for a particular sample size of difference rankings given that the population has an equal number of positive and negative rankings.

From the graph we can see that the observed *T* value obtained **from this particular sample** is the maximum value it could have obtained. Also the median value can be seen to be 10.5 as calculated earlier.

For reference, the R code used to produce this graph is given below. To produce a graph for a particular sample size you only need to change the first line of the code:

```
n<-6          change the 6 to the sample size you require

efsize<-n*(n+1)/2

x <- seq(0,efsize)

plot(x, dsignrank(x,n), type="h",

xlab="T (rank sum) value", ylab="p",

main=paste("Wilcoxon distribution (n =",n,")"))
```

20.9 Effect size – Clinical importance

Because we are dealing with ranks, and not using the normal distribution assumptions, an effect size measure is problematic (what is the pooled standard deviation here?). Often researchers just treat the ranks as data and use the rank equivalent to the standard deviation.

$$\text{Standard deviation} = \sigma_{T^+} = \sqrt{\frac{n(n+1)(2n+1)}{24}}$$

Using this expression as the denominator along with the numerator as the observed difference from that expected (i.e. value of T^+ minus the mean of T), were T is the sum of difference ranks. This actually gives us a standardised score:

$$z = \frac{T^+ - \mu_{T^+}}{\sigma_{T^+}} = \frac{T^+ - n(n+1)/4}{\sqrt{n(n+1)(2n+1)/24}}$$ for our example this becomes

$$z = \frac{21 - 6(7)/4}{\sqrt{6(7)(13)/24}} = \frac{10.5}{\sqrt{\frac{546}{24}}} = \frac{10.5}{4.769} = 2.201$$

We can alternatively ignore the denominator and as a result produce the **signed-rank statistic,** 'S' which can be considered as an unstandardized effect size measure:

$$S = \left| T^{+or-} - \frac{n(n+1)}{4} \right|$$ *For our example:* $S = \left|21 - \frac{6(7)}{4}\right|$ or $S = \left|0 - \frac{6(7)}{4}\right|$

S = 10.5

The vertical lines indicate that the result is a positive value | |.

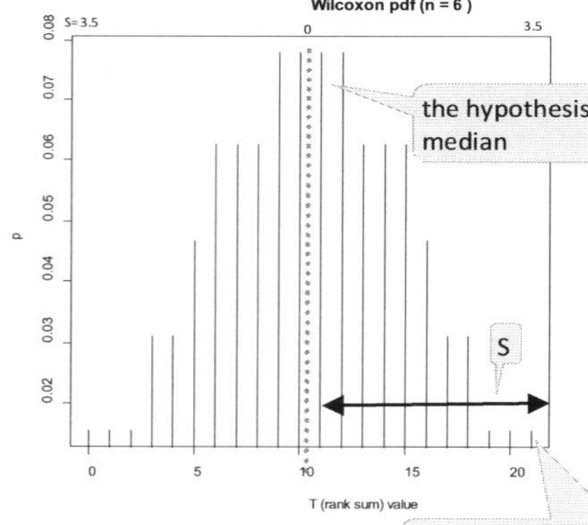

Because S is not a very intuitive measure lets consider some typical values it might take. S would equal zero when the positive and negative ranks balance one another. That is when it equals the median rank sum. We know for our sample this is 10.5 from the distribution graph. It reaches a maximum value when T either equals zero or is at its maximum value, which in this instance is 21. So our value here of 10.5 indicates a maximum effect size when n=6.

Because many readers are unfamiliar with the S measure it is a good idea to indicate that it is just the difference between the mean sum of difference ranks (σ_{T^+} =10.5) and the sum of either the positive of negative ranks (T^+ =21).

In addition, to contextualise this value, I would report how many units the difference scores median was from the expected median (0), which is +3.5 units for our sample. In fact, it is often sufficient to simply report this value as an indication of effect size for this type of data.

20.10 Finding and reporting ties

I have mentioned previously that it is important to report the number of ties, particularly zero ties, as these effectively reduce the sample size available for the analysis. This requires the use of R code which you can type into either the *R Console* or R Commander *Script* windows.

```
post<-c(7,6,8,7,6,9)
pre<-c(1,4,3,5,4,5)
difference <- post - pre
atable <- table(difference)
atable
percent <- (atable/length(difference))*100
percent
hist(difference, breaks=length(difference))
```

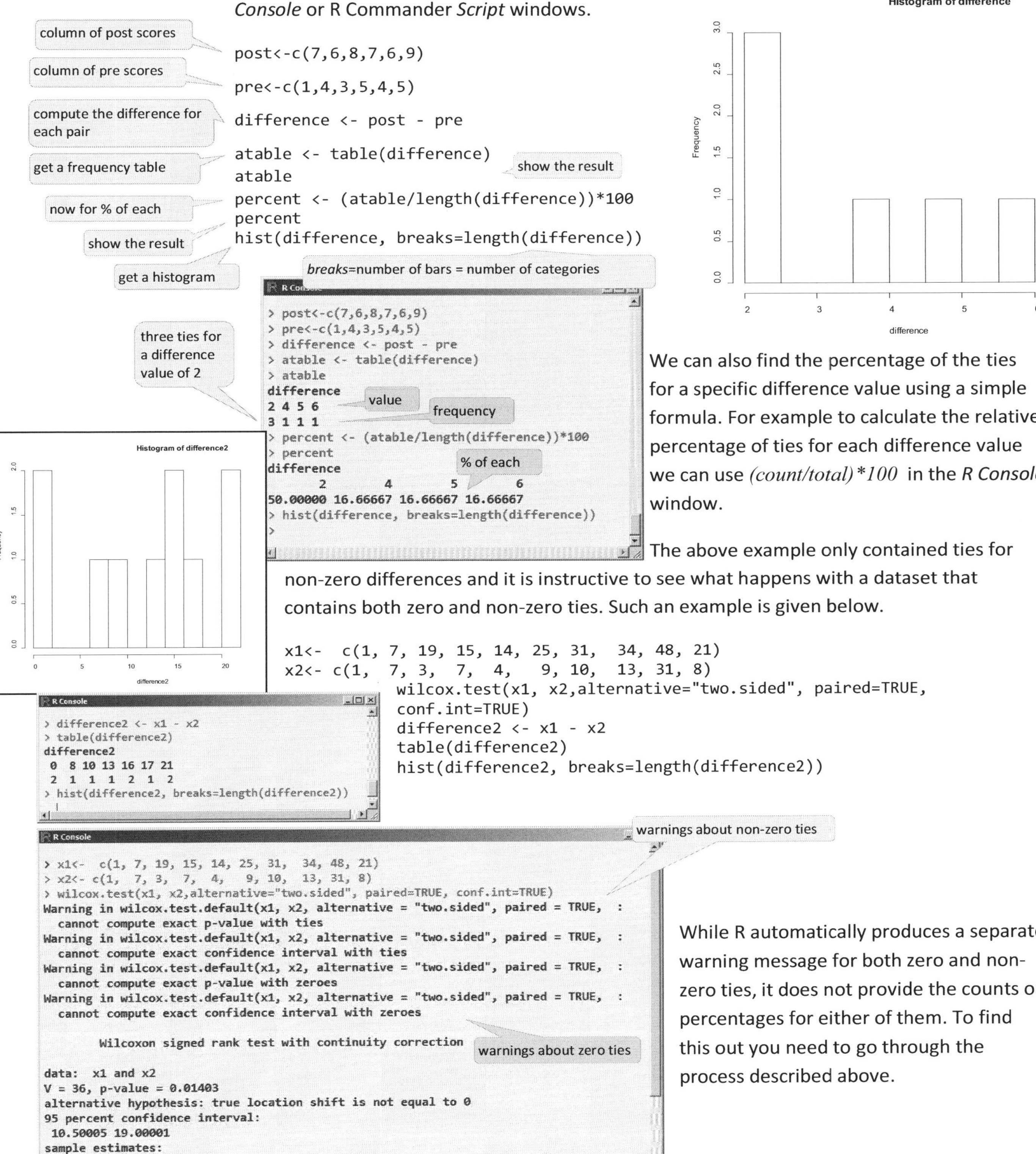

column of post scores
column of pre scores
compute the difference for each pair
get a frequency table
show the result
now for % of each
show the result
get a histogram
breaks=number of bars = number of categories
Histogram of difference
Frequency
difference
R Console
> post<-c(7,6,8,7,6,9)
> pre<-c(1,4,3,5,4,5)
> difference <- post - pre
> atable <- table(difference)
> atable
difference
2 4 5 6
3 1 1 1
> percent <- (atable/length(difference))*100
> percent
difference
2 4 5 6
50.00000 16.66667 16.66667 16.66667
> hist(difference, breaks=length(difference))
three ties for a difference value of 2
value
frequency
% of each
Histogram of difference2
Frequency
difference2
R Console
> difference2 <- x1 - x2
> table(difference2)
difference2
0 8 10 13 16 17 21
2 1 1 1 2 1 2
> hist(difference2, breaks=length(difference2))
R Console
> x1<- c(1, 7, 19, 15, 14, 25, 31, 34, 48, 21)
> x2<- c(1, 7, 3, 7, 4, 9, 10, 13, 31, 8)
> wilcox.test(x1, x2,alternative="two.sided", paired=TRUE, conf.int=TRUE)
Warning in wilcox.test.default(x1, x2, alternative = "two.sided", paired = TRUE, :
cannot compute exact p-value with ties
Warning in wilcox.test.default(x1, x2, alternative = "two.sided", paired = TRUE, :
cannot compute exact confidence interval with ties
Warning in wilcox.test.default(x1, x2, alternative = "two.sided", paired = TRUE, :
cannot compute exact p-value with zeroes
Warning in wilcox.test.default(x1, x2, alternative = "two.sided", paired = TRUE, :
cannot compute exact confidence interval with zeroes
Wilcoxon signed rank test with continuity correction
data: x1 and x2
V = 36, p-value = 0.01403
alternative hypothesis: true location shift is not equal to 0
95 percent confidence interval:
10.50005 19.00001
sample estimates:
(pseudo)median
15.49994
warnings about non-zero ties
warnings about zero ties

We can also find the percentage of the ties for a specific difference value using a simple formula. For example to calculate the relative percentage of ties for each difference value we can use *(count/total)*100* in the *R Console* window.

The above example only contained ties for non-zero differences and it is instructive to see what happens with a dataset that contains both zero and non-zero ties. Such an example is given below.

```
x1<-  c(1, 7, 19, 15, 14, 25, 31,  34, 48, 21)
x2<- c(1,  7, 3,  7,  4,   9, 10,  13, 31, 8)
wilcox.test(x1, x2,alternative="two.sided", paired=TRUE,
conf.int=TRUE)
difference2 <- x1 - x2
table(difference2)
hist(difference2, breaks=length(difference2))
```

While R automatically produces a separate warning message for both zero and non-zero ties, it does not provide the counts or percentages for either of them. To find this out you need to go through the process described above.

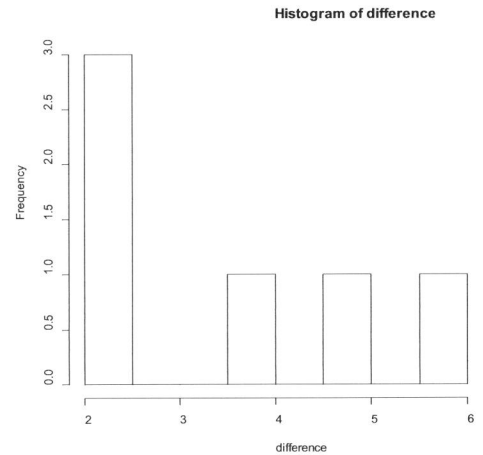

20.11 Writing up the results

From a larger group of newly enrolled members to an exercise gym 6 were randomly chosen and asked to indicate on a scale of one to ten how satisfied they were with their personal trainer both after an initial session and again after two months. All subjects demonstrated an increase in satisfaction (median improvement 3.5) with a Wilcoxon signed ranks statistic Exact p-value (2 tailed) = 0.031. Due to the small sample size (there were no ties) a confidence interval of only 80% is reported; 80% CI 2.00 to 4.99.

The effect size was measured using the signed-rank statistic (S) achieving a maximum value of 10.5 in contrast to an expected value of 0 if there had been equal numbers of positive and negative changes in the satisfaction ratings.

20.12 Tips and Tricks

It is often useful to compare the individual difference scores and you can use the graphical technique described in the chapter *Comparing pre post test means*, in the section *Digging deeper showing pre post changes.*

There is an extension to the Wilcoxon Matched Pairs statistic for when you have several repeated measures called the Friedman rank sum statistic.

Always remember that the focus of your attention is on the difference scores not the individual values.

For those of you who might be interested in seeing the code behind the Wilcoxon Matched Pairs function in R simply type *stats:::wilcox.test.default* into the R Console window. While much of it will be gobbledygook you will see several lines you can make sense of, and it is a great way to start to understand more about complex R code.

YouTube help for the Wilcoxon Matched Pairs statistic in SPSS, R Commander and R: http://youtu.be/Hq1Ogxa4mC8

21 Comparing 2 distributions: Mann-Whitney U Statistic

The Mann-Whitney U statistic assesses the degree of overlap of two samples. A large degree of overlap results in a high overall U value and a high associated p-value (i.e. **in**significant p-value). In contrast the situation where there is minimal or no overlap (in other words a high degrees of separation) results in a small or zero overall U value and a low associated p-value (i.e. significant p-value). The MWU (**M**ann-**W**hitney **U**) statistic works much the same way as the Wilcoxon Matched Pairs statistic (previous chapter), except this time the statistic is calculated from the combined ranks of all the scores from both groups.

In this chapter we will consider two example datasets to demonstrate these aspects:

- small sample, comparing distribution of 'high' ratings (N=10) for those who jogged compared to those who read an article
- large sample, comparing distribution of consultation satisfaction ratings (N=65) between those who visited a GP and a triage nurse

Control	Experimental
2	5
3	6
1	4
4	7
3	6

We will start with the small sample example. The data opposite was obtained from an experiment where 10 students were asked to rate, on a scale of one to ten, how 'high' they felt. The control group reading an article from the British Journal of Sports Medicine and the treatment group having just completed a five mile jog. The scores are therefore independent, each subject only providing a single score, and of ordinal **level of measurement**.

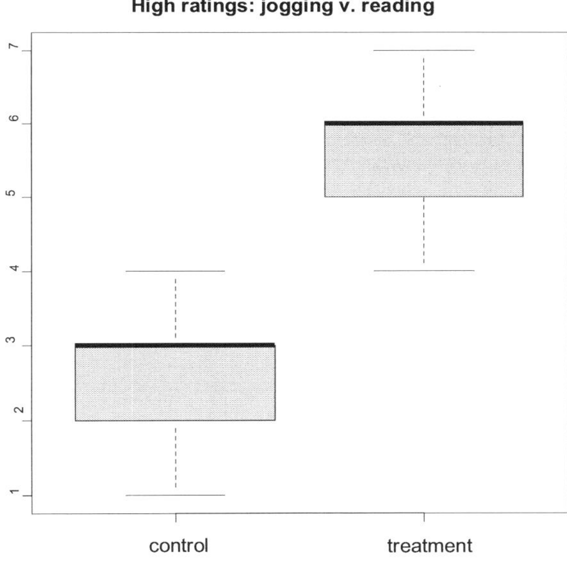

As usual we will first describe the data by way of boxplots, histograms and numeric values. How to do this is described later in the chapter.

The boxplots show the medians to be 3 and 6 for the two groups, with 50% of the scores occurring within a range of two values (the interquartile range) with no overlap. This is an important aspect to note as it suggests that we will obtain a high U value and a significant associated p-value indicating that it is unlikely for the two samples to come from the same population. Also both the median values are at the top of the boxes (difficult to see).

Because we are dealing with ordinal results we will only consider those descriptive statistics that do not require the data to have equal intervals ignoring statistics like the means, variance and standard deviation etc. focusing instead on the ranks and mean ranks for each group.

Ranks				
	group	N	Mean Rank	Sum of Ranks
result	control	5	3.10	15.50
	experiment	5	7.90	39.50
	Total	10		55

For the small dataset the first thing we note after working out the combined ranks are the ties for the 3s, 4s, and 6s. If both samples came from the same population it is sensible to assume that both groups will have a similar sum of the ranks. In other words the average of the ranks for each group will be similar, just like the positive and negative ones in the Wilcoxon statistic.

score	Rank (highest to lowest)	Group (1=control 2=treatment)
2	9	1
3	7.5	1
1	10	1
4	5.5	1
3	7.5	1
5	4	2
6	2.5	2
4	5.5	2
7	1	2
6	2.5	2

In contrast, from the table opposite we can see that both the mean rank and sum of ranks for each group are very different, reflecting the higher values for the treatment group. Thinking back to the Wilcoxon matched pairs statistic you will remember that the maximum value for the sum of ranks is equal to $n_{tot}(n_{tot}+1)/2$ which in this instance is (10x11)/2 = 55.

If we assumed there was no difference between the groups, half of this value (sum of ranks) would be allocated to each group (i.e. $n_{tot}(n_{tot}+1)/4 = 27.5$). However, the sum of ranks for the control group is 15.5 compared to 39.5 for the experimental group. There is a disproportionate number of the higher scores in the experimental group, if we think our assumption of no difference between groups is valid.

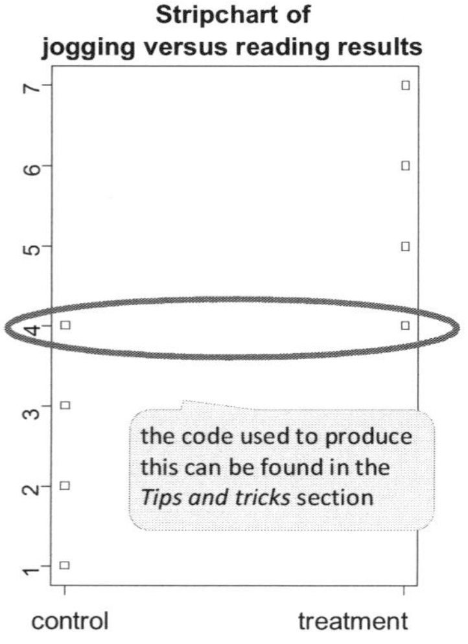

The MVU statistic uses a value called U which is the degree of overlap for each group, therefore there are two U values and the final U value reported is usually taken to be the smaller of the two. The degree of overlap can quite clearly be seen by producing a scatter plot, boxplot or stripchart (shown opposite) of the two sets of scores from which you can make a rough estimate of U. For the small dataset example show opposite, only one value overlaps with a tie which contributes half a unit value to U, so we will probably have a final U value of around 0.5. How this value is calculated is explained below.

21.1 Mann-Whitney U is the non-parametric equivalent to the t statistic – I think not!

To understand what the associated probability of U means it is necessary to first understand how U is exactly calculated and how it behaves, including its minimum and maximum values (this is true for most inferential statistics) unfortunately if you don't make the effort to understand this you can end up coming to the wrong conclusions.

A good example of this type of misunderstanding is how people often quote that the Mann-Whitney U statistic is the non-parametric equivalent to the two independent sample *t* statistic. This makes people think you are finding out the same thing from the Mann-Whitney U (MWU) statistic as you would from using the *t* statistic, but this time with ordinal or non-normally distributed data. In other words you are investigating the possible difference in medians (or even worse means – which have limited validity in ordinal data) between the groups. The meaning of U and its associated probability is different and much more subtle than that. Therefore, because of this common misunderstanding I have presented three explanations of U below.

in reality the U values go up in steps (they are discrete so the midpoint of U_1 and U_2 does not always cross at the exact half way point.

21.2 The meaning of U

21.2.1 Verbal explanation

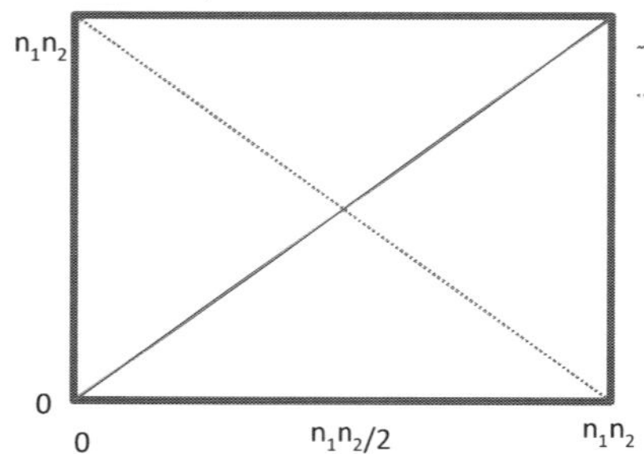

This is the usual verbal explanation; by considering the ranked observations from both groups combined, U is the minimum of the following two values. Where x_i is the *i*'th ranking in group 1 $=x$, and y_i is the *i*'th ranking in group $2 = y$:

- $U_{xy} = U_2$ is the number of pairs for which the value of $x_i<y_i$
- $U_{yx} = U_1$ is the number of pairs for which the value of $x_i>y_i$ Armitage, 1971, p.398).

Any pairs where $x_i=y_i$ count as ½ a unit towards both U_{xy} and U_{yx}

It can be shown that $U_{xy}+ U_{yx}= n_1n_2$, that is the two U values always add up to n_1n_2, therefore as one U value increases the other decreases as demonstrated in the form of a diagram opposite.

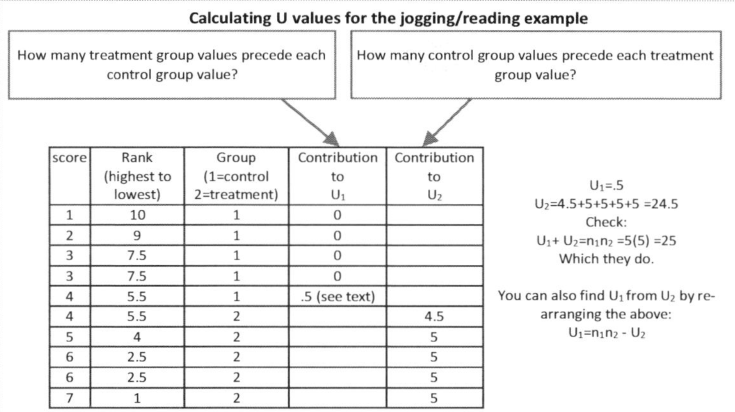

Calculating U values for the jogging/reading example

How many treatment group values precede each control group value? → Contribution to U_1

How many control group values precede each treatment group value? → Contribution to U_2

score	Rank (highest to lowest)	Group (1=control 2=treatment)	Contribution to U_1	Contribution to U_2
1	10	1	0	
2	9	1	0	
3	7.5	1	0	
3	7.5	1	0	
4	5.5	1	.5 (see text)	
4	5.5	2		4.5
5	4	2		5
6	2.5	2		5
6	2.5	2		5
7	1	2		5

U_1=.5
U_2=4.5+5+5+5+5 =24.5
Check:
U_1+ U_2=n_1n_2 =5(5) =25
Which they do.

You can also find U_1 from U_2 by re-arranging the above:
$U_1=n_1n_2 - U_2$

The above verbal definitions can be applicated to our small jogging/reading dataset (Bland, 2000, p.211), shown opposite.

Considering the first definition (i.e. U_{xy}); if the two groups were completely separate, one U value would be at its maximum and the other have a zero value. Also because the final U value is defined as the minimum of the two we would say here that the U value (no subscript) is equal to zero.

A fictitious dataset which is completely intermingled

Score rank	Group (1=control 2=treatment)	Contribution to U_1 (group 2 values before group 1 value)	Contribution to U_2 (group 1 values before group 2 value)
1	1		
2	2		1
3	1	1	
4	2		2
5	1	2	
6	2		3
7	1	3	
8	2		4
9	1	4	
10	2		5
		U_1=10	U_2=15

In contrast, when the two groups completely intermingle, both U_1 and U_2 have similar values each being around half the maximum value of n_1n_2 (i.e. $n_1n_2/2$). However, one of the U values will always be less than the other so we would still end up with a final unique minimum U value. The example opposite demonstrates this situation. Here the two U values are around the average (5x5)/2=12.5 resulting in a final U value of 10. This presents a very different situation to that of our small jogging versus reading dataset where we had a very small overall U value with very little overlap.

$$U = \min(U_1, U_2)$$

$$U_1 = n_1 n_2 + \frac{n_1(n_1+1)}{2} - R_1$$

$$U_2 = n_1 n_2 + \frac{n_2(n_2+1)}{2} - R_2$$

Rather than tabulating the various rankings there is a mathematical formula for both U values - shown opposite. R is the sum of ranks for each group; n is the number of observations, where n_1 is the number in group one and n_2 the number in the other group. Given this equation, it is possible to calculate the U values directly and calculate them for all possible arrangements of rankings for any two sample sizes (for further details see the *Re-sampling* chapter). This leads us to the next explanation of U.

21.2.2 Distribution explanation

The graph below is the distribution of all U values for two samples of 5 observations, given that they come from the same distribution. The x axis is the actual U value and the y axis the relative frequency (think here probability) of a particular value occurring given that both sample come from the dame distribution. We see that the U values nearer $n_1n_2/2$ (i.e. 12.5) occur most frequently and those with more extreme U values less frequently. We also see that the range of values is as expected from the previous discussion; a minimum value of zero and a maximum here of 25 (5x5). Note that the graph is for either U_1 or U_2 not the final U value as this is the minimum of the two.

the code used to produce this can be found in the *Tips and tricks* section

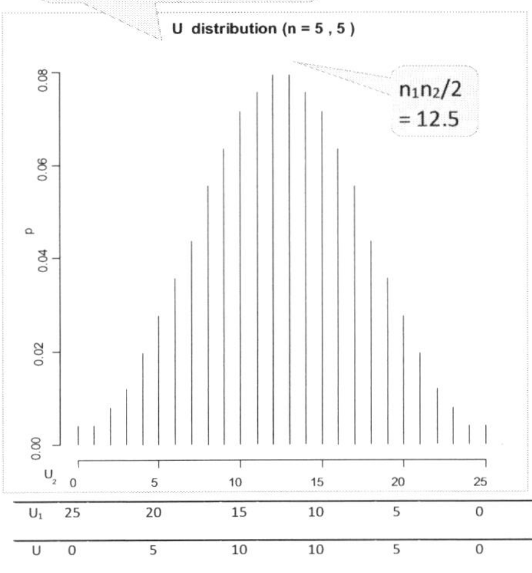

If you felt that the shape of the curve opposite looked a bit like the normal distribution (see **Standard Deviation**) you would be correct. As the sample sizes increase the similarity becomes even greater, and for samples larger than 30 the U and normal distribution values (z scores) provide good approximations to each other. The graph on the right demonstrates this. Also because there are so many U values now with (N=60) they appear like a continuous black filling.

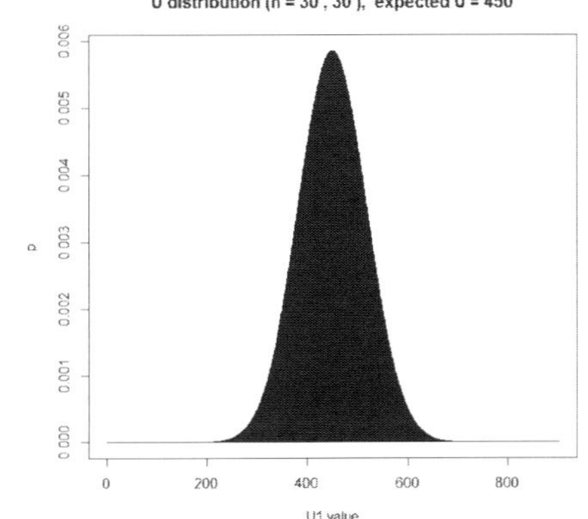

21.2.3 Degree of enfoldment/separation

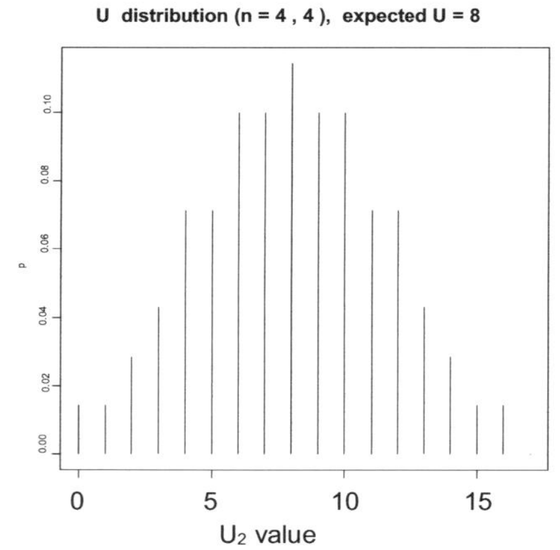

The diagram opposite provides the distribution of U values for two sample of size 4 where we can see that the U value ranges from 0 to 16 with the most frequent value being 8. While this diagram is useful it does not help most people to conceptualise how U indicates the degree of overlap or enfoldment between the two distributions U (Hart, 2001).

In contrast the diagram below does show clearly the degree of overlap that is associated with a particular U value for two samples of four observations.

The U_2 value increases as the groups increasingly overlap. However as U is the minimum value of the two U's, U=*min*(U_1,U_2), the actual value of U reaches a maximum, (4x4)/2 = 8, value and then decreases as the graph shows.

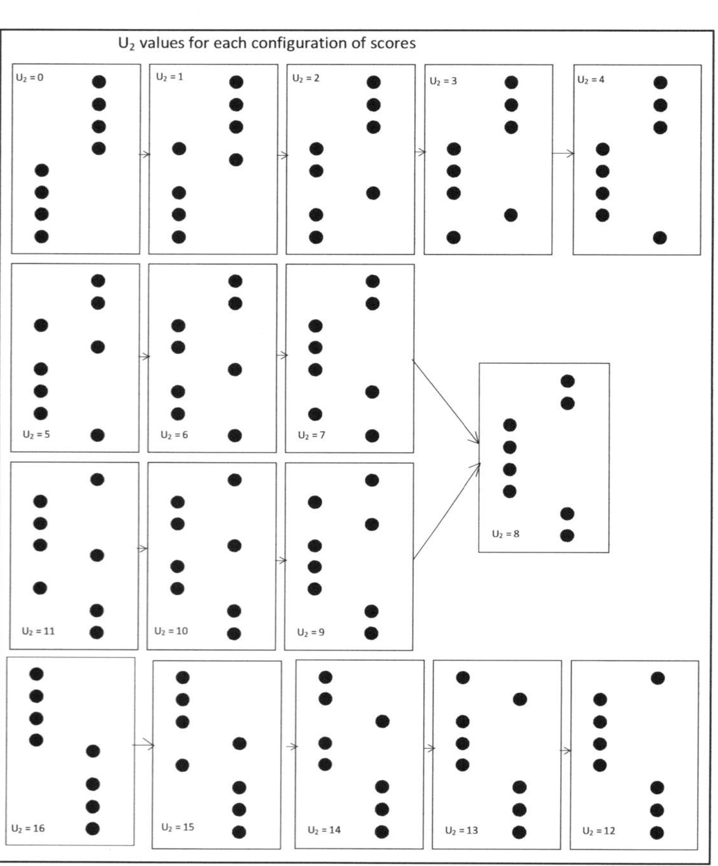

When U is at its maximum, one set of scores is completely enfolded by the other set, and furthermore the enfoldment is symmetrical. This is in contrast to when U_2=4 or U_2=12 (i.e. U_1=4) where again one set of scores is enfolded in the other but here it is not symmetrical.

Key points

- When U = 0 this would indicate that there is no overlap between the groups. A situation which is unlikely to happen if we assume the two samples are from a single population.
- When U= $n_1n_2/2$ - where n is the number of observations in each group - this is the maximum value for U and is when all the scores from one group are surrounded in the middle of the other group.

The rest of this chapter will concentrate on the practicalities of applying and interpreting the Mann-Whitney U statistic to a larger dataset using R Commander and R.

21.3 Data preparation

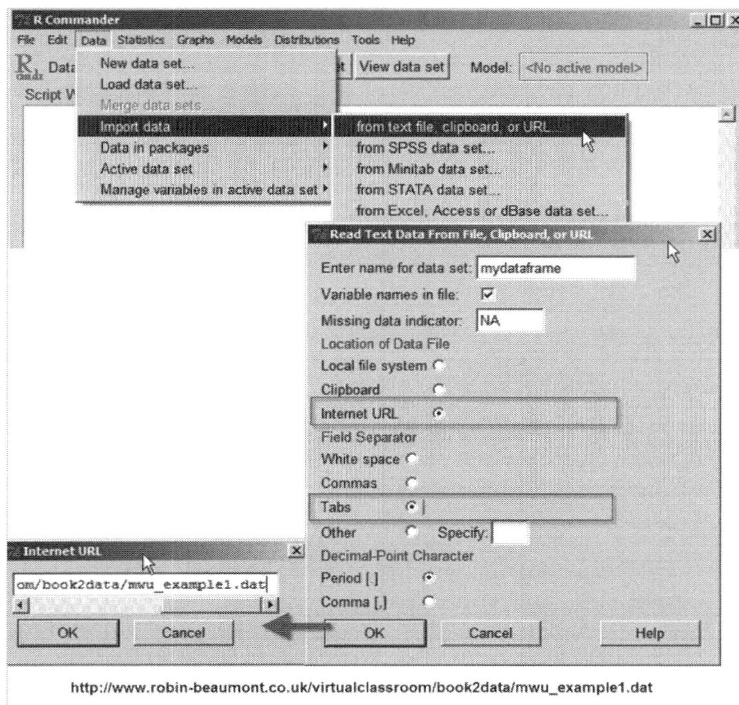

The basis for this example is from MSME3 describing a group of patients who either made emergency appointments with a GP or agreed to see the triage nurse. At the end of the consultation they rated their satisfaction with the consultation from 0 to 5; zero being not satisfied to 5 very satisfied, the sample consists of 64 patients in total.

To load R Commander type the following in the *R Console* window:

```
library(Rcmdr)
```

To load the sample dataset select the R Commander menu option:

Data-> Import data->from text file, the clipboard or URL

I have given the resultant dataframe the name *mydataframe*, also indicating that it is from a URL (i.e. the web) and the columns are separated by *tab* characters. Clicking on the **OK** button brings up the *Internet URL* box in which you need to type the following to obtain my sample data:

http://www.robin-beaumont.co.uk/virtualclassroom/book2data/mwu_example1.dat

Click **OK** to load the data.

21.4 Data description

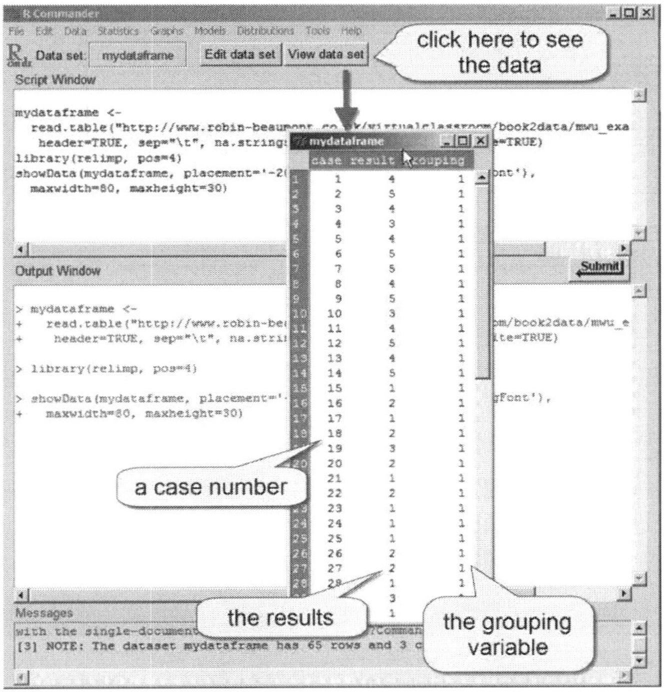

It is always a good idea to see the data and you can achieve this easily in R Commander by clicking on the **View data set** button.

Note that I have all the observations in a single column with another one indicating which group they belong to, such a variable is often called a **factor** or grouping variable.

Here 1=GP consultation and 2=Nurse triage consultation.

We note there are no missing values or silly values such as negative satisfaction scores or values above 5. Similarly all the satisfaction ratings are integer values reflecting that the participants could only provide whole values from 1 to five.

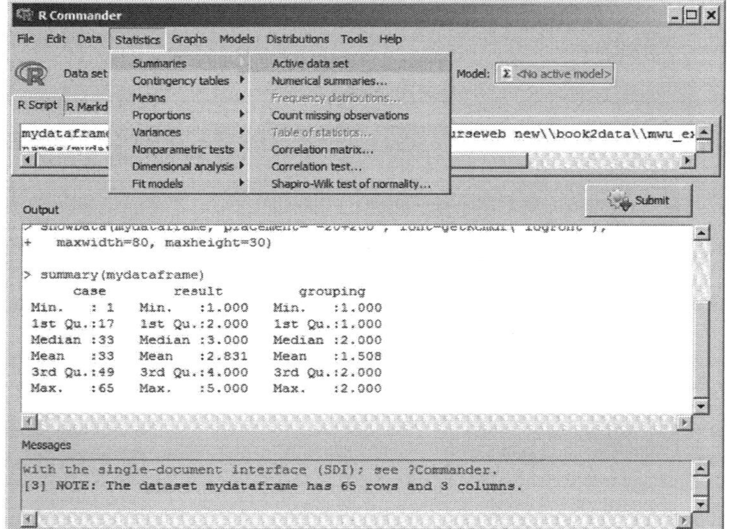

You can also see the summary statistics for each column by using the R Commander menu option:

Statistics->Active data set

Inspecting the *result* variable the mean satisfaction score is 2.8 out of 5 for the entire group.

Please note that incorrectly, the mean along with other values are reported for the *grouping* variable. This variable should be a factor and therefore R should only report the numbers at each level. To correct this we need to convert it to a factor.

21.4.1 Converting a grouping variable into a factor

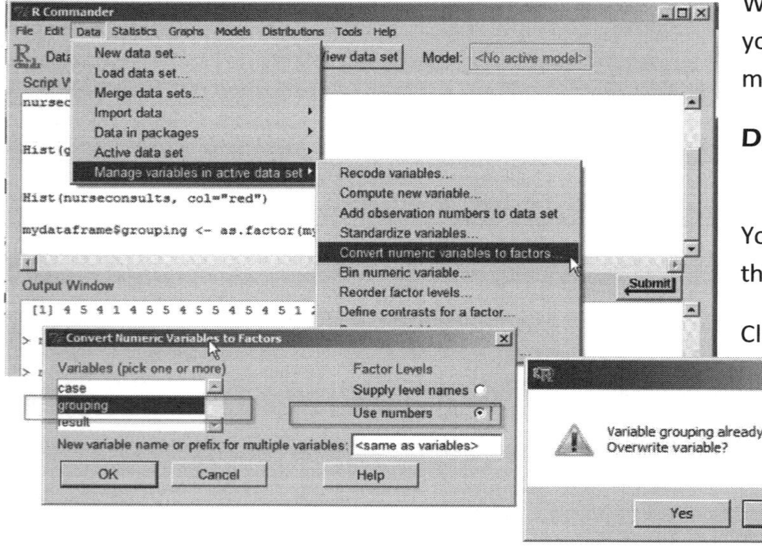

We can convert any column of data (i.e. variable) in your dataframe to a factor using the R Commander menu option:

Data-> Manage variables in active dataset -> convert numeric variables to factors

You then select the appropriate variable, *grouping* in this instance, and select the **Use numbers** option.

Click **OK** and again when a warning message appears.

It is instructive here to notice what R Commander has done under the bonnet by inspecting the *Script* window in R Commander:

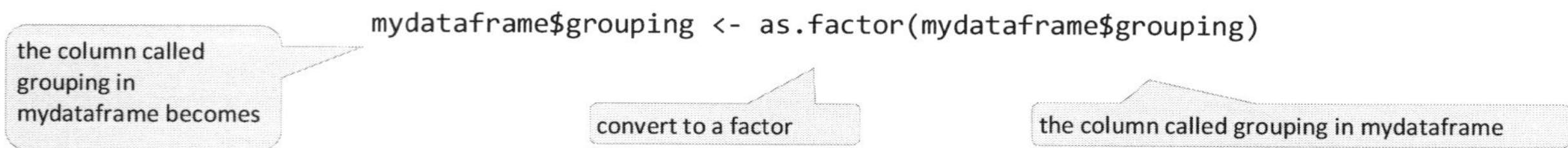

21.4.2 Viewing the distributions

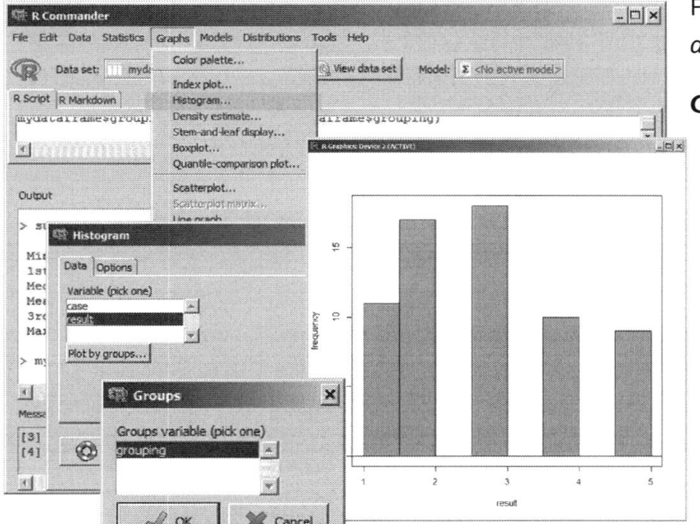

Following my own rule, *always graph the distribution of your data* select the R Commander option

Graphs-> Histogram

Select the *result* variable then click **OK**.

The result appears in a *R Graphics* window of the *RGui* window.

This suggests that the scores follow the bell shape, that is they are **normally distributed**, however what we need to consider is the distribution of scores for each group.

To plot the distribution of results for each group click on the **Plot by groups...** button and then in the *Groups* window select the *grouping* variable. Click on the **OK** button to close it.

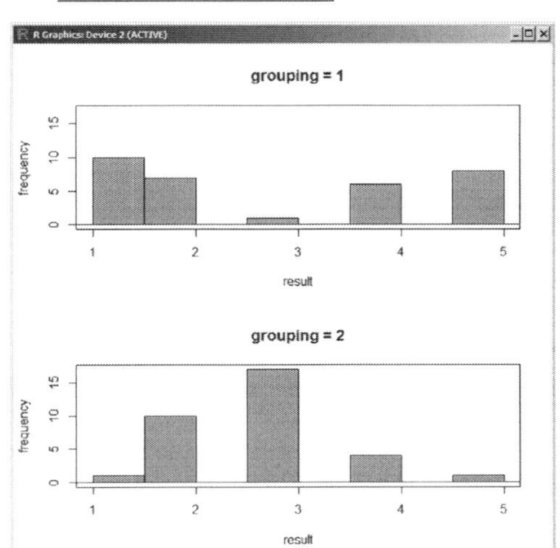

The button in the *Histogram* window then changes to indicate your selection: Plot by: grouping

From the two histograms opposite we can see that that satisfaction with the emergency GP consultations and those with the triage nurse have very different shapes: formally, we would say they follow very different distributions.

The *Tips and tricks* section at the end of this chapter describes how the above can be achieved using R code.

21.5 Table of counts for each group

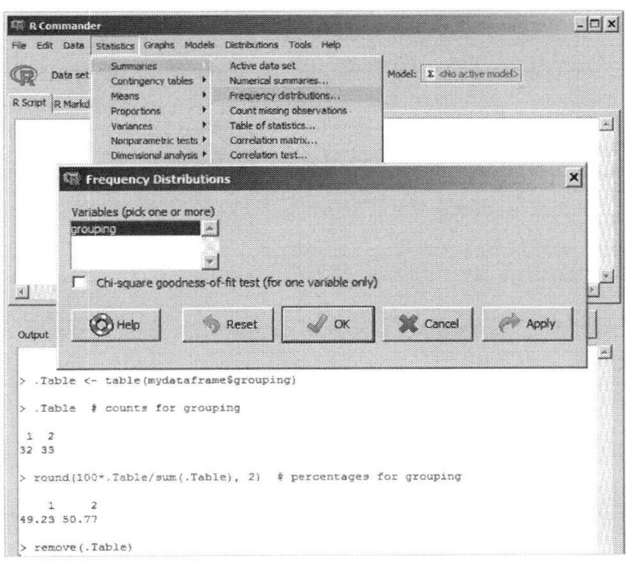

Now we have a factor we can see how many there are in each group. Select the R Commander menu option:

Statistics->Summaries->Frequency distributions...

Then complete the dialog box as shown opposite. Click on the **OK** button. We have 32 (49%) in the GP consultation group (=1) and 33 (51%) in the Nurse group. Although the MWU statistic copes well with unequal size groups it is good from a study design point of view to have two roughly equally sized groups with more than 30 subjects in each.

21.5.1 Boxplots

Having specified the grouping variable as a factor we can now obtain the boxplots using the R Commander menu option:

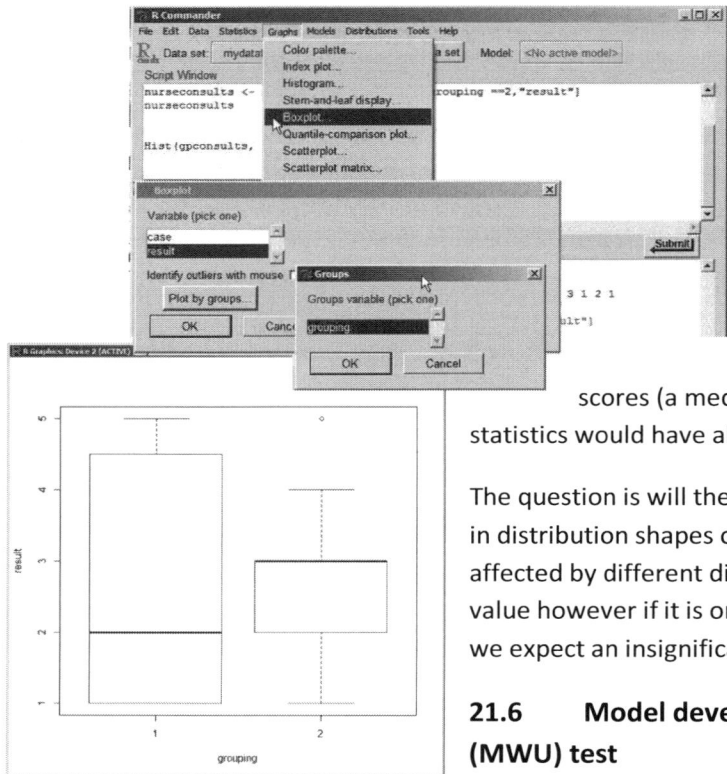

Graphs -> Boxplot

Select the *result* variable, and then indicate the grouping variable by clicking on the **Plot by groups . .** button and select the *grouping* factor. Finally click on both **OK** buttons.

We can see that the boxplots give virtually no indication of the very different distribution of the scores (a median of 2 versus 3). Similarly, looking at a set of summary statistics would have also left us in the dark.

The question is will the Mann-Whitney U statistic be affected by this difference in distribution shapes or will it just consider the degree of overlap? If it is affected by different distribution shapes we will be expecting a significant p-value however if it is only concerned with the degree of overlap, which is high, we expect an insignificant (large) p-value.

21.6 Model development / evaluation - the Mann-Whitney U (MWU) test

To obtain the MWU statistic use the R Commander menu option:

the MWU test is also known as the Two sample Wilcoxon test

Statistics-> Nonparametric tests -> Two sample Wilcoxon test

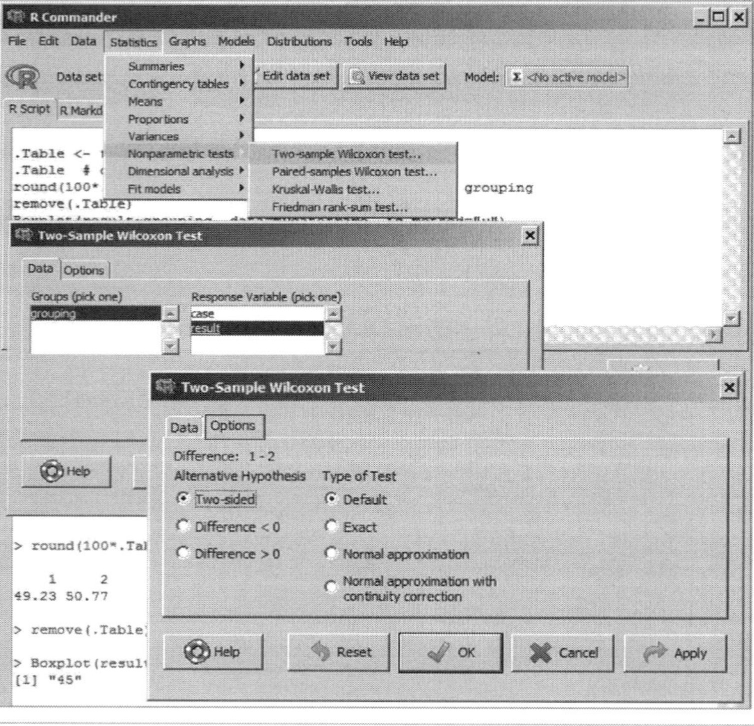

Then for the **Groups** variable select *grouping* and for the **Response Variable** select *result*. You can ignore the options tab.

Click **OK**.

The results then appear in the R Commander *Output* window.

```
> wilcox.test(result ~ grouping, alternative="two.sided", data=mydataframe)

        Wilcoxon rank sum test with continuity correction

data:  result by grouping
W = 506.5, p-value = 0.7774
alternative hypothesis: true location shift is not equal to 0
```

W is the same as U

Looking first at the U value of 506 it is useful to compare this to what would be expected if one sample was completely enfolded the other, being $(n_1 x n_2)/2$. For our sample this is $(32 \times 33)/2 = 528$. This value is very close to that reported of 506.5, indicating how completely entwined the samples are.

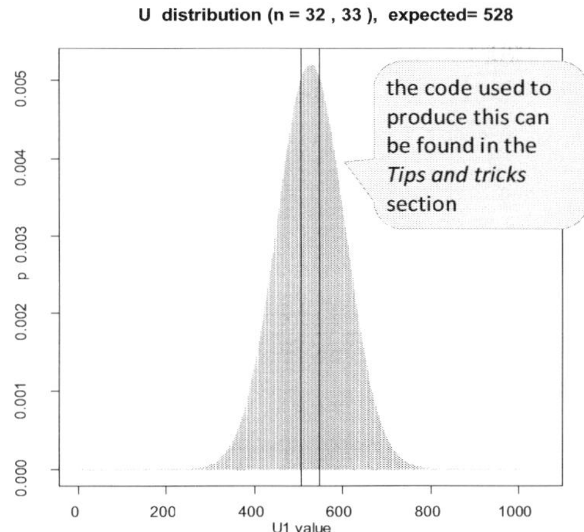
U distribution (n = 32 , 33), expected= 528
p
0.000
0.001
0.002
0.003
0.004
0.005
0
200
400
600
800
1000
U1 value
the code used to produce this can be found in the *Tips and tricks* section

For information I have produced the appropriate U distribution opposite, along with two lines indicating the observed U values (506.5 and its complement 549.494). Clearly the area outside the two values (i.e. more extreme) represents the vast majority of it hence the high p-value. Specifically the p-value of 0.77 indicates that we would obtain a U value equal to 506.5 or one more extreme seventy seven times in every hundred given that the samples come from a single population.

By taking the usual critical value of 0.05, because the observed p-value is greater than this we decide that the two samples do come from the same population.

From the discussion on the previous page it would appear that the Mann-Whitney U statistic is concerned with overlap and not the actual distribution shape of the samples. To obtain a confidence interval for the MWU statistic you need to use R code.

21.6.1 Obtaining a confidence interval: annotating R Commander code

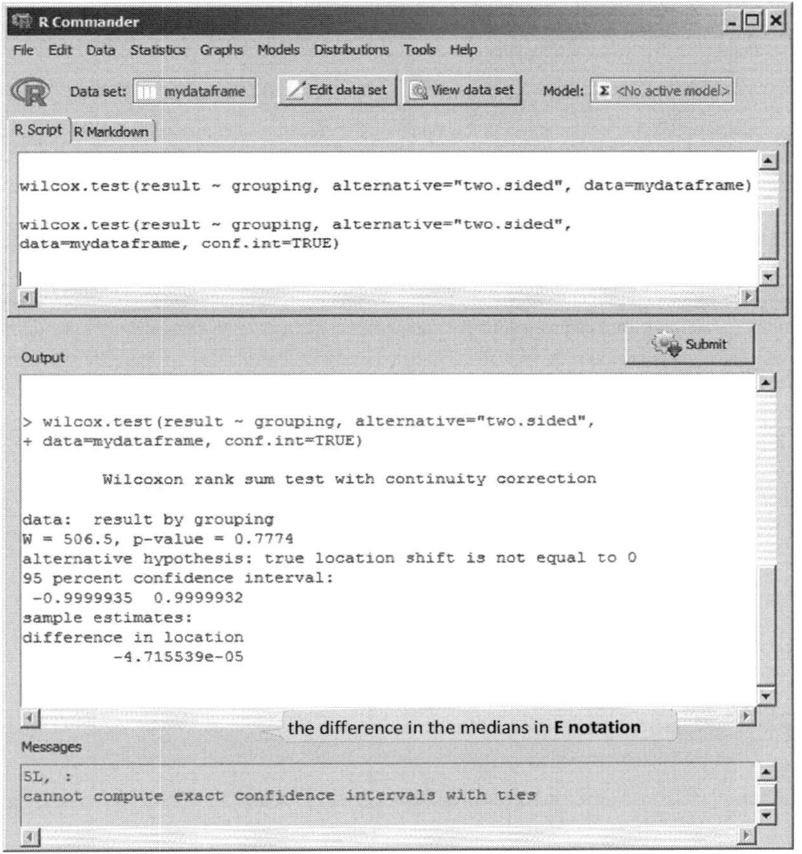
R Commander
File Edit Data Statistics Graphs Models Distributions Tools Help
Data set: mydataframe
Edit data set
View data set
Model: <No active model>
R Script
R Markdown
wilcox.test(result ~ grouping, alternative="two.sided", data=mydataframe)
wilcox.test(result ~ grouping, alternative="two.sided",
data=mydataframe, conf.int=TRUE)
Submit
Output
> wilcox.test(result ~ grouping, alternative="two.sided",
+ data=mydataframe, conf.int=TRUE)
Wilcoxon rank sum test with continuity correction
data: result by grouping
W = 506.5, p-value = 0.7774
alternative hypothesis: true location shift is not equal to 0
95 percent confidence interval:
-0.9999935 0.9999932
sample estimates:
difference in location
-4.715539e-05
the difference in the medians in **E notation**
Messages
5L, :
cannot compute exact confidence intervals with ties

The above output did not produce a confidence interval but we can annotate the R code R Commander produced to create one.

You need to add the following to the R code to that in the R Script window after *data=mydataframe*:

```
, conf.int=TRUE
```

Then select the code segment from *Wilcox.test(* to the closing bracket *)*. Click on the **Submit** button.

There is now a confidence interval in the output.

The very narrow confidence interval which straddles zero, reinforces our decision that the two samples come from the same population or at least they both have the same median value (**clinical significance**).

A detailed explanation of the R code is given below.

carry out the 2 independent samples Mann-Whitney U test

use the data in my dataframe

```
wilcox.test(result ~ grouping, alternative="two.sided", data=mydataframe, conf.int=TRUE)
```

use the result variable divided by the grouping variable

do a two sided test

request a confidence interval

21.7 Finding ties and mean rankings for each group

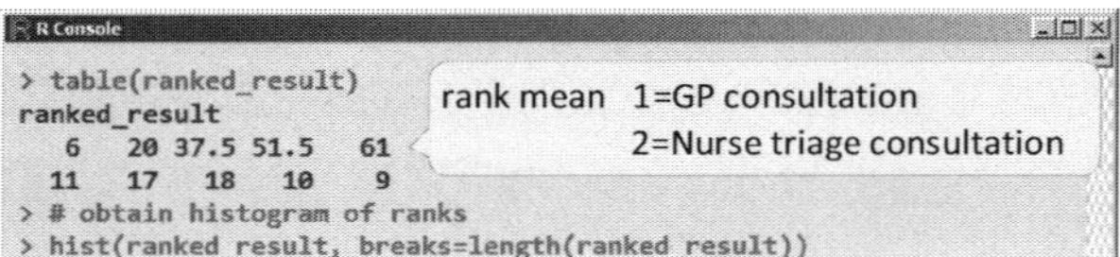

To investigate ties we can use a similar approach to that described in the *Wilcoxon Matched-pairs* chapter. Given the very high level of overlap and narrow range of scores (1 to 5) we will be expecting a large number of ties. The results confirm this with a minimum count of 9, for a ranking value of 61, representing 14% of the scores (9/65).

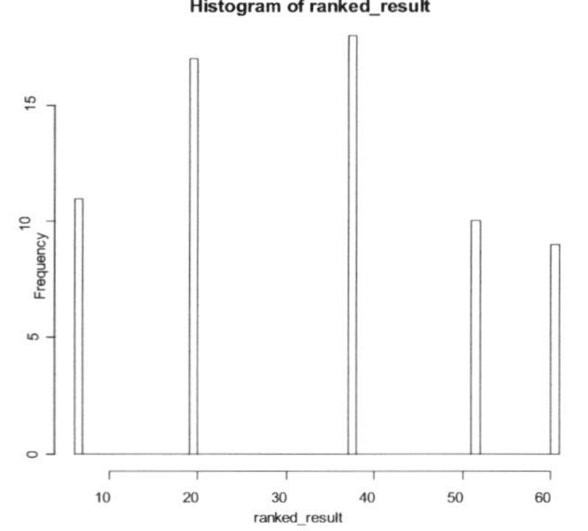

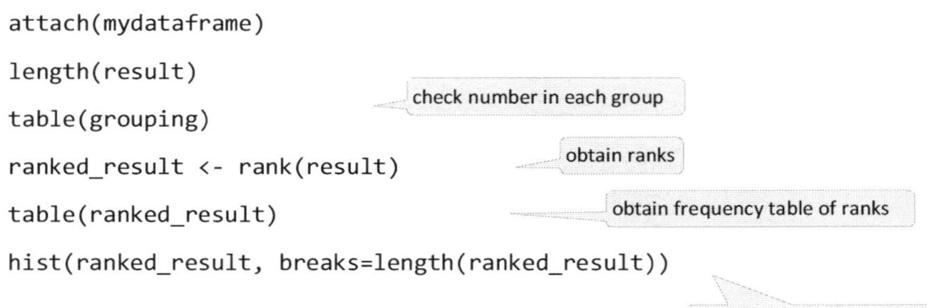

```
attach(mydataframe)
length(result)
table(grouping)
ranked_result <- rank(result)
table(ranked_result)
hist(ranked_result, breaks=length(ranked_result))
```


21.7.1 mean ranks

We can also find the mean ranks for each group, a value often required when submitting to journals.

From the output we can see how close these two values are. This contrasts to the big difference shown in the previous jogging versus reading example discussed at the start of the chapter.

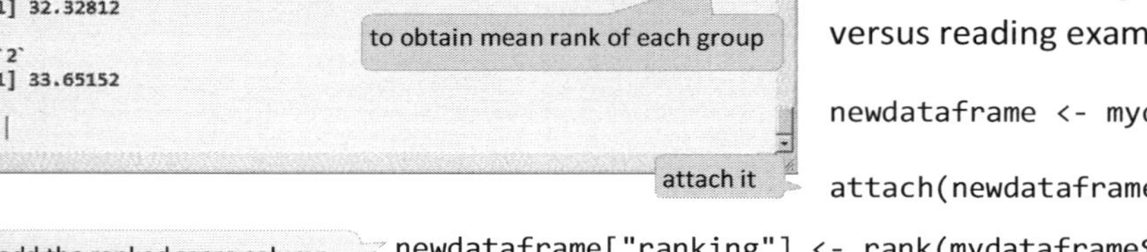

```
> newdataframe <- mydataframe
> attach(newdataframe)
The following objects are masked from newdataframe (position 3):

    case, grouping, result
> newdataframe["ranking"] <- rank(mydataframe$result)
> lapply((split(newdataframe$ranking, newdataframe$grouping)), mean)
$`1`
[1] 32.32812

$`2`
[1] 33.65152
```


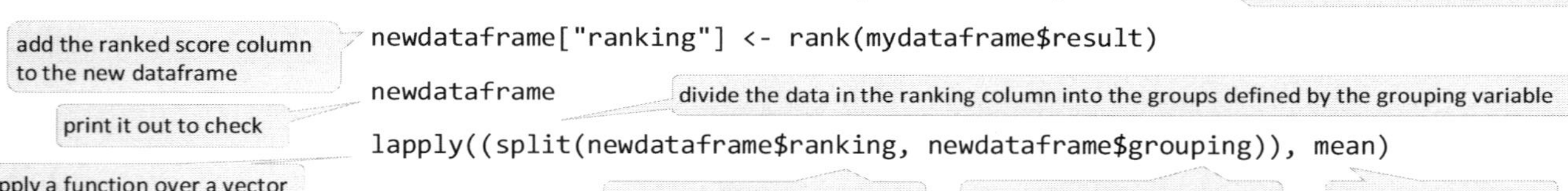

```
newdataframe <- mydataframe
attach(newdataframe)
newdataframe["ranking"] <- rank(mydataframe$result)
newdataframe
lapply((split(newdataframe$ranking, newdataframe$grouping)), mean)
```


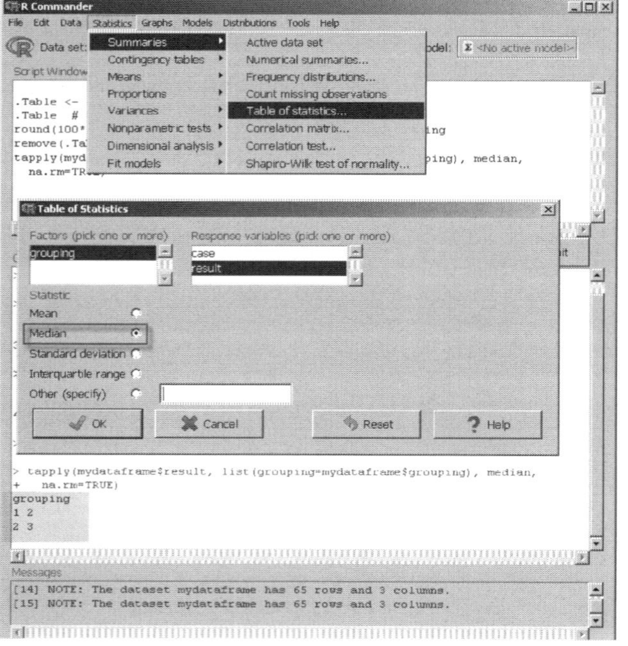

21.8 Effect size

One should always report the measure of the observed difference between the two sample medians. In this situation it is simply the difference in the medians between the two groups. You can obtain the median value for each group by using the R Commander menu option:

Statistics-> Summaries-> Table of statistics

Complete the dialog box as shown opposite.

The median satisfaction rating for the GP consultations is 2 compared to 3 for the nurse consultations, so the effect size is 3-2 = 1.

21.9 Writing up the results

Using the larger dataset:

64 satisfactions scores were obtained from patients who had visited a GP as an emergency consultation or alternatively a nurse triage consultation offered at the same GP practice. The scores for the GP consultations demonstrated a U shaped distribution while those for the Triage Nurse consultations demonstrated a more normal bell shaped distribution.

I would then insert the histograms.

Because of the nature of the data (non **normal distribution** and rating scores) a Mann-Whitney U (MWU) test was undertaken to assess the degree of overlap/separation of the two groups.

The satisfaction scores for the GP consultations (median=2 range 1 to 5) compared to the Nurse Triage consultations (median=1 range 1 to 5) demonstrated an effect size of 1 (median difference). A MWU test was performed (W=506.5, p-value=0.7774 two tailed, 95% Ci -.99 to +.99). As the p-value is above the critical value of 0.05 there is no evidence of a degree of separation that would indicate they came from different populations. However, given the very different observed distribution shapes and the very high level of ties, this result warrants further investigation, possibly with further data collection.

21.10 Tips and Tricks

The same comments apply here as given at the end of the *Two independent samples t test* chapter.

Remember when reporting non-parametric tests such as the Mann-Whitney U instead of reporting means and standard deviations, you give the non-parametric equivalents such as the median and interquartile range.

The investigation of the larger dataset emphasises the importance of not just relying upon one or two summary measures of the data. It is vitally important to inspect histograms and density plots to get a feel for the distribution of the scores in the various subgroups.

Youtube help: http://www.youtube.com/watch?v=K2Th7G9VYZg

21.10.1 R code for histograms

Instead of producing the histograms using the R Commander menus you can use R Code:

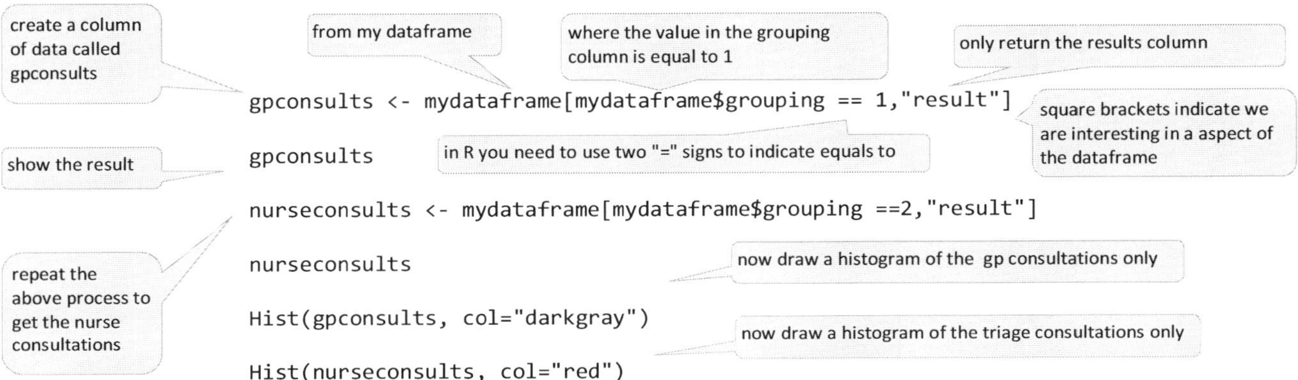
create a column of data called gpconsults
from my dataframe
where the value in the grouping column is equal to 1
only return the results column
gpconsults <- mydataframe[mydataframe$grouping == 1,"result"]
square brackets indicate we are interesting in a aspect of the dataframe
show the result
gpconsults
in R you need to use two "=" signs to indicate equals to
nurseconsults <- mydataframe[mydataframe$grouping ==2,"result"]
repeat the above process to get the nurse consultations
nurseconsults
now draw a histogram of the gp consultations only
Hist(gpconsults, col="darkgray")
now draw a histogram of the triage consultations only
Hist(nurseconsults, col="red")

21.10.2 R Code for when data is in a separate column for each group

```
control<- c(2,3,1,4,3); treatment <- c(5,6,4,7,6)

jogging_reading <- data.frame(control, treatment)

boxplot(control, treatment, data=jogging_reading,
names = c("control", "treatment"),
main="High ratings: jogging v. reading",
id.method="y", col = "gray90", medlwd=6)
```

make the median line thickness=6

Gray90 means 90% gray

The code used to produce the boxplot at the start of the chapter for the small dataset concerned with article reading versus jogging example is given opposite.

Here we have the results for each group in separate columns.

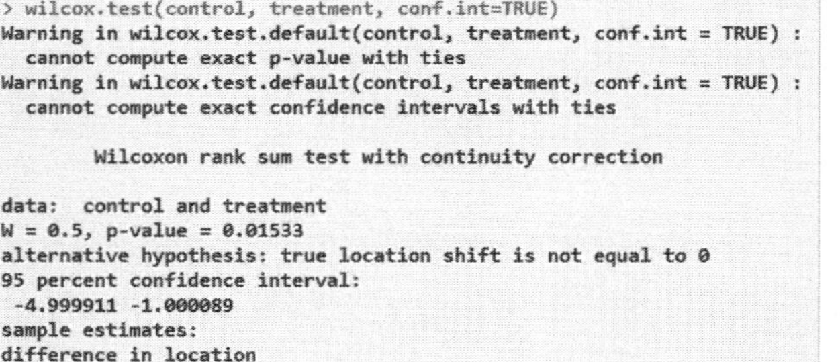

```
> wilcox.test(control, treatment, conf.int=TRUE)
Warning in wilcox.test.default(control, treatment, conf.int = TRUE) :
  cannot compute exact p-value with ties
Warning in wilcox.test.default(control, treatment, conf.int = TRUE) :
  cannot compute exact confidence intervals with ties

        Wilcoxon rank sum test with continuity correction

data:  control and treatment
W = 0.5, p-value = 0.01533
alternative hypothesis: true location shift is not equal to 0
95 percent confidence interval:
 -4.999911 -1.000089
sample estimates:
difference in location
                    -3
```

To carry out the Mann-Whitney U test you now need to indicate each column of data:

```
wilcox.test(control, treatment, conf.int=TRUE)
```

The result is shown opposite.

21.10.3 Stripcharts

A stripchart was used at the start of this chapter to show which specific value overlapped. You can produce basic stripcharts using the R Commander menu option:

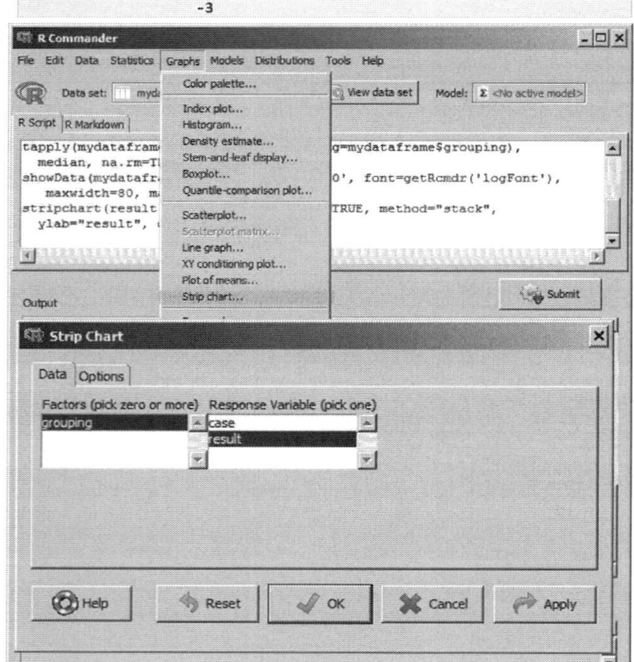

Graphs->Strip chart

However to gain finer control you need to use R code. For example the R code below produced the strip chart shown at the start of this chapter.

```
stripchart(list(jogging_reading$control,
                jogging_reading$treatment),
vertical = TRUE,
group.names= c("control","treatment"),
main = "stripchart of jogging versus reading results")
```

21.10.4 U distribution graphs

For those who are interested, the following R code produced the U distribution graphs displayed earlier in this chapter.

```
n1<-32  # group 1 size
n2<-33   # group 2 size
U_result <- 506.506
U_comp <- n1*n2 - U_result
U_comp
expect_u <- (n1*n2)/2
expect_u
efsize<-(n1*n2) + 1
efsize
x <- seq(0,efsize)
plot(x, dwilcox(x,n1,n2), type="h", xlab="U1 value", ylab="p", col="pink",
main=paste("U  pdf (n =",n1, ",", n2 ,"), ","expected U =",expect_u ))
abline(v=U_result) # one U value - optional
abline(v=U_comp)    # the other U value, they are symmetrical- optional
```

optional, you might want to indicate a particular U value

we can calculate one U value from the other U1=n1n2 - U2

expected U value is (n1 * n2)/2

now for the plot

22 Comparing an observed proportion to a population value: the Binomial test

Frequently we want to compare an observed (sample) proportion to an expected one (a population value). Taking a medical example, imagine we have 328 patients who suffered post-operative infections from a group of 1361 patients (i.e. 1033 non-infected patients) and we know that the country-wide post-operative infection rate is 25%. With this information, we want to know the probability of obtaining a rate like ours or one more extreme given that it is a random sample from a population with an incidence of 25%. The 'given' part of the previous sentence is important as it helps define our model. If we don't know the population value, or are unable to make a reasonable estimate of it, we can't develop a model. This chapter answers this type of question.

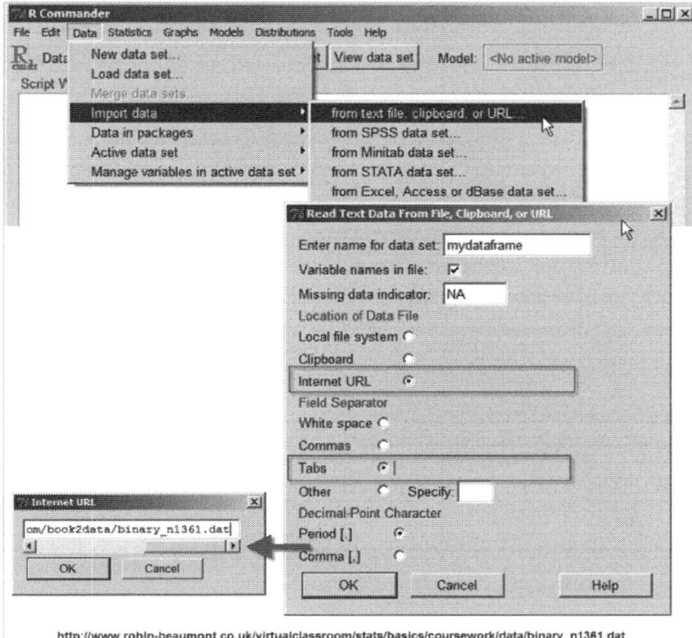

22.1 Data preparation

To load R Commander type the following into the *R Console* window:

```
library(Rcmdr)
```

To obtain the data directly from my website select the R Commander menu option:

Data-> from text, the clipboard or URL

I have given the resultant dataframe the name *mydataframe*, also indicating that it is from a URL (i.e. the web) and the columns are separated by *tab* characters.

Clicking on the **OK** button brings up the Internet URL box in which you need to type the following to obtain my sample data:

```
http://www.robin-beaumont.co.uk/virtualclassroom/book2data/binary_n1361.dat
```

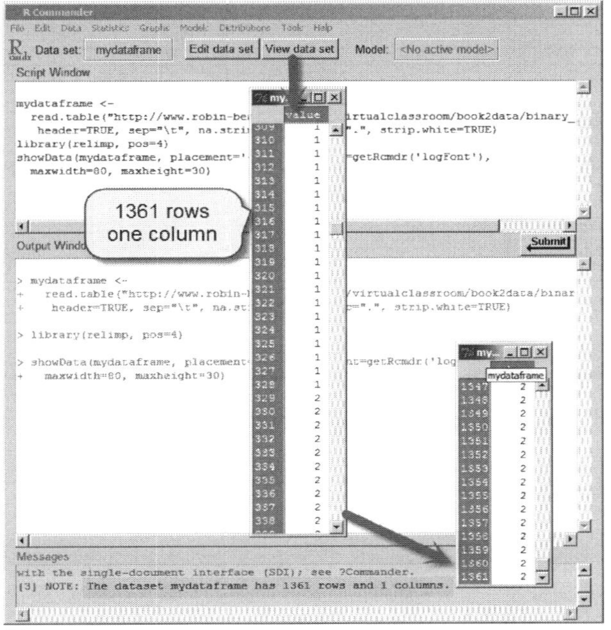

Click **OK**.

It is always a good idea to check the data and you can achieve this easily in R Commander by clicking on the **View data set** button.

You can see we have 1361 rows and one column called *value*, the column being filled with '1's (those infected) and '2's (those uninfected).

22.2 Converting a numeric variable to a factor

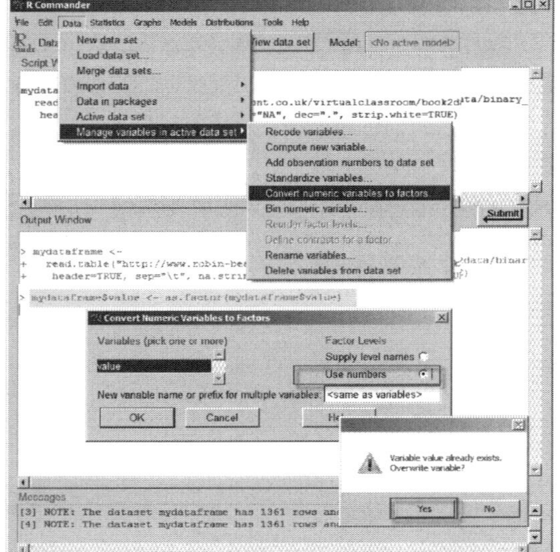

Unfortunately, R Commander needs to know the *value* column is a **factor** to be able to carry out the binomial test. To convert it use the menu option:

Data-> Manage variable in active dataset -> convert numeric variables to factors

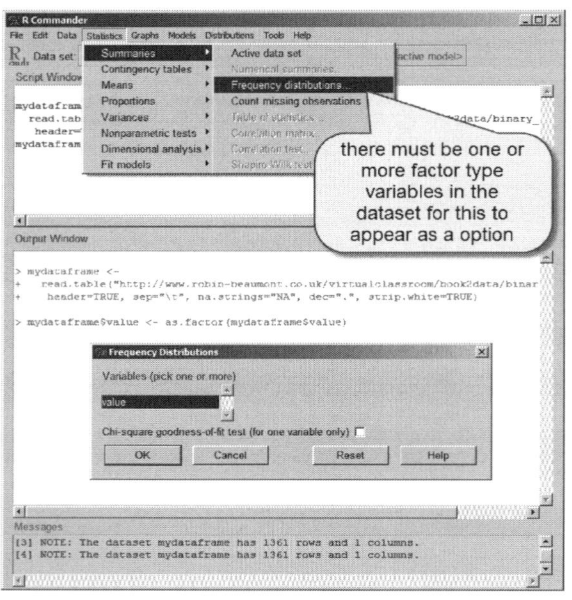

We can also create a table of the counts for each factor level by selecting the R Commander menu option:

Statistics-> summaries -> Frequency distributions

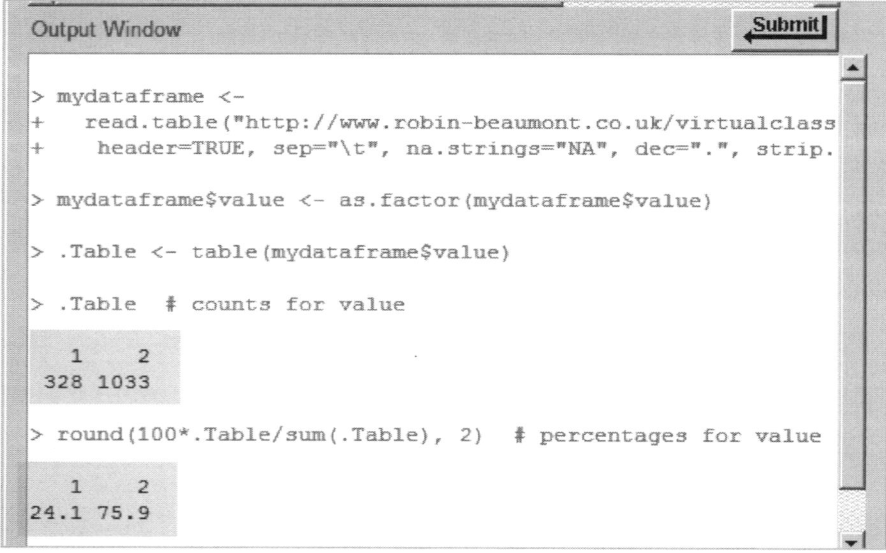

Output Window
Submit
> mydataframe <-
+ read.table("http://www.robin-beaumont.co.uk/virtualclass
+ header=TRUE, sep="\t", na.strings="NA", dec=".", strip.
> mydataframe$value <- as.factor(mydataframe$value)
> .Table <- table(mydataframe$value)
> .Table # counts for value
1 2
328 1033
> round(100*.Table/sum(.Table), 2) # percentages for value
1 2
24.1 75.9

The results indicate that 24.1% of the observed sample is infected compared to the national average of 25%.

22.3 Model development and evaluation

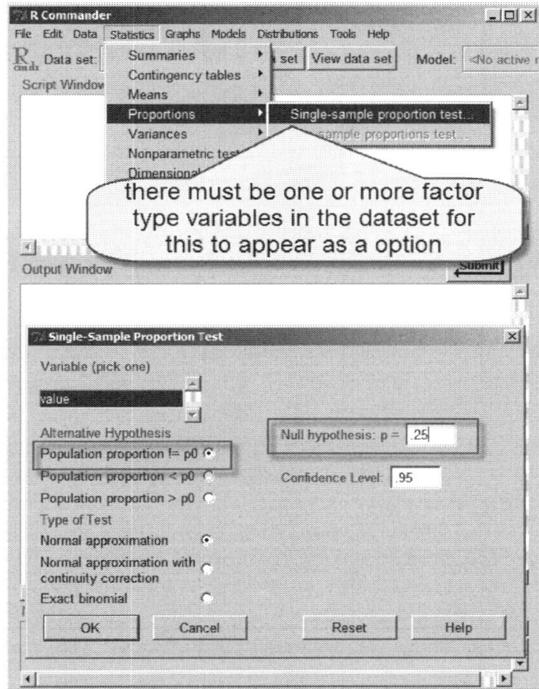

We now want to see how likely it would be to obtain an observed proportion such as ours or one even more extreme, given that the population value is 25%.

Select the R Commander menu option:

Statistics-> Proportions-> Single-sample proportion test

Complete the dialog box as show opposite. Making sure you have changed the hypothesised population proportion, called the **Null hypothesis** in the dialog box, (p=proportion) to 0.25

!= in R means 'not equal to'. This is known as a two sided **p-value** (see glossary entry for further details).

There are three options available to calculate the **p-value**, the most accurate is the **Exact binomial** test. In the screen shots below I have run the analysis using all three options and they only differ by 0.02 so in this instance, probably because we have a large sample size of 1361, the option makes very little difference. However, with small sample sizes the difference would be more obvious.

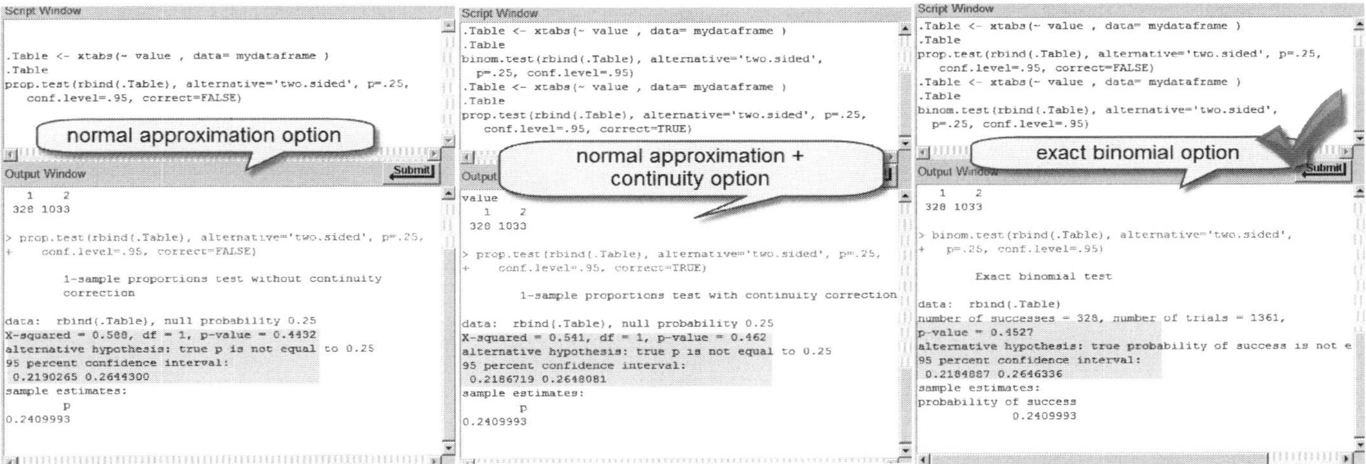

The confidence interval can be interpreted in the following way; we are 95% confident that the interval .21 to .26 contains the population proportion. As this interval contains the 25% hypothesised proportion we can say that the sample could have well come from it. An alternative interpretation is provided by the **p-value**. The most accurate way of interpreting this p-value of 0.24 is:

Given a population which has a proportion of 25% we would obtain a proportion of 24.1% or one more extreme in a sample of size 1361 twenty four times in every hundred.

We can also set an arbitrary value called a critical value, which is usually 0.05 (interpreted as one in twenty times) to create a decision rule. If our observed proportion or one more extreme than this, is predicted to occur more frequently than this (i.e. p-value greater than critical value) we accept that the sample comes from the hypothesised population. In contrast if our sample, produces a p-value which indicates it occurs less frequently than the critical value we assume the sample comes from a population with a different proportion.

As our p-value is above the critical value we come to the conclusion that our sample comes from a population with a proportion of 25% and the observed difference is just due to the vagaries of random sampling. Considering the p-value graphically I find helps understand this (see the *Showing the null hypothesis distribution and explaining the p-value* section below).

22.4 Counts instead of raw data – using R directly

In the script/output windows on the previous page you can see the R code generated by the various menu selections we made. Before the *prop.test()* or *binom.test()* commands there are several other lines of R code. These other lines of code convert our raw data into a table of frequencies which each subsequently uses. In R a dot is just another character and I find names like *.Table* confusing so I have slightly edited the code to make it a bit easier to understand.

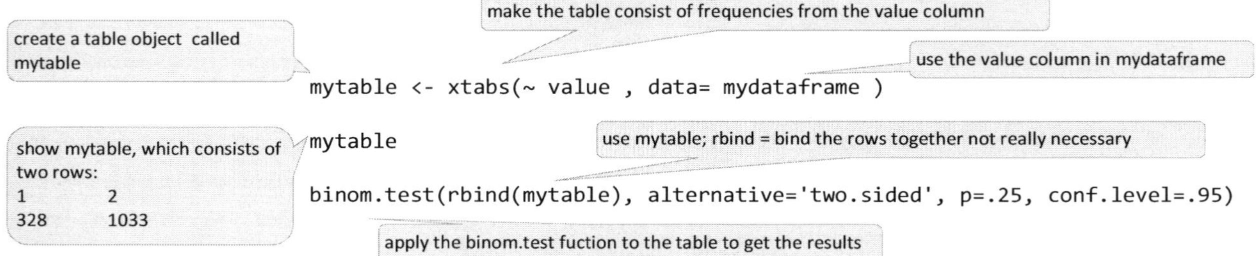
make the table consist of frequencies from the value column
create a table object called mytable
use the value column in mydataframe
mytable <- xtabs(~ value , data= mydataframe)
show mytable, which consists of two rows:
1 2
328 1033
mytable
use mytable; rbind = bind the rows together not really necessary
binom.test(rbind(mytable), alternative='two.sided', p=.25, conf.level=.95)
apply the binom.test fuction to the table to get the results

```
> mytable <- xtabs(~ value , data= mydataframe )
> mytable
value
   1    2
 328 1033
> binom.test(rbind(mytable), alternative='two.sided', p=.25, conf.level=.95)

        Exact binomial test

data:  rbind(mytable)
number of successes = 328, number of trials = 1361, p-value =
0.4527
alternative hypothesis: true probability of success is not equal to 0.25
95 percent confidence interval:
 0.2184887 0.2646336
sample estimates:
probability of success
             0.2409993
```

The important aspect the above analysis of the R code demonstrates is that the actual R command that carried out the binomial test uses the counts of the two categories rather than the raw data.

Using this knowledge we could have entered the counts directly and obtained our result in a single line of R code:

data consists of a single column with two values

```
binom.test(c(328, 1033), alternative='two.sided', p=.25, conf.level=.95)
```

one category | the other category | the population proportion

If we can't even be bothered to work out how many there are in the second category we can use:

```
binom.test(328, 1361, alternative='two.sided', p=.25, conf.level=.95)
```

one category | total number | the population proportion

22.5 Showing the null hypothesis distribution and explaining the p-value

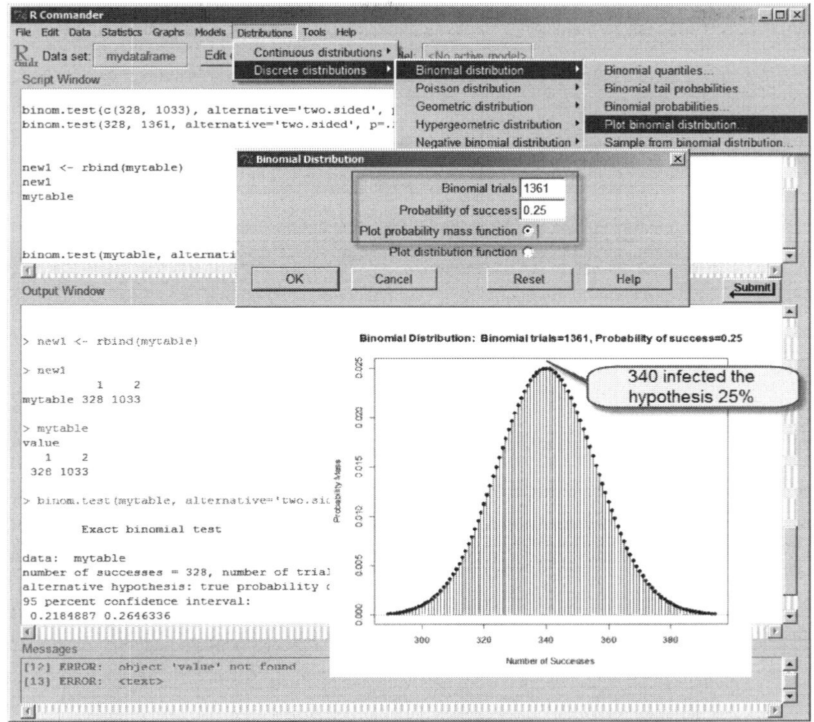

Selecting the menu option shown opposite and setting up the binomial distribution dialog box as shown produces a graph of the likely outcomes given that the proportion of successes (i.e. infected individuals) is equal to 25% (340 of 1361) in the population.

The observed rate of 328 is 12 points less than the expected value. However we considered the 'more extreme' situation implying both sides of the expected value so we have to consider both 328 and 352 (i.e. 340+12). We can annotate the diagram to show this by using the *abline()* function. Typing the code below into the R Commander *Script* window, then highlighting it and clicking on the **Submit** button adds the lines to the graph as shown below. Alternatively you can type it into the *R Console* window and then press the enter key (↵).

abline = add a line function

v=vertical line

lwd= line width

col=colour

```
abline(v=328, lwd=5, col="red")
```

make the colour red

```
abline(v=352, lwd=5, col="blue")
```

make it a vertical line (v) where x=352

make the line width=5. See glossary entry **lwd**

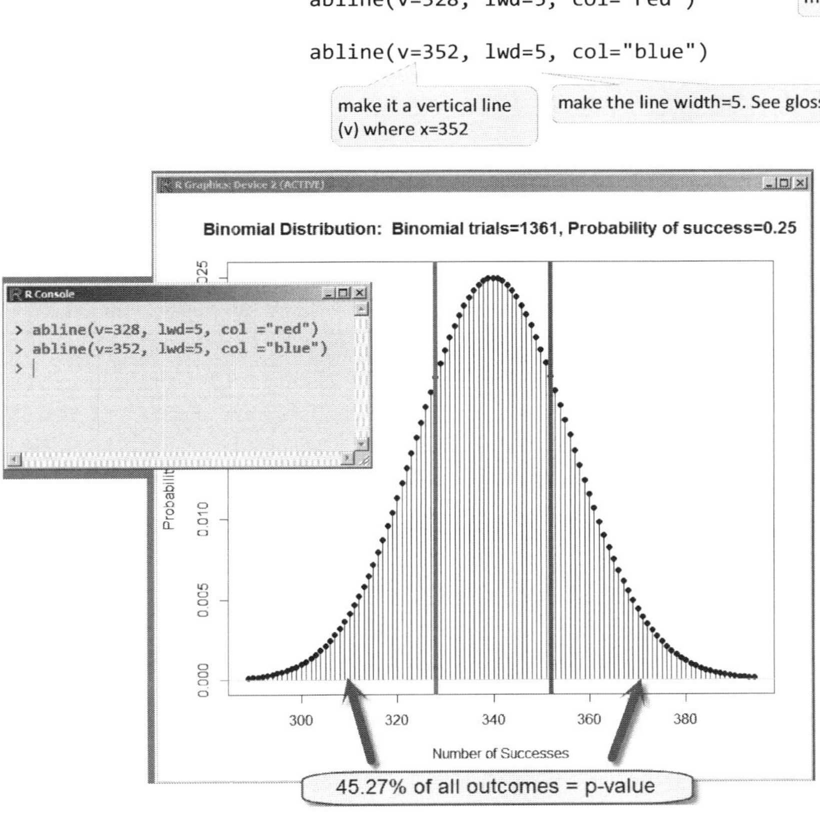

The tail areas under the curve being each side of our two lines is the p-value. It represents a percentage of all possible outcomes assuming the population proportion is 25%.

22.5.1 Adding Critical values

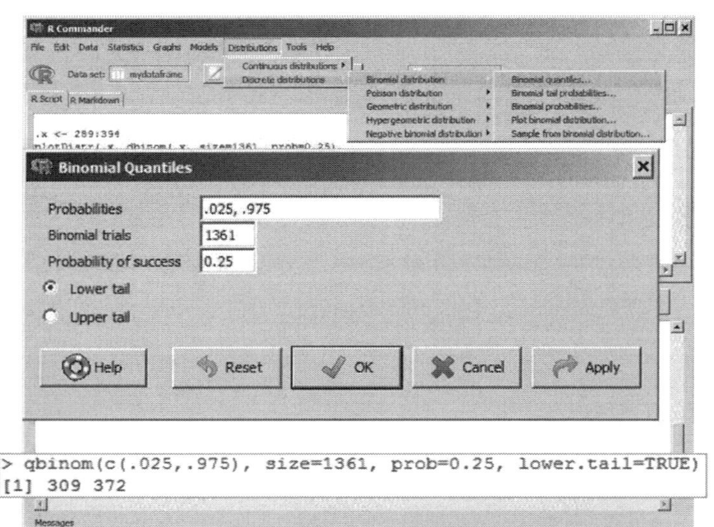

We could also consider the critical values discussed earlier and add these to the diagram. To do this we need to find the specific number of successes (i.e. the x values in the previous plot) which will divide the area up so that we have two tail areas each equal to the critical value divided by two. Taking the usual critical value of 0.05 means we need two tail areas each representing 2.5% of the total area. Considering the curve from the left (lower tail) this is 0.025 and 0.975 of the total area which we know to be 1.

The Binomial Quantile function, *qbinom()* returns the required x values and you can access it using the R Commander menu option:

Distributions->Discrete distributions->Binomial distribution->Binomial quantiles...

In the **Probabilities** option add *.025, .975* In the **Binomial trials** option add *1361* and in the **Probability of success** enter *0.25* We are considering the curve from left to right so make sure the **Lower tail** option is selected. Finally click on the **OK** button.

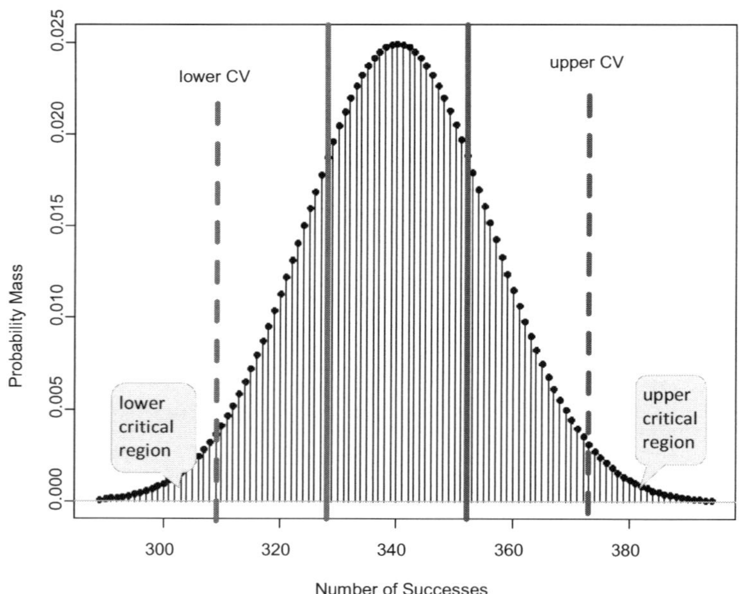

The results are 309 and 372 representing the x values we require. We can now use these values to add the additional lines to the plot where values more extreme than these are said to be within the critical or rejection regions: make the line dotted

```
abline(v=309, lwd=5, col="red", lty=2)

abline(v=372, lwd=5, col="blue", lty=2)

text(309,.02,"lower CV")

text(372,.02,"upper CV")
```

add some text

x coordinate y coordinate text to add

We can see now why our observed p-value of 0.45 stopped us rejecting the null distribution, it included values outside of the critical regions.

22.6 Tips and Tricks

It might seem like a lot of extra work to draw a graph and mark the points in R Commander. However, I always find it helpful to have a visual aid when interpreting the p-value even if it is a rough sketch on a scrap of paper.

23 Several independent proportions compared with the average: two way tables

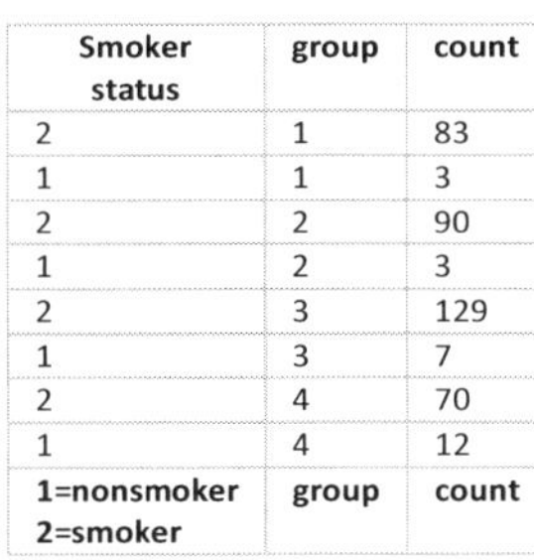

Smoker status	group	count
2	1	83
1	1	3
2	2	90
1	2	3
2	3	129
1	3	7
2	4	70
1	4	12
1=nonsmoker 2=smoker	group	count

We can extend the approach described in the previous chapter to compare a number of independent binary proportions. Say we have four groups of patients and we wish to compare the proportion of smokers in each of the four groups against the overall average.

	Smoker status	
group	smoker =2	nonsmoker =1
1	83	3
2	90	3
3	129	7
4	70	12
total	372	25

This time the statistic is called the chi-squared statistic (X^2), and follows a chi-squared (χ^2) distribution.

We can show the data in several ways, of which two equivalent tables are given above.

23.1 Data preparation

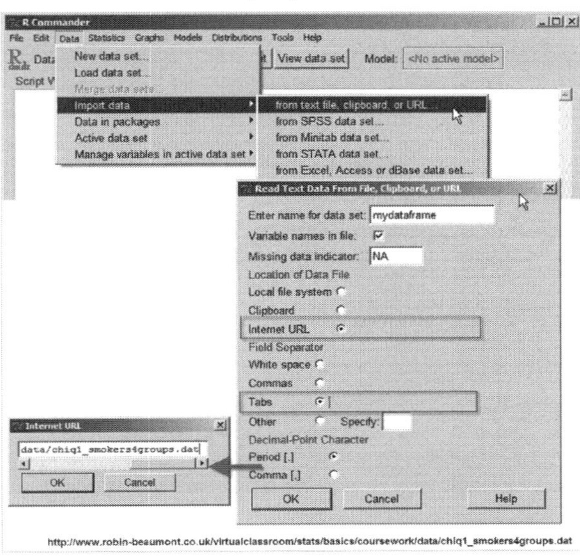

Start R Commander by typing the following into the *R Console* window:

```
library(Rcmdr)
```

You can obtain the data directly from my website by selecting the R Commander menu option:

Data-> from text, the clipboard or URL

I have called the dataframe *mydataframe*, also indicating that it is from a URL (i.e. the web) and the columns are separated by *tab* characters.

Clicking on the **OK** button brings up the *Internet URL* box, in which you need to type the following to obtain my sample data:

```
http://www.robin-beaumont.co.uk/virtualclassroom/book2data/chiq1_smokers4groups.dat
```

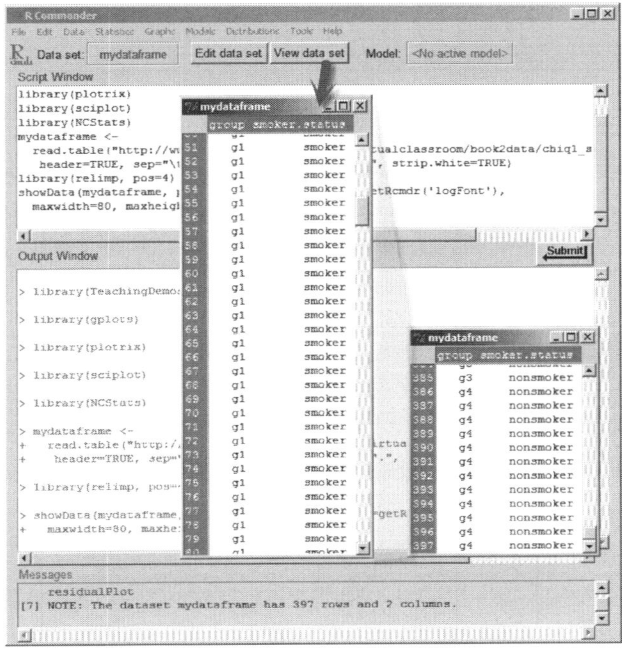

Click **OK**.

It is always a good idea to check the data and you can achieve this in R Commander by clicking on the **View data set** button. You can see we have 397 rows and two columns, the first column indicating which group they belong to and the second one their smoking status.

23.2 Model development and evaluation

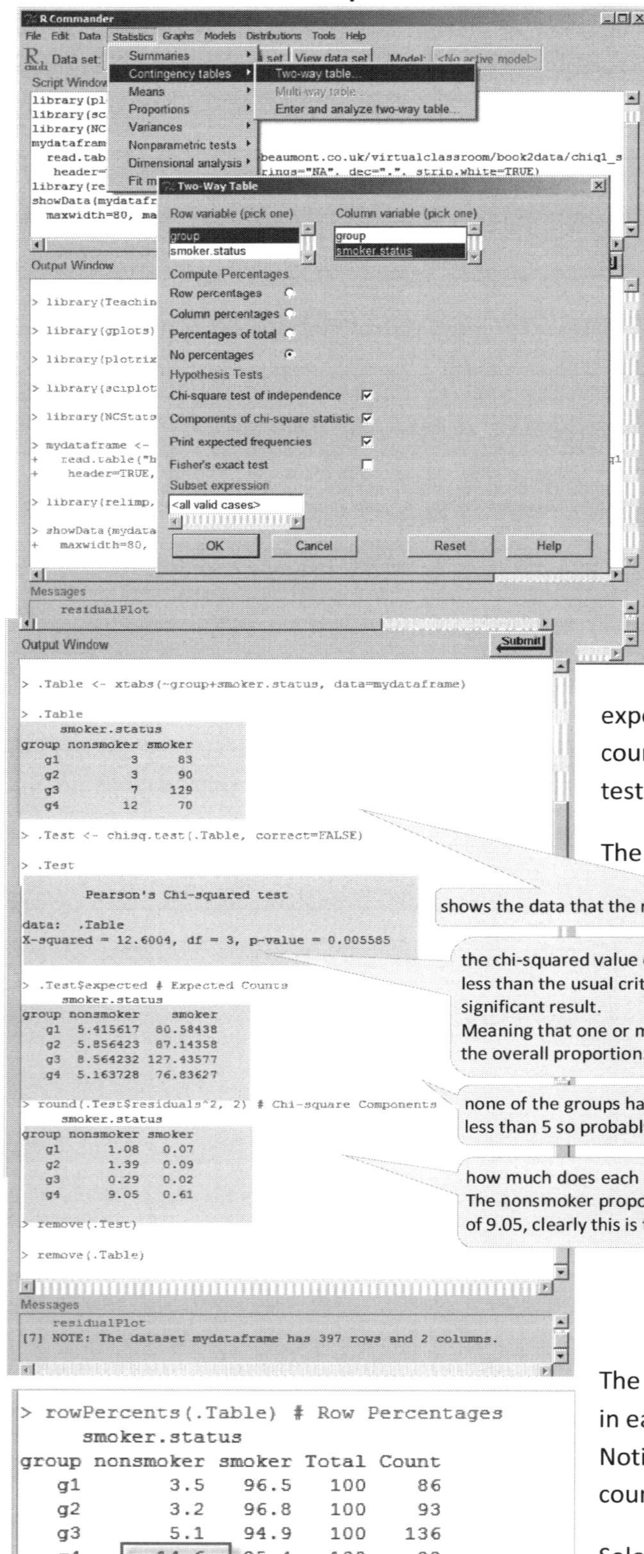

Two-Way Table
Row variable (pick one)
Column variable (pick one)
Compute Percentages
Row percentages
Column percentages
Percentages of total
No percentages
Hypothesis Tests
Chi-square test of independence
Components of chi-square statistic
Print expected frequencies
Fisher's exact test
Subset expression
OK
Cancel
Reset
Help
> .Table <- xtabs(~group+smoker.status, data=mydataframe)
> .Table
smoker.status
group nonsmoker smoker
g1 3 83
g2 3 90
g3 7 129
g4 12 70
> .Test <- chisq.test(.Table, correct=FALSE)
> .Test
Pearson's Chi-squared test
data: .Table
X-squared = 12.6004, df = 3, p-value = 0.005585
> .Test$expected # Expected Counts
smoker.status
group nonsmoker smoker
g1 5.415617 80.58438
g2 5.856423 87.14358
g3 8.564232 127.43577
g4 5.163728 76.83627
> round(.Test$residuals^2, 2) # Chi-square Components
smoker.status
group nonsmoker smoker
g1 1.08 0.07
g2 1.39 0.09
g3 0.29 0.02
g4 9.05 0.61
> remove(.Test)
> remove(.Table)
[7] NOTE: The dataset mydataframe has 397 rows and 2 columns.

We will now develop a model to see if any of our observed proportions are statistically significantly different from the overall proportion. We will set the critical value to 0.05, that is one in twenty.

Select the R Commander menu option:

Statistics-> Contingency tables-> two way table

Complete the dialog box as show opposite. Make sure you have selected both variables.

Besides the **chi-square test** option I have also selected both the **Components of the chi-square statistic** and the **Print expected frequencies** options.

Selecting the *Components of the chi square statistic* option allows us to assess, how much each proportion deviates from that expected. Viewing the *expected frequencies* allows the assessment of cells which might have an expected frequency of less than five. This is important as expected counts of <5 can effect the validity of the traditional chi-squared test result.

The results are shown opposite.

shows the data that the results are based on, always a good thing to check

the chi-squared value of 12.6 and the associated p-value 0.0055. The p-value is less than the usual critical value of 0.05. Therefore we have a statistically significant result.
Meaning that one or more of the proportions of smokers is different to that of the overall proportion.

none of the groups has a expected value of less than 5 so probably the result is valid

how much does each of the proportions differ from that expected?
The nonsmoker proportion for group 4 have a very high chi-squared component of 9.05, clearly this is the group.

```
> rowPercents(.Table) # Row Percentages
     smoker.status
group nonsmoker smoker Total Count
   g1       3.5   96.5   100    86
   g2       3.2   96.8   100    93
   g3       5.1   94.9   100   136
   g4      14.6   85.4   100    82
```

The expected count cells give details of the values that should be in each cell if each group had the same proportion of smokers. Notice that group 4 has 12 smokers compared to the expected 5.2 count.

Selecting the row percentages option and re-running the analysis shows that group 4 has 3 times the proportion of nonsmokers compared to the other groups.

We could collapse all the other groups data together and see if we still obtained a similar p-value as a final check.

23.2.1 Counts as data

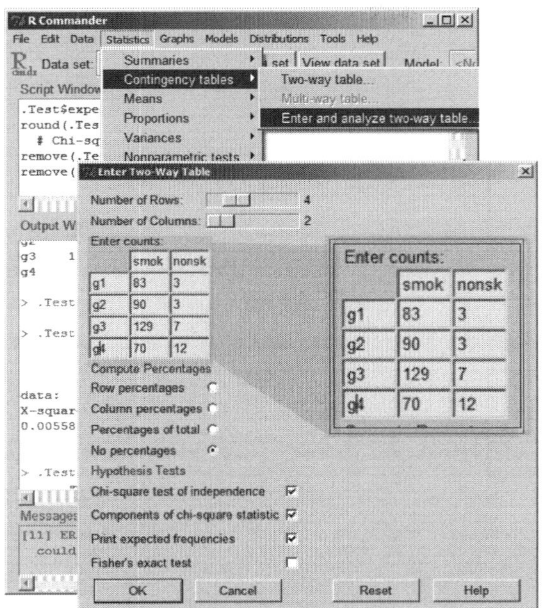

Often there is a need to analyse a set of counts rather than the raw data, which might not be available. In this situation we can make use of R Commander's table entry dialog box.

Select the R Commander menu option:

Statistics-> Contingency tables->
Enter and analyze two-way table

Complete the dialog box as show opposite making sure you fill in the cells and the descriptors for both the columns and rows

As in the previous example, besides selecting the chi-square test, I have chosen both the **Components of chi-square statistic** and the **Print expected frequencies** options.

Clicking on the **OK** button produces identical results to that obtained from the raw data.

23.2.2 R code

We can also carry out the above analysis directly using R code. Considering the two situations:

- **Raw data**

If you have not have imported the data from the previous exercise and given the dataframe the name mydataframe you need to do this first typing the following into either the *R Console* or R Commander *Script* window:

```
mydataframe <- read.delim("http://www.robin-beaumont.co.uk/virtualclassroom/book2data/chiq1_smokers4groups.dat",header=TRUE)

names(mydataframe)
```

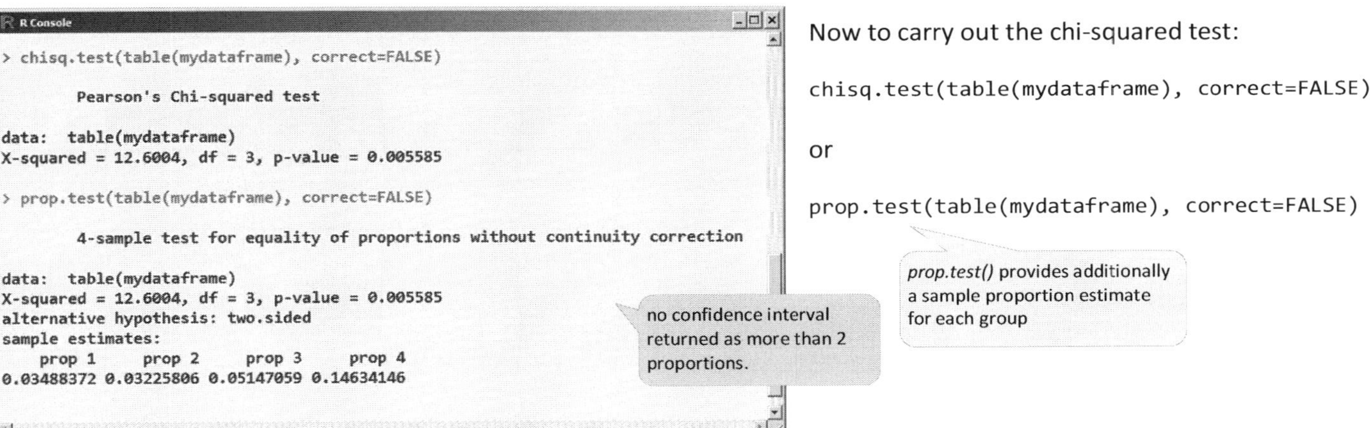

Now to carry out the chi-squared test:

```
chisq.test(table(mydataframe), correct=FALSE)
```

or

```
prop.test(table(mydataframe), correct=FALSE)
```

- **Counts**

Again this is just a few lines of R code:

```
smokers  <- c( 83, 90, 129, 70 ) #smokers
patients <- c( 86, 93, 136, 82 ) # total
prop.test(smokers, patients)
```

Notice here that the 'patient' group includes the smokers, as the *prop.test* behaves similarly to the *binom.test()* command (see *The binomial test* chapter).

23.3 Mosaic plots

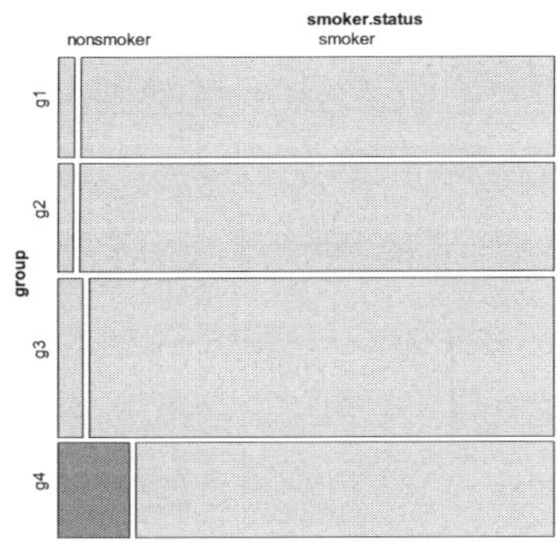

An extremely useful type of graph called a mosaic plot shows immediately those proportions that deviate from that expected assuming the null hypothesis is true, that is equal proportions across groups.

To obtain a mosaic plot you need to use the R Commander *Script* window or the *R Console* window. The first stage is to install and load the *vcd* (Visualizing Categorical Data) package.

```
# install the vcd library

install.packages("vcd", dependencies=TRUE)

# load the vcd library

library(vcd)
```

the mosaic plot works with proportions so if you have raw data you need to use the table command to create them on the fly

Now to create the mosaic plot:

```
mosaic(table(mydataframe), shade=TRUE, legend=TRUE)
```

We can see immediately from the mosaic plot the disproportionally large number of non-smokers in group 4. The Pearson residuals are a measure of how much each observed count deviates from that expected. Referring back to the R Commander output the Pearson residuals are the same as the chi-squared component values. R Commander provides the squared values, so for group four we have $3.01^2 = 9.05$ (approximately).

23.4 Writing up the results

4 groups of patients (n=86, 93, 136, 82) were investigated to compare the proportion of smokers in each. The incidence of smoking ranged from 96.8% to 85.4%. A chi-squared test was performed ($X^2 = 12.60$, df=3; all expected cell counts exceeded 5) producing a statistically significant p-value=0.0055 indicating that one or more of the groups had a different proportion of smokers/non-smokers to the overall average. Examination of the individual residuals and mosaic plot indicated that group 4 (n=82; 14.6% non-smokers) had a larger proportion of non-smokers compared to the other groups.

A mosaic plot would have been included in the results.

24 Comparing several independent categories: contingency tables

	Blood type				
Severity of condition	**O**	**AB**	**B**	**A**	***Total***
Severe	31	7	9	28	75
Mild	31	8	22	44	105
Absent	476	90	211	543	1320
Total	538	105	242	615	1500

Analysis of a contingency table allows the evaluation of possible differences between the distribution of frequencies across several nominal variables which can take more than two values, that is they have more than 2 categories (known as multinomial variables).

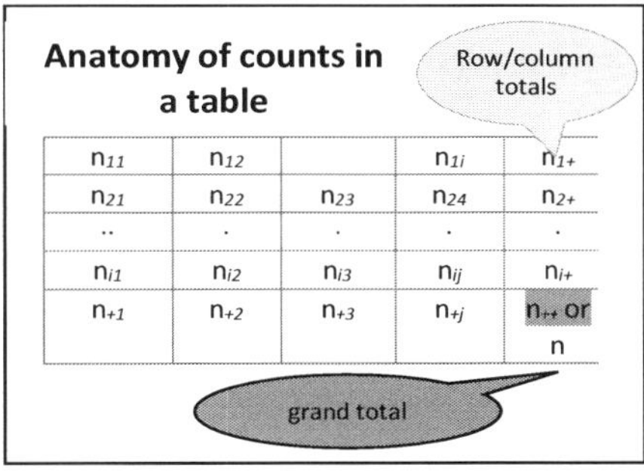

Anatomy of counts in a table
Row/column totals
n11 n12 n1j n1+
n21 n22 n23 n24 n2+
ni1 ni2 ni3 nij ni+
n+1 n+2 n+3 n+j n++ or n
grand total

Daniel, 1991 p.546 provides an example of a research team who collected data on 1500 patients with a particular condition which can be classified as Absent, Mild or Severe. Along with this the researchers also recorded the blood type of each person (A, B, AB, or O). Some readers may notice that the severity of the condition is really an ordinal rather than a nominal variable (see **levels of measurement**) thereby disregarding what I said in the previous paragraph about contingency tables consisting of nominal variables. However, often in contingency tables even interval/ratio variables are treated as nominal variables, such as grouping participants into several age bands. In this example we have proportions between the four blood types and the severity of the condition.

A little revision

From school you may remember that you refer to table cells a particular way, given opposite. $n_{1,1}$ is the value for the first row, first column and $n_{2,5}$ is the value of the second row 5th column. Alternatively if the maximum value for the row and column is clear and is less than 10 the comma can be omitted (n_{23} as in the table opposite). We often wish to show a summed value, say for a particular row or column which we do by using a + subscript, n_{i+} indicating the summation of all the values in the ith row and similarly n_{+j} indicates that it is the value for summing over all the values in the jth column. Unfortunately, this is not a standard, some texts using a dot instead (i.e. $n_{i.}$). A table with *I* rows and *J* columns is called an *I*× *J* or *I* by *J* table

24.1 The chi-squared test – how it works

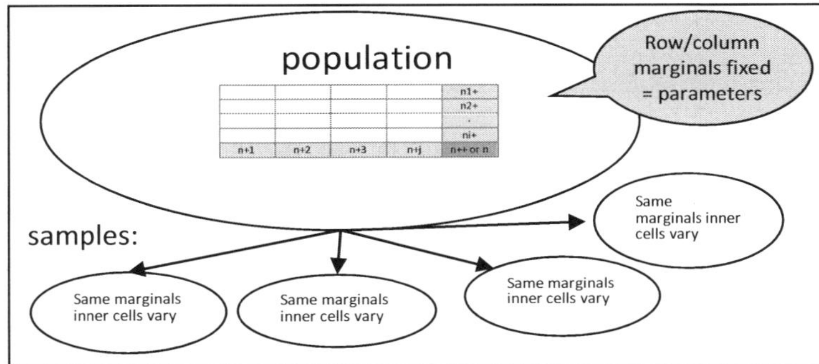

You might well ask how are we going to compare the above table of counts to a population and see how one or more rows or columns differ from the others beyond that due to random sampling. The secret lies in the shaded values, technically called the marginal totals, or just marginals. When we assume each of the inner cells is dependent (i.e. contingent) upon the fixed marginals there are only an infinite number of tables that can be produced. If this is the case we can calculate the probably of a particular table configuration and we have a p-value.

> The proportion in cell *ij* =
> Marginal proportion of category *i*
> Multiplied by
> Marginal proportion of category *j*

As we are concerned with proportions we can represent this in terms of probabilities or proportions, where we use the *pi* symbol to represent population proportions. Using this notation we can state the null distribution:

$$h_0 : \pi_{ij} = \pi_{i+}\pi_{+j} \text{ = null distribution}$$

This can, in turn be interpreted as the **proportions for each category for each variable is the same as the overall average across variables**. So the alternative is: $h_0 : \pi_{ij} \neq \pi_{i+}\pi_{+j}$

$$X^2 = \sum_{all\,cells} \frac{(Observed\ frequency - Expected\ frequency)^2}{Expected\ frequency}$$

$$= \sum_{all\,cells} \frac{(O-E)^2}{E} \sim \chi^2 - distribution\,with\,df = k-1$$

We can measure the deviation from the expected frequency for each cell and add these together. Specifically this is done using the chi-squared statistic (X^2) given opposite. Looking at the equation this value is the observed frequency minus the expected frequency squared for each cell divided by the expected frequency summed across all cells. The individual cell values are called the chi-squared (X^2) components and the square root of each the Pearson residual.

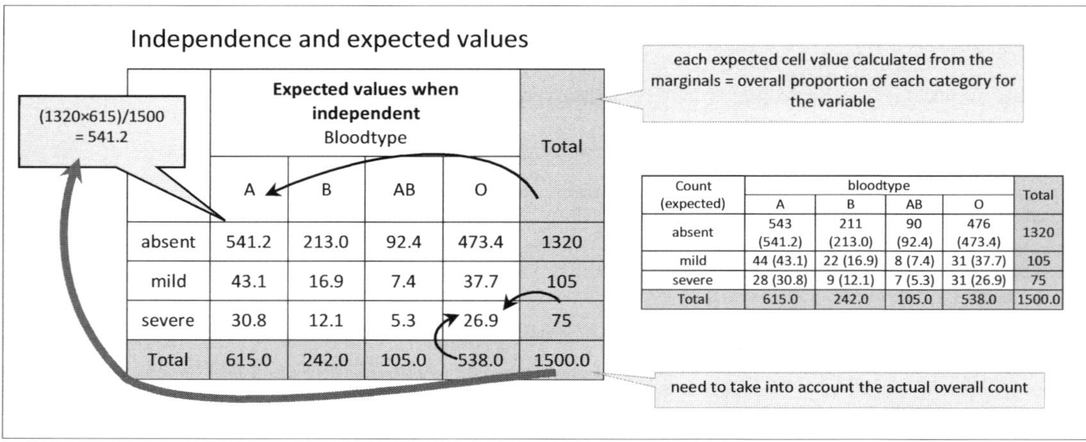

	Expected values when independent Bloodtype				Total
	A	B	AB	O	
absent	541.2	213.0	92.4	473.4	1320
mild	43.1	16.9	7.4	37.7	105
severe	30.8	12.1	5.3	26.9	75
Total	615.0	242.0	105.0	538.0	1500.0

Count (expected)	bloodtype				Total
	A	B	AB	O	
absent	543 (541.2)	211 (213.0)	90 (92.4)	476 (473.4)	1320
mild	44 (43.1)	22 (16.9)	8 (7.4)	31 (37.7)	105
severe	28 (30.8)	9 (12.1)	7 (5.3)	31 (26.9)	75
Total	615.0	242.0	105.0	538.0	1500.0

each of these is a chi-squared component

$$X^2 = \frac{(31-26.9)^2}{26.9} + \frac{(7-5.3)^2}{5.3} + \frac{(9-12.1)^2}{12.1} + \frac{(28-30.8)^2}{30.8} + \frac{(31-37.7)^2}{37.7} + \frac{(8-7.4)^2}{7.4} + \frac{(22-16.9)^2}{16.9} + \frac{(44-43.1)^2}{43.1} + \frac{(476-473.4)^2}{473.4} + \frac{(90-92.4)^2}{92.4} + \frac{(211-213.0)^2}{213.0} + \frac{(543-541.2)^2}{541.2}$$

$$X^2 = .0059 + .0180 + .0623 + .0138 + .0209 + 1.511 + .0574 + 1.177 + .2459 + .7942 + .5833 + .6249$$

$$X^2 = 5.1163$$

this is always a positive value

The chi-squared statistic (X^2) is never a negative value because the chi-squared components consist of squared values. This is different to the values of other statistics we have come across so far (*t* and normal) which can take negative values. This is demonstrated in the calculations for the chi-squared statistic (X^2) for the example table given above.

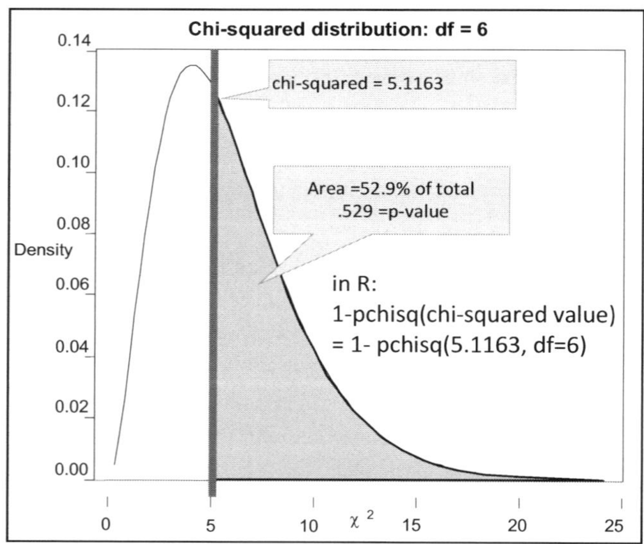

Given certain conditions the X^2 value follows approximately a chi-squared distribution (χ^2) the exact shape being defined by the number of rows and columns. Because the X^2 value consists of only positive values when we consider 'more extreme values', we only refer to the area to the right of the observed chi-squared distribution.

Taking the blood group and illness severity example we obtain a chi-squared value of 5.1163 (df=6) with an associated p-value=0.529 (details in the next section).

In other words we would expect a table with the observed counts or one which would result in a larger chi-squared value fifty three times in every hundred on average from a population defined by the marginal values. That is where all the category proportions equal the marginals within random sampling variability.

You may think I'm being rather measured in my interpretation of the chi-squared p-value but this is deliberate. What we have done is basically taken a whole set of counts and converted then into a single value.

However often the interest is in specific proportions in various categories, **all the chi-squared value gives us is an overall measure of fit**, it does not tell us to what degree specific categories proportions vary from the expected value. To find this out we need to carry out some type of graphical or residual analysis.

The chi-squared distribution is defined by a single parameter known as the Degrees of freedom (df) this is equal to:

(number of rows-1) X (number of columns-1); In this instance (3-1) x (4-1) = 6

24.1.1 Too good a fit

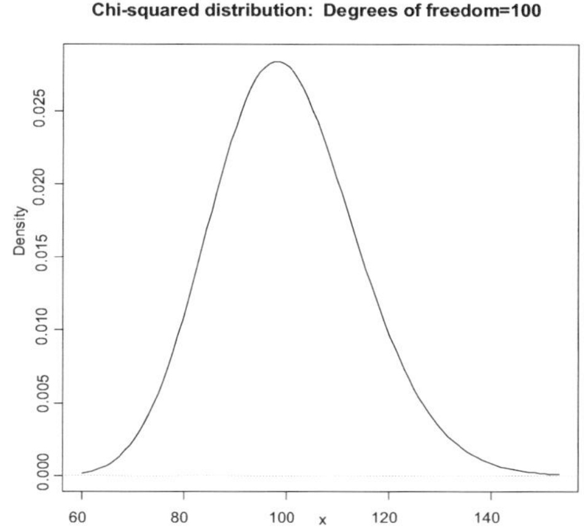

One aspect that often confuses people is the fact that given that the expected value of the chi-squared statistic (X^2) component is zero (when there is no difference between the expected and observed frequencies) why does not the chi-squared distribution bunch near zero regardless of how many rows and columns there are in the table. For example the above chi-squared distribution (for df=6) appears to reach its maximum around 4 to 5? In effect this shift from zero can be thought of as taking into account the danger of expecting a near perfect fit. (Yule & Kendall, 1953 p.469, Speigel, 1975 p.219). The degree of shift is related to the degrees of freedom (i.e. number of column and rows). The more cells we have the greater the deviation from zero. In fact 50% of the area under the curve lies each side of the degrees of freedom value. Also as the degrees of freedom increases the distribution becomes more symmetrical, and at large values approximates the normal distribution as shown opposite.

Yule & Kendall, 1953 explain it this way:

> We are just as unlikely to get very good correspondence between observed and expected values as we are to get very bad correspondence and, for precisely the same reasons, we must suspect the sampling technique if we do. In short, very close correspondence is too good to be true.
>
> The student who feels some hesitation about this statement may like to reassure himself with the following example. An investigator says that he threw a die 600 times and got exactly 100 of each number from 1 to 6. This is the theoretical expectation, X^2=0 and p-value=1, but should we believe him? We might, if we knew him very well, but we should probably regard him as somewhat lucky, which is only another way of saying that he has brought off a very improbable event. (p.470)

Key point
The final chi-squared value only tells you about overall fit not which cell(s) deviate the most or least.

You may be interested to know that when Pearson in 1900 introduced the chi-squared test he got the degrees of freedom wrong and it was R.A.Fisher in 1922 who corrected him (Agresti, 2002 p.79), so if you have problems calculating them you are in good company! Also interestingly in the 1900 article (quoted in Plackett, 1983 p.68) he had more trouble in evaluating X^2 for deviations which sum to zero. This aspect is the longest piece of algebra in the paper, suggesting that it was a discovery more recent than that of the distribution of X^2, which occupied only 16 lines.

We will now consider two examples, the first is a simple 2 x 2 table and the second one revisits the blood group illness severity table discussed above.

For the simple 2 x 2 table I will use the example from MSME3 describing the effect two different types of antibiotic had upon a group of 480 patients after five days. Improvement versus no improvement.

	amox	erythr
improv	144	160
Noimprov	96	80

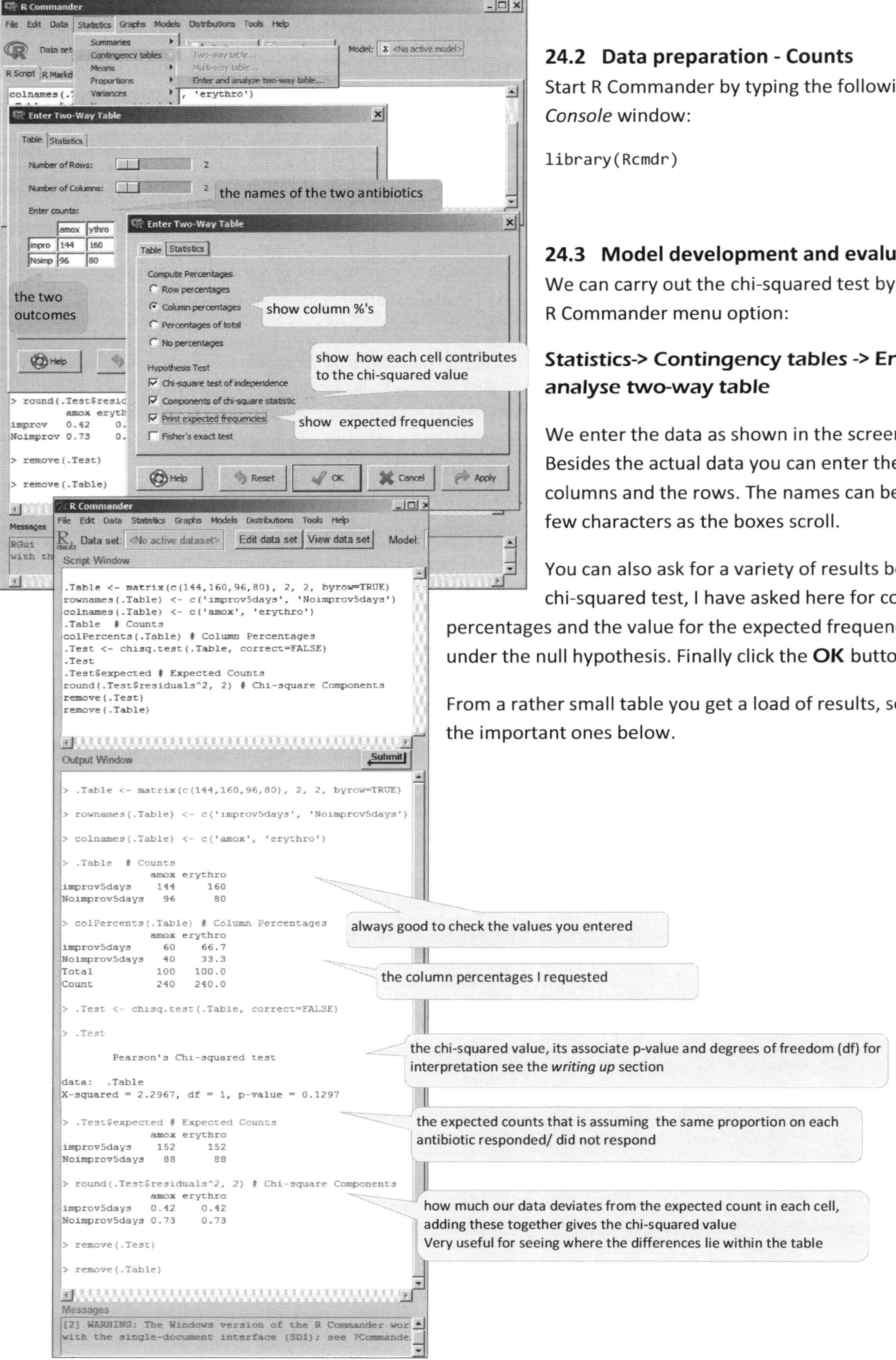

24.2 Data preparation - Counts

Start R Commander by typing the following into the *R Console* window:

```
library(Rcmdr)
```

24.3 Model development and evaluation

We can carry out the chi-squared test by selecting the R Commander menu option:

Statistics-> Contingency tables -> Enter and analyse two-way table

We enter the data as shown in the screenshot opposite. Besides the actual data you can enter the names for the columns and the rows. The names can be more than a few characters as the boxes scroll.

You can also ask for a variety of results besides the chi-squared test, I have asked here for column percentages and the value for the expected frequency in each cell under the null hypothesis. Finally click the **OK** button.

From a rather small table you get a load of results, so I've highlighted the important ones below.

24.3.1 Using raw data

Select the R Commander menu option:

Statistics-> Contingency tables-> two way table

Complete the dialog box as described in the previous chapter on page 112.

```
> mycol_names <- c("Amoxicillin", "Erythromycin")
> myrow_names <- c("improved@5days", "no_improvement@5days")
> dimnames(thedata) <- list(status = myrow_names, antibiotic = mycol_names)
> thedata
                      antibiotic
status                 Amoxicillin Erythromycin
  improved@5days               144          160
  no_improvement@5days          96           80
> myresult <- chisq.test(thedata)
> myresult

        Pearson's Chi-squared test with Yates' continuity correction

data:  thedata
X-squared = 2.0185, df = 1, p-value = 0.1554

> myresult$observed
                      antibiotic
status                 Amoxicillin Erythromycin
  improved@5days               144          160
  no_improvement@5days          96           80
> myresult$expected
                      antibiotic
status                 Amoxicillin Erythromycin
  improved@5days               152          152
  no_improvement@5days          88           88
> myresult$residual
                      antibiotic
status                 Amoxicillin Erythromycin
  improved@5days        -0.6488857    0.6488857
  no_improvement@5days   0.8528029   -0.8528029
```

Alternatively, you can carry out the analysis using R code using the following **function** in the *R Console* window:

`chisq.test(`x`, `y`)`

Where x and y are columns of raw data. However if your data is in the form of a matrix R thinks this is a table of counts. The example opposite demonstrates this, producing the same result as that previously achieved in R Commander.

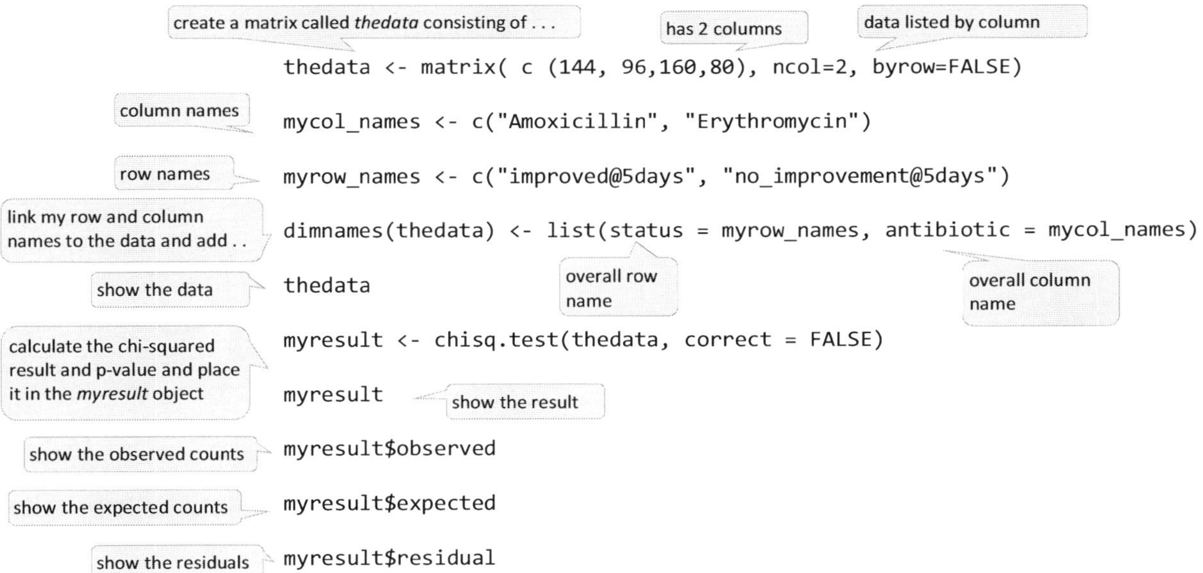
create a matrix called *thedata* consisting of . . .
has 2 columns
data listed by column
thedata <- matrix(c (144, 96,160,80), ncol=2, byrow=FALSE)
column names
mycol_names <- c("Amoxicillin", "Erythromycin")
row names
myrow_names <- c("improved@5days", "no_improvement@5days")
link my row and column names to the data and add . .
dimnames(thedata) <- list(status = myrow_names, antibiotic = mycol_names)
overall row name
overall column name
show the data
thedata
calculate the chi-squared result and p-value and place it in the *myresult* object
myresult <- chisq.test(thedata, correct = FALSE)
myresult
show the result
show the observed counts
myresult$observed
show the expected counts
myresult$expected
show the residuals
myresult$residual

24.4 Writing up the results

480 patients suffering from bronchopneumonia were randomly allocated to one of two antibiotic treatments, amoxicillin or erythromycin. After 5 days of treatment, any improvement was noted. 144 (60%) of those receiving Amoxicillin improved compared to 160 (67%) of those receiving Erythromycin. A chi-squared test (X^2=2.3, df=1, p-value = 0.13, all expected cell counts >5) produced a statistically insignificant result (setting the critical value at 0.05) forcing us to the decision that the observed difference in the proportions of patients that recovered on either antibiotic was due to random sampling. In other words they come from a single population with a single proportion.

24.5 Larger tables

We discussed at the beginning of this chapter a table that had two variables each with more than two categories, the Illness severity/ Blood group example and we will now carry out the analysis of this table

You can carry out the analysis in R Commander using either counts or raw data:

Using counts	**Using raw data** http://www.robin-beaumont.co.uk/virtualclassroom/book2data/chiq1_daniel_bg_condition.dat

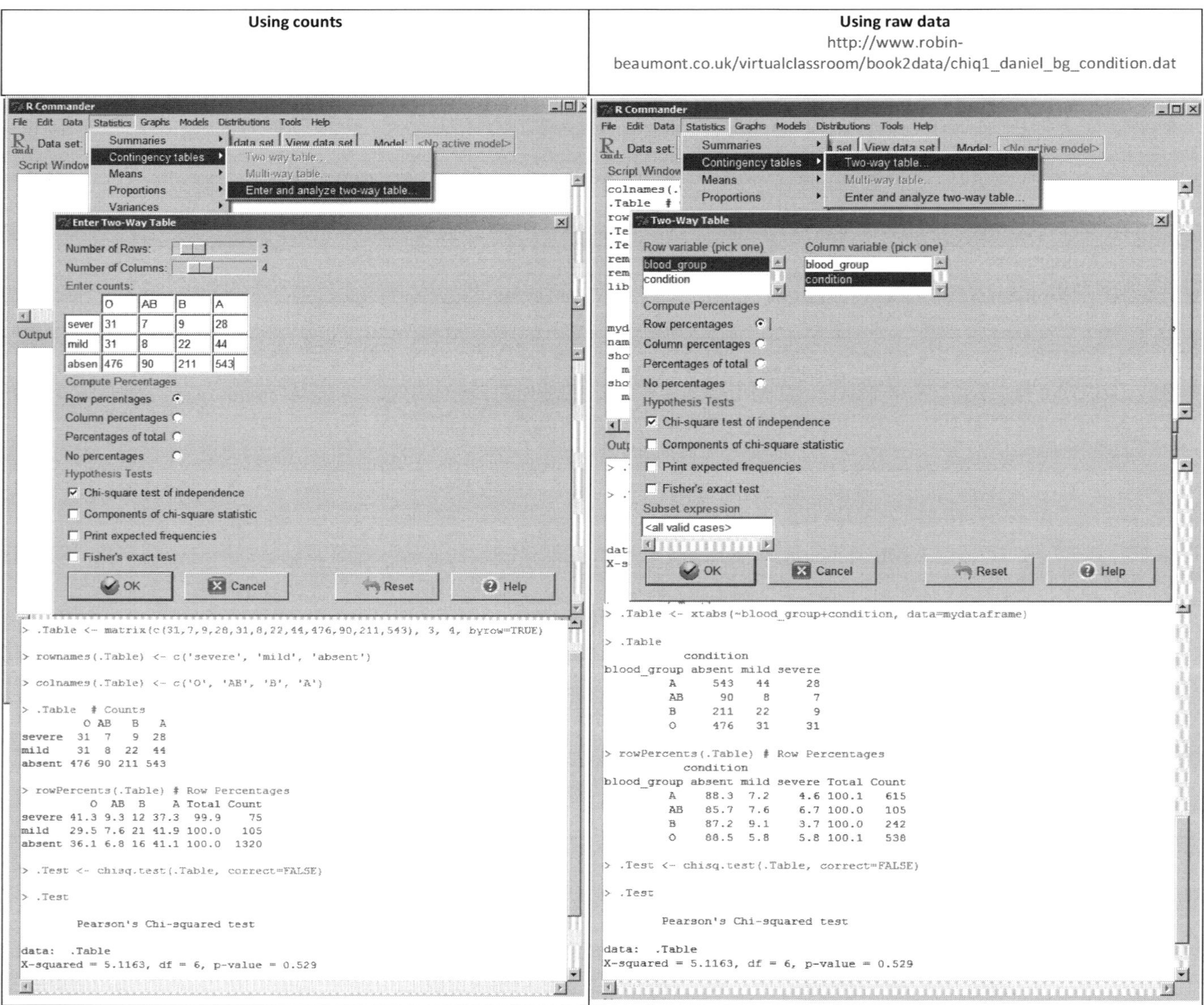

The R equivalent analysis is given below, first using counts:

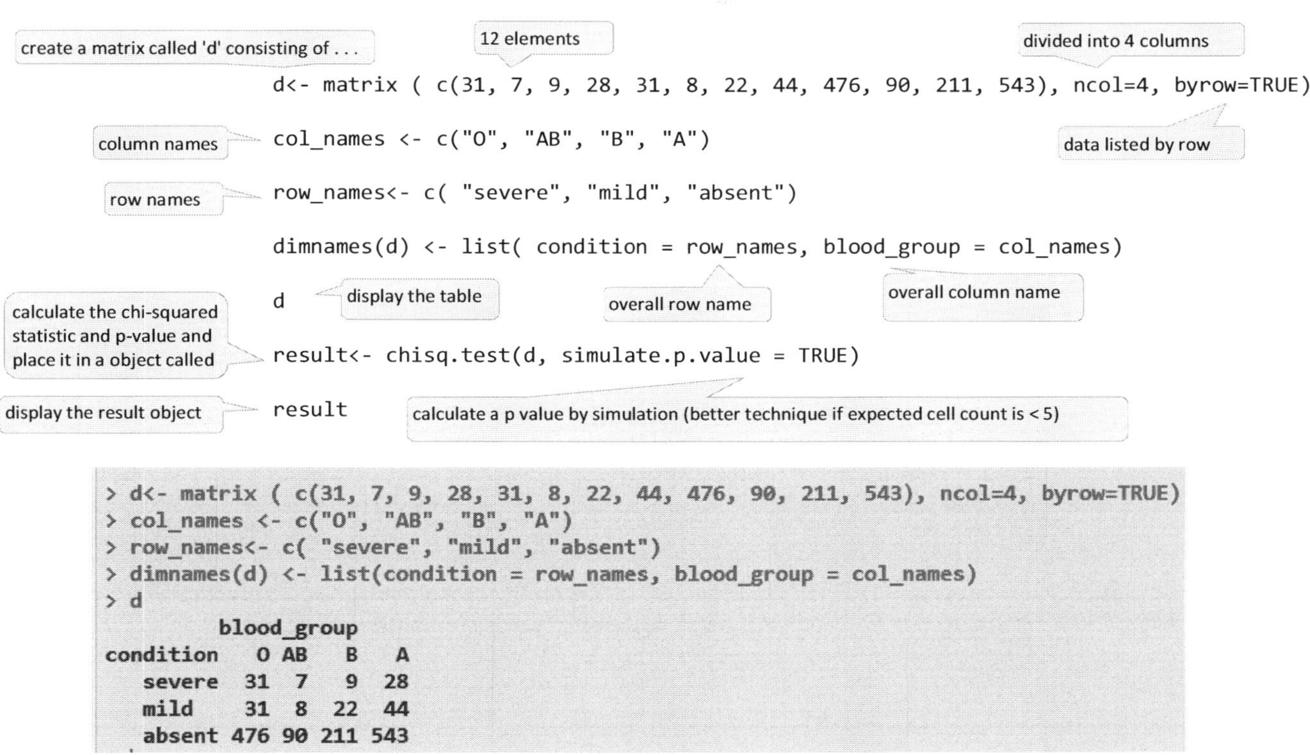
create a matrix called 'd' consisting of . . .
12 elements
divided into 4 columns
d<- matrix (c(31, 7, 9, 28, 31, 8, 22, 44, 476, 90, 211, 543), ncol=4, byrow=TRUE)
column names
col_names <- c("O", "AB", "B", "A")
data listed by row
row names
row_names<- c("severe", "mild", "absent")
dimnames(d) <- list(condition = row_names, blood_group = col_names)
d
display the table
overall row name
overall column name
calculate the chi-squared statistic and p-value and place it in a object called
result<- chisq.test(d, simulate.p.value = TRUE)
display the result object
result
calculate a p value by simulation (better technique if expected cell count is < 5)

```
> d<- matrix ( c(31, 7, 9, 28, 31, 8, 22, 44, 476, 90, 211, 543), ncol=4, byrow=TRUE)
> col_names <- c("O", "AB", "B", "A")
> row_names<- c( "severe", "mild", "absent")
> dimnames(d) <- list(condition = row_names, blood_group = col_names)
> d
         blood_group
condition   O AB   B   A
   severe  31  7   9  28
   mild    31  8  22  44
   absent 476 90 211 543
```

Along with a similar analysis using the raw data where the last line produces a graphical summary of the data called a mosaic plot described in detail in the following chapter.

```
mydataframe <- read.delim("http://www.robin-beaumont.co.uk/virtualclassroom/book2data/chiq1_daniel_bg_condition.dat", header=TRUE)

names(mydataframe)

mydataframe

myresult <- chisq.test(table(mydataframe), correct=FALSE)

myresult

myresult$observed

myresult$expected

myresult$residual

library(vcd)

mosaic(table(mydataframe),residuals_type = "pearson", gp = shading_Friendly )
```

get the data and place it in a dataframe object called *mydataframe*

use the raw data

the data has column names

display the names of the columns

display the data

do not apply Yates correction to the calculations

put the result into *myresult* object

display the result

carry out a chi-squared test

display observed, expected and residual values values

load the *vcd* library so that I can get a mosaic plot

calculate person residuals

use this shading style

24.6 Residual analysis – Extended Association plots

This topic is described in more detail in the *Mosaic and extended association plots* chapter.

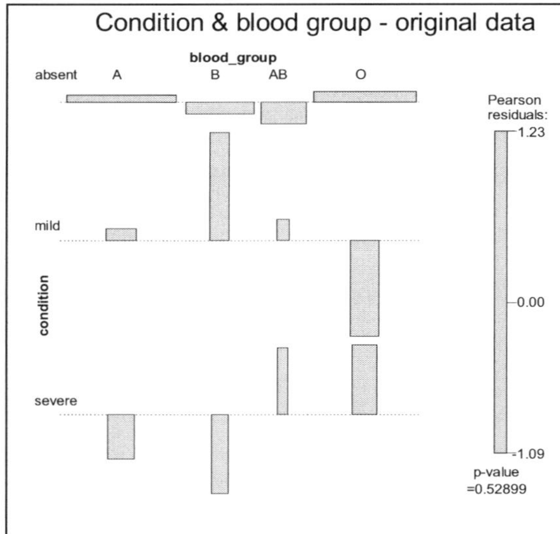

We have discussed the importance of looking at the residuals for each cell to get a feel for the data, and have done this so far by looking at the chi-squared components and mosaic plots.

An alternative to the mosaic plot, also provided in the R *vcd* package, which graphs the actual residuals is the **Extended Association Plot** (ASP) obtained by using the *assoc()* function. Any residuals with an extreme change in colour should be noted.

The illness versus blood group dataset shows little difference between the observed and expected values producing a high p-value and no large residuals (first extended association plot opposite). In contrast to demonstrate a table with a significant p-value I have edited the data slightly for the second extended association plot increasing the proportion of subjects with blood group A and the severe illness and reducing those with blood group O and the severe illness.

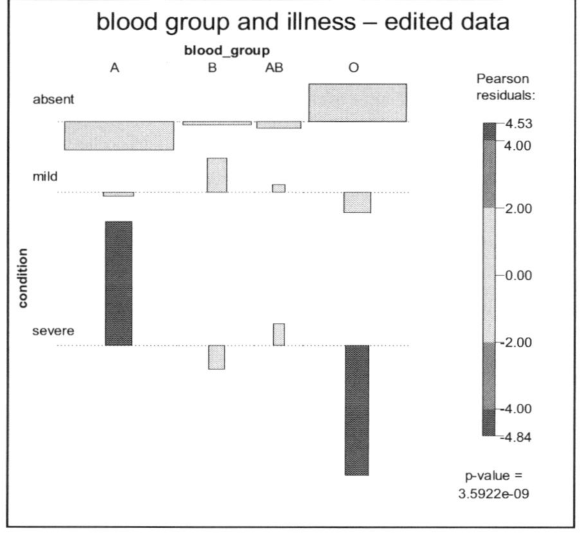


```
# matrix( c(col 1 data, col 2 data etc),  nrow = 3, ncol=4)
# define the names for each level
# name the rows and then the columns, give the severe group
# a very high proportion of blood group A
# and also the severe group have a very low
# proportion of blood group O
# If you don't have the vcd package installed
# un-comment the next line
# install.packages("vcd")
library(vcd)

thedata<-matrix(c(543, 44, 58, 211, 22, 9, 90, 8, 7, 476, 31, 1), nrow=3, ncol=4,
dimnames = list(condition= c("absent", "mild", "severe"),
blood_group =c("A"  , "B", "AB", "O")))
thedata
assoc(thedata, main = "blood group and illness", shade = TRUE)
```

From the extended association plot opposite you can see immediately the dramatic effect this has on the residuals. Also notice that the chi-squared p-value now equals 0.00000000359 (that is what the e-09 means), you would never report a value like this but write p<.0001. If we were hoping to demonstrate that those patients with blood group A had a higher incidence of the severe illness compared to the other blood groups we would be running off to celebrate now. Unfortunately, much research data looks more like the top plot, and any amount of data manipulation will not transform it.

24.7 When you have expected counts of less than five in a cell

When an expected frequency (not the observed frequency) for a cell is less than five the chi-squared associated p-value might be inaccurate. To correct this we can calculate an exact p-value using a resampling technique (the advanced chapter *Confidence intervals for effect sizes* describes this in more detail). To carry out a resampling technique, specifically a technique known as Monte Carlo simulation here, for the chi-squared test we need to use R code, annotating the first R code example in this chapter we add the option *simulate.p.value= TRUE* to it:

```
result<- chisq.test(d, simulate.p.value = TRUE)
```

```
> result<- chisq.test(d, simulate.p.value = TRUE, B= 10,000)
> result

        Pearson's Chi-squared test with simulated p-value
        (based on 10 replicates)

data:  d
X-squared = 5.1163, df = NA, p-value = 0.7273

> result<- chisq.test(d, simulate.p.value = TRUE, B= 10000)
> result

        Pearson's Chi-squared test with simulated p-value
        (based on 10000 replicates)

data:  d
X-squared = 5.1163, df = NA, p-value = 0.5347
```

You can also specify the number of replications you want the p-value based on using the $B = number$ option where the default is 2000. Note, do not use commas to separate large numbers as R interprets the commas as decimal points. You will see in the example opposite where I have entered $B = 10,000$, R has taken this to mean 10.

```
result<- chisq.test(d, simulate.p.value = TRUE, B= 10000)
```

Comparing the simulation method to the traditional method used to produce the p-value shows very little difference, 0.5357 compared to 0.529.

We would interpret this p-value to mean:

That we would obtain a chi-squared value equal to or more extreme that that observed around 53 times in a hundred if the proportions across blood groups and illness severity bands were the same.

Setting a critical value equal to 0.05 (one in twenty times) as this is greater than that threshold we accept that any observed differences in proportions across different illness levels or blood group is due to random sampling variability.

24.8 Why no confidence intervals?

There are no confidence interval values returned when you carry out a chi-squared test. This is basically because no one can agree on how to create valid measures, If you have a variable with two categories it is possible but often variables have more than two categories (i.e. multinomial) and this causes problems. For one solution see: http://www.rforge.net/doc/packages/NCStats/html/gofCI.html and also my youtube videos which describe a workaround.

24.9 Tips and Tricks

The chi-squared value is a summary measure for the whole table and once you obtain a statistically significant result it is very helpful to know which of the cells contributed to the greatest degree - always look at the residuals, see the mosaic and extended association plots chapter for more details.

To obtain a valid result it is necessary to have an expected count of more than 5 in each cell and when this expected cell count assumption is violated it is appropriate to obtain a p-value by using a process called Monte Carlo simulation (demonstrated in the last example).

Youtube help:

http://www.youtube.com/watch?v=n85QkIUPHy0

This is the first of 5 videos I have produced discussing the various types of chi-squared test.

25 Measuring the degree to which two variables co-vary: correlation

Correlation forms the basis of very many statistical procedures and I could write a whole book about it. Correlation can be defined as:

'The degree to which points cluster about the line of best fit'

Words in **bold** indicate a glossary entry

I would add 'for two variables that possess, ordinal, interval or ratio **level of measurement**'. The line of best fit I will discuss latter. The statement above assumes that the correlation is concerned with a 'straight' line in other words it is a *linear* relationship. While in this chapter we will only consider correlation measures that relate to a linear relationship it is possible to develop similar measures for curved or even oscillatory curves such as a sine wave.

The following sections provide details of the theory behind correlation. Please feel free to skip them if you just need the practical aspects.

25.1 That is correlation?

Sir Francis Galton (1822-1911) first proposed the term "co-relation" between two variables in 1888 when studying the height and length of forearms of 348 men (Snedecor & Cochran, 1980 p. 178). Much of statistics in the first decade of the 20th century was concerned with measuring and correlating variables within large data sets, just looking at the titles of the articles in the journal Biometrika from 1900 to 1920 is fascinating and by 1920 the famous Karl Pearson included an article entitled, "Notes on the history of Correlation" (Pearson, 1920).

25.1.1 The distribution when there is no relationship

It is important to realise that we are no longer thinking of two separate variables but a value that has two aspects, think of each value having an x and y part, equivalent to a **point** on a two dimensional plane (i.e. x and y axis). To plot the distribution of these points we need a third dimension to show how many of these points occur for each x, y value. This means we have moved from a flat (two dimensions) normal distribution to one with three dimensions. When the distribution of points has a zero correlation it looks like an upturned pudding basin.

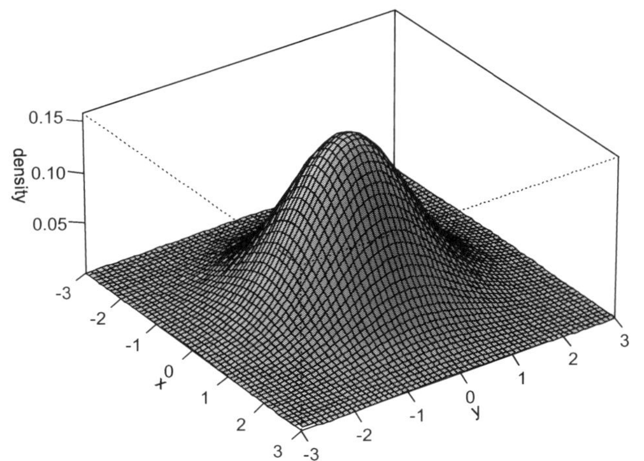

Because it is difficult to draw and conceptualise three dimensional graphs these are often converted to two dimensional scatterplots. We will start our discussion of correlation by considering these two dimensional graphs, and how they help assess the degree to which two ordinal/interval/ratio variables 'co-vary'.

25.1.2 Scatter Plots

Two examples of scatter plots are given below, first an appropriate use and then an inappropriate one.

Example a) appropriate use of scatter plot two ordinal/interval/ratio measurement level variables

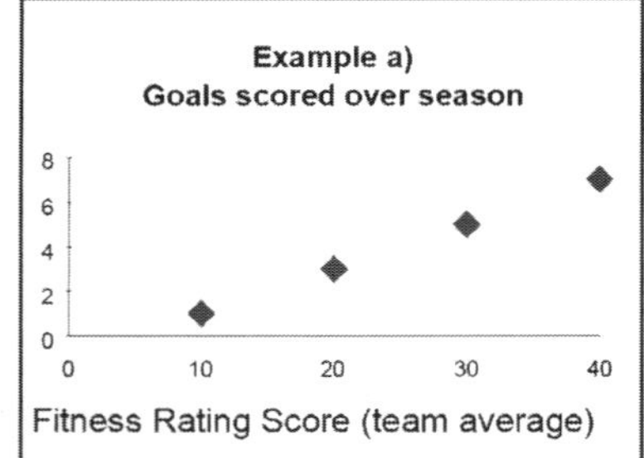

The scatter plot opposite provides details for two ratio variables, goals scored in a season along with the average mythic fitness rating score taken at the beginning of the season.

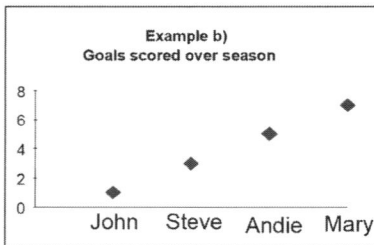

Example b)
Goals scored over season
8
6
4
2
0
John Steve Andie Mary

Example b) Misuse when one of the variables is nominal measurement level data

This scatter plot provides details of goals scored over a session by various players. It shows what, I think, someone might mistakenly call an association or relationship. Clearly, it is not possible to investigate the correlation between these two variables using the above definition as the x axis consists of levels of a nominal variable. Notice that the x axis values could have been placed in any order it just so happens that they have been placed in what looks like a 'correct' order. Data that only possesses a nominal level of measurement cannot be ordered, and a much better way of presenting such data is to use a bar chart or ideally a mosaic plot.

It is all very well looking at scatter plots (in fact it is an essential preliminary step) when considering the possibility that two variables might co-vary. However what we need is an actual measure of the degree to which the two sets of scores 'co-vary', luckily the covariance statistic provides this.

25.1.3 Covariance

Let us start by considering the **sample variance** (equation below far left) this is the average of the squared deviations from the mean for a single variable. Expanding this expression we end up with the equation beside it. We know this measure increases as the values for the variable becomes more dispersed. To calculate the covariance we change the second bracket to show the scores from the second variable (y_i).

$$\frac{\sum_{i=1}^{n}(x_i-\bar{x})^2}{n-1}=\frac{\sum_{i=1}^{n}(x_i-\bar{x})(x_i-\bar{x})}{n-1}$$

Looking at the equation it appears that the covariance is an average measure of the deviations from both means. By considering several scenarios we can understand more fully what this means.

$$Covariance = \text{cov}_{xy} = \frac{\sum_{i=1}^{n}(x_i-\bar{x})(y_i-\bar{y})}{n-1}$$

25.1.4 Scenario 1 – all points on line rising.

$$Covariance = \text{cov}_{xy} = \frac{\sum_{i=1}^{n}(x_i - \bar{x})(y_i - \bar{y})}{n-1}$$

$$= \frac{(1-2.5)(1-2.5)+(2-2.5)(2-2.5)+(3-2.5)(3-2.5)+(4-2.5)(4-2.5)}{4-1}$$

$$= \frac{(-1.5)(-1.5)+(-0.5)(-0.5)+(0.5)(0.5)+(1.5)(1.5)}{3}$$

$$= \frac{2.25+.25+.25+2.25}{3} = \frac{5}{3} = 1.666$$

```
> x<- c(1, 2, 3, 4)
> y <- c(1, 2, 3, 4)
> cov(x,y)
[1] 1.666667
```

Take the following pairs of scores *(x,y); (1,1), (2,2), (3,3), (4,4)* – if you make a quick sketch of the points they all fit on a straight line with an angle of 45% . Also both have a mean of *10/4=2.5* so using the formula opposite we can calculate the covariance as shown opposite. We can either do this the long hand way or using R with the *cov()* **function,**

What use is this value of 1.666? Well it just so happens to be equal to the product of the standard deviations for both of the groups. In this instance it is $s_x \cdot s_y = 1.29099 \cdot 1.29099 = 1.666$ where $\cdot$ = multiplication. Therefore, we can provisionally say that if all the points are on a line of 45% then the covariance value is equal to the product of the sample variances.

25.1.5 Scenario 2 – all points on line – falling.

$$\frac{\sum_{i=1}^{n}(x_i - \bar{x})(y_i - \bar{y})}{n-1} = \frac{(1-2.5)(4-2.5)+(2-2.5)(3-2.5)+(3-2.5)(2-2.5)+(4-2.5)(1-2.5)}{4-1}$$

$$= \frac{(-1.5)(1.5)+(-0.5)(0.5)+(0.5)(-0.5)+(1.5)(-1.5)}{3}$$

$$= \frac{-2.25-.25-.25-2.25}{3} = \frac{-5}{3} = -1.666$$

```
> x<- c(1, 2, 3, 4)
> y <- c(4, 3, 2, 1)
> cov(x,y)
[1] -1.666667
```

Take the following pairs of scores *(x,y); (1,4), (2,3), (3,2), (4,1)* –if you make a quick sketch of the points they would all fit on a straight line with an angle of -45% this time. Both variables again have the same mean 2.5 and also the same variance 1.29099 but this time the covariance is -1.666.

That is this time we get the same value except it is negative. Once again it is equal to the product of the two standard deviations.

In fact regardless of the angle of the line if all the points fall on it the covariance is equal to the sum of the standard deviations. Just to make sure you have got it:

When all the points are on a straight line the covariance = product of the standard deviations for each variable

$= s_x \cdot s_y$

There is one small proviso to the above; if either all the x or y values are the same then the variance of each respectively will be zero (s_x or s_y) so as zero times anything equals zero so will the covariance. Graphically this is the same as saying that if the line is either perfectly horizontal or vertical the covariance will be zero.

Given that the line is not either perfectly horizontal or vertical and when all the points are on the straight line, we know that the covariance is equal to the product of the two standard deviations. However, it would be nice to know how this descriptive statistic behaves when we do not get a perfect fit. Does it get larger or smaller, in other words are the values we have found so far the entire range of covariances, or just for a particular dataset? We will now investigate this by considering the lines of best fit.

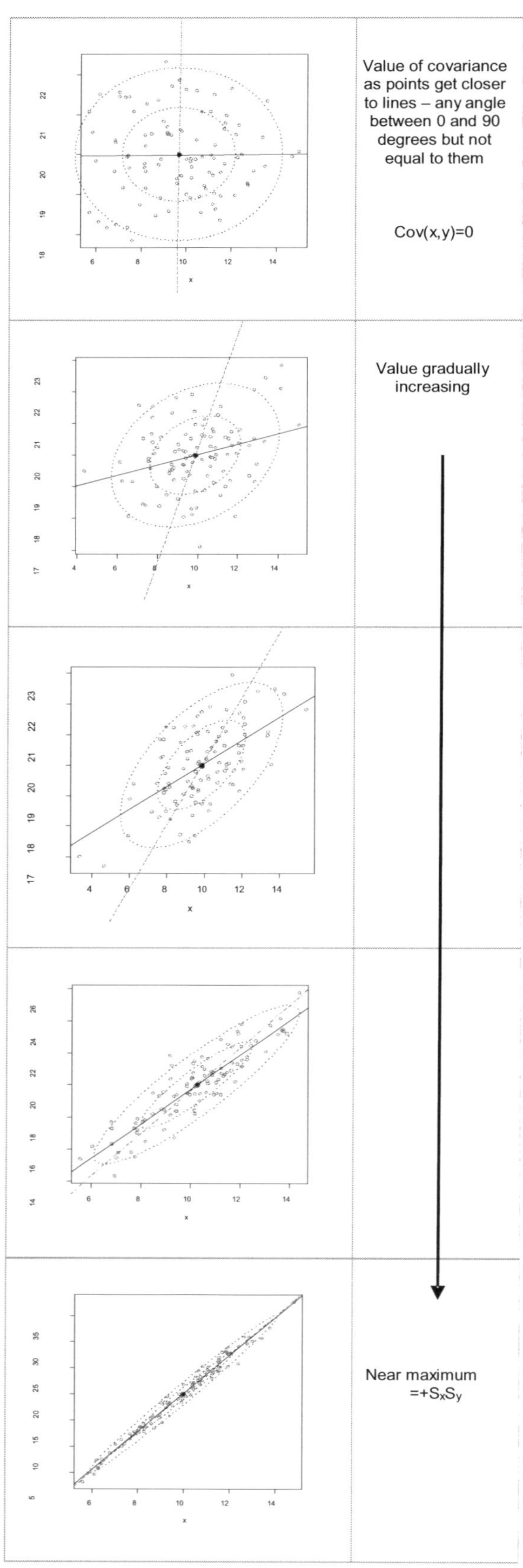

Value of covariance as points get closer to lines – any angle between 0 and 90 degrees but not equal to them
Cov(x,y)=0
Value gradually increasing
Near maximum =+SxSy

When the points do not cluster to any degree along the 2 lines the covariance is near zero. As the points cluster this gradually either increases or decreases in value until it is equal to the product of the standard deviations, given our proviso about horizontal and vertical lines. The diagrams opposite illustrates this for the situation with a rising line.

The maximum value of the covariance is problematic as it is specific to each dataset's standard deviations. It would be nice to be able to have a number that we can remember that tells us, for all datasets, how close the points cluster around a straight line. . . .

25.1.6 The standardised covariance = correlation coefficient

We now know that the maximum value our covariance can achieve is the product of the standard deviations. If we divide the observed covariance by this value we can see that the maximum value achievable for any dataset will be -1/+1. The new value is called the **Pearson Product-moment correlation coefficient** and we use the letter '*r*' to identify it.

$$r = \frac{\mathrm{cov}(x,y)}{S_x S_y} = \frac{\text{Covariance}}{\text{product of standard deviations}}$$

A coefficient is the mathematical word for a particular type of 'value'. A correlation coefficient, or value, is therefore another descriptive / summary statistic that is in this instance concerned with the degree of 'co-variation' between two continuous variables standardised to have a maximum value of 1 and a minimum value of -1.

25.1.7 *r* = Angle between the two lines of best fit

The scatterplots opposite each have two lines; one minimises the horizontal lines from the points to the line (dotted line- extended) and the other which minimises the vertical lines from the points to the line (solid line - the usual line of best fit). Now if you consider the cosine of the angle between them when the lines are at right angles the cosine (90 degrees)= 0 and when all the points are on the straight lines the cosine(0) =1, in fact the **Pearson Product-moment correlation coefficient is equivalent to this value.** We can demonstrate this in R using the *cos(radians)* function where 180 degrees= π radians. In R π=*pi*.

```
> # cos 90 degrees
> cos(pi/2)
[1] 6.123032e-17
> # cos 0 degrees
> cos(0)
[1] 1
```

R uses **E notation** where E-17 indicates the value is equal to a decimal point followed by 16 zeros, basically zero.

25.1.8 Standardised scores = gradient

If our dataset consisted of standardised (z) scores, the **gradient** (i.e. **slope**) of the standard line of best fit (minimising the vertical errors) is **equal to r** (see Rodgers & Nicewander, 1988, Pagano, 1990 p123). Standardised scores (i.e. *z* scores) are obtained by subtracting the variable mean from each of the raw scores and dividing by the standard deviation. It results in a set of scores with a mean of zero and a standard deviation of one.

25.1.9 Examples and interpretation

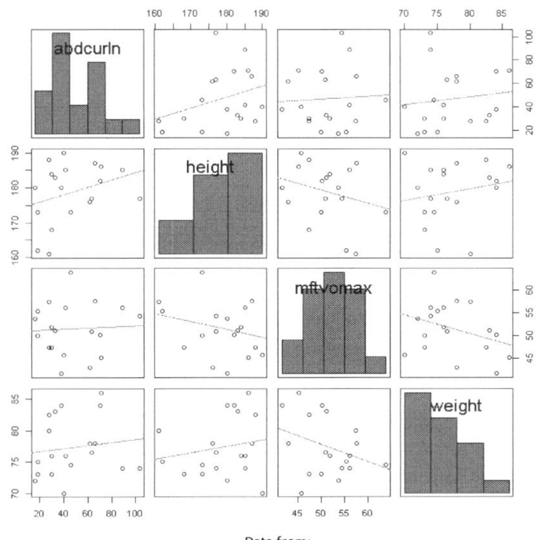

Data from:
http://www.robin-beaumont.co.uk/virtualclassroom/book2data/correlation_vo2max.dat

	abdcurln	height	mftvomax	weight
abdcurln	1.00000000	0.3154629	0.05612656	0.1254682
height	0.31546294	1.0000000	-0.26106278	0.1850520
mftvomax	0.05612656	-0.2610628	1.00000000	-0.3372503
weight	0.12546815	0.1850520	-0.33725030	1.0000000

We will now consider what correlations from an actual dataset look like. A group of 39 sports students provided data about themselves in an attempt to see how various variables correlated. The scatterplot matrix opposite shows the results for four variables; height, weight, number of abdominal curls they could manage within a single session and a measure of physical fitness, the maximum capacity to transport and use oxygen during incremental exercise known as the maximal oxygen consumption (VO_2 max) which is measured in litres of oxygen per minute (L/min).

The table opposite shows the correlations for each pair of variables. There is a correlation of 0.3154 for the number of abdominal curls and height. In contrast, we have a correlation of a much smaller value of 0.05612 for the number of abdominal curls and Vo_2Max. We also have two negative correlations, that between Vo_2Max and height and that between weight and Vo_2Max.

It is rather a sobering demonstration of what actual data looks like! The highest correlation is between weight and Vo_2Max of -0.3374 indicated that as weight goes up Vo_2Max has a tendency to drop. There are several explanations of these findings which any academic paper would thoroughly discuss such as; biased data collection, the small numbers of subjects, the mixture of males and females and a possible wide age range where age is known to affect Vo_2Max.

Variables that produce a correlation value close to +1 or -1 are said to be highly correlated. A value close to +1 is said to yield a high **positive correlation** and a value close to -1 a high **negative correlation**. When variables produce a correlation near to 0 they are said to be **uncorrelated**. Clearly all this is open to personal interpretation. The correlation value cannot be interpreted as a percentage. It being incorrect to say that a sample with a correlation of r=.67 has a relationship of 67% (Howell, 2007 p.238). In contrast, its squared value is a useful measure which we will discuss latter.

It must be stressed that the correlation 'value' only describes the degree of clustering of points near the straight line of best fit. Clearly this may happen, or not, by accident and obviously we would prefer a way of knowing if this is the case or not. In other words, we need to be able to infer from our sample to a population which either has zero correlation or has one identical to our sample estimate. Traditionally, and for most of this chapter, we will assume the population correlation coefficient is equal to zero. See the chapter *Confidence intervals for Effect sizes* for further details. Therefore, for now, we will assume the situation is like that of the paired *t* statistic where we assumed that the population from which our sample came had a mean of zero. In the correlation situation we need to go through a similar process to the one we did concerning the *t* statistic; develop a sample estimate of the population parameter, calculate the sampling distribution when the value is zero (the null distribution) and calculate a p-value given the null distribution and our observed value. Let us start by thinking about how to estimate the population correlation coefficient from our sample, in other words taking into account sampling variability.

25.1.10 Taking into account sampling variability - Adjusted r = rho (ρ).

Adjusted r

$$r_{adj} = \sqrt{1 - \frac{(1-r^2)(N-1)}{N-2}}$$

As with the other statistics we have discussed there is a version for the population along with a sample estimate. r, which we have been discussing so far, is a biased estimate of the population correlation coefficient (rho=ρ) **even though it is this one that is normally reported!** The most important aspect that affects the bias is sample size, so it is no surprise that a sample estimate of the population correlation coefficient is calculated with a 'N-something' added, where N is the number of points shown in the formulae opposite. The important thing to realise is that usually it is lower than the unadjusted correlation coefficient because we not taking into account the random sampling, with the discrepancy decreasing as the sample size increases. The example opposite is for a r value of .3345 (N=20) which when adjusted is reduced to .25.

```
> res <- .3345
> N<- 20
> adj_r <- sqrt(1-((1-res^2)*(N-1))/(N-2))
> adj_r
[1] 0.2501016
```

As with the population and sample values for the mean and variance we also represent the population correlation coefficient by a Greek letter, this time its rho (ρ).

25.1.11 Sampling distribution of rho(ρ) population correlation coefficient=0

Standard deviation for repeated samples:

$$SE_r = \sqrt{\frac{1-r^2}{N-2}}$$

To be able to provide a **p-value** for our correlation (r) we need to be able to define first the sampling distribution of r. Before we can do that we need to know if the statistic (r) has a standard error, that is the standard deviation of (r) across random samples from a population with a correlation coefficient ρ (rho) of zero. Luckily, it is possible to calculate the standard deviation (i.e. sampling error) for the distribution, as shown opposite.

$$t_{df=N-2} = \frac{\text{Observed correlation}}{SE_r}$$

$$= \frac{r}{\sqrt{\frac{1-r^2}{N-2}}} = \frac{signal}{noise}$$

$$= \frac{r\sqrt{N-2}}{\sqrt{1-r^2}}$$

So now we can create an expression analogous to the t statistic; the observed correlation divided by the variability one would expect due to random sampling if the correlation in the population were zero.

Luckily for us this statistic follows a *t* distribution (Fisher, 1915) so we can obtain a p-value.

When the number of points (N) is roughly over 30 there is little difference between the *t* distribution and the normal. The degrees of freedom (df) for the *t* distribution is in this instance the number of pairs -2.

25.1.12 Interpreting the p-value

```
> cor.test(mydataframe$mftvomax, mydataframe$weight)

        Pearson's product-moment correlation

data:  mydataframe$mftvomax and mydataframe$weight
t = -1.5062, df = 18, p-value = 0.1494
alternative hypothesis: true correlation is not equal to 0
95 percent confidence interval:
 -0.6768662  0.1267295
sample estimates:
       cor
-0.3345532
```

Here the p-value is interpreted, as 'one would obtain a correlation equal to or greater than that observed from a sample size of N points given that it came from a population where the correlation is equal to zero'.

Taking as an example the correlation between Vo_2Max and weight of -.3345 with 20 data points (df=18) here we obtained a *t* value of -1.5062 and an associated p-value of 0.1494. We can interpret this to indicate:

A sample of 20 pairs of data would produce a correlation of -.3345 or one more extreme 15 times (i.e. p-value=0.15) in a hundred if we had randomly sampled from a population with zero correlation.

We can also set a critical value which is usually 0.05 (one in twenty) and say that if our p-value is less than this we are prepared to reject this and our sample does not come from a population with a correlation of zero. One might argue that this must come from a population with a correlation value possibly equal to that observed but this is not good practice; a better way to proceed is to consider the confidence interval approach discussed next.

25.1.13 Confidence interval

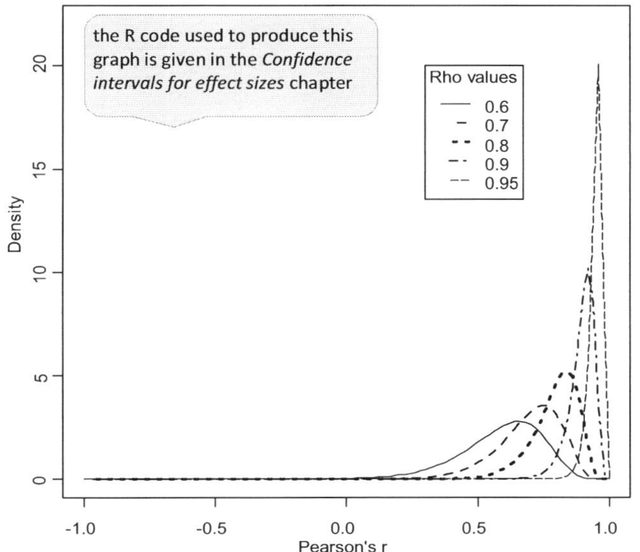

In past chapters we have calculated the confidence interval by using a measure of the standard deviation of the parameter estimate (i.e. the standard error). This time it is a bit trickier because the sampling distribution of the correlation only possesses a normal distribution when it is zero otherwise the distribution tends to get buffeted up against the minimum or maximum value. The diagram opposite demonstrates this.

To overcome this problem with the non-normal shape of these distributions three approaches have been taken, listed below.

Luckily most of the time the computer just produces the result for you, as we can see on the previous page. There we have a 95% confidence interval of -0.6768 to 0.1267 for the correlation between Vo_2Max and weight. You can interpret this to mean:

Approaches to obtaining a confidence interval for a correlation

- Transform the values so that they follow a standard normal distribution – This was the traditional method using what is called Fishers transformation, we can then use our good old z score to create the necessary confidence interval. This is the method R uses.
- Use the *t* statistic value above (this is the method SPSS uses to calculate the p-value).
- Computer intensive approaches – either create a permutation test or use the boot strapping procedure I describe in the re-sampling chapter.

We are confident that 95% of the time the population correlation values lies within the interval of -0.6768 to 0.1267. Which in this instance includes zero.

The advantage of the confidence interval value over the p-value is clear here. Using the p-value approach we only have information about the likelihood of obtaining the observed correlation value or one more extreme, given that the population value is zero. Whereas with the confidence interval approach we have obtained a range representing a level of certainty.

25.1.14 Effect size

$$d = \frac{2r}{\sqrt{1-r^2}} \text{ and } r = \frac{d}{\sqrt{d^2+4}}$$

While the correlation coefficient is classed as an effect size measure it is frequently converted to a value known as Cohen's *d*, the details of which are given opposite.

r	d
0.05	0.1001
0.1	0.2010
0.15	0.3034
0.2	0.4082
0.25	0.5164
0.3	0.6290
0.35	0.7473
0.4	0.8729
0.45	1.0078
0.5	1.1547
0.55	1.3171
0.6	1.5000
0.65	1.7107
0.7	1.9604
0.75	2.2678
0.8	2.6667
0.85	3.2271
0.9	4.1295
0.95	6.0849
1	Division by zero

The plot below provides details of the relationship between *d* and *r*. As *r* gets larger *d* more rapidly rises as we get closer to the maximum correlation value. For reference, the following R code produced the plot.

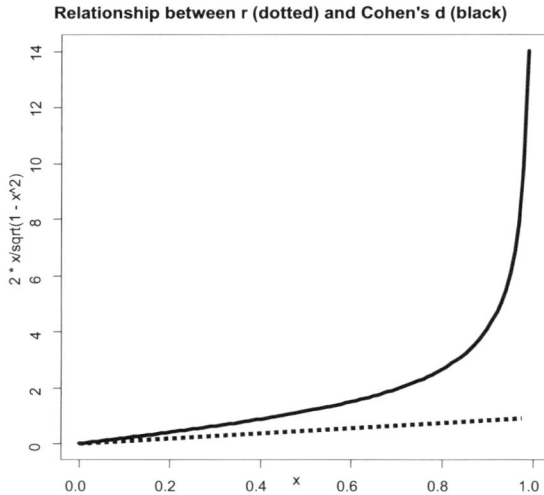

```
curve (2*x/sqrt(1 - x^2),  from="0", to="1",
col = "black", lwd=4,  main="Relationship
between r(dotted) and Cohen's d" )

curve (x*1, add = TRUE, col = "violet",lty=3)
```

A more widely used measure which can also be classed as an effect size measure is r^2 called the **coefficient of determination** or more commonly called **r squared**.

25.1.15 Coefficient of Determination (r^2)

The coefficient of determination, being the correlation coefficient squared, provides a readily understandable interpretation of the correlation.

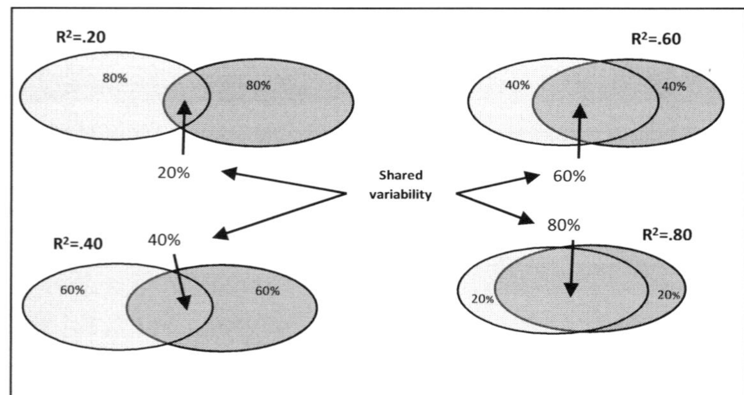

r^2 can be considered as a proportion, unlike *r*, as it is in effect measuring the proportion of explained variation compared to the total variation. Put another way it is the proportion of predicted to total variability. As a result, when all the observed variation is accounted for by the predicted portion (the line of best fit) it will equal one. You can show this relationship nicely by way of a Venn diagram.

25.1.16 Comparison between r^2 and *r*

r	r^2
0.1	0.01
0.2	0.04
0.3	0.09
0.4	0.16
0.5	0.25
0.6	0.36
0.7	0.49
0.8	0.64
0.9	0.81
0.95	0.90
0.98	0.96
0.99	0.98

Statisticians hold a wide range of views concerning which is the best measure to report either *r* or r^2 (Howell, 1992 p. 242). I prefer to use r^2. While most statistical applications tend to produce correlations, it is a simple matter of squaring the values to obtain r^2. It is instructive to examine a table of the r and associated r^2 values to see how large correlations can appear to mean a remarkably small amount if we re-interpret them as coefficients of determination. For example, a correlation of 0.5 means that only 25% of the variability is accounted for by the correlation.

25.1.17 Rank correlation

With variables that possess only an **ordinal level of measurement** we are restricted to only considering the ranks of the values and an equivalent to the Pearson correlation coefficient is computed. One such measure is the Spearman rho correlation, often indicated by the Greek letter ρ (rho).

The Spearman rho correlation basically applies the Pearson formula to the ranks. Unfortunately, this is obscured in most textbooks which provide what appears to be a very different formula, for each correlation type. In fact the two are identical, if you don't believe me look at Marques de Sá, 2007 p.69.

The problem with the Spearman correlation is the same as that of the other non-parametric statistics discussed elsewhere – Ties. When there are a large number of them the results tend to be unreliable, in which case, an alternative statistic called Kendall's tau (τ) coefficient can be used. Kendall's tau comes in three varieties called Kendall's tau-a, Kendall's tau-b and I think you can guess the third variety.

25.1.18 Common misconceptions about correlation

Five common misconceptions surrounding correlation are listed below.

Misconception	Description
Correlation implies causation	Just because two variables co-vary does not mean that one causes the other, however it can be said that correlation is a pre-condition for measuring causation. Read http://en.wikipedia.org/wiki/Correlation_does_not_imply_causation for further details
Correlation is only to do with monotonic / linear associations	We have discussed the particular type of association that Pearson and rank correlations measure. Non-linear relationships require special techniques not discussed in this chapter.
Non-homogeneous groups do not need to be considered	Some samples represent a number of distinct groups. For example, cholesterol levels may have been taken from both males and females and also from patients of different ages, where it is possible that each subgroup will provide a different correlation. There is also a danger that the sample becomes so heterogeneous that the various associations cancel out one another resulting in what appears to be a very low correlation. Two techniques are available for removing the influence of possible intervening variables, partial and semi-partial (also called part) correlation discussed in the *Multiple regression* chapter.
A significant p-value proves the observed correlation	It is important to remember that the associated p-value of a correlation is providing information concerning the likelihood of obtaining a correlation of the given value or one more extreme given that the correlation is assumed to be zero in the population. That is when we obtain a statistically significant correlation (i.e. p value less than the critical value) all we are saying is that the population correlation is NOT zero we are **definitely not saying that we have proved that the population value is equal to the observed correlation in our sample**. Confidence intervals provide more information.
A correlation value and its associated p-value are equivalent	Often people get muddled up between the correlation value and the associated p-value. While one is derived from the other they are distinct values with different meanings.

We will now see how to obtain correlation coefficients, associated confidence intervals and p-values in R Commander and R.

Here we will discuss two examples, the first of which is from MSME3 page 55 describing 12 HbA1c and fasting blood glucose measurements taken from 12 diabetic subjects.

25.2 Data preparation

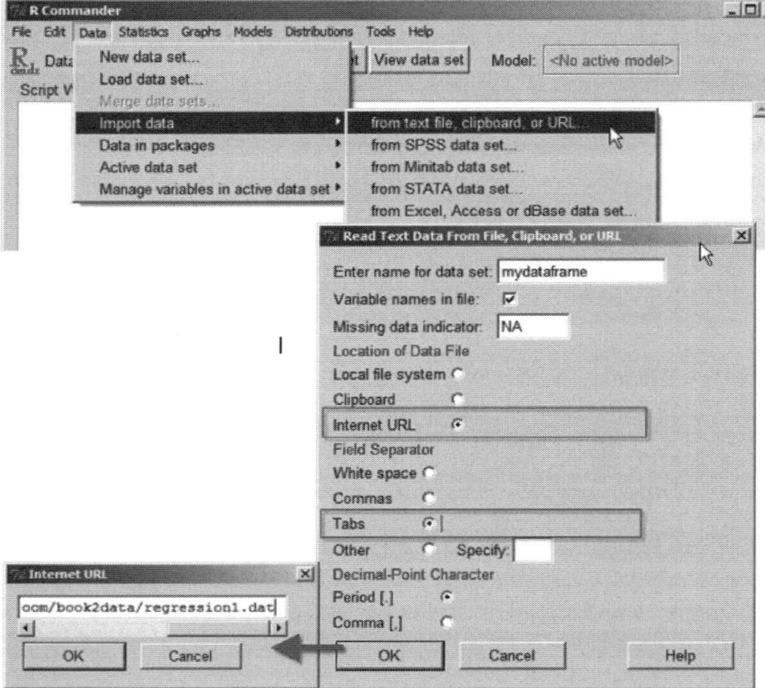

To load R Commander type the following into the *R Console* window:

```
library(Rcmdr)
```

Obtain the data directly from my website by selecting the R Commander menu option:

Data-> Import data->
from text, the clipboard or URL

I have given the resultant dataframe the name *mydataframe,* also indicating that it is from a URL (i.e. the web) and the columns are separated by *tab* characters.

Clicking on the **OK** button brings up the *Internet URL* box, in which you type the following to obtain my sample data:

```
http://www.robin-beaumont.co.uk/virtualclassroom/book2data/regression1.dat
```

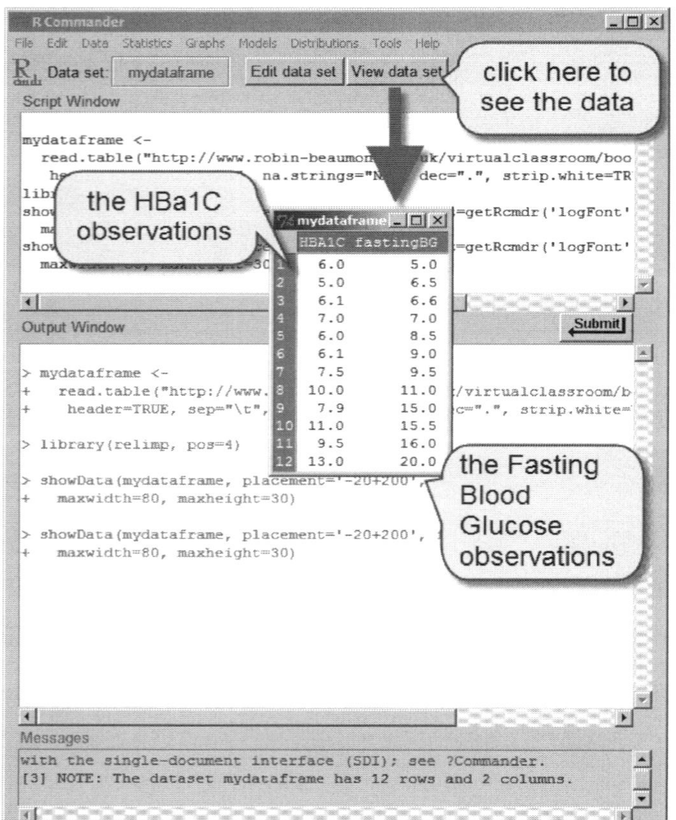

Click **OK**.

It is always a good idea to check the data and you can achieve this easily in R Commander by clicking on the **View data set** button.

We now have a dataframe containing the necessary data so we can now request the correlation values, confidence intervals and p-values.

25.3 Model development and evaluation

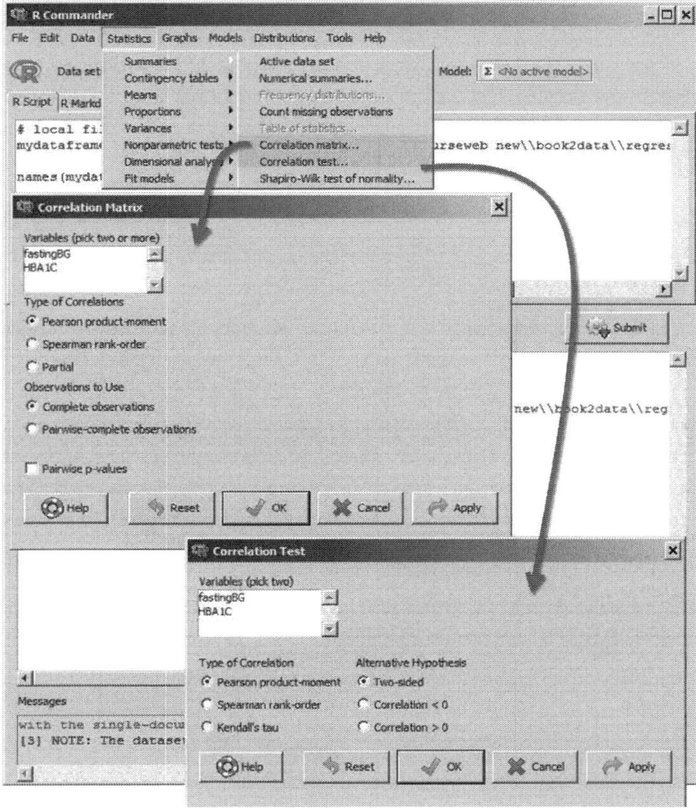

Using R Commander to obtain the values we have two relevant menu options.

Statistics->Summaries->Correlation matrix . . .

The correlation matrix option provides the correlation values (r) in the form of a table, the values reported to three decimal places (i.e. 0.889), along with a table of p-values underneath it (i.e. 0.0001).

The *Observations to use* option allows you to specify if you want to only use rows that are complete or consider them on a pairwise basis. For small datasets the two options can produce significantly different results. See the *Publication quality graphics* chapter for an example.

The *adjusted p-value* is a p-value that takes into account the fact that the more correlations you have in the table to more likely you are to find significant p-values so adjusts them accordingly. The other option is to use:

Statistics->Summaries->Correlation test . . .

This option provides more detail including a confidence interval, but has the downside that you can only investigate one pair at a time.

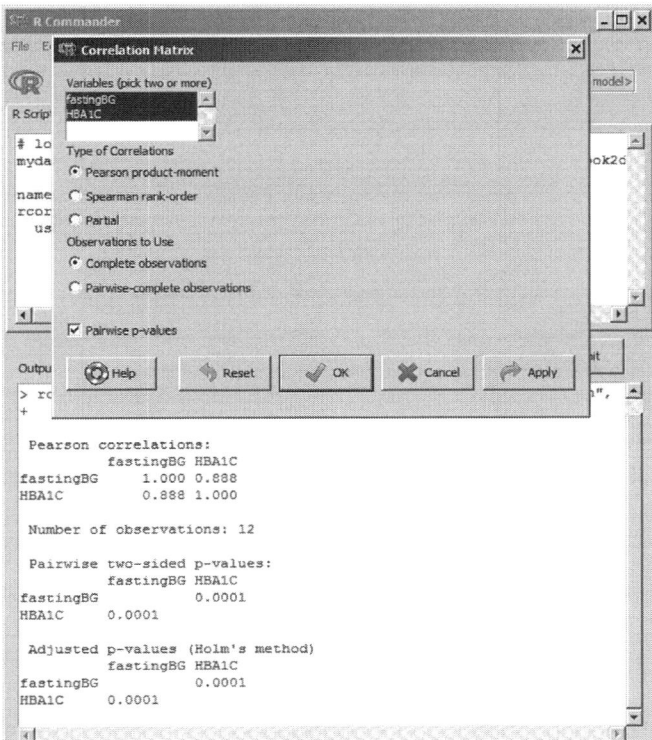

```
 Pearson correlations:
           fastingBG HBA1C
fastingBG      1.000 0.888
HBA1C          0.888 1.000

 Number of observations: 12

 Pairwise two-sided p-values:
          fastingBG HBA1C
fastingBG           0.0001
HBA1C     0.0001

 Adjusted p-values (Holm's method)
          fastingBG HBA1C
fastingBG           0.0001
HBA1C     0.0001
```

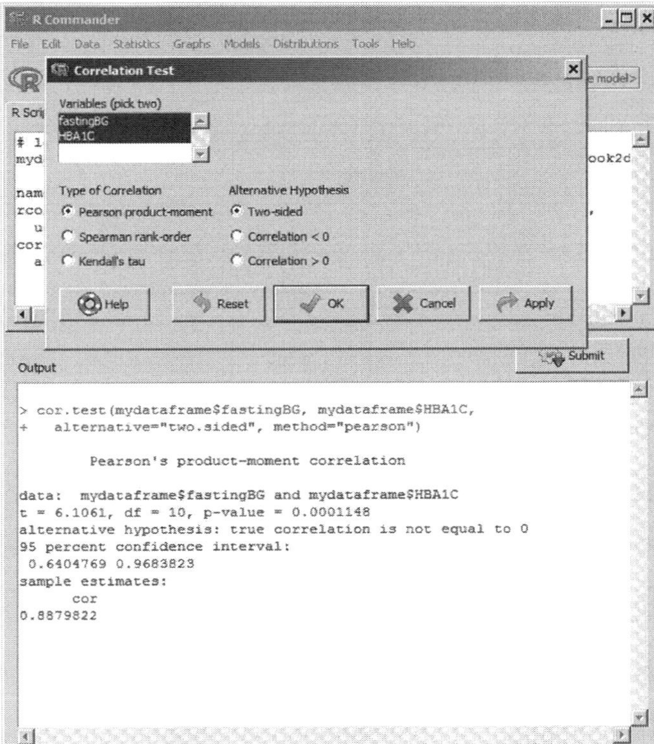

25.3.1 Doing it directly in R

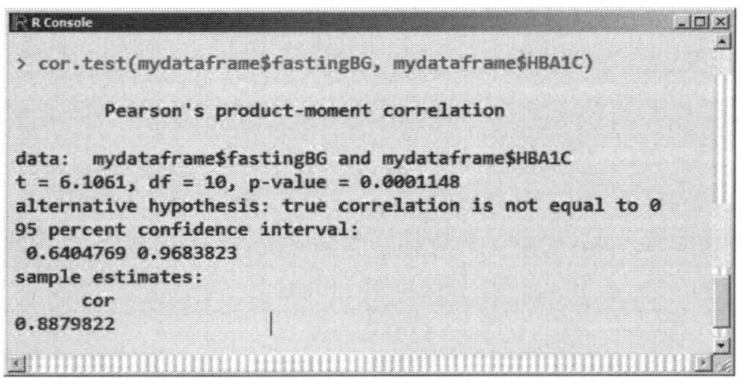
R Console
> cor.test(mydataframe$fastingBG, mydataframe$HBA1C)
Pearson's product-moment correlation
data: mydataframe$fastingBG and mydataframe$HBA1C
t = 6.1061, df = 10, p-value = 0.0001148
alternative hypothesis: true correlation is not equal to 0
95 percent confidence interval:
0.6404769 0.9683823
sample estimates:
cor
0.8879822

To obtain correlation values without p-values we simply enter the following command in the *R Console* window:

```
cor(mydataframe)
```

or we can specify a certain number of decimal places by

```
round(cor(mydataframe), 4)
```

will give *r* values to 4 decimal places

For a single correlation with a 95% CI and p-value enter:

```
cor.test(mydataframe$fastingBG, mydataframe$HBA1C)
```

25.4 I want to see a set of scatterplots and correlations?

Often journal articles have a series of small scatterplots and correlation values appearing together in a table like structure allowing assessment across a collection of variables. You can achieve this in R.

To demonstrate this I have extracted a dataset from the *DAAG* package called *ais*.

The *ais* dataframe consists of 202 observations on Australian athletes for the following 13 variables. The data were collected to study how various characteristics of the blood varied with sport, body size and the athlete's gender.

- rcc - red blood cell count, in $10^{12} l^{-1}$
- wcc - while blood cell count, in 10^{12} per liter
- hc - hematocrit, percent
- hg - hemaglobin concentration, in g per decaliter
- ferr - plasma ferritins, ng dl^{-1}
- bmi - Body mass index, kg $cm^{-2} 10^2$
- ssf - sum of skin folds
- pcBfat - percent Body fat
- lbm - lean body mass, kg
- ht - height, cm
- wt - weight, kg
- sex - a factor with levels: f, m
- sport - a factor with levels: B_Ball, Field Gym, Netball, Row, Swim, T_400m, T_Sprnt, Tennis, W_Polo

Notice that the last two variables I have grayed out as they would make little sense in correlations for them as they are factors.

load the required data

```
mydataframe <- read.delim("http://www.robin-beaumont.co.uk/virtualclassroom/book2data/ais_daag.dat", header=TRUE)
```

```
names(mydataframe)
```

check the names of the columns and attach the dataframe

```
attach(mydataframe)
```

```
library(psych)
```

load the psych package

```
pairs.panels(mydataframe)
```

now use the *pairs.panels* command in the psych package to get the required plot

```
pairs.panels(mydataframe[1:11])
```

above produces a diagram which includes the last two variables (columns 12 and 13) to omit those use this command:

I have used the simplest approach here. Some other options:
lm=True ; produces straight linear regression lines.
ellipses=FALSE; removes the red ellipses.
method="spearman" or "kendall" produces other types of correlations

The result is shown on the following page:

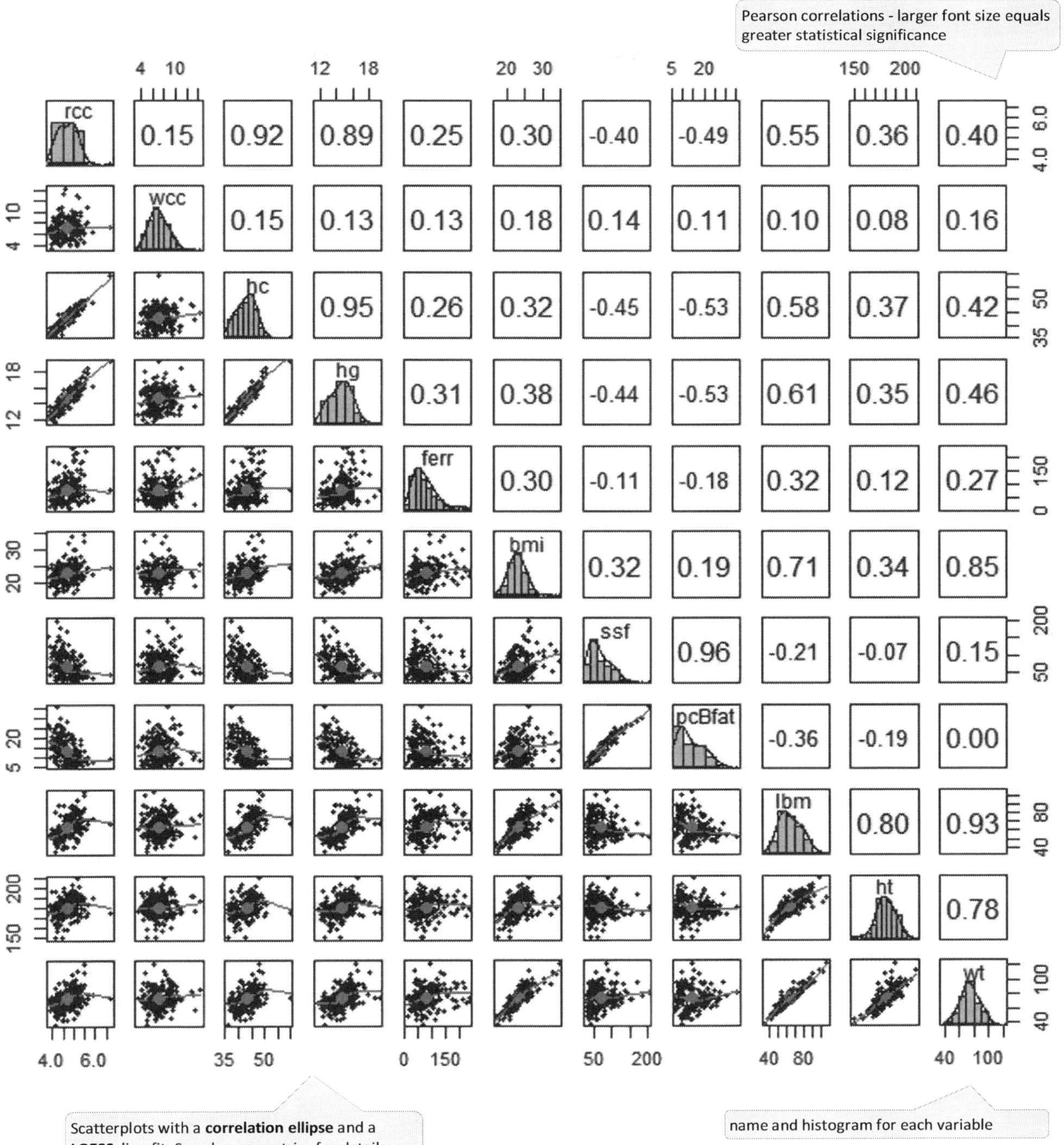

My YouTube video at the end of this chapter describes how you can obtain a result similar to the above in R Commander.

color=FALSE produces a grayscale plot. The default is 51 gray levels. To change them to say 20 add n=20. You can also add a title using: main="a title".

Another method of displaying a large number of correlations is to have a table where each cell's color reflects the correlation. This can easily be done using the psych package *cor.plot()* function, typing the following into the *R Console* window, produces the output below.

```
cor.plot(cor(mydataframe[1:11]), color=FALSE)
```

leave out if you want a colored one

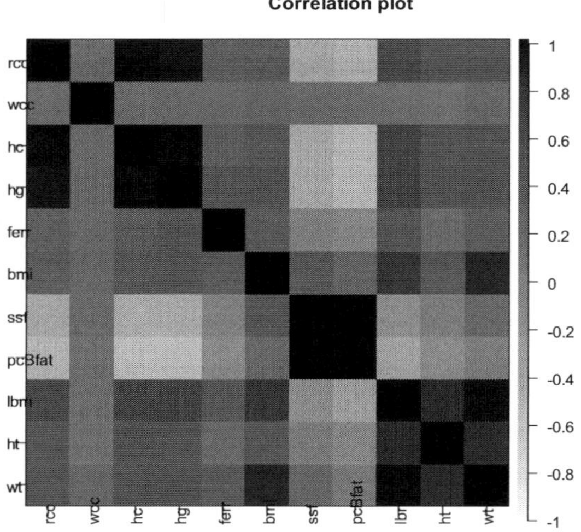

The key on the right hand side indicates the correlation value where white equals -1 and black equals 1. This graphical approach immediately highlights any clustering of correlations. A colored alternative ranges from dark red (-1) to blue (1) with white indicating a zero correlation.

It also highlights the danger of correlations with variables which are just combinations of other variables. For example, someone who did not know that BMI is calculated from height and weight would be excited by the dark shading around the three variables in the bottom right hand corner.

25.5 Writing up the results

Normally when reporting correlations you need to indicate the type of correlation; Pearson, Spearman or Kendall, the sample size (n which equals pairs), the actual correlation value and also its associated confidence interval or, failing that, the p-value. If you are using a rank order correlation measure you should also report the number (+percentage) of ties.

25.6 Tips and Tricks

While most correlations are for continuous or ordinal data, you can produce special types of correlations for binary or nominal data.

YouTube help:

http://www.youtube.com/watch?v=a9ndjEYAB10&feature=share&list=PL9F0EBD42C0AB37D0

25.7 Techie notes

Several diagrams have appeared in this chapter for purely pedagogical purposes. One specific diagram is that displaying both regression lines. While this is not required routinely those teaching this subject might find the code below useful.

```
install.packages("MethComp")
require(MethComp)
###  some data x and y are vectors
bothlines(x, y, Dem=TRUE, col=rainbow(3), lwd=2 )
# optionally add the line obtained
# by allowing errors in both variables
# Deming regression using the command Den-TRUE.
```

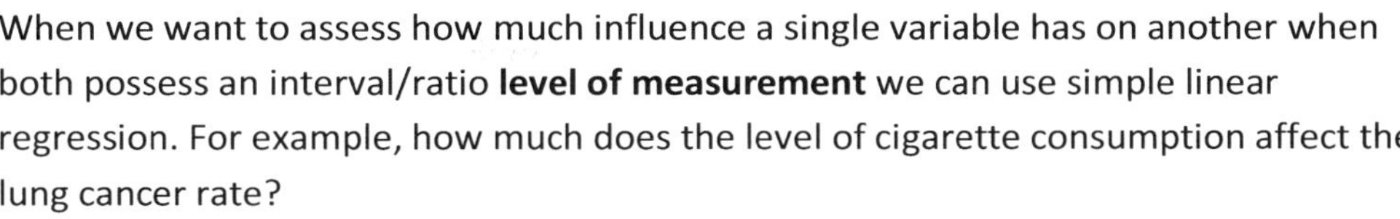

26 Measuring the influence of one variable on another: regression

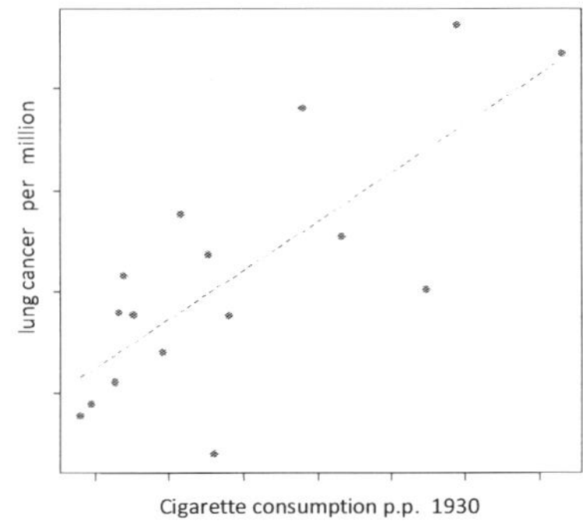

When we want to assess how much influence a single variable has on another when both possess an interval/ratio **level of measurement** we can use simple linear regression. For example, how much does the level of cigarette consumption affect the lung cancer rate?

In a previous chapter *Measuring the degree to which two variables co-vary: Correlation*, we discussed the lines of best fit by either minimising the horizontal or vertical lines (actually the squared values of them) from the predicted x or y value and the corresponding value on the line. We discovered that both lines coincided when the points all fell on the line with a resultant correlation of 1 or -1.

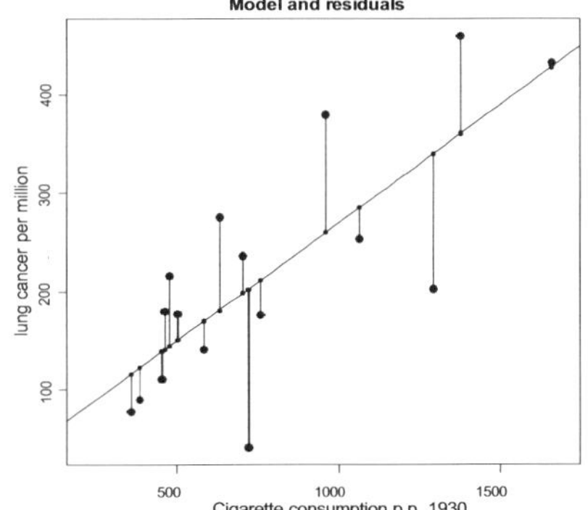

The vertical line between a particular observed y value and that which a regression line would predict is called the *residual* or *error*. The residual can be positive, negative or zero.

In this chapter we will only consider the line of best fit that minimises the squared vertical residuals. Many introductory statistics books automatically talk about the 'line of best fit' when they actually mean just this, the line that minimises these squared vertical residuals. For statisticians this has some important consequences. Primarily we are no longer considering the bivariate normal distribution (i.e. the upturned pudding basin shape) we used in the correlation chapter. One variable (the y value) is now considered to be what statisticians call a random variable, when what they mean is that it follows a specific distribution, while each value of the x variable is set or pre-determined in some way, and we say it is a *fixed variable*. In practice researchers often ignore this and both the x and y variables are considered to be random variables (Howell, 2007 p232, p250).

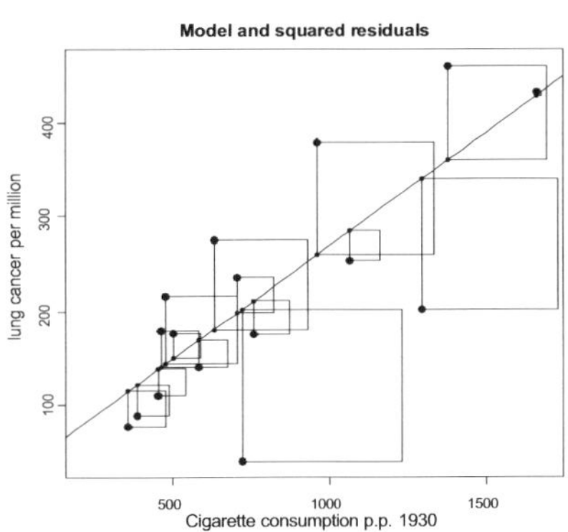

The terminology for the x and y variables used in regression depends upon the context, in an experimental design a *researcher manipulates an independent variable* and observes changes in a dependent variable. The independent variable might be the particular levels of calorie intake, exercise or drug dosage etc. where the level is allocated by the researcher to the subjects, and the aim is to investigate *causation* rather than just association. The dependent variable might be any measure that the researcher believes will be directly affected by the independent variable.

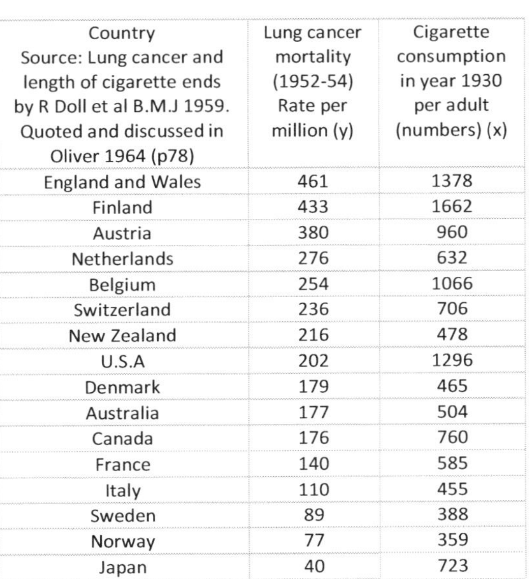

Country Source: Lung cancer and length of cigarette ends by R Doll et al B.M.J 1959. Quoted and discussed in Oliver 1964 (p78)	Lung cancer mortality (1952-54) Rate per million (y)	Cigarette consumption in year 1930 per adult (numbers) (x)
England and Wales	461	1378
Finland	433	1662
Austria	380	960
Netherlands	276	632
Belgium	254	1066
Switzerland	236	706
New Zealand	216	478
U.S.A	202	1296
Denmark	179	465
Australia	177	504
Canada	176	760
France	140	585
Italy	110	455
Sweden	89	388
Norway	77	359
Japan	40	723

In contrast, for the majority of *observational* (frequently *retrospective*) studies a number of measures are taken and the researcher uses their own knowledge of the area to decide which would be the independent and dependent variables. SAS1 uses the generic terms input and output variables which I particularly like.

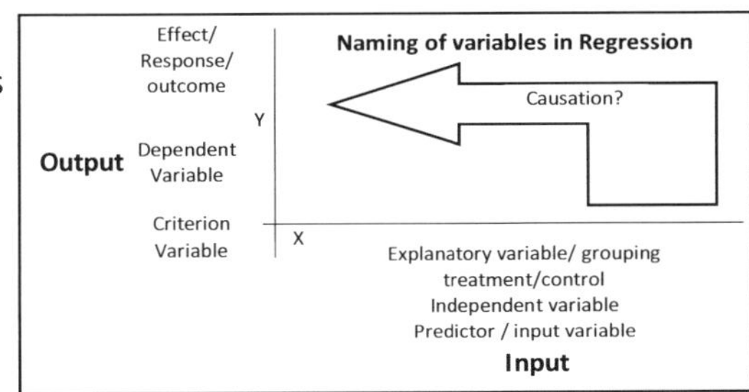

For example, in the above scatter diagrams the cigarette consumption for 1930 per adult person is considered to be the 'predictor' variable and the incidence of lung cancer per million as the 'criterion' (i.e. dependent) variable.

The term '**on**' is used to indicate if we have minimised the horizontal or vertical residues; if it is the vertical ones as above we say Y **on** X (or Y given X). Therefore, in the above scatterplots, we have regressed incidence of lung cancer on adult cigarette consumption for 1930 or we could say we have regressed incidence of lung cancer given adult cigarette consumption for 1930.

The line of best fit in the above diagrams, often simply called the regression line, is the focus of this chapter so let's begin by looking at the equation that defines the line.

26.1 $\hat{y}_i=a+bx_i$

You may be familiar with something similar to the above expression from your school days (called a *simple linear regression* equation), but just in case you have forgotten it, here is a quick review:

- $\hat{y}_i$ pronounced 'Y hat *i*', indicates that it is an estimated value being the value on the line produced by the equation. The i indicates it to be the ith value, where i takes the value of 1 to n where n is the number of points in the dataset. This variable is the 'output': mathematically it is the expectation of $\hat{y}_i$ given a particular x value x_i. The model is a function of x_i, a and b and can be formally written like this $E(\hat{y}_i|x_i)=f(x_i, a, b)$ which we will discuss below.
- $\hat{y}_i$,a and b are all ratio/interval level values (in later chapters we will consider other regression models where this is not the case).
- y_i, x_i is the observed y and x value for the ith point.
- $\bar{y}$ is the average y value.
- a is a parameter that we need to estimate. It is the value when x=0 and called the intercept.
- b is the second parameter that we need to estimate. It is the slope/gradient and represents the change in x divided by the change in y.

26.2 The regression model concept

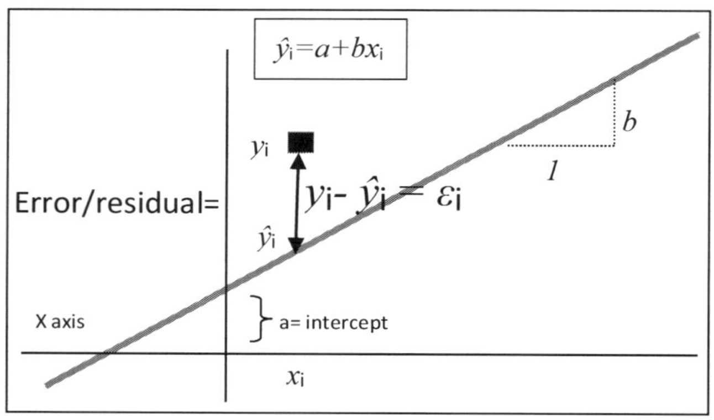

The above equation is an example of a general concept called a *model* – where all models produce expected values. All models are basically:

Observed Data = Model + Error

In this instance the model consists of two parameters (values to be estimated), so given our data and the above model we want to *find the values of the slope and intercept that make the data most likely.* To achieve this we use a technique called *maximum* **likelihood** *estimation* (Crawley, 2005 p.129). This technique finds the estimate(s) that minimises these errors, a process that should sound familiar to you. Have a quick look at the 'finding the centre' and 'spread' in the online high school revision chapters, if it seems a little hazy.

Most statistical applications compute the maximum likelihood estimates (i.e. the slope and intercept parameters in this instance), with a single mouse click, but before we perform this magic, I would like to describe in a little more detail what is going on exactly in terms of this model building idea.

While the equation given at the top of this section provides the estimated value of the outcome variable of a particular value a more correct, general equation is:

$$Y = \alpha + \beta X + \varepsilon$$

where Alpha (α) and beta (β) are unknown constants that we need to estimate (i.e. parameters) and epsilon (ε) is a variable with zero mean and variance $\sigma^2_{y.x}$ This is our error, also called the noise or disturbance term.

To summarize:
Aim of regression is to minimise the sum of the errors squared

= $\sum \varepsilon_i^2$

Technique used = maximum likelihood/Least Squares
Creates model parameters that maximises likelihood of observed data

Returning back to our function described previously, this can now more correctly state:

the expectation of y given x is a function of four values x, β, α and ε

$E(Y|x)=f(x,\beta,\alpha,\varepsilon)$

This function is now being used to imply both a linear regression model and also the error distribution, which is taken usually to be normally distributed.

26.2.1 Regression model assumptions: LINE

Daniel, 1991 provides an acronym – LINE - to list the assumptions a valid regression model dependents on:

- **L** – The relationship is a linear one
- **i** – X values are independent
- **n** – For each value of x the values follow a normal distribution
- **e** – The spread of values for each value of x is constant

These will be studied in more detail when working through the examples.

26.2.2 Why 'simple' regression?

In the above equation $\hat{y}_i=a+bx_i$ we have only one b, that is one predictor variable, however it is possible to have any number of them such as:

$$\hat{y}_i=a+b_1x_i+b_2x_i+b_3x_i+b_4x_i+b_5x_i\,..\,+b_nx_i$$

The above equation indicates we have n input (independent) variables attempting to predict $\hat{y}_i$. In this instance, where the outcome is predicted lung cancer incidence, the inputs might represent variables such as saturated fat intake, level of exercise, age and socioeconomic group etc. This is clearly a much more complex model than our '**simple**' one hence the term simple when we only have a single predictor variable. In contrast when there are multiple inputs modelled the term **multiple regression** is used, a situation which we will discuss in subsequent chapters.

26.2.3 Building a regression model

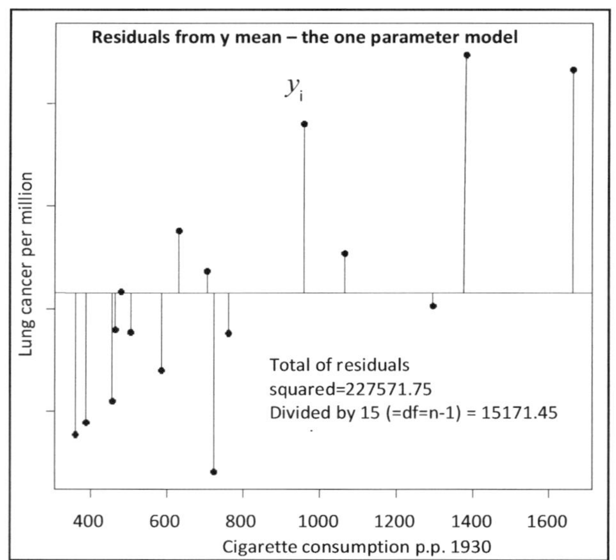

Lung cancer mortality (1952-54) Rate per million Descriptive Statistics					
	N	Minimum	Maximum	Mean	variance
death rate per million lung cancer	16	40.00	461.00	215.3750	15171.45

First, let us consider a very simple model to represent our lung cancer data, instead of having a line with a slope imagine that we just take a horizontal line along the average y value (shown as $\bar{y}$). Doing this our model simply becomes $\hat{y}_i = \bar{y} + e$, that is the estimated value of y_i (i.e. $\hat{y}_i$) is equal to the mean value of y plus the specific ith error where e equals this error. This error is the difference between the y mean and the individual y_i value.

The y mean value was chosen because it minimises the sum of squared errors/residuals from the mean (for details see the online chapter, measuring the centre, in the high school revision section). Squared values are used to stop the negative and positive values cancelling one another out and ending up with zero.

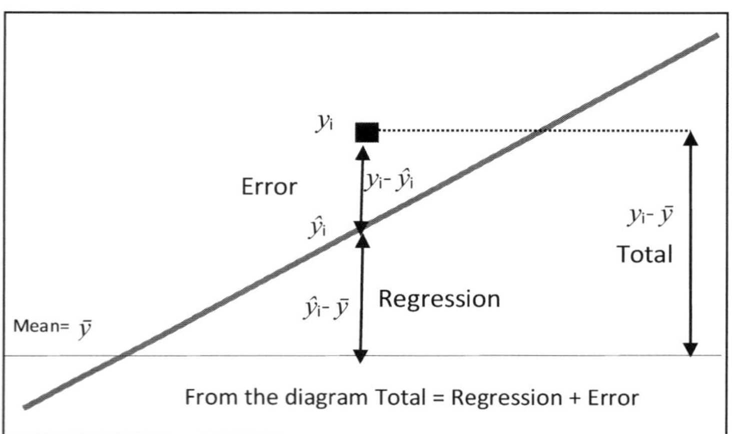

The above one parameter model is really pretty useless as it predicts a lung cancer value which is the same for any country. However, it is obvious from the scatterplot that the cancer rate increases as cigarette consumption increases. What would be more useful is to give the line the ability to have a slope as well. This is achieved by adding a slope parameter so now there is a unique predicted value for each level of cigarette consumption.

When a slope parameter is added the error term/residual is divided into two parts. That due to the regression (difference between actual and predicted y_i) and that due to error/residual (difference between mean and predicted).

For this dataset, by allowing the line to have a slope, it now follows more closely the actual values, consequently the errors (residuals) appear to be reduced, as shown in the diagram opposite. We will now investigate how these errors are calculated.

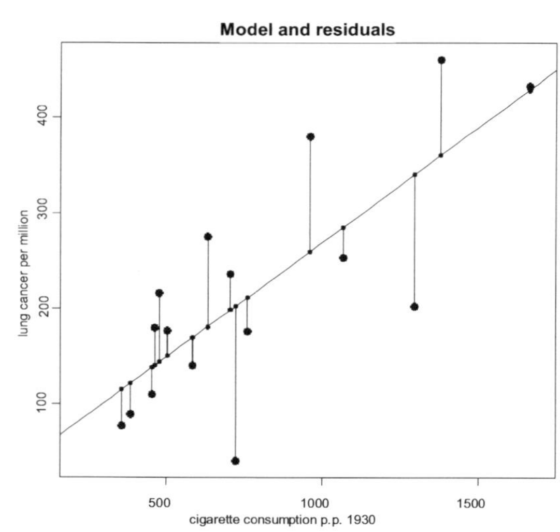

26.2.4 Total is equal to model plus error (SST=SSR+SSE)

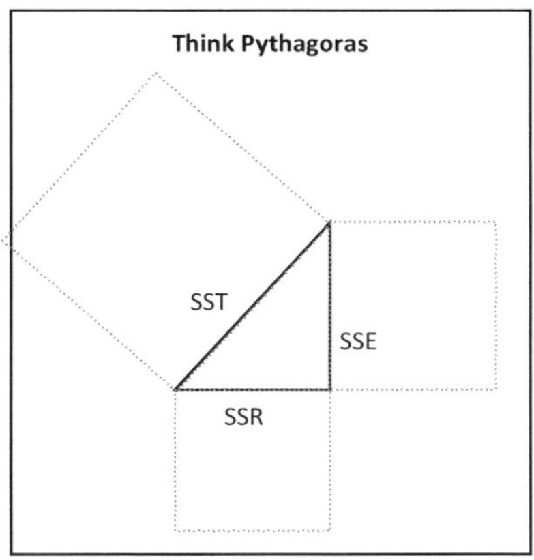

The above expression is just as important as the $\hat{y} = ax+b$ one when understanding regression. From the previous section we can add up all the squared values for all the errors, regression values and deviations from the mean Y value. That is all the $y_i - \hat{y}_i$, $\hat{y}_i - \bar{y}$, and $y_i - \bar{y}$ expressions, if you find the mathematical equations too much do not fear, if you understand the graphical explanation opposite that is fine.

SSE = Error sum of squares= $\sum_{i=1}^{n}(y_i - \hat{y}_i)^2$

SSR = Regression sum of squares = $\sum_{i=1}^{n}(\hat{y}_i - \bar{y})^2$

SST = Total sum of squares = $\sum_{i=1}^{n}(y_i - \bar{y})^2$

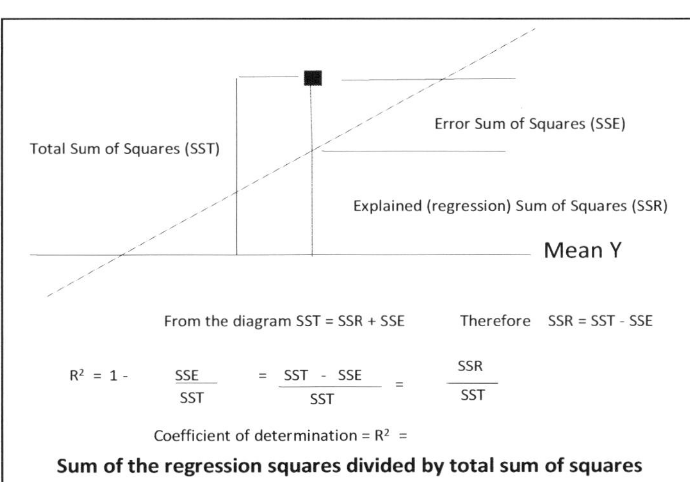

Error Sum of Squares (SSE)
Total Sum of Squares (SST)
Explained (regression) Sum of Squares (SSR)
Mean Y
From the diagram SST = SSR + SSE
Therefore SSR = SST - SSE
$R^2 = 1 - \frac{SSE}{SST} = \frac{SST - SSE}{SST} = \frac{SSR}{SST}$
Coefficient of determination = R^2 =
Sum of the regression squares divided by total sum of squares

26.2.5 R squared (not adjusted)

It is important to realise that the total sums of squares (SST), is the deviation from the Y mean.

In addition, it can be shown that our old correlation value squared is the same value as the regression sum of squares divided by the total sum of squares. So when we have a perfect squared correlation (i.e. r=+1) SSR = SST, this indicates that all the variability in our model can be explained by the regression; that is all the y_i's are on the regression line.

Conversely when none of the variability in the data can be attributed to the regression equation the correlation = 0 indicating that SSR=0 and therefore SST=SSE, meaning that the equation has no predictive value. The above explanations you may realise, are very similar to several paragraphs in the previous correlation chapter. In statistics you will find the same old ideas constantly reappearing and sums of squares, correlations and coefficients are such examples.

Check your understanding:
Say out loud 10 times "In regression the total sum of squares is equal to the sum of the Regression and Error sums of squares. SST=SSR+SSE"
Draw a rough scatterplot with some fake data and a possible regression line, mark on it SST, SSR and SSE for one or two of the points.

26.2.6 Other names for the sums of squares

Crawley (2005) (used in this chapter)	Field 2009	Norman & Streiner 2008 (also used by SPSS/SPSS)	Howell 2007	Miles & Shevlin 2001	R
SSY = Total	SST		SS_Y = Total	Total	
SSR = Regression	SSM = Model	$SS_{Regression}$ = Regression	$SS_{\hat{y}}$ = Regression	Regression	Variable name
SSE = Error	SSR = Residual	$SS_{Residual}$ = Residual	$SS_{Residual}$ = Residual	Residual	Residuals

Unfortunately, different authors use different names for the various sums of squares. The table opposite provides a typical list and should help if you start looking at any of these books, otherwise just ignore it.

26.2.7 Averaging the sums of squares – mean/adjusted SS

SSR = Regression sum of squares = $\sum_{i=1}^{n}(\hat{y}_i - \bar{y}_i)^2$

therefore $MS_{regression}$ = MSR = $\frac{1}{1}\sum_{i=1}^{n}(\hat{y}_i - \bar{y}_i)^2$

SSE = Error sum of squares = $\sum_{i=1}^{n}(y_i - \hat{y}_i)^2$

therefore $MS_{error/residual}$ = MSE = $\frac{1}{n-2}\sum_{i=1}^{n}(y_i - \hat{y}_i)^2$

To provide a mean value (also called the adjusted value) for the various sums of squares each needs to be divided by an appropriate value. We achieve this by dividing each sum of squares value by the associated degrees of freedom.

Traditionally (that is from the early 20th century when this approach was developed by Fisher), the various values have been laid out in a particular way called an ANOVA table. An ANOVA table is part of the standard output when you do a regression analysis. The mean of SSR becomes MSR and the mean of SSE becomes MSE. One important thing to note is that by doing this the value is now what is called the **variance** for the particular value, so by taking their square root you would obtain standard deviations of each.

The degrees of freedom are related by the equation $df_{total} = df_{residual} + df_{regression}$ where total is just the number of cases minus 1, and that for regression the number of parameter estimates ignoring the constant/intercept term, which is one in this case. Finally the degrees of freedom for the residual, for our current model (the two parameter, simple regression model), is the number of cases minus 2. So:

$df_{regression} = df_{total} - df_{residual}$ = 1= (n-1)- (n-2). While it is good to know the underlying formula, in practice all statistics programs will work this out for you.

26.2.8 ANOVA table - overall fit

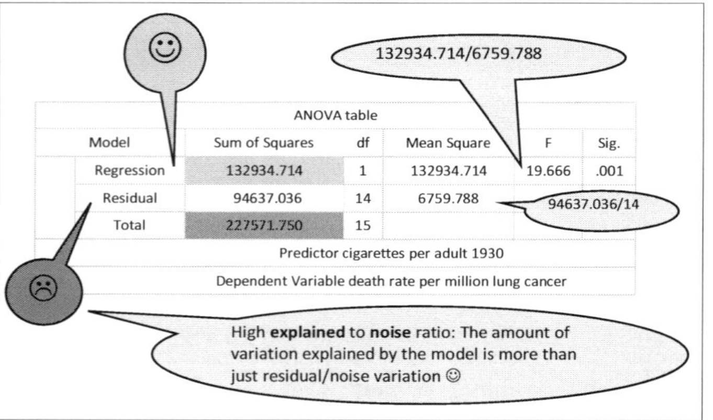

ANOVA table

Model	Sum of Squares	df	Mean Square	F	Sig.
Regression	132934.714	1	132934.714	19.666	.001
Residual	94637.036	14	6759.788		
Total	227571.750	15			

Predictor cigarettes per adult 1930

Dependent Variable death rate per million lung cancer

132934.714/6759.788
94637.036/14
High **explained** to **noise** ratio: The amount of variation explained by the model is more than just residual/noise variation ☺

How does our simple regression model compare to the simple one parameter ($\bar{y}$) model i.e. where there is no difference between the various countries' cigarette consumption and lung cancer? This question is answered in the ANOVA table and the R squared value.

The ANOVA (ANalysis Of VAriance) table opposite for our cigarettes/lung cancer data demonstrates the general layout used in simple regression.

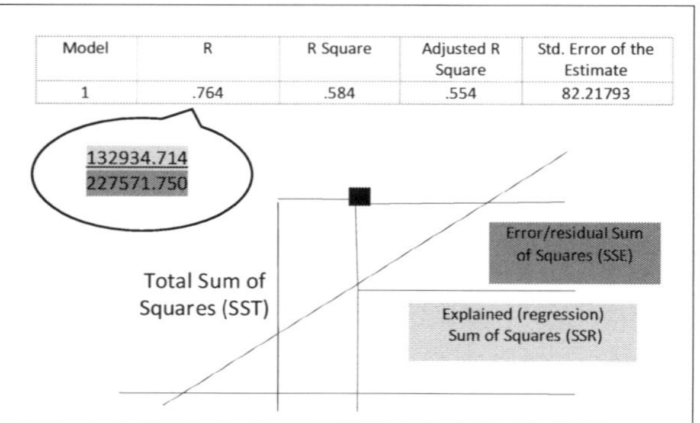

Model	R	R Square	Adjusted R Square	Std. Error of the Estimate
1	.764	.584	.554	82.21793

The regression sum of squares can be considered to be 'good', that is what we want, hence the smiley face opposite. In contrast the residual/error sums of squares are what we don't want. Taking the mean value of each of these and considering their ratio produces what is known as an F statistic. You can also think of this as the present model's fit divided by the previous one parameter model's fit.

When both the regression (numerator) and residual (denominator) are the same except for sampling error the F ratio follows an F distribution. This is equivalent to saying that the additional parameter in the simple regression model (b) is equal to zero in the population or that indeed Y is not related to X (Weisberg, 1985 p.18). That is the correlation (ρ) in the population is equal to zero. (Howell, 2007 p.255). It is pertinent to consider a few facts about the F distribution at this point.

26.2.9 The F ratio/ Distribution

$$F = \frac{MS_{regression}}{MS_{residual/error}} = \frac{signal}{noise/sampling} \sim f_{(\text{df\_numerator},\text{df\_denominator})}$$

The previous paragraph provided an interpretation of the F ratio. This value (also called the F statistic) follows a mathematically defined distribution, which is generally called a probability density function (pdf). We can show this in the equation above, using the ~ sign to mean "is distributed as" or "has the pdf".

Denominator df (when numerator df=1)	Mean value of F (df2/(df2-2))
6	1.5
7	1.4
8	1.333
9	1.285
10	1.25
50	1.041
100	1.020
1000	1.002

As with other distributions the F distribution shape is defined by certain values (i.e. parameters), 2 in this instance, simply called the numerator and denominator degrees of freedom. For simple regression these values become numerator =1 and denominator=(n-2) where n is the number of cases.

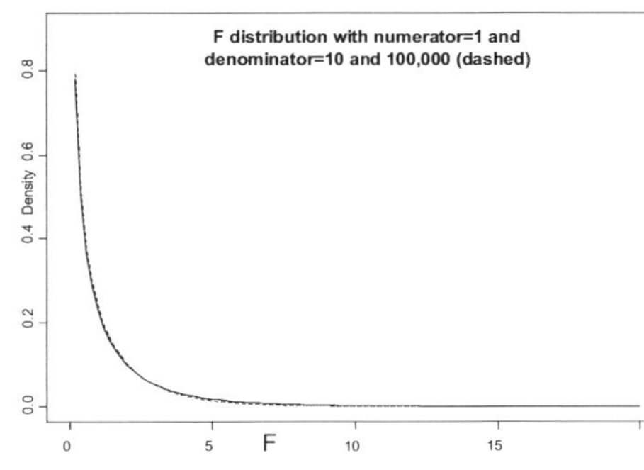

Conceptually it is rather difficult to understand the shape of an F distribution. Possibly the easiest way of gaining some understanding of it for the simple regression situation, is to note that the mean F value, that is where 50% of its area lies each side of it, is always near 1 regardless of the sample size. As the p-value is defined as the area under the curve to the right of a particular F value in the positive direction we can see that extremely high F values in this situation will equate to very small p-values.

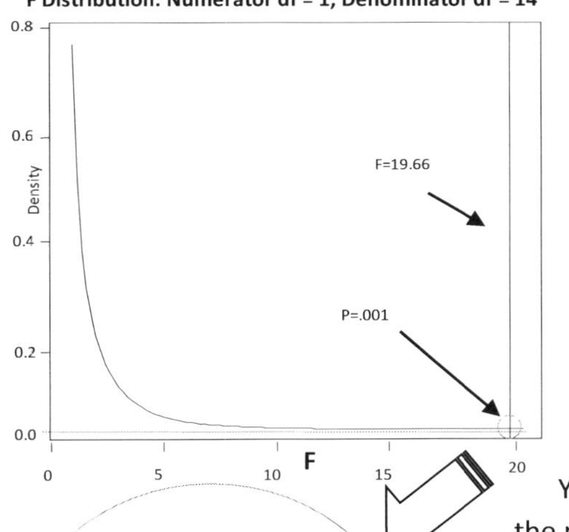

Given the above interpretation we can say the smaller the associated p-value, the more likely we will accept our simple regression model ('two parameter model') over the one parameter model.

Returning to our dataset the diagram opposite shows the $F_{(1,14)}$ distribution along with a line indicating our F value of 19.66; the area to the right of this value is the associated p-value and is equal to 0.001. If we were writing this up in an article we would indicate it thus ($F_{(1,14)}$=19.66, p-value=0.001). This value is a long way along its tail. Also notice that for the F distribution we only consider one end of it, they are the more extreme values only in the positive direction.

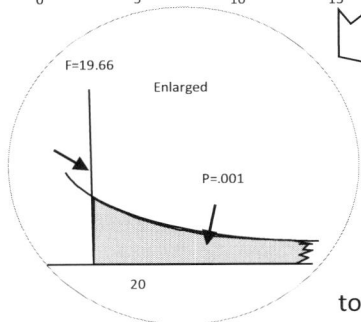

You can easily obtain the plot of the F distribution yourself using R Commander. The appropriate menu option is shown opposite.

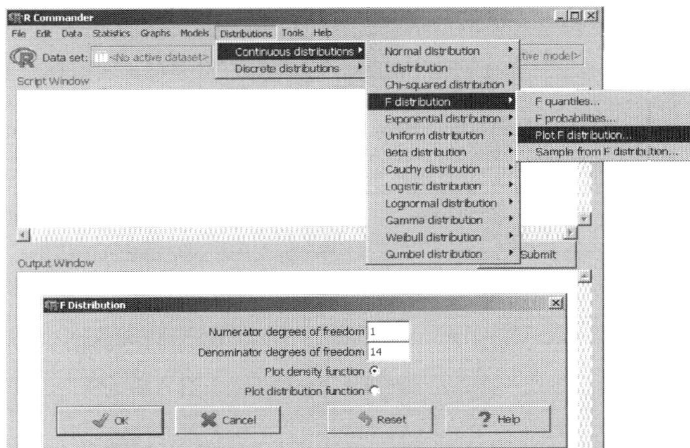

While considering the overall fit of the model it is equally important to consider each parameter in turn.

> ***Check your understanding:***
> Given that the F ratio is basically good (the model) divided by the bad (sampling error) which of the following F values would you be most happy with 2.3, 15.3, 5.76, 1, 0.3

26.2.10 Individual *b* evaluation

```
R Console
> cancer <- c(461, 433, 380, 276, 254, 236, 216, 202, 179, 177, 176, 140, 110, 89, 77, 40)
> cigs <- c(1378, 1662, 960, 632, 1066, 706, 478, 1296, 465, 504, 760, 585, 455, 388, 359, 723)
> cigs_mortality<-data.frame(cbind(cancer,cigs))
> names(cigs_mortality)
[1] "cancer" "cigs"
> # 'lm' means linear model y~x
> model<-lm(cancer ~ cigs)
> summary(model)

Call:
lm(formula = cancer ~ cigs)

Residuals:
    Min      1Q  Median      3Q     Max
-162.58  -33.50  -11.96   47.07  120.29

Coefficients:
            Estimate Std. Error t value Pr(>|t|)
(Intercept) 28.30877   46.92479   0.603 0.555971
cigs         0.24105    0.05436   4.435 0.000566 ***
---
Signif. codes:  0 '***' 0.001 '**' 0.01 '*' 0.05 '.' 0.1 ' ' 1

Residual standard error: 82.22 on 14 degrees of freedom
Multiple R-squared: 0.5841,     Adjusted R-squared: 0.5544
F-statistic: 19.67 on 1 and 14 DF,  p-value: 0.0005659

> summary.aov(model)
            Df Sum Sq Mean Sq F value    Pr(>F)
cigs         1 132935  132935  19.666 0.0005659 ***
Residuals   14  94637    6760
---
Signif. codes:  0 '***' 0.001 '**' 0.01 '*' 0.05 '.' 0.1 ' ' 1
> |
```

The b indicates what a one unit change in the input (independent) variable has upon the outcome (dependent) variable according to the model.

The coefficients table opposite in the R output provides the a and b values for the $\hat{y}_i = a + bx_i$ simple regression equation. So, where the middle dot (·) means multiplication, we now have:

$\hat{y}_i = 28.309 + 0.241 \cdot x_i$

According to the model an increase in cigarette consumption of one unit increases the lung cancer death rate per million by 0.241.

```
R Console
> var(cigs_mortality)
         cancer      cigs
cancer 15171.45  36766.18
cigs   36766.18 152528.06
> |
```

b is calculated as the covariance of xy divided by the variance of x, where x in the example is cigarette consumption per adult in 1930 and y is incidence of lung cancer per million.

We can easily check this in R using the *var()* function requesting the variance and covariance values for the variables in our dataset. The results shown opposite indicate that the variance of the cancer variable is 15171.45, and that of cigarette consumption 152528.06. The covariance between the two is 36766.18, which we can obtain either by using the *var()* or *cov()* R functions, so b is:

```
R Console
> cov(cancer,cigs)
[1] 36766.18
> b_check<- cov(cancer,cigs)/var(cigs)
> b_check
[1] 0.2410453
> |
```

$$b = \frac{\text{cov}(x, y)}{\text{var}(x)} = \frac{36766.18}{152528.06} = 0.24104$$

This value is identical to that R produced in the coefficients table, but you should now have some insight as to how it reflects a ratio of the co-variation in x and y along with the variability in x.

> **Key point**
> The *b* value is an effect size measure, and its standard error indicates its accuracy.

The standard (*std*) error column indicates the degree of accuracy for the parameter estimate (remember it is the **standard deviation** of its sampling distribution of the mean). Smaller sizes here, in relation to the parameter estimate are better, and often it is a good idea to note the relative size of the standard error to the parameter estimate. We can also use this relationship to form a test. Assuming that the value for *b* is zero in the population and considering our observed b value to be due to random sampling variability, we can create a *t* statistic and get an associated p-value.

$$t = \frac{b - \rho}{SE_b} = \frac{signal}{noise / sampling} \Rightarrow \frac{.241}{.054} = 4.435 \sim t^{4.435}_{(\text{df=n-2})} \Rightarrow t^{4.435}_{(14)} = 0.00056 \text{ (p-value)}$$

Interpreting the above p-value of 0.00056 we can say that given that the population value of b=0 we would obtain a t value equal to that observed or one more extreme 56 times in a hundred thousand. The last column of the above table labelled $Pr(>|t|)$ is the p-value.

Now you may think that it is merely coincidence that both the F value and t value have the same p-value (.00056) in the above example. Mathematically this is because $F_{(1,n-2)} = (t_{(n-2)})^2$ as can be seen in the above $(4.435)^2$=19.66. In the simple regression situation, as above, considering the overall fit of the model (ANOVA table and the multiple R) is equivalent to considering b. However in the situation with multiple regression where we have more than one predictor this is not the case.

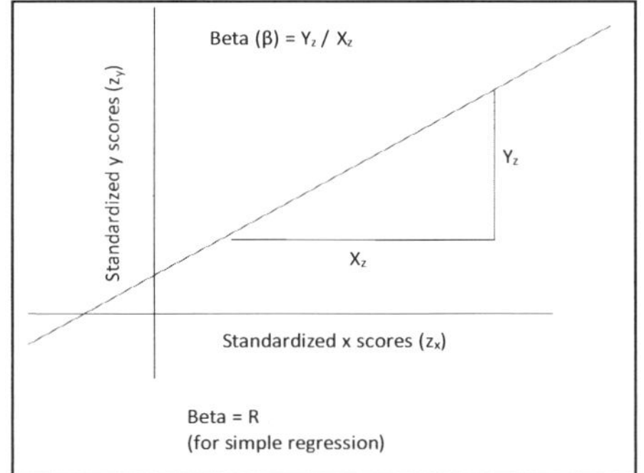

26.2.11 Beta (β)

A standardised score (z scores), is a variable that has been adjusted to have a mean of zero and a standard deviation of one. It is obtained by subtracting the mean from each score and dividing it by the standard deviation. If we do this with our scores and then carry out the regression, our b value becomes **Beta (β) and in the simple regression situation this is equivalent to the correlation**. If you look back at the correlation chapter the reason for this should become clear.

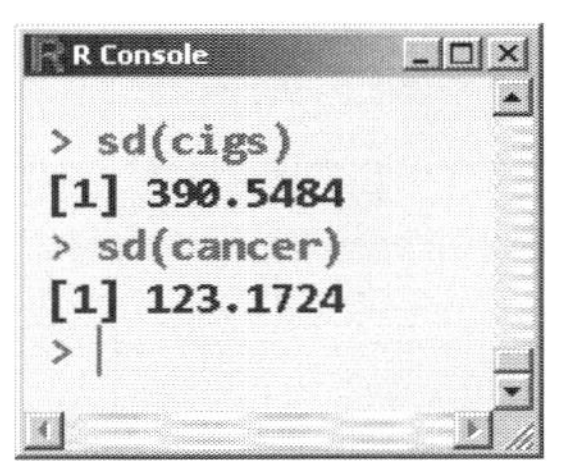

Beta (β) indicates what a one standard deviation change in the predictor (input) variable has upon the dependent (outcome) variable according to the model, in standard deviation units. This may be useful if you want to compare the relative effect of different predictors in a regression, but is not useful in simple regression. We can easily convert b's into Betas if you know the standard deviation or variance of the predictor and response variables. I have calculated these opposite. We can then apply the following formula: Beta is calculated from the b coefficient multiplied by the standard deviation of the raw scores for the variable (cigs) all divided by the standard deviation of the raw scores for the response variable (cancer):

$$\beta = \frac{b_i s_i}{s_y} = \frac{.24104 \bullet 390.5484}{123.1724} = .7642$$

Key point
The beta value is a standardised effect size measure.

This Beta (β) value indicates what a one standard deviation change in cigarette consumption will have upon the incidence of lung cancer, in standard deviation units. Therefore, we can say that a rise in one standard deviation in cigarette consumption will raise the incidence of lung cancer by three quarters of a standard deviation.

As in other situations where we have come across standardised scores this allows us to compare sets of scores more easily.

While the above approach using p-values provides us with a decision to either accept or reject the regression equation, it does not give us information about either how the equation relates to individual points, or any assessment of the assumptions of the regression equation fitting technique - an area we have deliberately ignored so far.

Confidence and prediction intervals are a good way of seeing how well the individual data points fit the model, so we will begin by investigating them.

26.2.12 Confidence and Prediction intervals

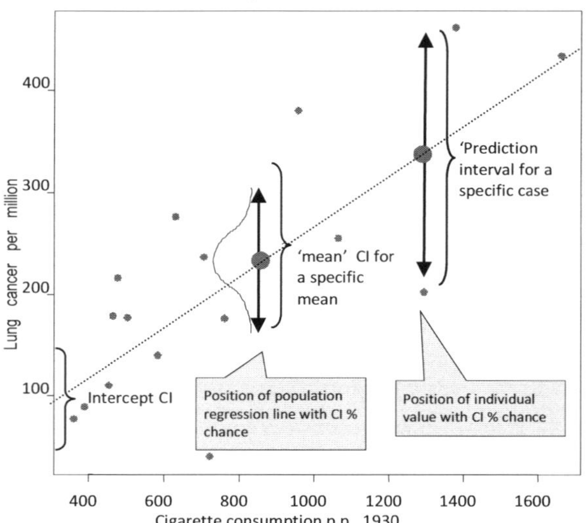

A confidence interval is a width with a lower and upper value. In our simple regression equation we can consider two different confidence intervals, that for the a in the equation $\hat{y}_i = a + bx_i$ that is the intercept value and the second one for a specific y value on the regression line. The point on the regression line at x_i is assumed to be the mean for all the possible y values at this x value. Obviously, in many situations this is theoretical, as we do not have a dataset which gives us many y values for a single x value. Mathematically we believe the set of y values for any specific x value is normally distributed with a mean equal to $\hat{y}_i$.

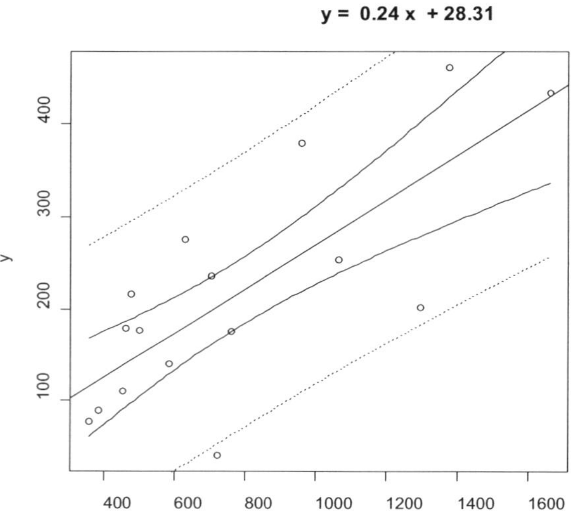

Additionally considering the parameter estimates also allows us to reflect on the uncertainty in $\hat{y}_i$ as a predictor of y for an individual case/subject. This range of uncertainty is called a *prediction* or *tolerance interval.* (Gardner & Altman, 1989 p.40). However while Gardner & Altman make this distinction between confidence intervals and prediction intervals, other authors and statistics applications are less clear and use different terms, for example SPSS call them prediction intervals for means or individuals.

An example of what these confidence and prediction intervals look like for the cigarette/ lung cancer data is given opposite. The narrower interval is that for the (mean) confidence interval and the wider one is the prediction interval, both are 90% limits. Notice that the width for the confidence interval increases at the extreme x values and is narrowest at the $\bar{y}$ (and coincidentally the mean x value) value.

While you can find the equations for the confidence and prediction intervals in Gardner & Altman I will concentrate on how to obtain and interpret these values in R. For example, from the plot opposite we can see one point is outside the wide prediction interval and it would be sensible to consider if this is a legitimate value.

This chapter has presented a lot of theory before starting to demonstrate how to carry out a regression analysis in R. This have been deliberate as it is relatively simple to do the button pressing, but requires a fair degree of knowledge to interpret the results and understand what is happening.

26.3 Carrying out simple regression

I have used here the example from MSME3 page 55 describing 12 HbA1c and fasting blood glucose measurements taken from 12 diabetic subjects which we used in the correlation chapter.

26.3.1 Data preparation

Carry out the steps described in the preliminaries section of the Correlation chapter, by the end of which you will have a dataframe called *mydataframe* consisting of 12 observations.

26.3.2 Model development: A basic regression analysis in R Commander

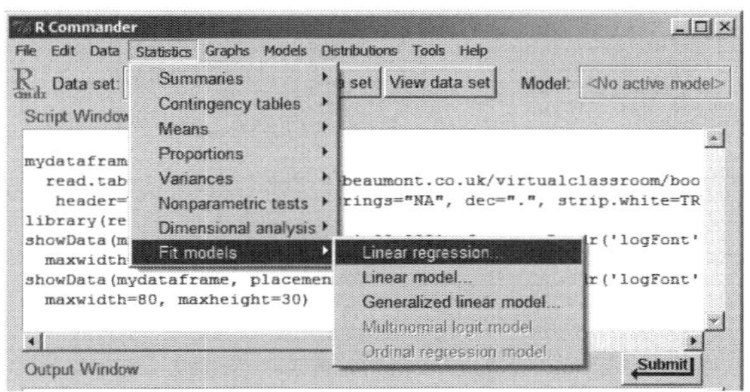

I will assume you have carried out a preliminary analysis of the data looking at each variable in turn. Now to carry out the regression analysis in R Commander. Choose the R Commander menu option:

Statistics-> Fit models -> Linear regression

In the *Linear regression* box you need to give the model a name and I have called it *hba1c_model,* you need also to select both the **Response variable** (outcome/dependent) and the **Explanatory variable** (input/independent).

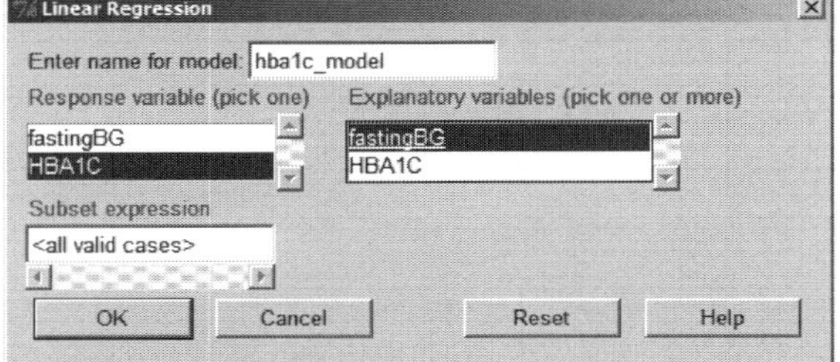

Click **OK.**

The results are shown below.

Please note that while the gradient value is the same as that reported in MSME3 p.55, the intercept value is slightly lower at 2.9 instead of 3.2. This is because I guessed the values from the scatter plot in MSME3.

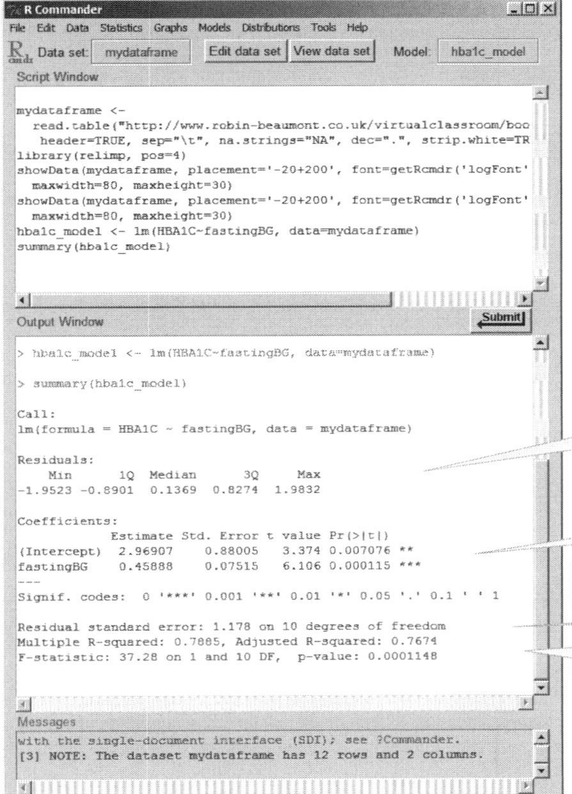

R Commander
Data set: mydataframe
Model: hba1c_model
hba1c_model <- lm(HBA1C~fastingBG, data=mydataframe)
summary(hba1c_model)
Call:
lm(formula = HBA1C ~ fastingBG, data = mydataframe)
Residuals:
Min 1Q Median 3Q Max
-1.9523 -0.8901 0.1369 0.8274 1.9832
Coefficients:
Estimate Std. Error t value Pr(>|t|)
(Intercept) 2.96907 0.88005 3.374 0.007076 **
fastingBG 0.45888 0.07515 6.106 0.000115 ***
Residual standard error: 1.178 on 10 degrees of freedom
Multiple R-squared: 0.7885, Adjusted R-squared: 0.7674
F-statistic: 37.28 on 1 and 10 DF, p-value: 0.0001148
[3] NOTE: The dataset mydataframe has 12 rows and 2 columns.

A residual is the difference between the observed and model calculated value, the smaller the value, the better the fit

The two values calculated for the model, the intercept and the gradient. Pr(>|t|) is the p-value

R squared value is considered by some to be the % of variability in the data that is accounted for by the model

F value = overall fit. Not so relevant in simple regression

26.3.3 Visualising the line of best fit

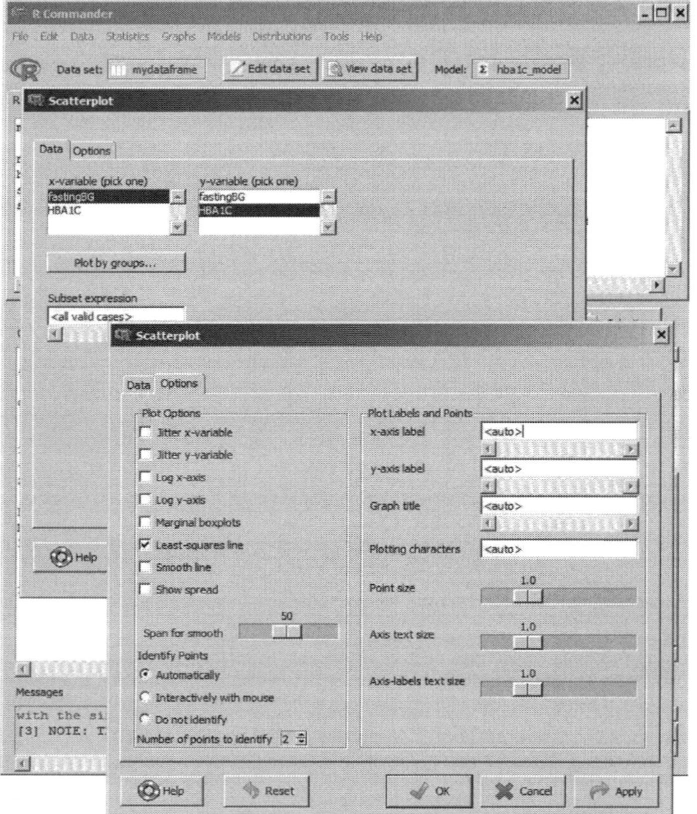

To see the actual data along with the line of best fit choose the R Commander menu option:

Graphs -> Scatterplot

Complete the scatterplot dialog box as shown opposite. If you wish to see the case number for each point select the 'identify points' option.

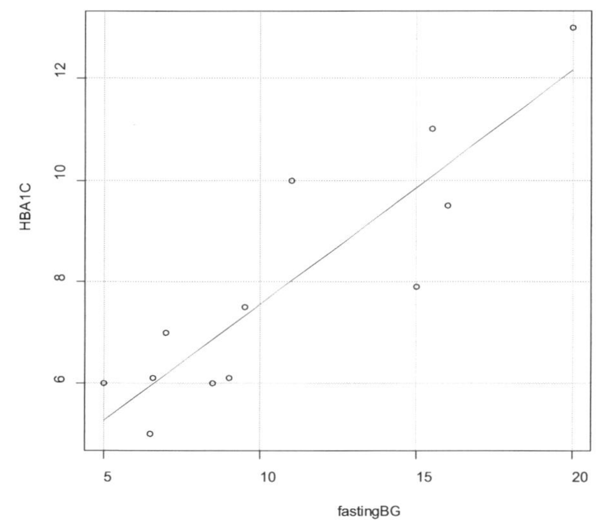

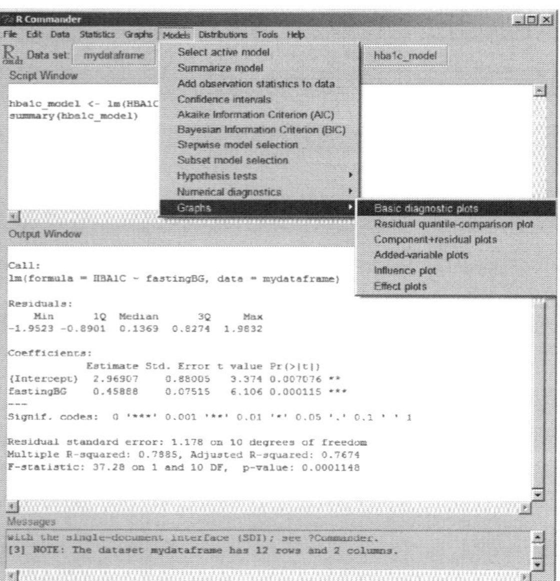

26.3.4 Diagnostics

While it is fine carrying out a regression analysis, the validity of the model is dependent upon many assumptions, such as the LINE ones mentioned earlier and R Commander provides a quick way of assessing some of these. Select the R Commander menu option:
Models -> Graphs-> Basic diagnostic plots

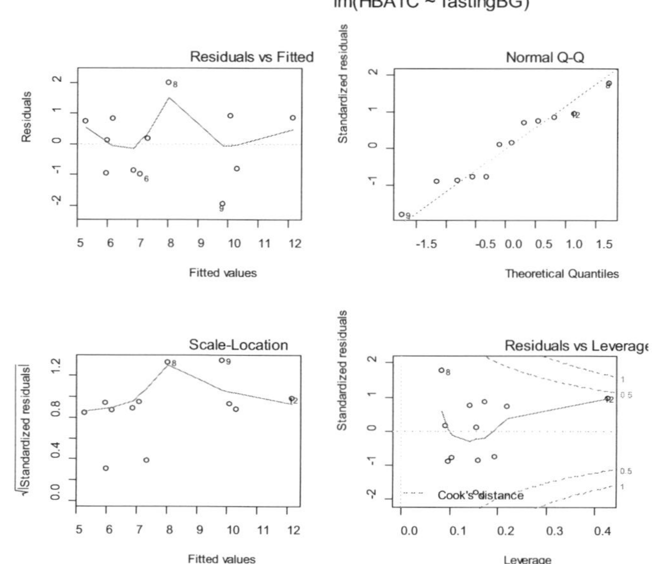

This results in a number of graphs, for details of which see the *Multiple Linear Regression* chapter.

26.3.5 Obtaining confidence intervals

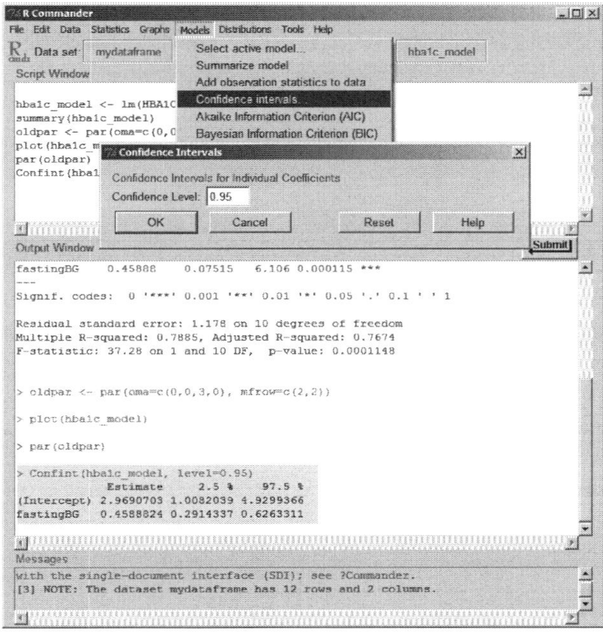

Many journals require the reporting of confidence intervals for any values (technically called parameters) produced from a model. In R Commander you obtain these by selecting the R Commander menu option:

Models -> Confidence intervals

The results are shown opposite.

It is best to avoid using the dash sign to indicate the lower/upper values, instead use the word 'to'. For example fastBG (95% CI 0.291 to 0.626) is better than fastBG (95% CI 0.291 - 0.626).

26.3.6 Confidence interval for R squared

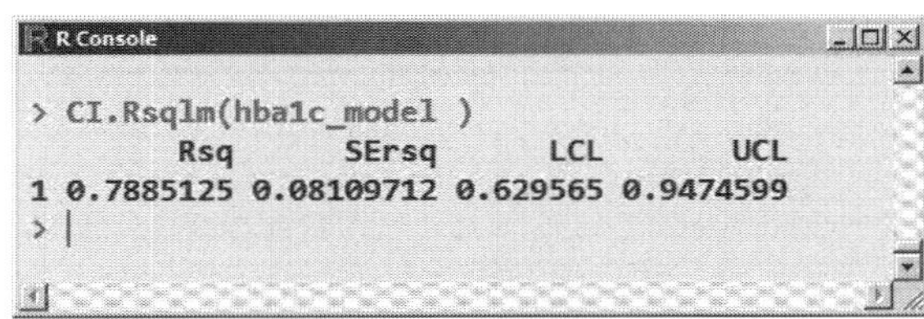

Although we have produced confidence intervals for the model values, we have not done so for the R squared value. To do this we need to use the *CI.Rsqlm()* **function** from the psychometric package, to install and load this package type the following into either the *R Console* or R Commander *Script* windows:

```
install.packages("psychometric", dependencies=TRUE)
library(psychometric)
CI.Rsqlm(hba1c_model )
```

26.3.7 Drawing confidence /prediction intervals

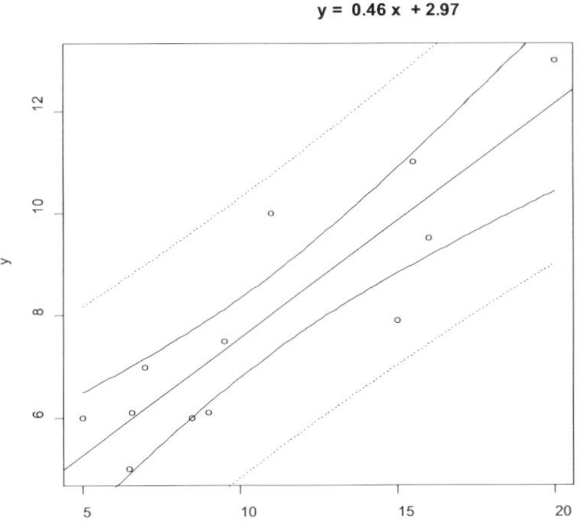

The easiest way to do this is to find an *R* package that does it for you and luckily there is one, *UsingR*. To install and load it type the following commands into the R *Console* window. The code is explained in *Packages: the apps* chapter.

```
install.packages("UsingR", dependencies=TRUE)
library(UsingR)
```

To produce a scatterplot with both confidence and prediction intervals type the following in the R Console window:

```
simple.lm(mydataframe$fastingBG, mydataframe$HBA1C, show.ci=TRUE)
```

The nearer ones are the confidence intervals and the wider ones are the prediction intervals.

One important thing to note from the documentation accompanying the psychometric package is that the limits are approximations. The accuracy of it is poor for sample sizes of less than 60 so we should be very careful about interpreting these values with our small sample of 12.

26.3.8 Influence measures

While we can get a visual idea of which particular point(s) affect the regression line, in the above analysis it would be nice to be able to have a more quantitative evaluation. To do this we need to be a bit more specific about what we are actually talking about! Norman & Streiner, 2008 p.155 give the following equation:

Influence = Distance X Leverage

Distance This can simply be the magnitude of the residual/error i.e. $y_i - \hat{y}_i$ but it can also be any expression that uses this in some way, such as the squared, or standardised values etc.

Leverage This is the degree to which a particular value can affect the regression line. Norman & Streiner express this as the degree of typicality of the values for the dependent variable(s), the less typical the value the more leverage the point has. Crawley, 2005 p.123 makes a very important point, **a highly influential point means one that forces the regression line close to it, and hence the influential point may have a very small residual**. A Leverage value for case i is often represented as h_i.

Leverage influence statistics		
R Name	**Long name**	**Explanation**
dfb.1	Standardised Dfbeta for y	Change in parameter estimate when case excluded
dfb.fsBG	Standardised Dfbeta for fastingbg	Change in parameter estimate when case excluded
dffit	Standardised Dffit	Difference between predicted y with case deleted and original predicted y. If no influence =0
cov.r	covariance ratios	A measure of how much a case affects the variance of the parameters. A value close to one indicates little influence.
cook.d	Cook's distances	Overall influence. A value >1 suggests too greater influence
Hat (leverage statistic)	hat matrix	Range between 0 (no influence) and 1 (complete influence). Average value is (k+1)/n where k=number of predictors; n=number of cases

So the actual influence of any one particular point is a combination of both Distance (from the predicted value) and Leverage. Let's get some influence measures for each of the points in our dataset.

Type the following command in the R Commander script window:

influence.measures(hba1c_model)

Highlight it and click on the submit button to produce the output shown opposite. Each of the six columns provides a value for each point.

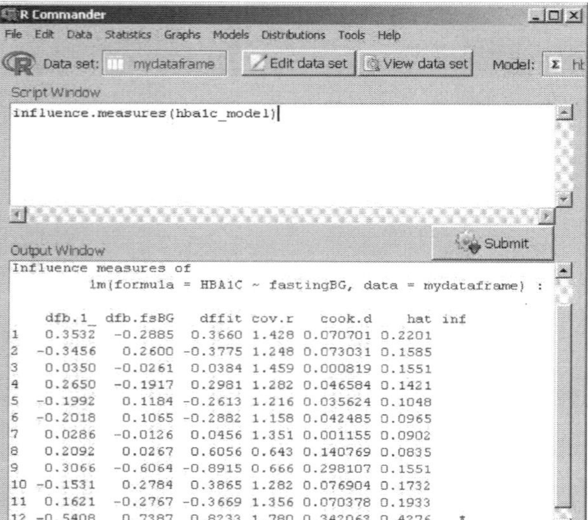

Using the criteria mentioned in the above table it would appear that no value exerts a undue influence. The largest Cook's distance (labelled *cook.d*) is .34 with a hat value of .42. The average value should be 2/12 = .166 so .42 is just over two and a half times that (.42/.166 =2.52). We would probably start to be seriously worried if it was above three times the average (.166 x 3= .49).

26.3.9 Residuals

We can also investigate how far each point is from the line of best fit with three options:

- residuals in the original units or
- normalize the residuals to have unit variance, or
- divide the residuals by their standard error

These are implemented in R with the following commands:

```
resid(hba1c_model)
rstandard(hba1c_model)
rstudent(hba1c_model)
```

non standardized residuals

standardized residuals 95% of points should be <|1.96|

Studentized residuals = residual divided by its standard error

You can create graphs of these results by wrapping them in the *plot()* command:

```
plot(resid(hba1c_model))
plot(rstandard(hba1c_model))
```

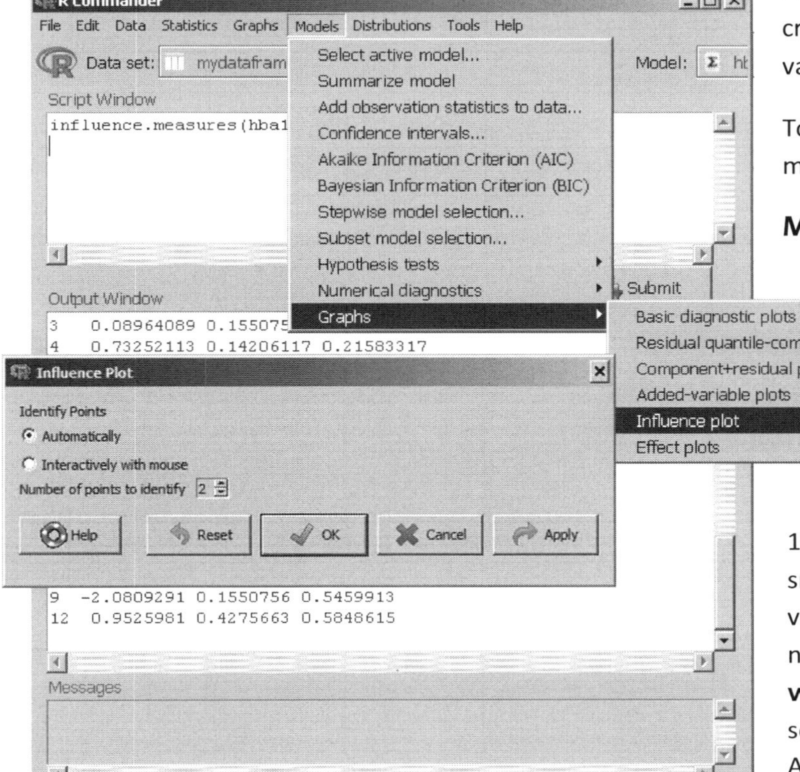

Using the R Commander menu option you can also create a plot of both an influence measure (hat values) along with residuals (studentized).

To produce the plot you select the R Commander menu option:

Models-> graphs -> influence plot

A message box them appears asking if you want to identify the points by clicking on them, select the ***Automatically*** option.

The resultant plot is shown below. It is immediately clear that while point 12 has the large hat value its residual is relatively small. In contrast points 9 and 8 have low hat values but large residuals. I discovered the case number for point 8 by selecting the **Interactively with mouse** option from the **Identify Points** selection when rerunning the above menu option. Also when requesting the plot at the start of the analysis you can request the case numbers to be displayed (see previous page).

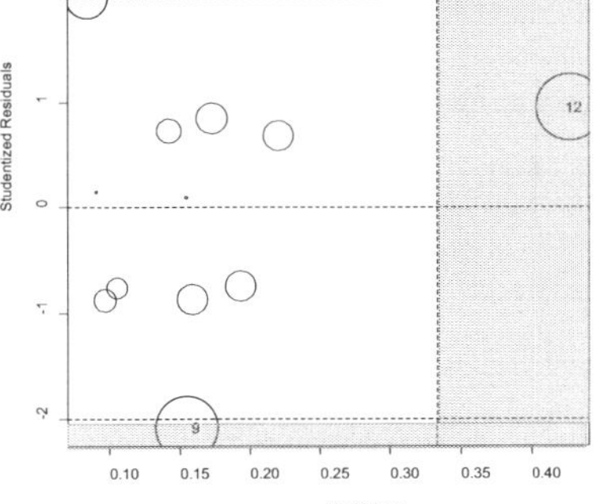

Note the graph has a vertical dotted line indicating where two times the average hat value is. Similarly, the two horizontal lines indicate where excessive studentized residual values occur (>|2|). In other words the two extreme top and bottom strips along with the far right (shaded) one. These are the regions of the graph where troublesome values might occur, that is where the model is least effective at accurately predicting values.

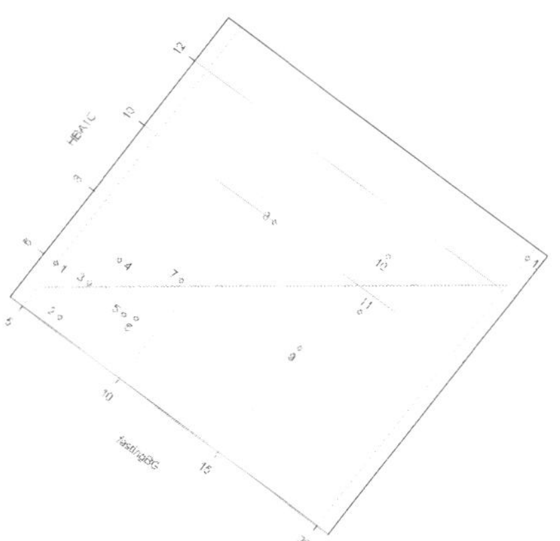

The relationship between the studentized residuals in the above plot and the distances from the line in the scatterplot can be seen if you rotate the scatterplot as shown opposite.

We will now consider the situation where you have identified a unusually influential point (in reality it could be a subgroup of them) that you feel might be atypical and would like to rerun the analysis without it. As an example I will consider case 12 to be the problem (although we know it is not).

26.3.10 Analysing a subset

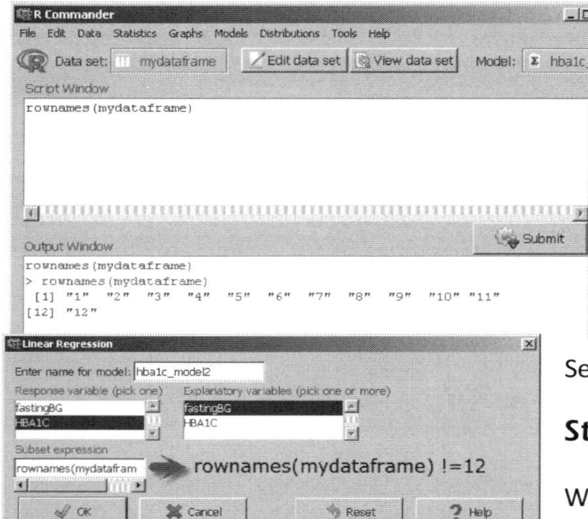

Although our present dataframe only appears to have two columns, there is a hidden one which contains the row numbers which you can see by typing in the R Commander script window:

```
rownames(mydataframe)
```

As shown opposite the rows have names that are numbered from 1 to 12, and we can make use of this to rerun the analysis leaving out row 12.

Select again the R Commander menu option:

Statistics-> Fit models -> Linear regression

We then enter a new name for the model *hba1c_model2.* This time we enter the expression *rownames(mydataframe) != 12* into the subset expression box informing R Commander to include all rows that do not equal 12 (i.e. the *!=* expression), in other words to omit row 12.

You could have avoided the use of the R Commander menu by typing in the script window:

```
hba1c_model2 <- update(hba1c_model1,
                subset=(rownames(mydataframe) != 12))

summary(hba1c_model2)
```

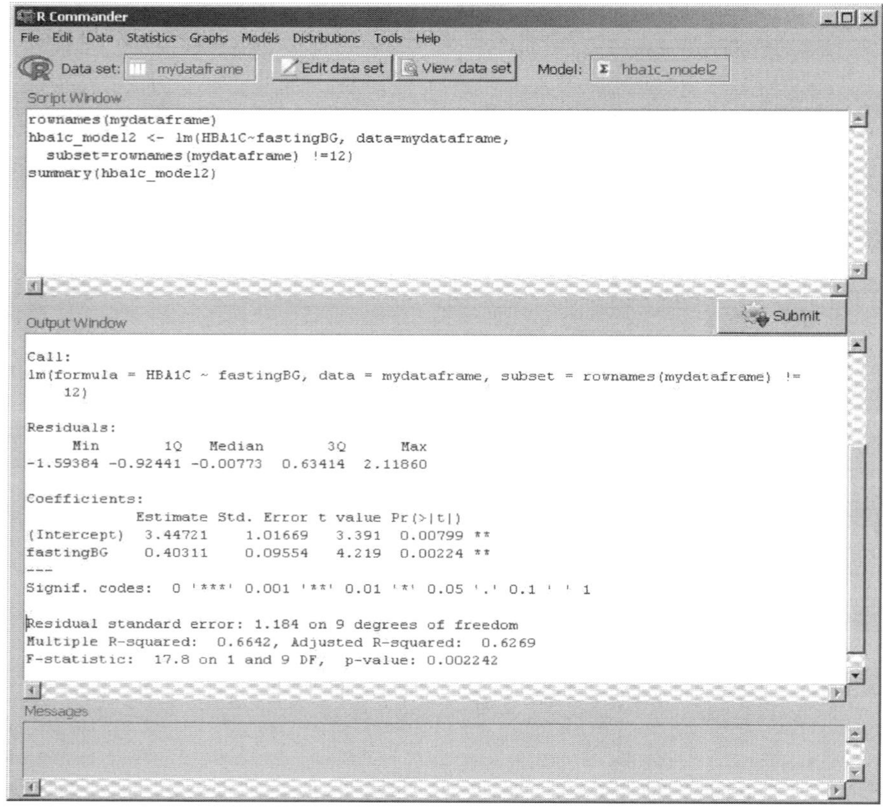

The results shown opposite indicate that the R squared value is now .6642 in comparison to .7885 in the previous analysis. Also the *fastingBG* parameter has gone from being three star significant to two star. So informally, it would appear that the extra case produced a better fitting model.

What would be useful to know is if the two models are quantitatively different as we did so previously with the one parameter versus two parameter models.

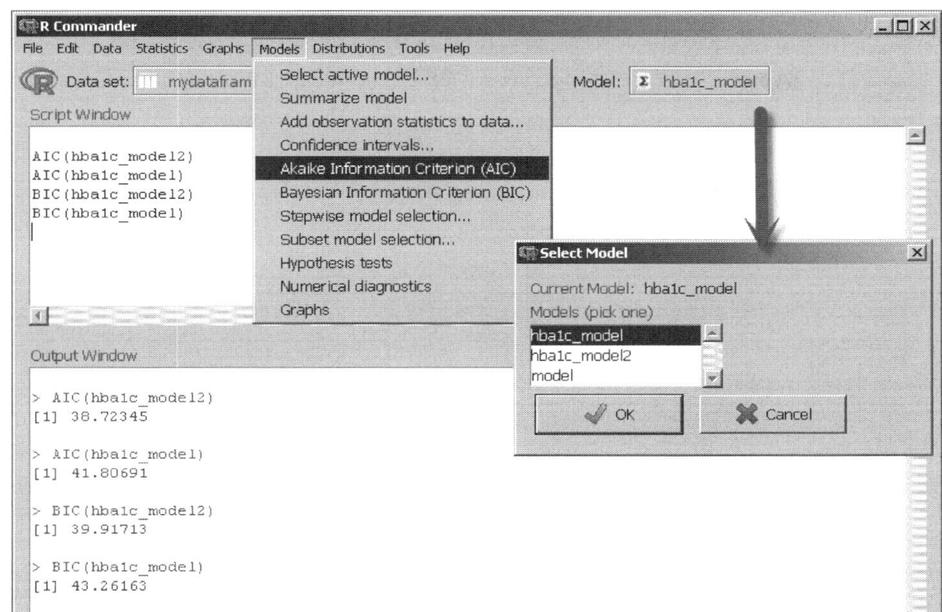

There are two measures of model fit, Akiake's Information Criterion (AIC) and the Bayesian Information Criterion (BIC) where lower values indicate a more parsimonious model. The values can be obtained for each model and you can quickly swap between the models by using the *Selection Model* box, which appears when you click on the space at the right of the **Model** name panel.

We can see that both the AIC and BIC are smaller for the original model. So given these results along with the R squared difference and the parameter estimate we come to the conclusion that the point should definitely stay unless there is some reason why the value might be considered to be invalid.

Clearly, I have demonstrated the above technique on a rather trivial scenario, but hopefully it has provided insight into model development. This process is developed in far greater depth in subsequent chapters.

26.4 Model development directly in R

As an alternative to using the R Commander menu options you can enter the R code directly. I will demonstrate this by considering some aspects of the analysis described in the previous section. You can load the dataset from my website using the following R Code:

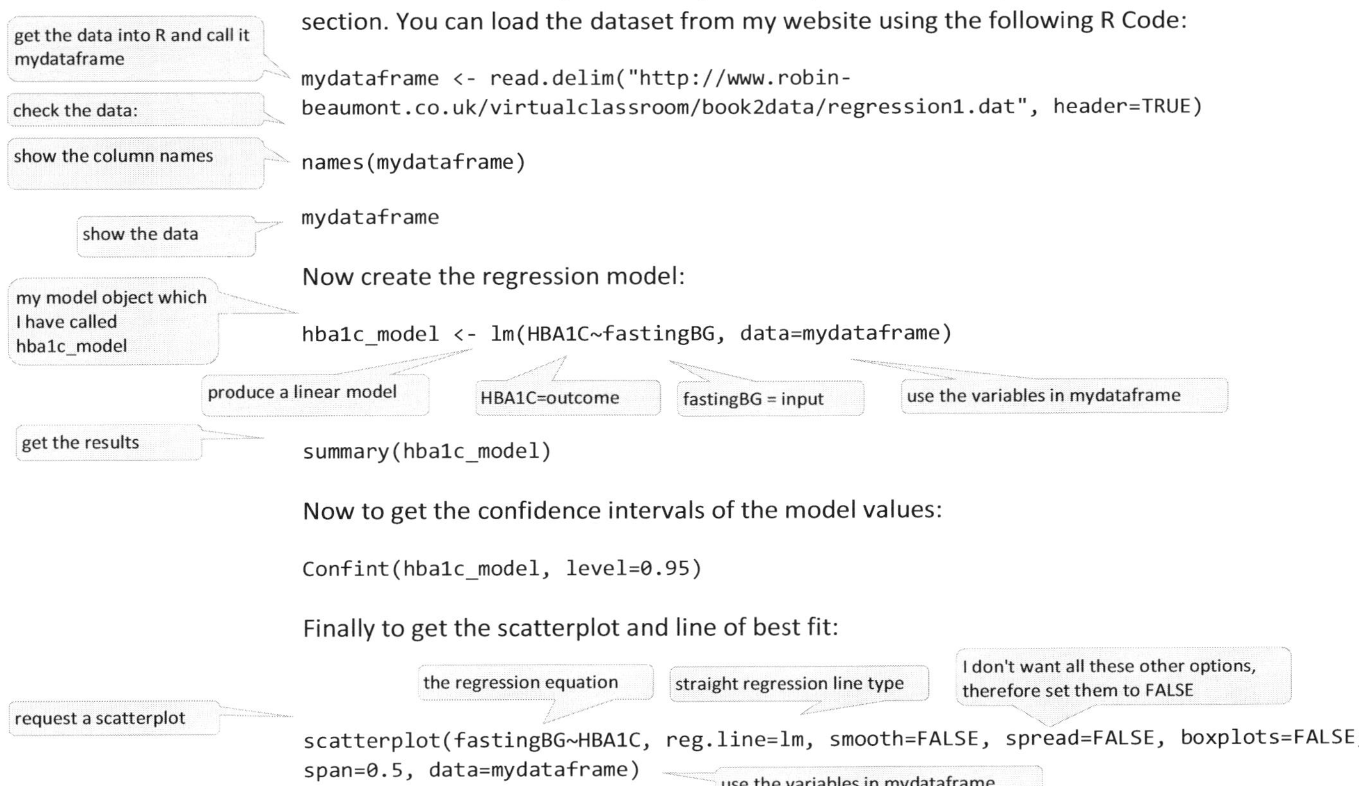


```
mydataframe <- read.delim("http://www.robin-
beaumont.co.uk/virtualclassroom/book2data/regression1.dat", header=TRUE)
```

```
names(mydataframe)
```

```
mydataframe
```

Now create the regression model:

```
hba1c_model <- lm(HBA1C~fastingBG, data=mydataframe)
```

```
summary(hba1c_model)
```

Now to get the confidence intervals of the model values:

```
Confint(hba1c_model, level=0.95)
```

Finally to get the scatterplot and line of best fit:

```
scatterplot(fastingBG~HBA1C, reg.line=lm, smooth=FALSE, spread=FALSE, boxplots=FALSE,
span=0.5, data=mydataframe)
```

The *lm()* **l**inear **m**odel, function in the above code produces a model **object** which contains all the information produced in the R output, but by wrapping it around the summary function we get only the most important aspects displayed.

If you do not wish to use the *UsingR* package to display the actual confidence intervals on the plot look at the professional graphics chapter. Alternatively, to do it with R code the R Ecology and Evolution blogspot has a good tutorial providing full details, http://r-eco-evo.blogspot.co.uk/2011/01/confidence-intervals-for-regression.html

26.5 Writing up the results

12 HbA1c and fasting blood glucose measurements were obtained from 12 diabetic subjects with the aim to produce a regression model predicting HbA1c level given Fasting blood glucose level. The regression equation produced an adjusted R squared value of .767 indicating that 76.7% of the variability in the Hba1c can be accounted for by variation in the fasting blood glucose level. The model produced the following results:

	R^2 (adjusted 95% CI)	B (95% CI)	SE B	P-value
	0.78 (.62 to .94)			
Constant (intercept)		2.96 (1.00 to 4.92)	0.88	.007
Fasting Blood Glucose		0.45 (0.29 to 0.62	0.07	<.001

I would also include a plot of the points along with the line of best fit and the CI and prediction intervals.

Things to note:

- <.001? - The p-value for the Fasting blood glucose variable I have given as ' <.001' indicating that it is less than 0.001. This is because in the results it is 0.000115. I have done this as it is normal to present results in a table to a consistent number of decimal places, usually 2 to 4. Here I have gone for two decimal places for the values and CI's and 3 for the p-values.

- The value of 0.45 for the Fasting blood glucose level indicates that from the model an increase of one mmol/l in this value predicts an increase of 0.45 in the HbA1c value. This can be confusing as you would think the value associated with it would relate directly to that variable, not the outcome.

- Standardised beta (β) values are often reported and you can obtain these values using the *lm.beta(model)* function within the *QuantPsyc* package, or the equation I provided earlier in this chapter. If there is only one input variable in the regression equation, the single standardised beta value is the same as the correlation between the input and output variable. You can also obtain it by simply square rooting the (multiple) R squared value, shown opposite producing a value of .8879. This is because standardising the scores makes the intercept vanish (i.e. equal zero).

26.6 Summary – a caution

Both SAS1, SAS2 and to a greater extent Norman & Streiner, 2008 contain sections within their chapters stressing the importance of assessing the assumptions that regression analysis makes along with the dangers of over-interpreting the predicted values you might calculate from the regression model. Always check:

- The predicted value is not something that is impossible i.e. a negative fasting blood sugar level.
- Possibility of several distinct subgroups within your sample that might have different regression equations, common subgroups that are mixed include; males/females, and specific age ranges.
- Are there sufficient numbers of observations across the whole range of values for both the input and output variables? Also is it sensible to make predictions on a range for which you have no observations?
- Is the relationship best modelled by a straight line or would a curved line be more appropriate. If this is the case transforming the data might result in a good straight line fit?

26.7 Tips and tricks

Youtube help:

http://www.youtube.com/watch?v=g05KDiXrVt0&feature=share&list=PL9F0EBD42C0AB37D0

27 Health statistics

This section focuses on various statistical techniques that are commonly found in the health literature but not exclusively so. For example the *Levels of agreement* chapter describes methods that are frequently used in psychology.

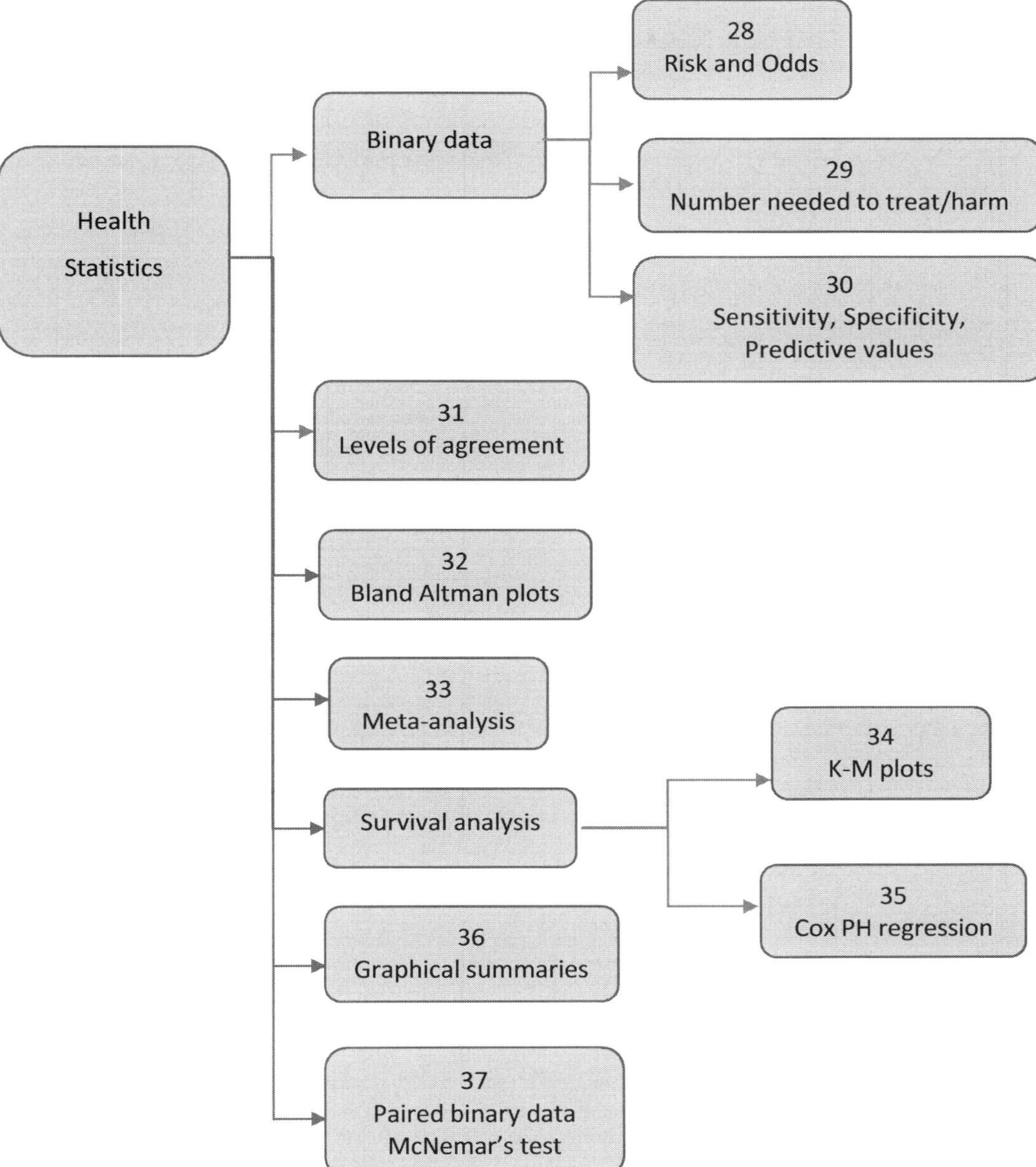

28 Risk and odds ratios

If you are unsure what risk, odds and odds ratios are check out the glossary entries for **Risk, Odds** and **Odds ratio** before reading this chapter.

For this chapter I have used the example from MSME3 page 41 describing the association between skiing (disease=outcome) and knee injury (exposure=risk factor). The cases (injured=exposed) and controls (healthy=unexposed) were matched for age and sex.

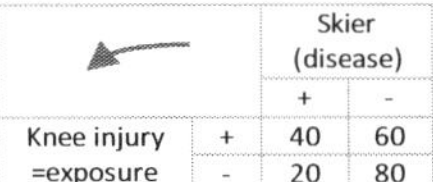

		Skier (disease)	
		+	-
Knee injury =exposure	+	40	60
	-	20	80

28.1 Using OpenEpi

R Commander does not provide a menu option for risk and odds ratios so instead I will demonstrate an excellent online application called *OpenEpi*. If you wish you can download it for free rather than use it online.

Open your web browser and go to the url: http://www.openepi.com/OE2.3/Menu/OpenEpiMenu.htm

Select the **Two by Two table** option on the left of the screen. You will then see a number of tabs across the top of the screen, including Start and Enter.

Select the **Enter** tab, now you can enter the data, and change the names of the columns/rows. Enter the data and edit the row column names, as shown opposite.

Click on the **calculate** button.

The results are shown below,

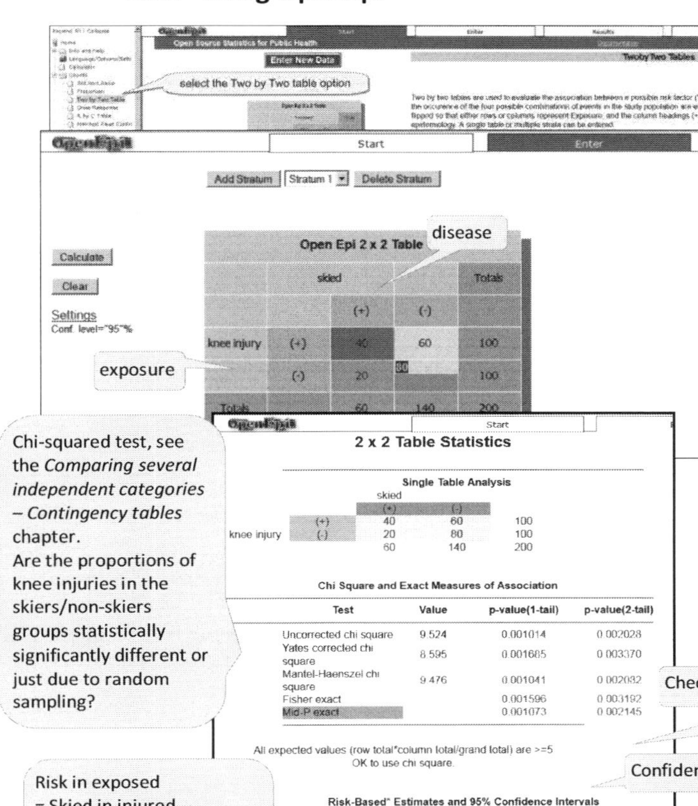

Chi-squared test, see the *Comparing several independent categories – Contingency tables* chapter. Are the proportions of knee injuries in the skiers/non-skiers groups statistically significantly different or just due to random sampling?

2 x 2 Table Statistics

Single Table Analysis

		skied (+)	skied (-)	
knee injury	(+)	40	60	100
	(-)	20	80	100
		60	140	200

Chi Square and Exact Measures of Association

Test	Value	p-value(1-tail)	p-value(2-tail)
Uncorrected chi square	9.524	0.001014	0.002028
Yates corrected chi square	8.595	0.001685	0.003370
Mantel-Haenszel chi square	9.476	0.001041	0.002032
Fisher exact		0.001596	0.003192
Mid-P exact		0.001073	0.002145

All expected values (row total*column total/grand total) are >=5
OK to use chi square.

Checks for us, the danger of invalid results if the expected cell counts are <5

Confidence intervals are important as they provide a level of certainty for our estimates

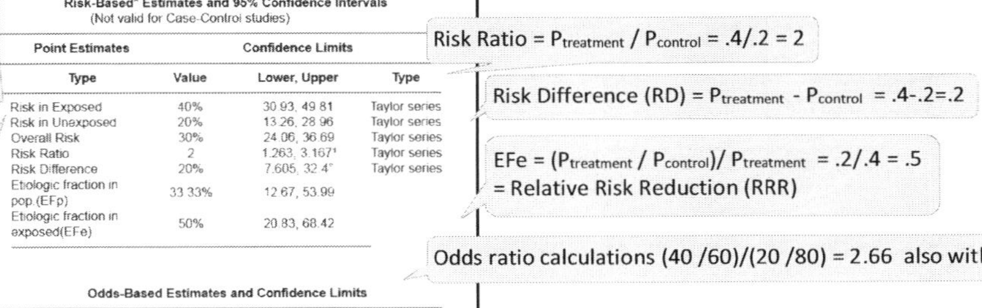

Risk-Based* Estimates and 95% Confidence Intervals
(Not valid for Case-Control studies)

Point Estimates		Confidence Limits	
Type	Value	Lower, Upper	Type
Risk in Exposed	40%	30.93, 49.81	Taylor series
Risk in Unexposed	20%	13.26, 28.96	Taylor series
Overall Risk	30%	24.06, 36.69	Taylor series
Risk Ratio	2	1.263, 3.167¹	Taylor series
Risk Difference	20%	7.605, 32.4°	Taylor series
Etiologic fraction in pop.(EFp)	33.33%	12.67, 53.99	
Etiologic fraction in exposed(EFe)	50%	20.83, 68.42	

Risk in exposed = Skied in injured $= P_{treatment} = 40/(40+60) = .4$ = EER (Experimental Event Rate)

Risk in unexposed = Skied in healthy $= P_{control} = 20/(20+80) = .2$ = CER (Control Event Rate)

Risk Ratio $= P_{treatment} / P_{control} = .4/.2 = 2$

Risk Difference (RD) $= P_{treatment} - P_{control} = .4 - .2 = .2$

EFe $= (P_{treatment} / P_{control}) / P_{treatment} = .2/.4 = .5$ = Relative Risk Reduction (RRR)

Odds ratio calculations (40 /60)/(20 /80) = 2.66 also with confidence intervals

Odds-Based Estimates and Confidence Limits

Point Estimates		Confidence Limits	
Type	Value	Lower, Upper	Type
CMLE Odds Ratio*	2.653	1.415, 5.07¹	Mid-P Exact
		1.358, 5.313¹	Fisher Exact
Odds Ratio	2.667	1.417, 5.02¹	Taylor series
Etiologic fraction in pop.(EFp|OR)	41.67%	19.18, 64.15	
Etiologic fraction in exposed(EFe|OR)	62.5%	29.41, 80.08	

*Conditional maximum likelihood estimate of Odds Ratio
(P)indicates a one-tail P-value for Protective or negative association, otherwise one-tailed exact P-values are for a positive association.
Martin,D; Austin,H (1991) An efficient program for computing conditional maximum likelihood estimates and exact confidence limits for a common odds ratio. Epidemiology 2, 359-362.
° ¹ 95% confidence limits testing exclusion of 0 or 1, as indicated
P-values < 0.05 and confidence limits excluding null values (0,1, or [n]) are highlighted.
LookFirst items: Editor's choice of items to examine first.

Results from OpenEpi, Version 2, open source calculator--TwobyTwo

Notice that we do not get the actual odds for each group, only the odds ratio. In the above example I have considered skiing as the disease/outcome and knee injury status as the risk factor, following MSME3. However, it would be more logical to have treated injury status as the disease and skiing as the risk factor. This would produce different risk based measures ($P_{treatment}$=.66; $P_{control}$=.429; RD=.238; RR=1.55) but the same Odds ratio (40/20)/(60/80)=2.6 (SAS1 p.31).

28.2 Doing it directly in R: the epitools and epibasix packages

In R you can obtain the above results using various epidemiology packages. I will look first at the *epitools* package. To install and load the package type the following into the *R Console* window.

```
install.packages("epitools", dependencies=TRUE)

library(epitools)
```

We now need to get the data into R, and I will create a matrix object with two columns and two rows:

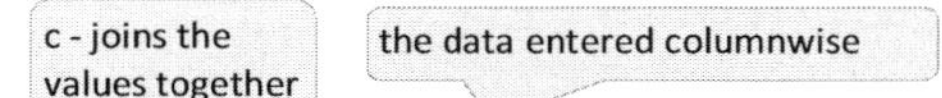

specify how the data is divided into rows and columns. i.e. here 2 rows and 2 columns

create a data object called *thedata* which is a matrix

```
thedata <- matrix(c(40, 20, 60, 80 ), nrow = 2 , ncol = 2)
```

create a matrix from . . .

defaults to column then row

	Injured knees	healthy
cases	40	60
controls	20	80

Now specify the row and columns names, the opposite way round to how we entered the data in the matrix command:

add row and column names to thedata

create a list consisting of . . .

```
dimnames(thedata) <- list("group" = c("case", "control"), "outcome" = c("injury", "no_injury"))
```

row heading

row names

column heading

column names

Now using the *epitab()* function in *epitools*, with the above data produces the odds ratio with a confidence interval:

```
> thedata
         outcome
group     injury no_injury
  case        40        60
  control     20        80
> epitab(thedata)
$tab
         outcome
group     injury        p0 no_injury        p1 oddsratio
  case        40 0.6666667        60 0.4285714  1.000000
  control     20 0.3333333        80 0.5714286  2.666667
         outcome
group        lower    upper     p.value
  case          NA       NA          NA
  control 1.416591 5.019874 0.003191817
```

```
epitab(thedata)
```

You can use additional commands to compare various methods of obtaining the confidence intervals, for the odds ratio there are five options:

```
oddsratio(thedata); oddsratio.midp(thedata)
oddsratio.fisher(thedata); oddsratio.wald(thedata)
oddsratio.small (thedata)
```

Similarly you can use the *riskratio()* function to find the risk ratios, including a bootstrap estimate.

However, before doing that, you can see in the above output that the case, rather than control, group has been used as the reference group (i.e. it has an odds ratio of 1) and it would be better when reporting risk to have the control group as the reference group. This is achieved by using the *rev="both"* option in the function:

```
> riskratio(thedata, rev = "both")
$data
         outcome
group     no_injury injury Total
  control        80     20   100
  case           60     40   100
  Total         140     60   200

$measure
         risk ratio with 95% C.I.
group     estimate    lower    upper
  control        1       NA       NA
  case           2 1.263006 3.167047

$p.value
         two-sided
group      midp.exact fisher.exact  chi.square
  control          NA           NA          NA
  case    0.002145336  0.003191817 0.002028231
```

```
riskratio(thedata, rev = "both")
riskratio.wald(thedata, rev = "both")
riskratio.small(thedata, rev = "both")
riskratio.boot(thedata, rev = "both")
or   riskratio(method = "boot", rev = "both")  etc.
```

```
> summary(epi2x2(thedata))
Epidemiological 2x2 Table Analysis

Input Matrix:
         outcome
group     injury no_injury
  case        40        60
  control     20        80

Pearson Chi-Squared Statistic (Includes Yates' Continuity Correction): 8.595
Associated p.value for H0: There is an association between exposure and outcome vs. HA: No associa$
p.value using Fisher's Exact Test (1 DF) : 0.003

Estimate of Odds Ratio: 2.667
95% Confidence Limits for true Odds Ratio are: [1.417, 5.02]

Estimate of Relative Risk (Cohort, Col1): 2
95% Confidence Limits for true Relative Risk are: [1.263, 3.167]

Estimate of Risk Difference (p1 - p2) in Cohort Studies: 0.2
95% Confidence Limits for Risk Difference: [0.066, 0.334]

Estimate of Risk Difference (p1 - p2) in Case Control Studies: 0.238
95% Confidence Limits for Risk Difference: [0.092, 0.384]

Note: Above Confidence Intervals employ a continuity correction.
```

If you are not interested in comparing different confidence interval estimation techniques you can use the *epibasix* package. The *epi2x2()* function gives the odds ratio, relative risk, and variations for both cohort (prospective) and case control (retrospective) studies, all with confidence intervals. As shown opposite wrapping it in the *summay()* function provides a neat output:

```
summary(epi2x2(thedata))
```

28.3 Writing up the results

Taking the example we have a odds ratio of 2.666 for cases/controls whereas 1/2.6666 = 0.3750 gives the odds ratio for controls/cases. This is because the two odds ratios are technically reciprocals.

So we can say that skiers have a 166% increase in the odds of a knee injury compared to those who do not ski, calculated from (2.66 -1)x100.

The above sentence is equivalent to saying that those who do not ski have a reduction in the odds of a knee injury of 63%, calculated from (1 - .37)x100.

A report would also include confidence intervals and emphasize the clinical important of these differences if any.

28.4 Relative risk versus odds ratios – study design

Retrospective Case control design Smokers and non-smokers among male cancer patients and controls (Doll & Hill 1950) quoted in Bland 2000			
	Smokers	**Non-smokers**	total
Lung cancer (admitted to hospital for ca lung)	647	2	649
Controls (admitted to hospital not for ca lung)	622	27	649
Smoker/non-smokers odds ratio= (647/2)/(622/27) =14.04			

While the relative risk and odds ratio are equivalent when the baseline risks are low, there is also another factor that needs to be taken into consideration, the manner in which the data was obtained. We can divide studies into ones that either look into the past (*retrospective*) where we know the outcome but not the exposure/treatment or *prospective* studies where we control who receives the treatment and wait to see the outcome. We can really only consider 'risk' in prospective studies, as looking back in retrospective studies usually only provides us with the number of those who have the disease/outcome (not those who did not). Usually, we don't know the risk in the population, but this is not always the case (see Langholz, 2010 for details).

The *Case control* approach, described at the beginning of the chapter, is often a retrospective study where outcome is measured after exposure (treatment) and the controls are selected on the basis of not having the outcome. In contrast, the Cohort study approach is usually prospective (where the outcome is measured before exposure/treatment), so we will have a truer reflection of the risk in the population.

Prospective cohort study (Doll & Hill 1956) Standardised death rate per 1,000 men (actually GPs) aged 35 53 months followup quoted in Bland 2000		
	Smokers	**Non-smokers**
Lung cancer	.90	.07
	Smoker/non-smokers odds ratio = .9/.07=12.9	

Bland (Bland, 2000 2nd ed. p.239; 3rd ed. p.242) provides a good example of the two types of approach, presenting the research by Doll & Hill concerning lung cancer and smoking, the early research used a retrospective case control design comparing smoking habits of those who were admitted to hospital with lung cancer to those admitted with some other condition.

Six years later they published results from a prospective cohort study consisting of 60% of all UK GPs following up the effect of smoking on their mortality after 53 months. I'm sure you can think of many reasons why the so called 'control' group in the 1950 research would not reflect the true incidence of lung cancer in all non smokers and Bland (Bland, 2000) discusses this at length – it's well worth a read if you have the time and inclination.

Cross sectional design 'association study' Hay fever and eczema in 11 year old children Campbell & Swinscow 2009 (p.30)			
	Hay fever present	No hay fever	total
Eczema present	141	420	561
No Eczema	928	13525	14453
Odds of hay fever given Eczema = 141/420 = .3357 Odds of hay fever given no eczema = 928/13525 =.06861 Odds **ratio** hayfever/Eczema = .3357/.06861 = 4.892 = eczema/hayfever ' A child with hayfever has approximately 5 times the odds of having eczema' and also because odds are reflective -> ' A child with eczema has approximately 5 times the odds of having hayfever ' The 'given' term in the above needs some care in interpreting as we are using it in the context of association. Normally we use it with regard to causation which is not reflective *A* causes *B* but *B* does not cause *A*			

A third type of study is the cross sectional study which obtains information at a single point in time, usually the present. This type of study has similar problems to the retrospective design. Campbell & Swinscow, 2009 p.28 also call these association studies and demonstrate how odds ratios for two different conditions can be analysed in these studies.

From the above we can conclude that:

- **Odds ratios should be quoted for all studies**
- **Relative risks should usually be quoted for Prospective studies only** (but see Langholz, 2010)

28.5 Tips and Tricks

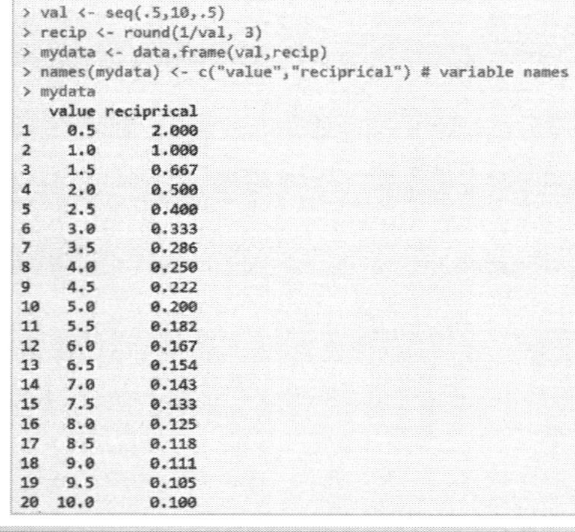

```
> val <- seq(.5,10,.5)
> recip <- round(1/val, 3)
> mydata <- data.frame(val,recip)
> names(mydata) <- c("value","reciprical") # variable names
> mydata
   value reciprical
1    0.5      2.000
2    1.0      1.000
3    1.5      0.667
4    2.0      0.500
5    2.5      0.400
6    3.0      0.333
7    3.5      0.286
8    4.0      0.250
9    4.5      0.222
10   5.0      0.200
11   5.5      0.182
12   6.0      0.167
13   6.5      0.154
14   7.0      0.143
15   7.5      0.133
16   8.0      0.125
17   8.5      0.118
18   9.0      0.111
19   9.5      0.105
20  10.0      0.100
```

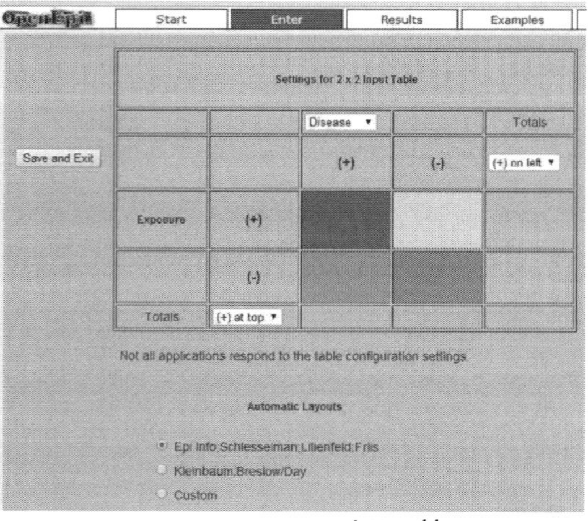

When calculating odds and risk ratios you need to be very clear about four things:

1. Which is the control (unexposed) and which is the intervention (exposed) group. Before entering my data I always read the openEpi example to remind myself.
2. When dealing with risks, concentrate on the groups with the least desirable outcome (technically we say 'suffered the event'). For instance in the above example we used the values from those who suffered knee injuries, in other situations it might be those who died or those who failed to respond to a particular treatment.
3. If you feel that you have the odds ratio the wrong way round consider this value over one (the reciprocal shown opposite), if this looks more like the value you were expecting you might indeed have the groups the wrong way round.
4. In OpenEpi you can change the table layout, that is which is the disease (outcome) and which are the Exposure categories by clicking on the **Settings** link on the **Enter** tab.

OpenEpi has great help resources including a pdf file which includes examples and full details of how the various results are calculated. It is available from: http://www.openepi.com/OE2.3/Documentation/TwoByTwoDoc.PDF

Youtube help:

http://www.youtube.com/watch?v=nFHL54yOnil&feature=share&list=PL9F0EBD42C0AB37D0

29 Number needed to treat/harm (NNT/NNH)

R Commander does not have a menu option for calculating NNT/NNH so again we are going to work through these analyses using Open Epi.

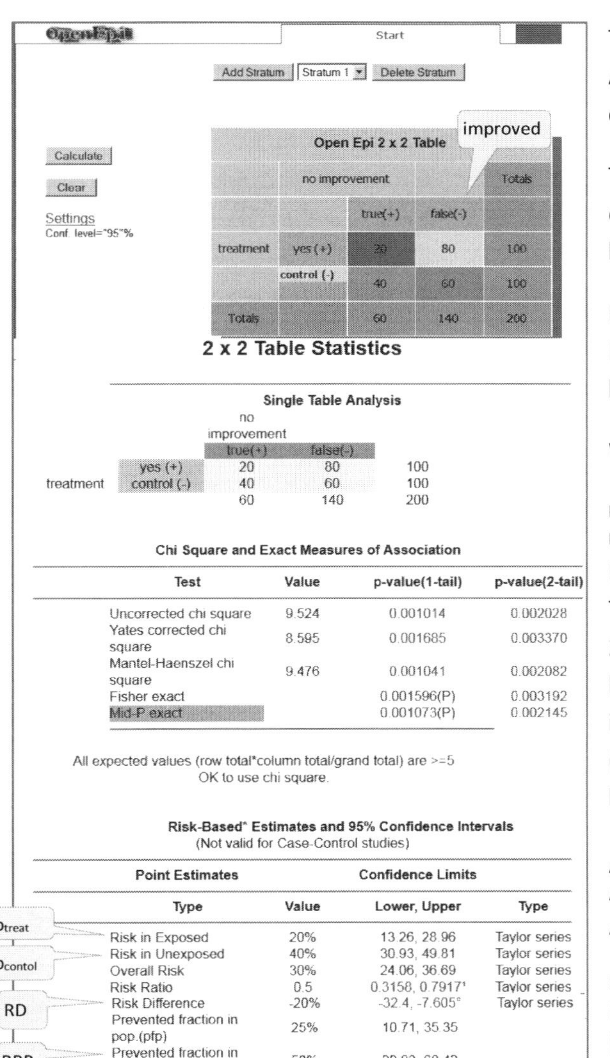

OpenEpi — Start

Add Stratum | Stratum 1 | Delete Stratum

Calculate | Clear | Settings — Conf. level="95"%

Open Epi 2 x 2 Table

		no improvement		Totals
		true(+)	false(-)	
treatment	yes (+)	20	80	100
	control (-)	40	60	100
Totals		60	140	200

2 x 2 Table Statistics

Single Table Analysis

		no improvement true(+)	false(-)	
treatment	yes (+)	20	80	100
	control (-)	40	60	100
		60	140	200

Chi Square and Exact Measures of Association

Test	Value	p-value(1-tail)	p-value(2-tail)
Uncorrected chi square	9.524	0.001014	0.002028
Yates corrected chi square	8.595	0.001685	0.003370
Mantel-Haenszel chi square	9.476	0.001041	0.002082
Fisher exact		0.001596(P)	0.003192
Mid-P exact		0.001073(P)	0.002145

All expected values (row total*column total/grand total) are >=5
OK to use chi square.

Risk-Based* Estimates and 95% Confidence Intervals
(Not valid for Case-Control studies)

Point Estimates Type	Value	Confidence Limits Lower, Upper	Type
Risk in Exposed	20%	13.26, 28.96	Taylor series
Risk in Unexposed	40%	30.93, 49.81	Taylor series
Overall Risk	30%	24.06, 36.69	Taylor series
Risk Ratio	0.5	0.3158, 0.7917¹	Taylor series
Risk Difference	-20%	-32.4, -7.605°	Taylor series
Prevented fraction in pop.(pfp)	25%	10.71, 35.35	
Prevented fraction in exposed(pfe)	50%	20.83, 68.42	

Odds-Based Estimates and Confidence Limits

Point Estimates Type	Value	Confidence Limits Lower, Upper	Type
CMLE Odds Ratio*	0.3769	0.1972, 0.7065¹	Mid-P Exact
		0.1882, 0.7366¹	Fisher Exact
Odds Ratio	0.375	0.1992, 0.7059¹	Taylor series
Prevented fraction in pop(PFpOR)	35.71%	12.9, 49.06	
Prevented fraction in exposed(PFeOR)	62.5%	29.41, 80.08	

*Conditional maximum likelihood estimate of Odds Ratio
(P)indicates a one-tail P-value for Protective or negative association; otherwise one-tailed exact P-values are for a positive association.
Martin,D; Austin,H (1991) An efficient program for computing conditional
...

> **Tip**
> The Risk difference (RD) in the OpenEpi output, provided in percentage format, can be used to get the ARD value simply by ignoring the plus or minus sign and converting it to a decimal value i.e. 20% = .2 etc.

The number needed to treat/harm (NNT/NNH) is derived from the ARD (Absolute Risk Difference). We just take the ARD and calculate its reciprocal (that is calculate its value over 1) = 1/ARD.

The table shown on the left is the example from MSME3 page 44 describing the effect of giving an oral antifungal (exposure group) had upon the incidence of vaginal candida (event=disease=risk).

Here I have treated the risk/event as the failure to improve. Interestingly we would get the same results in the calculations below if we have used the *improved* category instead.

We can use the formula in SAS1, page 62 to check the output:

Risk Difference RD. OpenEpi

RD = treatment event proportion (p_{treat}) - control event proportion ($p_{control}$)

RD = 20/100 - 40/100 = -.2 = -20%

This is a useful measure and tells us that the risk is reduced by 20% in the treatment group. The value is also useful indicating how to interpret the NNT calculated below. A negative value indicates that the risk has been reduced (☺) in the exposed (treatment) group whereas a positive value means that the risk has increased (☹).

Absolute Risk Difference ARD. Only reported in SAS1.

ARD = |control event proportion - treatment event proportion|

ARD = |40/100 - 20/100| = .2 = 20%

> '|' means absolute value. That is the non-negative value so |-.2|=.2= |.2|

Relative Risk Reduction RRR = Prevented fraction in exposed (pfe) in OpenEpi

RRR = (control event proportion - treatment event proportion)/control event proportion

RRR = ARD/$p_{control}$ = .2/.4 = .5 = 50%

> × 100 to get percentage

> tells us the percentage that did show improvement in the treatment group beyond that in the control group

Both NNT and NNH are calculated by:

1/ARD = 1/.2 = 5 or equivalently 100/20 = 5

Because the RD is negative, indicating that the risk in the control group is greater than that in the intervention/exposed group, the above value is called the NNT. A NNT of 5 indicates that 5 patients would need to be treated for one to benefit.

The ideal value for NNT is 1 or just above, but this is not always the case as in mass screening / vaccination contexts where we might be happy with values in the thousands. That is where the cost of the treatment might be trivial but the result of exposure lethal. This is the argument for low dose Aspirin as a preventative.

29.1 Confidence interval for NNT

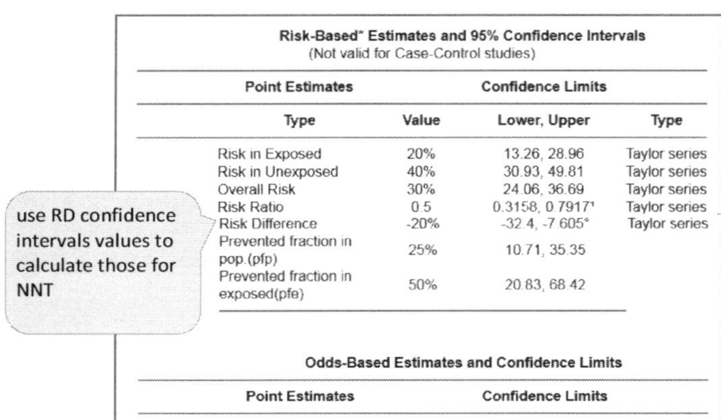

Risk-Based* Estimates and 95% Confidence Intervals
(Not valid for Case-Control studies)

Point Estimates		Confidence Limits	
Type	Value	Lower, Upper	Type
Risk in Exposed	20%	13.26, 28.96	Taylor series
Risk in Unexposed	40%	30.93, 49.81	Taylor series
Overall Risk	30%	24.06, 36.69	Taylor series
Risk Ratio	0.5	0.3158, 0.7917'	Taylor series
Risk Difference	-20%	-32.4, -7.605°	Taylor series
Prevented fraction in pop.(pfp)	25%	10.71, 35.35	
Prevented fraction in exposed(pfe)	50%	20.83, 68.42	

Odds-Based Estimates and Confidence Limits

Point Estimates		Confidence Limits	
Type	Value	Lower, Upper	Type
CMLE Odds Ratio*	0.3769	0.1972, 0.7065'	Mid-P Exact
		0.1882, 0.7366'	Fisher Exact
Odds Ratio	0.375	0.1992, 0.7059'	Taylor series
Prevented fraction in pop(PFpOR)	35.71%	12.9, 49.06	
Prevented fraction in exposed(PFeOR)	62.5%	29.41, 80.08	

*Conditional maximum likelihood estimate of Odds Ratio
(P)indicates a one-tail P-value for Protective or negative association; otherwise one-tailed exact P-values are for a positive association.
Martin,D; Austin,H (1991) An efficient program for computing conditional maximum likelihood estimates and exact confidence limits for a common odds ratio. Epidemiology 2, 359-362.
° ' 95% confidence limits testing exclusion of 0 or 1, as indicated
P-values < 0.05 and confidence limits excluding null values (0,1, or [n]) are highlighted.
LookFirst items: Editor's choice of items to examine first.

use RD confidence intervals values to calculate those for NNT

The output from OpenEPi produces a confidence interval for ARD (labelled Risk Difference in the OpenEpi output) and given the correct circumstances, can be used to form a confidence interval for the NNT estimate.

What I mean by the right circumstances is that the Risk difference is statistically significant, shaded opposite and coloured blue in the OpenEpi output. Given that our Risk difference is shaded opposite, indicating statistical significance we will develop a confidence interval for this NNT.

Observing the confidence intervals for the Risk difference in the OpenEpi output and ignoring the minus sign results in 32.4 to 7.605 finally converting the percentages to decimals and taking their reciprocals:

1/.324 = 3.08 and 1/.07605= 13.14

Remembering from the previous page the NNT = 5 our confidence interval is 3 to 13

Showing these values along a line we have:

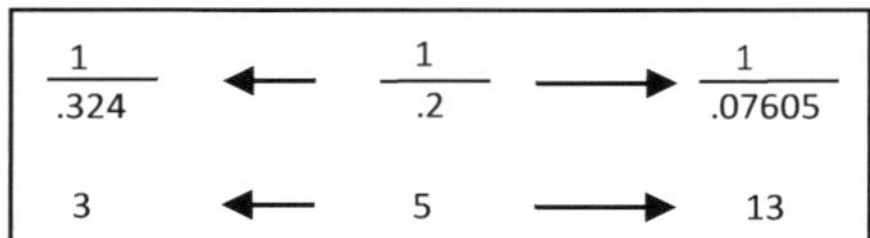

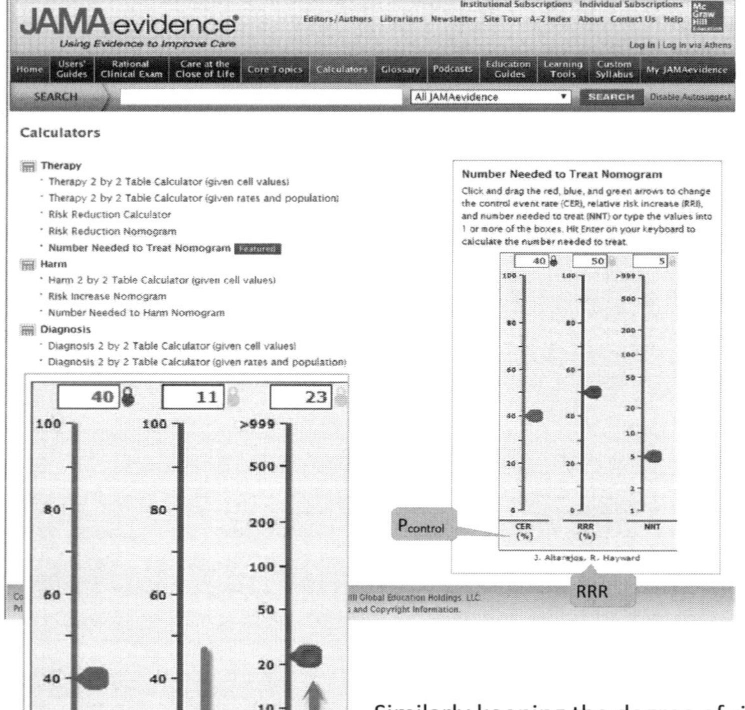

The NNT is related to the; CER=Control group Event Rate (prevalence here of *non-improvement*), and the RRR (Relative Risk Reduction). When one goes up the other must go down for the NNT to stay the same. This aspect is nicely demonstrated using the *Number needed to Treat* Nomogram available on the JAMAevidence website shown opposite. Available from:

http://jamaevidence.com/calculators

Similarly keeping the degree of risk in the control group ($P_{control}$ = CER) constant but reducing the RRR (the effectiveness of the treatment) the NNT increases, as shown opposite.

29.2 Number needed to Harm (NNH)

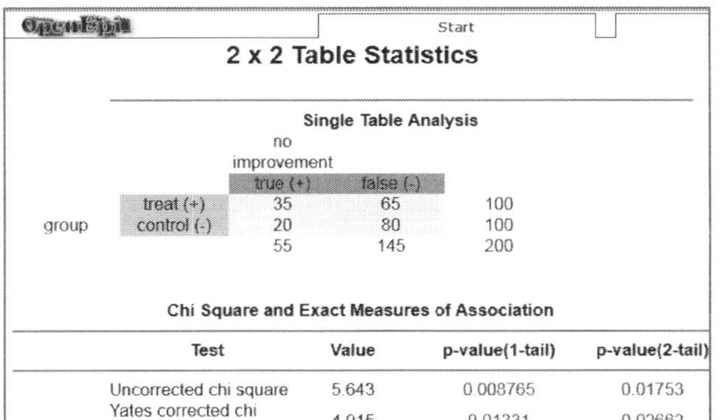

OpenEpi

Start

2 x 2 Table Statistics

Single Table Analysis

		no improvement		
		true (+)	false (-)	
group	treat (+)	35	65	100
	control (-)	20	80	100
		55	145	200

Chi Square and Exact Measures of Association

Test	Value	p-value(1-tail)	p-value(2-tail)
Uncorrected chi square	5.643	0.008765	0.01753
Yates corrected chi square	4.915	0.01331	0.02662
Mantel-Haenszel chi square	5.614	0.008907	0.01781
Fisher exact		0.01306	0.02612
Mid-P exact		0.009257	0.01851

All expected values (row total*column total/grand total) are >=5
OK to use chi square.

Risk-Based* Estimates and 95% Confidence Intervals
(Not valid for Case-Control studies)

Point Estimates		Confidence Limits	
Type	**Value**	**Lower, Upper**	**Type**
Risk in Exposed	35%	26.35, 44.76	Taylor series
Risk in Unexposed	20%	13.26, 28.96	Taylor series
Overall Risk	27.5%	21.77, 34.08	Taylor series
Risk Ratio	1.75	1.089, 2.812[1]	Taylor series
Risk Difference	15%	2.8, 27.2[0]	Taylor series
Etiologic fraction in pop.(EFp)	27.27%	5.15, 49.4	
Etiologic fraction in exposed(EFe)	42.86%	8.177, 64.44	

use RD to calculate NNT

Odds-Based Estimates and Confidence Limits

Point Estimates		Confidence Limits	
Type	**Value**	**Lower, Upper**	**Type**
CMLE Odds Ratio*	2.146	1.135, 4.125[1]	Mid-P Exact
		1.087, 4.325[1]	Fisher Exact
Odds Ratio	2.154	1.136, 4.082[1]	Taylor series
Etiologic fraction in pop.(EFp|OR)	34.09%	9.103, 59.08	
Etiologic fraction in exposed(EFe|OR)	53.57%	12, 75.51	

*Conditional maximum likelihood estimate of Odds Ratio
(P)indicates a one-tail P-value for Protective or negative association, otherwise one-tailed exact P-values are for a positive association.
Martin,D; Austin,H (1991) An efficient program for computing conditional maximum likelihood estimates and exact confidence limits for a common odds ratio. Epidemiology 2, 359-362.
[0] [1] 95% confidence limits testing exclusion of 0 or 1, as indicated
P-values < 0.05 and confidence limits excluding null values (0,1, or [n]) are highlighted.
LookFirst items: Editor's choice of items to examine first.

Results from OpenEpi, Version 2, open source calculator--TwobyTwo

http://www.openepi.com/OE2.3/TwobyTwo/TwobyTwo.htm
Source file last modified on 09/21/2010 03:10:42

In the above example we had negative Risk difference (☺) however we can have a situation where the risk increases in the exposed (i.e. treatment) group (☹). When this occurs the same formula (1/ARD) produces a value which we call the Number needed to Harm ((NNH).

Adjusting the values in the previous example so that we have a risk in the treatment group of .35 and only .20 in the control group we have:

Risk Difference RD. OpenEpi
= treatment event proportion - control event proportion
= 35/100 - 20/100 = -.15 = +15%
This positive value (☹) indicates that the risk is 15% greater in the exposed (treatment) group compared to the control (unexposed) group.

Absolute Risk Difference ARD. Only reported in **SAS1**
= |control event proportion - treatment event proportion|
= |20/100 - 35/100| = .15 = 15%

And therefore 1/ARD = 1/.15 = 6.66

Because the RD is positive this value is a NNH - that is the risk in the control group is greater than in the intervention/exposed group. Taking this to the nearest integer, 7, we can say that for every 7 patients treated 1 additional patient fails to show any improvement compared to those receiving the control treatment.

29.3 Confidence interval for NNH

We can use the same approach as described for developing the NNT confidence interval, with the same proviso that we have a statistically significant Risk difference, which we do in the output opposite (note the shaded text).

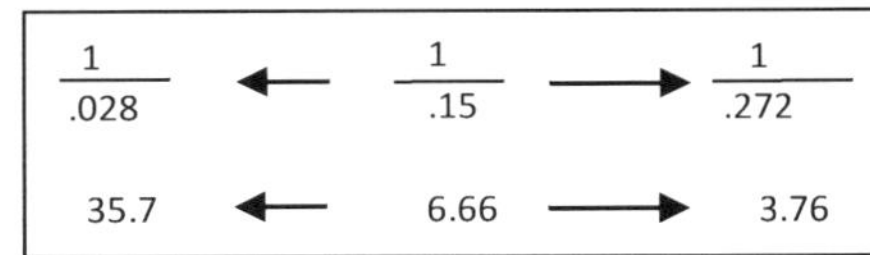

29.4 Tips and Tricks

In the above examples I used the Risk Difference (RD) value to decide if the value from the 1/ARD formula was a NNT or a NNTH. However, I could just as easily have used the Risk Ratio value where a value of less than one indicates a NNT value and a value of more than one indicates a NNTH value from our equation.

If you want to create a Confidence interval for a NNT estimate which is not statistically significant it requires manipulating the data further. See Altman, 1998 for details.

30 Sensitivity, specificity, predictive values and likelihood ratios

Sign =
healthcare worker observation

Symptom =
something reported by a patient not necessarily important!

Test =
a action performed that produces a result

In most areas of healthcare making or excluding a particular diagnosis is an essential activity. When a test, sign or symptom is indicative of a specific disease, it is said to be a *pathognomonic* finding. Two such examples are a particular rash called Erythema chronicum migrans found in Lyme disease, and Levine's sign (hand clutching of chest) in Angina pectoris. Unfortunately, the absence of a pathognomonic sign does not necessarily mean the absence of the specific disease. Where the absence of a particular finding does exclude a particular disease it is known as a *sine qua non*. An example of a sine qua non is a vaginal pH of less than 4.5 which excludes bacterial vaginosis.

Unfortunately, pathognomonic findings and sine qua non are not universal. Many signs, symptoms and test results only provide a certain level of reassurance when making or excluding a diagnosis. To quantify this uncertainty specific values such as; Sensitivity (True Positive Rate, TPR), Specificity (True Negative Rate, TNR) and Likelihood ratios, along with several other measures, have been developed. This chapter investigates these.

30.1 Sensitivity (TPR), Specificity (TNR) & Predicted values

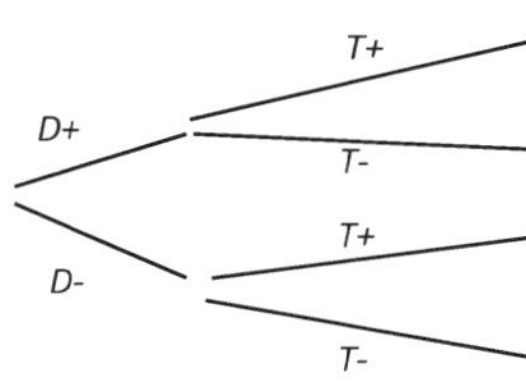

The terms sensitivity, specificity and predictive value, are all ratios of a test result to a particular diagnosis. Many people find these terms confusing and while both MSME3 and SAS1 make excellent jobs of explaining the values using tables and narrative descriptions, I find it easier to understand them by way of a tree diagram. I will discuss the example described in MSME3 page 64 in this format below.

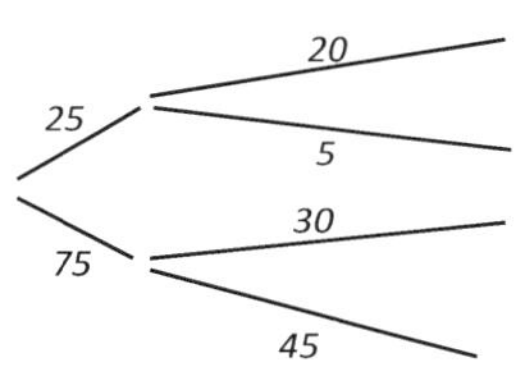

		Gastric Cancer (*D*)		
		Present	Absent	total
Blood test (*T*)	Positive	20	30	50
	Negative	5	45	50
total		25	75	100

The example opposite shows the blood results from 100 patients who were tested for Gastric cancer along with actual findings based on biopsy.

We can represent this as a tree diagram, where D= disease and T = test with the +/- indicating the result.

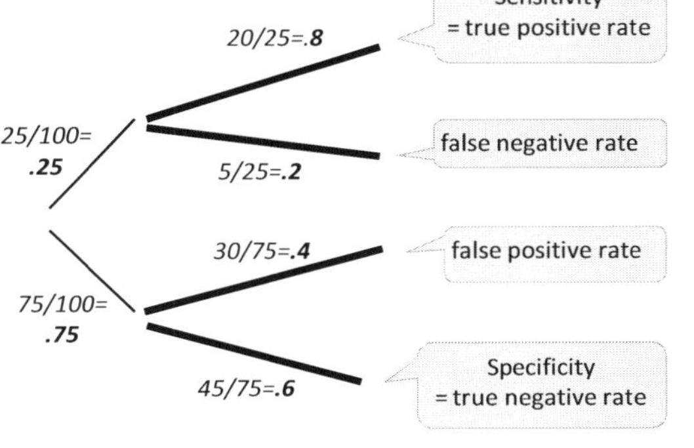

We can easily work out the probabilities at each stage by dividing the number on the branch by those at that particular stage, this also conveniently provides the sensitivity and specificity values:

- True Positive Rate = Sensitivity = true +/all + = 0.8
- True Negative Rate = Specificity = true -/all - = 0.6
- False Positive Rate = false +/all+ = 0.4
- False Negative Rate = false -/all - = 0.2

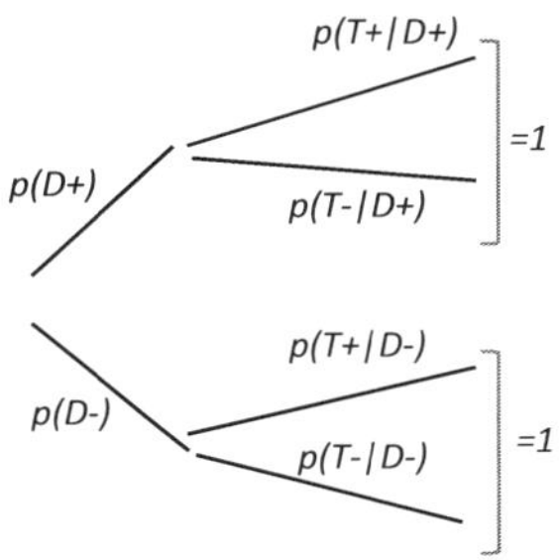

The tree diagram on the left shows the associated probabilities for each branch, that is p(*T+/D+*) indicates the probability of a positive test result *given that* the person has gastric cancer. Note that the sub-branches add up to one. In practical terms, it would be more informative to know the probability of someone having gastric cancer given that the person has a positive test result. Such a probability will now be calculated.

The Positive Predictive Value (PPV) is defined as being the probability of the patient having the condition (*D+)* GIVEN THAT they have a positive test (*T+*). This can be expressed as the probability p(*D+*/*T+*). This is the thing both you, and the patient usually wants to know.

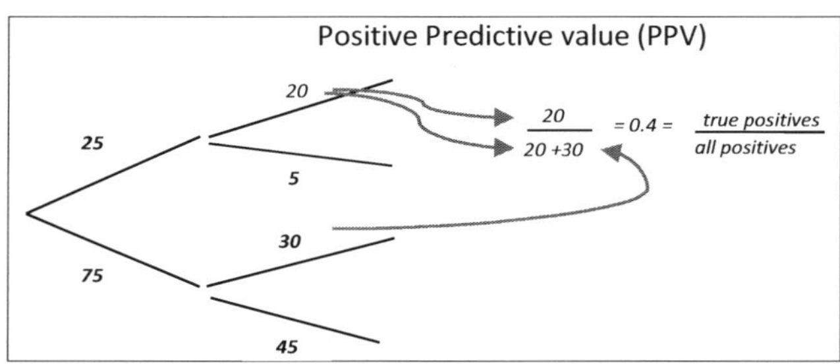

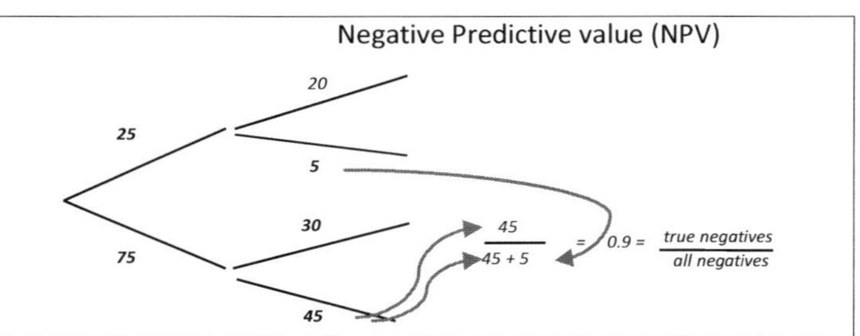

note that p(D+/T+) =0.4 does not equal p(T+/D+)=0.8

conditional probabilities are not like ordinary numbers here a×b ≠ b × a

We can take the same approach to finding the Negative Predictive Value, also shown above.

The Negative Predictive Value (NPV) is defined as being the probability of the patient NOT having the condition (*D-)* GIVEN THAT they have a negative test (*T-*). This can be expressed as the probability p(*D-*/*T-*).

From the above we can say that patients have a 40% chance of having gastric cancer if they have a positive test. However, if they have a negative test they have a 90% chance of not having cancer. While the above values provide useful information, something that is tailored more to the individual is even more useful; this is where likelihood ratios help.

30.2 Likelihood ratios

relationship between odds and probabilities.

see the glossary entry **Odds** for further details.

A likelihood ratio is one probability divided by another. In this situation we can consider two possible likelihood ratios; that is for a positive (LR+) or that for a negative test (LR-) result. Taking the positive test first, here we have the ratio of having the disease to not having it where the positive likelihood ratio (LR+) indicates how much the **odds** of the condition changes when the test is positive (this is not the same as the probability). It acts as a modifier permitting for more individually tailored care.

```
> probs <- seq(0,1, 0.05)
> odds <- round(probs/(1-probs) , 3)
> table <- data.frame(probs, odds)
> table
   probs   odds
1   0.00  0.000
2   0.05  0.053
3   0.10  0.111
4   0.15  0.176
5   0.20  0.250
6   0.25  0.333
7   0.30  0.429
8   0.35  0.538
9   0.40  0.667
10  0.45  0.818
11  0.50  1.000
12  0.55  1.222
13  0.60  1.500
14  0.65  1.857
15  0.70  2.333
16  0.75  3.000
17  0.80  4.000
18  0.85  5.667
19  0.90  9.000
20  0.95 19.000
21  1.00    Inf
```

probability odds

```
Positive likelihood ratio= probability of positive test given disease /
                           probability of a positive test in those with no
disease
= True Positive Rate / False Positive Rate  = Sensitivity / (1 - Specificity)
=  0.8 / 0.4 = 2 = odds of having gastric cancer have doubled
```

this is not the same as saying that the probability has doubled

For the situation with a negative result the specificity and sensitivity values are swapped around so the LR- likelihood ratio now becomes:

```
= (1 - Sensitivity) / Specificity
= False Negative Rate / True Negative Rate = 0.2 /0.6  = .33
= odds of NOT having gastric cancer are reduced by a third
```

this is not the same as saying that the probability has reduced by a third

In other words sometimes a positive result and a negative result set the balance to a different degree. It is essential to realise that a trebling of odds is not the same as the trebling of probabilities. For example, in the table opposite, if we consider the baseline probability of 0.25 we can see the comparable odds is 0.333. However, if we treble the odds to 0.999 this equals a probability of only 0.50, similarly taking, a baseline probability of 0.75 (odds = 3) trebled in terms of odds = 9 but has a probability of 0.9.

30.3 Fagan's nomogram: individualised care

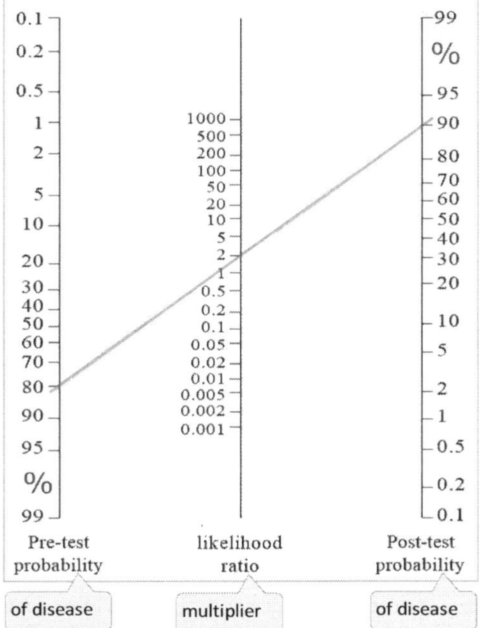

We can use the likelihood ratio along with a pre-test probability of a positive test to develop individualised probabilities of a positive result. The easiest way to do this is by using a diagram proposed originally by a UK GP, Terrence Fagan, in 1975 (Glasziou, 2001). Using Fagan's nomogram the odds are automatically converted to probabilities, this is why a pre-test probability of .8 and likelihood ratio of 2 results in a probability of approximately 0.9 (=2×4=8 odds =probability of .89 from table on previous page). Fagan's original nomogram has been improved upon in several ways and I will demonstrate one such improvement known as the 2 step Fagan nomogram which you can download from: http://www.adelaide.edu.au/vetsci/research/pub_pop/2step-nomogram/

How you obtain the pre-test probabilities and the likelihood ratios for a given test for a specific patient is a thorny issue. The pre-test probability is often either a value taken from large 'population' type studies or expert knowledge by the physician using the nomogram.

Specific likelihood ratios are even more difficult to pin down. There are two basic approaches; calculate the value from known specificity and sensitivity values, or alternatively search the literature/web for such values. The KT (Knowledge Translation) Clearinghouse website is one such resource, funded by the Canadian Institute of Health Research (CIHR) and the University of Toronto, Faculty of Medicine. The following link provides tables of likelihood ratios for several common tests: http://ktclearinghouse.ca/cebm/glossary/lr

Among the likelihood ratios listed on the KT page are three for Dementia; the clock drawing test, the Hopkins verbal learning test and the Mini-Mental status examination (taking 26/27 as cut-off point). The prevalence of dementia in the UK for the 70-79 year age group is 1 in 25 =0.04. (source UK Alzheimer's society http://www.alzheimers.org.uk/).

However, it is unlikely for a physician to consider testing for dementia unless they felt there was a reasonable (?) chance the patient was suffering from it. We will therefore consider the pre-test probability of a positive test result to be 40% (=0.4), that is you believe the particular patient is ten times more likely than the general population of 70-79 year olds to be suffering from dementia. We can then calculate the post-test probability of a positive test indicating the presence of dementia, as shown opposite.

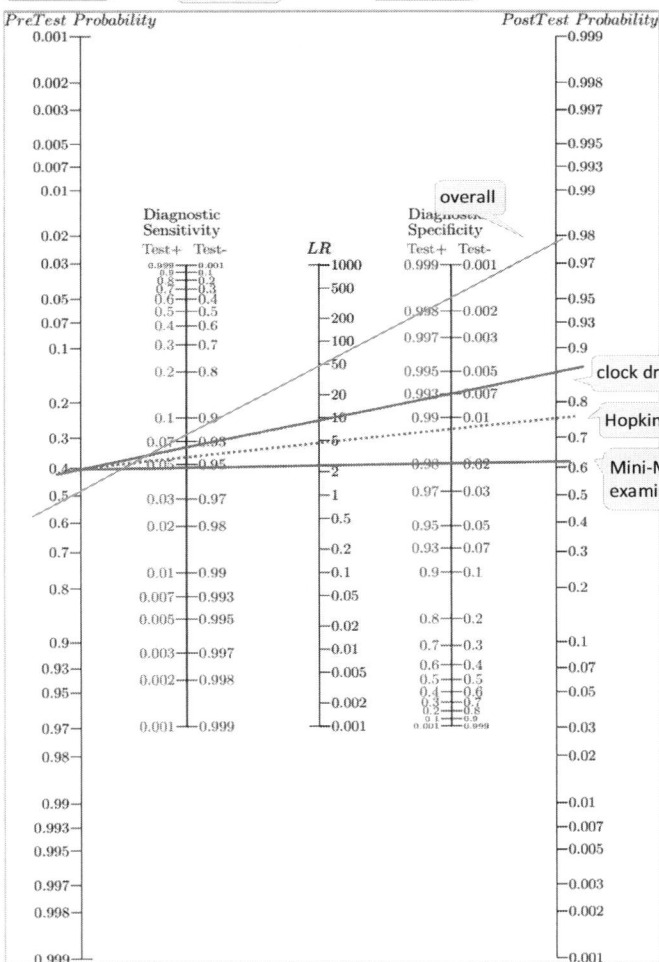

I have used the 2 step Fagan nomogram opposite to show the results. The lines have been drawn from the pre-test probability to the likelihood ratio (LR) scale and then extended to the post-test probability scale. Ignore the other scales for now.

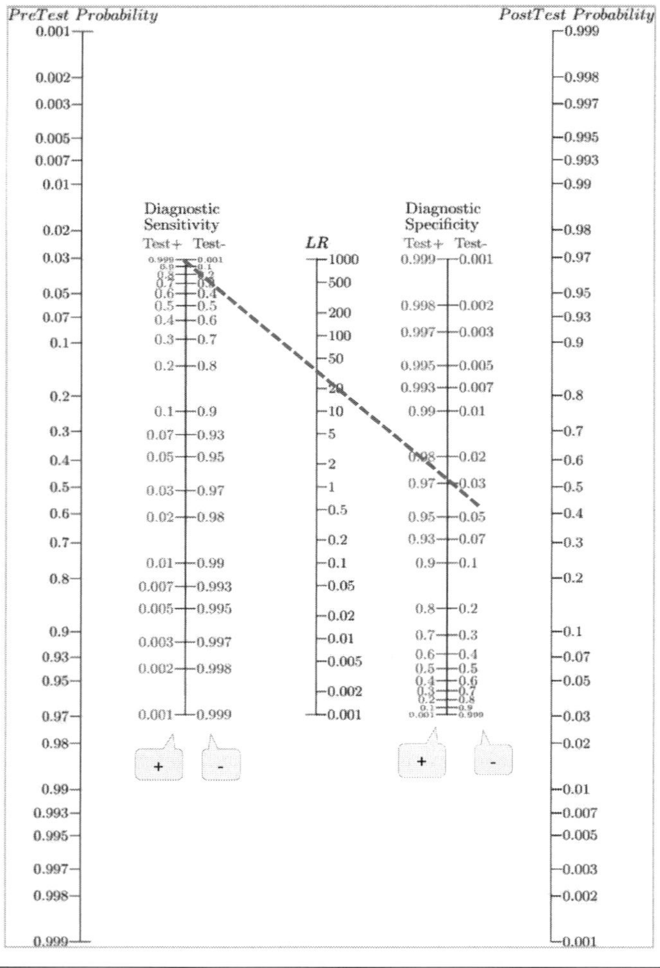

There is also the possibility of considering the three tests together where *pre-test odds × LR1 × LR2 × LR3 = post-test odds* so for our test battery we have for the overall likelihood ratio, 9.6 ×4.8×2.5= 115.2. This assumes that the tests do not influence each other, that is they are independent. This might not be the case in an elderly patient if they were subjected to the tests sequentially without adequate rest periods between them!

If you do not have a likelihood ratio (LR) for a specific test it might be possible to obtain the specificity and sensitivity values from which the LR can be calculated, for which you can use the 2 step Fagan nomogram. For a LR+ use the left hand sensitivity/specificity scales and for the LR- calculation use the right hand sensitivity/specificity scales (marked opposite). For example, the most optimistic reports for mammograms give a specificity of 0.97 and sensitivity of 0.95, this gives a LR+ of around 30 in the diagram opposite (values taken from the online family practice notebook at http://www.fpnotebook.com/Prevent/Epi/LklhdRt.htm). Alternatively you could have used the LR+ equation given in the previous section *=sensitivity/(1-specificity) = .95/(1-.97)=.95/.03 = 31*

The LR+ value can be used as an indicator of a prospective test's usefulness and can be used in conjunction with a minimum pre or post-test probability value to set a treatment threshold (TT).

Interpreting LR+

- LR+ over 5 indicates the test will be useful in verifying the disease, where greater the value the more use it will be.
- LR+ near 1 (i.e between 5 to 0.2) indicates the test result will not add anything. A waste of time unless you just want to confirm what you know!
- LR+ below 0.2: indicates the test will be useful in ruling out the disease, where smaller the value the more use it will be.

So far we have only considered tests which possess a binary scale (+/1) however the same approach can be applied to test results that follow an interval/ratio **level of measurement**. The webpages for the online Evidence-Based Medicine Course (OMERAD) for the MSU Primary Care Faculty, at Michigan State University written by Mark Ebell, and Henry Barry provides a good example, http://omerad.msu.edu/ebm/Diagnosis/Diagnosis8.html, taken from Guyatt, Oxman, & Ali et al. 1992.

Guyatt, Oxman, & Ali et al. 1992 considered the diagnosis of iron deficiency anaemia (IDA) from the serum ferritin level. Whereas laboratories generally report a single cutoff of around 65 mmol/l, lower values more strongly suggest a diagnosis of iron deficiency anaemia. Using all values below 65 mmol/l as a "positive" test, the LR+ is 6 = (59+22+10)/(1.1+4.5+10) =91/15.6 and the LR- is 0.12 = (1-0.91)/(1-.156)=0.09/0.844. However considering the various bands produces a more graduated level of evidence for IDA.

Serum ferritin (mmol/l)	Number with IDA (% of total=TPR)	Number without IDA (% of total=FPR)	LR	Comment IDA=iron deficiency anaemia
Original paper Guyatt, Oxman, & Ali et al. 1992				
< 15	474 (59%)	20 (1.1%)	52 (=59/1.1)	Strong evidence for IDA
15-34	175 (22%)	79 (4.5%)	4.8 (=22/4.5)	Moderate evidence for IDA
35-64	82 (10%)	171 (10%)	1 (=10/10)	No evidence either way
65-94	30 (3.7%)	168 (9.5%)	0.39 (=3.7/9.5)	Weak evidence against IDA
> 94	48 (5.9%)	1332 (75%)	0.08 (=5.9/75)	Strong evidence against IDA

The above interpretation of a diagnostic test allows for individually tailored care, but unfortunately, there appear few published likelihood ratios and I have never seen Fagan's nomogram on a GP's wall or computer yet!

In the academic domain when reporting the above likelihood ratios in a paper it is necessary to also provide confidence intervals which are discussed below.

30.4 Confidence intervals

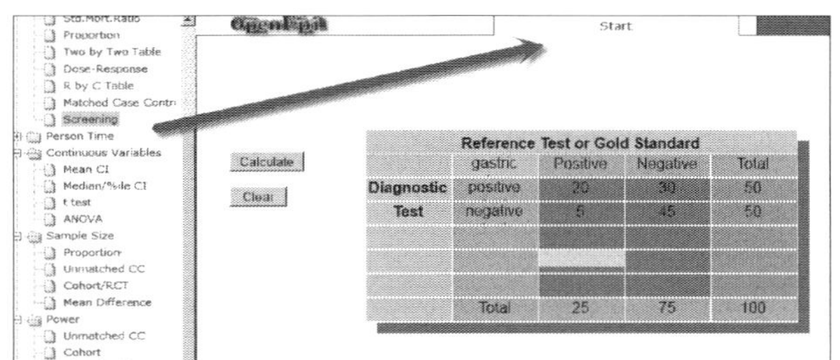

To demonstrate the various confidence intervals we will now return back to the 100 patients who were tested for gastric cancer along with actual findings based on biopsy discussed at the start of the chapter.

The easiest way to obtain confidence intervals for the above estimates is to use the *OpenEpi* online application, found at:

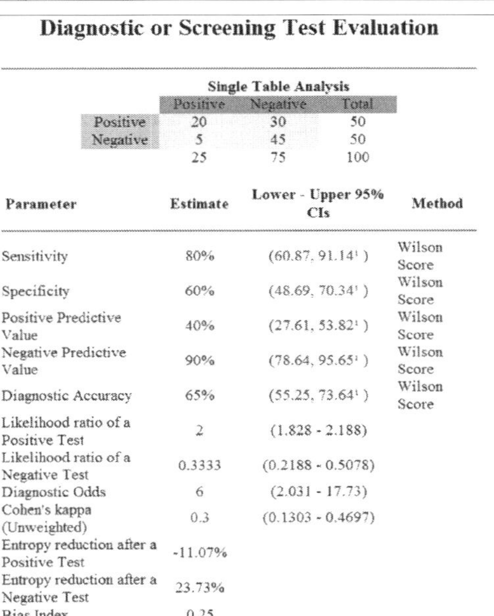

Diagnostic or Screening Test Evaluation

Single Table Analysis

	Positive	Negative	Total
Positive	20	30	50
Negative	5	45	50
	25	75	100

Parameter	Estimate	Lower - Upper 95% CIs	Method
Sensitivity	80%	(60.87, 91.14[1])	Wilson Score
Specificity	60%	(48.69, 70.34[1])	Wilson Score
Positive Predictive Value	40%	(27.61, 53.82[1])	Wilson Score
Negative Predictive Value	90%	(78.64, 95.65[1])	Wilson Score
Diagnostic Accuracy	65%	(55.25, 73.64[1])	Wilson Score
Likelihood ratio of a Positive Test	2	(1.828 - 2.188)	
Likelihood ratio of a Negative Test	0.3333	(0.2188 - 0.5078)	
Diagnostic Odds	6	(2.031 - 17.73)	
Cohen's kappa (Unweighted)	0.3	(0.1303 - 0.4697)	
Entropy reduction after a Positive Test	-11.07%		
Entropy reduction after a Negative Test	23.73%		
Bias Index	0.25		

http://www.openepi.com/OE2.3/Menu/OpenEpiMenu.htm

Select the screening option on the left of the screen. You will then see a number of tabs across the top of the screen, including Start and Enter.

Select the **Enter** tab, and enter the data, you can also change the names of the columns/rows to make the output more understandable. Enter the data and edit the row column names, as shown above. Click on the **Calculate** button to produce the results as shown opposite. We see that the $LR+$ 95% CI value is relatively narrow (1.8 to 2.2). Referring back to the *Interpreting LR+* table on the previous page interpreting the LK+ values this result indicates that the test adds nothing to the diagnostic process.

An alternative online resource can be found at:

http://www.medcalc.org/calc/diagnostic_test.php

30.5 Prevalence

The prevalence of a disease is the proportion of people diagnosed by the gold standard (**SAS1** p.103). In our example, it is 25/100 = 25% representing the proportion of those diagnosed by biopsy. Clearly this value does not represent an accurate population value and Campbell (SAS1 p.105) highlights the importance of differentiating between sample and population values.

- Sensitivity, Specificity, Likelihood ratios = Sample values which are not dependent upon prevalence values.
- PPV, NPV = population values, dependent upon prevalence values. Taking into account fully the population prevalence values requires more complex calculations than discussed in this chapter, and beyond the books remit. For further details see the R statistics *bdpv* library website.

30.6 Tips and Tricks

```
>
> fagan.plot(0.8, 2, "+")
>
```

The *TeachingDemos* R package contains the *Fagan.plot(probs.pre.test, LR, test.result="+")* function. Specifying "+" provides a LK+ value and "-" the LK- value. The equivalent to the first Fagan nomogram presented in this chapter is therefore:

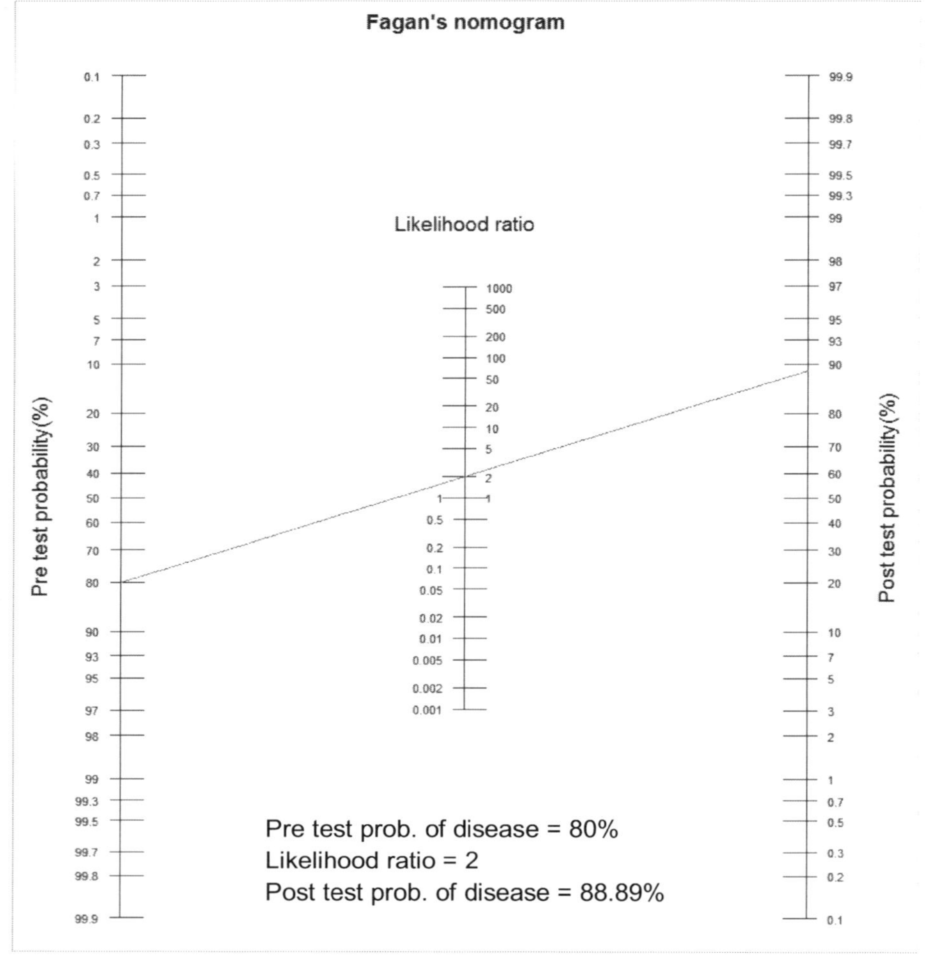


```
fagan.plot(0.8, 2, "+" )
```

Most of the techniques discussed in this chapter have been around a number of years along with many more sophisticated approaches such as Clinical Decision Support Systems (CDSS). However, these have not been taken up by the healthcare profession to a significant degree despite much research effort having gone into understanding the detailed reasons why this is the case.

In this chapter I provided one extension to Fagan's nomogram and another is provided by Marasco, Doerfler & Roschier, 2011, who also provide an additional nomogram for rare diseases.

You might be interested to know how p(*D+*/*T+*) can be obtained from the tree diagram.

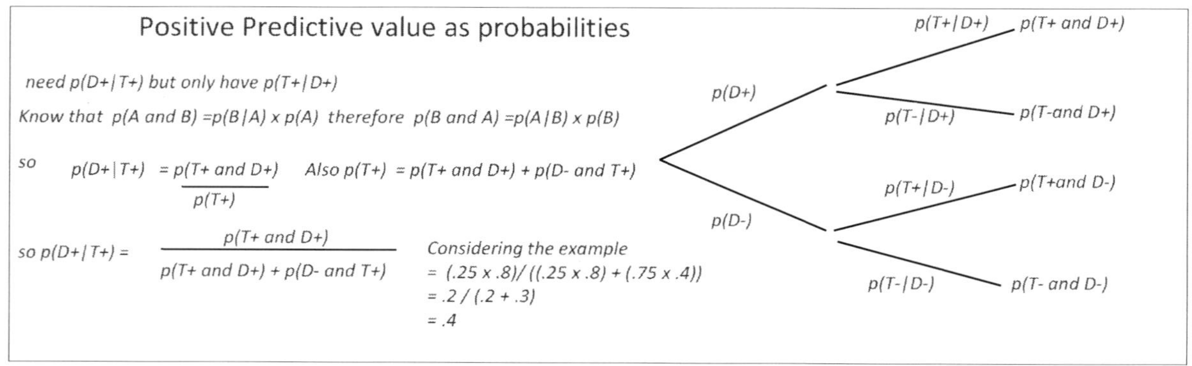

Positive Predictive value as probabilities
need p(D+|T+) but only have p(T+|D+)
Know that p(A and B) =p(B|A) x p(A) therefore p(B and A) =p(A|B) x p(B)
so p(D+|T+) = p(T+ and D+) / p(T+)
Also p(T+) = p(T+ and D+) + p(D- and T+)
so p(D+|T+) = p(T+ and D+) / p(T+ and D+) + p(D- and T+)
Considering the example
= (.25 x .8)/((.25 x .8) + (.75 x .4))
= .2 / (.2 + .3)
= .4
p(D+)
p(D-)
p(T+|D+)
p(T-|D+)
p(T+|D-)
p(T-|D-)
p(T+ and D+)
p(T-and D+)
p(T+and D-)
p(T- and D-)

31 Levels of agreement: Kappa, Krippendorff and the ICC

Rater
Is a person who estimates a rating. Rating scales can be of ordinal, interval or ratio **level of measurement.**

A common clinical rating scale is the question, how painful is this on a scale of 1 to 10?

Measuring the level of agreement between two raters at first appears a simple task, just take the percentage they agree. However, this approach is fraught with problems and the simple question 'how do we measure the level of agreement between two or more raters beyond that due to chance' has produced a large volume of literature with little consensus as to what is the best measure.

In fact a large proportion of the literature focuses on what is wrong, rather than what is right with the current measures. I will ignore most of this and present both the commonly used traditional measures alongside some more recently developed ones.

When talking about any measures of agreement we should always consider the following three important points:

bold words in the text indicate entries in the glossary providing more information

1. Is there a gold standard with near perfect reliability with which we can compare our raters' judgements or are we just comparing them (with each other)?
2. What is the **level of measurement** used for the ratings; nominal, ordinal or interval/ratio?
3. Are there just two raters or more?

Krippendorff (Hayes & Krippendorff, 2007; Krippendorff, 2004) has attempted to develop a measure that takes all these aspects into account, but first we will look at the more traditional measure developed by Cohen (1960) called Kappa (κ).

31.1 Kappa - two raters nominal data

Kappa measures the level of agreement beyond that due to chance, for nominal data for two raters, with two or more categories. A variation called weighted kappa can be used for ordinal **measurement level** data (Krippendorff, 2004).

Kappa and strength of agreement	
value	strength
<0.00	Less than even due to chance
0.00 to 0.20	Slight
0.21 to 0.40	Fair
0.41 to 0.60	Moderate
0.61 to 0.80	Good
0.81 to 1.00	Very good

The kappa value is often interpreted as strength of agreement (table opposite) however this should be treated with a level of scepticism, as both the number of observations and the number of categories can affect the value.

The example opposite is from of a detailed study looking at the particularly dubious practice of applied kinesiology (Lüdtke, Kunz, Seeber & Ring, 2001). This is where practitioners diagnose various allergens and food intolerances by rating muscle strength after the allergen is usually laid on the skin.

Level of agreement between two raters using applied kinesiology to access possible allergens. (translated from the German) Lüdtke, Kunz, Seeber, Ring 2001

Allergen	N	Kappa	95%-Confidence interval
Lactose	23	0.16	-0.21 - 0.52
House dust mite	23	0.11	-0.31 - 0.52
Cat hair	22	0.31	-0.09 - 0.70
Grass pollen	23	-0.02	-0.42 - 0.38
Acetylsalicylic acid	23	-0.16	-0.50 - 0-19
Wasp venom	23	-0.14	-0.54 - 0.27
Bee venom	23	0.47	0.13 - 0.81
Candida albicans	22	0.25	-0.15 - 0.66
Sodium chloride solution	23	0.42	0.01 - 0.82
Histamine	23	0.21	-0.18 - 0.59
total	228	0.17	0.04 - 0.30

In the list of possible allergens opposite Sodium chloride solution was added as a placebo. We can see that overall the Kappa value was very low (0.17). It must be remembered here that it is measuring the level of agreement between the raters not the level of correct agreement. This was also assessed in the original article and also found to be no better than chance.

We will now consider another dataset with which we can calculate Kappa ourselves.

		Judge A		
	Category	1	2	3
Judge B	1	.25	.13	.12
	2	.12	.02	.16
	3	.03	15	.02

		Judge A		
	Category	1	2	3
Judge B	1	88	14	18
	2	10	40	10
	3	2	6	12

We will use the fictitious data from the original article by Cohen, 1960. Cohen presented the data as both proportions and counts (shown opposite) assuming a total of 200 observations across two raters, he also suggested that one could think of the categories as being those with schizophrenia, neuroticism and brain damage, ignoring the questionable validity of these groupings.

I have converted these original counts into a set of raw data which you can download from my website using the following commands in the R Console.

```
cohen_1960data <- read.delim("http://www.robin-beaumont.co.uk/virtualclassroom/book2data/cohen_1960.dat", header=TRUE)

cohen_1960data
```

```
> cohen_1960data <- read.delim("http://www.robin-beaumont.co.uk/virtualclassroom/book2data/cohen_1960.dat", header=TRUE)
> cohen_1960data
        judge.A      judge.B
1  schizophrenic schizophrenic
2  schizophrenic schizophrenic
3  schizophrenic schizophrenic
4  schizophrenic schizophrenic
5  schizophrenic schizophrenic
```

From the above you can see there are two columns labelled judge.A and judge.B, scrolling down the window shows 200 rows representing the 200 patients.

31.1.1 Using the psych and irr packages to find Kappa

You can use the *irr* package to obtain Cohen's Kappa where the two commands below typed into the R Console install and load the necessary *irr* R package.

```
> kappa2(cohen_1960data)
 Cohen's Kappa for 2 Raters (Weights: unweighted)

 Subjects = 200
   Raters = 2
    Kappa = 0.492

        z = 9.46
  p-value = 0
```

```
install.packages("irr", dependencies=TRUE)

library(irr)
```

Type the following command to obtain Kappa.

```
kappa2(cohen_1960data)
```

The result, shown above, provides the Kappa value along with the number of observations and raters.

```
R Console
> library(psych)
> cohen.kappa(cohen_1960data)
Call: cohen.kappa1(x = x, w = w, n.obs = n.obs, alpha = alpha)

Cohen Kappa and Weighted Kappa correlation coefficients and confidence boundaries
                 lower estimate upper
unweighted kappa  0.39     0.49  0.59
weighted kappa    0.32     0.45  0.58

 Number of subjects = 200
```

The *psych* package provides a comparable function called *cohen.kappa()*, demonstrated opposite, which has the advantage of giving both the unweighted and weighted version of kappa (explained below) along with a confidence interval.

31.1.2 Setting the data up for the analysis

The above method made it appear relatively easy to obtain Kappa because both variables had been defined as factors. This was because in the actual text file the values were in quote marks (" or ') therefore R assumed the values were **factor** levels rather than numbers. Also the data was in a structure where the columns represented the results from the different raters and the rows represented the rating. Finally, these were the only columns in the dataframe.

Assuming that our dataframe did possess a third column, to use only the second and third columns for the *kappa2()* function you would write the following:

```
kappa2(cohen_1960data[,2:3])
```

Similarly it might be the first and fourth columns you wish to use, in which case you would write:

```
kappa2(cohen_1960data[,c(1,4)])
```

31.2 Krippendorff - the do it all statistic

The *irr* package also contains the *kripp.alpha()* function which calculates Krippendorff's alpha statistic. The *kripp.alpha()* function takes two values, the dataframe we wish to analyse along with a indication of the level of measurement used by the raters.

```
kripp.alpha(x, method=c("nominal","ordinal","interval","ratio"))
```

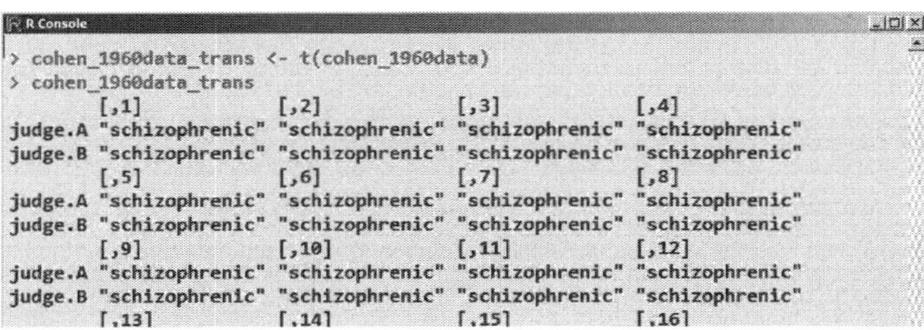

```
> cohen_1960data_trans <- t(cohen_1960data)
> cohen_1960data_trans
        [,1]           [,2]           [,3]           [,4]
judge.A "schizophrenic" "schizophrenic" "schizophrenic" "schizophrenic"
judge.B "schizophrenic" "schizophrenic" "schizophrenic" "schizophrenic"
        [,5]           [,6]           [,7]           [,8]
judge.A "schizophrenic" "schizophrenic" "schizophrenic" "schizophrenic"
judge.B "schizophrenic" "schizophrenic" "schizophrenic" "schizophrenic"
        [,9]           [,10]          [,11]          [,12]
judge.A "schizophrenic" "schizophrenic" "schizophrenic" "schizophrenic"
judge.B "schizophrenic" "schizophrenic" "schizophrenic" "schizophrenic"
        [,13]          [,14]          [,15]          [,16]
```

To use this function we need to swap the columns for the rows as it expects; Rows = raters, column = case .

In R this swapping, technically called a transposition, is achieved by using a single letter *t* instructing R to produce a new dataframe which is the transposed version of the old one. The result is shown opposite.

```
cohen_1960data_trans   <- t(cohen_1960data); cohen_1960data_trans
```

```
> kripp.alpha(cohen_1960data_trans)
 Krippendorff's alpha

 Subjects = 200
   Raters = 2
    alpha = 0.488
Warning message:
In kripp.alpha(cohen_1960data_trans) : NAs introduced by coercion
> |
```

Now we can obtain Krippendorff's alpha for our data frame by typing into the R Console:

```
kripp.alpha(cohen_1960data_trans,
"nominal")
```

The result is shown opposite, whereas we had 0.492 for Kappa we have 0.488 for Krippendorff's alpha, little difference.

31.2.1 Ordinal level of measurement variables

```
> kripp.alpha(cohen_1960data_trans, "ordinal")
 Krippendorff's alpha

 Subjects = 200
   Raters = 2
    alpha = 0.494
Warning message:
In kripp.alpha(cohen_1960data_trans, "ordinal") :
  NAs introduced by coercion
>  kappa2(cohen_1960data, "squared")
 Cohen's Kappa for 2 Raters (Weights: squared)

 Subjects = 200
   Raters = 2
    Kappa = 0.455

        z = 6.75
  p-value = 1.5e-11
```

For ordinal **level of measurement** variables there are two methods; either use the weighted Kappa function *kappa2()*, or Krippendorff's alpha indicating you have the ordinal data:

*kripp.alpha(*cohen_1960data_trans, *"ordinal")*

kappa2(cohen_1960data, "squared")

Again the two values only differ by a small amount.

31.2.2 Interval/ ratio data level of measurement variables

```
R Console
>
>
> data(anxiety)
> icc(anxiety, model="twoway", type="agreement")
 Single Score Intraclass Correlation

   Model: twoway
   Type : agreement

   Subjects = 20
     Raters = 3
   ICC(A,1) = 0.198

 F-Test, H0: r0 = 0 ; H1: r0 > 0
 F(19,39.7) = 1.83 , p = 0.0543

 95%-Confidence Interval for ICC Population Values:
   -0.039 < ICC < 0.494
>
```

```
> anxiety_trans <- t(anxiety)
> kripp.alpha(anxiety_trans, "interval")
Krippendorff's alpha

Subjects = 20
  Raters = 3
   alpha = 0.163
> kripp.alpha(anxiety_trans, "ratio")
Krippendorff's alpha

Subjects = 20
  Raters = 3
   alpha = 0.135
```

In the *irr* package you can either use the intraclass correlation coefficient (ICC) or once again Krippendorff's alpha this time using *kripp.alpha(dataframe,"interval")* or *kripp.alpha(dataframe,"ratio")* remembering to transpose the dataframe first.

The *icc()* **function** in the *irr* package contains many options as there are many varieties of Intraclass correlation coefficient (**ICC**). The example opposite, taken from the *irr* package manual, uses the build-in dataset *anxiety* with the *icc()* function. You can see that besides getting an ICC value (0.198) a confidence interval, *F*-test and p-value is given. The exact meaning of the **p-value** depends upon which variety of *icc()* you have chosen. For further details see the *irr* online manual at http://cran.r-project.org/web/packages/irr/irr.pdf.

The anxiety dataset provides anxiety ratings of 20 subjects, rated by three raters with values ranging from 1 (not anxious at all) to 6 (extremely anxious). Comparing the ICC value opposite with Krippendorff's, remembering to transpose the dataframe first, once again gives little difference in the three estimates.

31.2.3 More than 2 raters?

You may have noticed in the above example that we had three raters and there are various extensions in the *irr* package to cope with this situation as demonstrated above. The online *irr* manual provides further details.

31.3 Tricks and tips

- **Two raters binary variables** If you have only two raters and binary data use openEpi. An example of using this software in provided in the Numbers needed to treat/harm chapter, study the OpenEpi output.

- **GUI front end** The package KappaGUI provides a front end to *irr* but is rather limited only coping with a datafile in *csv* format. Once downloaded and installed, to start the KappaGUI type *StartKappa()* into the R Console.

- **Power calculates and required sample size** The *irr* package provides functions allowing you to calculate the required minimum sample size for various designs, again details are provided in the online *irr* manual from the link given above.

32 Bland - Altman plots

Bland-Altman plots (Bland & Altman, 1999), also called Tukey mean-difference plots, have become very popular for assessing the reliability of two or more supposedly equivalent methods of measuring continuous data, for example different methods of taking temperature or blood pressure.

The MethComp (Functions for analysis of Method Comparison studies) R package provides a number of functions that produces these graphs. In addition, the main developer of the package, Bendix Carstensen, at the Department of Biostatistics, University of Copenhagen, has an excellent set of resources concerned with the package along with teaching medical statistics in general http://www.bendixcarstensen.com/.

32.1 Installing a R package from a local zip file

In contrast to the usual method presented in this book of installing a package, this time we will investigate another method, downloading a zipped version of the library which is then installed locally using R's Console window.

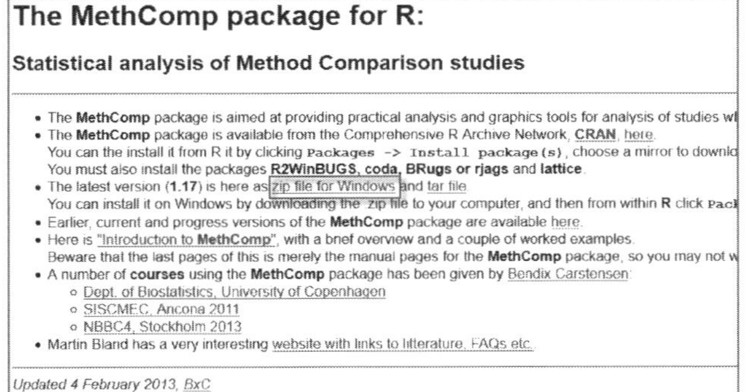

The MethComp package for R:
Statistical analysis of Method Comparison studies
Updated 4 February 2013, BxC

Go to the *MethComp* webpage, shown opposite, at: http://bendixcarstensen.com/MethComp/

Click on the **zip file for Windows** link and save the file locally.

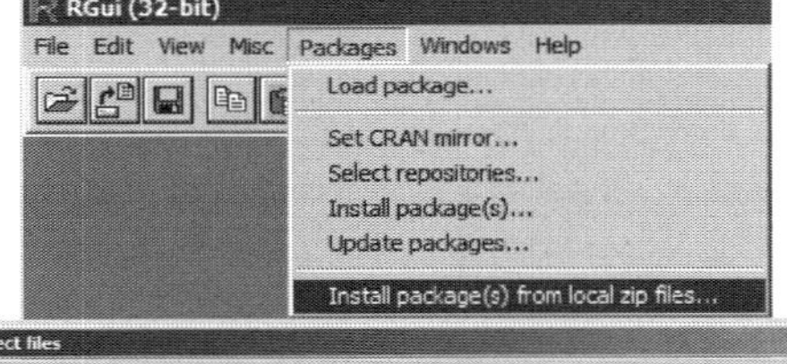

Within R, in the *R Console* window select the menu option

Packages-> install package(s) from local zip files

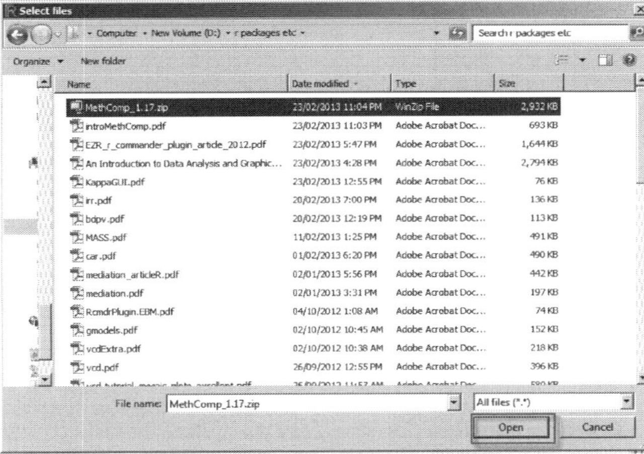

A *Select files* dialog appears, select the file you just downloaded and click on the **Open** button.

```
R Console
> utils:::menuInstallLocal()
package 'MethComp' successfully unpacked and MD5 sums checked
> library(MethComp)
Loading required package: nlme
>
```

A message appears informing you the installation was successful. Remember, to use the package you need to first load it by typing in the command:

```
library(MethComp)   # note the upper case M and C
```

32.2 Creating Bland-Altman plots using the MethComp package

```
> data(ox)
> ox
   meth item repl    y
1    CO    1    1 78.0
2    CO    1    2 76.4
3    CO    1    3 77.2
4    CO    2    1 68.7
5    CO    2    2 67.6
6    CO    2    3 68.3
7    CO    3    1 82.9
8    CO    3    2 80.1
9    CO    3    3 80.7
10   CO    4    1 62.3
11   CO    4    2 65.8
12   CO    4    3 67.5
13   CO    5    1 75.8
14   CO    5    2 73.7
15   CO    5    3 76.3
16   CO    6    1 78.0
17   CO    6    2 78.8
18   CO    6    3 77.3
19   CO    7    1 86.3
20   CO    7    2 84.3
```

```
345 pulse   54    3 77.0
346 pulse   55    1 92.0
347 pulse   55    2 85.0
348 pulse   55    3 88.0
349 pulse   56    1 84.0
350 pulse   56    2 82.0
351 pulse   56    3 84.0
352 pulse   57    1 92.0
353 pulse   57    2 90.0
354 pulse   57    3 82.0
355 pulse   58    1 84.0
356 pulse   58    2 87.0
357 pulse   58    3 89.0
358 pulse   59    1 94.0
359 pulse   59    2 89.0
360 pulse   59    3 91.0
361 pulse   60    1 79.0
362 pulse   60    2 77.0
363 pulse   60    3 80.0
364 pulse   61    1 72.0
365 pulse   61    2 70.0
366 pulse   61    3 68.0
```

The following is very much an adaption of the material in the *MethComp* manual.

The *ox* dataset, which is part of the *MethComp* package, provides details of the measurement of arterial oxygen saturation (SaO_2) in 61 children (Children's Hospital in Melbourne), either using a chemical method by analysing gases in the blood (*CO*) or with a pulse oximeter measuring transcutaneously (*pulse*). The two methods were performed at the same time and each measure was repeated three times.

Normal limits & effects of decreased oxygen saturation	
SaO_2	**Effect**
96 to 99%	Normal level
85% to 90% (hypoxic)	No evidence of impairment
65% and less	Impaired mental function on average
55% and less	Loss of consciousness on average

We can see the data by typing in the R Console window

```
data(ox); ox
```

The first variable *meth* takes either the value *co* or *pulse* indicating the measurement method. *Item* is the child unique identifier. *repl* represents the three measurements that were taken in rapid success using the same method (technically called replications), finally *y* is oxygen saturation (%).

The following command provides a basic Bland-Altman plot:

```
BA.plot( ox )
```

The *y* axis represents the difference between the two measures, so if both were perfectly equivalent all the points would be on the y=0 line. The y=2.4 line represents the mean difference and is the estimated bias. This indicates that the pulse oximeter method consistently provides a slightly higher value by 2.4 units of O_2 saturation.

The two lines at 14 and -9 indicate tolerance levels. They represent 95% confidence limits (average difference ± 1.96 standard deviation of the difference). If the majority of actual differences lie between these, that is +14 to -9, the two methods may be used interchangeably. However, considering the above table of normal limits and the effects of decreased oxygen saturation, a range of 23 units covers normal to hypoxic. The confidence intervals might be significantly reduced if a subsequent analysis considered values only above 70% as the larger discrepancies occur where O_2 saturation is below this value.

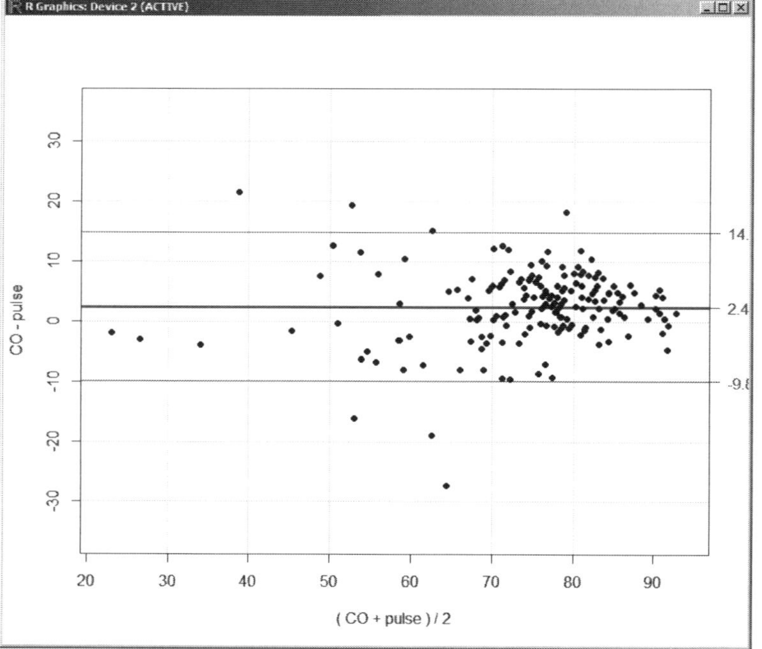

Such a funnel shaped distribution would indicate what is called *proportional bias* suggesting that the bias depends upon the actual value rather than being spread equally over the range. See Bland & Altman, 1999 for details of how you might further analyse such data.

The x axis provides the average value of the two methods for each specific measure. We can see that the majority of the measures are around the 70 to 95% range as would be expected.

In a remarkably small number of lines of code we can produce sophisticated Bland - Altman plots:

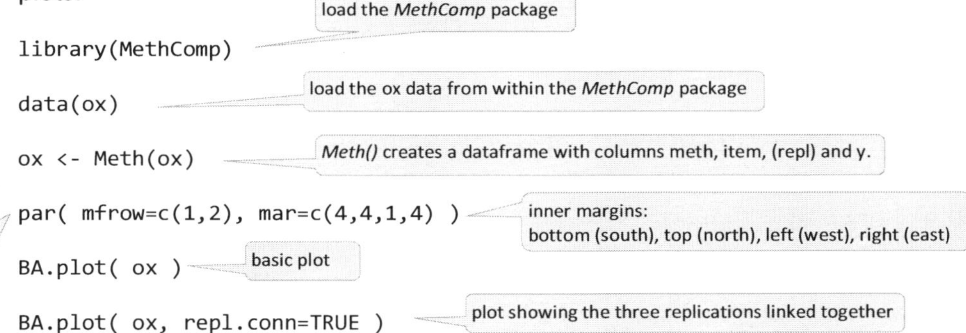


```
library(MethComp)
data(ox)
ox <- Meth(ox)
par( mfrow=c(1,2), mar=c(4,4,1,4) )
BA.plot( ox )
BA.plot( ox, repl.conn=TRUE )
```

The spider like lines indicate the three replications for each reading.

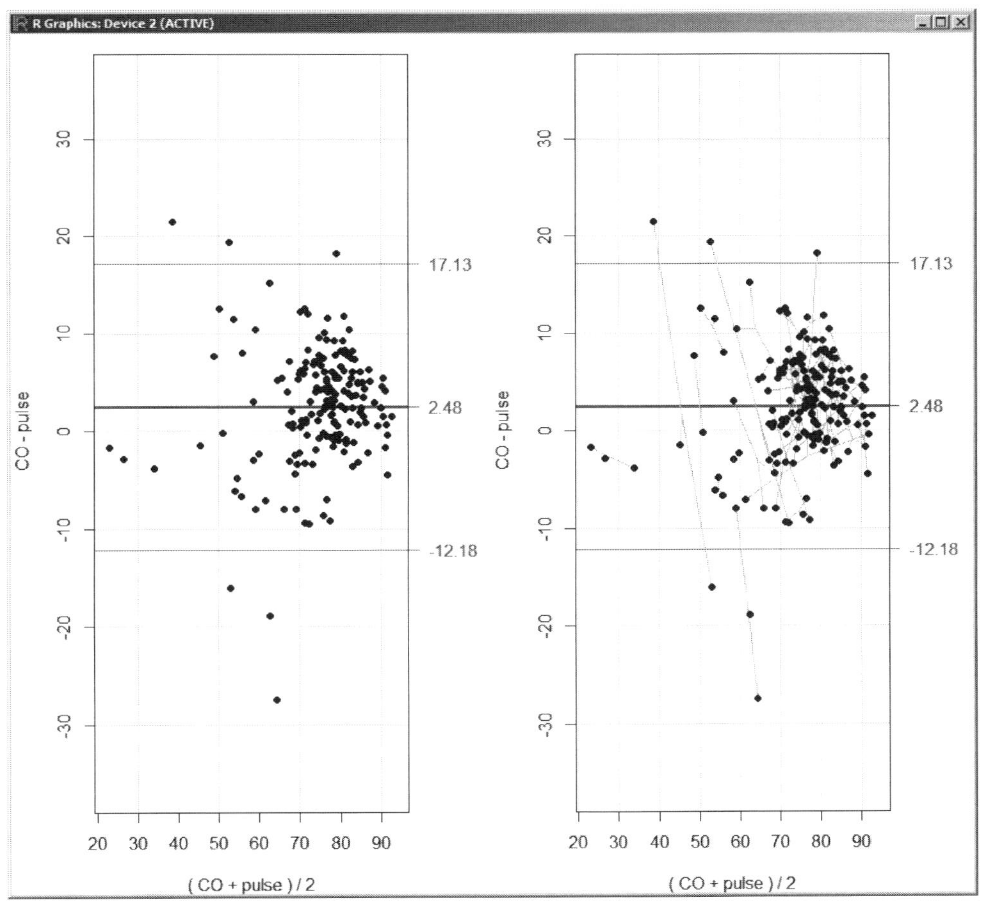

32.3 Tips and Tricks

If you are thinking about including a Bland-Altman plot in a publication I should emphasise I have only touched on this subject and I strongly recommend that you look at Bendix Carstensen's tutorial concerning these plots before you create a final version. You can find it at http://bendixcarstensen.com/MethComp/introMethComp.pdf. There are also extensions of the plot for multiple comparisons detailed in the *MethComp* material.

33 Meta-analysis: the basics

A meta-analysis allows you to aggregate the results (usually some type of effect size) from two or more studies to hopefully obtain a more accurate estimate. You can use it for results from several experiments done by yourself or, as is more commonly done, to compare a number of published results.

To introduce the concepts associated with meta-analysis we will consider the plot most commonly associated with it, the forest plot. The forest plot below (produced with R code discussed later) uses the data given in Cumming, 2012 (p.181 – 262) whose book I would thoroughly recommend for a non-mathematical, but very practical and insightful introduction to the subject.

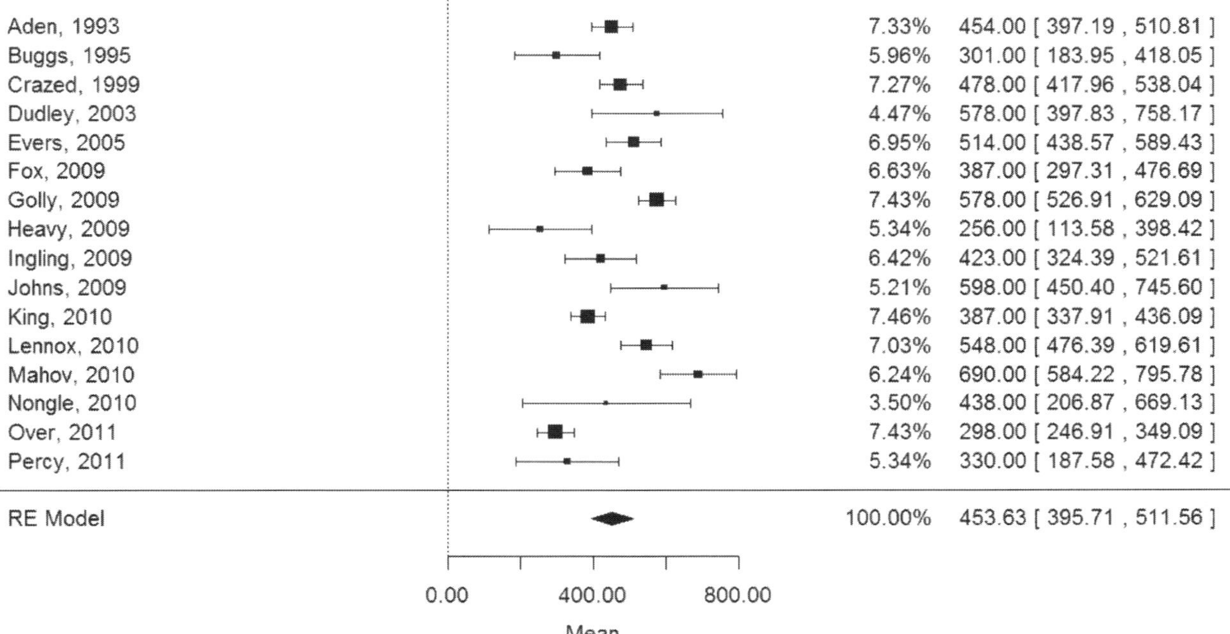

The above forest plot summarises the results from 16 studies concerned with measuring the mean reaction time (a continuous variable) along with an overall estimate taking into account all those studies (Cumming, 2012).

The far left column lists the names and years of the studies. This area of the plot is called the 'slab' in the R package *metafor*, which we will be using to undertake meta-analysis later.

The horizontal line for each study represents a confidence interval (usually 95% as in this case) of the outcome/effect size, that is the mean reaction time here.

The shaded square associated with each trial has a size proportional to its sample size. For example study 14 only contributes 3.5% of the values to the total dataset so has a very small square but study one contributes over twice as much so has a much larger box. The relative contribution of each study is shown in the first column to the right of the confidence intervals along with the mean and a 95% confidence interval.

The first column on the right (this is not standard plot layout) provides the weighting each study gives to the final estimate. In the above, they appear very uniform, varying by only 2 % (5% to 7%).

At the bottom of the forest plot is what meta-analysis is all about. You will see that we have a summary measure of all the individual studies, showing that the overall mean is estimated to be 453.6 with a 95% confidence interval of 395.71 to 511.56, which is also shown as a diamond shape at the bottom.

RE stands for Random effects. There are many ways of developing the model and particularly how to estimate the model parameters (i.e. the overall mean here). We can either think of each parameter as a fixed value or as a normal distribution with a mean and variance/standard deviation, called a random effect. For example, in the above forest plot we have an overall estimated mean of 453.63. We can either think that each of the observed study means represents an exact deviation from this mean, modelled by what is known as a fixed effect parameter or alternatively, we can think that the actual parameter estimate can vary itself following a normal distribution, called random. In other words, we are now no longer thinking of an exact value from a fixed value for each observed study mean but a more elastic measure that varies for each study. There are two important consequences of adopting a random effects approach:

- The accuracy of the estimates is reduced as we are now taking into account possible differences between studies resulting in wider confidence intervals.
- Conceptually we are no longer thinking of just the selected studies but a population of studies from which they came. In other words, we are now extrapolating from the studies in the current meta-analysis to a population of similar studies.

Traditionally the focus was on estimating fixed effects; however, it has now become the norm to consider the outcome/effect size measure to be modelled as a random effect. Remember by random effect I mean one with a normal distribution, it is definitely not random.

One of the most common aspects of meta-analysis is measuring and attempting to reduce the level of heterogeneity across the studies. Originally, all studies included in a meta-analysis were required to be homogenous as there was little point in summarising disparate things, and this, broadly speaking, is still true. However now it is possible to take into account conceivable moderators, that is those variables that may affect the outcome/effect size measure. There are several related measures of heterogeneity used in meta-analysis.

Heterogeneity in this instance is taken to indicate the measure of variability between the studies, for our example here, the variability in means across them. There are several related heterogeneity measures, each of which we will now investigate.

The simplest measure is called *tau* (τ) which is the standard deviation of the estimate of the result measure over the studies in the meta-analysis. Often tau squared (τ^2) is given representing the variance. If *tau*, or *tau squared*, equals zero then there is no heterogeneity between the studies and they are said to be homogeneous.

The next measure in terms of simplicity is Cochran's Q, which is calculated as the sum of squared differences (weighted) between individual study outcome (effect size) measures and the pooled value across studies. Q is useful because when you have sufficient studies which

> **Q is not always Q**
> You have a variety of ways of calculating Q in the *metafor* package in R. Therefore the reported values may not appear to respect the formulae given here.

are homogeneous it has a chi-square distribution with k (number of studies) minus 1 degrees of freedom (=df). Unfortunately, the number of studies involved in the meta-analysis unduly influences the value so the critical value is usually set at 0.1. This is because in this situation Q has very low power (high probability of missing heterogeneity when it is present) with less than 10 studies. As a result of this problem, along and others, emphasis is now placed on other measures discussed below (Higgins, Thompson, Deeks & Altman, 2003).

$$I^2 = 100 \times \frac{Q - df}{Q}$$

where df = number of studies - 1

A slight manipulation of Cochran's Q measure, which takes into account the number of studies, results in a measure known as Higgins *I* squared (I^2). This describes the variation across studies that is due to an actual possible difference in outcome (heterogeneity) rather than that accounted for by random sampling. Often this value is multiplied by a hundred producing a percentage value and described as an '**in**consistency of studies' measure. I have highlighted the 'in' in inconsistency to emphasize that lower values are desirable (i.e. greater homogeneity). I^2 is interpreted in a rather qualitative way where less than 25% indicates a low level of heterogeneity, values between 26-74% moderate heterogeneity, and those above 75% as possessing a high level of heterogeneity. It is a similar concept to the intraclass correlation (Higgins & Thompson, 2002) introduced in *the levels of agreement* chapter.

$$H^2 = \frac{Q}{df}$$

where df = number of studies - 1

H^2 is another measure developed by Higgins and related to I^2. It estimates the ratio of the total amount (or unaccounted amount in mixed-effects models) of variability in the effect size estimates to the amount of sampling variability (Higgins & Thompson, 2002). H^2 is not a percentage measure like I^2 but works similarly with high values indicating more heterogeneity.

There is a danger with the above guidelines that they become set in stone, quoting Higgins himself:

> **Heterogeneity**
> No universal rule could cover definitions for 'mild', 'moderate' or 'severe' heterogeneity, but it would seem that H^2 values exceeding 1.5 might induce considerable caution and values below 1.2 might cause little concern. These correspond to values of I^2 of 56 per cent and 31 per cent. Thus, mild heterogeneity might account for less than 30 per cent of the variability in point estimates, and notable heterogeneity substantially more than 50 per cent.
>
> However, these suggestions are tentative, not least because the practical impact of heterogeneity in a meta-analysis also depends on the size and direction of treatment effects (Higgins & Thompson, 2002 p.1553).

```
tau^2 (estimated amount of total heterogeneity): 11082.8418 (SE = 5624.2855)
tau (square root of estimated tau^2 value):      105.2751
I^2 (total heterogeneity / total variability):   87.41%
H^2 (total variability / sampling variability):  7.94

Test for Heterogeneity:
Q(df = 15) = 119.1178, p-val < .0001
```

The results from a typical meta-analysis, using the data from the previous forest plot produces tau, I^2, H^2 and Q measures as shown opposite.

Working upwards:

- p-value associated with Q is .0001 indicating that the results show a statistically significant level of heterogeneity between studies.
- H^2 is way above 1.5 indicating heterogeneity.
- I^2 Indicates that a large percentage, 87% of the total variability, can be accounted for by heterogeneity between studies rather than random variability.
- tau (τ^2) squared is also large indicating that the set of studies show a high level of heterogeneity.

To sum up; we have both statistically significant and an actual high level of heterogeneity in our batch of studies ☹.

Obviously, we would either conclude that the overall estimate from these studies should be treated with a great deal of caution or use some method to try to reduce this heterogeneity which we do later. However, first let us begin by replicating the meta-analysis described above.

33.1 Data preparation

author	year	mean	sd	number	method na= unspecified
Aden	1993	454	142	24	alloc
Buggs	1995	301	158	7	alloc
Crazed	1999	478	137	20	random
Dudley	2003	578	260	8	random
Evers	2005	514	144	14	na
Fox	2009	387	165	13	alloc
Golly	2009	578	125	23	random
Heavy	2009	256	218	9	alloc
Ingling	2009	423	225	20	random
Johns	2009	598	213	8	random
King	2010	387	112	20	alloc
Lennox	2010	548	155	18	alloc
Mahov	2010	690	179	11	random
Nongle	2010	438	312	7	alloc
Over	2011	298	125	23	na
Percy	2011	330	218	9	alloc

The table opposite shows the fictitious results with a typical meta-analysis style layout for data used to produce the forest plot at the start of this chapter. We have the following columns of data:

author = the author of the paper

year = date of publication

mean = the outcome variable, the result. This does not need to be a continuous variable.

sd = standard deviation

number = number that provided data in the study

method = method used to select the subjects. That is either they have been randomly selected from a larger group, or allocated which implies they were selected based on some characteristic not specified in the paper. *NA* indicates that no details were provided, its R's missing value indicator.

Several points to note concerning the above dataset; each of the 16 studies are considered to be independent that is we do not expect to use the same dataset more than once and if this occurs the results are invalid. The number is for those in each study that provided a result, not the number that may have initially enrolled. Finally, we have a variable, *method*, that could potentially affect the outcome (i.e. mean). We might additionally include several such variables, called *moderators*, which the meta-analysis can take into account when computing the overall estimate, more about that later.

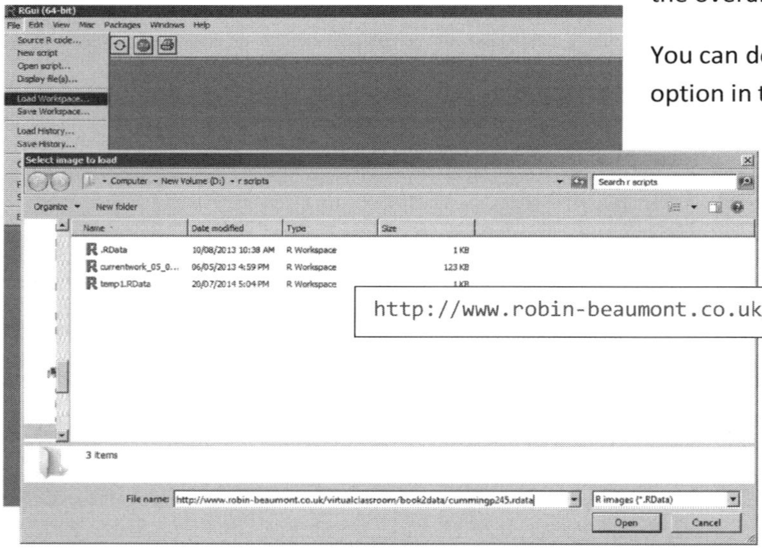

You can download and load the dataset by selecting the menu option in the RGui window:

File->Load Workspace

Then in the *Select image to load* dialog box copy the following into the **File name** text box:

```
http://www.robin-beaumont.co.uk/virtualclassroom/book2data/cummingp245.rdata
```

We now have a dataframe called *cummingp245* loaded into R. Typing the following two lines of code means that we can refer directly to the columns in the *cummingp245* dataframe and using the names function displays what the columns are called.

```
attach(cummingp245)

names(cummingp245)
```

```
> attach(cummingp245)
> names(cummingp245)
[1] "author" "year"   "mean"   "sd"     "number" "method"
```

We can also check the levels for the method variable/column/vector by typing into the R Console:

```
levels(cummingp245$method)
```

This indicates that the method variable has three levels; *"alloc", "na" and "random".*

As is usual, it is a good idea to inspect the dataframe by viewing a simple set of summary statistics using the summary command. Type the following into the R Console.

```
summary(cummingp245)
```

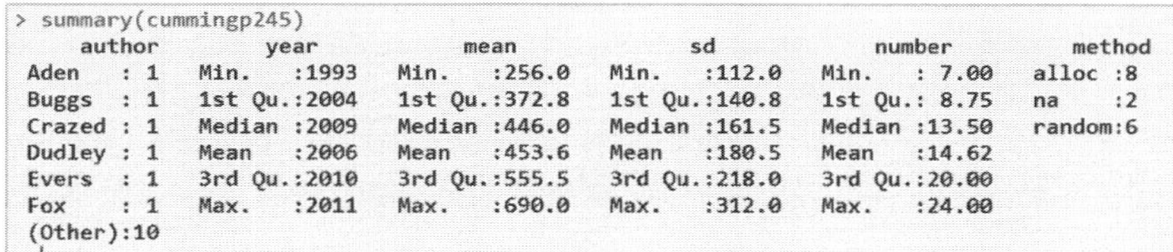

```
> summary(cummingp245)
    author        year           mean             sd            number          method
 Aden   : 1   Min.   :1993   Min.   :256.0   Min.   :112.0   Min.   : 7.00   alloc :8
 Buggs  : 1   1st Qu.:2004   1st Qu.:372.8   1st Qu.:140.8   1st Qu.: 8.75   na    :2
 Crazed : 1   Median :2009   Median :446.0   Median :161.5   Median :13.50   random:6
 Dudley : 1   Mean   :2006   Mean   :453.6   Mean   :180.5   Mean   :14.62
 Evers  : 1   3rd Qu.:2010   3rd Qu.:555.5   3rd Qu.:218.0   3rd Qu.:20.00
 Fox    : 1   Max.   :2011   Max.   :690.0   Max.   :312.0   Max.   :24.00
 (Other):10
```

Several summary values make no sense, such as author, and year. We can see that the mean is 453.6. However, remember this does not take into account the size of each study, which as we can see from the numbers variable varies between 7 and 24. We can also see the numbers in each of the three levels of the method variable/column.

Now onto the meta-analysis proper.

33.2 Meta-analysis in R using the metafor package

To be able to carry out the meta-analysis in R we now need to install and load the metafor package http://www.metafor-project.org/doku.php by typing the following line into the R Console window

```
install.packages("metafor", dependencies=TRUE)
```

We will be asked to select a site from which to download it – if you are not sure which is the closest select *o-cloud.*

Once installed we need to load the library into R which is achieved by typing:

```
library(metafor)
```

Within the *metafor* package there is a function called *rma()* which carries out meta-analysis on appropriately structured data, which ours is. By appropriately structured data, I mean that the datafile contains at least an outcome measure (*mean*), a measure of variability (*sd*), and the number in each study (*number*). Type the following two lines of code below.

```
res <- rma(measure ="MN", mi=mean, sdi=sd,  ni=number, method="DL", data=cummingp245)

summary(res)
```

Some comments are required here:

rma() is the meta-analysis function in the metafor package. It can take a large number of different types of outcome/effect size measures.

Measure = "MN" indicates that the outcome is a measure consisting of raw means, and the actual mean value is stored in the column called *mean*. The sample size for each study contributing to this mean is stored in a column called *number.* The standard deviation value for each study is stored in a column called *sd.* All three columns are in the dataframe called *cummingp245*. As with most statistical procedures there are several ways of computing the results and I have chosen the *DerSimonian-Laird* method indicating we want a random effects model. Rather than discussing the results immediately, I will request a forest plot using the following command:

```
forest(res, showweight=TRUE)
```

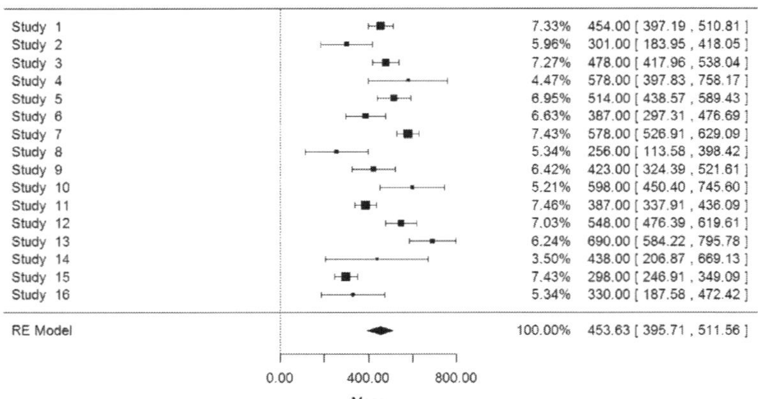

Setting the *showweight* option to *TRUE* makes the shaded squares be proportional to the sample sizes.

The first column, stating study 1 to study 16, is not really that helpful and can easily be replaced with a panel (slab) of more appropriate text which I have done in the code below. The slab now consists of two columns, the values from the *author* along with the *year* vectors separated by a comma. The *paste()* command sticks the values together shown in the parentheses and you can have as few or as many as you want.

```
forest(res, showweight=TRUE, slab=paste(author, year, sep=", "))
```

This results in the same forest plot as presented at the start of the chapter. To obtain the model results we type into the R Console window

```
summary(res)
```

```
> res <- rma(measure ="MN", mi=mean, sdi=sd,  ni=number, method="DL", data=cummingp245)
> summary(res)

Random-Effects Model (k = 16; tau^2 estimator: DL)

  logLik  deviance       AIC       BIC
-99.3061   45.6395  202.6122  204.1574

tau^2 (estimated amount of total heterogeneity): 11082.8418 (SE = 5624.2855)
tau (square root of estimated tau^2 value):      105.2751
I^2 (total heterogeneity / total variability):   87.41%
H^2 (total variability / sampling variability):  7.94

Test for Heterogeneity:
Q(df = 15) = 119.1178, p-val < .0001

Model Results:

estimate       se     zval     pval    ci.lb    ci.ub
453.6340  29.5553  15.3486   <.0001 395.7066 511.5614
```

What has happened is that R has performed what is known as a random effects model (a type of regression), with a single parameter estimate given at the bottom of the output. These values are equivalent to the values given at the bottom of the forest plot. We have already discussed the various heterogeneity measures.

You may have a sense of Déjà vu here as several of the headings in the above output you will have seen in other chapters in this book discussing different types of regression. This is because mathematically meta-analysis is just a type of regression, in fact reading this book you may feel that all statistical techniques are just forms of regression or correlation which is not far from the truth.

33.3 Publication bias

The next aspect that most meta-analysis researchers consider is the possible publication bias of the published reports. This term means that only statistically significant results get published and is a problem when you are trying to get a representative range of results in a meta-analysis. Another plot called the funnel plot allows you to assess this qualitatively. In the funnel plot the y axis is the effect/outcome measure and the x axis is a measure of precision.

We can obtain a funnel plot of the result by typing in the *R Console* window:

```
funnel(res)
```

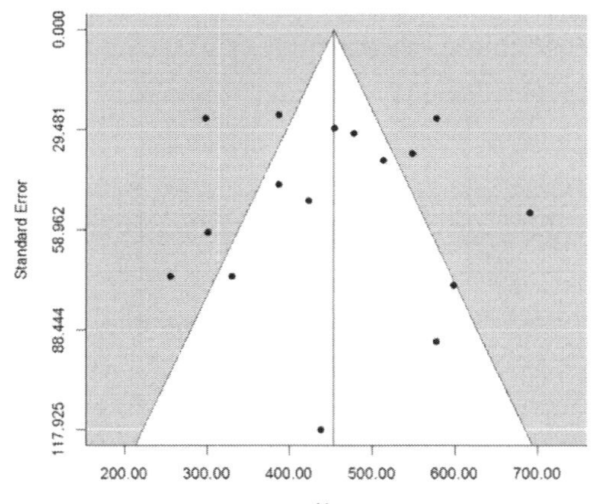

A funnel plot that would display possible publication bias would possess dots with a remarkably asymmetrical distribution. In contrast, the funnel plot opposite seems very symmetrical indicating no evidence of publication bias. The metafor package also has an ingenious function called *trimfill()* which allows you to get metafor to plot on the funnel plot what studies might be missing to make the plot more symmetrical (Viechtbauer, 2010 p.31-32, 2013).

33.4 Accounting for heterogeneity

Previously we noted the very high level of heterogeneity of the results and now we will consider adding two possible moderator variables, that is possible variables that might affect the outcome/effect size measure and therefore reduce the heterogeneity. We have two such variables in the dataset: random, indicating the way the subjects were selected and also the year of publication.

We will start by seeing if taking into account the method of patient selection (a factor), reduces the heterogeneity of the results. To do this we carry out the meta-analysis twice each time just including a particular factor level of the method variable. There is no need to consider the *na* factor level as it indicates that no information was provided.

```
res_alloc <- rma(measure ="MN", mi=mean, sdi=sd,  ni=number, method="DL",
data=cummingp245,
subset=(method=="alloc"), slab=paste(author, year, sep=",    "))
# need to specify slab content in above as subset will not work below
summary(res_alloc)
forest(res_alloc, showweight=TRUE, main="studies using NONrandom allocation")
res_random <- rma(measure ="MN", mi=mean, sdi=sd,  ni=number, method="DL",
data=cummingp245,
subset=(method=="random"),  slab=paste(author, year, sep=",    "))
summary(res_random)
forest(res_random, showweight=TRUE, main="studies using random allocation")
```

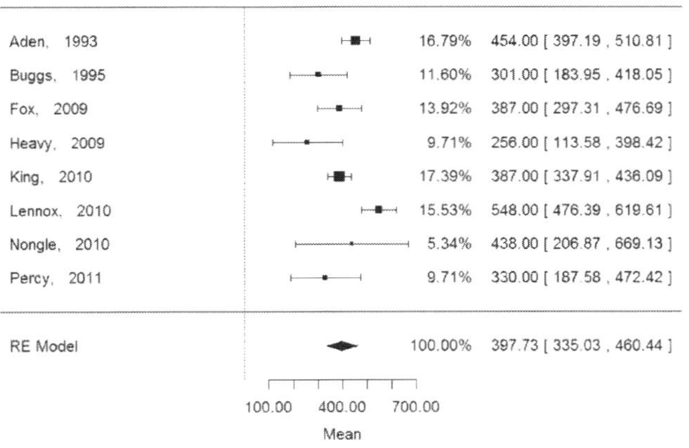

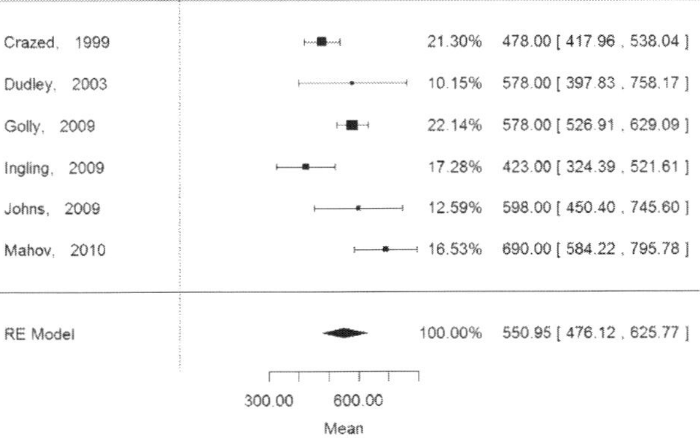

We can see that there is a large difference in the estimated mean; between those studies that used non-random allocation compared to those that used random allocation. The estimated means being 397 compared to 550.

We can also see what using either of these two different subsets has upon the various model statistics:

```
> summary(res_alloc)

Random-Effects Model (k = 8; tau^2 estimator: DL)

  logLik  deviance       AIC       BIC
-47.3665   17.1488   98.7330   98.8919

tau^2 (estimated amount of total heterogeneity): 5257.6706 (SE = 4473.3929)
tau (square root of estimated tau^2 value):      72.5098
I^2 (total heterogeneity / total variability):   73.58%
H^2 (total variability / sampling variability):  3.79

Test for Heterogeneity:
Q(df = 7) = 26.4977, p-val = 0.0004

Model Results:

estimate       se     zval     pval    ci.lb    ci.ub
397.7323  31.9941  12.4314   <.0001 335.0251 460.4395      ***

---
Signif. codes:  0 '***' 0.001 '**' 0.01 '*' 0.05 '.' 0.1 ' ' 1
```

```
> summary(res_random)

Random-Effects Model (k = 6; tau^2 estimator: DL)

  logLik  deviance    .  AIC       BIC
-35.4841   13.0772   74.9683   74.5518

tau^2 (estimated amount of total heterogeneity): 5903.1002 (SE = 5781.1131)
tau (square root of estimated tau^2 value):      76.8316
I^2 (total heterogeneity / total variability):   75.11%
H^2 (total variability / sampling variability):  4.02

Test for Heterogeneity:
Q(df = 5) = 20.0910, p-val = 0.0012

Model Results:

estimate       se     zval     pval    ci.lb    ci.ub
550.9483  38.1760  14.4318   <.0001 476.1246 625.7719      ***

---
Signif. codes:  0 '***' 0.001 '**' 0.01 '*' 0.05 '.' 0.1 ' ' 1
```

Higgins I^2 has come down from 87% to 73% and 75% ☺ but is still classed as a high value.

> **AIC and BIC**
> You can find out more about these in the, *Logistic regression* chapter, section: *Model development and evaluation.*

The model fit statistics AIC and BIC, where smaller values are better, have come down from 202.6 and 2204 to around 98 and 74 respectively. ☺

We can see how statistically significant adding the method moderator variable is to the model by using the *mods* option in the *rma()* function. But first we need to create a temporary dataframe, which I have called *data_reduced.* This dataframe does not include the two studies where the method variable is equal to *NA*, that is we want to include all studies where the method variable value is NOT equal to *NA*. In R we use the expression *!=* to indicate not equal to.

Unfortunately even if the dataframe does not contain any rows with a particular factor level the actual factor levels are stored away in its structure so you use the *droplevels()* function to remove the *NA* level entirely.

```
data_reduced <- cummingp245[cummingp245$method !="na",]

data_reduced <- droplevels(data_reduced)

res2 <- rma(measure ="MN", mi=mean, sdi=sd,  ni=number, method="DL", mods = ~method, data=data_reduced )

summary(res2)
```

```
> summary(res2)

Mixed-Effects Model (k = 14; tau^2 estimator: DL)

  logLik  deviance       AIC       BIC
-82.8472   30.2191  171.6943  173.6115

tau^2 (estimated amount of residual heterogeneity):     5521.0446 (SE = 3547.3602)
tau (square root of estimated tau^2 value):             74.3037
I^2 (residual heterogeneity / unaccounted variability): 74.24%
H^2 (unaccounted variability / sampling variability):   3.88
R^2 (amount of heterogeneity accounted for):            38.65%

Test for Residual Heterogeneity:
QE(df = 12) = 46.5887, p-val < .0001

Test of Moderators (coefficient(s) 2):
QM(df = 1) = 9.5993, p-val = 0.0019

Model Results:

              estimate       se     zval    pval     ci.lb     ci.ub
intrcpt       397.3559  32.5523  12.2067  <.0001  333.5545  461.1572  ***
methodrandom  153.3251  49.4872   3.0983  0.0019   56.3321  250.3182   **

---
Signif. codes:  0 '***' 0.001 '**' 0.01 '*' 0.05 '.' 0.1 ' ' 1
```

We now have three additional values in the output.

R squared (*R^2*) indicates how much of the heterogeneity the model takes into account, you can think of this value as a measure of model fit, 38% being a relatively small amount. Because it is a pretty unreliable measure it is referred to as a pseudo R squared value.

The test of the moderators section works like the overall F test in regression where the null hypothesis is that all the regression weights are zero. Given that we have a value very close to 0.001 we can reject this assertion.

The third addition to the output is the *methodrandom* row which represents the parameter estimate for those studies that used random subject selection. Notice that the actual value is 397 + 153 = 550 as shown in the previous individual group analysis. We can see that both random and non-random levels are statistically significant.

We could also add another moderator to the model; *publication year*, which is a continuous variable in this instance.

```
res3 <- rma(measure ="MN", mi=mean, sdi=sd,  ni=number, method="DL", mods = ~method +
year, data=data_reduced )

summary(res3)
```

```
> summary(res3)

Mixed-Effects Model (k = 14; tau^2 estimator: DL)

  logLik  deviance        AIC        BIC
-82.8248   30.1744   173.6497   176.2059

tau^2 (estimated amount of residual heterogeneity):     6699.0313 (SE = 4383.4362)
tau (square root of estimated tau^2 value):             81.8476
I^2 (residual heterogeneity / unaccounted variability): 76.03%
H^2 (unaccounted variability / sampling variability):   4.17
R^2 (amount of heterogeneity accounted for):            25.56%

Test for Residual Heterogeneity:
QE(df = 11) = 45.8996, p-val < .0001

Test of Moderators (coefficient(s) 2,3):
QM(df = 2) = 8.8276, p-val = 0.0121

Model Results:

               estimate         se     zval    pval        ci.lb       ci.ub
intrcpt      -3866.8106  8619.6565  -0.4486  0.6537  -20761.0269  13027.4057
methodrandom   153.1454    53.2931   2.8736  0.0041     48.6928    257.5979  **
year             2.1257     4.2983   0.4945  0.6209     -6.2989     10.5503

---
Signif. codes:  0 '***' 0.001 '**' 0.01 '*' 0.05 '.' 0.1 ' ' 1
```

Inspecting the R squared value - indicating model fit we can see that it has dropped from 38% to 25% which is not good. Also the year moderator is not statistically significant, indicating that it does not influence the effect size/outcome measure. Clearly, it should be dropped from the model.

We can obtain a graphical summary of any model by simply typing in the R Console *plot(model)*, for example typing in *plot(res2)* produces the four plots shown below.

The first two plots have already been discussed. The two residual plots provide a method of assessing how well each of the studies fit the model. We can see that for both levels the distribution of residuals is almost identical. Inspecting the standardised residuals shows that studies 7, 8, 11 and 12 (ringed) have larger residuals than the others. We might now want to go back and see if their studies were any different to the others.

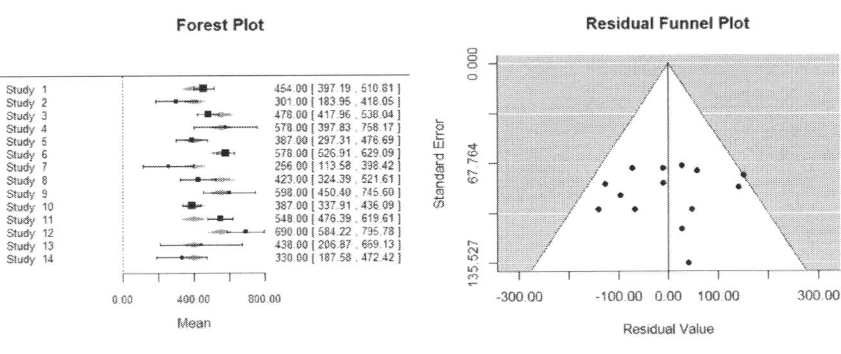

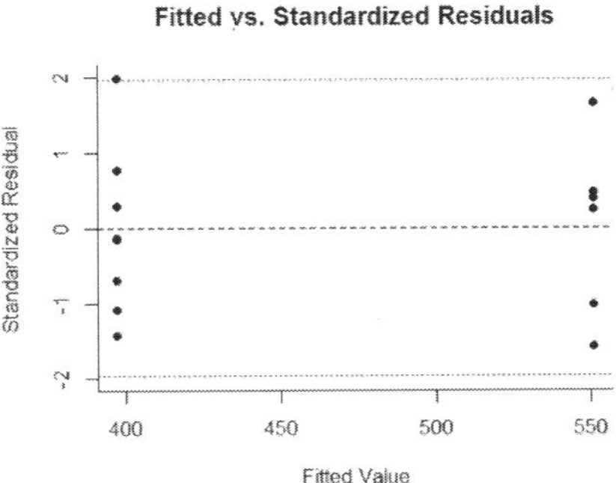

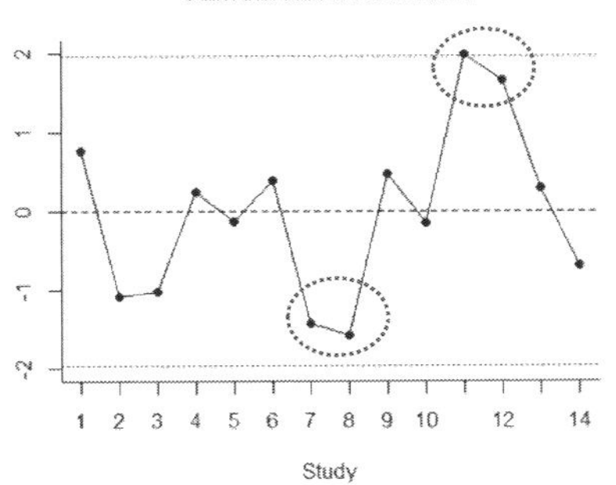

If we had used the slab option when we specified the *res2* model, we would have obtained the actual study authors' names here rather than just a study number which equates to the row number in the dataframe.

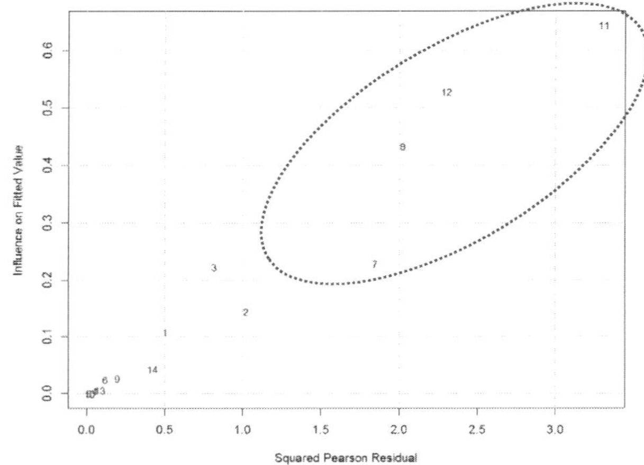

An alternative to the residual plots is the Baujat plot indicating the level of influence each study has on the model. We can obtain a Baujat plot for the *res2* model by typing into the R Console:

```
baujat(res2)
```

The Baujat plot reflects the values previously presented in the standardized residuals plot. Here the higher the influence value on the Baujat plot the larger the residual is on the standardized residuals plot. Therefore we have again the same information identifying the four studies 7, 8, 12 and 11 as the most influential.

A summary of the findings is provided below.

33.5 Writing up the results

A random effect meta-analysis was carried out on a group of 16 studies published between 1990 and 2003 concerned with measuring mean reaction time. Each study contributed between 5.2% and 7.4% of the weighting to the estimate. The initial analysis demonstrated a high level of heterogeneity (I^2=87.4%) with a highly statistically significant value for Cohen's Q (q=119, df=15, p-value<.0001). Subsequent analysis considered two moderators; the method used to select the subjects and the year of publication. Adding the year of publication to the model was not statistically significant (95% ci -6.29 to 10.55) and reduced the pseudo-R squared value down to 25% whereas including the method used to select the subjects increased the model fit and was statistically significant (QM= 9.59, df=1, p-value=0.0019).

The final estimate for the studies that used non-random selection was a mean reaction time of 397 (95% ci 335 to 460) and for those that used random selection a mean reaction time of 551 (95% ci 476 to 626). It should be noted that both these subsets retain a high level of heterogeneity ((I^2=73% and 75% respectively 74% overall) and a relatively low pseudo-R squared value 38.65%.

Visual inspection of the funnel plot displayed a high level of symmetry and therefore no evidence of publication bias.

33.6 Enhancing the forest plot

You can enhance the forest plot with various additions, several of which I will now demonstrate using the Ebola outbreak information available from the WHO website. Here the outcome measure is the proportion surviving. I have created an R dataset from this information which you can download from my website by typing the following into the R Console window

```
load(url("http://www.robin-beaumont.co.uk/virtualclassroom/book2data/ebola.rdata"))
```

```
> names(ebola)
[1] "outbreak" "year"     "xi"       "ni"       "strain"
[6] "country"
```

```
names(ebola)
```

xi here presents the number of deaths in each outbreak and *ni* the total number of cases.

```
> res <- rma(measure = "PR", xi=xi, ni=ni, method="DL", data=ebola)
> summary(res)

Random-Effects Model (k = 25; tau^2 estimator: DL)

  logLik  deviance       AIC       BIC      AICc
  8.5996   87.9053  -13.1991  -10.7614  -12.6537

tau^2 (estimated amount of total heterogeneity): 0.0321 (SE = 0.0165)
tau (square root of estimated tau^2 value):      0.1792
I^2 (total heterogeneity / total variability):   97.28%
H^2 (total variability / sampling variability):  36.74

Test for Heterogeneity:
Q(df = 24) = 881.7438, p-val < .0001

Model Results:

estimate      se     zval     pval    ci.lb    ci.ub
  0.6381  0.0385  16.5842   <.0001   0.5626   0.7135      ***

---
Signif. codes:  0 '***' 0.001 '**' 0.01 '*' 0.05 '.' 0.1 ' ' 1
```

We then, as in the previous example, load the *metafor* package and use the *rma()* function, requesting it to provide summary information:

```
library(metafor)
```

PR = Proportion Raw as the outcome measure

```
res <- rma(measure = "PR", xi=xi, ni=ni, method="DL", data=ebola)

summary(res)
```

We interpret these values as we did previously. We could also consider adding *strain* as a moderator. However, the focus here is to demonstrate an enhanced forest plot where I have requested a credibility/prediction interval (similar to a confidence interval) along with column headings.

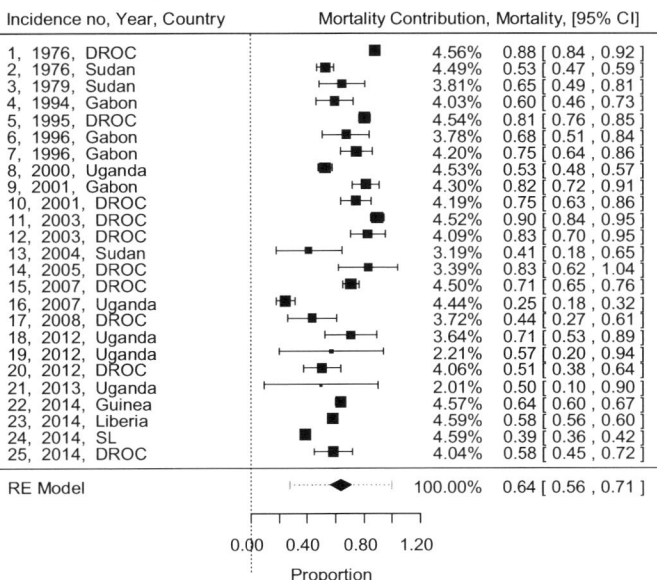

```
> forest(res, showweight=TRUE,
+ slab = paste(outbreak, year, country, sep=",  "),
+ main = "Ebola outbreaks Random-Effects Model",
+ addcred=TRUE)
> text(-1.8, 27, "Incidence no, Year, Country", pos = 4)
> text(.75, 27, "Mortality")
> text(2.9, 27, "Contribution, Mortality, [95% CI]", pos = 2)
```

```
attach(ebola)
```

necessary so that the slab function knows about the variables

```
forest(res, showweight=TRUE,

slab = paste(outbreak, year, country, sep=",  "),

main = "Ebola outbreaks Random-Effects Model", addcred=TRUE)
```

add a credibility/prediction interval

Ebola outbreaks Random-Effects Model

Incidence no, Year, Country	Contribution	Mortality, [95% CI]
1, 1976, DROC	4.56%	0.88 [0.84 , 0.92]
2, 1976, Sudan	4.49%	0.53 [0.47 , 0.59]
3, 1979, Sudan	3.81%	0.65 [0.49 , 0.81]
4, 1994, Gabon	4.03%	0.60 [0.46 , 0.73]
5, 1995, DROC	4.54%	0.81 [0.76 , 0.85]
6, 1996, Gabon	3.78%	0.68 [0.51 , 0.84]
7, 1996, Gabon	4.20%	0.75 [0.64 , 0.86]
8, 2000, Uganda	4.53%	0.53 [0.48 , 0.57]
9, 2001, Gabon	4.30%	0.82 [0.72 , 0.91]
10, 2001, DROC	4.19%	0.75 [0.63 , 0.86]
11, 2003, DROC	4.52%	0.90 [0.84 , 0.95]
12, 2003, DROC	4.09%	0.83 [0.70 , 0.95]
13, 2004, Sudan	3.19%	0.41 [0.18 , 0.65]
14, 2005, DROC	3.39%	0.83 [0.62 , 1.04]
15, 2007, DROC	4.50%	0.71 [0.65 , 0.76]
16, 2007, Uganda	4.44%	0.25 [0.18 , 0.32]
17, 2008, DROC	3.72%	0.44 [0.27 , 0.61]
18, 2012, Uganda	3.64%	0.71 [0.53 , 0.89]
19, 2012, Uganda	2.21%	0.57 [0.20 , 0.94]
20, 2012, DROC	4.06%	0.51 [0.38 , 0.64]
21, 2013, Uganda	2.01%	0.50 [0.10 , 0.90]
22, 2014, Guinea	4.57%	0.64 [0.60 , 0.67]
23, 2014, Liberia	4.59%	0.58 [0.56 , 0.60]
24, 2014, SL	4.59%	0.39 [0.36 , 0.42]
25, 2014, DROC	4.04%	0.58 [0.45 , 0.72]
RE Model	100.00%	0.64 [0.56 , 0.71]

Mortality

0.00 0.40 0.80 1.20

Proportion

x y

text to right of x

```
text(-1.8, 27, "Incidence no, Year, Country", pos = 4)

text(.75, 27, "Mortality")
```

text centred by default

text to left of x

```
text(2.9,27,"Contribution, Mortality, [95% CI]",pos= 2)
```

The positioning of the text is by trial and error.

33.7 Summary

This introductory meta-analysis chapter has presented what appears to be a relatively easy process, but in reality, it is often far from that. Collecting the data, formatting and manipulating the outcome/effect size measures can be very time consuming and frustrating.

In this chapter we have only considered the single group design, with two types of outcome measures; proportions and means, but meta-analysis has been developed to cope with most research designs and many outcome measures. Furthermore, it is a very lively research area, where one might say that there are now as many people carrying out meta-analysis, as there are actively undertaking original research.

Some enthusiastic people are suggesting that a meta-analysis should form part of every literature review to provide a quantitative review of the research.

33.8 Tips and tricks

There are many sources of information concerning meta-analysis and below I've listed a few I have found useful.

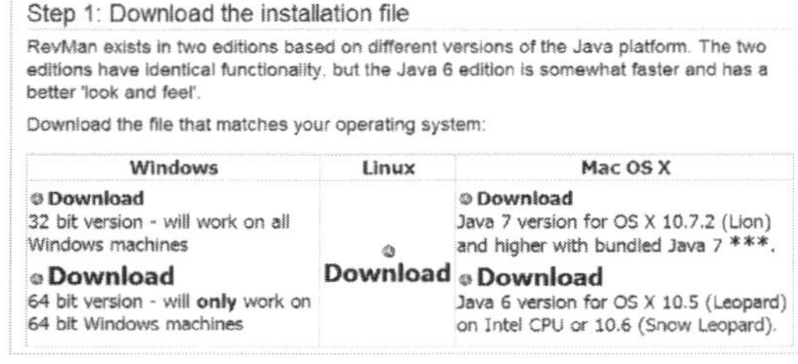

The Cochrane Collaboration website contains many meta-analysis examples, teaching material and free software, called *RevMan* which allows you to carry out meta-analysis and Cochrane reviews http://tech.cochrane.org/Revman. *RevMan* is available for Windows, Linux and the Mac. You can see the formula for many of the statistics used in meta-analysis by looking at the document on the site called *Standard Statistical Algorithms in Cochrane Reviews*.

The book by Geoff Cumming (2012) is highly recommended for an understanding of the meta-analysis process along with the interpretation of results.

The metafor website http://www.metafor-project.org/doku.php/metafor provides examples from several articles concerned with meta-analysis replicating the analysis using R code and the metafor package.

Fast becoming the definitive reference is Borenstein, Hedges & Higgins et al., 2009. The book also introduces their own software called comprehensive meta-analysis (CMA) which you can download a free trial of at http://www.meta-analysis.com/ .

An interesting approach has been to develop software based on various R packages including *metafor* which then provides a graphical front end. One such package is *OpenMeta[Analyst]*. To the end user they are interacting with a windows point and click interface while underneath the various graphical windows are calls to R functions. You can download *OpenMeta[Analyst]* from http://www.cebm.brown.edu/open_meta

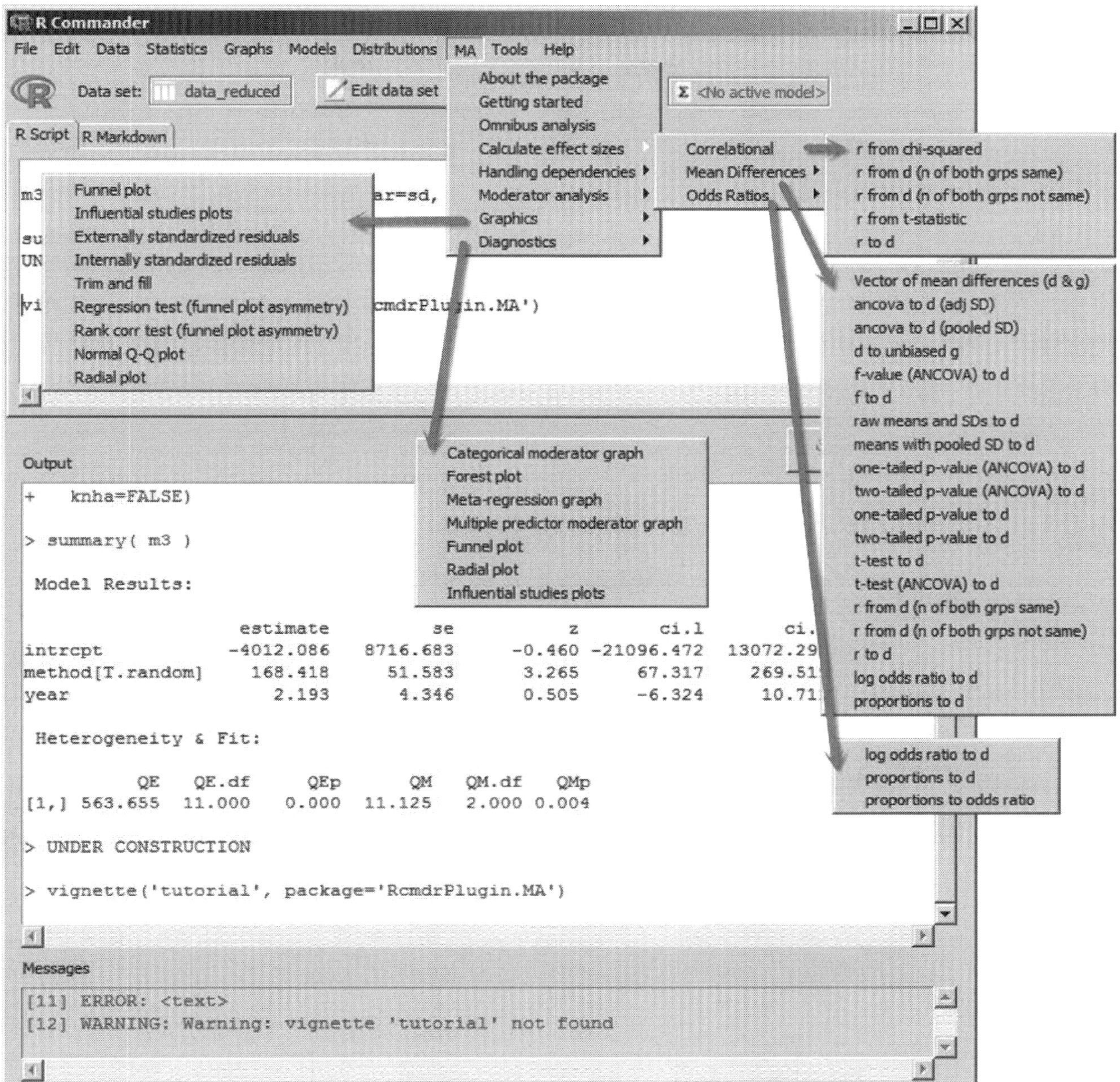

An R Commander plugin for meta-analysis has also being developed, called *RcmdrPlugin.MA*. Once the package has been downloaded and installed it adds a extra menu option to R Commander **MA.** The additional menu options allow the calculation of a multitude of effect size measures along with most of the options resulting from a meta-analysis (see the screenshot above). Unfortunately, the plugin is rather unstable, and provides less flexibility in actual model specification than when using R code directly.

Being able to interpret the results from a meta-analysis along with an assessment of their validity is an essential aspect of evidence based medicine (EBM). To help clinicians gain a working understanding of these aspects *BMJ* contains a regular feature called *Endgames* which always includes a statistical question along with excellent explanations. Several *Endgames* have focused on meta-analysis:

The British Medical Journal (BMJ) provides a regular series of high quality articles entitled endgames several of which have considered aspects of meta-analysis. **Detail:**	**Outcome measure**
Sedgwick P. 2011a Meta-analyses I BMJ, 342:d45 (Published 11 January 2011)	relative risk
Sedgwick P. 2011b Meta-analyses II BMJ, 341:d229 (Published 19 January 2011)	measure relative risk
Sedgwick P. 2011c Meta-analyses III BMJ, 342:d244 (Published 26 January 2011)	measure relative risk
Sedgwick P. 2011d Meta-analyses IV BMJ, 342:d540 (Published 02 February 2011)	mean difference
Sedgwick P. 2011e Meta-analyses V BMJ, 342:d686 (Published 09 February 2011)	standardised mean differences
Sedgwick P. 2011f Meta-analyses VI BMJ, 342:d937 (Published 16 February 2011)	standardised mean differences and subgroups
Sedgwick P. 2011g Meta-analyses VII BMJ, 342:d1108 (Published 23 February 2011)	mean differences and publication bias
Sedgwick P. 2011h Meta-analyses: funnel plots. BMJ, 343:d5372 (Published 31 August 2011)	
Sedgwick P. 2012a How to read a forest plot. BMJ, 345:e8335 (Published 7 December 2012)	
Sedgwick P. 2012b Meta-analyses: tests of heterogeneity. BMJ, 344:e3971 (Published 13 June 2012)	
Sedgwick P. 2013 How to read a funnel plot. BMJ, 346 – (Published 1 March 2013)	

34 Plotting survival over time: K-M (Kaplan-Meier) plots

Survival analysis is concerned with looking at how long it takes for an **event** of some sort to happen. The event is usually something that you do not want to happen such as death or failure, however it might be a positive thing such as 'recovery' or healing or a specific treatment state such as remission. Campbell, 2009 p.141 provides the example of an exercise stress test where the event is the point at which the subject cannot carry on any longer on the machine.

In certain cases, some subjects will not reach the event during the course of the study. If this happens, they are said to be **censored**. Censoring can happen for many reasons. For example, in the case of the exercise stress test above, the equipment might fail before the subject can complete the test (event). There are several different types of censoring and more details can be found in Lee & Wang, 2003 p.4; Cox & Oates, 1984 p.5. The degree of censoring can affect the reliability of the results and there are recommendations for the maximum % of censoring allowable in a group along with sample size.

For many years analysis of such data needed the help of a statistician and a mainframe computer. When I undertook survival analysis of various types of renal patient in the late 1980's I needed to use a mainframe computer and a very unfriendly statistical package called BMDP, this has all changed now and you can easily carry out complex analyses of survival data on your laptop. Survival analysis has become a major area of medical statistical research with the UK leading the way, with one of the most widely used and influential models being the Cox regression model developed by professor D R Cox at Oxford University in the 1970's (http://en.wikipedia.org/wiki/David_Cox_(statistician).

Why do we need to consider the analysis of 'survival' data differently from other data? Well there are two reasons; a proportion of the data is Censored and also the fact that time to the event does not follow a normal distribution, instead tending to follow an exponential one (╰).

The easiest way to get some understanding of what an analysis of survival data entails is to consider how you might graph a typical dataset. The most common graph is the Kaplan-Meier product-limit (K-M) graph which estimates what is called the survival function $S(t)$ against time. The next section looks in detail at how the plot is made, but if you wish just move onto the practical aspect sections.

34.1 How the K-M plot is made

Survival time (weeks)	Status 1= completed 0=censored
6	1
19	1
32	1
42	1
42	1
43	0
94	1
126	0
169	0
207	1
211	0
227	0
253	1
255	0
270	0
310	0
316	0
335	0
346	0
[survival1.sav]	

The following are a set of survival times (in days from entry to a trial) for patients with stage 3 diffuse hystiocytic lymphoma (McKelvey & Gottlieb, 1976). The graph on the right is the Kaplan—Meier product-limit (K-M) graph of the data, commonly called the **K-M plot**, the vertical dashes represent the censored items, showing how the majority of them are at the far right of the graph as would be expected.

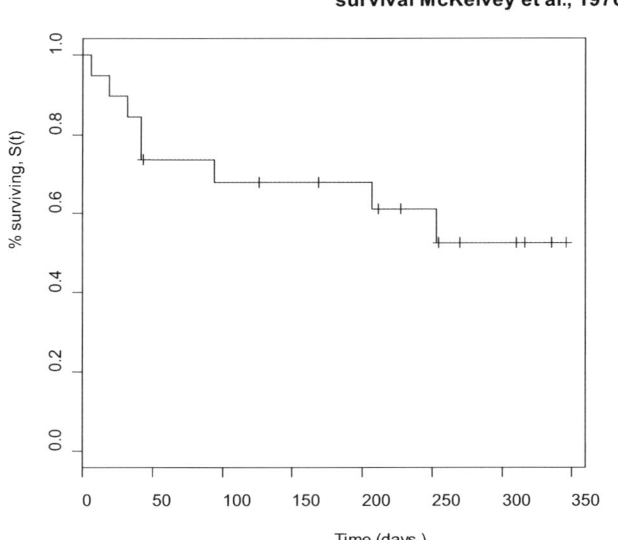

In the table on the left the second column 'status' indicates the status (1=completed; 0=censored) for each subject. Alternatively often censored observations are indicated by a star, asterisk or + symbol. Notice that in our dataset, 8 out of the 19 observations are not censored - representing 50%. This is usually a good quick check to ensure you have used the censored / uncensored values the right way round, as the censored observations do not affect the % surviving. Unfortunately here 50% does not help matters much as coding either way we will end up with a line at 50%, subsequent examples demonstrates this more successfully. Obviously, the effective sample size decreases as we move from the left to the right of the K-M plot which also affects the accuracy of the estimates. Some authors provide a table along the bottom of the x-axis or a separate one indicating the number of non-censored observations at several time points.

The lines on the K-M plot relate to a set of x,y values and investigating how these are calculated provides insight into what $S(t)$ means, that is the values on the y axis. The table below shows how this is carried out. First let's consider some points of nomenclature.

i = A time point, numbered on ordered survival times (t_i), from 0 to p, where p is equal to the number of cases/observations.

t_i = Survival times. As more than one case might fail at the same time (t_i) there may be duplicate entries (see table below).

d_i = Number of cases failing at time t_i i.e. suffer an event.

n_i = Number at risk just before time i. Notice the downward pointing arrows in table below.

r_i= Number alive just before time i.

$(n_i-d_i)/n_i$ = Proportion surviving interval i =probability of surviving i .
Notice the horizontal arrows in table below.

$S(t)$ = Probability of surviving from start (i=0) to t_i = cumulative survival probability = Kaplan-Meier Product limit estimator.

failure time Ranked from shortest to largest, then uncensored, censored (*) t_i	***i***	**events** d_i = No. failing at time t_i	**no. at risk** n_i = no. at risk r_i= no. alive just before *i*	**proportion surviving interval *i*** =$(n_i-d_i)/n_i$ *=probability of surviving i*	***Survival function S(t)*** = probability of surviving from start (i=0) to t_i = cumulative survival probability = Kaplan Meier Product limit estimator
0	0	-	19	-	1
6	1	1	19	(19-1)/19= 0.9474	0.947
19	2	1	19-1=18	(18-1)/18= 0.9444	0.895
32	3	1	18-1=17	(17-1)17= 0.9412	0.842
42	4	2	17-1=16	(16-2)/16= 0.8750	0.737
42	5				
43*	6	0	16-2=14	1	
94	7	1	16-3=13	(13-1)/13 0.9231	0.680
126*	8	0	12	1	
169*	9	0	11	1	
207	10	1	11-1=10	(10-1)/10 0.9000	0.612
211*	11	0	9	1	
227*	12	0	8	1	
253	13	1	8-1=7	(7-1)/7 0.8571	0.525

current survival time i

up to time i

There are three important aspects to note about the above:

- The proportion surviving in the penultimate column relates to the **current** survival time *i* –it is a conditional probability (Norman & Streiner, 2008 p.278)
- The *S(t)* function relates to **all time intervals up** to time *i*
- The failure time t_i is a value not an interval. This needs to be measured accurately, requiring diligent follow-up and avoidance of relying upon recall from participants.

$$S(t_{13}) = \prod_{i=1}^{13}\left(\frac{n_i - d_i}{n_i}\right)$$

The last column uses the term 'product' when describing the *S(t)* function. This is because it is calculated by multiplying (i.e forming the product of) all the previous proportions surviving. Taking our 253 survival time (the largest non censored observation).

or more generally

$S(t_{13})$ = 0.9474 x .9444 x .9412 x .8750 x .9231 x .9000 = .8571 (notice I have missed out the censored observations as x1 does nothing).

$$S(t_j) = \prod_{i=1}^{j}\left(\frac{n_i - d_i}{n_i}\right)$$

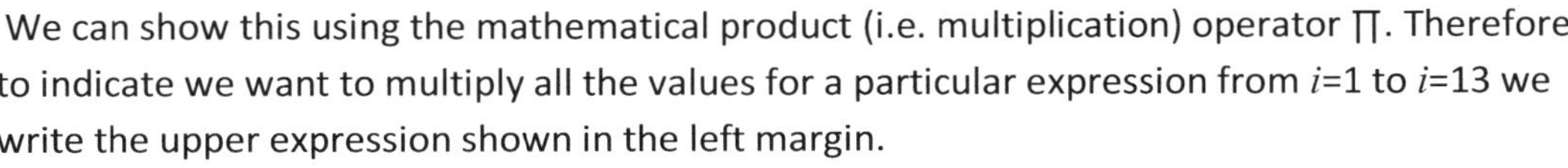

We can show this using the mathematical product (i.e. multiplication) operator $\prod$. Therefore to indicate we want to multiply all the values for a particular expression from *i*=1 to *i*=13 we write the upper expression shown in the left margin.

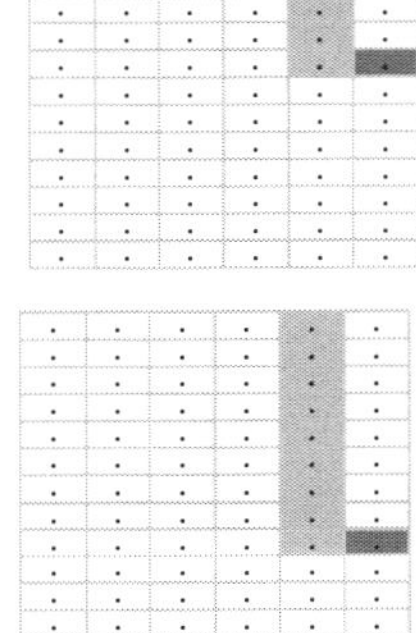

Sometimes what appears to be a very frightening equation can be something quite simple underneath! All we are doing is multiplying the values for each period up to the time we require, it's even simpler when we look at it graphically, considering the value for the 5th and 9th rows. If you have a knowledge of probability, you may have realised that this is just the same as using the 'and' rule, for example the probability of throwing a 6 and then throwing another 6 with the usual dice is 1/6 x 1/6 = 1/36. This illustrates another interesting fact about this model, the survival to time *i* is simply the product of all the individual conditional survival rates for each previous period. You as the researcher must decide if you think this adequately models your data.

34.2 Producing Kaplan-Meier (K-M) plots

To demonstrate K-M plots I'm using information from 26 ovarian cancer chemotherapy patients treated with cyclophosphamide (monotherapy group), or a mixture of cyclophosphamide and adriamycin (combined therapy group) taken from Collett, 1994 p.291. The dataset also includes details of their ages and outcome.

To produce K-M plots you need to first load the necessary package before you load the data.

34.2.1 K-M plots in R Commander

We must first install the *RcmdrPlugin.survival* package, from within the R Console window type the following. As usual you will be asked from which location you wish to download it:

```
install.packages("RcmdrPlugin.survival", dependencies=TRUE)
```

Now close down R and re-start it.

Back in R, load R Commander by typing the following command in the *R Console* window:

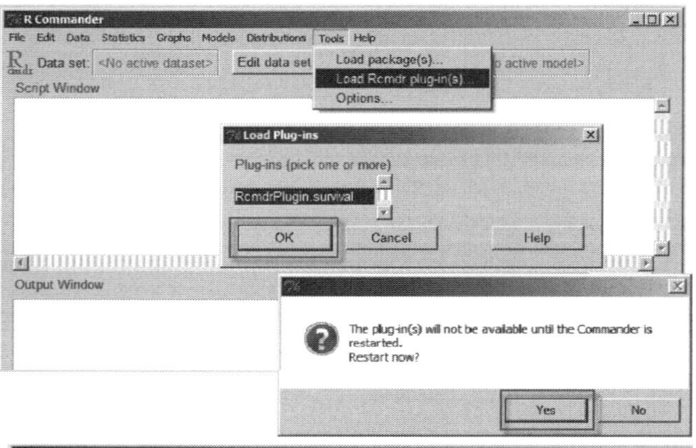


```
library(Rcmdr)
```

From within R Commander select the menu option:

Tools-> Load Rcmdr plug-in(s) . . .

Then select the **RcmdrPlugin.survival** option and click **OK**.

You are warned that R Commander will close and re-open, click **YES**.

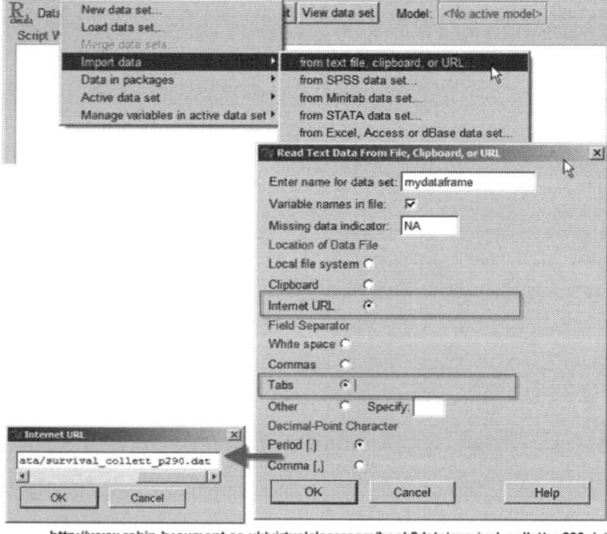

We are now ready to load the dataset described above. You can do this by selecting the R Commander menu option:

Data->Import data->from text, the clipboard or URL

Give the new dataframe the name *mydataframe*, indicating that it is from a URL (i.e. the web) and the columns are separated by *tab* characters.

Clicking on the **OK** button brings up the internet URL box, type the following in it to obtain my sample data:

```
http://www.robin-beaumont.co.uk/virtualclassroom/book2data/survival_collett_p290.dat
```

Click **OK**.

We can see the data from the 26 cases opposite.

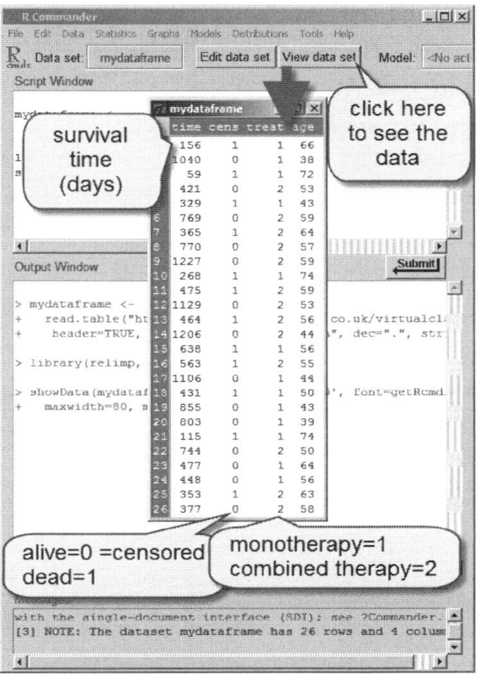

Before we can graph this data we need to let R, via R Commander, know that this dataset has a particular 'survival' structure. That is the dataset must include as a minimum, a variable indicating the length of time to the event or end of the trial period; along with an event indicator.

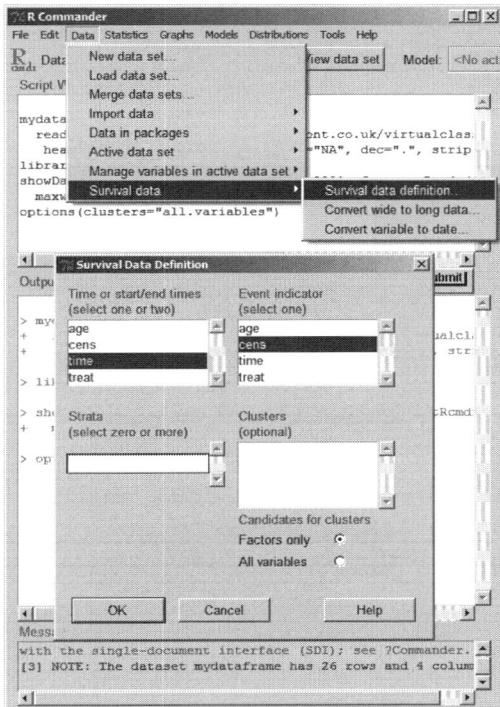

Select the R Commander menu option:

Data -> Survival data-> Survival data definition

Complete the *Survival Data Definition* dialog box as shown opposite.

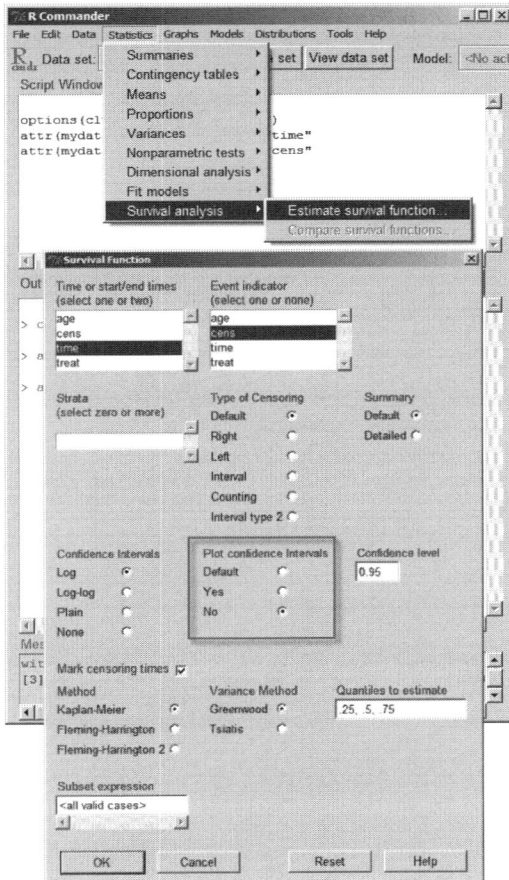

Now we have defined a survival data structure we can make use of it to draw the K-M plot, do this by selecting the menu option:

Statistics -> Survival analysis -> Estimate survival function

Complete the *Survival Function* dialog box as shown opposite.

Note I have deliberately de-selected **plot confidence intervals**.

The K-M plot appears in the main R window.

While the K-M plot opposite is useful, it would be better to see the data divided by treatment group. To achieve this we need to convert the treat variable to a factor.

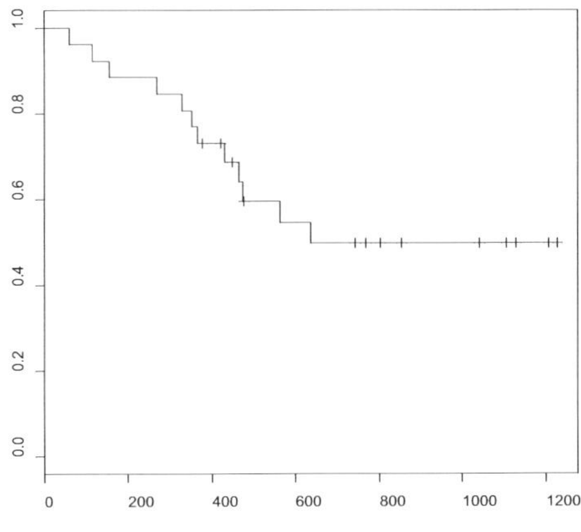

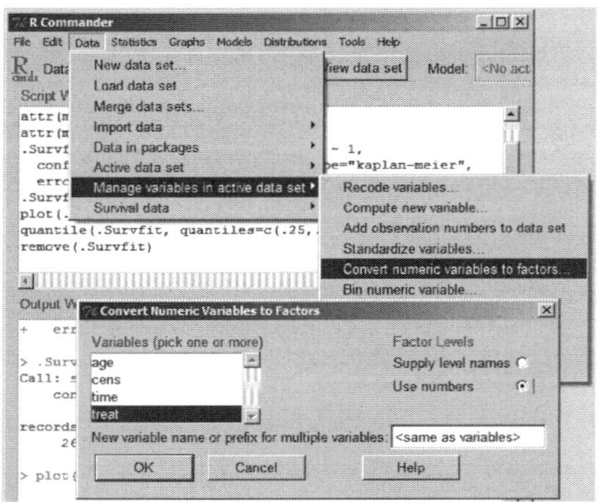

To convert the treat variable to a factor select the R Commander menu option:

Data -> Manage variables in active data set->

Convert numeric variables to factors . .

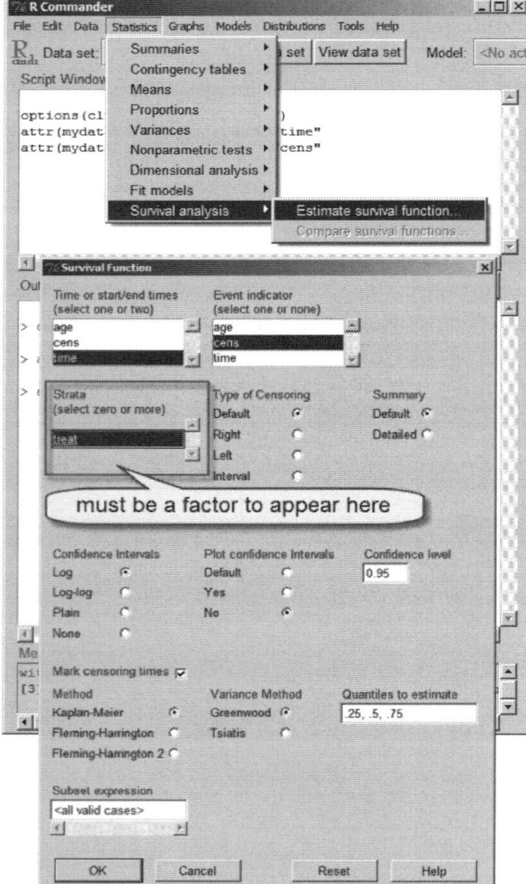

Select the *treat* variable and for the type of **Factor Levels** option select *Use numbers*. A warning message, which you can ignore, will appear so click **OK** in response to it.

When you now select the menu option

Statistics -> Survival analysis -> Estimate survival function

The *treat* variable appears in the **Strata** list box, select it.

Click **OK**.

The Summary option provides a printout of the estimates (including 95% CI's) for each time an event occurred, however if only one event occurred at a specific time the width of the CI's will be zero.

The K-M plot below suggests that the monotherapy group (i.e. group=1) do worse at any stage of followup. Later we will consider this more formally with the Logrank or Gehan-Breslow statistics.

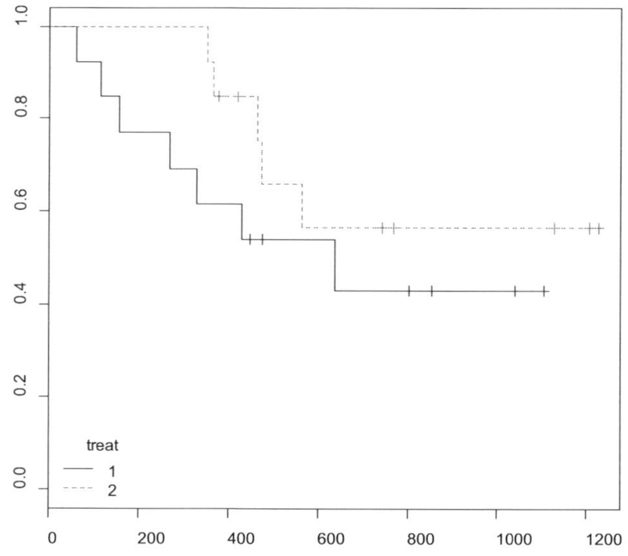

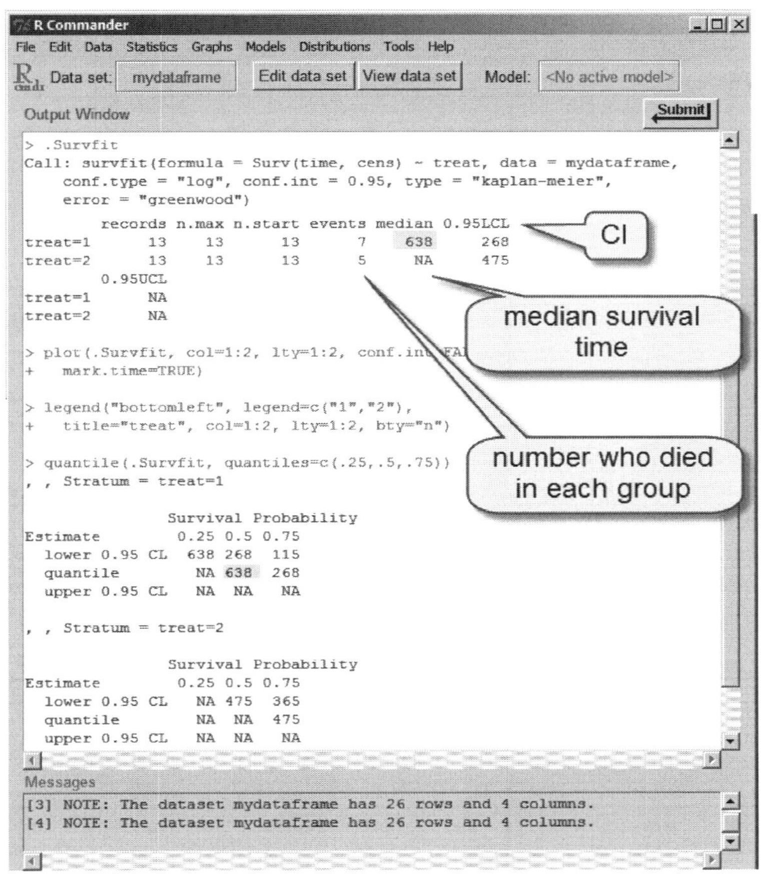

Besides the actual K-M plot, a number of results appear in the R Commander Output Window.

The output includes the Median survival time and 95% Confidence Intervals. Note that there are several with the value NA, meaning they are **N**ot **A**vailable. This is because the group 2 observations fail to reach the 50% mortality level, a good thing for the participants but not for the statistics.

As with most statistical output there is much repetition such as the median survival times which is shaded.

34.3 Logrank and Breslow Tests

These statistics along with their associated p-values allow the quantitative comparison of survival curves. The logrank statistic is formed by calculating the expected failures assuming that the groups are the same, which is then added up. The statistic then compares the observed against the expected values and carries out a chi-squared test on them.

The problem with the Logrank statistic is that it does not take into account where the difference is in the survival curve, in other words it does not matter if there is a difference that changes over time, or may even reverse. In contrast, the Gehan-Breslow statistic gives more weight to earlier differences in the dataset. In R you simply indicate which statistic you want with the "$rho=$" statement as will now be demonstrated. R provides a slightly modified version of the Gehan-Breslow statistic known as the Peto and Peto modification (Peto, Pike, Armitage, Breslow, Cox, et al., 1977).

34.3.1 Logrank and Breslow statistics in R Commander

To obtain the Logrank and Breslow statistics select the R Commander menu option:

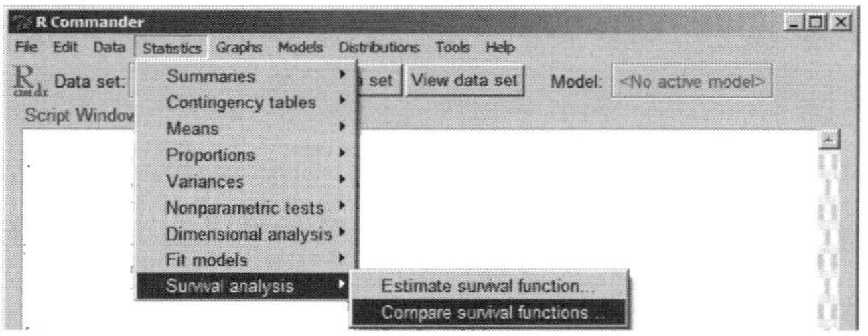

Statistics-> Survival Analysis -> Compare survival functions . . .

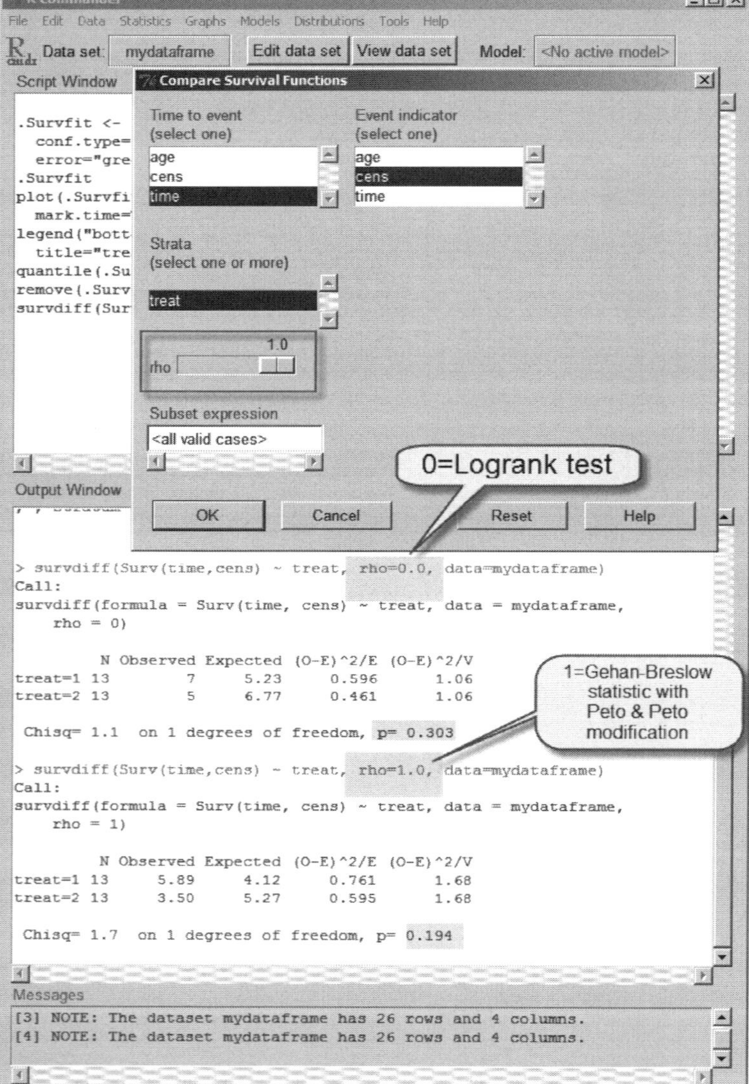

Complete the dialog box as shown opposite, when you select **rho**=*0* you obtain the Logrank statistic and its associated **p-value**. Alternatively, if you set it to 1 you obtain the Breslow equivalent.

Focusing on the Logrank statistic, this indicates that we would have obtained data that would have produced the observed Logrank statistic value or that producing a more extreme Logrank statistic value just over thirty times in every hundred, assuming the curves are just random samples from a single population.

Similarly, for the Gehan-Breslow statistic we can say that we would have obtained data that would have produced the observed Gehan-Breslow statistic value or one more extreme, just over nineteen times in every hundred assuming the curves are just random samples from a single population.

If you take the statistic test approach you can then take this further by setting a critical value (usually 0.05) and decide to reject the assumption that they are just random samples from a single population (the null distribution) when the observed p-value is less than the critical value.

Using this approach, considering both associated probabilities, we accept the null hypothesis and come to the conclusion that the two curves are indeed just displaying random sampling variability, and therefore come from the same population.

34.3.2 K-M plots directly in R

If you have not worked through the previous R Commander session, install and load the survival package with the following command. This may not be necessary if you have done so in the past:

```
install.packages("survival", dependencies=TRUE)
```

Now load the dataset into R with the following command:

get the data into R and call it mydataframe

```
mydataframe <- read.delim("http://www.robin-beaumont.co.uk/virtualclassroom/book2data/survival_collett_p290.dat", header=TRUE)
```

check the data by:

show the column names

```
names(mydataframe)
```

show the data

```
mydataframe
```

Now create the regression model:

now load the survival library (have assumed you have installed it)

```
library(survival)
```

attach mydataframe so that I can refer to column names directly

```
attach(mydataframe)
```

# indicates a comment line

```
# now create a survival object/dataset, note the upper case S
```

mysurv is made to be a survival object containing the time and status columns

another comment line

```
mysurv <- Surv(time, cens)

# now create a survival curve object, note the lower case S
```

myKMest is made to be a survival curve object containing the survival object divided by each group

don't show the confidence interval - this would only happen if we asked for a single group anyway

```
myKMest <- survfit(mysurv~treat, conf.int=FALSE )
```

plot my survival curve with:

line type. 1=solid (default), 3=dotted,

line colour

another comment line (partial this time)

```
plot(myKMest,
lty =c(1,3),
col=c("blue","red"))      # end of the plot command
```

main title =

x axis label =

y axis label =

show line colours

show line type

```
title(main= "Survival ovarian cancer n=26",
xlab="Time (days )",
ylab=" % surviving, S(t)" ) # end of the plot command
```

legend =

show labels; montherapy=1, combined=2

```
legend(x=10,y=0.2, legend=c("monotherapy","combined"), col=c("blue","red"), lty=c(1,3))
```

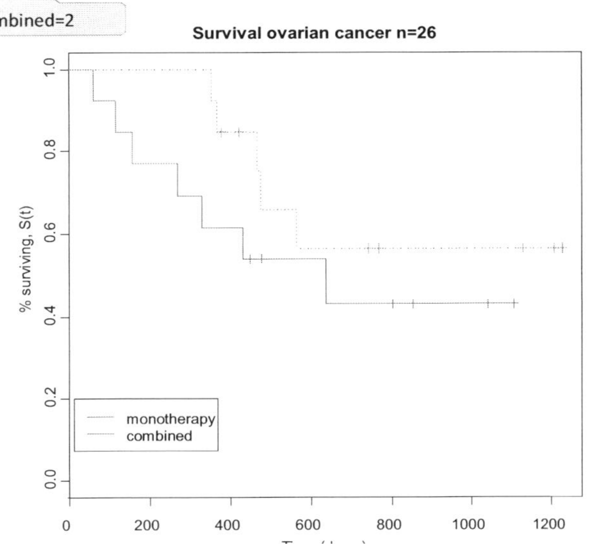

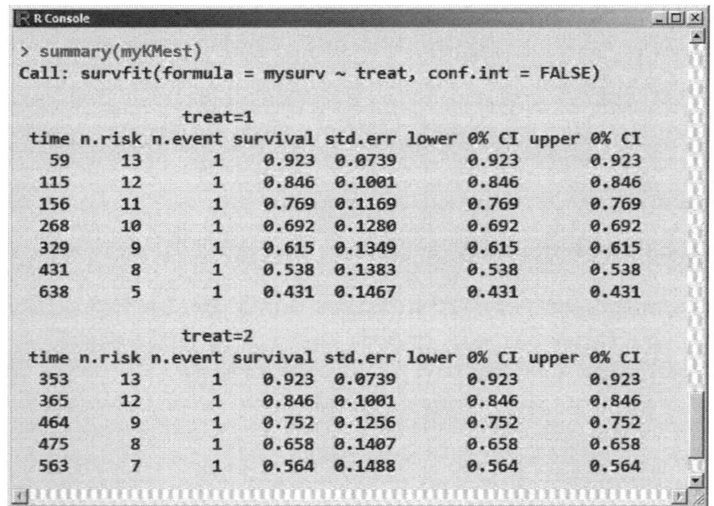

```
R Console
> summary(myKMest)
Call: survfit(formula = mysurv ~ treat, conf.int = FALSE)

                treat=1
 time n.risk n.event survival std.err lower 0% CI upper 0% CI
   59     13       1    0.923  0.0739       0.923       0.923
  115     12       1    0.846  0.1001       0.846       0.846
  156     11       1    0.769  0.1169       0.769       0.769
  268     10       1    0.692  0.1280       0.692       0.692
  329      9       1    0.615  0.1349       0.615       0.615
  431      8       1    0.538  0.1383       0.538       0.538
  638      5       1    0.431  0.1467       0.431       0.431

                treat=2
 time n.risk n.event survival std.err lower 0% CI upper 0% CI
  353     13       1    0.923  0.0739       0.923       0.923
  365     12       1    0.846  0.1001       0.846       0.846
  464      9       1    0.752  0.1256       0.752       0.752
  475      8       1    0.658  0.1407       0.658       0.658
  563      7       1    0.564  0.1488       0.564       0.564
```

We can obtain a list of the estimates for each time a event occurred by typing:

```
summary(myKMest)
```

34.4 Producing K-M plots for publication

Pocock, Clayton & Altman, 2002 made a number of recommendations concerning the display of K-M plots, and many journals, including the BMJ, have adopted them. The most important modification compared to the ones we have previously produced is the incorporation of a table along the bottom showing the number still in the cohort over time. Several people have written R code to produce such a plot and the *survplot()* function within the *rms* R package does this. However, here I have chosen to use an alternative technique, additionally demonstrating how you can make use of R code from another computer.

Tatsuki Koyama at the Department of Biostatistics at Vanderbilt University has developed an R function, *kmplot()*, that takes a survival curve object and returns the plot we require. In a previous chapter we have made use of libraries to get new functions into R but Tatsuki has just provided the raw function. So we need to take a different tack this time to use it.

The R code below uses the keyword source, along with the name of the file on my web server which happens to contain Tatsuki's R code. Typing the code below into the *R Console* the R code is added to your session, so in effect you have silently typed it.

```
source("http://www.robin-beaumont.co.uk/virtualclassroom/book2data/r_code/TatsukiRcodeKMplot.r")
```

You now have installed the additional **function** called *kmplot()*.

To produce a K-M plot with the sample size details included we will now use the *kmplot()* function with the survival curve object we previously created which we named *myKMest*.

marks for events if you want a line put '|'

xaxis.at specifies where 'n at risk' will be computed and printed.

specifies what will be printed at xaxis.at.

line type. 1=solid (default), 2=dotted,

the group names

draw a grid on the plot

draw a legend on the plot

specify the font for the title 1 corresponds to plain text (the default), 2 to bold face, 3 to italic and 4 to bold italic.

add a title

```
kmplot (myKMest, mark='', simple=FALSE,
xaxis.at=c(0,200, 400, 600, 800, 1000, 1200),
xaxis.lab=c(0,200, 400, 600, 800, 1000, 1200),
lty.surv=c(1,2), lwd.surv=1, col.surv=c(1,4),     # survival.curves
col.ci=0, # confidence intervals not plotted
group.names=c('monotherapy','combined'),
group.order=c(1,2), # order of appearance in the n.risk.at table and legend.
extra.left.margin=6, label.n.at.risk=TRUE, draw.lines=TRUE,
cex.axis=.8, xlab='days', ylab='Survival Probability', # labels
grid=TRUE, lty.grid=1, lwd.grid=1, col.grid=grey(.9),
legend=TRUE, loc.legend='bottomleft',
cex.lab=.8, xaxs='r', bty='L', las=1, tcl=-.2  )
title(main='Chemotherapy for Ovarian cancer', adj=.1, font.main=1, line=0.5, cex.main=1)
```

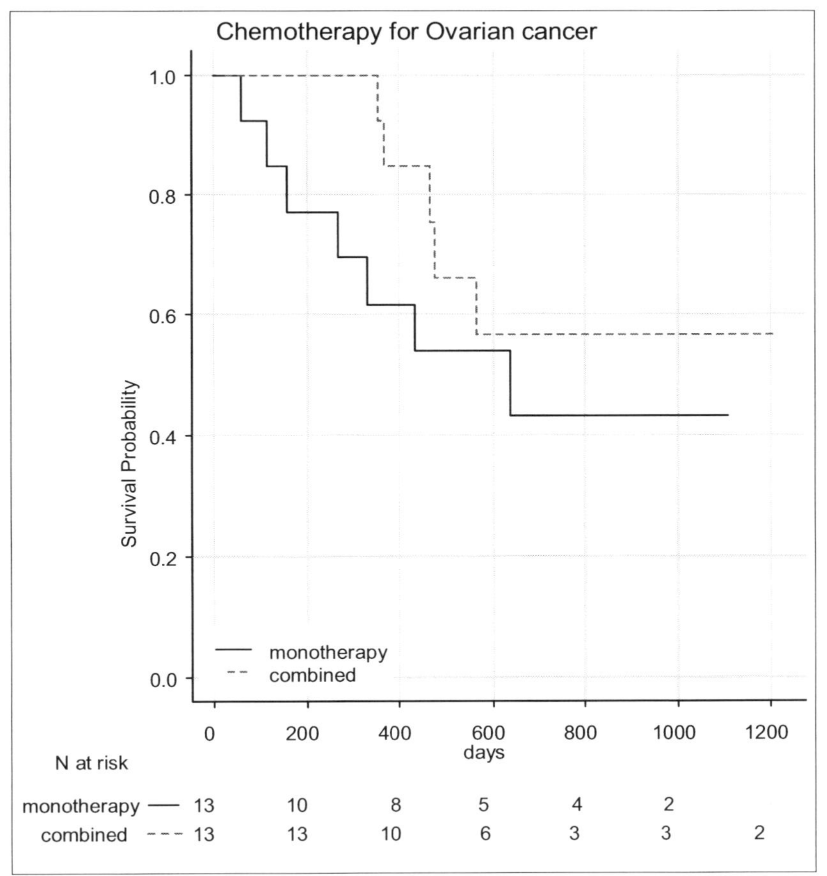

The comments boxes above give details of the various options you need to set.

Often the times of the events are not shown on survival curves when you have a table underneath it. If you do want to show them as in the previous K-M plots simply change *mark=' '* to *mark='|'*.

The second and third lines specify when and what to show along the x axis. To decide upon the places you want the values in your table just draw a simple K-M plot using R Commander, as we did before, first to give you an idea.

The *survplot()* function within the *rms* package allows many more options than those of the *kmplot()* function above but requires greater study.

34.5 Tips and Tricks

K-M plots are very useful but often there are very few left in the cohort at the end of the trial, this is why it is so useful to see the numbers at each stage under the plot.

Curves can cross over indicating that mortality rates change over time for different groups. If you suspect the differences are due to some characteristic of the subgroup you should discuss this with a statistician.

Often continuous variables, such as age, are dichotomised to allow the plotting of survival curves for different age ranges.

To obtain the Logrank and Breslow statistics using R code is straightforward:

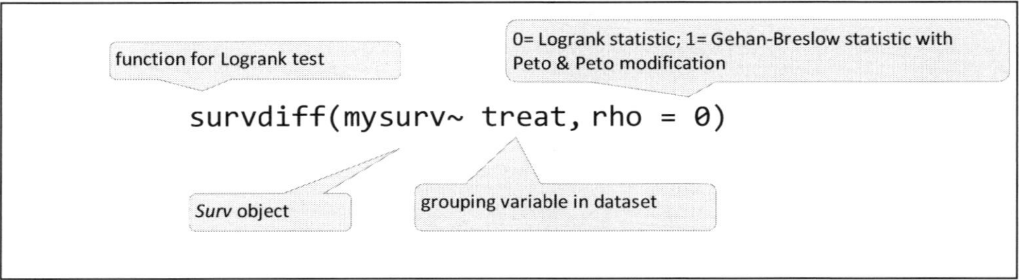

Youtube help:

http://www.youtube.com/watch?v=CNYOWZmwOIA&feature=share&list=PL05FC4785D24C6E68

35 Investigating effects upon survival over time: Cox PH regression

Analysing what influences survival is the focus of this chapter using the Cox Proportional Hazards Regression model. It follows on from the previous chapter discussing the Kaplan-Meier (K-M) plot where we looked at the survival curves for the two different chemotherapy régimes but we did not take into account the affect that age might also have had upon outcome which we will now investigate.

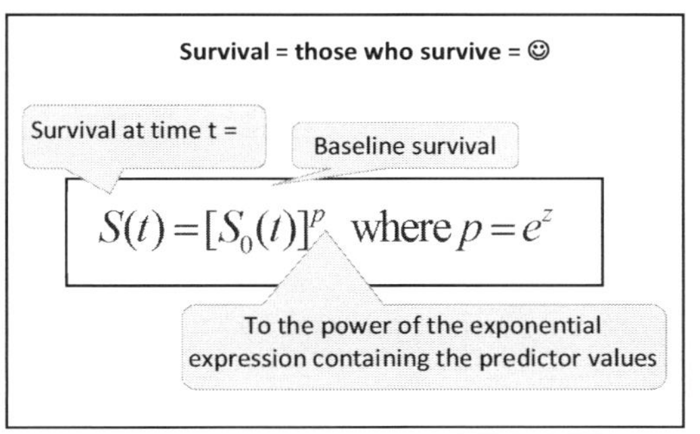

Survival = those who survive = ☺
Survival at time t =
Baseline survival
$S(t)=[S_0(t)]^p$ where $p=e^z$
To the power of the exponential expression containing the predictor values

The Cox regression model is just one of several approaches that attempts to evaluate survival curves taking into account other variables that may affect the survival, such variables are called; confounders, covariates or predictors.

The Cox model can be expressed a number of ways, and we will first look at the approach that defines it in terms of the cumulative survival function (*S(t)* i.e. the y axis on the K–M plot). Here we can say that the predicted survival at time *t* is equal to the baseline survival taken to the power of an expression that contains the various predictors we might want to include.

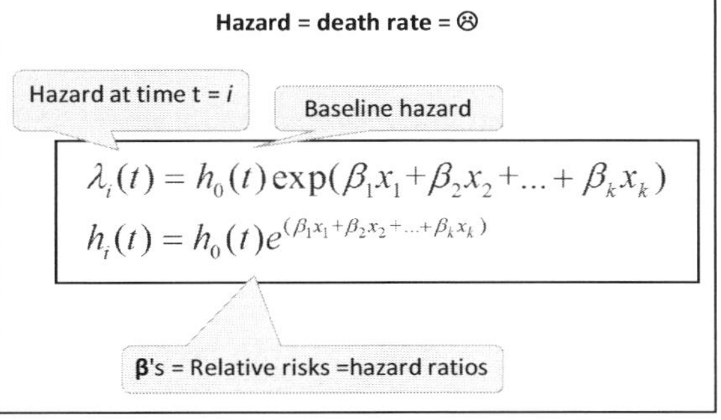

While this format provides a conceptual framework back to the survival function the most common way of defining the model is by considering the **Hazard function** (represented by *h* or the lambda character λ) as the predicted value. The hazard is the risk per unit time, and can vary between zero and infinity so the Cox model becomes that shown above.

Ovarian cancer data (Collett 1994 p.290) [survival_collett_p290.sav]				
Subject no	time	Cens [censored = 0] [died = 1]	Treat [1=single; 2=combination]	Age [years]
1	156	1	1	66
2	1040	0	1	38
3	59	1	1	72
4	421	0	2	53
5	329	1	1	43
6	769	0	2	59
7	365	1	2	64
8	770	0	2	57
9	1227	0	2	59
10	268	1	1	74
11	475	1	2	59
12	1129	0	2	53
13	464	1	2	56
14	1206	0	2	44
15	638	1	1	56
16	563	1	2	55
17	1106	0	1	44
18	431	1	1	50
19	855	0	1	43
20	803	0	1	39
21	115	1	1	74
22	744	0	2	50
23	477	0	1	64
24	448	0	1	56
25	353	1	2	63
26	377	0	2	58

Survival is the optimistic side of the coin whereas the Hazard (i.e. death rate) is the flip side. Note that the expression within the exponentiation function *exp()* looks like a regression equation, which it is. In the Cox model the betas (β's) represent the **relative hazards (RH)** which are also called the **hazard ratios (HR)**. If you have come across the logistic regression equation in another chapter you will see similarities, there the betas (β's) represented log odds ratios. An example of the Cox model will make things clearer.

The data opposite (Collett, 1994 p.291) represents information from 26 ovarian cancer chemotherapy patients treated with cyclophosphamide (the monotherapy group) or a mixture of cyclophosphamide and adriamycin (the combined therapy group), along with their age and outcome. The K-M plot in the previous chapter suggested that the observed differences in survival between the two treatment groups was merely due to random sampling.

Looking at the above dataset it seems reasonable to suggest that both the type of therapy and also age might affect survival. In this context we call such variables **covariates**, and luckily the Cox regression model just happens to provide us with a valid model if the data obey certain rules.

$$\lambda_i(t) = h_0(t)\exp(\beta_{treat}x_i + \beta_{age}x_i)$$

$$h_i(t) = h_0(t)e^{(\beta_{treat}x_i + \beta_{age}x_i)}$$

In other words we want to consider a Cox regression model that looks like the one shown opposite.

> **Iterative methods**
> For further details, see the glossary entry **likelihood and Maximum likelihood,** the *Logistic regression* chapter and the techie section at the end of this chapter.

Whereas in ordinary regression we have a direct method of estimating the β's, this is not possible in Cox regression. Here Instead, the computer searches for values that minimise the difference between the estimated and observed values from the model, using an iterative process to produce partial likelihood values (the betas).

The *Age* covariate is just the typical interval/ratio variable, but *Treat* is a bit more complex to use as it represents a variable that divides up the data. *Treat* is a factor with two levels.

In Cox regression analysis, you generally number the different levels of a variable sequentially. In the example above we used {0,1} for *Cens* and {1,2} for *Treat*. Using {0,2} or {-1, +1} would produce a different coefficient value. Let's now create a Cox model from the data discussed above (we'll look at the output in more detail in a while).

35.1 Creating a Cox model

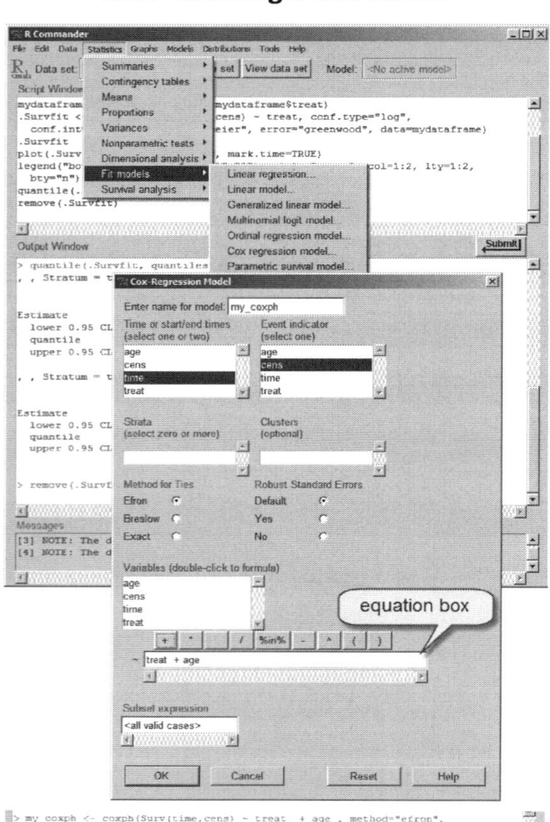

First consider how we can do this in R Commander. I assume you have just completed the steps described in the previous chapter that is you have:

- installed the *RcmdrPlugin.survival* package
- loaded the *survival_collett_p290.dat* file into R Commander

Do not convert the treat variable to a factor.

Now select the R Commander menu option:

Statistics -> Cox regression model

Complete the *Cox Regression Model* dialog box as shown opposite.

Pay particular attention to the actual formula in the equation box beginning with ~. You create an equation by selecting each variable required in turn and adding each by clicking on the + button. Finally click **OK**.

If you don't have a treat variable listed but only a *treat[factor]* variable, select this one then remove the '*[factor]*' part by editing the text in the equation box.

```
> my_coxph <- coxph(Surv(time,cens) ~ treat  + age , method="efron",
+   data=mydataframe)

> summary(my_coxph)
Call:
coxph(formula = Surv(time, cens) ~ treat + age, data = mydataframe,
    method = "efron")

  n= 26, number of events= 12

          coef exp(coef) se(coef)      z Pr(>|z|)
treat -0.79593   0.45116  0.63294 -1.258  0.20857
age    0.14657   1.15786  0.04585  3.196  0.00139 **
---
Signif. codes:  0 '***' 0.001 '**' 0.01 '*' 0.05 '.' 0.1 ' ' 1

      exp(coef) exp(-coef) lower .95 upper .95
treat    0.4512     2.2165    0.1305     1.560
age      1.1579     0.8637    1.0583     1.267

Concordance= 0.794  (se = 0.091 )
Rsquare= 0.456   (max possible= 0.932 )
Likelihood ratio test= 15.82  on 2 df,   p=0.00036[illegible]
Wald test            = 13.55  on 2 df,   p=0.001145
Score (logrank) test = 18.61  on 2 df,   p=9.1e-05
```

insignificant therefore remove from model ☹

significant therefore keep in model ☺

95% CI for age – does not include 1 ☺

significant therefore better than null model ☺

The results are given opposite. Setting our critical value as 0.05 means that the treat variable is not statistically significant while the age variable is. This means that we would probably drop the treat variable from the model but definitely keep the age variable. It would appear that age rather than the form of treatment given predicts survival in this dataset. Note that we have not considered how good the fit is which we will discuss latter.

35.1.1 Doing it directly in R

If you have not done any of the previous survival analysis examples, install the survival package with this code:

```
install.packages("survival", dependencies=TRUE)
```

Now load the dataset into R with the following command:

get the data into R and call it *mydataframe*

```
mydataframe <- read.delim("http://www.robin-beaumont.co.uk/virtualclassroom/book2data/survival_collett_p290.dat", header=TRUE)
```

check the data by:

show the column names

```
names(mydataframe)
```

show the data

```
mydataframe

# Now load survival library

library(survival)
```

now load the survival library. I have assumed you have already installed it (see page 17)

attach mydataframe so that I can refer to column names directly

```
attach(mydataframe)

# Now create the cox regression model:
```

# indicates a comment line

my_coxph is made to be a Cox PH object containing the survival object and 2 covariates, treat and age

```
my_coxph<- coxph(Surv(time,cens) ~ treat + age, data=mydataframe)

summary(my_coxph)
```

display the results

```
####################

# alternatively if you have carried out the previous chapters r code you
# will have a Surv object called mysurv from the line:
# mysurv <- Surv(time, cens)
# so the above formula could be changed to:
# my_coxph<- coxph(mysurv ~ treat + age, data=mydataframe)
####################
```

several lines of comments

R Console

```
> my_coxph<- coxph(Surv(time,cens) ~ treat + age, data=mydataframe)
> summary(my_coxph)
Call:
coxph(formula = Surv(time, cens) ~ treat + age, data = mydataframe)

  n= 26, number of events= 12

          coef exp(coef) se(coef)      z Pr(>|z|)
treat -0.79593   0.45116  0.63294 -1.258  0.20857
age    0.14657   1.15786  0.04585  3.196  0.00139 **
---
Signif. codes:  0 '***' 0.001 '**' 0.01 '*' 0.05 '.' 0.1 ' ' 1

      exp(coef) exp(-coef) lower .95 upper .95
treat    0.4512     2.2165    0.1305     1.560
age      1.1579     0.8637    1.0583     1.267

Concordance= 0.794  (se = 0.091 )
Rsquare= 0.456   (max possible= 0.932 )
Likelihood ratio test= 15.82  on 2 df,   p=0.0003666
Wald test            = 13.55  on 2 df,   p=0.001145
Score (logrank) test = 18.61  on 2 df,   p=9.1e-05

> |
```

The output is identical to that produced from R Commander on the previous page.

I feel a little more explanation of the above process might help. After having created the *Surv* object called *mysurv* for displaying the K-M plot we simply used it as the outcome variable in the formula for the Cox regression model:

```
my_coxph <- coxph(formula = mysurv ~ treat + age, data=mydataframe)
summary(my_coxph)
```

The left hand side of the ~ represents the outcome and the right hand side of it lists the input (predictor) variables. You also need to specify the dataframe you are using, *mydataframe* in this instance. The summary command applied to the Cox object gives the β's along with the exponential values (i.e. relative risks RR's = hazard ratios) and the 95% CIs of the RR's.

At the end there is what is known as a pseudo R-Squared value which can be interpreted like a rather unreliable R squared value in ordinary linear regression. Concordance is a similar measure and when given for continuous covariates, like age, is equivalent to Kendall's tau. A value of 0.5 indicates an agreement that is no better than chance, we have a value of .794 indicating good agreement between the predicted values produced by the model and the observed ones. You can obtain a table of concordance, discordant pairs and ties using the r code, *survConcordance(survobject)*. A third measure, the likelihood ratio test will be discussed below.

35.2 Log Likelihood ratio test

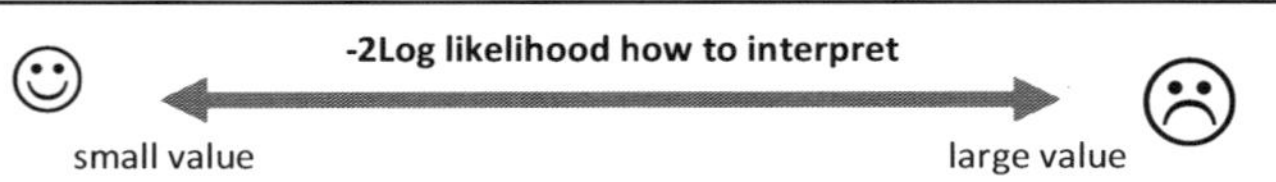

In the above output, the likelihood ratio test value provides a measure of model fit interpreted as shown opposite.

```
> my_coxphnull <- coxph(formula = mysurv ~ 1, mydata)
> summary(my_coxphnull)
Call:  coxph(formula = mysurv ~ 1, data = mydata)

Null model
  log likelihood= -34.98494
  n= 26
> my_coxphnull$loglik
[1] -34.98494
> # to get the null model -2LL  multiply by -2
> -2*my_coxphnull$loglik
[1] 69.96988
```

Furthermore comparing the -2 log likelihood values of different models provides a method of assessing their relative merits. The value of 15.82 given in the above output is the difference between the null model and the model we specified with two predictors, because we have added two β's (i.e. parameters) the DF is 2. The p-value is significant (i.e. 0.05) indicating that our currently fitting model is better than the null model. It is easily calculated:

```
-2(log likelihood of current model - log likelihood of previous model)
```

We can obtain the -2log likelihood (-2*LL*) value for the null model by specifying a new model which has a constant instead of predictor variables; again see the analogy to the situation in simple linear regression where we had a null model by fitting the data to the mean y value. In R we type the following

```
my_coxphnull <- coxph(formula = mysurv ~ 1, mydataframe)

summary(my_coxphnull)

my_coxphnull$loglik

-2*my_coxphnull$loglik
```

to get the null model -2LL multiply by -2

Multiplying the value by -2 gives the -2log likelihood value – then it is simply a case of subtracting one from the other to perform the log likelihood ratio test.

So the original (null model) -2*LL* is 69.96988 and the model we have has a change in -2*LL* from the previous model of 15.148, so the -2*LL* of our new model is 69.96988 – 15.148 = 54.82188. The above process is not necessary, but I wanted to show you how the value is worked out.

The value is also provided as part of the output from the *anova()* function which shows the change in log likelihood as each input variable is added to the model, simply type:

```
anova(my_coxph)
```

```
> anova(my_coxph)
Analysis of Deviance Table
 Cox model: response is mysurv
Terms added sequentially (first to last)

       loglik   Chisq Df Pr(>|Chi|)
NULL  -34.985
treat -34.459  1.0515  1  0.3051727
age   -27.074 14.7709  1  0.0001214 ***
---
Signif. codes:  0 '***' 0.001 '**' 0.01 '*' 0.05 '.' 0.1 ' ' 1
```

adding treat to the model has no affect

adding age to the model improves it

35.3 Interpreting the betas (β)

The Hazard Ratio (HR) is *exp(*β*)* and is the relative hazard corresponding to a unit change in the associated predictor. In this instance you can think of a hazard as a death rate, so greater the number the worse the outcome. *exp()* is the antilog function, also called the exponential function.

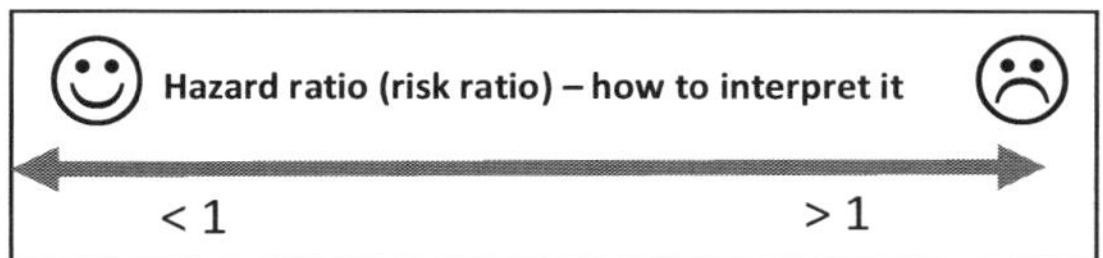

Considering each of the predictors in the *my_coxph* model separately we have (taken from previous page):

age: looking at the *exp(coef)* column 1.158 is the HR of death in a subject of age *a* + 1 years relative to that of a woman of age *a*. We can also consider multiple years, such as $(1.158)^5$ = 2.082298 so if you are five years older your odds of dying are twice that of that someone 5 years your junior. Similarly if you are 10 years older it goes up to $(1.158)^{10}$ = 4.335963, that is your odds of dying are over 4 times that of someone 10 years younger. These techniques also apply to the logistic regression model discussed in that chapter.

treat: here, 0.451 is the HR of death in a subject in group 2 (combined therapy) relative to that of a subject on monotherapy (group=1). This is because for nominal predictor variables (covariates) one category is the 'reference group' which means that it has a coefficient of 1, in other words the single therapy has a hazard ratio of 1. **In reality we would probably remove the 'group' predictor from the equation as the associated p-value is not statistically significant.**

We have not discussed the baseline hazard ratio $h_0(t)$*, which is the value when all the* β*'s* equal zero. Some statistical applications produce estimates of this value while others do not. In real world applications this is generally irrelevant - we are much more interested in the relative risks between subjects with differing coefficient values, not those mythological ones where all the β*'s* equal zero.

35.4 Hazard ratios, odds and probabilities

$$odds = \frac{p_i}{1-p_i} \Rightarrow p_i = \frac{odds}{1+odds}$$

In this context:

$$HR = \frac{p_i}{1-p_i} \Rightarrow p_i = \frac{HR}{1+HR}$$

Two common misinterpretations of the Hazard Ratio (HR) are:

- comparing HR values
- relating the HR values to the time to get to the event.

For example incorrectly thinking that a HR of 2 versus 1 means that the subject has twice the chance (i.e. probability) of succumbing to the event or that the time to get to the event is halved, neither of these interpretations is correct. See the excellent clear article by Spruance & Reid et al., 2004 for full details.

HR	probability
0.5	0.33
0.6	0.38
0.7	0.41
0.8	0.44
0.9	0.47
1	0.50
1.1	0.52
1.2	0.55
1.3	0.57
1.4	0.58
1.5	0.60
2	0.67
3	0.75
4	0.80
5	0.83
6	0.86
7	0.88
8	0.89
9	0.90
10	0.91

The best way to conceptualise Hazard Ratios is to think about odds, which we can convert to probabilities. The odds in this instance are HR's, as shown opposite.

The table opposite allows the comparison between Hazard Ratios and probabilities. We can see that a hazard ratio of 2 therefore corresponds to a 67% greater chance of reaching the event compared to the other group, and a hazard ratio of 3 corresponds to a 75% chance of reaching the event first. Clearly this can be either a good or bad thing depending upon what the event is! For our ovarian cancer dataset we have a HR of 2.08 for an increase in age of 5 years and 4.33 for a 10 year increase in age. Using the table opposite, as a rough and ready estimation, we see that this corresponds to an increased probability of death of 67% for five years and 80% for ten years.

35.5 Finding individual HR scores

As we have done in other regression models, now we know the values of the β's, we can substitute the values for a specific subject and get a predicted hazard ratio value back. However, we do no know the value of the baseline hazard, and this is not calculated via the Cox model, so all we have is the relative hazard to some theoretical baseline risk group, luckily, this is not usually a problem. Taking our Ovarian cancer data, let's consider two scenarios, the hazard for a 30 yr old receiving mono-therapy (=1) and a 60 year old receiving combination therapy (=0).

Taking the 30 yr old receiving mono-therapy (=1):

$$\lambda_i(t) = h_0(t)\exp(0.451x_{treat\_i} + 1.158x_{age\_i}) \Rightarrow h_i(t) = h_0(t)(e^{0.451x_{treat\_i}})(e^{1.158x_{age\_i}})$$

$$h_i(t) = h_0(t)(e^{(0.451)(1)})(e^{(1.158)(30)}) \Rightarrow h_0(t)(e^{(0.451)})(e^{34.74}) \Rightarrow h_0(t)e^{.451+43.74} \Rightarrow h_0(t)e^{44.191}$$
$$= h_0(t)(3.788)$$

Therefore the hazard for our 30 year old on mono-therapy is just under 4 times the baseline risk. You may be wondering why the '+'s become multiplication. This is because one of the rules of exponents state that $x^{(a+b+c)}=x^a x^b x^c$. Think about it this way $2^2=(2)(2)$ and also $2^3 =(2)(2)(2) = 2^{(1+1+1)}$ Also incidentially $x^0= 1$, so any covariate that has a zero value for beta means that the value in effect represents a multiplier of one so it have no effect on the baseline hazard as would be expected. Our 3.788, using the HR/probability table is equivalent to an increase in risk of 80%.

Taking the 60 year old receiving combination therapy (=0):

$$h_i(t) = h_0(t)(e^{(0.451)(0)})(e^{(1.158)(60)}) \Rightarrow h_0(t)(e^{(0.451)})(e^{34.74}) \Rightarrow h_0(t)e^{0+69.48} \Rightarrow h_0(t)e^{69.48}$$
$$= h_0(t)(4.241)$$

So the hazard for our 60 year old on combination therapy is just over 4 times the baseline risk and once again equals an increase in risk of around 80%. You can also produce survival curves for particular subjects, described in the next section.

35.6 Checking the model

While articles tend not to mention checking the various characteristics required of the data to produce a valid Cox PH model this is usually carried out behind the scenes.

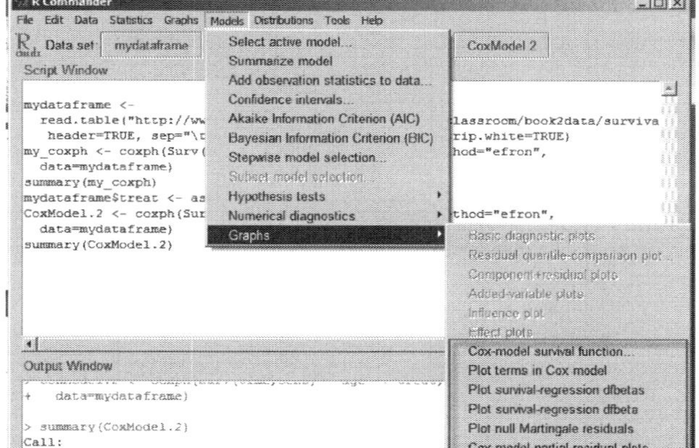

One way to do this is to inspect various graphs which can easily be obtained from the R Commander menu, shown opposite.

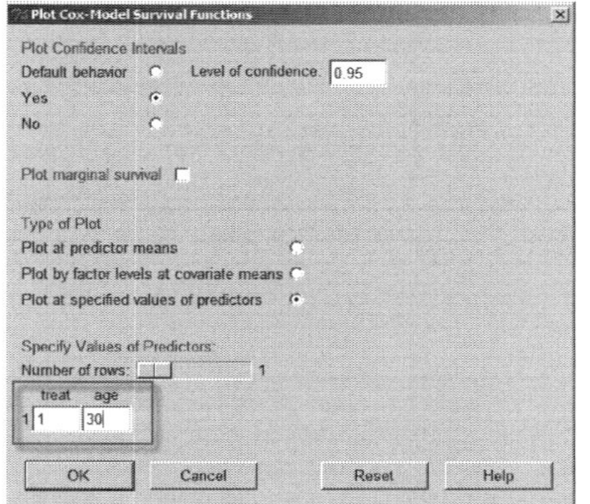

You can also investigate the modelled survival rates for individuals with different values for the various predictors, (covariates). For example taking two separate individuals, both belonging to treatment group 1 (=monotherapy) one having an age of 30 and the second an age of 60 can be used to produce specific survival curves. Also by requesting 95% confidence intervals we can obtain the graphs shown below. The second result shows a very wide confidence interval because we have very small sample sizes.

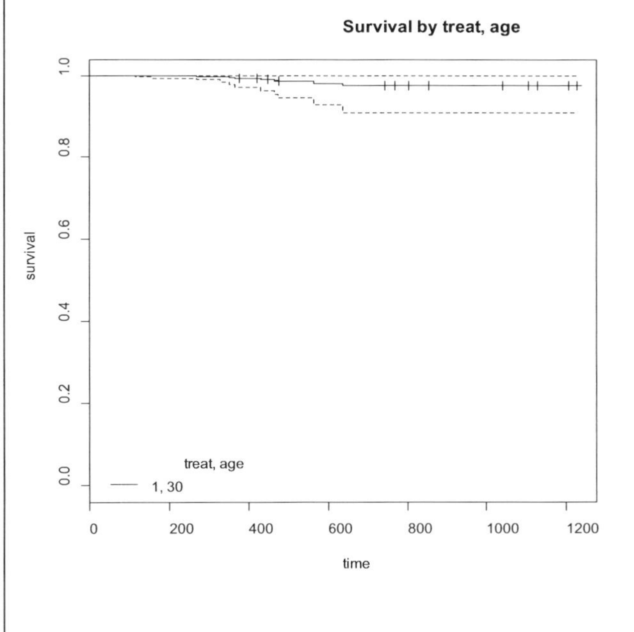

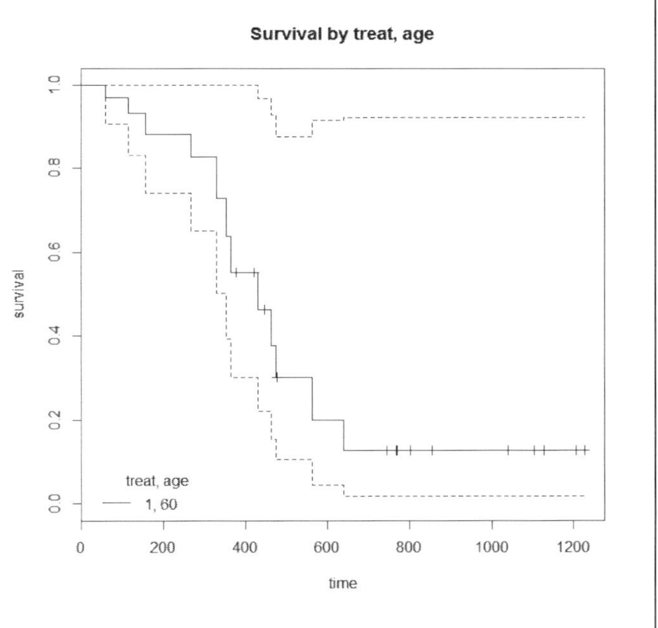

35.7 Assumptions, dangers and assessment of Cox regression

For Cox regression to produce valid results, there are several assumptions that the model makes, these are described well in both Campbell, 2006 and Norman & Streiner, 2009. Peat, Burton and Elliott, 2009 provide two checklists, one concerning the assumptions for carrying out a valid Cox regression (provided below) and another one listing issues to consider when reviewing an article regarding survival analysis.

Assumptions for standard survival analysis (from Peat, Barton & Elliott, 2009 p.131, adapted)

- Observations are independent, each person is only included once
- Survival prospects remain constant over the study period (although this can be modelled used advanced techniques)
- Censored observations have the same survival prospects as the non-censored participants

Also more generally:

- Degree of censoring and sample size needs to be considered for each time point.

35.8 Tips and Tricks

- For a first model always use the R Commander survival extension
- Always double check that you have coded the censored indicator the correct way round (event = 1)
- For covariates that are factors check that the level you consider to be the 'baseline' is the one you assume to have the least risk

In this chapter many aspects of survival analysis have not been considered:

- Many covariates with interaction. My website provides an example that considers 6 covariates along with interactions
- Modelling hazards that vary over time, known as time dependent covariates. Fox & Carvalalho, 2012 provide an example using the R Commander survival extension.
- Sometimes the Cox model might not fit the survival curve. If this is the case, a typical alternative is the frailty model (also called the Gaussian random effects model). Three packages in R provide methods for this type of model; *coxme* (Mixed Effects Cox Models), *frailtypack* and *frailtyHL*. The last package provides a model comparison fit measure (AIC) which can be useful in model selection.

Youtube help: http://youtu.be/wTLsw-Ckfvw

35.9 Techie notes: showing iteration details

I mentioned at the start of this chapter that the beta values are obtained by an iterative method. You can see this process by using using the *iter=* option in the *coxph()* function (Terry Therneau, personal communication, August 20, 2014).

Firstly, you can find the initial beta values by setting *iter=0*:

```
> fit0 <- coxph(formula = mysurv ~ treat + age, data=mydataframe, iter=0)
> fit0
Call:
coxph(formula = mysurv ~ treat + age, data = mydataframe, iter = 0)

      coef exp(coef) se(coef) z p
treat    0         1   0.6289 0 1
age      0         1   0.0377 0 1

Likelihood ratio test=0  on 2 df, p=1  n= 26, number of events= 12
```

```
fit0 <- coxph(formula = mysurv ~ treat + age, data=mydataframe, iter=0); fit0
```

This indicates that the initial values for both betas are set to 0.

To find the beta values for the first iteration we use the *iter=1* setting:

```
>   fit1 <- coxph(formula = mysurv ~ treat + age, data=mydataframe, iter=1)
> fit1
Call:
coxph(formula = mysurv ~ treat + age, data = mydataframe, iter = 1)


        coef exp(coef) se(coef)     z       p
treat -1.583     0.205   0.6978 -2.27 0.02300
age    0.158     1.171   0.0463  3.41 0.00064

Likelihood ratio test=14.2  on 2 df, p=0.000843  n= 26, number of events= 12
```

```
fit1 <- coxph(formula = mysurv ~ treat + age, data=mydataframe, iter=1); fit1
```

We see now that the initial estimates for both betas are slightly different from their final values of; *treat*=.7959; *age*=.1465 given in the previous section.

You can also compare the difference in the likelihood ratio test between specific iterations by setting the initial beta values to equal specific values using the *init=* option. The code below allows us to compare the likelihood between the initial value and the first iteration.

```
> fit2 <- coxph(formula = mysurv ~ treat + age, data=mydataframe, iter=1, init=fit1$coef)
>     fit2
Call:
coxph(formula = mysurv ~ treat + age, data = mydataframe, init = fit1$coef,
    iter = 1)

        coef exp(coef) se(coef)     z      p
treat -0.684     0.505   0.6299 -1.09 0.2800
age    0.145     1.156   0.0459  3.15 0.0016

Likelihood ratio test=1.63  on 2 df, p=0.443  n= 26, number of events= 12
```

```
fit2 <- coxph(formula = mysurv ~ treat + age, data=mydataframe, iter=1, init=fit1$coef)

fit2
```

The default number of iterations used to try to achieve convergence for the *coxph()* function is 20 which you can override by setting the *iter.max=* option. Here I have set it to 50.

```
> fit3 <- coxph(formula = mysurv ~ treat + age, data=mydataframe, iter.max = 50)
>     fit3
Call:
coxph(formula = mysurv ~ treat + age, data = mydataframe, iter.max = 50)


        coef exp(coef) se(coef)     z      p
treat -0.796     0.451   0.6329 -1.26 0.2100
age    0.147     1.158   0.0459  3.20 0.0014

Likelihood ratio test=15.8  on 2 df, p=0.000367  n= 26, number of events= 12
```

```
fit3 <- coxph(formula = mysurv ~ treat + age, data=mydataframe, iter.max = 50); fit3
```

```
> fit2$iter
[1] 1
> fit3$iter
[1] 5
```

You can find the number of iterations it took to achieve the required level of convergence using either using the structure function, *str()* and looking at the *iter* attribute which provides the number used, or extracting it directly using the expression *cox_model$iter*, substituting *cox_model* for your cox model object, for example *fit1$iter*.

```
> str(fit3)
List of 18
 $ coefficients     : Named num [1:2] -0.796 0.147
  ..- attr(*, "names")= chr [1:2] "treat" "age"
 $ var              : num [1:2, 1:2] 0.40061 0.00491 0.00491 0.0021
 $ loglik           : num [1:2] -35 -27.1
 $ score            : num 18.6
 $ iter             : int 5
```

We see that it took 5 iterations to achieve the required level of convergence. The level of convergence required is set by the epsilon option, *eps =* where the default is 1e-09.

The above code can be adapted by wrapping it in a control loop and obtaining the log likelihood value for each of the iterations:

setup the base model

repeat the loop for the number of times it took to converge in the original model

run the model for a single iteration

concatenate the; initial, new and difference between the log likelihood for each iteration and then add a line return character

save the model so you can use to the betas from it the next iteration

```
fit_test <- fit0

for (i in 0:my_coxph$iter) {

fit_test2 <- coxph(formula = mysurv ~ treat + age, data=mydataframe, iter=1,
init=fit_test$coef)

cat(fit_test2$loglik[2],"    ", fit_test2$loglik[1],"    ",fit_test2$loglik[2] -
fit_test2$loglik[1],"    ", "\n")

fit_test <- fit_test2

}
```

```
> fit_test <- fit0
> for (i in 0:my_coxph$iter) {
+ fit_test2 <- coxph(formula = mysurv ~ treat + age, data=mydataframe, iter=1, init=fit_test$coef)
+ cat(fit_test2$loglik[2],"    ", fit_test2$loglik[1],"    ",fit_test2$loglik[2] - fit_test2$loglik[1],"    ", "\n")
+ fit_test <- fit_test2
+ }
-27.90688      -34.98494      7.078063
-27.0923      -27.90688      0.8145755
-27.07377      -27.0923      0.01853367
-27.07377      -27.07377      1.394223e-06
-27.07377      -27.07377      3.197442e-14
-27.07377      -27.07377      3.552714e-15
```

difference gets smaller until it reaches epsilon

The *coxph.details(cox_model)* function provides additional information from the *cox_model* object, details of which can be find in the survival package manual.

36 Graphical summaries of data: aggregation

Often an analysis process involves more manipulation of the data than actual statistical analysis. An example of this is creation of aggregated datasets that consist of summary values from other datasets. In this chapter we will consider a typical problem: we want to plot hourly wage against years working at a health institution and have the data in the following format.

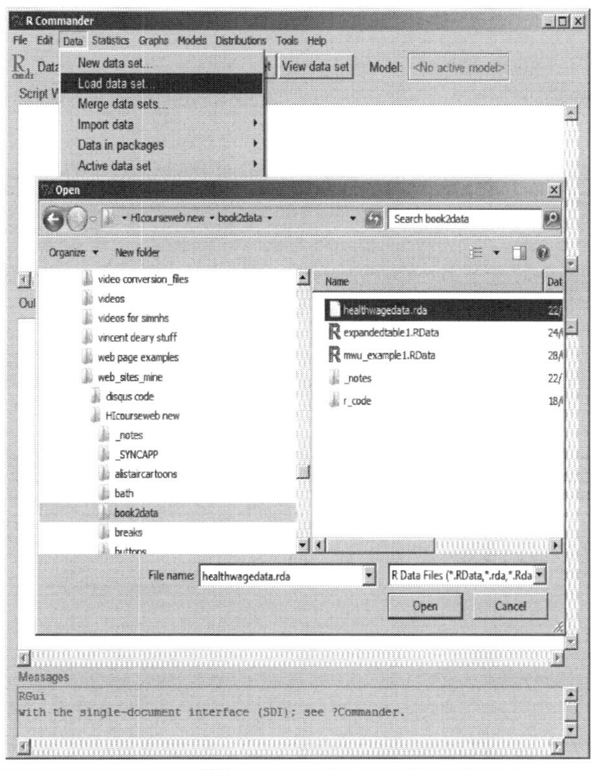

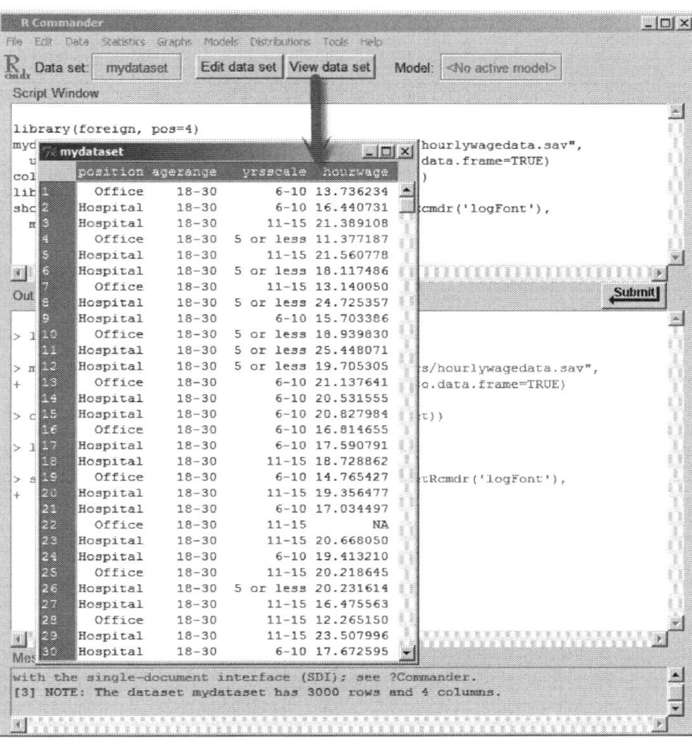

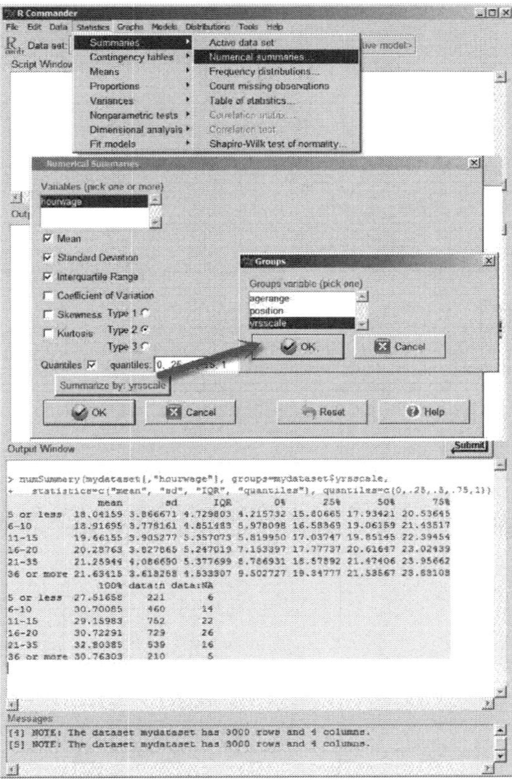

The first stage in the solution is to obtain the data; get either the *healthwagedata.sav* **or** the *healthwagedata.rda*, file from one of the urls below and store it on your local device.

http://www.robin-beaumont.co.uk/virtualclassroom/book2data/healthwagedata.rda
or
http://www.robin-beaumont.co.uk/virtualclassroom/book2data/healthwagedata.sav

The above left screenshot shows how to load the rda file.
We see there are many entries for each *yrsscale* (time worked with institution). While the *hourwage* shows the average hourly wage (top right).

Before we do anything let's check what the summary values are for each level of employment time using the menu option **statistics -> summaries -> numeric summaries** and setup the dialog box as shown opposite. Clearly the mean and median hourly rate go up with years employment, from 18 to 21.63.

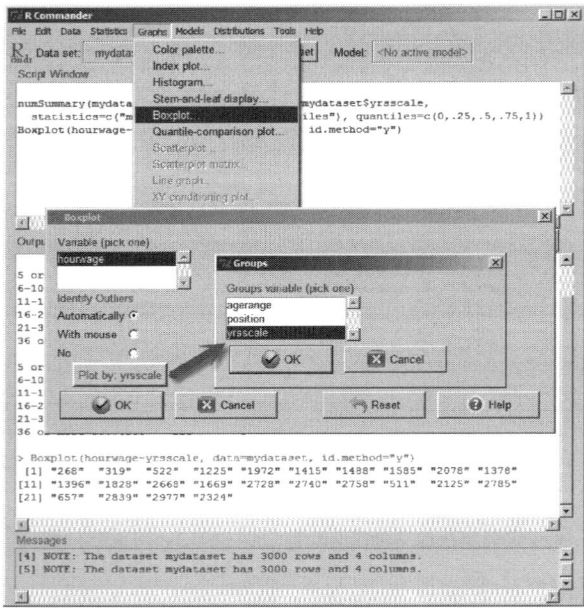

Because of the multiple hourly wage values for each level of employment (*yrsscale*), time a scatter plot of the raw data is not appropriate but we have two options:

- produce a series of boxplots or means for each group or
- aggregate the data, for example find the mean at each hourly wage against employment time and then plot these values.

We can easily produce a boxplot of the above findings. Select the menu option:

Graphs-> Boxplot

Setting the **Identify Outliers** option to *Automatically* produces a boxplot with the case numbers of extreme values marked.

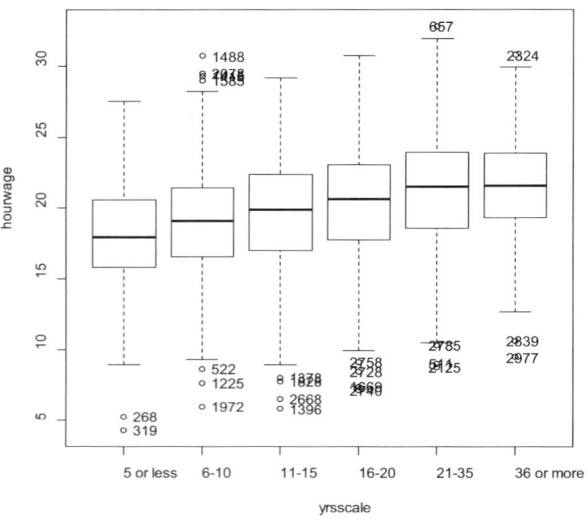

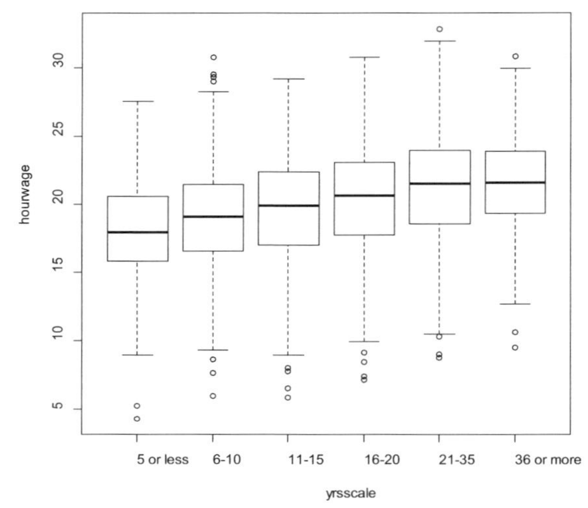

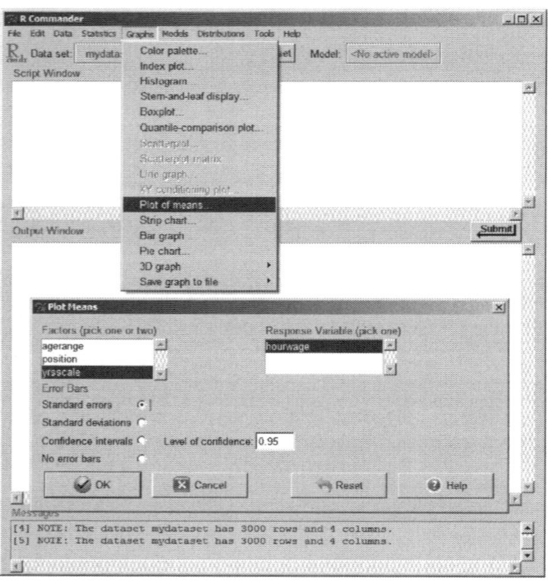

In contrast selecting the **Identify Outliers** option to *No* we now have a clearer, but possibly less useful, graph (above right).

Asking the question 'what do the many outliers suggest?' would require knowledge of the context in which the data was collected. For example, they might be miscoded values or a particular distinct subset of employees such as consultants. A definitive answer needs detailed knowledge of the environment from where the data was collected. Ignoring the outliers and assuming that the data are normally distributed at each *number of years* employment level we can produce a graph of means at each level:

Graphs->plot of means

We can also request the display of **Error Bars,** selecting the *standard errors* option we can see the estimated accuracy of the mean for each group.

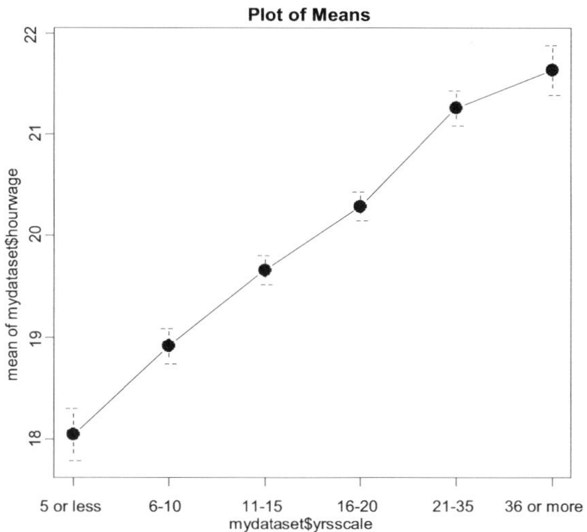

I feel that presenting the data like this does it a disservice as it now appears very clean giving no indication of those very low and high paid workers!

Notice that the x categories are in the correct order but this is not always the case, the *.rda* and *.sav* files contained additional information specifying the factor level order. However if we had used a plan text file (i.e. *.dat* or *.txt*) you would have needed to reorder the factor levels by using the R Commander menu option:

Data ->Manage variables in active dataset ->Reorder factor levels. . .

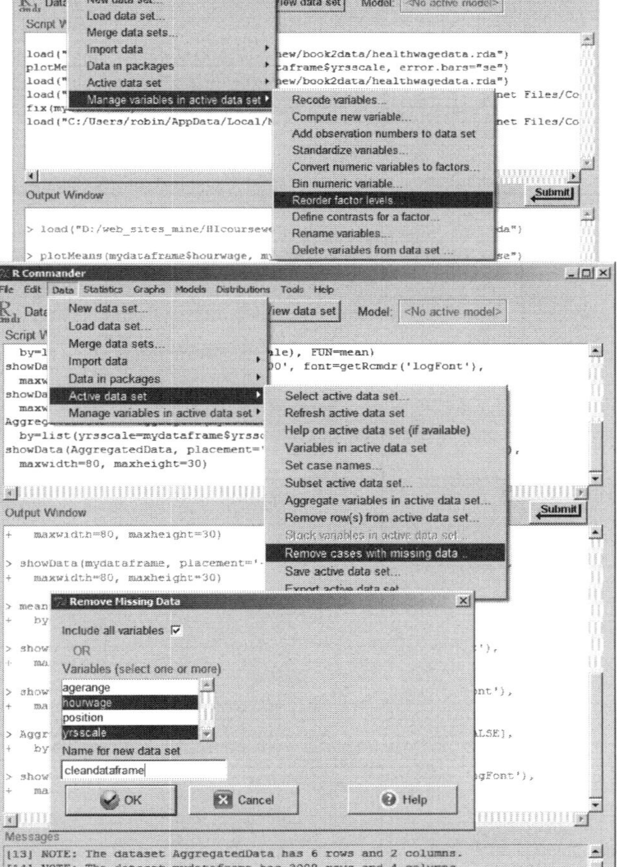

36.1 Removing missing values

The alternative strategy to the above is to produce a new dataframe which only consists of the summary values. To do this we first need to remove all those rows which have empty values for either the *hourwage* or *yrsscale* variables.

Select the menu option:

Data->Active data set->Remove cases with missing data

I have called the new dataframe *cleandataframe* which is automatically loaded and has 89 fewer records now.

36.2 Aggregating data

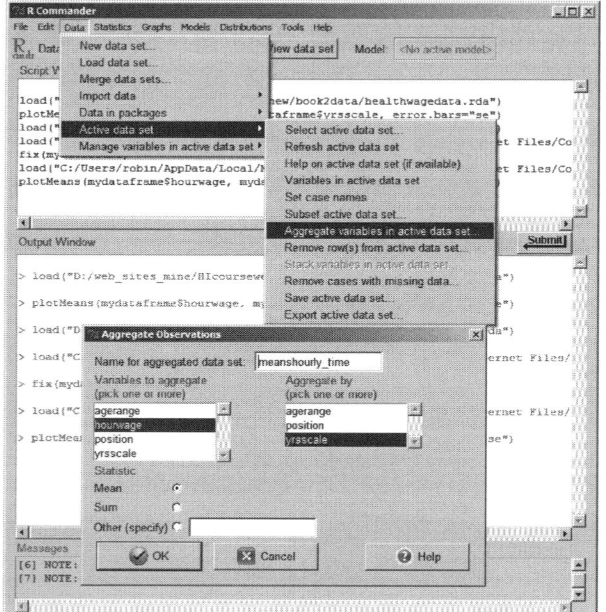

Aggregating data is the process of creating a new dataset from summary values of a parent dataset. This is a common occurrence with large datasets and this scenario provides a good example.

Having removed all the cases with missing data we can now create a new dataframe with just the aggregated data (i.e. the means) by selecting the menu option:

Data->Active data set

->Aggregate variables in active data set

Complete the dialog box as shown opposite.

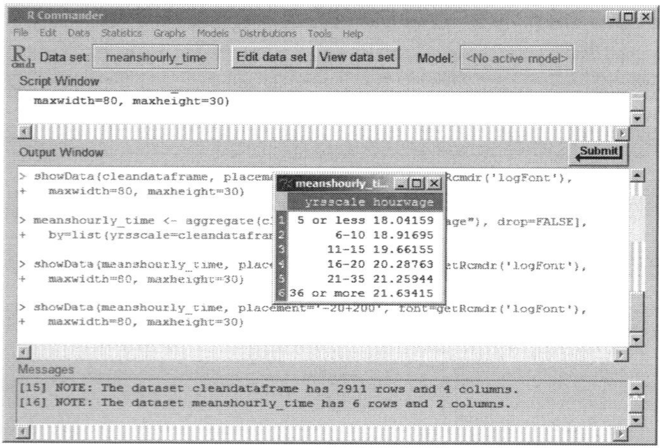

Notice that the new dataframe is automatically loaded.

The new dataframe has 6 records.

Often aggregated values are divided by factor levels as is the case with the *yrsscale* factor opposite. This can cause problems. Referring back to the original question; the need to plot hourly wage against years working. We therefore need the *yrsscale* variable to be an ordinal/ratio level variable, that is to be a numeric variable in R. This is because it is not possible to produce a scatterplot of a factor level against another variable.

We will use the Data Editor window to edit and convert the *yrsscale* variable.

Click the **Edit data set** button we can edit the new dataframe.

When you have finished make sure you close it by clicking on the X button on the top right hand side of the window.

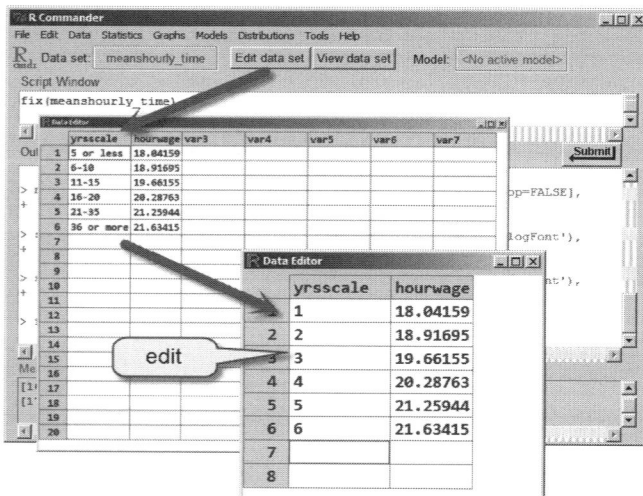

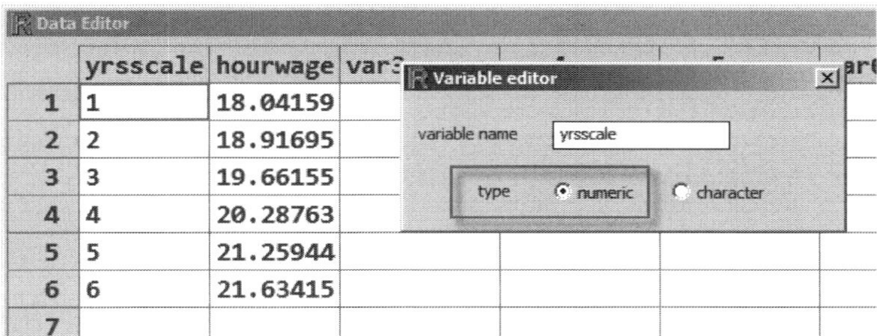

Once again click on the **Edit data set** button this time selecting the top of the *yrsscale* column and change the variable to numeric.

When you have finished make sure you close both the variable editor and the data editor windows with the X button.

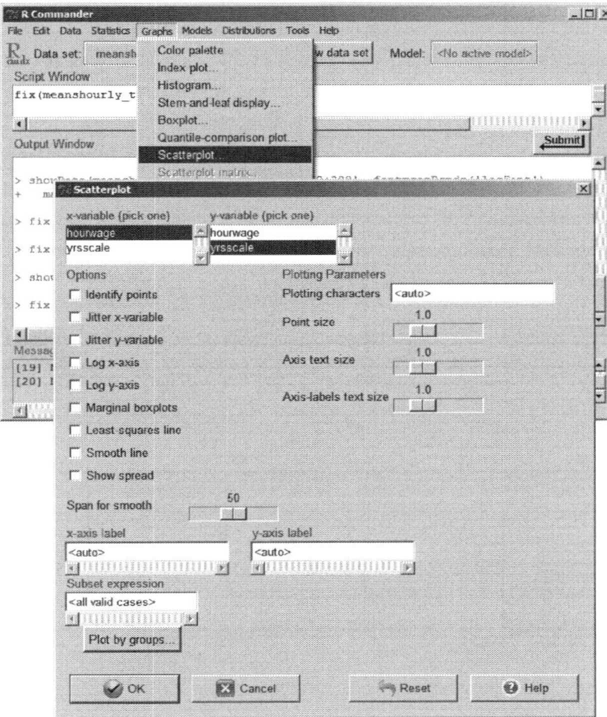

Now we can produce the scatterplot.

Select the menu option:

Graphs->Scatterplot

Complete the dialog box as shown opposite.

The result is shown below. Depending upon your needs you might feel this graph is inferior to the set of boxplots produced earlier. While I would agree, a group of managers might prefer to use the very strong message this one provides over the more messy boxplots.

36.3 Tips and Tricks

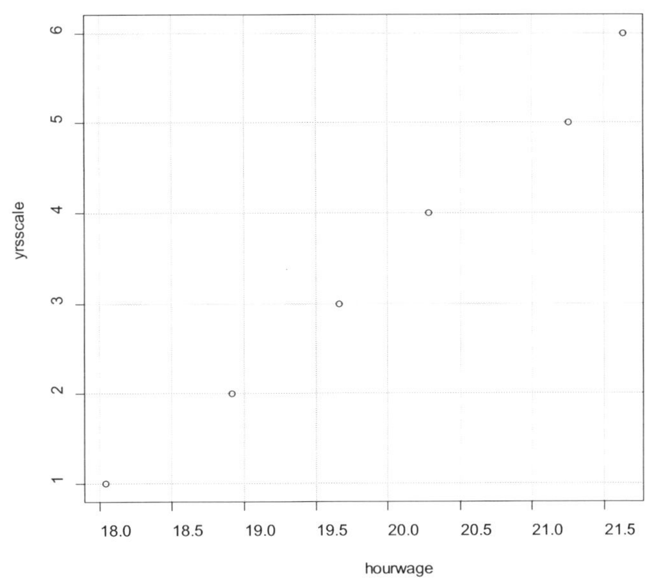

This chapter has worked through a typical problem which involved little statistical knowledge but a fair level of skill in data manipulation.

Often manipulation of data requires several stages and it is a good idea to draw some type of flow chart to help structure your thoughts and what you want the outcome to be.

37 Paired nominal data: comparing proportions using McNemar's test

Paired design

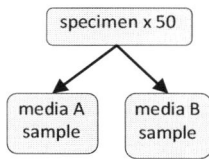

specimen x 50

media A sample | media B sample

I have included this test because it easily allows analysis of a frequently met research design, which I will demonstrate by using an example from Armitage, 2002 p.121.

Consider the situation where you have fifty specimens of sputum and you culture them on two different media A and B. You wish to compare the ability of the two media to detect tubercle bacilli. You will measure this simply as an increase, equivalent or decrease in growth for the media. We are in effect considering each specimen as a pair. Notice that we have actually 100 measurements (samples) and 50 data pairs (specimens). For various reasons we are only interested where there is a change represented in the shaded rows/cells in the tables below. You can think of the other values as ties and left out of the analysis, as they were in the various rank statistics we considered in previous chapters.

	Sample		
outcome	1	2	Number of pairs
1	*A*	*A*	*k*
2	*A*	Not *A*	*r*
3	Not *A*	*A*	*s*
4	Not *A*	Not *A*	*m*

	Medium		
outcome	1	2	Number of sputa pairs
1	+	+	*20*
2	+	-	*12*
3	-	+	*2*
4	-	-	*16*
			50

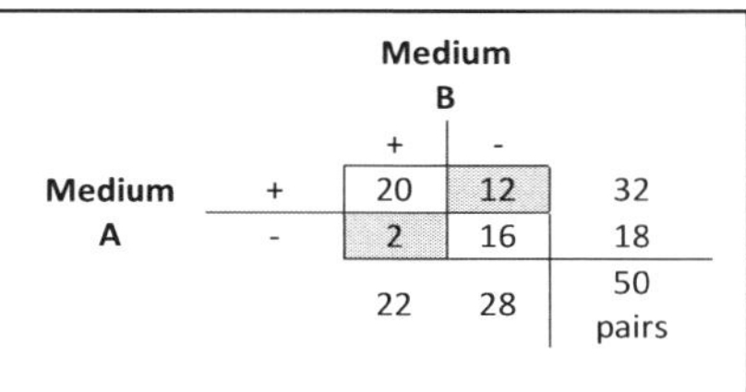

		Sample 2		
		A	Not A	
Sample 1	A	k	r	k + r
	Not A	s	m	s + m
		k + s r + m		N

		Medium B		
		+	-	
Medium A	+	20	12	32
	-	2	16	18
		22	28	50 pairs

If there was no difference in the two culture mediums one would expect roughly equal numbers in both the shaded cells, in other words π_0=0.5 and this is our null hypothesis a proportion of .5. So the expected number is simply (r+s)/2 which is 14/2 = 7 in each shaded cell. Inspecting our data, it seems to deviate a long way from this, indicating that one culture medium is better at detecting tubercle bacilli than the other. What we need is a p-value and/or confidence interval to allow us a method of assessing this deviation.

37.1 Carrying out the analysis in R

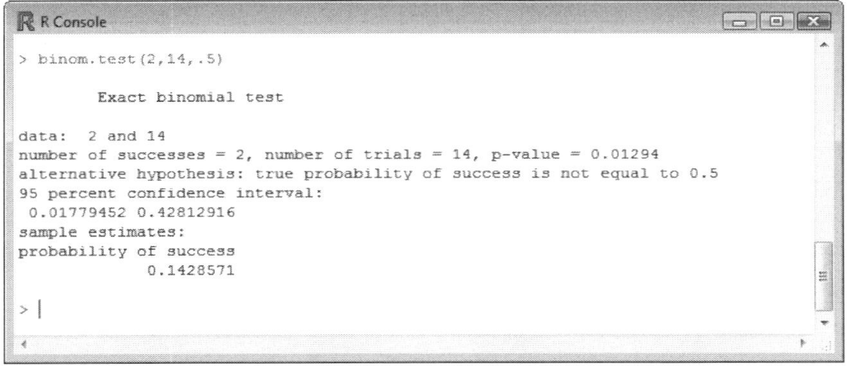

```
R Console
> binom.test(2,14,.5)

        Exact binomial test

data:  2 and 14
number of successes = 2, number of trials = 14, p-value = 0.01294
alternative hypothesis: true probability of success is not equal to 0.5
95 percent confidence interval:
 0.01779452 0.42812916
sample estimates:
probability of success
             0.1428571

> |
```

In essence we have what is known as the binominal test and you can easily carry this out in R using the single function, *binom.test(observed, total, expected proportion)* as shown opposite. This provides an exact p-value (Everitt & Hothorn, 2010). Note that the second value is the total number expected.

Notice that you also get the 95% confidence interval.

For large sizes (counts of more than 30) the sampling distribution approximately follows the chi-squared distribution and you can use the actual McNemar's test in R (Marques de Sá, 2007 p.205-6). The screenshot opposite demonstrates how to do this.

```
> x<- array(c(20,2, 12, 16), dim=c(2,2))
> mcnemar.test(x)

        McNemar's Chi-squared test with continuity correction

data:  x
McNemar's chi-squared = 5.7857, df = 1, p-value = 0.01616
```

```
x<- array(c(20,2, 12, 16), dim=c(2,2))

mcnemar.test(x)
```

37.2 Effect size

From the above example we can calculate the difference in the proportions by:

```
2/50 - 12/50 =   -0.2          or          12/50 - 2/50  = 0.2

(medium A decrease and B increase)/total - (medium A increase and B decrease)/total
or

(medium A increase and B decrease)/total  - (medium A decrease and B increase)/total
```

So we can see that the effect size between the two mediums is 0.2. Multiplying .2 by 100 gives the percentage of 20% indicating the difference between the proportions of growth in each of the mediums.

We can also calculate a value called the **odds ratio** which is: 12/2= 6 or 2/12 = 0.166. For further details about the odds ratio see the glossary entry. Taking the first odds ratio:

```
12/2 =  (medium A increase and B decrease)/(medium A decrease and B increase) = 6
```

Therefore, the odds of medium A increased and B decreased growth is 6 times that of medium A decreased and B increased growth. You can convert odds to probabilities by using the formula:

		Medium B +	Medium B -	
Medium A	+	20	12	32
	-	2	16	18
		22	28	50 pairs

```
Probability =  odds/(odds+1)
```

An odds of 6 is equivalent to a probability of 6/(6+1) = 6/7 = 0.8571. Similarly an odds of 0.166 has a probability of .166/(0.166+1) = 0.1423 indicating we have a much higher probability (read higher odds) of the situation occurring, that is medium A increased growth and medium B decreased growth. The probability conversions are sometimes difficult to interpret, so it is often best just to stick with the odds.

37.3 Tips and tricks

For something that appears so simple people often find McNemar's test difficult to understand, for a very good alternative explanation try:

http://stats.stackexchange.com/questions/76875/what-is-the-difference-between-mcnemars-test-and-the-chi-squared-test-and-how

For confidence intervals use the *mcnemar.exact(x)* function in the *exact2x2* library.

```
> mcnemar.exact(x)

        Exact McNemar test (with central confidence intervals)

data:  x
b = 12, c = 2, p-value = 0.01294
alternative hypothesis: true odds ratio is not equal to 1
95 percent confidence interval:
  1.335744 55.197091
sample estimates:
odds ratio
         6
```

```
install.packages("exact2x2");  library(exact2x2)

mcnemar.exact(x)
```

There are two extensions to McNemar's test allowing the analysis of repeated measure nominal data with more than two categories. The Stuart-Maxwell statistic (also known as Stuart's W_0) was the first one proposed and the *mh_test()* function in the coin package performs it. The *mh_test()* function requires a dataframe so we use the *as.table()* function to convert our array first.

```
> tablex <- as.table(x)
> mh_test(tablex)

        Asymptotic Marginal-Homogeneity Test

data:  response by groups (Var1, Var2)
         stratified by block
chi-squared = 7.1429, df = 1, p-value = 0.007526
```

```
install.packages("coin"); library(coin)
tablex <- as.table(x)
mh_test(tablex)
```

Alternatively, the same statistic is provided in the *irr* package, which does not require the data to be converted into a dataframe:

```
> stuart.maxwell.mh(x)
 Stuart-Maxwell marginal homogeneity

 Subjects = 50
   Raters = 2
    Chisq = 7.14

 Chisq(1) = 7.14
  p-value = 0.00753
```

```
install.packages("irr"); library(irr)

stuart.maxwell.mh(x)
```

The *irr* package also provides an improved version of the Stuart-Maxwell statistic known as the *marginal homogeneity statistic W of Bhapkar*. You can obtain the result using the *bhapkar(x)* function. Unfortunately, unlike the *stuart.maxwell.mh(x)* function it only takes raw data rather than an array of counts. Alternatively, there is an R script available from the links below which does work with a table of counts:

http://www.karlin.mff.cuni.cz/~omelka/Soubory/nmsa331/cviceni-10/bhapkar.test.R

or

http://www.robin-beaumont.co.uk/virtualclassroom/book2data/bhapkar_test_R.txt

```
> bhapkar.test(x)
Bhapkar's chi-squared test of marginal homogeneity in a squared table
Data:  x

W = 8.333333,  df = 1,  p-value = 0.003892417

Estimated marginal probabilities:
    category:  1    2
         row:  0.64  0.36
      column:  0.44  0.56
  difference:  -0.2  0.2
```

Select the R code on either of these pages and then paste it into the *R Console* window. You now have a function called *bhapker.test()* available.

If you just want W it is easily derived from the Stuart-Maxwell statistic (*W_0*) where the formula is:

```
> 7.14/(1-7.14/50)
[1] 8.329445
```

```
W = W_0 / (1 - W_0 / n) # n is the number of pairs
```

This provides the same value as that from the *bhapkar.test()* function.

38 Managing your data and R

The data management techniques we have discussed in the book so far are fine for the occasional user, or one who wishes to use R once for a specific project. However, if you plan to use R regularly for a whole range of statistical tasks it is necessary to develop further skills and this is the purpose of this section. Three specific aspects are considered; data management (both at the file and individual datum level), the R environment and using two tools to help with code development.

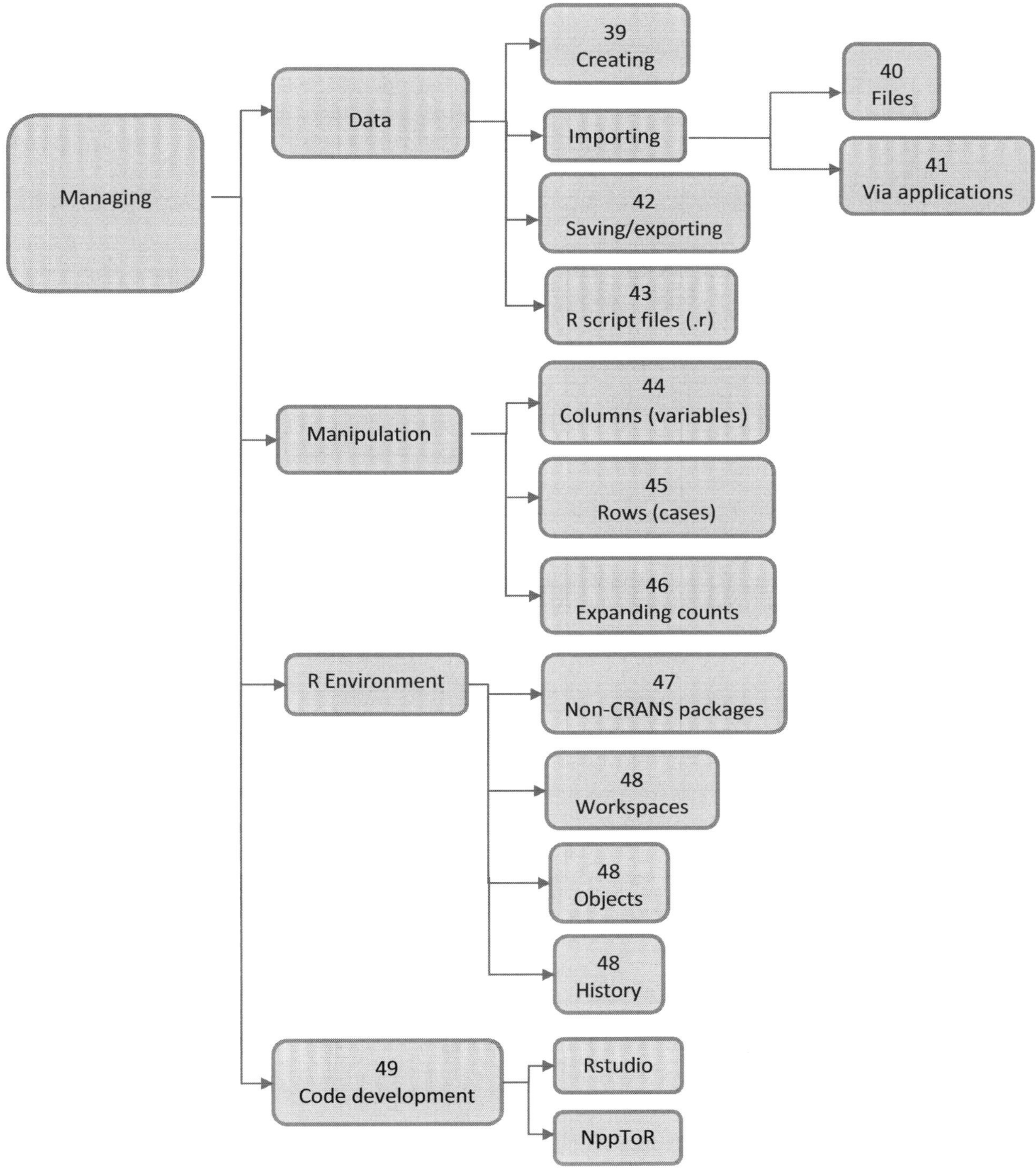

39 Creating datasets and distributions in R Commander and R

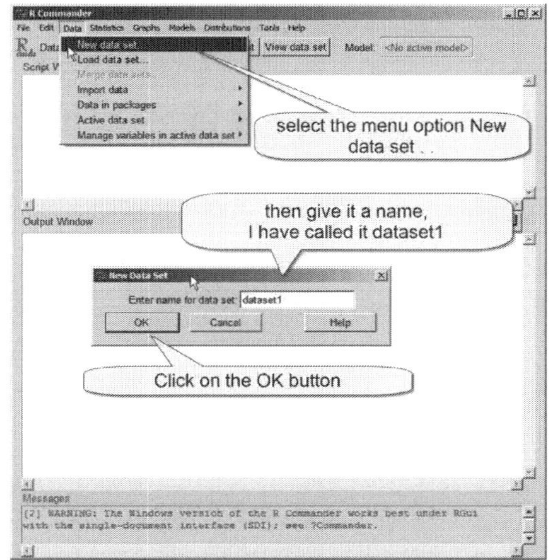

In most of the chapters in this book the datasets had already been created for you. If you want to create your own dataset (strictly an R dataframe), you can do so by using a menu option in either R Commander or R, or by using R code.

39.1 R Commander

To create a dataset using R Commander select the menu option:

Data->New dataset

You just need to provide a name for the dataset, anything will do and I often just call it *dataset1*.

The *Data Editor* window then appears. This being the same window you see if you had clicked on the **Edit data set** button.

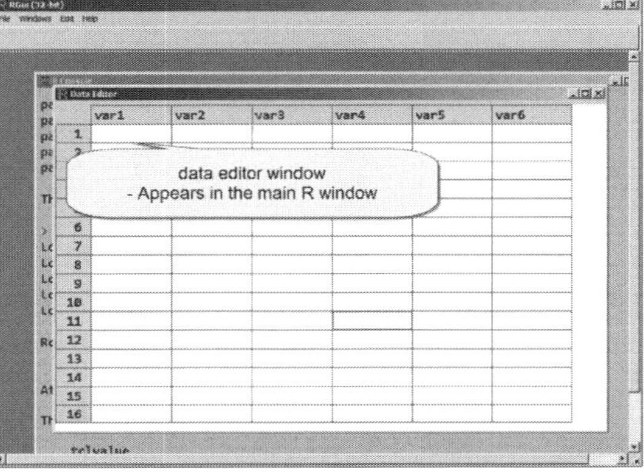

The window appears as a cross between a very basic version of *Excel* and *SPSS*. If you click on the top of the columns you can add a name to each, which you can then refer to in the various menu options.

Although you have given the dataset a name you must enter some values before it is recognised as such.

Once you have entered the necessary values. Remember to save your data, see the chapter *Saving and exporting your work and data* for details.

39.2 In R

From within R itself there are several ways to create a dataset. . .

39.2.1 Edit->Data editor menu option

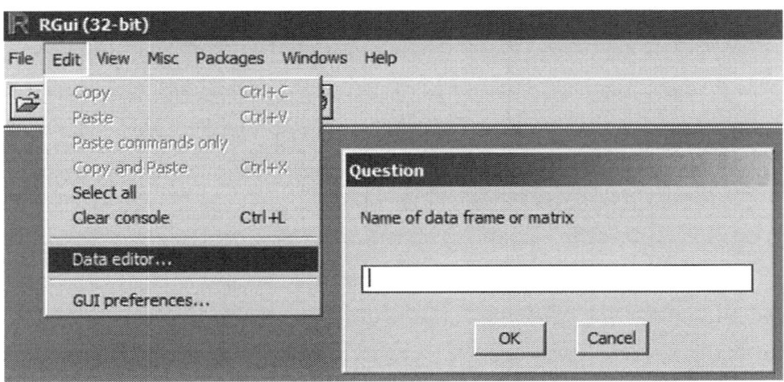

The following menu option brings up the dataset editor window, equivalent to the above R Commander screenshot. As in R Commander if there is no active dataset (dataframe) a dialog box, this time called *Question,* appears requesting a name for the new dataset.

Edit->Data editor . . .

As mentioned above, remember to save your newly created data.

39.2.2 R Code

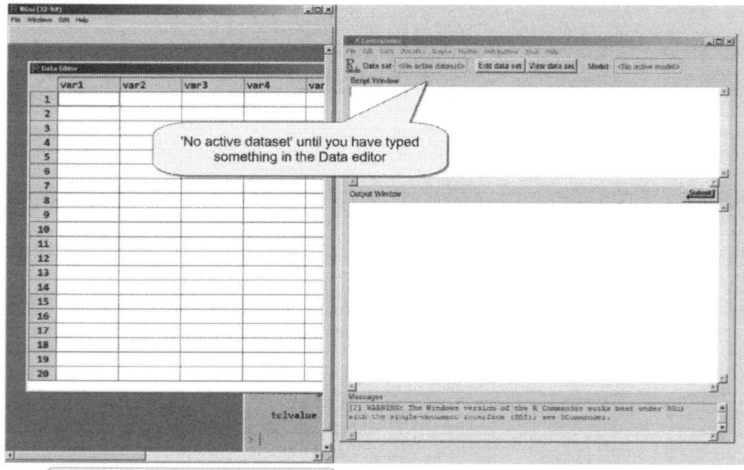

To create a dataframe in R directly is slightly more complex. As an example, assume you want to create a new dataframe called *test1*, consisting of two columns, one called group with three factor levels and a second one called *var2* with numeric values. To achieve this you would type the following into the R Console window. Remember, if you put the values in quotes, this instructs R to create a factor.

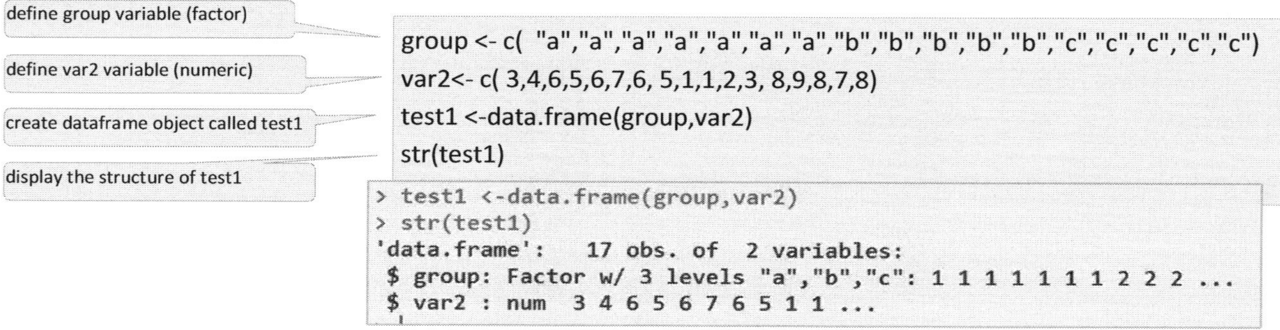

define group variable (factor)
define var2 variable (numeric)
create dataframe object called test1
display the structure of test1
group <- c("a","a","a","a","a","a","a","b","b","b","b","b","c","c","c","c","c")
var2<- c(3,4,6,5,6,7,6, 5,1,1,2,3, 8,9,8,7,8)
test1 <-data.frame(group,var2)
str(test1)
> test1 <-data.frame(group,var2)
> str(test1)
'data.frame': 17 obs. of 2 variables:
$ group: Factor w/ 3 levels "a","b","c": 1 1 1 1 1 1 1 2 2 2 ...
$ var2 : num 3 4 6 5 6 7 6 5 1 1 ...

39.3 Creating distributions

Sometimes it is desirable to create a whole set of scores from a theoretical distribution; this can be achieved several ways. R Commander provides menu options for most distributions – what follows is a demonstration for a normal distribution.

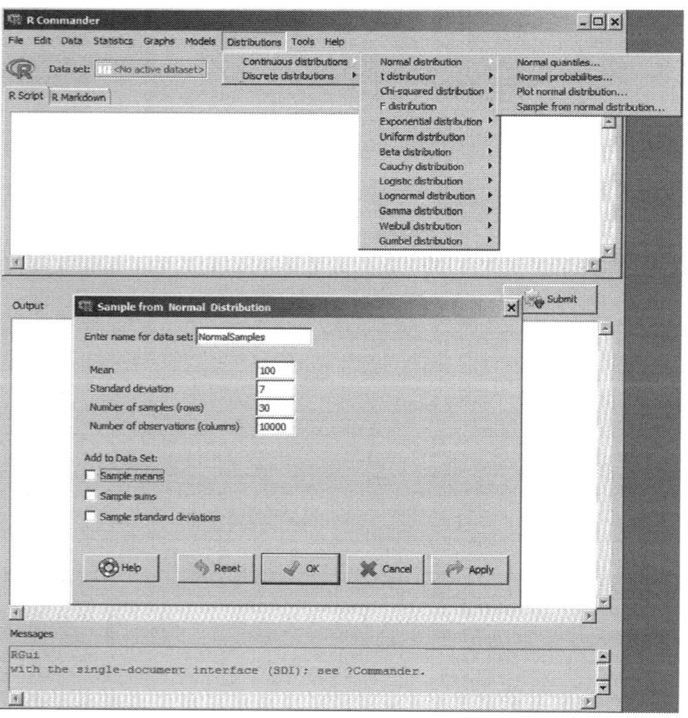

39.3.1 R Commander

To produce samples from a normal distribution, from within R Commander select the menu option:

Distributions-> Continuous distributions
->Normal distributions
-> sample from normal distribution

You can then specify the number of samples and size of each sample required. There is also an option to request a summary value for each sample.

I have requested 30 samples each consisting of 10,000 observations from a normally distributed population with a mean of 100 and a standard deviation of 7. I have given the resultant dataframe the name *NormalSamples*.

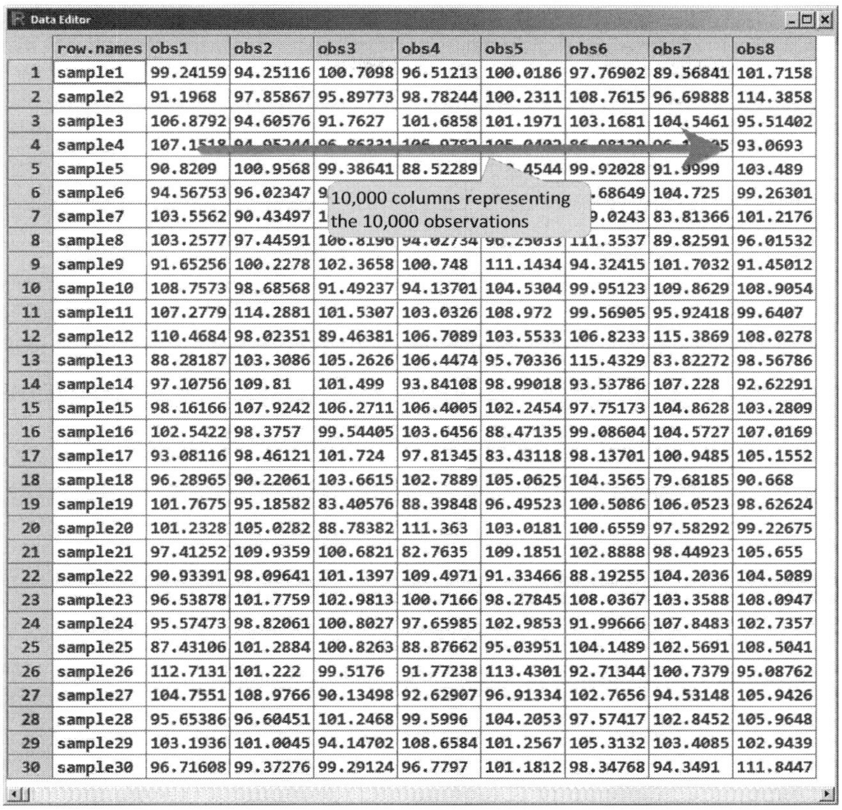

Data Editor

	row.names	obs1	obs2	obs3	obs4	obs5	obs6	obs7	obs8
1	sample1	99.24159	94.25116	100.7098	96.51213	100.0186	97.76902	89.56841	101.7158
2	sample2	91.1968	97.85867	95.89773	98.78244	100.2311	108.7615	96.69888	114.3858
3	sample3	106.8792	94.60576	91.7627	101.6858	101.1971	103.1681	104.5461	95.51402
4	sample4	107.1518	[illegible]	[illegible]	[illegible]	[illegible]	[illegible]	[illegible]	93.0693
5	sample5	90.8209	100.9568	99.38641	88.52289	[illegible]	99.92028	91.9999	103.489
6	sample6	94.56753	96.02347	[illegible]	[illegible]	[illegible]	[illegible]	104.725	99.26301
7	sample7	103.5562	90.43497	[illegible]	[illegible]	[illegible]	[illegible]	83.81366	101.2176
8	sample8	103.2577	97.44591	[illegible]	[illegible]	[illegible]	111.3537	89.82591	96.01532
9	sample9	91.65256	100.2278	102.3658	100.748	111.1434	94.32415	101.7032	91.45012
10	sample10	108.7573	98.68568	91.49237	94.13701	104.5304	99.95123	109.8629	108.9054
11	sample11	107.2779	114.2881	101.5307	103.0326	108.972	99.56905	95.92418	99.6407
12	sample12	110.4684	98.02351	89.46381	106.7089	103.5533	106.8233	115.3869	108.0278
13	sample13	88.28187	103.3086	105.2626	106.4474	95.70336	115.4329	83.82272	98.56786
14	sample14	97.10756	109.81	101.499	93.84108	98.99018	93.53786	107.228	92.62291
15	sample15	98.16166	107.9242	106.2711	106.4005	102.2454	97.75173	104.8628	103.2809
16	sample16	102.5422	98.3757	99.54405	103.6456	88.47135	99.08604	104.5727	107.0169
17	sample17	93.08116	98.46121	101.724	97.81345	83.43118	98.13701	100.9485	105.1552
18	sample18	96.28965	90.22061	103.6615	102.7889	105.0625	104.3565	79.68185	90.668
19	sample19	101.7675	95.18582	83.40576	88.39848	96.49523	100.5086	106.0523	98.62624
20	sample20	101.2328	105.0282	88.78382	111.363	103.0181	100.6559	97.58292	99.22675
21	sample21	97.41252	109.9359	100.6821	82.7635	109.1851	102.8888	98.44923	105.655
22	sample22	90.93391	98.09641	101.1397	109.4971	91.33466	88.19255	104.2036	104.5089
23	sample23	96.53878	101.7759	102.9813	100.7166	98.27845	108.0367	103.3588	108.0947
24	sample24	95.57473	98.82061	100.8027	97.65985	102.9853	91.99666	107.8483	102.7357
25	sample25	87.43106	101.2884	100.8263	88.87662	95.03951	104.1489	102.5691	108.5041
26	sample26	112.7131	101.222	99.5176	91.77238	113.4301	92.71344	100.7379	95.08762
27	sample27	104.7551	108.9766	90.13498	92.62907	96.91334	102.7656	94.53148	105.9426
28	sample28	95.65386	96.60451	101.2468	99.5996	104.2053	97.57417	102.8452	105.9648
29	sample29	103.1936	101.0045	94.14702	108.6584	101.2567	105.3132	103.4085	102.9439
30	sample30	96.71608	99.37276	99.29124	96.7797	101.1812	98.34768	94.3491	111.8447

10,000 columns representing the 10,000 observations

The data appears as shown opposite with each sample representing a row.

The R code produced from the above menu option is given below.

39.3.2 R Code

The R code produced by R Commander can be used as the basis for further development, for example you could change the parent distribution from normal *rnorm()* to Poisson *rpois()*. The original code is shown below.

convert the matrix to a dataframe

values from a normal population

produce 30 X 10,000 values

fit the values into a matrix with 10,000 columns

```
NormalSamples <- as.data.frame(matrix(rnorm(30*10000, mean=100, sd=7),   ncol=10000))
```

give the dataframe row names

```
rownames(NormalSamples) <- paste("sample", 1:30, sep="")
```

sample1 . . . . to sample30

give the dataframe column names

```
colnames(NormalSamples) <- paste("obs", 1:10000, sep="")
```

obs1 . . . . to obs10000

there are around 20 distributions in R you can use, for details type, **?Distributions()** in the R Console window.

By editing the various options in the above code such as; the number of values, population type and row/column names, many situations can be investigated. The online chapter *Poisson distribution bootstrapping* provides an example of this approach where graphs of confidence intervals and other statistics are produced as part of the process.

While R provides access to around 20 theoretical distributions, other packages extend this. For example the *ecodist* library contains a function *corgen()* which either returns a single column of the required correlation value to that of a supplied column or produces two columns of the specified correlation.

```
> library(ecodist)
> xval <- rnorm(30, mean=100, sd=7)
> len <- length(xval);  r= 0.8;
> yval <- corgen(len, xval, r, epsilon = 0.001)$y
> yval
           [,1]
 [1,] -0.01490344
 [2,] -0.70712702
 [3,] -2.13456610
 [4,] -0.13853892
 [5,] -1.33403511
 [6,]  0.52799321
```

In the code opposite the *ecodist* library is loaded then a column (vector) of thirty values from a normally distributed population (mean=100; SD=7) is produced. I then request a second column of the same length with a correlation to the first of 0.8 (+/- 0.001). The *epison* value controls the tolerance of the value of the correlation. Setting an *epison* value below 1 instructs R to continue resampling the population until a sample is found +/- the *epison* value.

```
install.packages("ecodist")
library(ecodist)
xval <- rnorm(30, mean=100, sd=7)
len <- length(xval);  r= 0.8;
yval <- corgen(len, xval, r, epsilon = 0.001)$y
yval
xy <- corgen(len=30, r=.8, epsilon=0.01)
xy
```

install and load the *ecodist* library

obtain 30 random values from a normal distribution (mean=100, sd=7)

specify the length and population correlation

just return the y column

correlation of r 0.80 +/- 0.001

you can also ask for the function to create both columns

```
> xy <- corgen(len=30, r=.8, epsilon=0.01)
> xy
$x
 [1] -0.90074942  0.31938997  1.99758852  0.18948864 -0.67045611  0.94449146
 [7]  0.16314379  1.04282274  0.17867388  0.68032412  0.22405276 -0.35796641
[13]  0.21524168 -1.87619738  1.70299285 -1.35362362  1.15832844  0.63027843
[19]  0.18906543 -0.83749727  0.67244005 -1.10705137 -2.08694478 -0.48407978
[25]  0.82052761 -0.05732552  0.76251285 -0.24428337 -1.58891834 -0.32626982

$y
 [1] -1.74297031  0.24384837  3.32990265 -0.84589878 -0.61697475  1.79270714
 [7] -1.94860661  1.14652685 -0.51525616 -0.38119805  1.10545581 -0.05519711
[13] -1.24684768 -1.92609458  3.18239040 -0.98517722  3.53580827  0.71950494
[19] -0.19637997  1.06747953 -0.04608036 -1.41063058 -2.48374507 -1.88881943
[25]  2.00881029  1.24744639  0.70973559 -0.90918939 -2.43784253 -0.45270765
```

To obtain a random sample from a population of the specified correlation use the option *population=TRUE.*

```
xysample<-corgen(len=30, r=-0.8, population=TRUE)
xysample;
cor(data.frame(xysample))
```

obtain the sample correlation

The above code, slightly modified, was used to produce the various plots demonstrating different correlation values in the *Measuring the Degree to which two variable covary: correlation* chapter.

39.4 Tips and hints

The following link on the CRAN site provides encyclopaedic knowledge of the distribution functions available in the various R packages:
http://cran.r-project.org/web/views/Distributions.html

Changing the active dataset - In R Commander, if you already have a dataset in R, to make the new one the active dataset use the menu option **Data-> Active dataset -> select active dataset**. If you have only one dataset when you attempt this you get the following rather cryptic error message:

Warning: There is only one dataset in memory

Which you can just ignore.

Unless you are working with a small dataset, usually you import the data into R Commander or R rather than creating it within R.

40 Importing your data into R

In this book we have imported a variety of text files and opened native R **.rda* files. While in the subsequent chapter, *Cutting and Pasting from Excel/Word to the R Data editor* I describe a quick and dirty way of importing small sets of data, this chapter's focus is on the two more common methods; using R Commander or R code directly. First considering the R Commander method.

40.1 R Commander

R Commander makes it easy for you to import data with the menu option:

Data -> import data

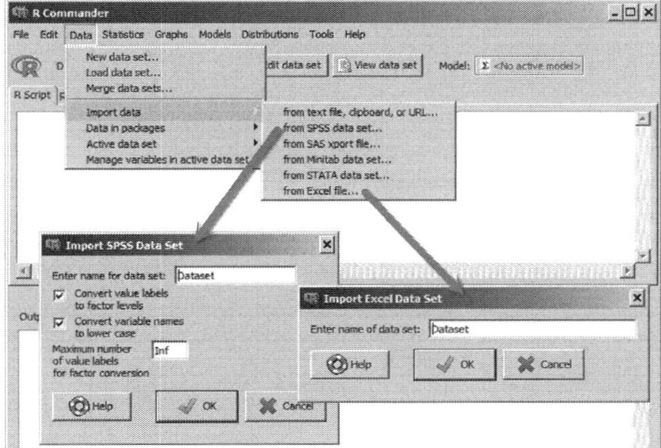

This provides opportunities to import data from textfiles, the clipboard, the web, SPSS, SAS, Minitab, STATA and Excel.

Importing text files and files from the web has been demonstrated in many of the previous chapters.

Each Import option asks a different set of questions about the data you are importing, for example if you select the *Excel* option and the file contains several sheets you will then be asked to specify which sheet you want to import.

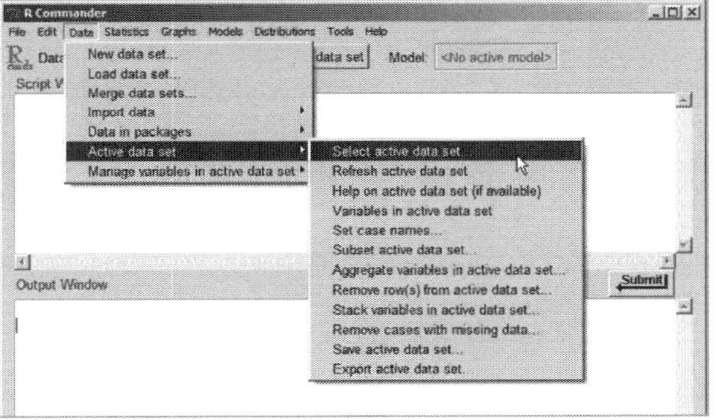

Any data you want to import must be allocated a dataset name (actually a dataframe), and you can think of this as an internal name of the data for use within R. It can be anything and I usually call it *dataset1*.

You can have more than one dataset within R at any one time but, if you do, then you need to specify which is the active dataset. The first dataset you open/import is assumed to be the active dataset.

You can change the active dataset (i.e. dataframe) anytime using the R Commander menu option:

Data->active data set->select active data set

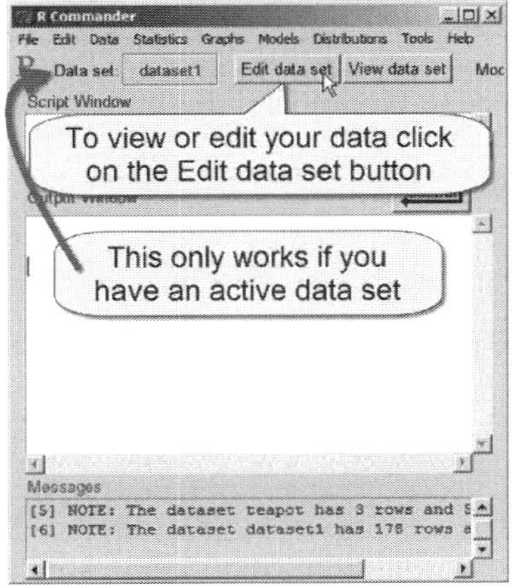

You can view or carry out simple edits on the data set by clicking on the **Edit data set** button.

40.2 R Code

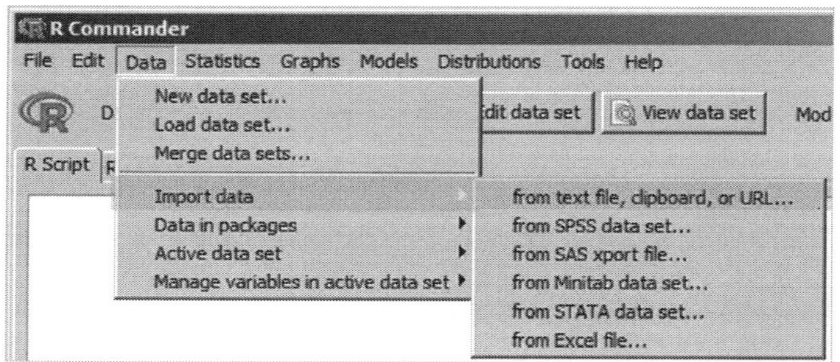

The R Commander **Data>Import data->from text file** menu option equates to the *read.table()* function discussed below.

The table below demonstrates some typical situations. If you wished to import a *tab* delimited file you would use the *read.delim()* function. Similarly, if you wished to import a comma delimited file the *read.csv()* function could be used. Both these functions assume each column starts with a column name (variable name). In fact both functions simply call the *read.table()* function but set certain values on the way, saving you the effort. The *sep* column in the table below refers to the character used to separate each column (i.e. field).

Default values for file import commands Simplified from Adler 2009				**File export equivalents**
Function	**header**	**sep**	**quote**	
read.table()	FALSE		\" or \'	Because write.table has col.names=TRUE we need to specify: write.table(dataframe, file = "myfile.csv", col.names = FALSE row.names = FALSE)
read.csv()	TRUE	,	\"	write.csv(dataframe, file = "myfile.csv", row.names = FALSE)
read.delim()	TRUE	\t	\"	write.csv(dataframe, file = "myfile.csv", row.names = FALSE, sep = "\t")

The official R documentation provides details of all the possible values you can pass to the functions, along with their defaults. I have included these below as occasionally setting a less well known option can be useful. For example, sometimes the missing value might not be the usual NA value:

you will never need to use the majority of these options but you never know.

```
read.table(file, header = FALSE, sep = "", quote = "\"'", dec = ".", row.names,
col.names, as.is = !stringsAsFactors, na.strings = "NA", colClasses = NA,
nrows = -1, skip = 0, check.names = TRUE, fill = !blank.lines.skip,
strip.white = FALSE, blank.lines.skip = TRUE, comment.char = "#",
allowEscapes = FALSE, flush = FALSE, stringsAsFactors = default.stringsAsFactors(),
fileEncoding = "", encoding = "unknown", text)
```

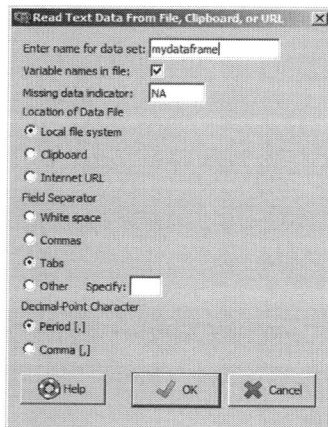

Shown below is the R code generated by R Commander from the dialog setup given on the left:

```
mydataframe <-read.table("D:/myrfiles/correlation_vo2max.dat",   header=TRUE,
sep="\t", na.strings="NA", dec=".", strip.white=TRUE)
```

The full details of the other two read functions are provided below:

```
read.csv(file, header = TRUE, sep = ",", quote="\"", dec=".", fill = TRUE,
comment.char="", ...)
```

this is the default value – a comma character

```
read.delim(file, header = TRUE, sep = "\t", quote="\"",)
```

this is the default value, indicating the first row is made up of the column names

this is the default value – a tab character

You can also import a number of proprietary file types using the foreign package. For example to import an *SPSS* file from the web use:

```
library(foreign)

campbell_mult_reg <-

read.spss("http://www.robin-beaumont.co.uk/virtualclassroom/book2data/campbell_mlr_p12.sav",

use.value.labels=TRUE,

max.value.labels=Inf,

to.data.frame=TRUE )
```

None of the above functions is designed for importing very large files (i.e. >100 megabytes), in which case you might want to try the *scan()*, further details of which can be found in the official *R data import/export manual* at:

http://cran.r-project.org/doc/manuals/r-release/R-data.html#Using-scan-directly

Some more examples of the *scan()* function are provided at:

http://wiener.math.csi.cuny.edu/Statistics/R/simpleR/stat022.html

40.3 Tips and hints

Remember . . .

- You can't carry out any analysis until you have specified an active dataset.
- You can always save your active dataset in R's format (*.rda) any time, see the chapter *Saving and exporting your work and data*.
- You can read R's native data file format (*.rda) using the R Commander menu option **Data->Load data set** or the *RGui* window menu option **File->Load Workspace**.
- As shown in the example above, you can replace a local file name with a url. Make sure you include the *http://* prefix or else you get an error:

```
> read.delim("www.robin-beaumont.co.uk/virtualclassroom/book2data/regression1.dat")
Error in file(file, "rt") : cannot open the connection
In addition: Warning message:
In file(file, "rt") :
  cannot open file 'www.robin-beaumont.co.uk/virtualclassroom/book2data/regression1.dat': No such file or directory
```

41 Cutting and pasting from Excel/Word to the R data editor

You can quickly cut and paste a selection of cells from an Excel table or a table in Word to create an R dataframe. It is a three stage process:

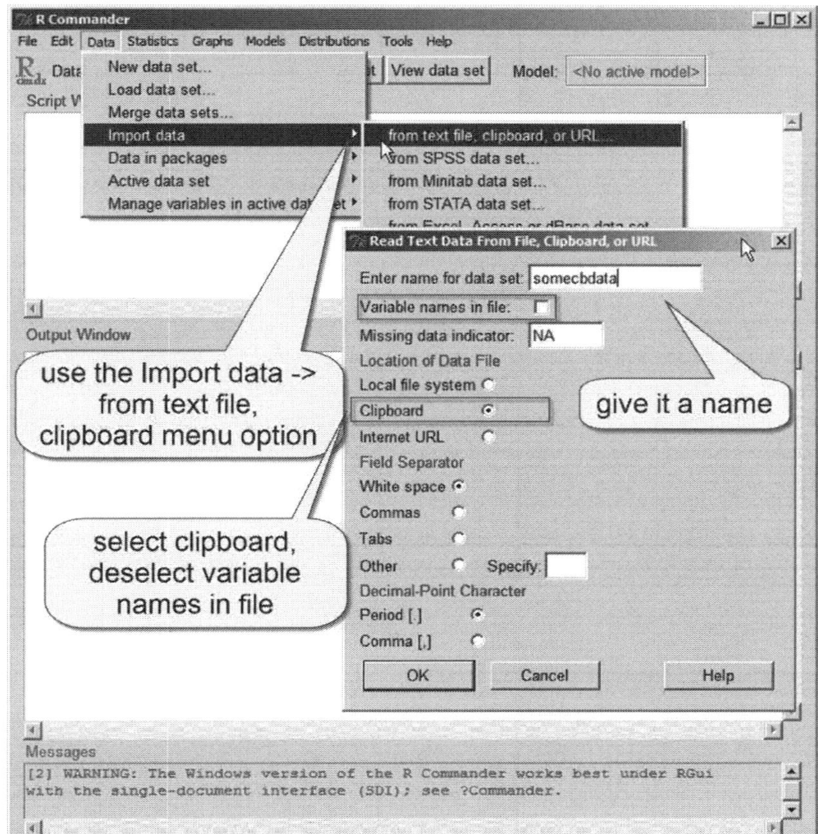

- within Word/Excel copy the relevant cells by first highlighting them and then selecting the menu option **Edit->Copy** (or *Ctrl+c*).
- within R Commander select the menu option: **Import data -> from text file, clipboard, or URL..**
- complete the dialog box shown opposite; Select the **clipboard** option and deselect the ***Variable names in file*** option, also give the new dataset a name.

41.1 Tips and hints

To make this new dataframe the active one use the menu option **Data-> Active dataset -> Select active dataset**

An exercise:

Create a table in Word or a sheet in Excel, containing the data shown below on the left.

Follow the instructions above, to paste this into R.
Your final result should look like that on the right.

34	1
45	3
34	2
24	4
27	2

Data Editor

	V1	V2	var3
1	34	1	
2	45	3	
3	34	2	
4	24	4	
5	27	2	
6			
7			
8			
9			

42 Saving and exporting your work and data

Saving your data is an important aspect of any analysis process. In complex analyses you will find that you save your data in many forms, including the native **.rda* file type. You can also save data in a variety of other formats using either R or R Commander.

42.1 R Commander

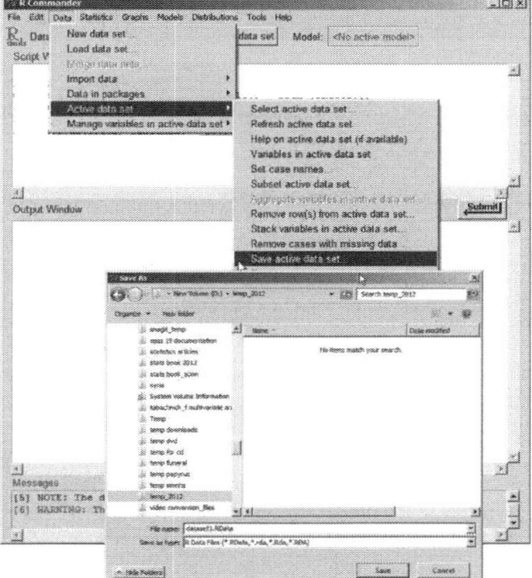

R Commander has two menu options for saving the active dataset, the first one, shown below, saves it in the binary R data file format **.rda*:

Data ->Active data set -> Save Active dataset

This saves it as an R data file which unfortunately can't be read by other programs, however it can be opened again by R or R Commander.

Note that you have the opportunity to save it under a different name in the *Save As* dialog box. This is useful if you wish to save the dataframe each time you manipulate it during a process.

The R Commander menu option **File->Save R Workspace** does the same thing.

42.2 Exporting your data

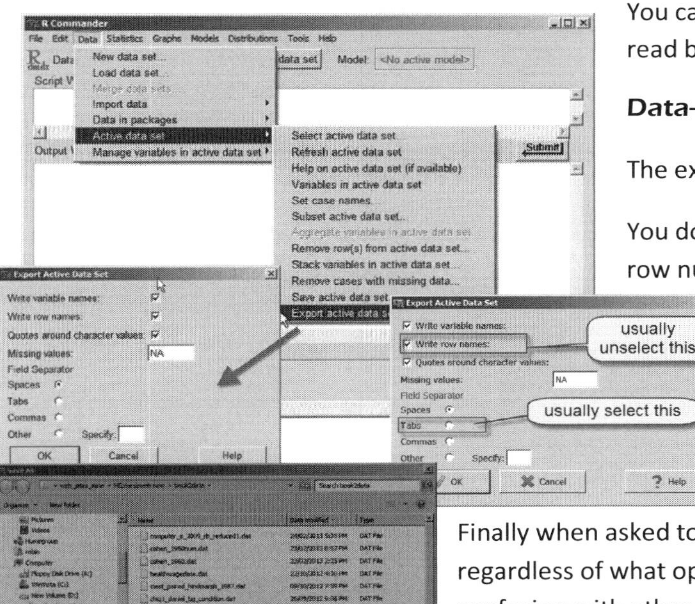

You can export your active **dataframe** to a text file which can be read by most programs by using the R Commander menu option:

Data-> Active data set-> Export active data set

The export active dataset has several options.

You do not usually want an additional column created with the row numbers so I unselect the 'write row names' option.

I also tend to use a *tab* character to separate each column and therefore click on the 'tabs' option for the field separator.

Finally when asked to provide a name for your file the extension is always .txt regardless of what options you have chosen which unfortunately can cause confusion with other programs. To avoid this you might want to use the following rule which I find helps:

Tab field separator = **.DAT* extension.

Comma field separator = **.csv* extension.

42.3 R Code

The above R menu options produce various R code fragments, shown below, which you could have typed directly into the *R Console* or R Commander *R Script* windows.

42.3.1 To save the active dataset:

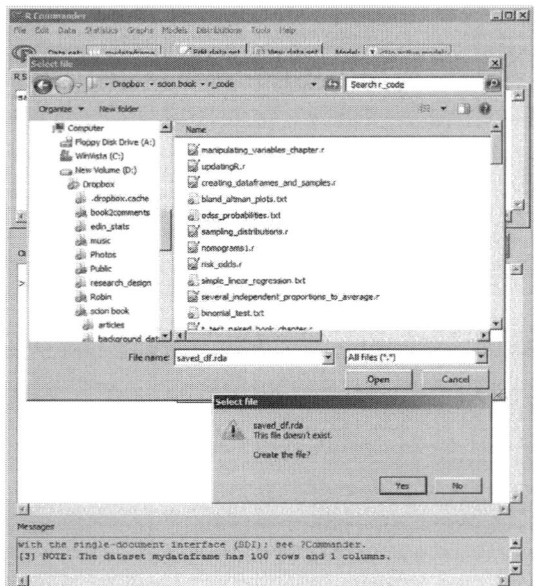

You specify the name of the dataframe (no quotes) and then where you want to save it (using quotes):

```
save(mydataframe, file="c:/mydata.txt")
```

If you don't know exactly where the folder is you can do it interactively using the *file.choose()* function.

```
save(mydataframe, file=file.choose())
```

Ignore the *Select file* window title and just think of it as a save file dialog box, also selecting **Yes** to the *Create the file?* dialog box.

42.3.2 To export the active dataset

To export a dataframe use the *write.table()* function.

```
write.table(mydataframe, "c:/mydata.txt", sep="\t")
```

For a full specification of all the options available see below:

```
write.table(x, file = "", append = FALSE, quote = TRUE, sep = " ", eol = "\n", na =
"NA", dec = ".", row.names = TRUE, col.names = TRUE, qmethod = c("escape", double"),
fileEncoding = "")
```

Once again you can use the *file.choose()* function to specify the file and folder name.

42.4 Tips and tricks

You can also export dataframes to excel and other programs using various R packages but I find that it is far easier to just export using the *write.table()* function and then import it directly into these programs.

43 R Script files (.r)

Why can't I use a word processor to create and edit R Script files?

Most word processors, like *Microsoft Word*, save files in their own proprietary format which R can't read.

An R script file is a text file which contains a set of **R commands**. This allows you to keep your common R commands in one place so that you can run them as many times as you want without the hassle of typing them in each time. An R script file can vary in size from a single line to many thousands.

R script files are identified by the *.r* extension.

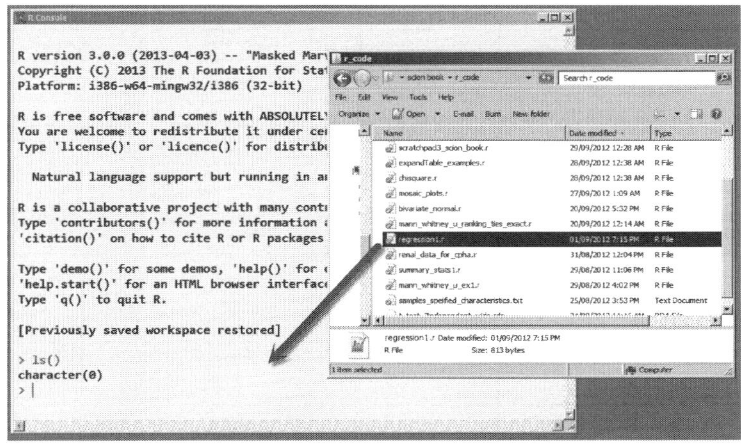

When you create/edit an R script file it is important not to use a word processor such as *Microsoft Word*, instead you must use a text editor such as *Notepad++*. All the R commands I present in each of the chapters in this book have been saved to a set of R script files. For example the R commands described in the simple regression chapter I have in a file called *regression1.r*

To instruct R to run an R script file such as the *regression1.r* file you simply find it in your file explorer program outside of R and drag the file to the *R Console* window. R then automatically runs all the commands that are in the file.

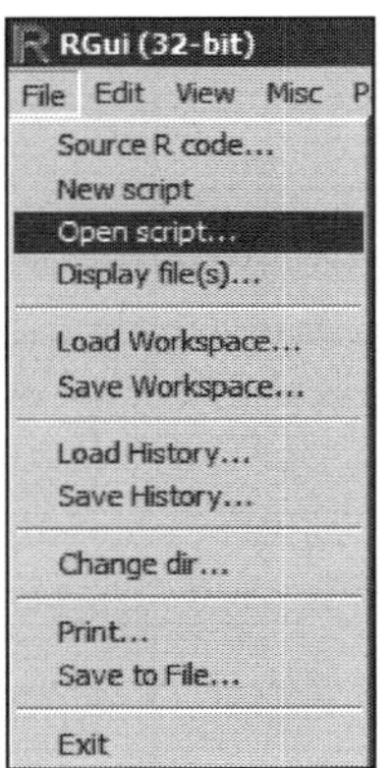

For a more subtle approach you can use the RGui window menu **File->Open** script which allows you to select the file which is subsequently opened up in an R script editor window. Within the R script window you can then use the right mouse button to bring up a popup menu which allows you to select the entire file or you can use the mouse to select one or more lines.

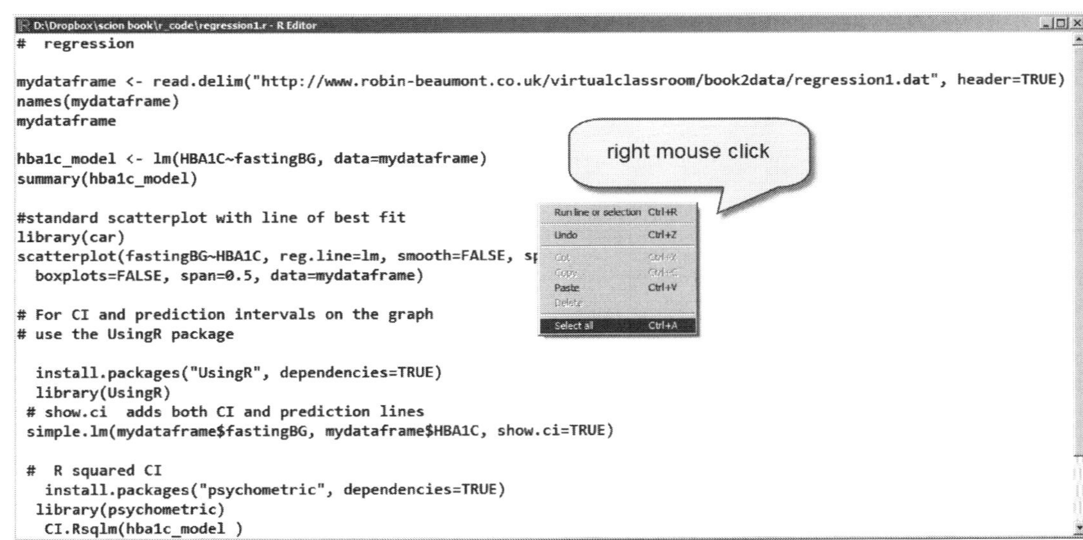

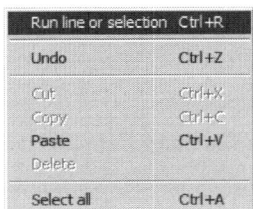

```
# regression

mydataframe <- read.delim("http://www.robin-beaumont.co.uk/virtualclassroom/book2data/regression1.dat", header=TRUE)
names(mydataframe)
mydataframe

hba1c_model <- lm(HBA1C~fastingBG, data=mydataframe)
summary(hba1c_model)

#standard scatterplot with line of best fit
library(car)
scatterplot(fastingBG~HBA1C, reg.line=lm, smooth=FALSE, sp
  boxplots=FALSE, span=0.5, data=mydataframe)

# For CI and prediction intervals on the graph
# use the UsingR package

 install.packages("UsingR", dependencies=TRUE)
 library(UsingR)
 # show.ci  adds both CI and prediction lines
 simple.lm(mydataframe$fastingBG, mydataframe$HBA1C, show.ci=TRUE)

 #  R squared CI
  install.packages("psychometric", dependencies=TRUE)
 library(psychometric)
  CI.Rsqlm(hba1c_model )
```

Run line or selection	Ctrl+R
Undo	Ctrl+Z
Cut	Ctrl+X
Copy	Ctrl+C
Paste	Ctrl+V
Delete	
Select all	Ctrl+A

To run the selection, select from the same popup menu the **Run line or selection** option or use the shortcut key combination *Ctrl+R*. Either will then run the commands with the results appearing in the usual *R Console* window.

43.1 Tips and Tricks

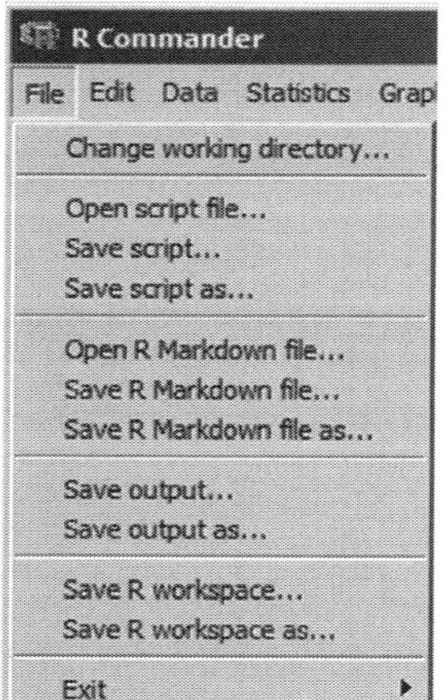

You can save the R commands you have entered, or R commander generated, during your R session using the *savehistory(file = "filename.Rhistory")* command described in the *Workspaces,* objects and history files chapter. This you can then edit in something like *Notepad++* saving it to a new file with the extension *.r* (dot r). You can then use the methods described in this chapter to run the commands.

There are two ways from within R Commander to create a file of R commands. The R Commander menu options **File->save script** or **File->Save output**. R Commander also has an option to save what is called a markdown file which saves both the output and the commands formatted as html. To enable this you will need to select the R Commander startup option **Use knitr**. Accessed from the R Commander menu option **Tools->Options** (shown below).

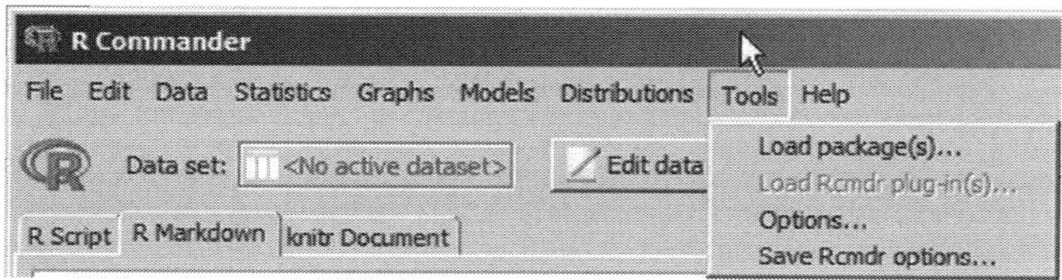

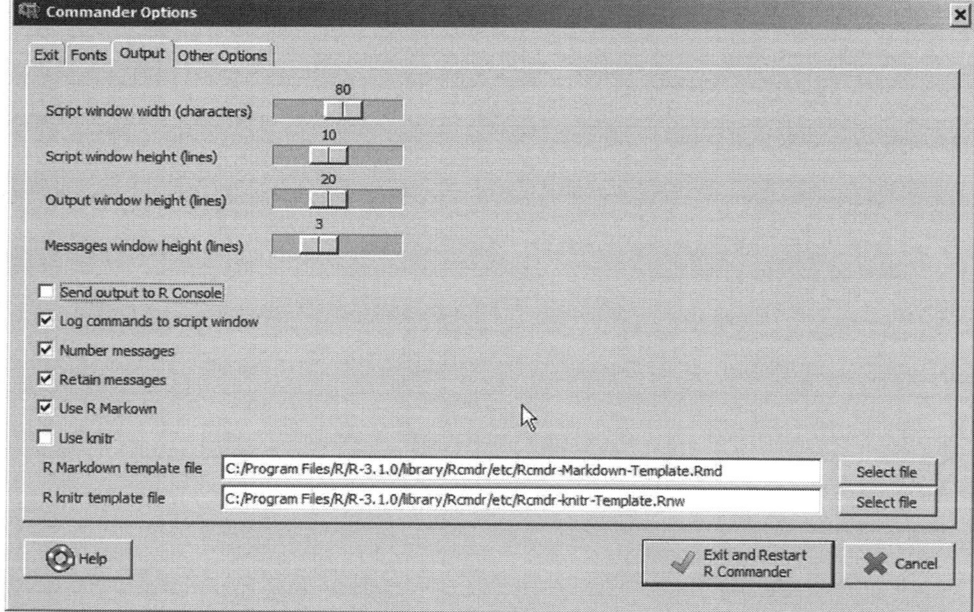

44 Manipulating variables (columns) in R Commander and R

var1	var2
2	2
4	5
3	4
5	5
6	6
5	2
6	3
4	4
NA	NA
1	3

Often you need to calculate new variables, select subgroups of rows or columns or delete them, all of which can be achieved from within R Commander using various menu options. This chapter will specifically consider the manipulation of columns (variables). For each menu option I give the R code equivalent for those of you who wish to dig deeper.

44.1 Recoding a variable

Given the data opposite consider, let's consider how it might be manipulated.

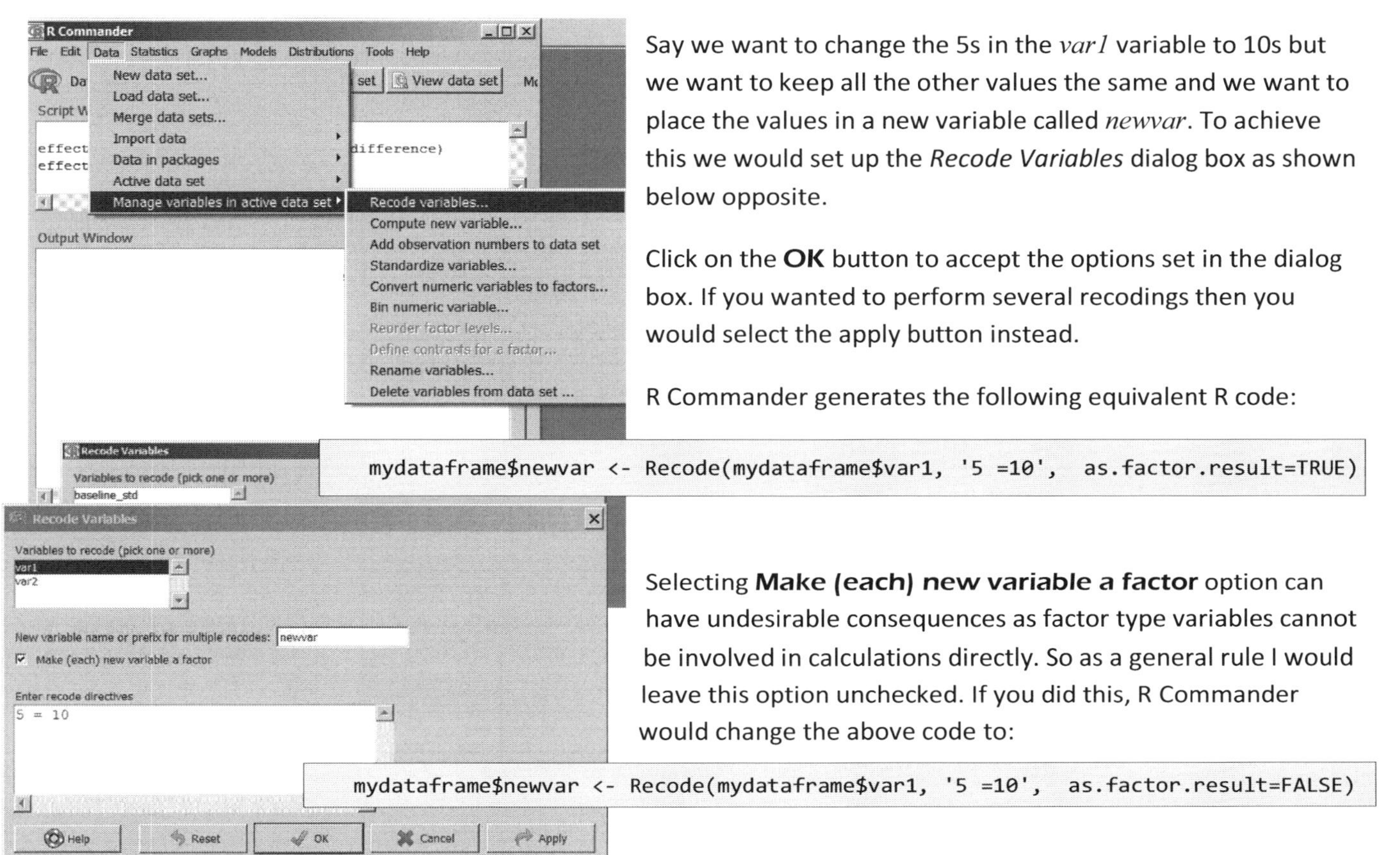

Say we want to change the 5s in the *var1* variable to 10s but we want to keep all the other values the same and we want to place the values in a new variable called *newvar*. To achieve this we would set up the *Recode Variables* dialog box as shown below opposite.

Click on the **OK** button to accept the options set in the dialog box. If you wanted to perform several recodings then you would select the apply button instead.

R Commander generates the following equivalent R code:

```
mydataframe$newvar <- Recode(mydataframe$var1, '5 =10',  as.factor.result=TRUE)
```

Selecting **Make (each) new variable a factor** option can have undesirable consequences as factor type variables cannot be involved in calculations directly. So as a general rule I would leave this option unchecked. If you did this, R Commander would change the above code to:

```
mydataframe$newvar <- Recode(mydataframe$var1, '5 =10',  as.factor.result=FALSE)
```

There are several options you have in the **Enter recode directives** box, taking some examples from R Commander help:

- one or more single values on separate lines, for example, *"missing" = NA*. and on the following line *2 = log(2)*.
- Making several values separated by commas equal one value.
  For example, *7,8,9 = "high"*.
- Making a range of values indicated by a colon equal one value. For example, *7:9 = "high"*. The special values *lo* and *hi* may appear in a range. For example, *lo:10=1*. Note that these values are unquoted.

44.2 Computing a new variable

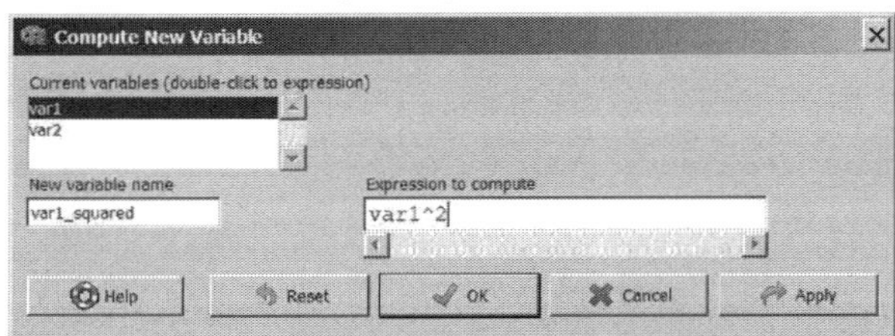

You can also specify a new variable which is based on a current one. If you double click on a variable in the **Current variables** box it will appear in the **Expression to compute** box; some examples of expressions are given below. If you wish to create an empty new variable simply put *NA* in the **Expression to compute** box.

The above example creates a new variable called *var1_squared* which is the squared value of the *var1* variable. The R code produced from the above uses the *with()* function on the right hand side, indicating which dataframe to use and the *dataframe$column* name format on the left, indicating in which dataframe to put the new variable.

Example expressions (taken from Karp 2010)			
Operators	Function	Example 1	Example 2
x + y	Addition	Variable 1 + Variable 2	Variable 1 + 25
x - y	Subtraction	Variable 1 – Variable 2	35 - Variable 1
x * y	Multiple	Variable 1*Variable 2	100*Variable 1
x / y	Division	Variable 1/Variable 2	Variable 1 / 63
x ^ y	X to the power of Y	Variable 1 ^ Variable2	Variable1^10
log10(x)	Log10 transformation	Log10(Variable 1)	
log(x, base)	Log transformation to a specified base	Log(Variable 1, 2)	

```
mydataframe$var1_squared <- with(mydataframe, var1^2)
```

44.3 Standardized Variables

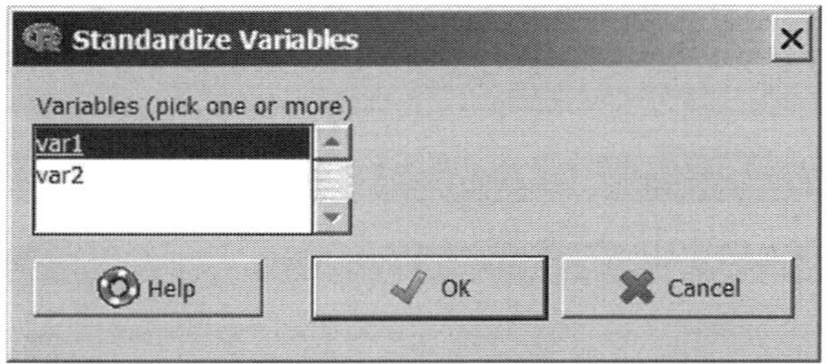

It is often desirable to convert a variable so that its mean value is zero and its Standard Deviation is 1. Such a transformation is often called a z score transformation, producing a standardized variable.

The *Standardize Variables* dialog box does not give you the opportunity to specify a name for the standardized variable, it automatically creating a new variable with a *Z.* prefix added to the original name. This is shown in the R code produced, but please note that this code is rather complex for a beginner (as you have the option to standardize more than one variable). The *scale()* function takes a variable and returns its standardized values.

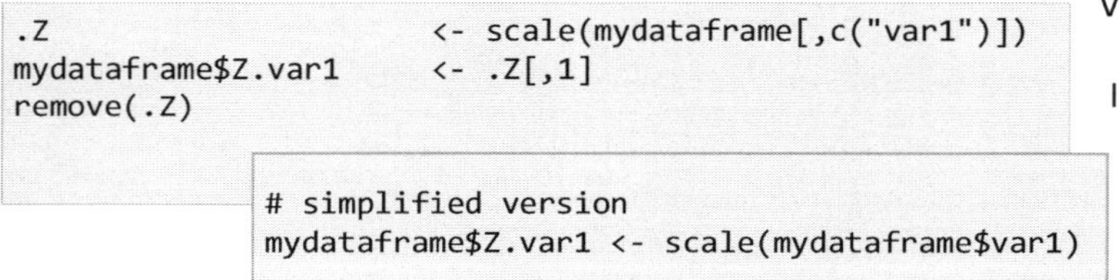

```
.Z                    <- scale(mydataframe[,c("var1")])
mydataframe$Z.var1    <- .Z[,1]
remove(.Z)
```

```
# simplified version
mydataframe$Z.var1 <- scale(mydataframe$var1)
```

I have simplified the original code opposite to produce the same result if you had only selected the *var1* variable.

var1	var2	var1_squared	Z.var1
2	2	4	-1.154701
4	5	16	0
3	4	9	-0.5773503
5	5	25	0.5773503
6	6	36	1.154701
5	2	25	0.5773503
6	3	36	1.154701
4	4	16	0
NA	NA	NA	NA
1	3	1	-1.732051

The results are shown opposite.

I would advise you to not use this menu option but R code directly as I find the '.' (i.e. the full stop as part of a variable name) confusing. You can edit the code to the following, which can easily be adapted:

```
mydataframe$Zvar1 <- scale(mydataframe$var1)
```

44.4 Converting numeric variable(s) to factors with level values

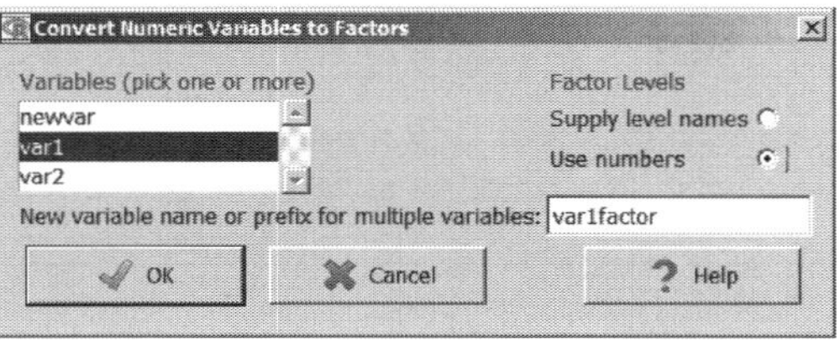

To convert a numeric variable to a **factor** (necessary to allow access to many of the menu options) you can use this menu option to indicate which variable to convert, optionally a new variable name and how you want the new variable to reflect the old one in terms of factor levels.

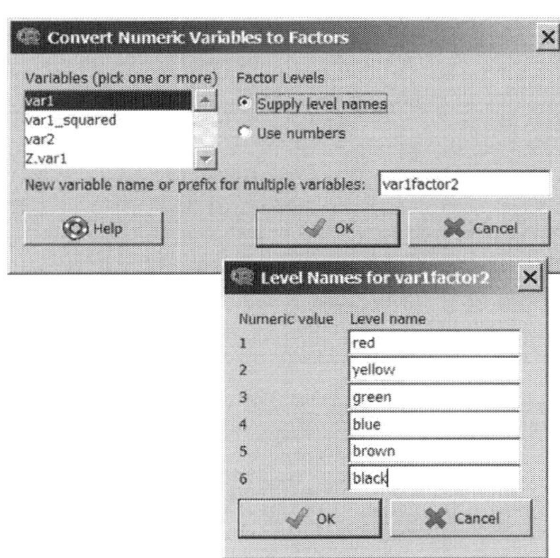

var1	var2	var1_squared	Z.var1	var1factor	var6
2	2	4	-1.154701	2	
4	5	16	0	4	
3	4	9	-0.5773503	3	
5	5	25	0.5773503	5	
6	6	36	1.154701	6	
5	2	25	0.5773503	5	
6	3	36	1.154701	6	
4	4	16	0	4	
NA	NA	NA	NA		
1	3	1	-1.732051	1	

If you are not worried about the names of the individual factor levels you can allow R Commander to just use numbers by selecting the use numbers option. Under the bonnet R Commander uses the *as.factor()* function to convert the variable.

```
mydataframe$var1factor <- as.factor(mydataframe$var1)
```

You may think this does nothing to your dataset as the new variable looks identical to the old numeric variable. However if you check the new variables properties in the edit window you will see that it is now a character variable (i.e. a factor).

44.5 Converting numeric variable(s) to factors with level names

The factor level values can be supplied by yourself, which is very useful if you wish to give them useful names. Also R Commander is clever enough to provide the correct number of boxes of level names you need to fill in depending upon the number of unique values in the original variable.

In this instance R Commander provides 6 level slots for the new factor variable *var1factor2* having found 6 unique values in the variable. I have given them arbitrary level names to demonstrate the approach. If you have a variable with many different unique values and wish to convert it to a factor, it is a good idea to first reduce the number by using the recode variable option described above. Alternatively use the 'bin' option described on the following page.

The R code produced by R Commander shows that in contrast to the above example we now have a *labels()* function which provides the level names. The *as.factor()* and *factor()* functions are equivalent.

```
mydataframe$var1factor2 <- factor(mydataframe$var1,  labels=c('red','yellow','green','blue','brown','black'))
```

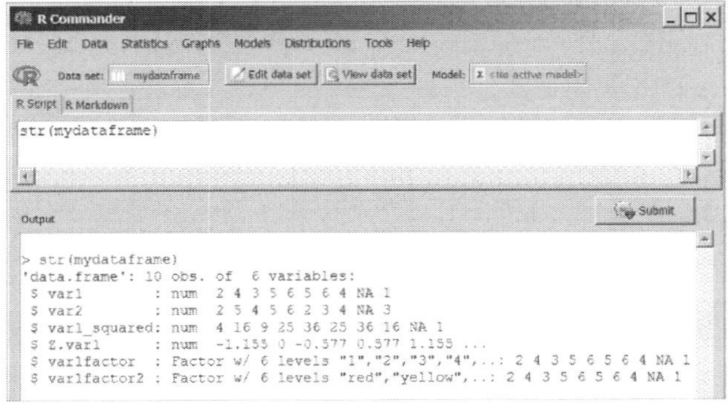

44.6 Which of my variables are factors?

From within R Commander you can check to see which of your variables are factors in two ways:

- menu option: **Statistics -> Summaries -> Frequency Distribution** where only the factors will be listed.
- select the **Edit data set** button then click on each row header which will tell you if it is numeric/categorical (i.e. factor).

From within R you can make use of the structure *str()* function as shown opposite. Once you know a variable is a factor typing *levels(factorvar)* lists the levels.

44.7 Converting and grouping a numeric variable to a factor with level values: 'binning'

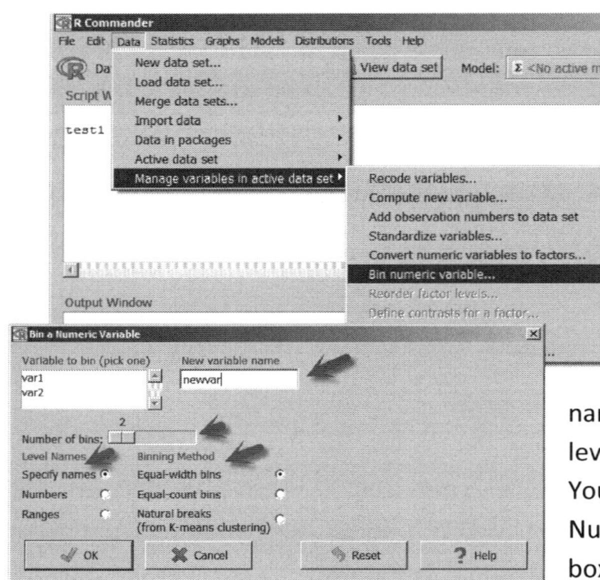

Binning numeric variables is very useful if you have a continuous variable such as *weight* or *age* and want to divide your data up into a certain number of groups (called bins).

This option allows you to select the variable you wish to create the groups from called the **Variable to bin**. You need to provide a name for the new variable along with the number of groups (i.e. factor levels) you wish to create (from 2 to 20 on the **Number of bins** slider). You can also specify how you want to provide the factor level names. Numbers are provided automatically or you can provide names (dialog box shown below). Finally, you can specify the method you want the computer to use to create the bins. Clicking on the **Help** button provides details on each of these methods.

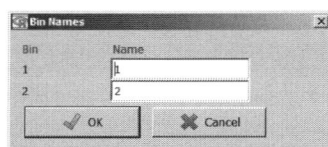

44.8 Reorder factor levels

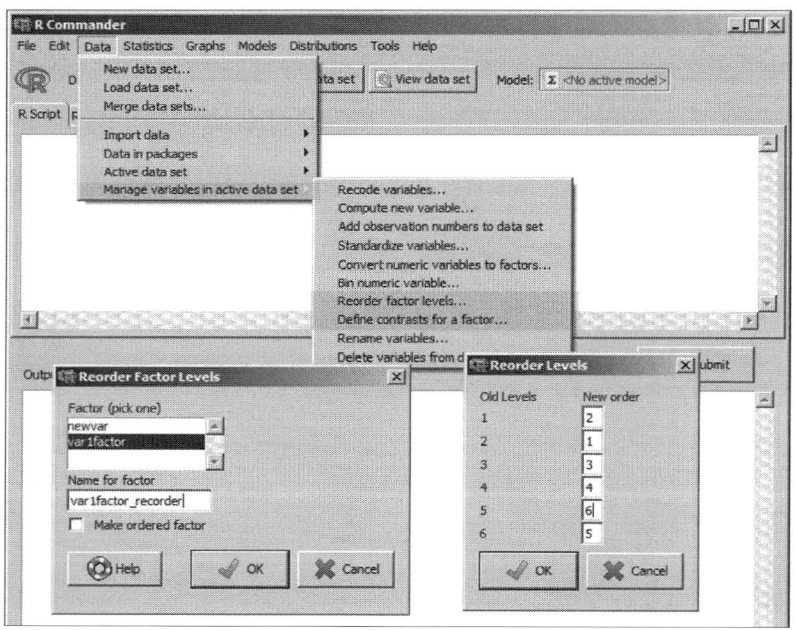

Sometime you need to reorder the levels of a factor either because you want to represent the data in some type of ordinal scale, or wish the levels to be displayed in a particular order in a bar chart etc.

There are two menu options in the **Manage variables in active data set** menu which achieve this. As with other menu options they only become available when you have one or more factors in the active dataframe.

In the **Reorder Factor Levels** dialog box you select the old factor and specify a name for the new factor, I have called it *var1factor_reorder*. You also have the option of indicating that the new factor is an ordered factor, selecting this option affects the results of some statistics and also the order in which the factor levels are displayed in graphs. The *Reorder Levels* box allows you to specify the new order, where I have swapped levels 1 and 2 and also levels 5 and 6.

To find out about the **Define contrasts for a factor** menu option see the glossary entry **Contrasts**. The resultant R code for both options is shown below.

```
mydataframe$var1factor_reorder <- factor(mydataframe$var1factor, levels=c('2','1','3','4','6','5'))
mydataframe$var1factor_ordered <- factor(mydataframe$var1factor, levels=c('2','1','3','4','6','5'), ordered=TRUE)
```

44.9 Renaming one or more variables

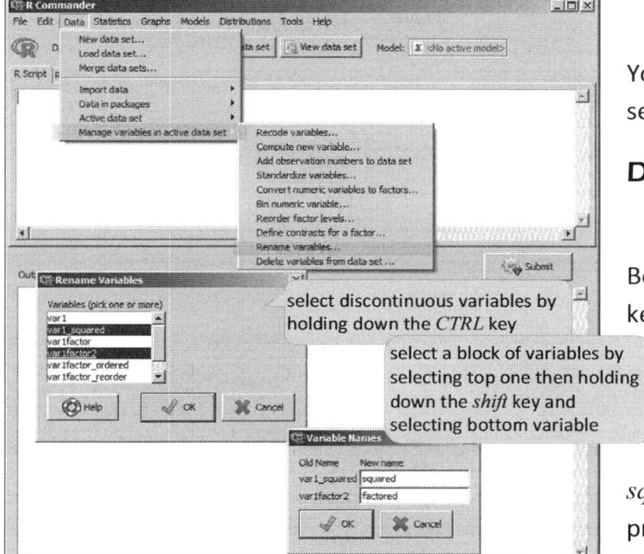

You can easily rename one or more variables (columns) by selecting the menu option:

Data-> Manage variables in active data set -> Rename variables. . .

Besides selecting a single variable you can hold down the *Ctrl* key to select several or a block by first selecting the top one, then holding down the *shift* key and then clicking on the bottom of the block of variables you are interested in.

Assuming that we are giving the variables the new names, *squared* and *factored*, the options shown opposite would produce the following R code:

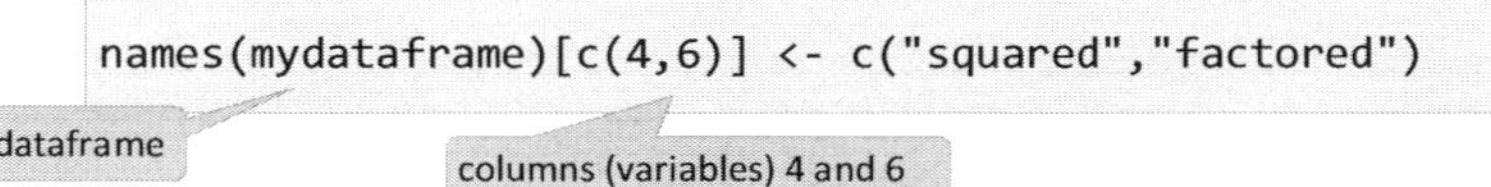

```
names(mydataframe)[c(4,6)] <- c("squared","factored")
```


44.10 Deleting one or more variables

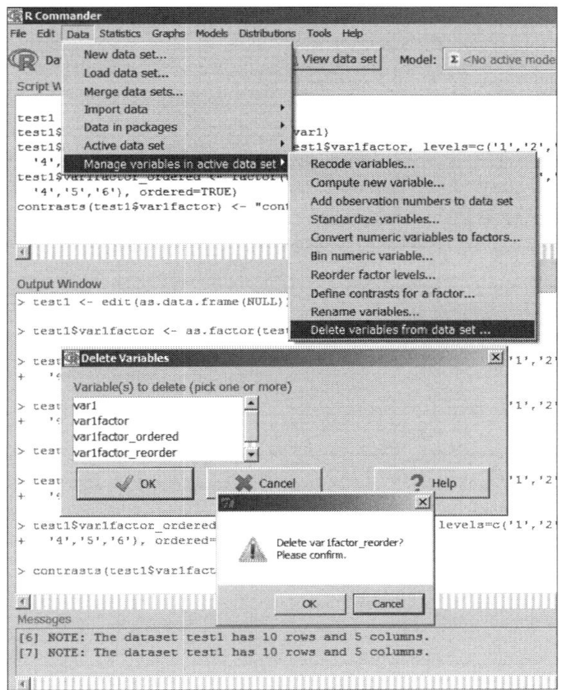

If you need to delete a column, you do so by selecting the menu option:

Data-> Manage variable in active data set -> Delete variables from data set . . .

After you have selected one or more variables using the technique described above, you are asked to confirm your decision - as it can't be undone.

Supposing that you wished to delete the *var1factor_order* variable in the dataframe, the resultant code would look like:

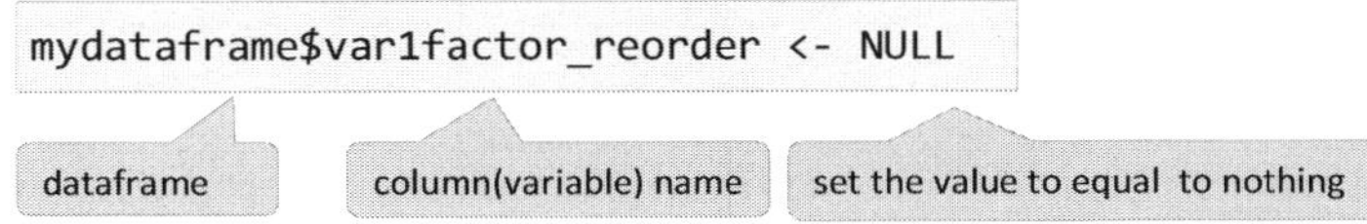

```
mydataframe$var1factor_reorder <- NULL
```

44.11 Tips and tricks

There are many freely available introductions to R which cover the topics presented in this chapter but few describe in detail the R Commander options, I think this is because most people realise that by planning their analysis carefully the actions described in this chapter can often be saved as a set of R commands. The collection of commands are easy to maintain and run *en masse* preventing the repetitive use of many dialog boxes. The chapter *R Script files* explains how to manage such files.

45 Manipulating cases (rows) in R Commander and R

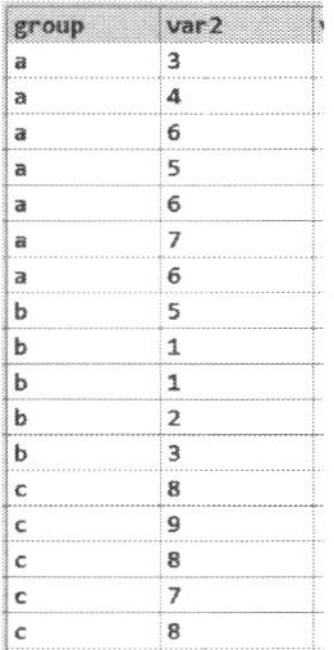

group	var2
a	3
a	4
a	6
a	5
a	6
a	7
a	6
b	5
b	1
b	1
b	2
b	3
c	8
c	9
c	8
c	7
c	8

For most analyses the columns represent variables and the rows cases. A common situation is that one or more variables (columns) allow you to group the data in some way. Taking the example opposite, the group variable indicates that the value associated with the *var2* variable belongs to either the *a, b* or *c* group. Technically the group variable is a **factor** with three levels.

You can recreate the dataframe shown opposite, called *test1* by copying the R code given in the section *R code* in the chapter *Creating datasets and distributions in R Commander and R.*

45.1 Copying a complete dataframe

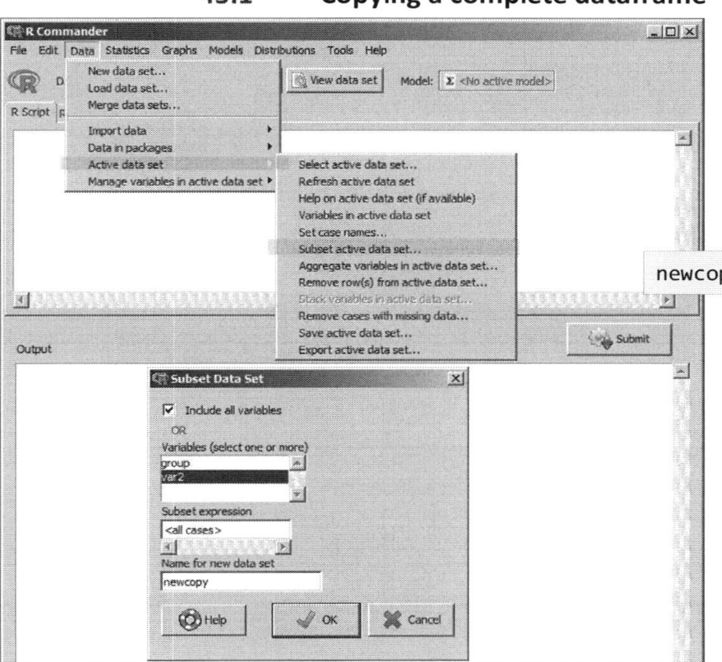

To copy a complete dataframe select the **Include all variables** option, making the **Subset expression** box blank and provide a name for the new data set, I have called it *newcopy*.

The R code this option generates is:

```
newcopy <- subset(test1, subset=)
```

A more elegant solution is simply:

```
newcopy <- test1
```

If you do not wish to lose the new dataframe remember to save it before exiting R Commander.

45.2 Copying a subset of rows (cases)

You can achieve this by typing in an expression into the **Subset expression** box such as:

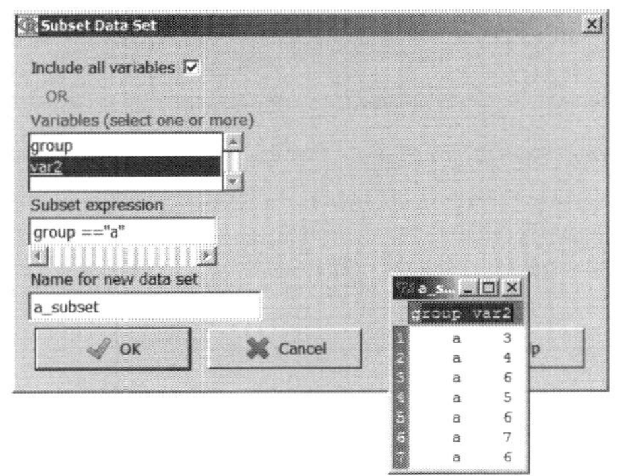


```
group == "a"
```

The code this produces is shown below. Note the double equals sign"==" to mean equals to.

```
a_subset <- subset(a_subset, subset=group =="a")
```

You can also create expressions that are much more complex - such as

```
group == “a” & var2 <= 5
```

The table opposite lists the common operators used in building expressions.

Mathematical operators you can use (adapted from Karp, 2010).

Symbol/code	Name	Use
==	equality	used to indicate the variable should equal
!=	Inequality	used to indicate the variable should not equal
&	And	used to combine multiple expressions
|	Or	used to combine multiple expressions
is.na(varname)		Include the missing values of a variable
!is.na(varname)		Exclude the missing values of a variable
>	Greater than	
<	Less than	
>=		More than or equal to
<=		Less than or equal to

45.3 Obtaining summary values for a subset of rows - Aggregation

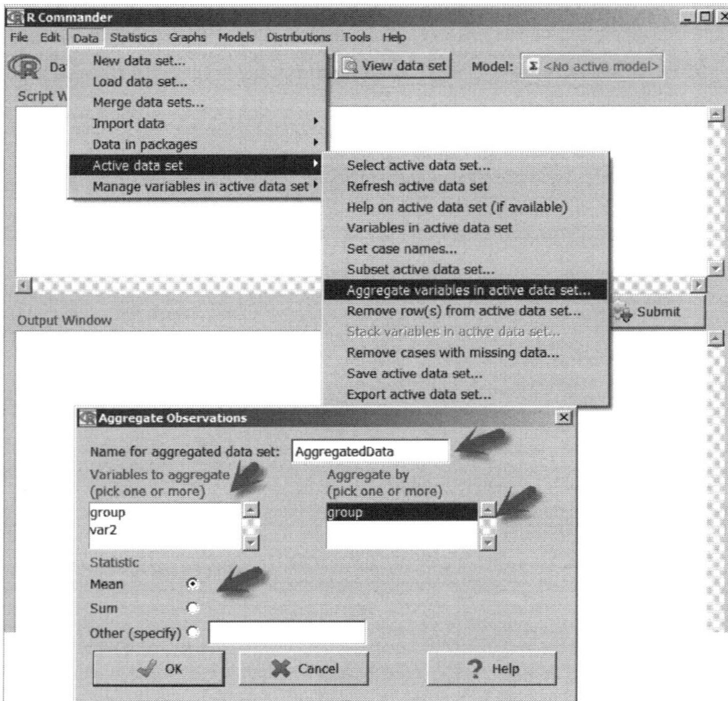

Often it is useful to get a set of summary statistics for a dataframe. For example you might want to find the mean values for all the cases in the a, b, or c groups for a particular variable, and one way of doing this is to use the menu option ***statistics-> summaries->Table of statistics***. However, this does not save the results - which you often want to do, particularly if you need to produce graphs of summaries.

You can achieve this by using the menu option
Data-> Active dataset
-> Aggregate variables in active data set.

In the resultant dialog box you specify a name for the dataframe produced, the variables to aggregate along with which variable you want to group by. You also need to indicate which summary statistic you want to use.

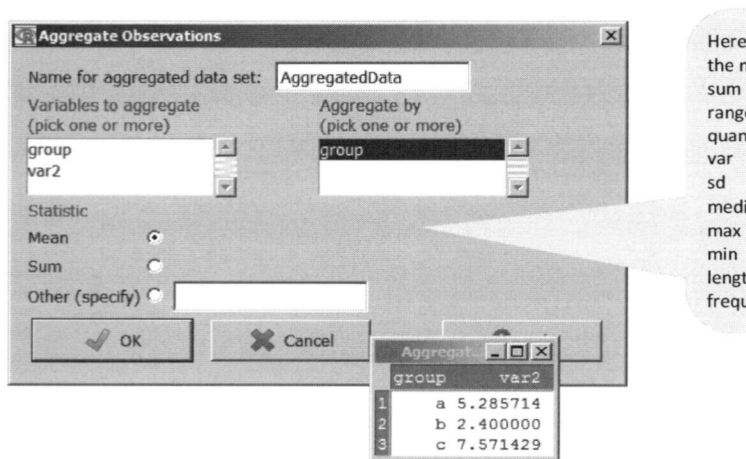

Here the summary value is the mean, it could also be:
sum
range
quantile
var
sd
median
max
min
length (useful if you want frequencies)

	group	var2	var3
1	a	3	5
2	a	4	6
3	a	6	7
4	a	5	5
5	a	6	6
6	a	7	9
7	a	6	8
8	b	5	10
9	b	1	4
10	b	1	3
11	b	2	5
12	b	3	2
13	c	8	3
14	c	9	6
15	c	8	8
16	c	7	7
17	c	8	5
18	c	6	8
19	c	7	7

The above dialog box options generates the following R code, which you can then subsequently edit and adapt as you wish.

```
AggregatedData <- aggregate(test1[,c("var2"), drop=FALSE], by=list(group= test1$group), FUN=mean)
```

45.4 Correlation matrices

Several statistical procedures, such as factor analysis and structural equation modelling, only require a set of summary values from a dataset rather than the original data, and it is often useful to save such a set of values from an analysis.

For this example I'll use a dataset described in the following table.

184 respondents completed a *computer system satisfaction* questionnaire. I have anonymised and edited the data removing many variables. The following variables are included:

Field name	**description**
q1	How often do you have problems logging onto the system
q2	overall response time
q3	menu system
q4	Ease of navigation between screens
q5	Contributes to job satisfaction?
q6	reduces productivity
q7	easy to use
q8	source of stress
q10	Supporting tasks in job
q11	overall how happy with the system
q6_invert	q6 scores inverted
q8_invert	q8 scores inverted

All marked on 5 point likert scales 1 = grossly inadequate/strongly disagree to 5 excellent/strongly agree

Q6_invert and q8_invert are these fields with inverted values for q6 and q8 respectively, to facilitate certain analyses.

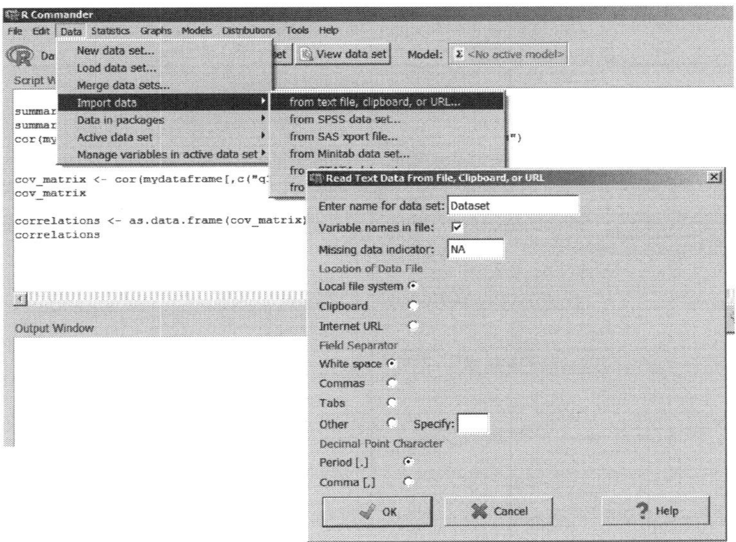

You can obtain the data directly from my website by selecting the R Commander menu option:

Data-> from text, the clipboard or URL

I have given the resultant dataframe the name *mydataframe*, also indicating that it is from a URL (i.e. the web), and that the columns are separated by *tab* characters.

Clicking on the **OK** button brings up the *Internet URL* box, in which you need to type the following to obtain my sample data:

http://www.robin-beaumont.co.uk/virtualclassroom/book2data/computer_s_2009_rb_reduced1.dat

Click **OK**.

Alternatively, you can type the following directly into the R Console window:

create a dataframe called mydataframe

```
filename <- "http://www.robin-
beaumont.co.uk/virtualclassroom/book2data/computer_s_2009_rb_reduced1.dat"

mydataframe <- read.delim(filename, header=TRUE)

mydataframe
```

display the dataframe

each column starts with the column name

45.4.1 Editing R Commander code to obtain a correlation matrix

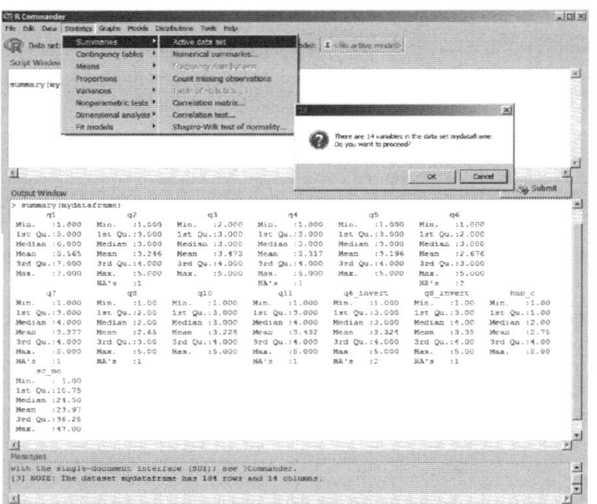

Before producing a correlation matrix it is always advisable to check the variables you have in the dataframe and note the number and percentage of missing values.

Using the R Commander menu option **statistics-> summaries active dataset** produces the output shown opposite, from which we can see that variable *q2* has one missing value (i.e. *NA*) as does *q4, q7, q8, q11 and q8_invert*. Both *q6* and *q6_invert* have two missing values. Given that we have 184 cases (note output from bottom of window) this is a desirably very low level of missing values. We can also check the ranges of the variables, indicating there are no inappropriate values.

Assume we wish to create a correlation matrix with variables *q1* to *q6*. We will start by producing the correlations using the R Commander menu option:

Statistics->summaries-> correlation matrix

Then select the required variables and click on the **OK** button. The result is shown opposite.

The important thing to note is the R code generated in the window:

```
cor(mydataframe[,c("q1","q2","q3","q4","q5","q6")],
use="complete.obs")
```

There are several things to note:

use="complete.obs" means that for each case if either of the pair has a missing value (i.e. *NA*) then the case is not included in the analysis.

[,c("q1","q2","q3","q4","q5","q6")] indicates that we wish to use the entire columns (note the ',' at the start), labelled *q1* etc.

```
cov_matrix <-   cor(mydataframe[,c("q1","q2","q3","q4","q5","q6")],use="complete.obs")
cov_matrix
```

To produce a matrix of these correlations all we need do is to copy and paste the R Commander generated code and modify it to that shown above.

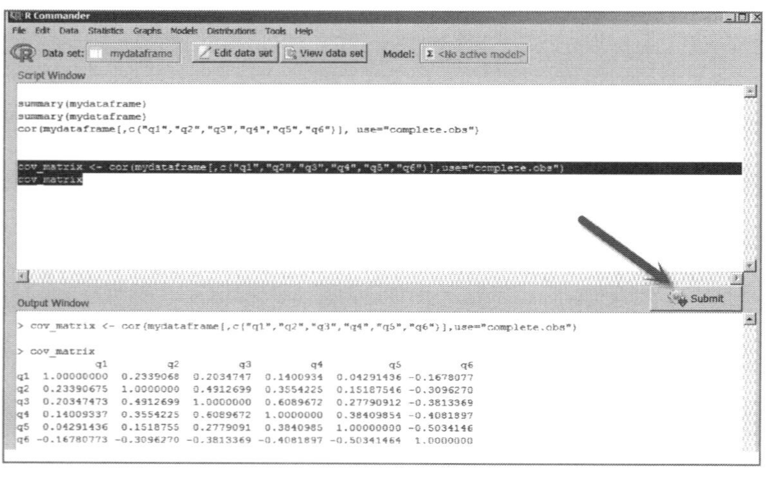

The screenshot opposite shows the result of running our edited code. What we now have is a matrix structure which we need to convert to a dataframe so that we can save /export it. This is achieved by using the *as.data.frame()* function.

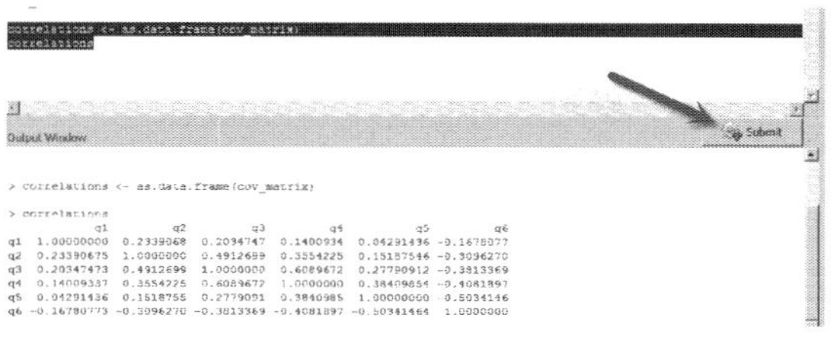

Typing into the R Commander syntax window the following two lines, highlighting it and then clicking on the submit button produces the output shown opposite:

```
correlations <- as.data.frame(cov_matrix)

correlations
```

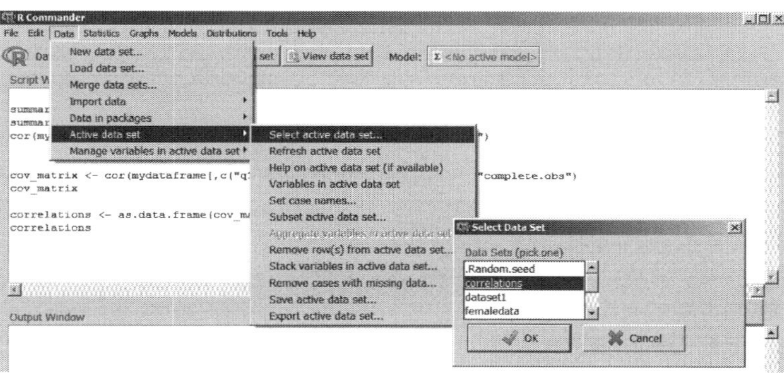

Now we have our set of correlations as a dataframe we need to make it the active dataset so that we can save/export it. This is done using the menu option shown opposite.

The process of saving a dataframe or exporting it as a *tab* or *comma* delimited file is covered in the chapter *Saving and exporting your work and data.*

45.5 Tips and Tricks

We could have combined two of the above steps into a single command thus:

```
correlations <- as.data.frame(cor(mydataframe[,c("q1","q2","q3","q4","q5","q6")],use="complete.obs"))
correlations
```

We could have written the matrix directly as output but there is no such R Commander option. To do this would require you to write more R code such as:

```
write.matrix(cov_matrix, file ="c:\temp.csv", sep = "\t")
```

The following code also seems to work using the *file.choose()* dialog box to specify the name of the file you want to save.

```
write.matrix(cov_matrix, file =file.choose())
```

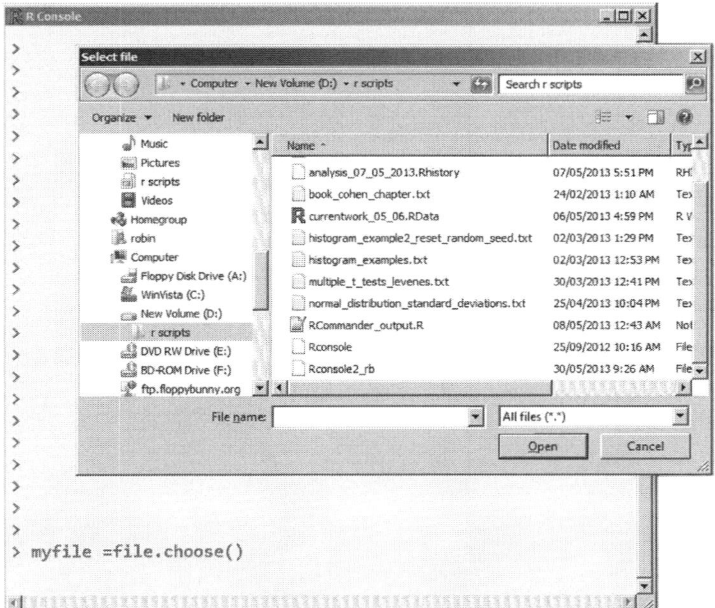

46 Expanding tables of counts into flat files

Sometimes it is necessary to convert a table of counts back into raw data. This can be done in R Commander/R and there is a **function** in two separate packages that allows you to do this:

- *expand.table()* in the *epitools* package
- *expandTable()* in the *NCStats* package, which I find slightly easier.

I will provide several examples using the *expandTable()* function in the NCStats package.

46.1 Preliminaries

The *NCStats* package is a bit tricky to install, see the chapter *Installing non CRANS packages*. Once you have it installed load the package:

if you have problems installing or loading it see the Tricks & Tips section at the end of this chapter for a easy workaround

```
library(NCStats)
```

Now we will consider some examples.

46.2 Example 1 - A single variable.

Here I am using the example from the chapter, *Comparing an observed proportion to a population value: the binomial test,* where we have 328 post operative infections from a group of 1361 patients (i.e. 1033 non infected patients). The code below uses the counts to generate the raw data. You can enter it into either the *R Console* window or the R Commander script window:

these are the factor levels if you want to specify factor level names use "level2" etc instead

```
d<- matrix( c(328, 1033), nrow=2, byrow=TRUE)

rownames(d) <- c(1, 2)

rawd <- expandTable (d, "value")

rawd
```

a two row, single column matrix

show the data

note the capital T

the counts

the name of the new variable

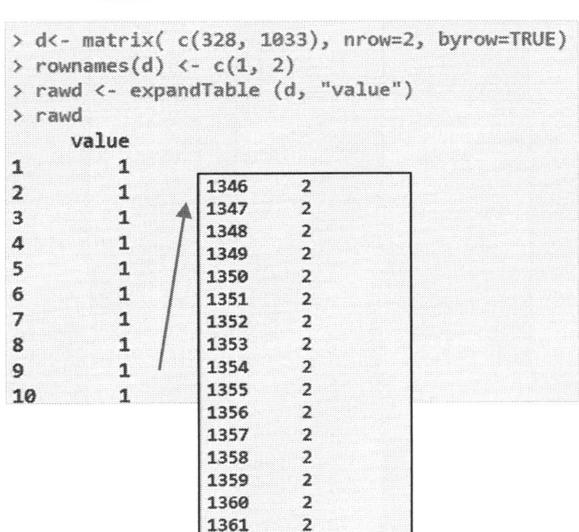

```
> d<- matrix( c(328, 1033), nrow=2, byrow=TRUE)
> rownames(d) <- c(1, 2)
> rawd <- expandTable (d, "value")
> rawd
   value
1      1
2      1
3      1
4      1
5      1
6      1
7      1
8      1
9      1
10     1

1346   2
1347   2
1348   2
1349   2
1350   2
1351   2
1352   2
1353   2
1354   2
1355   2
1356   2
1357   2
1358   2
1359   2
1360   2
1361   2
```

You can then save the new dataframe. See the chapter *Saving and exporting your work and data* for details.

```
R Console
> d<- matrix ( c( 83, 3, 90, 3, 129, 7, 70, 12), nrow = 4, byrow=TRUE)
> colnames(d) <- c("smoker", "nonsmoker")
> rownames(d)<- c( "g1", "g2", "g3", "g4")
> rawd<- expandTable(d, c("group", "smoker status")) #  rownames; colnames
> rawd
   group smoker status
1     g1        smoker
2     g1        smoker
3     g1        smoker
4     g1        smoker
5     g1        smoker
6     g1        smoker
7     g1        smoker
8     g1        smoker
9     g1        smoker
10    g1        smoker
11    g1        smoker
12    g1        smoker
13    g1        smoker
14    g1        smoker
15    g1        smoker
16    g1        smoker
17    g1        smoker
18    g1        smoker
19    g1        smoker
```

46.3 Example 2 - two variables

Here I'm going to use the example given in the chapter *Several independent proportions compared with the average,* where you can also find a table layout of the data. The code below was used to generate the raw data.

columns → smoker, nonsmoker, smoker, nonsmoker etc

```
d<- matrix ( c( 83, 3, 90, 3, 129, 7, 70, 12), nrow = 4, byrow=TRUE)
```

rows → group 1 | group 2 | group 3 | group 4

values for the 2nd factor →

```
colnames(d) <- c("smoker", "nonsmoker")
```

values for the 1st factor →

```
rownames(d)<- c( "g1", "g2", "g3", "g4")

rawd<- expandTable(d, c("group", "smoker status")) #  rownames; colnames

rawd
```

the counts

name for the first new column filled with values from rownames

name for the second new column filled with values from colnames

when this variable name is exported as a dat file the space is replaced with a dot; to avoid this use a '_' in place of spaces

```
> d<- matrix ( c( 83, 3, 90, 3, 129, 7, 70, 12), nrow = 4, byrow=TRUE)
>
> colnames(d) <- c("smoker", "nonsmoker")
> rownames(d)<- c( "g1", "g2", "g3", "g4")
> rawd<- expandTable(d, c("group", "smoker status")) #  rownames; colnames
> rawd
   group smoker status
1     g1        smoker
2     g1        smoker
3     g1        smoker
4     g1        smoker
5     g1        smoker
6     g1        smoker
```

```
380  g3  nonsmoker
381  g3  nonsmoker
382  g3  nonsmoker
383  g3  nonsmoker
384  g3  nonsmoker
385  g3  nonsmoker
386  g4  nonsmoker
387  g4  nonsmoker
388  g4  nonsmoker
389  g4  nonsmoker
390  g4  nonsmoker
391  g4  nonsmoker
392  g4  nonsmoker
393  g4  nonsmoker
394  g4  nonsmoker
395  g4  nonsmoker
396  g4  nonsmoker
397  g4  nonsmoker
```

Here is another example using the data presented in table format in Cohen's 1960 paper introducing the Kappa measure of agreement between raters, used in the *Levels of agreement* chapter:

```
d<- matrix ( c( 88,14,18, 10,40,10, 2,6,12), nrow = 3, byrow=TRUE)

colnames(d) <- c("schizophrenic", "neurotic", "brain damage")

rownames(d)<- c( "schizophrenic", "neurotic", "brain damage")

rawd<- expandTable(d, c("judge A", "judge B")) #  rownames; colnames

rawd

write.table(rawd, "cohen_1960.dat", sep="\t", row.names=FALSE)
```

```
> d<- matrix ( c( 88,14,18, 10,40,10, 2,6,12), nrow = 3, byrow=TRUE)
> colnames(d) <- c("schizophrenic", "neurotic", "brain damage")
> rownames(d)<- c( "schizophrenic", "neurotic", "brain damage")
> rawd<- expandTable(d, c("judge A", "judge B")) #  rownames; colnames
> rawd
        judge A       judge B
1  schizophrenic schizophrenic
2  schizophrenic schizophrenic
```

46.4 Tips and Tricks

The process described in this chapter is something that is not often done and can prove extraordinarily time-consuming to figure out if you don't know the above shortcuts.

Always check the result by obtaining a table of counts, for information see the chapter *Several independent proportions compared to an average – two way tables.*

The *NCStats* package I found tricky to install and if you only want to use the *expandTable()* function in it I suggest instead you simply copy the source code of the **function**, given below, into the *R Console* and then refer to it. I have renamed it *expandTable_RB().* Therefore to run the previous examples using this replace *expandTable* with *expandTable_RB.*

```
expandTable_RB <- function (x, var.names = NULL, ...)
{
    Freq <- NULL
    nr <- nrow(x)
    nc <- ncol(x)
    if (nr == 1 | nc == 1) {
        if (nr == 1)
            x <- t(x)
        df <- data.frame(rep(rownames(x), x))
        if (length(var.names) > 1)
            stop("Too many var.names given.", call. = FALSE)
        names(df) <- var.names
    }
    else {
        x <- as.data.frame.table(x)
        df <- sapply(1:nrow(x), function(i) x[rep(i, each = x[i,
            "Freq"]), ], simplify = FALSE)
        df <- subset(do.call("rbind", df), select = -Freq)
        for (i in 1:ncol(df)) {
            df[[i]] <- type.convert(as.character(df[[i]]), ...)
        }
        rownames(df) <- NULL
        if (!is.null(var.names)) {
            if (length(var.names) < 2)
                stop("Too few var.names given.", call. = FALSE)
            else if (length(var.names) > 2)
                stop("Too many var.names given.", call. = FALSE)
            else names(df) <- var.names
        }
    }
    df
}
```

I obtained the souce code for the *expandTable()* function by typing into the *R Console* window *NCStats::expandTable* once I had installed and loaded the *NCStats* package.

47 Installing non-CRANS packages

Sometimes it is necessary to install packages that are not on the official R repository, that is the *Comprehensive R Archive Network* (CRANS) site. One such alternative repository is RForge. Say we wanted to install a package called *NCStats* - which is on the RForge site, then we would type into the *R Console* window:

```
install.packages("NCStats",,"http://www.rforge.net/",dependencies=TRUE,type="source")
```

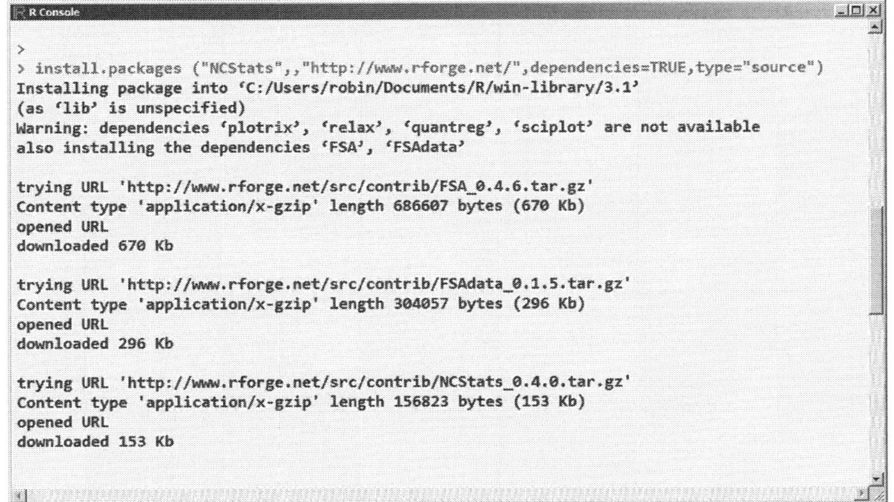

The *dependencies=TRUE* option means I have requested all dependent packages to be installed alongside the specified package above. However this does not always seem to happen when installing from a non-CRANS site which does create another additional level of complexity and associated errors. This frequently means that you will receive a series of errors about specific missing packages which you then need to install before the original package you installed works.

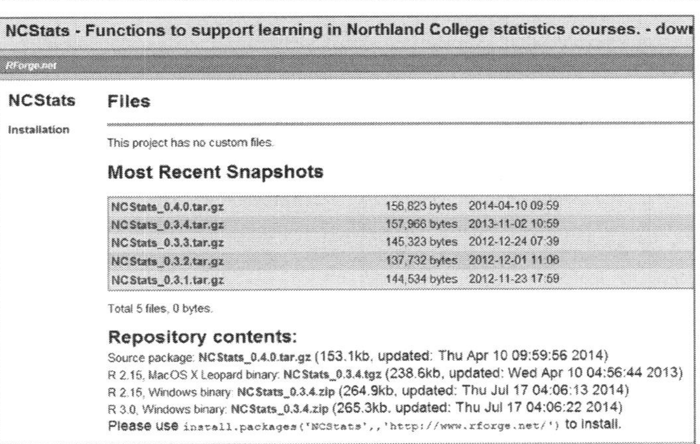

The *type ="source"* option overrides the default binary option for windows systems. Sometimes this does not produce a successful install and an alternative is to download a zipped (zip file) version of the library. For example, you can download a zipped version of the *NCStats* package here:

http://www.rforge.net/NCStats/files/

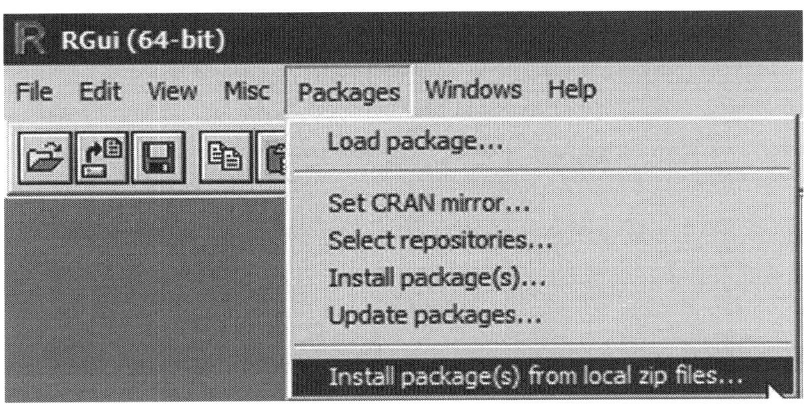

To install downloaded zipped R libraries use the *R Console* menu option:

Packages->Install packages(s) from local zip files...

Select the zipped file you saved earlier and the package should install itself. Remember to be able to use it you still need to load it using the *library(NCStats)* function.

48 Workspaces, objects and history files

There is much going on in R under the simple looking Console window and this chapter provides some details of these hidden aspects, consider it equivalent to looking under the bonnet of the car.

When you use R you are in a particular "workspace" which is your current R working environment. This environment contains all the dataframes you have, along with any other objects you may have created or accidentally accumulated on the way.

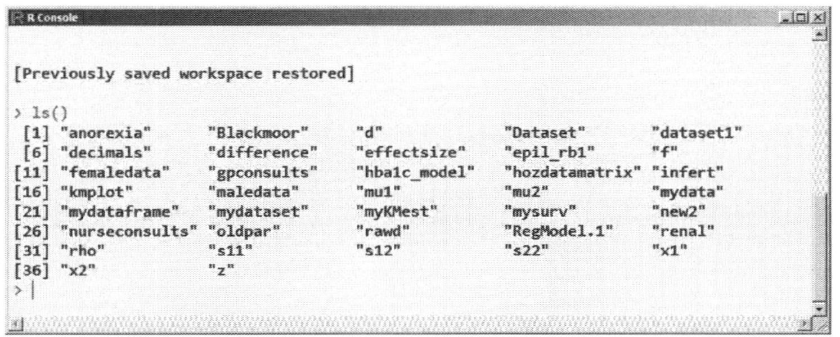

To list the objects you have in the workspace you type in the *R Console* window the command:

```
ls()
```

note the lower case L

In the example opposite I typed this command immediately after starting R, and as you can see I have 37 objects already available in my workspace. The number of objects loaded depends upon your own setup.

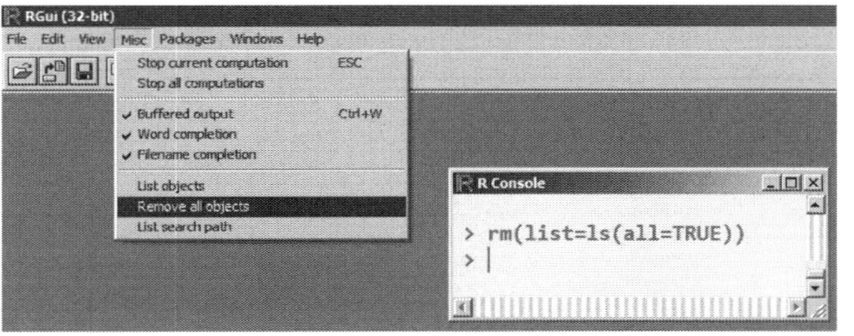

To remove all the objects in your workspace either select the RGui window menu option *Msc -> Remove all objects* or type the command:

note the lower case R

```
rm(list = ls(all=TRUE))
```

You can also use the *rm()* function to remove specific objects such as a specific dataframe, for example to remove just the *mydataframe* dataframe you would type *rm(mydataframe)*.

```
> search()
[1] ".GlobalEnv"        "package:stats"     "package:graphics"
[4] "package:grDevices" "package:utils"     "package:datasets"
[7] "package:methods"   "Autoloads"         "package:base"
>
```

You can also see all the packages you have loaded using the *search()* function.

To find out more about a particular object you can use the *str(object_name)* function (shorthand for structure), replacing the *object_name* value with your chosen one. In the example below I have applied the *str()* function to a dataframe object called *mydataframe*. I had previously assigned *mydataframe*, to the data described in the *Several independent proportions compared with the average: Two way tables* chapter. We see that it consists of two variables both of which are factors, one with two levels and the other with four.

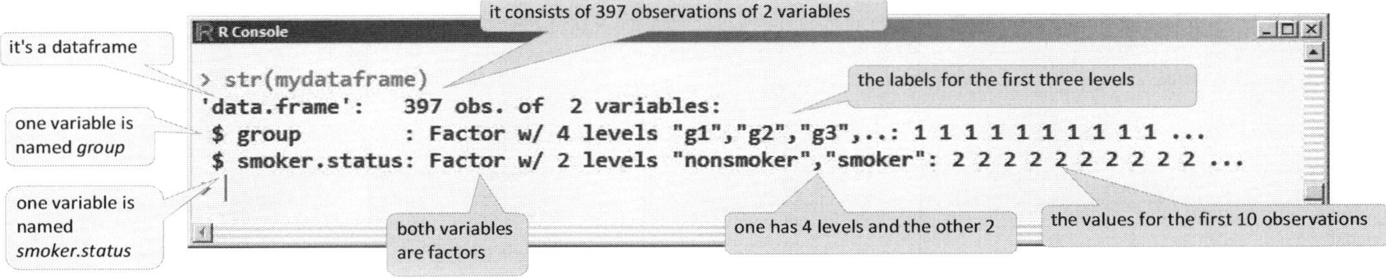

48.1 Saving and loading your workspace

The workspace is your immediate R working environment it keeps details of all the datasets and packages you have loaded.

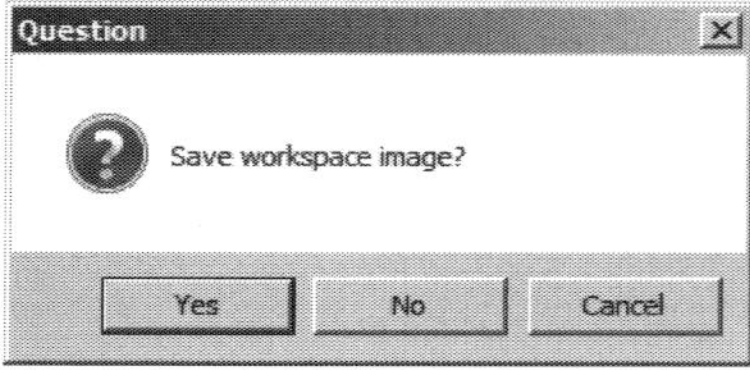
Question
Save workspace image?
Yes
No
Cancel

At the end of an R session, that is when you close R down, you are automatically asked if you wish to save the current workspace (called an image) and if you select the **yes** option this is then automatically reloaded the next time R is started.

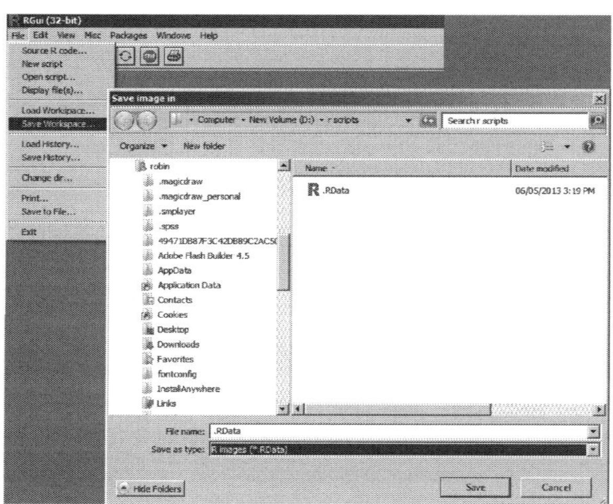

To save the current workspace to a specific file use the *RGui* window menu option **File-> Save workspace** or using the R function *save.image()* substituting *d:/mycurrentwork.rdata* for your own path and filename.

```
save.image("d:/mycurrentwork.rdata")
```

note the *rdata* extension this is important, however you could also use *rda* instead

To load a previously saved workspace, either use the *RGu*i menu option **File->Load workspace** or the R function:

```
load("d:/mycurrentwork.rdata") or load(file=file.choose())
```

You can also load a workspace from the web by wrapping the url in the *url()* function:

```
load(url("http://www.robin-beaumont.co.uk/virtualclassroom/book2data/cummingp245.rdata"))
```

48.2 Saving and loading your R commands - the history file

R Console
> savehistory(file = "d:/r scripts/analysis_07_05_2013.Rhistory")

Once you have carried out an analysis you can export all the commands you typed in R into a text file known as the history file. You do this either by selecting the **RGui** menu option **File->Save History** or the R command below replacing *filename* with your required path/filename.

```
savehistory(file = "filename.Rhistory")
```

R Console
Up /down arrows show previous commands stored in the history file
> library(car)

When you load a history file into R it does not run the commands, what it does is place the commands in memory. These you can see, and run, by pressing the **up** and **down arrow** keys.

48.3 Deleting the history file

To do this you need to find the specific history file that R opens by default when it starts up. This file is called *.Rhistory*, in other words it just has the extension and no first part to the file name. The file is usually located in the default working directory which you can find by typing the command *getwd()* after you start R.

Once you have found the directory, quit R, because if you try to delete the file while R is running R will just recreate it from memory when you close R down.

Once you have quit R use your usual file browser to navigate to the correct directory, select the *.Rhistory* file and then delete it. The next time you start R you will not have any commands from the previous session available to you when you use the arrow keys.

48.4 Tips and Tricks

Saving your workspace can be very useful if you want to have a break during an analysis.

The workspace, also called the global environment, is expanded in three ways; by loading packages, adding dataframes; or adding saved workspaces to it.

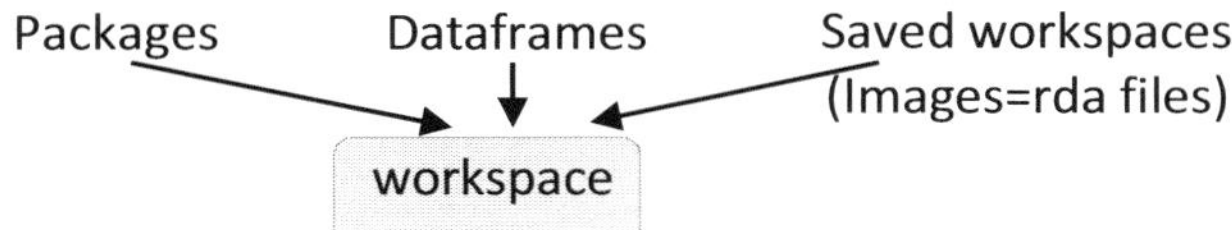

At first sight the saving of a history file may seem pretty useless as you can't simply run the R commands within it, however because it is a text file you can easily change it to an R script file which you can run in R. This is the topic of the next chapter.

To remove all objects in your workspace except the one called *nasty_one* type:

```
rm(list=setdiff(ls(), "nasty_one"))
```

49 Developing R Code: Rstudio and NppToR

In this book we have developed several small R code fragments using the *R Console* window or the *R Script* window in R Commander. If you want to produce larger pieces of R code it is a good idea to use what is called an *Integrated Development Environment* (IDE). IDEs allow you to; monitor and structure your code, produce R packages and pinpoint problems within it - a process commonly called debugging. As I intend this chapter to be a very basic introduction to IDEs I will focus on the debugging process.

RStudio, which is free, claims to be "the premier integrated development environment (IDE)" for R and is used by many R developers. The latest version is around 36 megabytes and also needs R to be installed for it to work.

49.1 Downloading and installing

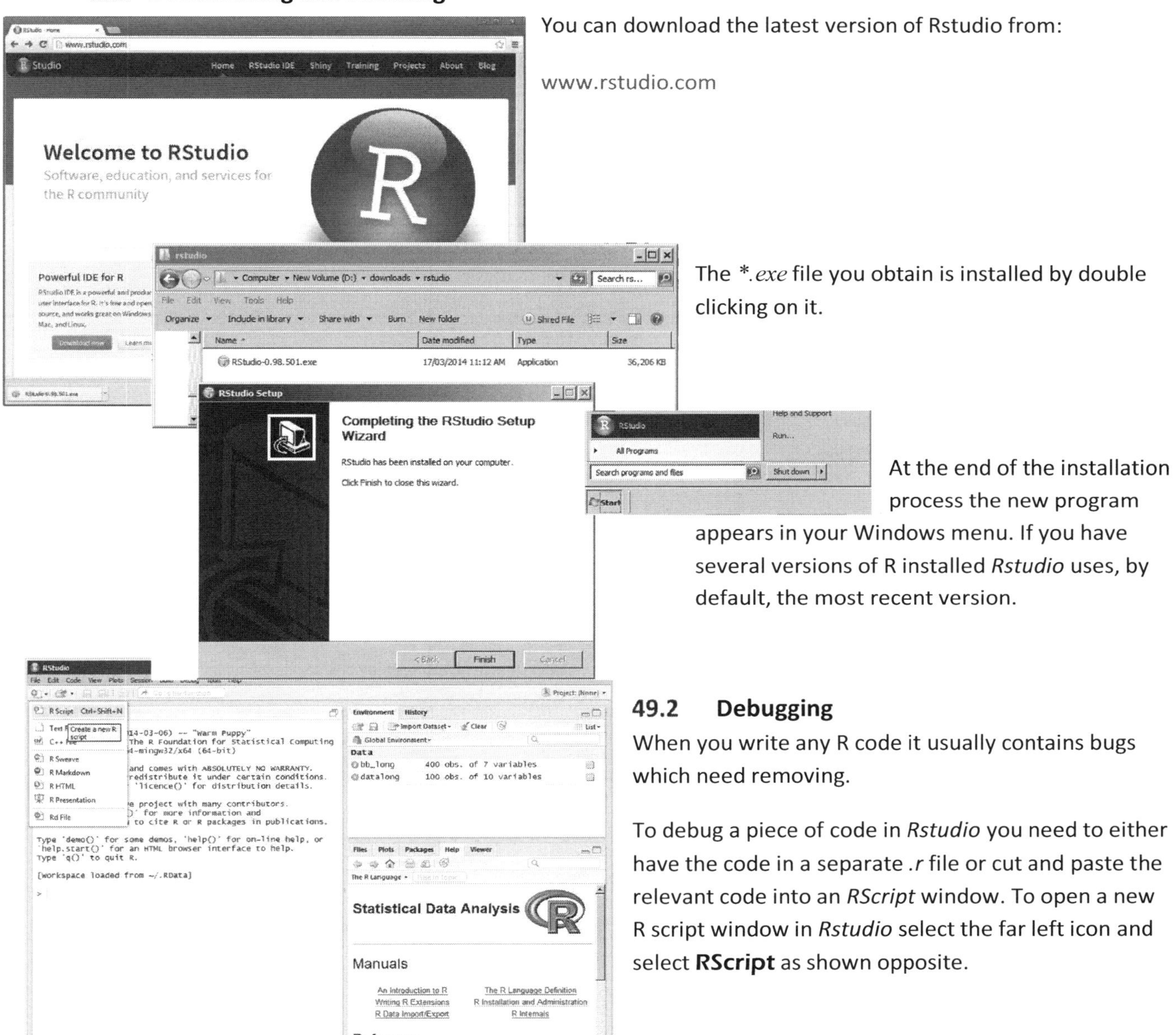

You can download the latest version of Rstudio from:

www.rstudio.com

The *.*exe* file you obtain is installed by double clicking on it.

At the end of the installation process the new program appears in your Windows menu. If you have several versions of R installed *Rstudio* uses, by default, the most recent version.

49.2 Debugging

When you write any R code it usually contains bugs which need removing.

To debug a piece of code in *Rstudio* you need to either have the code in a separate *.r* file or cut and paste the relevant code into an *RScript* window. To open a new R script window in *Rstudio* select the far left icon and select **RScript** as shown opposite.

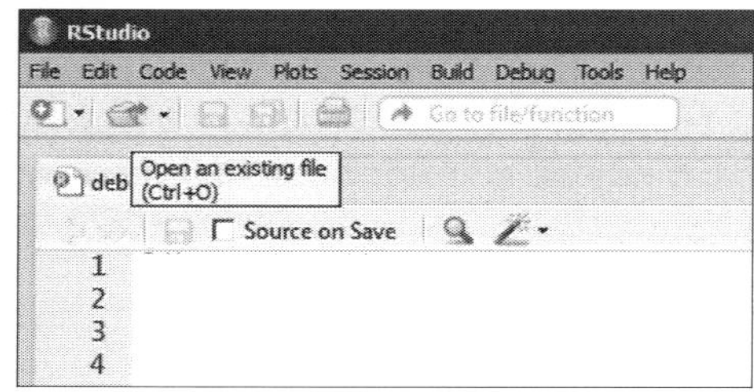

Alternatively, I have opened an existing piece of R code which contains a bug, using the **Open an existing file** icon (the second icon along the menu bar).

You can download the file from:

http://www.robin-beaumont.co.uk/virtualclassroom/book2data/rcode/debug_example1.r

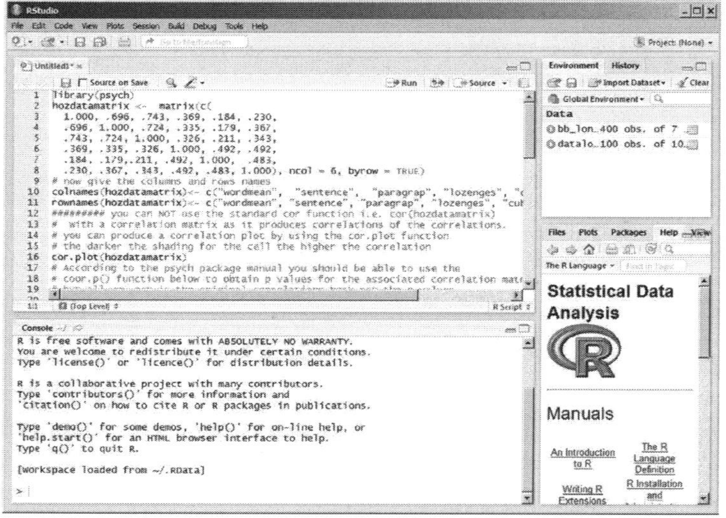

The code then appears in an *Untitled1* window, as shown opposite.

Before you can debug the code you need to ensure two conditions are fulfilled:

- the code is either in a file (as we have done), or you need to save the new code into a file first. If it is not currently saved you will see *untitled1.** shown at the top of the window.
- The code is 'sourced' which means that it can be executed in the Console panel, to set the file to sourced you need to:

Click on the Source icon and select the **Source** option.

If you have not sourced the file and try to debug it you get the error: *Breakpoints will be activated when the file is sourced.*

left mouse click here to create a breakpoint at line 24

```
   ## given below I have called it corr.p_temp
23 corr.p_temp <-function(r, n, adjust="holm") {
24   cl <- match.call()
25   if(missing(n)) stop("The number of subjects mus
26   t <- (r*sqrt(n-2))/sqrt(1-r^2)
27   p <- 2*(1 - pt(abs(t),(n-2)))
28   p[p>1] <- 1
29   if (adjust !="none") {
30     if(isSymmetric(p)) {sym <- TRUE
31                        lp <- upper.tri(p)   #the case of a symmetric matrix
32                        pa <- p[lp]
33                        pa <- p.adjust(pa,adjust)
34                        p[upper.tri(p,diag=FALSE)] <- pa
35     } else {
36       p[] <- p.adjust(p ,adjust)   #the case of an asymmetric matrix
37       sym <- FALSE}
38   }
39   result <- list(r = r,n=n,t=t,p=p,sym=sym, Call=cl)
40   class(result) <- c("psych", "corr.test")
41   return(result)
42 }
43
44 aresult1<- corr.p_temp(r = hozdatamatrix, n= 73, adjust="holm")
45 aresult1
46 aresult2<-corr.p_temp (r = hozdatamatrix, n= 73)
47 aresult2
48 aresult3<- corr.p_temp (r = hozdatamatrix, n= 73, adjust="none")
49 aresult3
50 aresult4<- corr.p_temp (r = hozdatamatrix, n= 73, adjust="holm")
51 aresult4
```

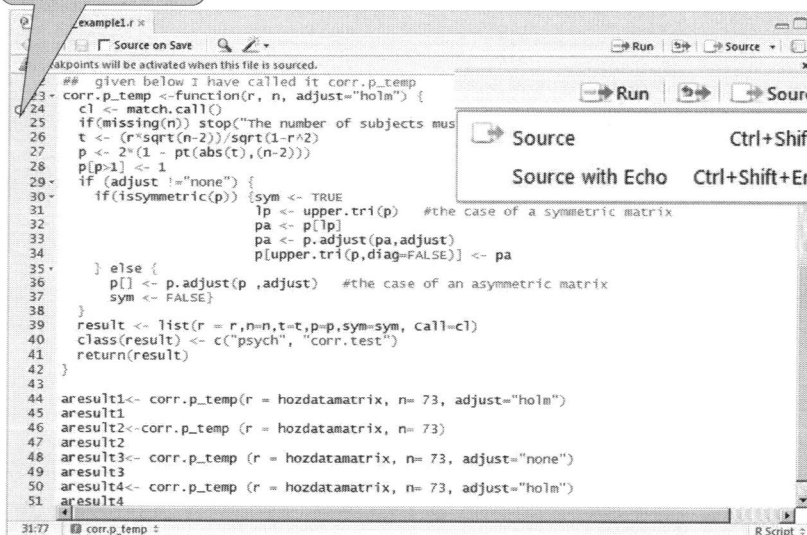

Once we have the code in a saved file which is sourced we can then debug it.

One of the basic techniques of debugging is to step over, which means to follow line by line, areas of the code, at the same time monitoring any changes that occur in the data items for each line.

Because the code might be very long and we don't want to spend time looking through all of it you can set a breakpoint which tells *Rstudio* when you want to start stepping over.

The specific bug I have in this piece of code is that when I use the *corr.p_temp()* function in it with the option *adjust='none'* I receive an error. So I want to see what happens for each line of code within the function when I send it this specific option.

To set a break point (●) click to the left of the line number in an R script, or press *Shift+F9*. I have created one in the script at line 24 which is at the beginning of the *corr.p_temp()* function.

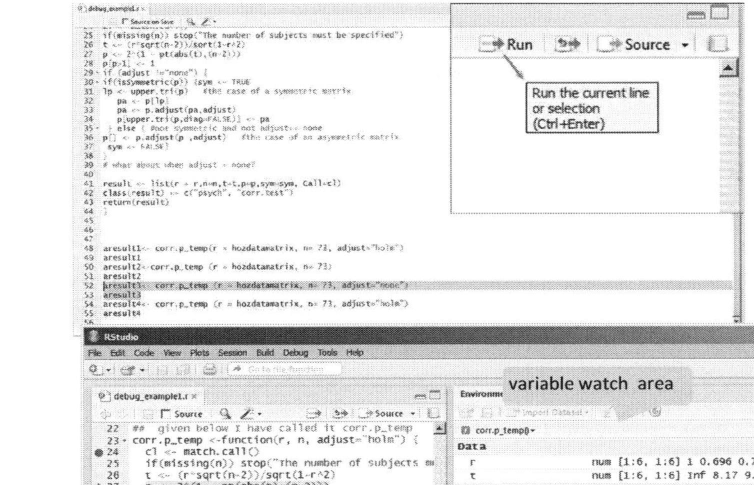

Now the breakpoint is set we need to instruct *Rstudio* which part of the code to run. As I'm interested in seeing what happens when I set the *adjust='none'* option I have highlighted lines 52 and 53, by left mouse clicking on the lines and dragging.

To run the selected code either click on the run icon or use the key combination *ctrl+enter*.

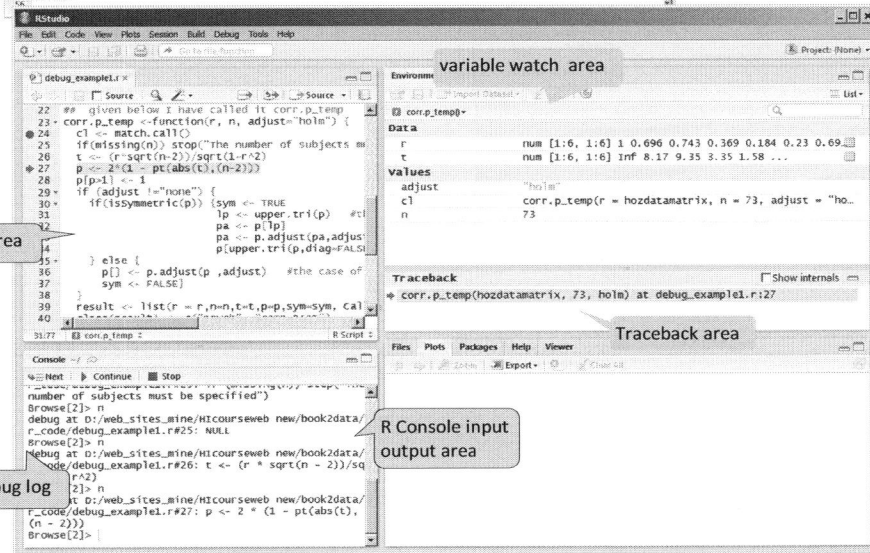

When the program gets to the breakpoint all stops. To move to the next line of executed code (i.e. to step through) you need to press:

F10

If you have seen enough before getting to the end of the selected code simply press:

shift + F8

There are several important aspects to take note of while stepping over the code:

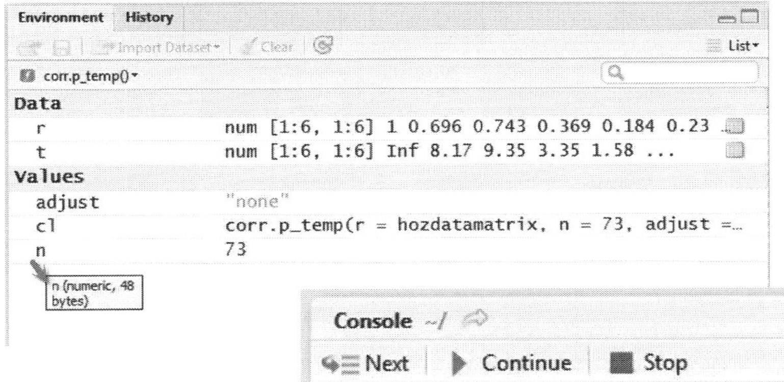

variable watch area – provides the current values for the various variables you have defined in your code as well as any data you have defined. Hovering over a particular value tells you how it is stored on the computer, while clicking on one of the listed data items creates another tab in the code area for you to be able to see more detail.

Console window + debug log area – provides both what you would see in the usual R Console window along with the debugging commands issued by *Rstudio*. This is because *Rstudio* makes use of the built in debugger in R so also echoes the commands it is sending to R.

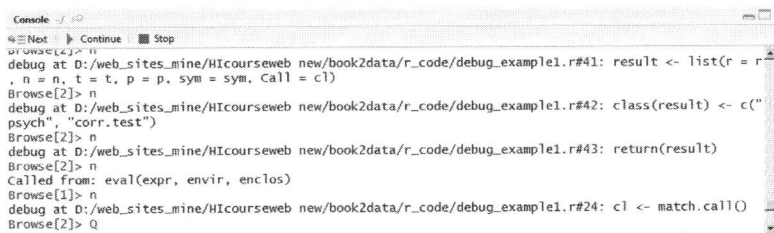

Traceback area – displays a list of the functions that are currently running, with the most recent function on top. By clicking on any function you can see the expression currently being evaluated.

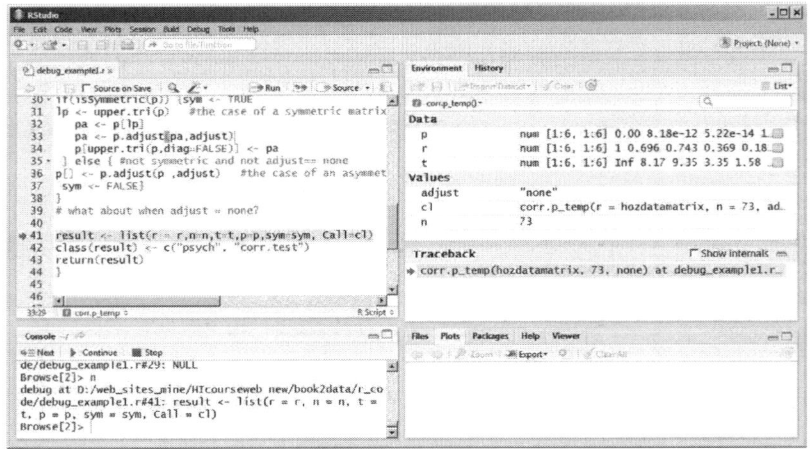

By stepping through the function we see that by the time we get to line 42 all the return variables have values allocated except for *sym* which has never been set when we use the option *adjust='none'*.

The next stage would be to reconsider the logic within the function and rewrite it. This simply might involve giving a default setting to the *sym* object at the start of the function or alternatively adding some more lines of code with more detailed branching.

Alternatively, in this instance, an email to the author of the function (within the *psych* package) produced a fix within a few hours – the wonders of the internet and the generosity of the many developers of the thousands of R packages continues to astound me.

49.3 Tips and tricks

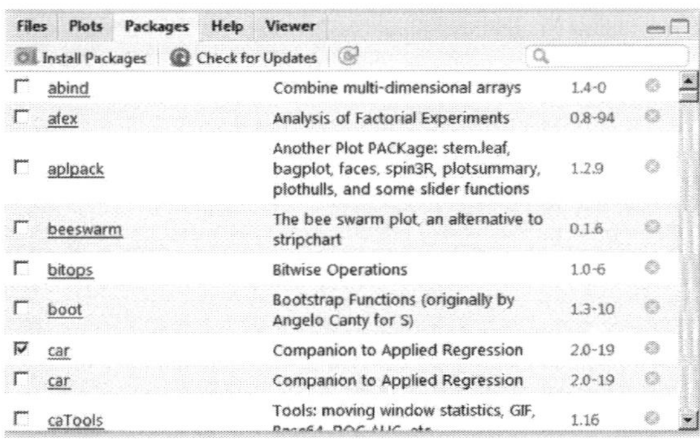

Rstudio is not just a debugger, and in fact some people recommend its use instead of R Commander for newbies to R. However, I am not as keen because it lacks a specific statistics menu structure.

You can find out which packages you currently have loaded by clicking on the Global Environment (shown below). Alternatively, as shown opposite, you can see which packages are loaded (ticked boxes) compared to those that are simply installed by inspecting the packages tab in the bottom right panel.

As shown below *Rstudio* works with R Commander. To call up R Commander from within *Rstudio* type *library(Rcomdr)* in the R Console panel.

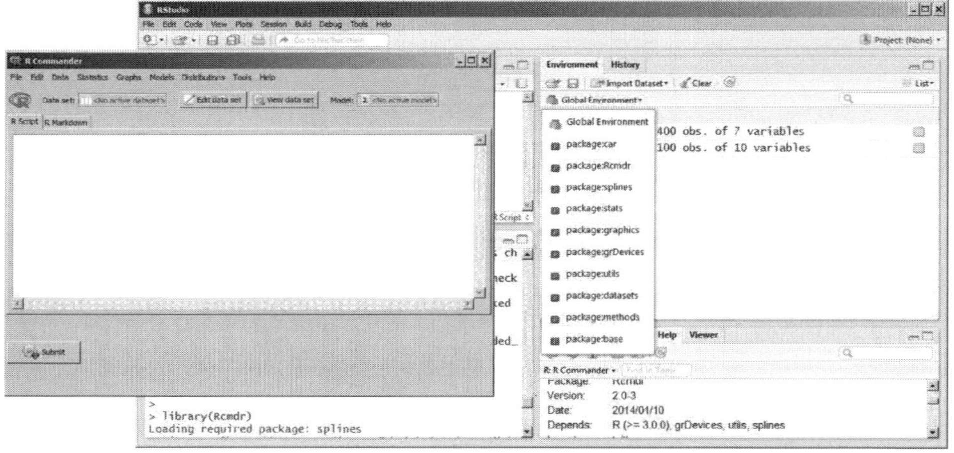

If you have wondered, how R packages are created there is an excellent anonymous video (author=trestletech?) describing the process in Rstudio at: http://youtu.be/9PyQlbAEujY

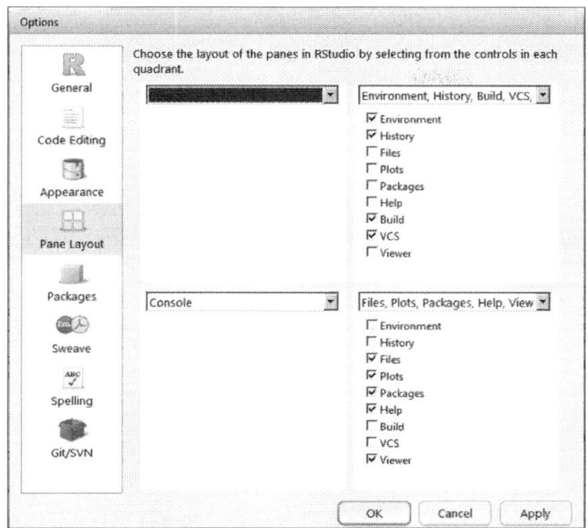

The Rstudio web site has good documentation including tutorials and there are many free videos available on youtube.

You can configure *Rstudio* a myriad of ways using the menu option:

Tools->global options

The Pane layout options, shown opposite, allow you to configure exactly what you have in each panel.

49.3.1 NppToR: linking Notepad++ to R

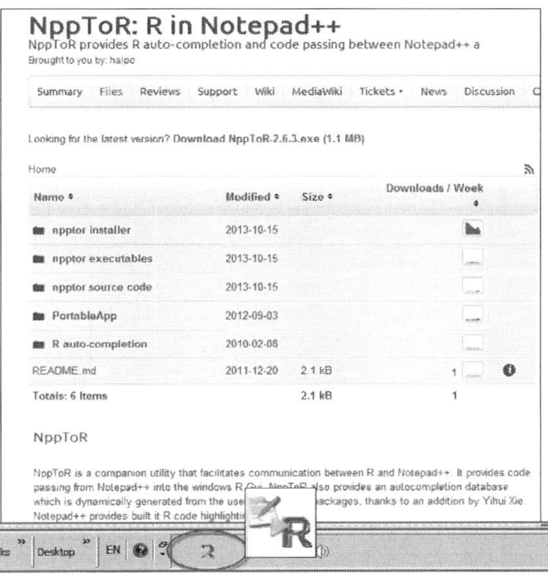

R studio is not the only IDE available (Versani 2011). For example you can get a free program that links *Notepad++* to R called *NppToR*. *NppToR* was developed by Andrew Redd who is a Bio-statistician and assistant professor at the University of Utah School of Medicine. *NppToR* is nowhere as near as sophisticated as *Rstudio* but still is very useful. You can download it for free from:

http://sourceforge.net/projects/npptor/files/

Once you have installed *NppToR* you can access it from the taskbar at the bottom of the screen, right clicking on the icon also offers many options, including starting up *notepad++* and setting automatic completion of R function names.

To allow the autocompletion function to work you also need to set the relevant options in *Notepad++* shown below.

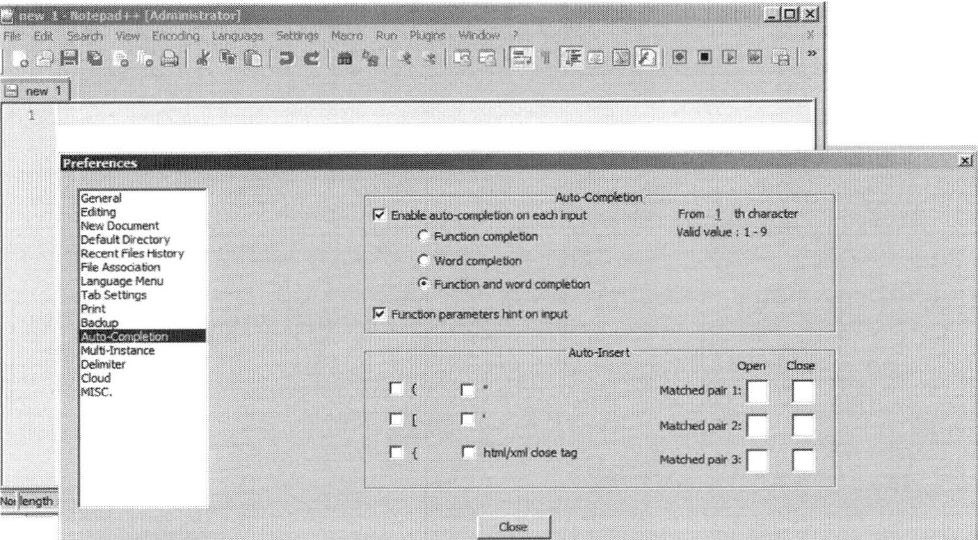

Then when typing R code in *Notepad++* you simply select the code and press the *F8* key to see the selected code executed in an R Console window. The author, Andrew Redd has produced a presentation available from:

http://www.r-project.org/conferences/useR-2010/slides/Redd.pdf

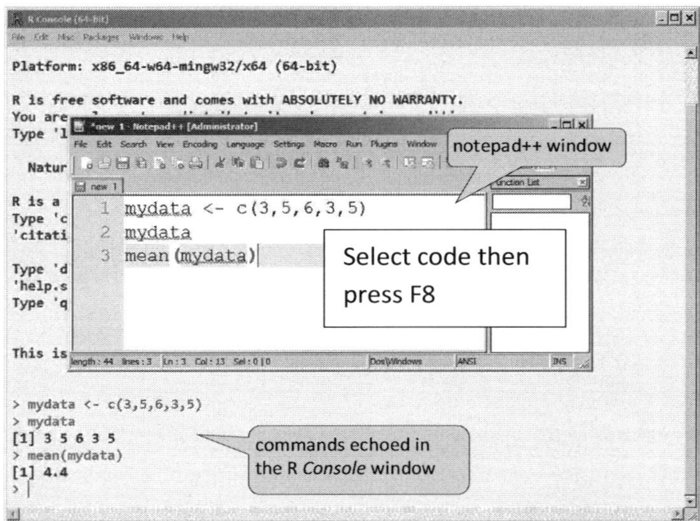

50 More ways of analysing your data

This section introduces several methods increasingly used to analyse and present data. The section starts and ends with two chapters investigating various graphical techniques; first looking at an alternative, more appropriate method than bar charts for displaying nominal data, and ending by considering the various approaches to developing publication quality graphics.

The statistical techniques chapters (darker shading) in this section are those that have blossomed since computer power, along with the appropriate applications, have become available. Such techniques include the creation of (virtual) populations from samples and allowing the computer to search for the best solution from a huge number of possibilities. Such approaches have provided practical solutions to several problems that appeared intractable in the past (i.e. multilevel models, confidence interval estimation etc.).

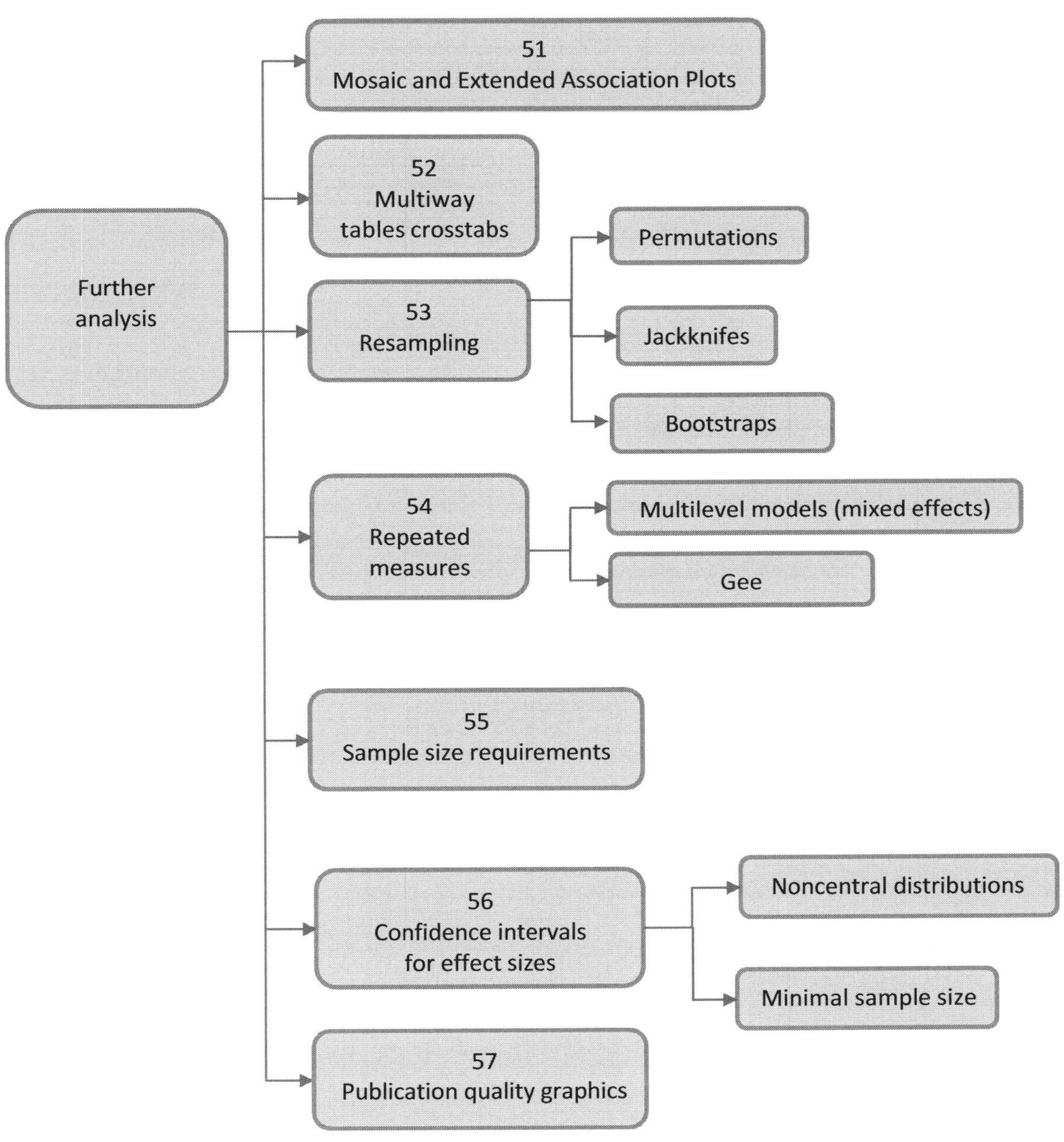

51 Mosaic and extended association plots

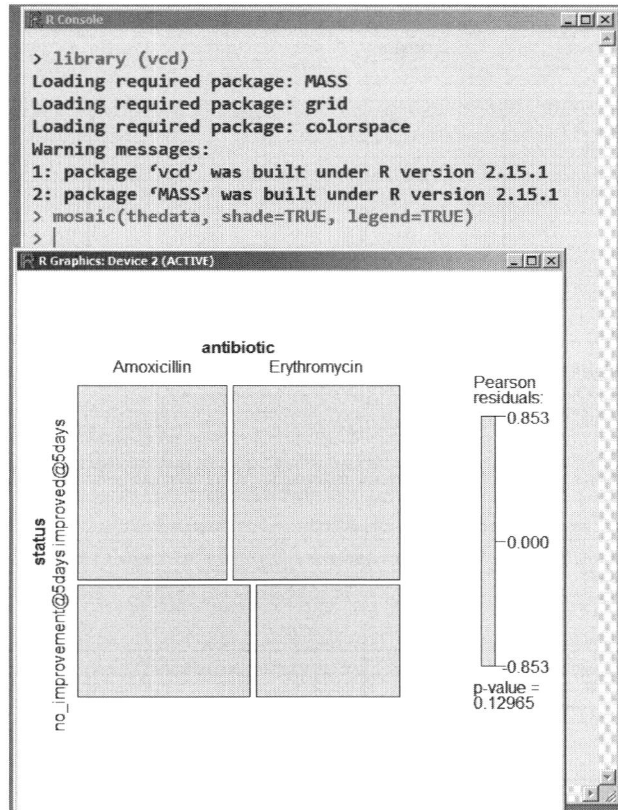

Mosaic and extended association plots allow the presentation of table data in the form of a diagram which immediately highlights those cells that contribute significantly to the chi square result. Mosaic and extended association plots were introduced in two chapters; *Several independent proportions compared with the average* and *Comparing several independent categories: Contingency tables.*

51.1 Two way tables

The following example makes use of the Amoxicillin/Erythromycin data described in the chapter *Comparing several independent categories,* section *Equivalent analysis directly in R.* Copy the R code in that section to produce the data. To create the Mosaic plot you need to install and load the *vcd* R library:

```
install.packages("vcd", dependencies=TRUE)

library(vcd)
```

the mosaic plot works on counts so if you have raw data you need to use the table function to create them on the fly.

```
mosaic(thedata, shade=TRUE, legend=TRUE)
```

All the squares opposite are the same color, indicating that none of the cells deviates much from the expected null model cell count.

However if our data had produced a statistically significant chi-squared p-value one or more of the cells would have changed shading/color.

```
> thedata <- matrix( c (110, 120,160,80), ncol=2, byrow=FALSE)
> mycol_names <- c("Amoxicillin", "Erythromycin")
> myrow_names <- c("improved", "no_improve")
> dimnames(thedata) <- list(status = myrow_names, antibiotic = mycol_names)
> thedata
            antibiotic
status       Amoxicillin Erythromycin
  improved           110          160
  no_improve         120           80
> myresult <- chisq.test(thedata)
> myresult

        Pearson's Chi-squared test with Yates' continuity correction

data:  thedata
X-squared = 16.2922, df = 1, p-value = 5.429e-05

> mosaic(thedata, shade=TRUE, legend=TRUE)
```

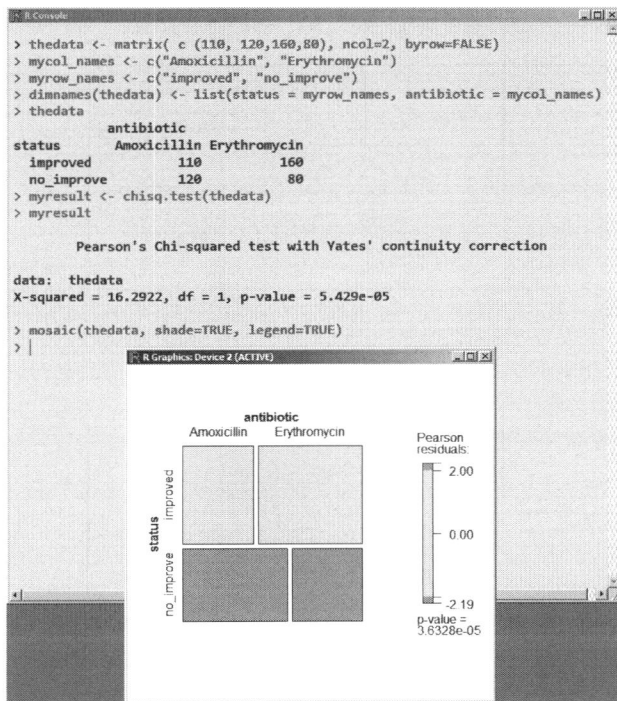

I will demonstrate this by editing the data, and shortening the row names to stop them colliding.

```
thedata <- matrix( c (110, 120,160,80), ncol=2,
byrow=FALSE)

mycol_names <- c("Amoxicillin", "Erythromycin")

myrow_names <- c("improved", "no_improve")

dimnames(thedata) <- list(status = myrow_names,
antibiotic = mycol_names)

thedata

myresult <- chisq.test(thedata)

myresult
```

Now with a highly statistically significant result we can see that the Amoxicillin group had far fewer numbers of patients improving at day five.

The default shading is not set to be particularly sensitive but this can be changed.

51.2 Mosaic plot options - shading and residual types

You can request other shading schemes and also other types of residual.
Quoting one of the people who developed the mosaic plot, Michael Friendly of York University Toronto "In contrast, one of the other shading schemes, from Friendly (1994) (use: *gp = shading_Friendly*), uses cutoffs of 2; 4, to shade cells which are individually significant at approximately the *p-value= 0:05* and *p-value = 0:001* levels, respectively".

To use these options we simply specify them in the *mosaic()* function:

```
mosaic(thedata, residuals_type = "pearson", gp = shading_Friendly )

# default; using components of Pearson's chi-squared

mosaic(thedata, residuals_type = "deviance", gp = shading_Friendly )

# using components of the likelihood ratio chi-squared

mosaic(thedata, residuals_type = "FT", gp = shading_Friendly )

# using Freeman-Tukey residuals
```

51.3 Mosaic plots versus bar charts

Hopefully you can see from the above discussion why it is useful to graph the counts to get an understanding of their distribution.

By far the best graphical method is the mosaic plot, and by using the Friendly style shading allows us to see immediately deviations that are statistically significant. In contrast, the bar chart is a poor substitute, not providing this level of sophistication.

51.4 Extended Association Plot

The extended association plot displays the residuals in the way shown below.

Things to note about the **Extended Association Plot**:

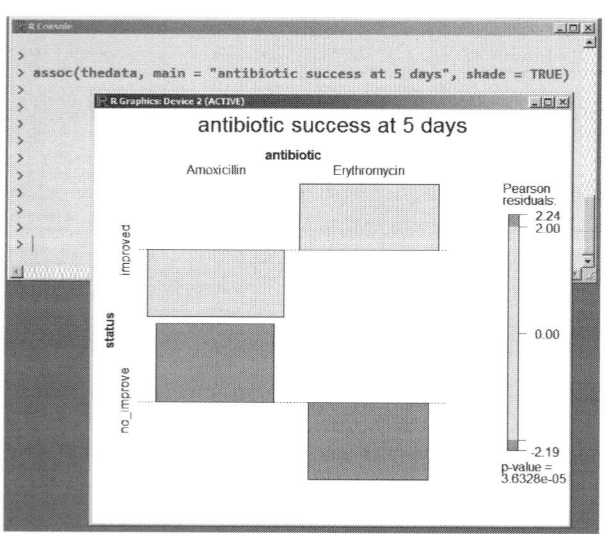

- the p-value associated with the overall chi-squared value is provided in the bottom right corner.
- the legend on the right indicates the Pearson residuals (default)
- if there are any which are extreme, a change in shading/color is noted.
- the heights of the bars represent the Pearson residual described in the *Comparing several independent categories: Contingency tables* chapter.

51.5 Tips and Tricks

It is unlikely that you will have come across the plots described in this chapter before. However, I am sure that in future they will become a more common feature in medical journals.

Remember if you have raw data you need to wrap the data in the *table()* function:

```
mosaic(thecountdata,  residuals_type = "pearson", gp = shading_Friendly )
```

or

```
mosaic(table(therawdata),  residuals_type = "pearson", gp = shading_Friendly )
```

The following chapter, *Multiway tables and Crosstabs* provides an example of a Mosaic plot produced from a multiway table.

Both the mosaic and the ASP were proposed by Michael Friendly who also developed the *vcd* and *vcdExtra* packages. Details of his approach can be found in his book, *Visualizing Categorical Data* (2000). Alternatively, you can consult his excellent tutorial available to download from the CRAN site at:

http://cran.us.r-project.org/web/packages/vcdExtra/vignettes/vcd-tutorial.pdf

There is an add-on package to R Commander called *RcmdrPlugin.mosaic* which is purported to allow the creation of mosaic plots via the R Commander menu structure, however I found it very temperamental and difficult to get to work. Please let me know if you have more success.

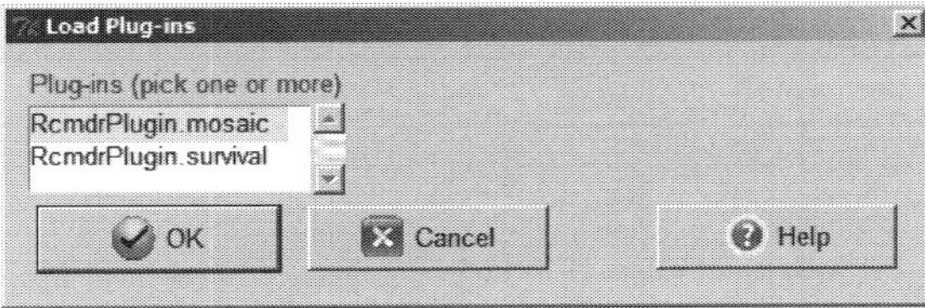

52 Multiway tables and crosstabs

	case	result	grouping	illness	sex
1	1	4	1	1	1
2	2	5	1	3	1
3	3	4	1	2	1
4	4	1	1	3	1
5	5	4	1	3	1
6	6	5	1	2	1
7	7	5	1	2	1
8	8	4	1	3	1
9	9	5	1	1	1
10	10	5	1	1	2
11	11	4	1	2	2
12	12	5	1	3	2

Multiway tables and crosstabs allow the presentation of table data which consists of more than two dimensions. An example should help understand this.

In the dataset opposite I have added two more variables to the dataset discussed in the *Summary statistics* chapter. There is now an *illness* index which grades each patient's condition from 1 to 3 as well as a *sex* variable that indicates if the patient was male (1) or female (2).

You can either download the data from the url below as described at the start of the *Summary statistics* chapter, remembering to replace the file name with the one given below :

```
http://www.robin-beaumont.co.uk/virtualclassroom/book2data/crosstabs1.dat
```

Alternatively you can type the following into the *R Console* window:

```
mydataframe <- read.table("http://www.robin-
                    beaumont.co.uk/virtualclassroom/book2data/crosstabs1.dat",header=T)

str(mydataframe)
```

The *tapply()* function was introduced in the *Summary statistics* chapter where we used it with two variables. You can also use it with more than two; say you wanted to find the mean values of your satisfaction scores divided by both the *sex* and *illness* status of the patients, you could do this by typing the following **R command** into the *R Console* window:

```
tapply(mydataframe$result,  mydataframe [c('sex','illness')],mean)
```

```
> tapply(mydataframe$result,  mydataframe [c('sex','illness')],mean)
   illness
sex        1        2        3
  1 2.333333 2.933333 2.882353
  2 3.333333 2.400000 2.600000
> |
```

We now have mean satisfaction scores for each of the patient illness groups divided into males and females. We can take this one step further by adding a further grouping variable which indicates if they were seen by a GP or triage nurse. To achieve this in R we simply add the *grouping* variable to the command:

```
tapply(mydataframe$result,  mydataframe [c('sex','illness','grouping')],mean)
```

```
> tapply(mydataframe$result,  mydataframe [c('sex','illness','grouping')],mean)
, , grouping = 1

   illness
sex        1     2        3
  1 2.000000 3.125 2.777778
  2 3.833333 2.250 1.500000

, , grouping = 2

   illness
sex        1        2        3
  1 2.666667 2.714286 3.000000
  2 2.833333 2.500000 3.333333

> |
```

Here we can see the mean satisfaction scores for each of the illness categories divided by both the sex of the patient and the type of interviewer.

In the above examples we have requested a summary function (i.e. the mean) to be calculated for each cell in the table. Alternatively we might want the count for each cell; this is achieved by using the *length* R command:

```
tapply(mydataframe$result,  mydataframe [c('sex','illness','grouping')],  length )
```

```
R Console
> tapply(mydataframe$result,  mydataframe [c('sex','illness','grouping')], length)
, , grouping = 1

   illness
sex 1 2 3
  1 3 8 9
  2 6 4 2

, , grouping = 2

   illness
sex 1 2 3
  1 3 7 8
  2 6 6 3

> |
```

Such tables are more useful if you can either see the percentages of each count in each cell (i.e. SPSS style crosstabs) or if you use mosaic plots, both of which are described below.

52.1 SPSS style crosstabs

If you are familiar with SPSS crosstabs tables and like them, you can produce the equivalent in R. Crossstab is short for cross tabulation, which is another word for a table of counts.

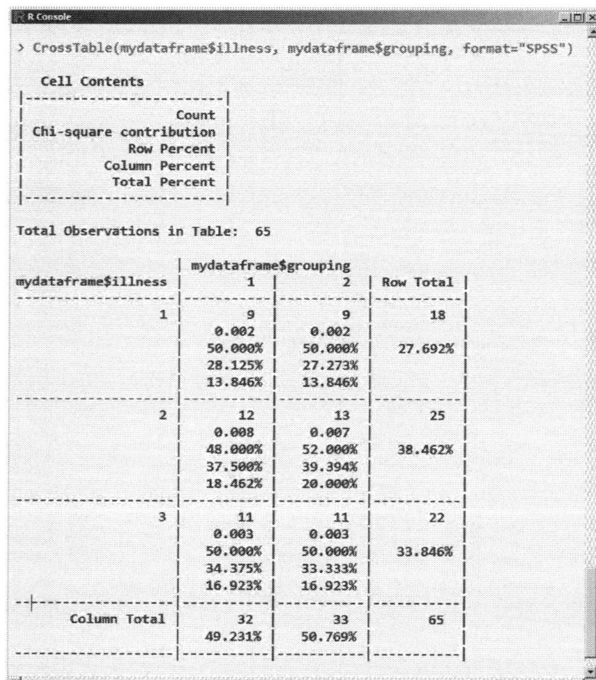

```
R Console
> CrossTable(mydataframe$illness, mydataframe$grouping, format="SPSS")

   Cell Contents
|-------------------------|
|                   Count |
| Chi-square contribution |
|             Row Percent |
|          Column Percent |
|           Total Percent |
|-------------------------|

Total Observations in Table:  65

                    | mydataframe$grouping
mydataframe$illness |         1 |         2 | Row Total |
--------------------|-----------|-----------|-----------|
                  1 |         9 |         9 |        18 |
                    |     0.002 |     0.002 |           |
                    |   50.000% |   50.000% |   27.692% |
                    |   28.125% |   27.273% |           |
                    |   13.846% |   13.846% |           |
--------------------|-----------|-----------|-----------|
                  2 |        12 |        13 |        25 |
                    |     0.008 |     0.007 |           |
                    |   48.000% |   52.000% |   38.462% |
                    |   37.500% |   39.394% |           |
                    |   18.462% |   20.000% |           |
--------------------|-----------|-----------|-----------|
                  3 |        11 |        11 |        22 |
                    |     0.003 |     0.003 |           |
                    |   50.000% |   50.000% |   33.846% |
                    |   34.375% |   33.333% |           |
                    |   16.923% |   16.923% |           |
--------------------|-----------|-----------|-----------|
       Column Total |        32 |        33 |        65 |
                    |   49.231% |   50.769% |           |
--------------------|-----------|-----------|-----------|
```

To obtain SPSS style Crosstabs in R use the *CrossTable()* function in the gmodels package. Use the R commands given below to install and then load the package. The final line requests a crosstab of cell counts for illness category against interviewer type (GP or nurse). Because the *CrossTable()* function offers several styles of format I have used the command *format = "SPSS"* to let R know that I want the SPSS style.

```
install.packages("gmodels", dependencies=TRUE)

library(gmodels)
```

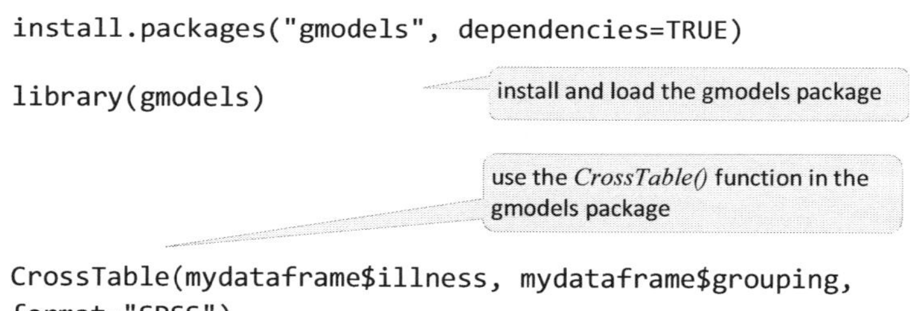
install and load the gmodels package
use the CrossTable() function in the gmodels package

```
CrossTable(mydataframe$illness, mydataframe$grouping,
format="SPSS")
```

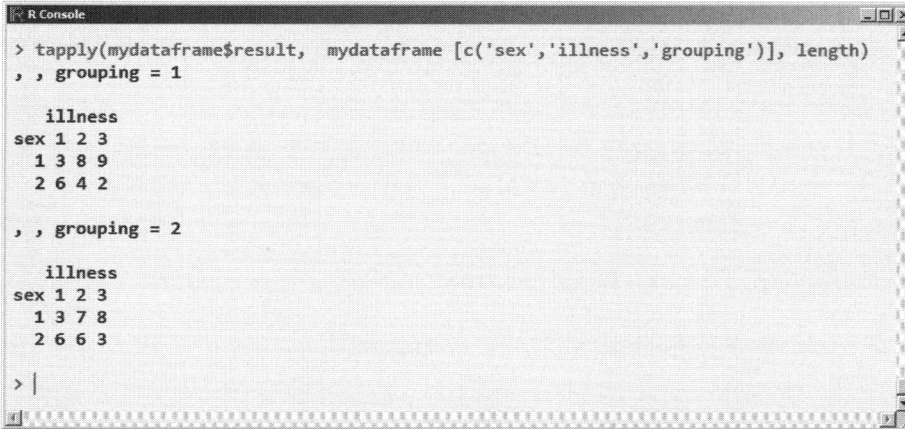

set the format option to SPSS

52.2 Mosaic plots for Multiway tables

You can also produce mosaic plots for the proportion of counts in each category. For example, for the number of participants in each of the illness categories divided by both their sex and the type of interviewer (grouping variable) we would use the following code:

```
install.packages("vcd", dependencies=TRUE)
library(vcd)
mosaic(~ grouping + illness+ sex, data = mydataframe,
main = "Proportions by group for illness & sex", shade = TRUE, legend = TRUE)
```

install and load the *vcd* package

divide by *grouping*, *illness* and *sex* counts

add a title

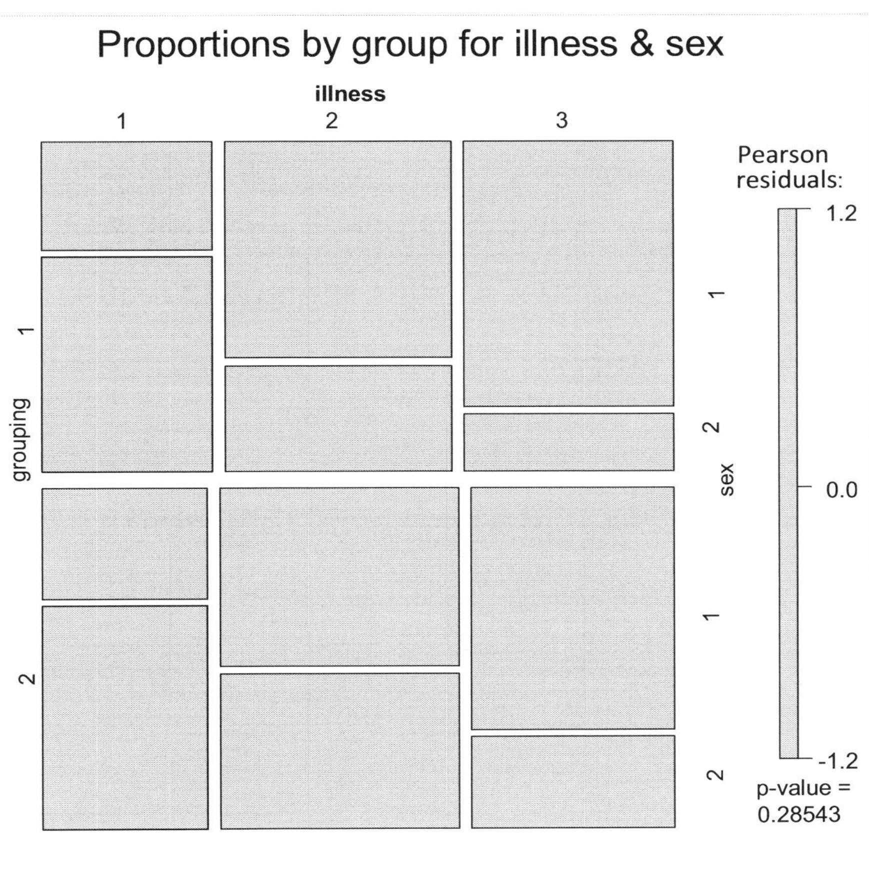

52.3 Tips and Tricks

Producing complex tables can take a considerable amount of time and, when created by hand, often contain errors. Therefore it is important that you always double-check them.

53 Resampling: permutations, jackknives and bootstraps

This chapter considers the ways we can create populations from the actual sample we have at hand.

There are three main methods; Permutation tests, Bootstrapping and Jackknifing. All three methods involve calculating the summary statistic (usually a mean or *t* value) many times. Permutation tests swap around group membership, while the other techniques focus on resampling from the original samples, although each has been adapted to allow group membership to be also taken into account.

Because all these techniques, other than for very small samples, require the use of a computer they are often also called computer intensive techniques. They can be used to provide estimates of values for which no formula exists, such as a confidence interval for the median. There is one very important rule to remember:

Resampling will provide valid results from a typical sample only if it is of sufficient size to include an adequate range of values

Because all these techniques sample from the original sample none of the samples produced will contain values that did not exist in the original one. Therefore it is essential that the original sample includes an adequate range of values. For example if we know that respondents could have provided a pain rating from 1 to 10 it is inappropriate to use a sample which only contained two values, unless this were a large sample and there was some good reason for such a restricted set of responses. The following sections will provide details and uses for each of the resampling techniques.

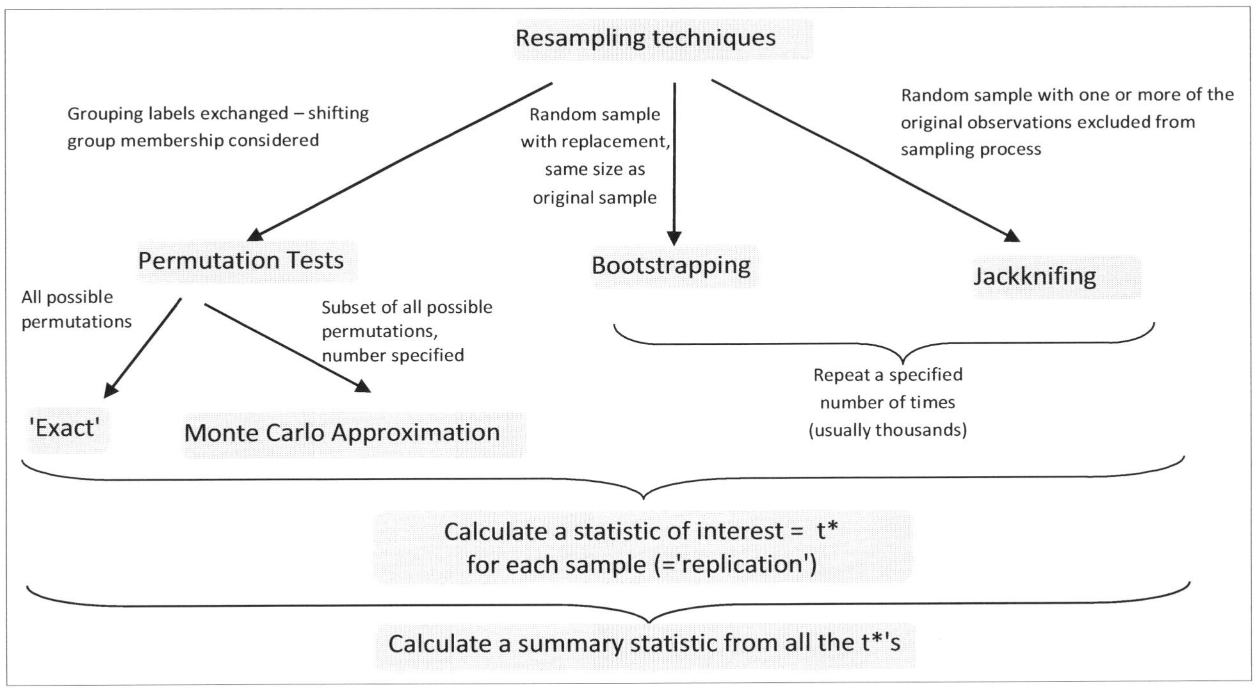

53.1 Permutation tests

Permutation tests, also called randomization, re-randomization, or exact tests allow the randomization aspect of an experiment or clinical trial to be assessed.

As an example of the permutation approach, take the two independent samples design and make the assumption that the treatment (designated by the group label) has no effect on the outcome measure. If this is the case, any value for the outcome measure obtained in either group can be exchanged with any subject's value in the other group. In other words we can legitimately swap values between the groups. For every possible exchange we can then calculate a summary value such as the mean difference score, *t* or U value. Repeating this process for all possible exchanges a frequency distribution of the statistic is then produced (technically called a conditional null distribution). Such a distribution is called a *conditional* distribution because it has been formed purely from the data. Here we are making no assumptions about the distribution of the data only that the grouping makes no difference to the outcome. This sample specific distribution of the statistic is then used to obtain a p-value or confidence interval which is then evaluated in the usual way.

	value	label
Control:	2	A
	3	B
	1	C
	4	D
	3	E
Experimental:		F
	5	G
	6	H
	4	I
	7	J
	6	k

Now consider the small sample example in the *Comparing 2 distributions: Mann-Whitney U Statistic* chapter with the ten observations shown opposite. Assuming that group membership is irrelevant for each score, we want to discover all the possible combinations of two groups of 5 observations from this set and calculate a summary statistic for each. Mathematically this is the same as discovering all possible combinations for a group of 5 items from a group of 10 items. We need also to take into account the duplicate values of which we have several, such as 6, for within this analysis we treat each duplicate as unique and we designate different labels to them.

The formula below indicates the number of ways (called combinations as order is NOT important here) in which we can divide n objects into two groups, one containing k members, and the other containing $n - k$ members. Feinstein, 2002 (chapter 12).

$$\frac{n!}{k!(n-k)!} = \frac{N!}{n_1!n_2!} = \frac{(\text{total number})!}{(\text{no. in group 1})!(\text{no. in group 2})!}$$

The exclamation mark means factorial so 3!=3x2x1 and 4!=4x3x2x1 etc. Also 0!=1 (just believe me).

Taking our example of 10 scores divided between two groups, the formula indicates there are 252 permutations:

$$\frac{N!}{n_1!n_2!} = \frac{10!}{5!5!} = \frac{10x9x8x7x6}{5x4x3x2} = \frac{3x2x7x6}{1} = 252$$

Therefore a single combination will have a probability of 1/252=0.003968254. To be able to produce the conditional null distribution we therefore need a list of all the combinations along with the value of a statistic calculated from each, a task which is easily accomplished in R.

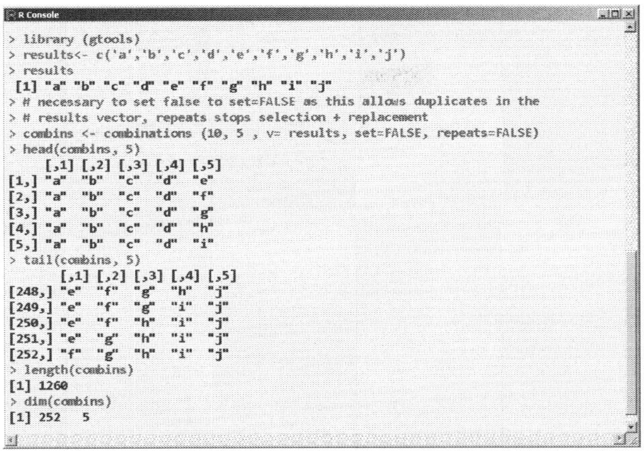

```
R Console
> library (gtools)
> results<- c('a','b','c','d','e','f','g','h','i','j')
> results
 [1] "a" "b" "c" "d" "e" "f" "g" "h" "i" "j"
> # necessary to set false to set=FALSE as this allows duplicates in the
> # results vector, repeats stops selection + replacement
> combins <- combinations (10, 5 , v= results, set=FALSE, repeats=FALSE)
> head(combins, 5)
     [,1] [,2] [,3] [,4] [,5]
[1,] "a"  "b"  "c"  "d"  "e"
[2,] "a"  "b"  "c"  "d"  "f"
[3,] "a"  "b"  "c"  "d"  "g"
[4,] "a"  "b"  "c"  "d"  "h"
[5,] "a"  "b"  "c"  "d"  "i"
> tail(combins, 5)
       [,1] [,2] [,3] [,4] [,5]
[248,] "e"  "f"  "g"  "h"  "j"
[249,] "e"  "f"  "g"  "i"  "j"
[250,] "e"  "f"  "h"  "i"  "j"
[251,] "e"  "g"  "h"  "i"  "j"
[252,] "f"  "g"  "h"  "i"  "j"
> length(combins)
[1] 1260
> dim(combins)
[1] 252   5
```

Before I demonstrate the easy way of doing this I will demonstrate the stages R goes through under the bonnet to provide deeper understanding.

The first stage is to produce all the possible combinations and to do this it is necessary to install and load the *gtools* package.

install the gtools package

```
install.packages("gtools", dependencies=TRUE)
update.packages(ask=FALSE)
```

update any packages it depends upon

load the gtools package

```
library (gtools)
results<- c('a','b','c','d','e','f','g','h','i','j')
```

create a column (vector) of the labels of our 10 observations

display them

```
results
```

place all the combinations in a matrix called combins

use the combinations function

v= the sample to use

indicates if duplicates should be removed from v. default=TRUE

```
combins <- combinations (10, 5 , v= results, set=FALSE, repeats=FALSE)
```

length of the v vector

select 5 observations each time from v

indicates if an individual combination can resample from v. default=FALSE

display first five rows

```
head(combins, 5)
```

display last five rows

```
tail(combins, 5)
```

display number of elements in combins

```
length(combins)
```

display number of rows and columns of combins

```
dim(combins)
```

Now it is possible to calculate a summary value for each combination, frequently referred to as the *statistic.* In this instance to emulate the Mann-Whitney U result we will consider the difference between the sum of the values for each above combination and that which would be obtained in the other group. Rather than directly calculating the combinations for the other group we can derive this value by realising that it is just the sum of the total for all the observations (2+3+1+4+3+5+6+4+7+6 = 41) minus the value for the combination in the combins column. We can then produce a frequency table along with a histogram of all the values, the conditional null distribution. For our observed combination this value is 13 – (41-13) = -15

We can also consider extreme values for the statistic and the number of combinations that have this value or one more extreme to produce what is called an exact p-value.

The following page provides the R code along with comments to carry out the above steps.

```
combinN <- combins
combinN[combinN=='a']<-2
combinN[combinN=='b']<- 3
combinN[combinN=='c']<- 1
combinN[combinN=='d']<- 4
combinN[combinN=='e']<- 3
combinN[combinN=='f']<- 5
combinN[combinN=='g']<- 6
combinN[combinN=='h']<- 4
combinN[combinN=='i']<- 7
combinN[combinN=='j']<- 6
# mode(combinN) <- 'numeric'
combinN<- apply(combinN, 2, as.numeric)
head(combinN)
resultsnum<- sum(c(2,3,1,4,3,5,6,4,7,6))
resultsnum
combinsum <- apply(combinN, 1, sum)
combinsum
statistic <-  combinsum - (resultsnum -  combinsum)
statistic
table(statistic)
hist(statistic)
```

the new matrix

copy the combinations

change 'a' to 2

recode the labels to the original numeric values

change 'b' to 3

change 'c' to 1

R still thinks the elements of the array are characters which would prevent us from performing calculations on them so it is necessary to convert them.

two possible ways to convert the elements I use the second, first one commented out

apply to row/columns 1 indicates rows, 2 indicates columns, c(1, 2) indicates rows and columns

display the first 5 rows

apply the as.numeric function to elements in combinN

a suitable statistic to emulate the MWU is to use the sum of the scores. Have the sums of the combinations for the first group for the second the value is the sum of the total minus that of the first group.

display the sum of all the observations

place the result in a column called combinsum

apply the sum function to each row

display combinsum

display statistic

display a frequency table

display a histogram

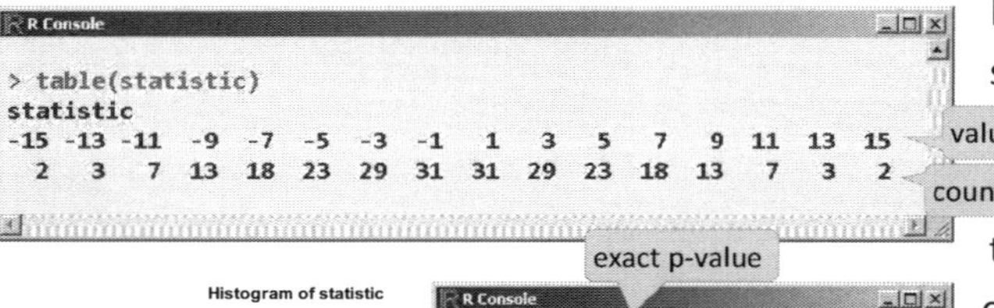

```
> table(statistic)
statistic
-15 -13 -11  -9  -7  -5  -3  -1   1   3   5   7   9  11  13  15
  2   3   7  13  18  23  29  31  31  29  23  18  13   7   3   2
```



```
> 4 *0.003968254
[1] 0.01587302
```

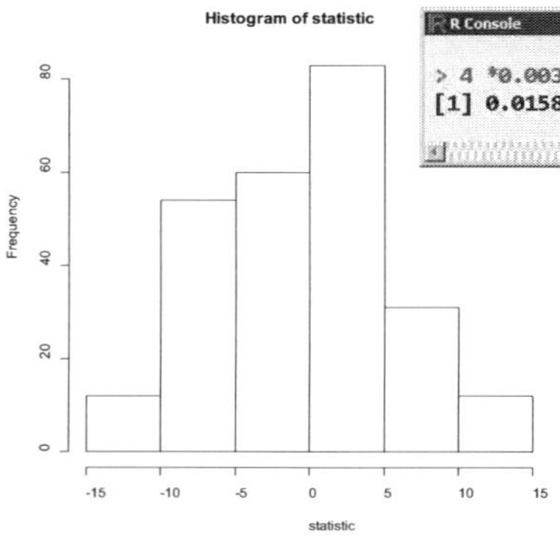

Referring back to the previous page we know that our observed sample produces a statistic of – 15 and we can see from the frequency table opposite that there are 2 combinations that have values that are equal to or more extreme than this value (2x 15 and 2x-15). We also know that the probability of each individual outcome is 0.003968254 so the total probability of these four outcomes is 4 x 0.003968254 which is 0.01587 which is very close to the p-value of 0.01533 obtained using the standard R Command when using the normal approximation.

Looking at the histogram we can see that the conditional null distribution is not symmetrical but does focus around zero. In reality we would require a much larger sample to produce a valid distribution but this has been done to demonstrate the principle.

53.1.1 Performing permutation tests in R - small samples

The above rather long description provided a step-by-step approach to carrying out a permutation test in R but luckily the R *coin* package can do this and more with a single command.

```
R Console
>
> results<- c(2,3,1,4,3,5,6,4,7,6)
> group <- factor(c(1,1,1,1,1,2,2,2,2,2))
> rst<- wilcox_test(results ~ group, distribution="exact", conf.int=TRUE)
> rst

        Exact Wilcoxon Mann-Whitney Rank Sum Test

data:  results by group (1, 2)
Z = -2.5298, p-value = 0.01587
alternative hypothesis: true mu is not equal to 0
95 percent confidence interval:
 -5 -1
sample estimates:
difference in location
                    -3
```

The code below performs the permutation test equivalent to the Mann-Whitney U test.

```
install.packages("coin", dependencies=TRUE)
library(coin)
results<- c(2,3,1,4,3,5,6,4,7,6)
group <- factor(c(1,1,1,1,1,2,2,2,2,2))
rst<- wilcox_test(results ~ group, distribution="exact",
conf.int=TRUE)
rst
```

install the *coin* package

load the *coin* package

column of results

grouping variable column

wilcox_test function, because there is a group variable it is equivalent to a MWU

produce a confidence interval

calculate the exact permutation distribution

display the result

R Console
> # note unable to request a CI
> rst<- wilcoxsign_test(pre ~ post, distribution="exact")
> rst
Exact Wilcoxon-Signed-Rank Test
data: y by x (neg, pos)
stratified by block
Z = -2.226, p-value = 0.03125
alternative hypothesis: true mu is not equal to 0
> |

We can also perform the equivalent permutation test for the Willcoxon matched pairs example, here producing a p-value of 0.03125 compared to 0.03401 in the chapter describing the test.

```
library(coin)
post<-c(7, 6,8,7,6,9)
pre<-c(1,4,3,5,4,5)
rst<- wilcoxsign_test(pre ~ post, distribution="exact")
rst
```

you can NOT request a confidence interval for this statistic

```
R Console
>
> qperm(rst, 0.95)
[1] 1.801996
```

You can't request a confidence interval but you can obtain a 95% quartile value using the *qperm()* function. The result opposite indicates that 95% of the statistic values are below 1.8.

53.1.1.1 Performing permutation tests in R - large samples

N	n_1	n_2	combinations
8	4	4	70
9	5	4	126
10	5	5	252
11	6	5	462
12	6	6	924
13	7	6	1716
14	7	7	3432
15	8	7	6435
16	8	8	12870
17	9	8	24310
18	9	9	48620
19	10	9	92378
20	10	10	184756
21	11	10	352716
22	11	11	705432
23	12	11	1352078
24	12	12	2704156
25	13	12	5200300
26	13	13	10400600
27	14	13	20058300
28	14	14	40116600
29	15	14	77558760
30	15	15	155117520

The above examples contained very small samples and in reality are rarely encountered in practice. I having used such small samples to demonstrate the mechanics of permutation tests. The table opposite shows that the number of possible combinations increases very rapidly beyond a sample size of 20. To apply permutation test techniques to large sample sizes, say above 100, you either need a super-computer, or to adopt one of the following alternatives:

- The Monte Carlo simulation approach considers a sample of all possible combinations to produce an approximate distribution. In the *coin* package you request this technique by setting the option *distribution="approximate"*.

- The asymptotic approach makes use of the fact that for large samples the distribution of the statistic obtained from the combinations follows a normal distribution. This asymptotic normal approximation gets closer to the real value as the sample size increases. In the *coin* package you request this method by setting the option *distribution=" asymptotic".*

```
R Console
> library(coin)
> model1<-  oneway_test(values ~ ind, distribution=approximate(B=10000))
> model1

        Approximative 2-Sample Permutation Test

data:  values by ind (control, treatment)
Z = -1.9938, p-value = 0.0455
alternative hypothesis: true mu is not equal to 0
```

We will consider these two approaches using the example consisting of 200 observations discussed in the *Two independent samples t test* chapter.

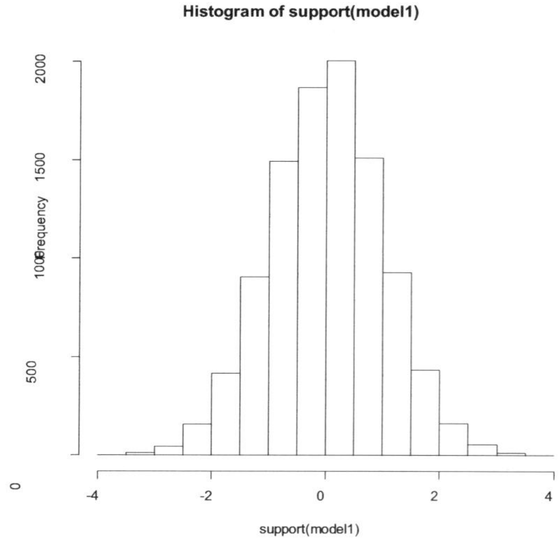

First the Monte Carlo approach, here the distribution is formed from a subset of all possible combinations, the subsequent distribution provides an almost identical p-value of 0.0455 compared to that of 0.0459 given in the *Two independent samples t test* chapter. The *coin* package equivalent to the 2 independent samples *t* test is *oneway_test()*:

```
library(coin)

model1<- oneway_test(values ~ ind,
distribution=approximate(B=10000))
```

form the distribution from the results of 10,000 combinations

You can also obtain a histogram of the distribution by typing, the following:

create and display a histogram

```
hist(support(model1))
```

the support function when used with an object produced by most of the functions in the *coin* package provides a column of the statistics value.

```
R Console
> model2<- oneway_test(values ~ ind)
> model2

        Asymptotic 2-Sample Permutation Test

data:  values by ind (control, treatment)
Z = -1.9938, p-value = 0.04618
alternative hypothesis: true mu is not equal to 0

>
> qperm(model2, 0.05)
[1] -1.644854
> qperm(model2, 0.95)
[1] 1.644854
> |
```

We can see that 10,000 results, compared to the 252 used in the previous example, produces a distribution which follows much more closely the bell shaped curve.

To repeat the exercise using the asymptotic approach we simply leave out the distribution command because this is the default. Once again it is not possible to obtain a 95% confidence interval so we request the 5% and 95% quartiles:

```
model2<- oneway_test(values ~ ind)

model2

qperm(model2, 0.05)

qperm(model2, 0.95)
```

We can see from the above that the p-values produced using the permutation approach are very similar to those produced using the usual approach but this should not be assumed and permutation test are considered to be useful when the data is decidedly not normally distributed. The table opposite lists the various permutations tests provided by the *coin* package.

Permutation tests offered in the *coin* package
(taken from the coin documentation)

oneway_test two- and K-sample permutation test
wilcox_test Wilcoxon-Mann-Whitney rank sum test
normal_test van der Waerden normal quantile test
median_test Median test
kruskal_test Kruskal-Wallis test
ansari_test Ansari-Bradley test
fligner_test Fligner-Killeen test
chisq_test Pearson's _2 test
cmh_test Cochran-Mantel-Haenszel test
lbl_test linear-by-linear association test
surv_test two- and K-sample logrank test
maxstat_test maximally selected statistics
spearman_test Spearman's test
friedman_test Friedman test
wilcoxsign_test Wilcoxon-Signed-Rank test
mh_test marginal homogeneity test (Maxwell-Stuart).

The fact that the p-values are almost identical to those obtained above using the more traditional approach provides yet another way of interpreting the p-values obtained. We can think of them as presenting the probably of obtaining a particular statistic or one more extreme given that the group allocation process had no effect on the outcome.

Using the exact approach, when it was practicable, we would obtain the same result each time on the same dataset as we do with the default asymptotic option. In contrast, the Monte Carlo approach only takes a random sample of all possible combinations meaning that we would only obtain a consistent result if the sample was sufficiently large. This idea of randomness, technically called a stochastic or non-deterministic method, is also a feature of the next approach, Bootstrapping - where we turn our attention from the group randomisation process to the sample itself.

53.2 Bootstrapping

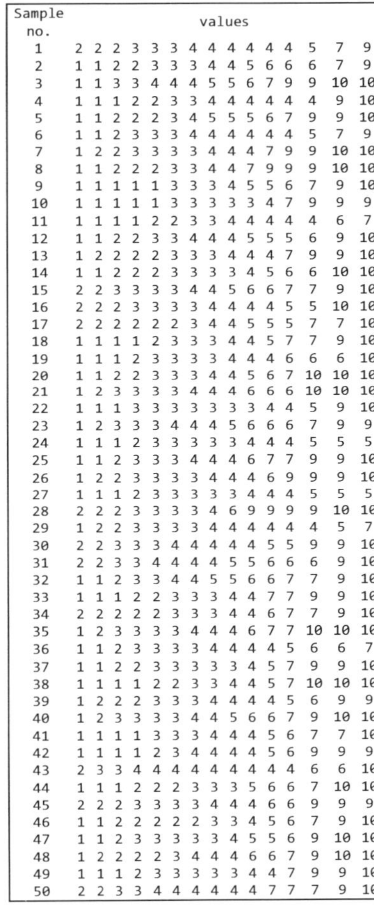

Sample no.	values														
1	2	2	2	3	3	3	4	4	4	4	4	4	5	7	9
2	1	1	2	2	3	3	3	4	4	5	6	6	6	7	9
3	1	1	3	3	4	4	4	5	5	6	7	9	9	10	10
4	1	1	1	2	2	3	3	4	4	4	4	4	4	9	10
5	1	1	2	2	2	3	4	5	5	5	6	7	9	9	10
6	1	1	2	3	3	3	4	4	4	4	4	4	5	7	9
7	1	2	2	3	3	3	3	4	4	4	7	9	9	10	10
8	1	1	2	2	2	3	3	4	4	7	9	9	9	10	10
9	1	1	1	1	1	3	3	3	4	5	5	6	7	9	10
10	1	1	1	1	1	3	3	3	3	3	4	7	9	9	9
11	1	1	1	1	2	2	3	3	4	4	4	4	4	6	7
12	1	1	2	2	3	3	4	4	4	5	5	5	6	9	10
13	1	2	2	2	2	3	3	3	4	4	4	7	9	9	10
14	1	1	2	2	2	3	3	3	3	4	5	6	6	10	10
15	2	2	3	3	3	3	4	4	5	6	6	7	7	9	10
16	2	2	2	3	3	3	3	4	4	4	4	5	5	10	10
17	2	2	2	2	2	2	3	4	4	5	5	5	7	7	10
18	1	1	1	1	2	3	3	3	4	4	5	7	7	9	10
19	1	1	1	2	3	3	3	3	4	4	4	6	6	6	10
20	1	1	2	2	3	3	3	4	4	5	6	7	10	10	10
21	1	2	3	3	3	3	4	4	4	6	6	6	10	10	10
22	1	1	1	3	3	3	3	3	3	3	4	4	5	9	10
23	1	2	3	3	3	4	4	4	5	6	6	6	7	9	9
24	1	1	1	2	3	3	3	3	3	4	4	4	5	5	5
25	1	1	2	3	3	3	4	4	4	6	7	7	9	9	10
26	1	2	2	3	3	3	3	4	4	4	6	9	9	9	10
27	1	1	1	2	3	3	3	3	3	4	4	4	5	5	5
28	2	2	2	3	3	3	3	4	6	9	9	9	9	10	10
29	1	2	2	3	3	3	3	4	4	4	4	4	4	5	7
30	2	2	3	3	3	4	4	4	4	4	5	5	9	9	10
31	2	2	3	3	4	4	4	4	5	5	6	6	6	9	10
32	1	1	2	3	3	4	4	5	5	6	6	7	7	9	10
33	1	1	1	2	2	3	3	3	4	4	7	7	9	9	10
34	2	2	2	2	2	3	3	3	4	4	6	7	7	9	10
35	1	2	3	3	3	3	4	4	4	6	7	7	10	10	10
36	1	1	2	3	3	3	3	4	4	4	4	5	6	6	7
37	1	1	2	2	3	3	3	3	3	4	5	7	9	9	10
38	1	1	1	1	2	2	3	3	4	4	5	7	10	10	10
39	1	2	2	2	3	3	3	4	4	4	4	5	6	9	9
40	1	2	3	3	3	3	4	4	5	6	6	7	9	10	10
41	1	1	1	1	3	3	3	4	4	4	5	6	7	7	10
42	1	1	1	1	2	3	4	4	4	4	5	6	9	9	9
43	2	3	3	4	4	4	4	4	4	4	4	4	6	6	10
44	1	1	1	2	2	2	3	3	3	5	6	6	7	10	10
45	2	2	2	3	3	3	3	4	4	4	6	6	9	9	9
46	1	1	2	2	2	2	2	3	3	4	5	6	7	9	10
47	1	1	2	3	3	3	3	3	4	5	5	6	9	10	10
48	1	2	2	2	2	3	4	4	4	6	6	7	9	10	10
49	1	1	1	2	3	3	3	3	3	4	4	7	9	9	10
50	2	2	3	3	4	4	4	4	4	4	7	7	7	9	10

This is when many samples of the original sample (sometimes called a pseudo population in this context) are taken. Each generated sample, which is called a bootstrapped sample, is then used to produce one or more statistics of interest (the *boot* package calls these t star = t*). From this distribution of t star values further summary statistics are produced. Often the number of samples taken is above 10,000 to produce a realistic distribution.

It may appear strange that the bootstrapped samples are the same size as the original one, surely this would imply that the samples are identical? The answer is no as sampling is "with replacement" meaning that each time a value is taken from the original sample it is immediately replaced before selecting the next value from it. A demonstration should make this clearer.

Assuming we have a dataset of the following values, noting that it contains one '10' and two '1's:

```
data1 <- c(3,5,4,7,6,2,4,3,9,10,2,1,1,3,4)
```

I have created 50 typical bootstrapped samples shown opposite, which I have sorted. Notice that many do not contain all the original values, with some having a minimum value of 2, therefore omitting the two '1' values while others have a maximum value of 5 omitting the '9' and '10' values.

To undertake bookstrapping in R you usually need to use the *boot* package and also know how to code R functions. See the Glossary entry **Functions**.

53.2.1 Bootstrapping functions

To write functions that are required by the *boot* package they need to conform to a specific structure which involves adding an indicator variable (*indices*) which is used by the *boot()* function.

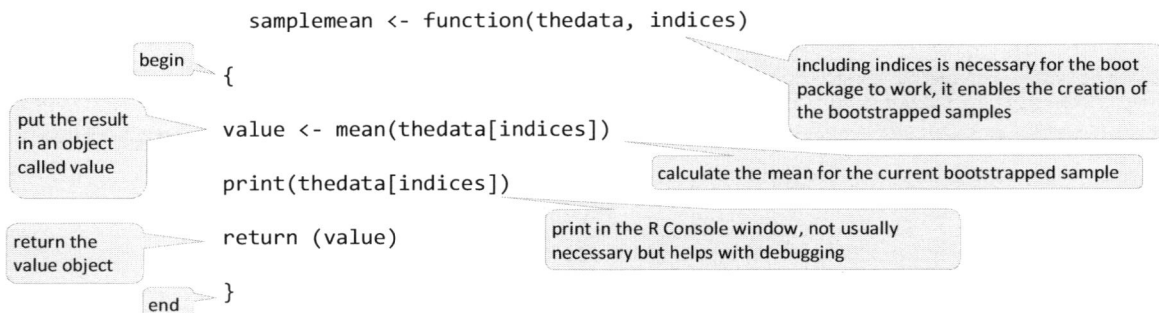

```
samplemean <- function(thedata, indices)
{
value <- mean(thedata[indices])
print(thedata[indices])
return (value)
}
```

begin

put the result in an object called value

return the value object

end

including indices is necessary for the boot package to work, it enables the creation of the bootstrapped samples

calculate the mean for the current bootstrapped sample

print in the R Console window, not usually necessary but helps with debugging

The indicator variable is a column the same length as the sample with numbers 1 to n where n is the length of the sample, the *boot()* function in the *boot* package then selects with replacement, the required number of values from the indicator variable. For example taking the *data1 <- c(3,5,4,7,6,2,4,3,9,10,2,1,1,3,4)* sample, suppose the *boot()* function produces the following indicator variable for one particular bootstrap sample:

```
Index <- c(6, 11, 11, 1, 8, 14, 3, 7, 15, 3, 3, 7, 2, 4, 9)
```

Applying the index to the sample data gives:

```
2, 2, 2, 3, 3, 3, 4, 4, 4, 4, 4, 4, 5, 7, 9
```

R Console

```
>
> index <- c(6, 11, 11, 1, 8, 14, 3, 7, 15, 3, 3, 7, 2, 4, 9)
> data1 <- c(3,5,4,7,6,2,4,3,9,10,2,1,1,3,4)
> data1[index]
 [1] 2 2 2 3 3 3 4 4 4 4 4 4 5 7 9
```

I have engineered it so that the above vector is the same as the first row of the 50 samples given on the previous page to emphasis the process.

Note that the value, or values returned from our function is the value called *t star* (t^*) in the subsequent output from the *boot* package but it can be anything we want it to be.

The above information now allows us to create all the necessary R code to produce 100 bookstrapped samples, obtain the mean of each and produce a histogram of these.

install and load the boot package

```
install.packages("boot", dependencies=TRUE)
library(boot)
```

my samplemean function. Indented to help see it

have removed the Print function as I do not want to see the 100 samples. Now just return the value

```
        samplemean <- function(thedata, indices)
        {
        value <- mean(thedata[indices])
        return (value)
        }
```

the sample data

```
data1 <- c(3,5,4,7,6,2,4,3,9,10,2,1,1,3,4)
```

produce 100 replicates = bootstrap samples

create a new boot object called bootmean

```
bootmean <-  boot(data1, samplemean, R = 100)
```

display the result

```
bootmean
```

provide boot with: the data; the function that calculates the summary value(s)

```
R Console
> bootmean <-  boot(data1, samplemean, R = 100)
> bootmean

ORDINARY NONPARAMETRIC BOOTSTRAP


Call:
boot(data = data1, statistic = samplemean, R = 100)


Bootstrap Statistics :
    original      bias    std. error
t1* 4.266667 -0.08266667   0.7075745
> plot(bootmean)
```

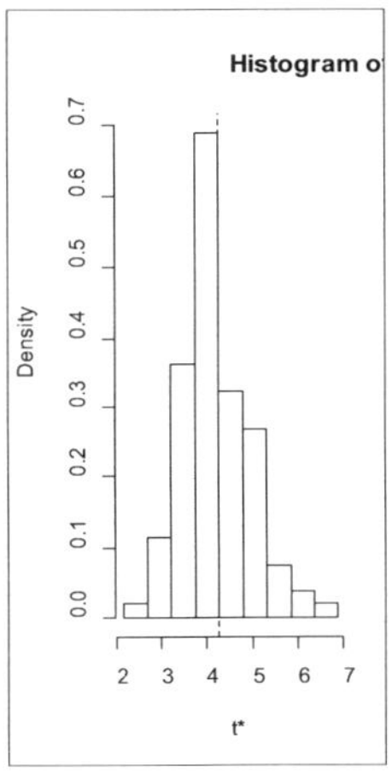

The results include the mean value of all the individual means calculated from the 100 samples $t1^*$ along with their standard error (i.e. the standard deviation of the means).

Typing *plot(bootmean)* produces a histogram of the statistic. It is always sensible to inspect the plot to see the distribution shape observing how close it is to a normal distribution.

Histogram o

Density

0.0 0.1 0.2 0.3 0.4 0.5 0.6 0.7

2 3 4 5 6 7

t*

You can also obtain confidence intervals for the statistic with the command:

```
boot.cl(bootobject, type = "all")
```

So for the above example we can use:

```
boot.ci(bootmean, type = "all")
```

```
R Console
> boot.ci(bootmean, type = "all")
BOOTSTRAP CONFIDENCE INTERVAL CALCULATIONS
Based on 100 bootstrap replicates

CALL :
boot.ci(boot.out = bootmean, type = "all")

Intervals :
Level      Normal              Basic
95%   ( 2.828,  5.528 )   ( 2.771,  5.467 )

Level     Percentile            BCa
95%   ( 3.067,  5.762 )   ( 3.041,  5.716 )
Calculations and Intervals on Original Scale
Some basic intervals may be unstable
Some percentile intervals may be unstable
Some BCa intervals may be unstable
Warning message:
In boot.ci(bootmean, type = "all") :
  bootstrap variances needed for studentized intervals
```

This gives four different confidence intervals using different approaches. The *Normal* method makes a normal distribution approximation to the bootstrap statistic.

The *percentile* approach uses the 5 and 95 quartiles, whereas the *BCa* (adjusted bootstrap percentile) method uses a more complex method.

This approach of obtaining confidence intervals is extremely useful when the sample is sufficiently large and representative.

You can use the bootstrapping technique to obtain valid estimates of almost any summary statistic, and therefore confidence limits as well - which is where they come into their own.

This example uses the data presented in the correlation chapter and produces a confidence interval for it, by way of 1000 bootstrapped samples.

load the boot package

```
library(boot)
```

my boot_corr function. Indented to help see it

start the code for the function

note how the index is used to create the sample

note the comma ',' to indicate to select not only rows equal to index but also ALL the columns

traditional correlation function between the HBAIC and the fastingBG columns

return the statistic

end the code for the function

```
    boot_corr <- function (thedata, index)
    {
    thesample <- thedata[index,]
    value <- cor(thesample$HBA1C , thesample$fastingBG)
    return(value)
    }
```

use the dataset from the correlation chapter

display the names of the two column

attach the datafram so I can refer to the columns directly

display the correlation produced using the traditional approach

now get a bootstrap equivalent and place it in the boot object called bootcorr

display the result

display the confidence intervals

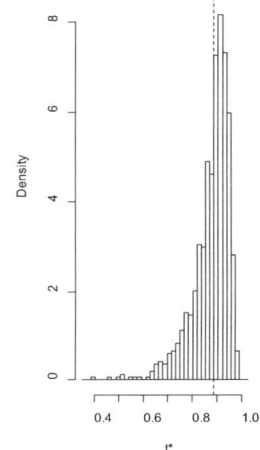

```
mydataframe <- read.delim("http://www.robin-
beaumont.co.uk/virtualclassroom/book2data/regression1.dat", header=TRUE)
names(mydataframe)
#  "HBA1C"     "fastingBG"
attach(mydataframe)
cor(HBA1C, fastingBG)
bootcorr <-  boot(mydataframe, boot_corr, R = 1000)
bootcorr
plot(bootcorr)
boot.ci(bootcorr, type = "all")
```

```
R Console
Call:
boot(data = mydataframe, statistic = boot_corr, R = 1000)


Bootstrap Statistics :
     original      bias    std. error
t1* 0.8879822 -0.01164426  0.07557479
> plot(bootcorr)
> plot(bootcorr)
> boot.ci(bootcorr, type = "all")
BOOTSTRAP CONFIDENCE INTERVAL CALCULATIONS
Based on 1000 bootstrap replicates

CALL :
boot.ci(boot.out = bootcorr, type = "all")

Intervals :
Level      Normal              Basic
95%   ( 0.7515,  1.0478 )   ( 0.8096,  1.0973 )

Level     Percentile            BCa
95%   ( 0.6786,  0.9664 )   ( 0.6476,  0.9600 )
Calculations and Intervals on Original Scale
Warning message:
In boot.ci(bootcorr, type = "all") :
  bootstrap variances needed for studentized intervals
>
```

Comparing the confidence intervals produced opposite to that using the usual method (see the correlation chapter) which produced values of .6404 to .9683, these equate closely to the *BCa* estimate. Also note that both the *Normal* and *Basic* estimate produce impossible values of 1.04 and 1.09, respectively for the higher limits when the maximum is 1.

One should always check that confidence intervals are actually possible. Some writers cut off the limits if they are not possible so the values of 1.04 and 1.09 would become 1.00. I would recommend that if such an adjustment is made, it is always reported.

53.3 The Jackknife

In contrast to a bootstrapped sample, a jackknife sample does not include all the original data in the possible sample. Frequently a specific observation is left out of the resampling process (called delete-1 jackknife). By leaving out a specific observation each time and repeating the resampling allows us to see how each item influences the calculated statistic.

Within R the *boot* package a plot called the *Jack after boot* is available. This plot provides information about the effect each observation has on the estimates by running jackknifing on each of the bootstrapped samples and then comparing them. Bearing in mind the *bootmean* object we created earlier we can obtain a *Jack after boot* plot using the following command:

```
jack.after.boot(bootmean, index = 1)
```

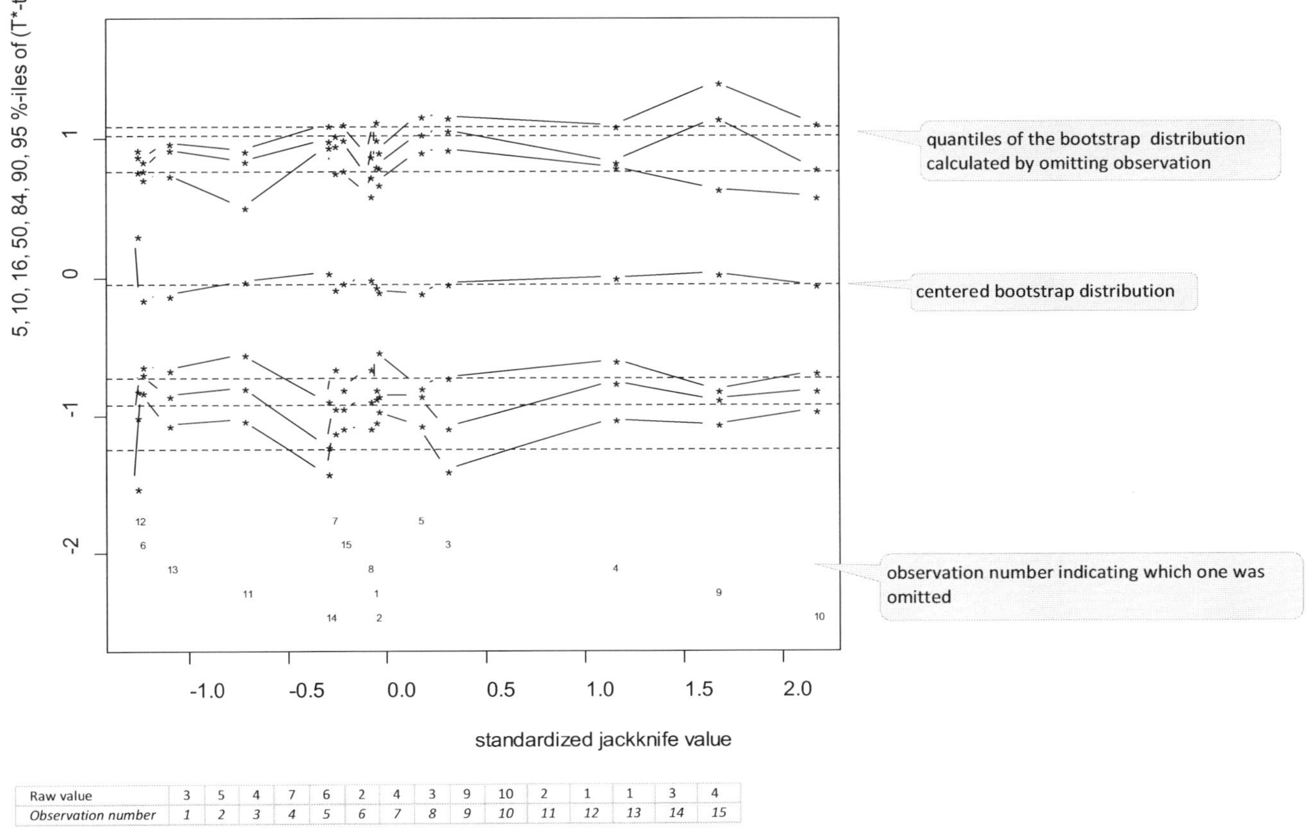

Raw value	3	5	4	7	6	2	4	3	9	10	2	1	1	3	4
Observation number	*1*	*2*	*3*	*4*	*5*	*6*	*7*	*8*	*9*	*10*	*11*	*12*	*13*	*14*	*15*

In the above focusing our attention on the centred distribution, we can see that omitting observation number 12, with the value '1' changes the bootstrap estimate the most, but even that is less than one standardised unit of the estimated statistic, in this case the mean.

Permutation tests, bootstrapping and jack after boot approaches are very useful but you can run into problems when using them, as they frequently require large amounts of computer memory. An issue we will now turn to.

53.4 Memory management

The error shown opposite often appears when; Permutation tests with larger datasets using the exact approach is specified or, bootstrapping a very large number of replications.

This error message indicates a failure to allocate memory, according to R help either because the size exceeded the address-space limit for a process or, more likely, because the system was unable to provide the memory. On a 32-bit system even if there is enough free memory available, there may well not be a large enough contiguous block of memory space for it.

```
R Console
> oneway_test(values ~ ind, distribution="exact")
Error: cannot allocate vector of size 79.5 Mb
> |
```

32 bit

result in Mb

```
> memory.size(FALSE)
[1] 13.57
> memory.size(TRUE)
[1] 14.44
```

```
> memory.limit()
[1] 3583
```

64 bit

```
> memory.size(FALSE)
[1] 26.33
> memory.size(TRUE)
[1] 26.62
```

```
> memory.limit()
[1] 8191
```

This is because the amount of memory allocated to R depends on both the type of system you have (32 versus 64 bit) and the setting within R. You can find the amount of current free memory available to R by typing *memory.size(FALSE)*. Typing *memory.size(TRUE)* gives the maximum possible memory available at the start of an R session. The screen shots opposite demonstrate this.

32 bit

```
> memory.limit()
[1] 3583
> memory.limit(size=3590)
[1] 3590
> memory.limit()
[1] 3590
> memory.limit(size=4000)
[1] 4000
> memory.limit(size=5000)
Error in memory.limit(size = 5000) :
  don't be silly!: your machine has a 4Gb address limit
```

64 bit

```
> memory.limit(size=8000)
[1] 8000
> memory.limit()
[1] 8000
```

In R the maximum possible memory allocation for a 32-bit system is 4095 Mb and for a 64-bit system 8Tb. According to R help you can find the maximum available for your system by typing *memory.limit()*, however from the screenshot opposite it would appear that you can allocate up to 4000Mb (4Gb) for a 32-bit setup even when the system reports a lower value of 3583 . Running the 32-bit version of R limits the availability of memory even if you have more than 4Gb installed. To increase the amount of memory allocated to R use the *memory.limit(size=MB)* function setting the *MB* value to that required. If you specify a value that exceeds the max *memory.limit()*, the action is different depending if you are running the 32-bit or 64-bit version of R. In the 32-bit version R simply creates extra virtual memory on the hard disk (it is said to be paged off) if you have your system setup for paged virtual memory to a maximum of 4Gb. Trying to set a value above this limit results in an error message - shown opposite. In contrast running R 64-bit on a machine with 8Gb of memory and windows 64-bit there appears no upper limit for setting this value? However, I assume that setting the value near to, or above the actual installed ram dramatically increases processing time as you will be using the hard disk as virtual memory.

From the above rather convoluted description it is obvious that memory management is rather a complex topic. One quick fix is to simply abandon the R session and restart R rather than resetting the limit. This works by removing all the other dataframes etc. you may have loaded in your R session.

Several specific R packages are available to help support very large datasets and develop parallel R processes to speed up processing. See Rosario, 2010 for details.

53.5 Tricks and Tips

Remember that with all resampling techniques the results are only as good as the sample you provide. The sample needs to be sufficiently large to be representative meaning very simply it includes a decent range of the values you expect and does not appear to be biased in any way.

The number of replications should usually be greater than 10,000. Specified by the $R=$ value in the *boot* package and the $B=$ *value* in the *coin* package.

Although the numbers produced to create the random samples are 'random' they follow a sequence starting at a random seed number. By specifying this number before you run the resampling method you can duplicate the results. Before specifying the random seed it is always a good idea to save the current one R is using by typing

```
save.seed <- .Random.seed  # to save your current random seed number

set.seed(number)  # to set the random seed number to number
```

If you want to find out retrospectively what random seed number was used by the boot object type:

summary(bootobject)

Memory management in windows is a very complex subject and I would advice you to stay clear of it. If you do want to begin learn about it I would recommend Mark Russinovich's technical blog covering topics, such as Windows troubleshooting, and memory management at:

http://blogs.technet.com/b/markrussinovich/archive/2008/11/17/3155406.aspx

54 Repeated measures: mixed models and Gee

Repeated measures are commonly used in all types of research. It is a form of analysis where the outcome is repeatedly measured over time. Each subject therefore provides multiple values for the same variable. In statistical terms, we generally use repeated measures in one of two ways:

- to see if there are differences between various groups (such as those that are male or female, or a specific treatment compared to a placebo)
- to assess any correlation or variance in the repeated measures themselves.

We can also take a combined approach to take both the group differences and the repeated measures correlation/variance into account.

The attempt to provide a robust way of analysing repeated measures has a history of several blind alleyways. From the 1940's until the late 1990's it was standard practice to use repeated measures ANOVA techniques which often required complex manipulation of various sums of squares and also made many demands on the data. Ideally, the gap between the repeated measurements needed to be equal; similar numbers were required in each subgroup, and the data needed to possess quite stringent characteristics concerning variance over time which, if violated, meant that the results were invalid. All this meant that the analysis of repeated measures was a nightmare for the amateur. In the 1990's two alternative approaches were developed:

- Generalised estimating equations (GEE)
- Mixed effects models

Generalised estimating equations (GEE) are often considered the quick and dirty approach; they provide estimates of the average value ("population-averaged") over the repeated measures by moving the focus away from the repeated measures at the individual level. This is partly achieved by the user defining a *working correlation/covariance matrix*, which is a matrix specifying the expected correlation between each repeated measure along with the variance. Such a model is called a marginal model and there are no random effects in these models.

An alternative way to account for the nature of the repeated measures is to fit a particular model known as a mixed effects model. Such models allow the individual level estimation of intercepts and slopes resulting in what's called a conditional model. The parameters of a conditional model have a subject-specific interpretation rather than the population-averaged interpretation of a marginal model.

For more details about -2log likelihood , AIC's and BIC's see the chapter, Logistic regression: a binary outcome.

Mixed effect models are evaluated using a variety of maximum likelihood techniques and produce -2log likelihood, AIC's and BIC's which can be used to compare models. Unfortunately this is not possible with GEE models and assessment of the suitability of a particular *working correlation matrix* is an area of ongoing research, Hin & Wang, 2009.

54.1 An example – Beat the blues

The dataset for this chapter is a complex set of repeated measures data, taken from Landau & Everitt, 2004 (p. 196), repeated in Everitt & Hothorn, 2006, quoting them:

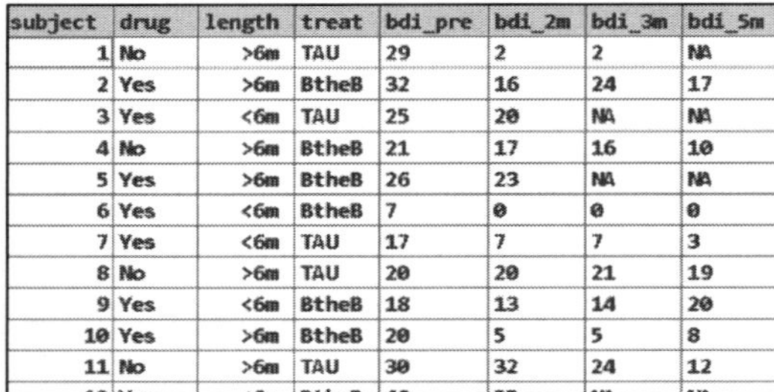

subject	drug	length	treat	bdi_pre	bdi_2m	bdi_3m	bdi_5m
1	No	>6m	TAU	29	2	2	NA
2	Yes	>6m	BtheB	32	16	24	17
3	Yes	<6m	TAU	25	20	NA	NA
4	No	>6m	BtheB	21	17	16	10
5	Yes	>6m	BtheB	26	23	NA	NA
6	Yes	<6m	BtheB	7	0	0	0
7	Yes	<6m	TAU	17	7	7	3
8	No	>6m	TAU	20	20	21	19
9	Yes	<6m	BtheB	18	13	14	20
10	Yes	>6m	BtheB	20	5	5	8
11	No	>6m	TAU	30	32	24	12

"The trial was designed to assess the effectiveness of an interactive program using multi-media techniques for the delivery of cognitive behavioral therapy for depressed patients and known as *Beating the Blues* (BtheB). In a randomized controlled trial of the program, patients with depression recruited in primary care were randomized to either the BtheB program, or to *Treatment as Usual* (TAU). The outcome measure used in the trial was the *Beck Depression Inventory II* (Beck et al., 1996) with higher values indicating more depression. Measurements of this variable were made on five occasions. [bdi_pre, bdi_2m, bdi_3, bdi_5m, bdi_8m]"

The dataset (in wide format) consists of the variables shown opposite.

Besides the details provided above there are several other variables in the above dataset: the drug variable indicates if the subject was receiving antidepressants; and the length variable indicates if the current episode of depression was less or more than six months. You can think of the variables in terms of two levels, level two being the subject and level one the repeated measure.

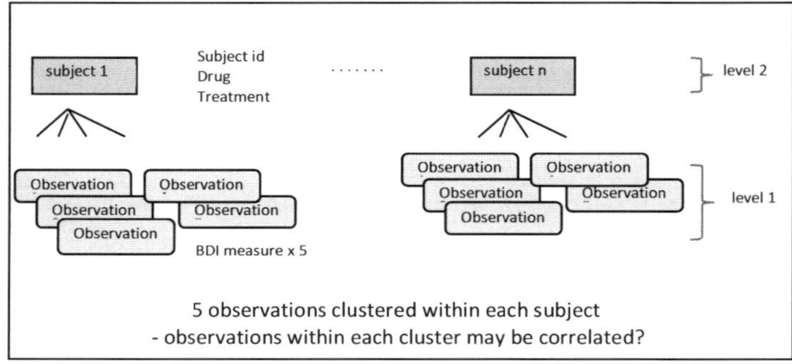

Notice that I have classed the Drug and Treatment variables to be at level two in the above. This is because in this design they have only been measured once and are assumed to be invariant. If they had been repeatedly measured, for example say each subject had a period of both *TAU* and *BtheB* and/or a period with and without antidepressant medication, they would be at level one.

For much of the analysis in this chapter we will concentrate on a subset of the above variables, that is the treatment (*Tau* versus *BtheB*) and the repeated *BDI* measures.

54.1.1 Obtaining and loading the data

Please click on the following link and select to save the file locally

http://www.robin-beaumont.co.uk/virtualclassroom/stats/statistics2/data/lmm3/bb_wide.rdata

Take a note of where you saved the file then use the following code to load the file in R.

```
load(file=file.choose())
```

54.1.2 Very basic summary statistics

Type the following to produce simple summary statistics

```
> summary(bb_wide)
    subject    drug       length     treat
       1: 1   No :56   <6m:49   TAU  :48
       2: 1   Yes:44   >6m:51   BtheB:52
       3: 1
       4: 1
       5: 1
       6: 1
 (Other) :94
    bdi_pre          bdi_2m          bdi_3m
 Min.   : 2.00   Min.   : 0.00   Min.   : 0.00
 1st Qu.:15.00   1st Qu.: 8.00   1st Qu.: 6.00
 Median :22.00   Median :15.00   Median :13.00
 Mean   :23.33   Mean   :16.92   Mean   :14.81
 3rd Qu.:30.25   3rd Qu.:23.00   3rd Qu.:20.00
 Max.   :49.00   Max.   :48.00   Max.   :53.00
                 NA's   :3       NA's   :27
     bdi_5m          bdi_8m
 Min.   : 0.00   Min.   : 0.00
 1st Qu.: 3.00   1st Qu.: 3.00
 Median :10.00   Median :10.50
 Mean   :12.76   Mean   :11.13
 3rd Qu.:20.00   3rd Qu.:15.25
 Max.   :47.00   Max.   :40.00
 NA's   :42      NA's   :48
>
```

```
summary(bb_wide)
```

We can see the 100 cases are equally divided across each treatment group. Also the median value for Becks Depression Inventory score for the five repeated measures appears to go steadily down from 22 pre to 10.5, for the last measure.

In any repeated measures analysis the first thing to do is to get a feel for the data and I find the best way to do this is to graph the outcome variable (i.e. the *BDI* measure) against any variable I sense that might affect it. In this instance the particular treatment they were given is the focus so we will consider this now.

54.2 Trend over time for each treatment group

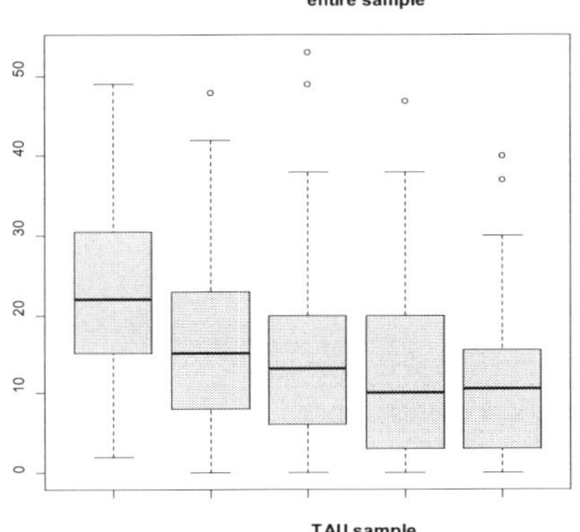

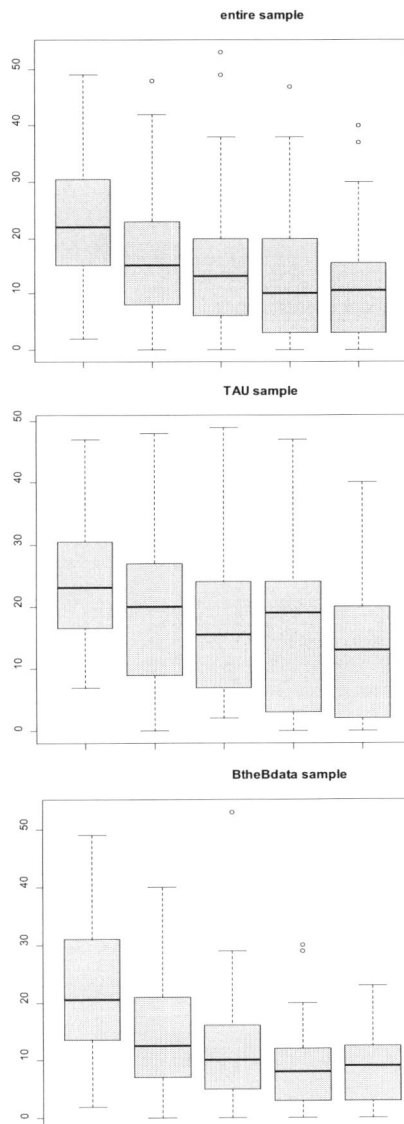

We need to inspect the change over time for the depression scores (*BDI* measure) for the two treatment groups. The traditional way of doing this is to draw a set of error bar charts, but I have chosen to use boxplots instead as these also show any outliers. The R code to produce the boxplots opposite is given below, along with comments, using the hash (#) character.

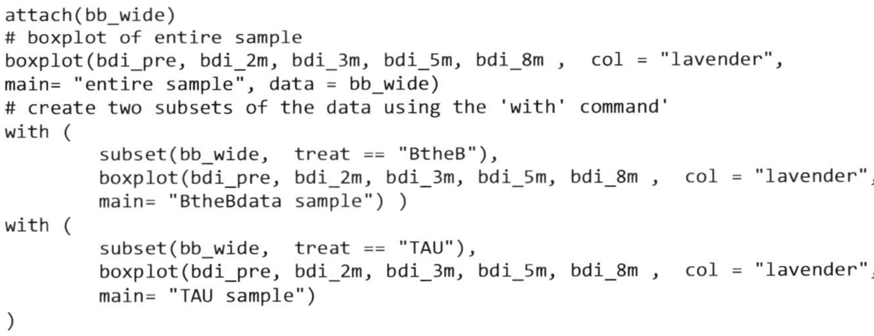

```
attach(bb_wide)
# boxplot of entire sample
boxplot(bdi_pre, bdi_2m, bdi_3m, bdi_5m, bdi_8m ,  col = "lavender",
main= "entire sample", data = bb_wide)
# create two subsets of the data using the 'with' command'
with (
        subset(bb_wide,  treat == "BtheB"),
        boxplot(bdi_pre, bdi_2m, bdi_3m, bdi_5m, bdi_8m ,  col = "lavender",
        main= "BtheBdata sample") )
with (
        subset(bb_wide,  treat == "TAU"),
        boxplot(bdi_pre, bdi_2m, bdi_3m, bdi_5m, bdi_8m ,  col = "lavender",
        main= "TAU sample")
)
```

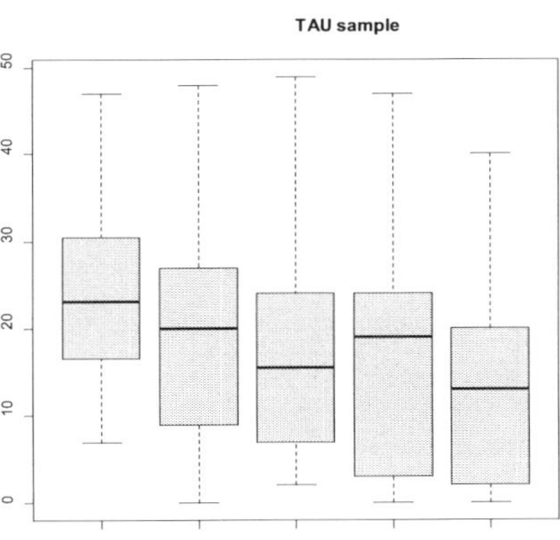

BtheBdata sample

From the boxplots opposite, we see that the median values for the traditional treatment group (*TAU*) do not fall as rapidly as the beating the blues (*BtheB*) group.

While the boxplots opposite do provide a graphical summary of the two groups over time they do not show the individual trajectories, in other words they do not take into account the correlated nature of the repeated measures. To do this we need to construct individual profile plots, but unfortunately to produce these we need to get all the *dbi* observations in a single column.

The next section describes how to change the data from the original wide format to a specific long format with all the *dbi* observations in a single column.

54.3 Converting from wide to long format

To convert the dataset from wide to long format we need to install and load the *reshape2* package done using the two lines of code below. If you're not sure whether you have installed it already, you can simply reinstall it.

```
install.packages("reshape2")

library(reshape2)
```

The *melt()* function in the *reshape2* package stacks a number of columns into a single one. It also creates a new variable indicating the orginal column from which each observation originated. The *id.vars* option indicates which variables are to stay as separate columns while the *measure.vars* option indicates which are to be stacked. You can also indicate the name of the two new variables melt creates.

In the R code below, I have included the *bdi_pre* variable in both the *id.vars* and *measure.vars* lists. This was done so that the *bdi_pre* values can be used in two specific ways; to draw profile plots and also to include as a covariate in the model development process.

```
bb_long<- melt(bb_wide, id.vars = c("subject","drug","length","treat", "bdi_pre"),

measure.vars= c("bdi_pre", "bdi_2m","bdi_3m","bdi_5m","bdi_8m" ),

variable.name= "time", value.name="bdi"  )

bb_long
```

```
> bb_long
   subject drug length treat bdi_pre    time bdi
1        1   No    >6m   TAU      29 bdi_pre  29
2        2  Yes    >6m BtheB      32 bdi_pre  32
3        3  Yes    <6m   TAU      25 bdi_pre  25
4        4   No    >6m BtheB      21 bdi_pre  21
5        5  Yes    >6m BtheB      26 bdi_pre  26
6        6  Yes    <6m BtheB       7 bdi_pre   7
7        7  Yes    <6m   TAU      17 bdi_pre  17
8        8   No    >6m   TAU      20 bdi_pre  20
9        9  Yes    <6m BtheB      18 bdi_pre  18
10      10  Yes    >6m BtheB      20 bdi_pre  20
```

The results are shown opposite.

To produce profile plots there needs to be a time type variable for the x axis. To achieve this the time variable is converted to a new one called month which represents the time in months (0,2,3,5, or 8) from the beginning of the trial to when the particular bdi measurement was taken.

```
> head(bb_long)
  subject drug length treat bdi_pre    time bdi month
1       1   No    >6m   TAU      29 bdi_pre  29     0
2       2  Yes    >6m BtheB      32 bdi_pre  32     0
3       3  Yes    <6m   TAU      25 bdi_pre  25     0
4       4   No    >6m BtheB      21 bdi_pre  21     0
5       5  Yes    >6m BtheB      26 bdi_pre  26     0
6       6  Yes    <6m BtheB       7 bdi_pre   7     0
```

```
# now create a new variable month
bb_long$month[bb_long$time=="bdi_pre"] <- 0
bb_long$month[bb_long$time=="bdi_2m"] <- 2
bb_long$month[bb_long$time=="bdi_3m"] <- 3
bb_long$month[bb_long$time=="bdi_5m"] <- 5
bb_long$month[bb_long$time=="bdi_8m"] <- 8
attach(bb_long)
head(bb_long)
names (bb_long)
```

```
> bb_long <- bb_long[order(bb_long$subject,bb_long$month)
> bb_long
    subject drug length treat bdi_pre    time bdi month
1         1   No    >6m   TAU      29 bdi_pre  29     0
101       1   No    >6m   TAU      29  bdi_2m   2     2
201       1   No    >6m   TAU      29  bdi_3m   2     3
301       1   No    >6m   TAU      29  bdi_5m  NA     5
401       1   No    >6m   TAU      29  bdi_8m  NA     8
2         2  Yes    >6m BtheB      32 bdi_pre  32     0
102       2  Yes    >6m BtheB      32  bdi_2m  16     2
202       2  Yes    >6m BtheB      32  bdi_3m  24     3
302       2  Yes    >6m BtheB      32  bdi_5m  17     5
402       2  Yes    >6m BtheB      32  bdi_8m  20     8
3         3  Yes    <6m   TAU      25 bdi_pre  25     0
103       3  Yes    <6m   TAU      25  bdi_2m  20     2
203       3  Yes    <6m   TAU      25  bdi_3m  NA     3
303       3  Yes    <6m   TAU      25  bdi_5m  NA     5
403       3  Yes    <6m   TAU      25  bdi_8m  NA     8
```

The data also needs to be ordered on the subject and month variables, which will facilitate the GEE analysis latter on.

```
bb_long <-
bb_long[order(bb_long$subject,bb_long$month),]

bb_long
```

We now have a dataframe which is in the appropriate format to produce some profile plots.

54.4 Individual profiles by treatment group

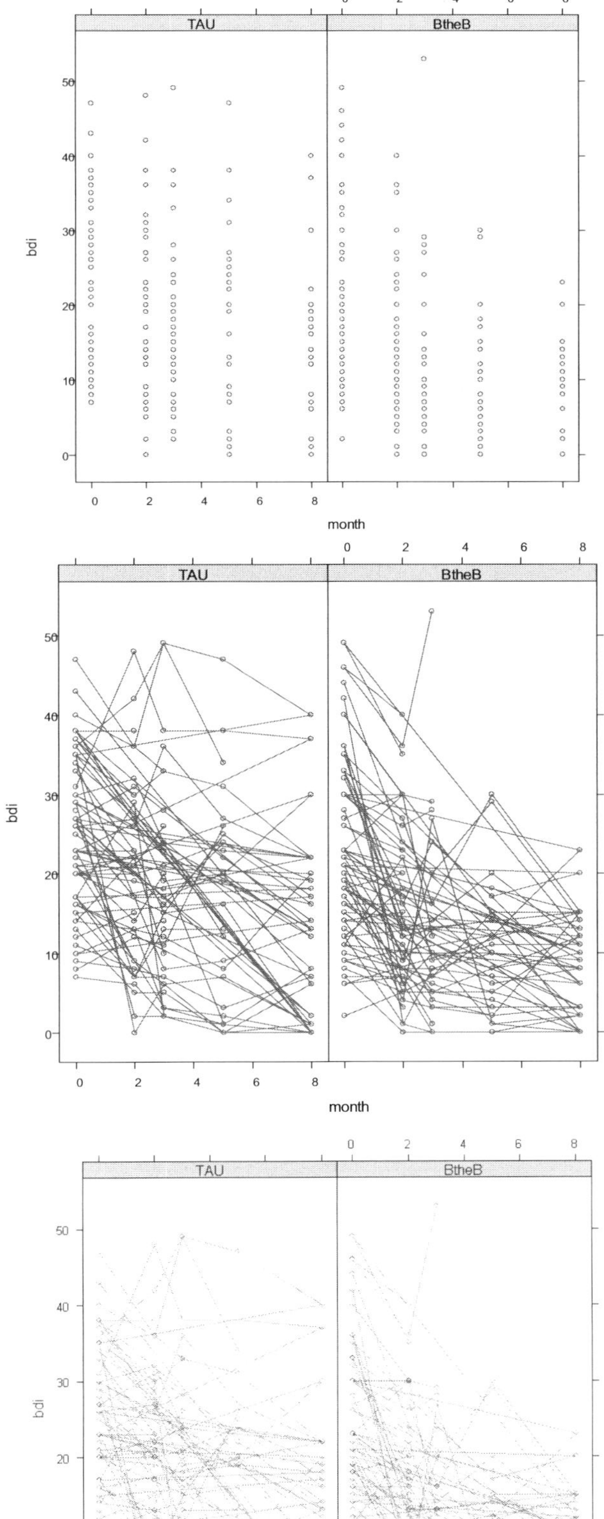

To produce the individual profile plots by treatment group shown opposite we need to install and load the *lattice* package achieved with the two lines of code below. If you're not sure whether you have installed it already, you can simply reinstall it.

```
install.packages("lattice")
```

```
library(lattice)
```

We can draw a simple profile plot divided by the two treatment groups using the following command.

```
xyplot(bdi ~ month | treat, data = bb_long)
```

You can think of the expression bdi~month | treat as meaning bdi by month divided by treat levels

We can also request that the individual points are joined up by adding type="b" to the code.

```
xyplot(bdi ~ month | treat, data = bb_long, type="b")
```

If you require lines with no markers use the lower case L, "l" for the type option. Finally, if you have a large amount of data, is it useful to see where the lines are most dense which is done by setting a level of transparency using the alpha option as shown below:

```
xyplot(bdi ~ month | treat, data = bb_long, type="l",
col= "grey",  alpha = 0.2)
```

Looking at the above plots the two groups look very similar. Possibly the *BtheB* group seem to have baseline levels of *bdi* levels higher than those in the other group and possibly end up at month 8 with slightly lower scores, in other words they have a greater reduction in *bdi* scores, but there appears more similarity than difference between the two groups. One thing we can be sure of is that if any model indicates there is a difference between the two groups it will be a subtle one. We will start this process by looking as the relationship across the repeated measures.

54.5 Reporting the correlations across repeated measures by group

```
> corr.test(bb_wide[bb_wide$treat == "TAU",5:9], adjust="none")
Call:corr.test(x = bb_wide[bb_wide$treat == "TAU", 5:9], adjust = "none")
Correlation matrix
        bdi_pre bdi_2m bdi_3m bdi_5m bdi_8m
bdi_pre    1.00   0.61   0.63   0.50   0.41
bdi_2m     0.61   1.00   0.82   0.76   0.72
bdi_3m     0.63   0.82   1.00   0.86   0.80
bdi_5m     0.50   0.76   0.86   1.00   0.85
bdi_8m     0.41   0.72   0.80   0.85   1.00
Sample Size
        bdi_pre bdi_2m bdi_3m bdi_5m bdi_8m
bdi_pre      48     45     36     29     25
bdi_2m       45     45     36     29     25
bdi_3m       36     36     36     29     25
bdi_5m       29     29     29     29     25
bdi_8m       25     25     25     25     25
Probability values (Entries above the diagonal are adjusted for multiple tests.)
        bdi_pre bdi_2m bdi_3m bdi_5m bdi_8m
bdi_pre    0.00      0      0   0.01   0.04
bdi_2m     0.00      0      0   0.00   0.00
bdi_3m     0.00      0      0   0.00   0.00
bdi_5m     0.01      0      0   0.00   0.00
bdi_8m     0.04      0      0   0.00   0.00
```

TAU group

```
> corr.test(bb_wide[bb_wide$treat == "BtheB",5:9], adjust="none")
Call:corr.test(x = bb_wide[bb_wide$treat == "BtheB", 5:9], adjust = "none")
Correlation matrix
        bdi_pre bdi_2m bdi_3m bdi_5m bdi_8m
bdi_pre    1.00   0.63   0.54   0.56   0.41
bdi_2m     0.63   1.00   0.71   0.76   0.58
bdi_3m     0.54   0.71   1.00   0.59   0.40
bdi_5m     0.56   0.76   0.59   1.00   0.66
bdi_8m     0.41   0.58   0.40   0.66   1.00
Sample Size
        bdi_pre bdi_2m bdi_3m bdi_5m bdi_8m
bdi_pre      52     52     37     29     27
bdi_2m       52     52     37     29     27
bdi_3m       37     37     37     29     27
bdi_5m       29     29     29     29     27
bdi_8m       27     27     27     27     27
Probability values (Entries above the diagonal are adjusted for multiple tests.)
        bdi_pre bdi_2m bdi_3m bdi_5m bdi_8m
bdi_pre    0.00      0   0.00      0   0.03
bdi_2m     0.00      0   0.00      0   0.00
bdi_3m     0.00      0   0.00      0   0.04
bdi_5m     0.00      0   0.00      0   0.00
bdi_8m     0.03      0   0.04      0   0.00
```

BtheB group

```
> corr.test(bb_wide[,5:9], adjust="none")
Call:corr.test(x = bb_wide[, 5:9], adjust = "none")
Correlation matrix
        bdi_pre bdi_2m bdi_3m bdi_5m bdi_8m
bdi_pre    1.00   0.61   0.57   0.51   0.38
bdi_2m     0.61   1.00   0.79   0.78   0.70
bdi_3m     0.57   0.79   1.00   0.82   0.72
bdi_5m     0.51   0.78   0.82   1.00   0.81
bdi_8m     0.38   0.70   0.72   0.81   1.00
Sample Size
        bdi_pre bdi_2m bdi_3m bdi_5m bdi_8m
bdi_pre     100     97     73     58     52
bdi_2m       97     97     73     58     52
bdi_3m       73     73     73     58     52
bdi_5m       58     58     58     58     52
bdi_8m       52     52     52     52     52
Probability values (Entries above the diagonal are adjusted for multiple tests.)
        bdi_pre bdi_2m bdi_3m bdi_5m bdi_8m
bdi_pre    0.00      0      0      0   0.01
bdi_2m     0.00      0      0      0   0.00
bdi_3m     0.00      0      0      0   0.00
bdi_5m     0.00      0      0      0   0.00
bdi_8m     0.01      0      0      0   0.00
```

entire dataframe

We have now considered the distributions of the results, by way of boxplots, and individual profile plots, divided by treatment group. Both suggest that the *beating the blues* group has a more rapid reduction in depression scores compared to that of the treatment group over the time period. We now need to start to consider the development of a mathematical model to evaluate these hunches. Two such models being the Linear Mixed Model (LMM) and GEE.

However, as a preface to developing these it is useful, the reasons for which we will see later, to consider the variance and correlation at each time point for the dependent variable. We have in effect considered the variance at each time point already with the boxplots, so now we need to consider the correlations.

To obtain values for each subset of data based upon whether they received Treatment as Usual (*TAU*) or beating the blues (*BtheB*) we will use the following command.

bb_wide[bb_wide$treat == "BtheB",5:9]

The above indicates we want rows where the treat valuable is equal to *BtheB. B*y changing the value in the above (shown below) to *TAU* we can also produce the correlations for the other group.

The easiest way to obtain a number of correlations is to use the *corr.test()* function in the *psych* package. Both times we request only columns 5 to 9 which is necessary to stop the function throwing an error, because all the columns sent to the *corr.test()* function must be numeric.

```
# psych library
# install.packages("psych")
library(psych)
corr.test(bb_wide[bb_wide$treat == "TAU",5:9],
adjust="none")
corr.test(bb_wide[bb_wide$treat == "BtheB",5:9], adjust="none")
corr.test(bb_wide[,5:9], adjust="none")
```

From the last table in each above output, labelled *Probability values* (=p-values), we can see that all the measures have large correlation values and every one is statistically significant. We will use this information when considering the working correlation/error covariance matrix for the model.

Take home message: **the repeated BDI measures are highly correlated both taking the whole sample and within each treatment group. In addition, the correlations get weaker as the measurements gets further apart. The variance at each time point (from the boxplots) appears to remain constant.**

54.6 Naive – wrong analysis: Ignoring the subject

Usually you would not disregard the repeated nature of the bdi measurements at the person level (i.e. subject clustering). However, for demonstration purposes it is instructive to carry out this inappropriate analysis as it is often wrongly undertaken. A warning, ignoring the dependent (i.e. correlated) nature of repeated measures can lead to invalid conclusions, as we will see.

```
> summary(bb_wrong)

Call:
lm(formula = bdi ~ bdi_pre + month + treat + drug + length, data = bb_long[bb_long$month !=
    0, ])

Residuals:
    Min      1Q  Median      3Q     Max
-23.678  -5.418   0.015   5.327  27.269

Coefficients:
            Estimate Std. Error t value Pr(>|t|)
(Intercept)    7.323      1.728    4.24  3.1e-05 ***
bdi_pre        0.574      0.055   10.44  < 2e-16 ***
month         -0.938      0.236   -3.97  9.4e-05 ***
treatBtheB    -3.322      1.101   -3.02   0.0028 **
drugYes       -3.569      1.147   -3.11   0.0021 **
length   >6m   1.711      1.111    1.54   0.1246
---
Signif. codes:  0 '***' 0.001 '**' 0.01 '*' 0.05 '.' 0.1 ' ' 1

Residual standard error: 8.67 on 274 degrees of freedom
  (120 observations deleted due to missingness)
Multiple R-squared:  0.395,     Adjusted R-squared:  0.384
F-statistic: 35.8 on 5 and 274 DF,  p-value: <2e-16
```

When carrying out the analysis we want to use the baseline bdi measurement as a separate variable rather than part of the treat column, but at the same time omit the rows where this value is repeated when the month variable is equal to zero. To achieve this we use the following expression, where *!=* means not equal to:

```
bb_long[bb_long$month != 0,])
```

We now create a regression model with *dbi* as the dependent (outcome) variable and *bdi_pre*, *month*, *treat*, *drug* and *length* as the independent (input) variables. So ignoring the possible subject clustering the regression equation is:

```
bb_wrong <- lm(bdi ~ bdi_pre + month +treat + drug + length, data=bb_long[bb_long$month != 0,])

bb_wrong
```

```
> library(nlme)
> bb_wrong2 <- gls(bdi ~ bdi_pre + month +treat + drug + length,
+ data=bb_long[bb_long$month != 0,], method ="ML", na.action= na.omit)
> summary(bb_wrong2)
Generalized least squares fit by maximum likelihood
  Model: bdi ~ bdi_pre + month + treat + drug + length
  Data: bb_long[bb_long$month != 0, ]
       AIC      BIC    logLik
  2012.294 2037.738 -999.1471

Coefficients:
                Value Std.Error   t-value p-value
(Intercept)  7.322903 1.7276495  4.238651  0.0000
bdi_pre      0.573963 0.0549747 10.440500  0.0000
month       -0.937843 0.2365022 -3.965471  0.0001
treatBtheB  -3.322536 1.1006886 -3.018597  0.0028
drugYes     -3.568663 1.1471681 -3.110846  0.0021
length   >6m 1.710675 1.1105646  1.540365  0.1246

 Correlation:
            (Intr) bdi_pr month  trtBtB drugYs
bdi_pre     -0.595
month       -0.548  0.036
treatBtheB  -0.357  0.136  0.020
drugYes     -0.007 -0.295 -0.038 -0.318
length   >6m -0.153 -0.316 -0.057  0.028  0.149

Standardized residuals:
         Min           Q1          Med           Q3          Max
-2.759640361 -0.631426947  0.001760626  0.620831656  3.178154212

Residual standard error: 8.580085
Degrees of freedom: 280 total; 274 residual
```

It would appear that treatment is a statistically significant predictor of outcome as well as the baseline score, length of time and if they were receiving drugs. However, notice the low R squared value of .38, which squared is .14 indicating that the model only accounts for 14% of the variability in the data.

If we had failed to realise that the above analysis ignored the repeated nature of the bdi measurements we would assume that there is a statistically significance difference between the bdi measures and the group to which they belonged – we will subsequently find this is erroneous.

To allow comparison between this inappropriate model and those which we are going to develop requires that we use the Maximum Likelihood (ML) approach to calculate the parameter estimates rather than the least squares approach as in the above *lm()* function. To do this we need to use the *gls()* function in the *nlme* package.

```
# install.packages("nlme") # remove the hash sign if you do not have it installed
library(nlme)
bb_wrong2 <- gls(bdi ~ bdi_pre + month +treat + drug + length,
data=bb_long[bb_long$month != 0,], method ="ML", na.action= na.omit)
summary(bb_wrong2)
```

```
> bb_intercept_only <- gls(bdi ~ 1,
+ data=bb_long[bb_long$month != 0,], method ="ML", na.action= na.omit)
> summary(bb_intercept_only)
Generalized least squares fit by maximum likelihood
  Model: bdi ~ 1
  Data: bb_long[bb_long$month != 0, ]
       AIC      BIC    logLik
  2143.015 2150.285 -1069.508

Coefficients:
               Value Std.Error  t-value p-value
(Intercept) 14.43214 0.6604231 21.85288       0

Standardized residuals:
       Min         Q1        Med         Q3        Max
-1.3082974 -0.7643876 -0.1751519  0.5953870  3.4962396

Residual standard error: 11.03124
Degrees of freedom: 280 total; 279 residual
```

The estimated parameter values, standard errors and p-values are identical using the ML approach. The Maximum Likelihood (ML) approach has also resulted in several measures of model fit being produced, AIC, BIC and logLik which can be used to compare models.

Another model that is often created is that which uses just one parameter, the intercept. In other words how well does the data fit a model consisting of a straight horizontal line of y= 14.4 which is the mean of the repeated *bdi* measurements, excluding the baseline ones.

We create such a model using the following code:

```
bb_intercept_only <- gls(bdi ~ 1,
data=bb_long[bb_long$month != 0,], method ="ML", na.action= na.omit)
summary(bb_intercept_only)
```

In the above analysis we have two inappropriate models, the first ignored the dependency of the repeated measures and the second assumed the data could be modelled by a single straight horizontal line. We will now consider several, more appropriate models for analysing repeated measures. To do this we need to install a package called *lme4*.

54.7 Installing lme4 and other packages to obtain p-values (if you must!)

lme4 Techie note: Prior to 2014, the most recent version of *lme4* was to be found on the non-crans r-forge server. However, this is no longer the case and you can now install it in the usual way.

To carry out an appropriate analysis of the repeated measures we need to download and install the *lme4* package in the usual way:

```
install.packages("lme4")
library(lme4)
```

The developers and maintainer of the *lme4* package believe it unwise to produce p-values for the parameter estimates (providing confidence intervals instead). If you do really need p-values use either the *car* or *lmerTest* packages demonstrated later in this chapter. Further details are also provided in the documentation to the *lme4* package at http://cran.r-project.org/web/packages/lme4/lme4.pdf

Why no p values?

See: http://stats.stackexchange.com/questions/22988/significant-effect-in-lme4-mixed-model and from the author of lme4: https://stat.ethz.ch/pipermail/r-help/2006-May/094765.html

54.8 Random Intercept models

> **MANOVA**
> Landau & Everitt, 2004 p.214 make the interesting observation that the random intercept model is similar in some ways to a (M)ANOVA. However that approach only uses complete cases whereas the random intercept analysis uses all cases including incomplete ones.

We will begin by adding a random intercept, being a conceptual necessity for repeated measures analysis (Twisk, 2006). This simply means we are now allowing each subject's baseline score to be taken into account along with the dependent nature of the repeated measures at the subject level.

A random intercept model for our data can be represented by the following R model code.

```
bdi ~ bdi_pre + month +treat + drug + length + (1|subject)
```

bdi being to the left of the tilde symbol (~) is defined as the output variable (dependent) while those variables to the right of the tilde are the input (independent) variables.

```
> bb_rintercept <- lmer(bdi ~ bdi_pre + month +treat + drug + length + (1|subject),
+ data=bb_long[bb_long$month != 0,],
+ REML = FALSE, na.action= na.omit)
> bb_rintercept
Linear mixed model fit by maximum likelihood  ['lmerMod']
Formula: bdi ~ bdi_pre + month + treat + drug + length + (1 | subject)
   Data: bb_long[bb_long$month != 0, ]
     AIC      BIC   logLik deviance df.resid
1887.492 1916.570 -935.746 1871.492      272
Random effects:
 Groups   Name        Std.Dev.
 subject  (Intercept) 6.984
 Residual             5.014
Number of obs: 280, groups: subject, 97
Fixed Effects:
 (Intercept)      bdi_pre        month   treatBtheB      drugYes  length  >6m
      5.5924       0.6397      -0.7048      -2.3291      -2.8250       0.1971
> |
```

The term *1|subject* allows the modelling of a random effect for each subject. In other words, we now have a set of **parallel** regression lines which are normally distributed about the mean value of the *bdi* values excluding the baseline values. In fact, the model produces a separate intercept estimate for each subject which we can and will subsequently plot. We will discuss these results after creating another model to allow comparison.

```
library(lme4)
bb_rintercept <- lmer(bdi ~ bdi_pre + month +treat + drug + length + (1|subject),
data=bb_long[bb_long$month != 0,],
REML = FALSE, na.action= na.omit)
bb_rintercept
```

54.8.1 Extracting additional information about the model

It is often necessary to find out more about a particular model than that displayed by R and there are three very useful functions you can apply to *lme4* model objects to achieve this. Applying the three functions; *coef(), ranef(), fixef()* to our present model gives:

Random and fixed estimates	Random effects estimates	Fixed effects estimates
coef(bb_rintercept)	**ranef(bb_rintercept)**	**fixef(bb_rintercept)**
> coef(bb_rintercept) $subject (Intercept) bdi_pre month treatBtheB 1 -10.7699718 0.6396746 -0.7047725 -2.329117 2 6.7586590 0.6396746 -0.7047725 -2.329117 3 7.3413305 0.6396746 -0.7047725 -2.329117 4 4.9526969 0.6396746 -0.7047725 -2.329117 5 10.3058349 0.6396746 -0.7047725 -2.329117	> ranef(bb_rintercept) $subject (Intercept) 1 -16.36241068 2 1.16622014 3 1.74889159 4 -0.63974204 5 4.71339601 6 -1.54554152	> fixef(bb_rintercept) (Intercept) bdi_pre month treatBtheB 5.5924389 0.6396746 -0.7047725 -2.3291166 drugYes length >6m -2.8249742 0.1971235

Note that for subject one the intercept value is equal to -10.76 which is the sum of the fixed effect intercept and the random effect i.e. -10.77 = 5.59 + (-16.36). In contrast to the random effect parameter, *(Intercept)* the fixed effects parameters have the same value for all the subjects.

54.9 Confidence intervals.

To obtain confidence intervals for the parameters there is a different method for the fixed and random effects. Confidence intervals for fixed effects are obtained by using the profile function, and then applying the *confint()* function to it.

```
> confint(pr01)
                   2.5 %     97.5 %
.sig01         5.8930767  8.3177622
.sigma         4.5440746  5.5699854
(Intercept)    1.1418580 10.0143601
bdi_pre        0.4860493  0.7943727
month         -0.9953219 -0.4167332
treatBtheB    -5.6254192  0.9905566
drugYes       -6.2360435  0.6009748
length    >6m -3.0622690  3.4270318
```

note that the name of the variable is (Intercept)

```
pr01 <- profile(bb_rintercept); confint(pr01)
```

In the latest version of *lme4* you can simply type:

```
confint(bb_rintercept)
```

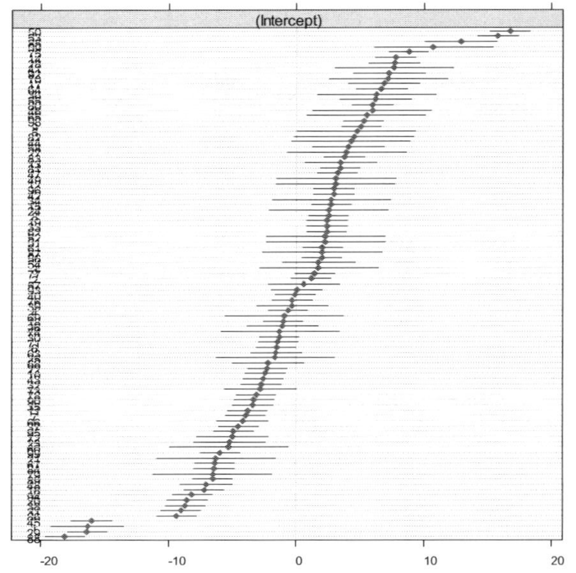

The confidence intervals obtained require the model to be re-calculated many times so the process might take some time. We can also use these values to indicate which of the parameters would have p-values of less than 0.05, that is the ones that do not cross the zero mark. Therefore, we have a statistically significant estimate for the intercept, *bdi_pre* and *month* parameters.

To obtain confidence intervals for the random effects (only one in the current equation) we use two functions, the first *ranef()* with the option *condVar = TRUE*. This produces a list of the required values which we can then apply to the *dotplot()* function in the lattice package to obtain a graph.

```
dotplot(ranef(bb_rintercept, condVar = TRUE))
```

The graph shows that most confidence intervals are around the zero mark as a good model would behave, however there are several that have extreme negative or positive values, possibly due to lack of data for particular subjects.

note that the name of the variable is (Intercept)

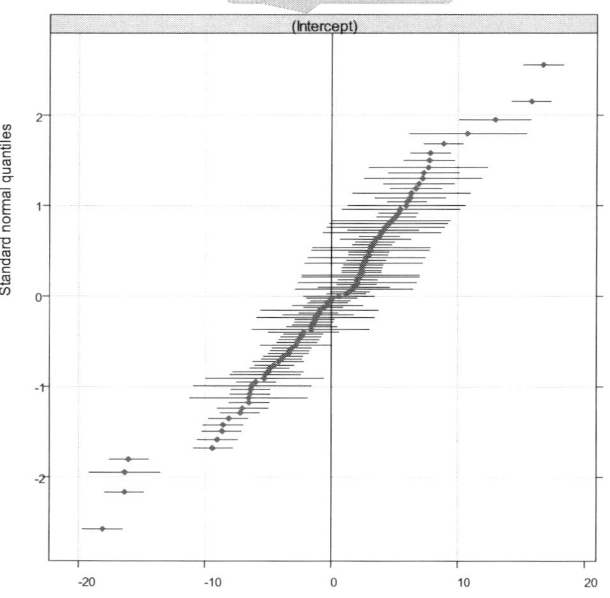

The outliers can be seen more clearly by using the *qqmath()* function to plot the result. This time the confidence intervals are ordered along a standardised normal axis where we see that four subjects have confidence intervals greater than or less than 2.

```
qqmath(ranef(bb_rintercept, condVar=TRUE))
```

There are several ways we can identify these outliers, firstly using a graphical technique which also demonstrates some additional *dotplot()* function options. You can specify the size of the font for the axes by using the scales option which, I have set to 40% of its normal size and changed its colour to blue in the code below.

```
dotplot(ranef(bb_rintercept, condVar = TRUE),
scales=list(cex=.4, col="blue"))
```

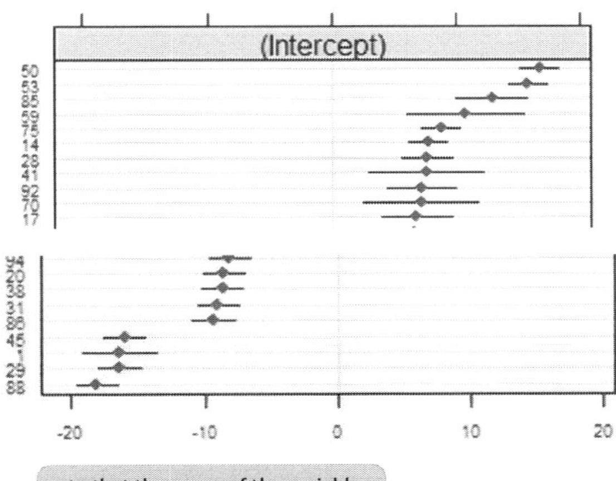

It is now possible to see that subjects 50, 53, 85 and 59 have excessively high random effect values and subjects 45, 1, 29 and 88 have excessively low ones.

You could also find the excessive values by using the object returned from the *ranef()* function, requesting the subject values list and converting it to a dataframe.

note that the name of the variable is (Intercept) including the brackets

```
ran_cc<- as.data.frame(ranef(bb_rintercept, postVar = TRUE)[["subject"]])
```

```
> ran_cc
   (Intercept)
1 -16.36172336
2   1.16611186
3   1.74883028
```

Now we have a dataframe with the relevant details in a column called *(Intercept)* we can request subsets of its rows based on certain criteria. For our purposes I request all rows where the random intercept value *(Intercept)* is either greater than +12 or less than -12, as this looks like a reasonable cut off value from the above plots. This is the same as setting the absolute value to more than 12.

```
ran_cc[abs(ran_cc$'(Intercept)') > 12,]  # '(intercept)' is the default name!

ran_cc[(ran_cc$'(Intercept)' > 12 | ran_cc$'(Intercept)' < -12) ,]  # equivalent to above
```

```
> ran_cc[(ran_cc$'(Intercept)' > 12 | ran_cc$'(Intercept)' < -12) ,]
[1] -16.36172 -16.38156 -16.01664  16.70161  15.77002  12.88172 -18.09959
> |
```

The above returns the same values we identified visually.

```
> dim(ran_cc)
[1] 97  1
```

Another aspect, which will be more important when manipulating the data later, is the number of random intercept estimates the model has created. You might be expecting one for each subject, however this is not the case, as the *ran_cc* object only has 97 rows so we do not have estimates for three of the subjects, the reason for which will become clear later. To spot the subjects with the missing random effects estimates just look at the *ran_cc* object by typing the name in the R Console window.

The next stage is to assess visually how well the model's regression line fits each subject's data.

54.9.1 Plotting subject regression lines based on random intercepts for the data

This is a complex process and I will start by plotting just the original data for each subject. This is done using the lattice package:

```
library(lattice)

xyplot( bdi ~ month | subject )
```

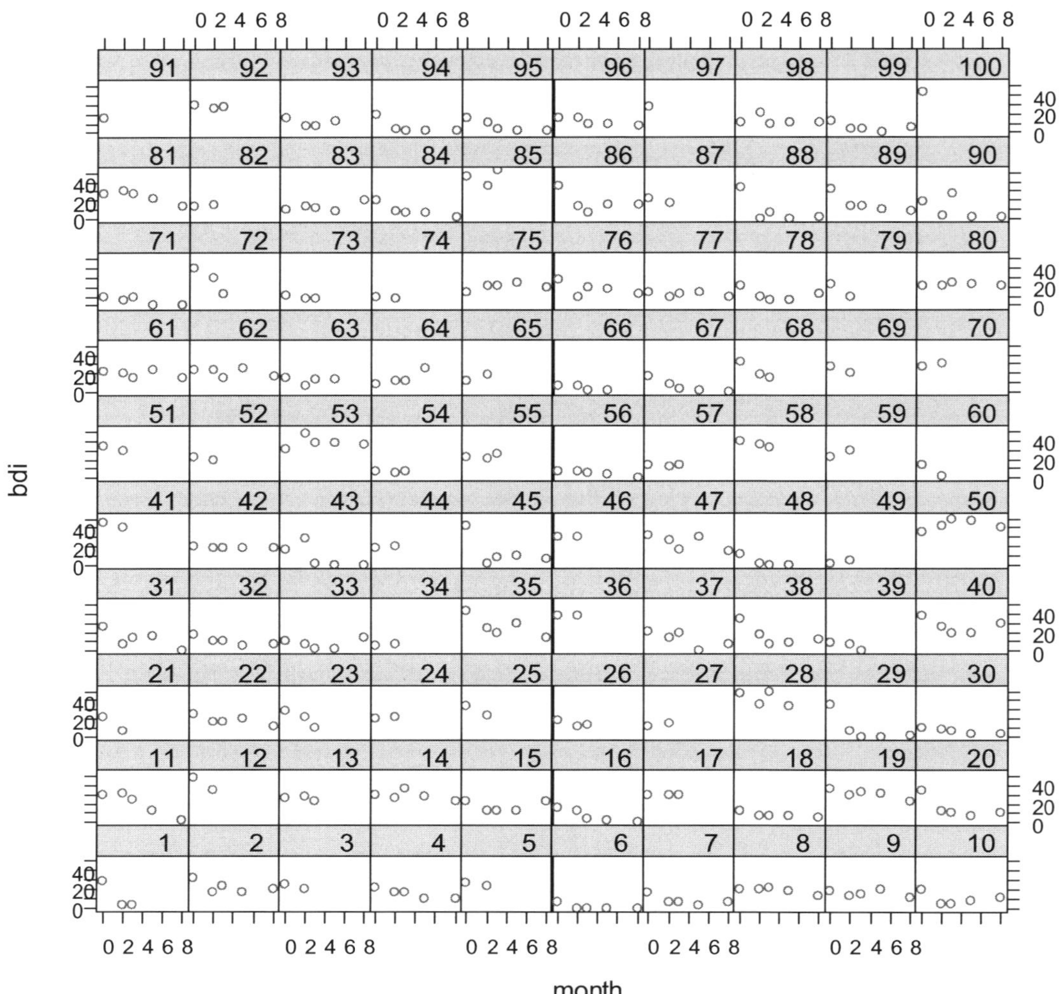

Things to note:

- the values at the top of each panel represent the subject. The y axis is the *bdi* measures and the x axis is the time in *month*s. Note that the baseline measure is also included
- three subjects (91, 97 and 100) only have a single value and similarly several others only have two points. Could these three subjects with the single value be those that the *lmer()* function failed to produce random effect estimates for?
- the values, like the profile plots, seem to follow many different trajectories

Now we want to annotate the above plot to include the regression line represented by our equation where the intercept varies for each subject but always has the same slope is also displayed. We will make use of the values produced from the *coef()* function shown earlier. We collect these values, save then in an object called *vals1* and convert it to a dataframe.

```
vals1 <- coef(bb_rintercept)[["subject"]]
vals1 <- as.data.frame(vals1)
```

```
> vals1
   (Intercept)   bdi_pre     month treatBtheB   drugYes length   >6m
1 -10.7693289 0.6396762 -0.704765  -2.329083 -2.824953    0.1970805
2   6.7585063 0.6396762 -0.704765  -2.329083 -2.824953    0.1970805
3   7.3412248 0.6396762 -0.704765  -2.329083 -2.824953    0.1970805
```

The vals1 object is now a dataframe with columns representing each of the equation parameters.

Now we add a hidden column, the rownames attribute of vals1, which happens to contain the subject id to a new object which I have called *yslope*.

```
yslope = as.data.frame(cbind(vals1[,1],vals1[,3],  as.integer(rownames(vals1))))
```

```
> yslope
    intercept     month subid
1 -10.7693289 -0.704765     1
2   6.7585063 -0.704765     2
3   7.3412248 -0.704765     3
4   4.9526745 -0.704765     4
5  10.3056666 -0.704765     5

90   2.2742278 -0.704765    90
91  12.8542655 -0.704765    92
92   5.6234268 -0.704765    93
93  -2.5511040 -0.704765    94
94   0.6991288 -0.704765    95
95   8.5167709 -0.704765    96
96  10.8407517 -0.704765    98
97  -0.3753925 -0.704765    99
```

We then give the new columns sensible names, and remove the '()' and capital 'I' from the intercept column name. The result is shown opposite

```
colnames(yslope) <-c("intercept","month",  "subid")
```

We still only have 97 rows for 100 subjects and converting this to a 100 row long dataframe is the final stage in the data manipulation. First we create a object called *temp1* which is a dataframe consisting of a single column called *subid* with values of 1 to 100

```
temp1 <- data.frame(subid= 1:100)
temp1
```

```
> yslope <- merge(yslope, temp1 , by="subid", all = TRUE)
> yslope
    subid   intercept     month
1       1 -10.7693289 -0.704765
2       2   6.7585063 -0.704765
3       3   7.3412248 -0.704765

90     90   2.2742278 -0.704765
91     91          NA        NA
92     92  12.8542655 -0.704765
93     93   5.6234268 -0.704765
94     94  -2.5511040 -0.704765
95     95   0.6991288 -0.704765
96     96   8.5167709 -0.704765
97     97          NA        NA
98     98  10.8407517 -0.704765
99     99  -0.3753925 -0.704765
100   100          NA        NA
```

We now merge the *temp1* and *yslope* objects using the *subid* column to provide the match in such a way that where there are no values for a row an *NA* value appears.

```
yslope <- merge(yslope, temp1, by="subid",all=TRUE)
yslope
```

Finally, we change the *NA* values to zeros, as such *NA* values can cause problems with graphing:

```
yslope[is.na(yslope)] <- 0
yslope
```

Note that it has taken nearly a page of explaining to get the data in the required format for the graph, an alternative would have been to remove those subjects from the data set where only a single *dbi* value exists or even just include in the analysis subjects that provided three or more measurements. This would have been most easily achieved when the data was in the original wide format using something like the following code.

```
bb_wide[complete.cases(bb_wide),]   #leaves only complete cases
bb_wide[complete.cases(bb_wide[,5:8]) ,] #leaves only cases where columns 5 to 8 are not empty
```

I feel it is useful to see the complete data in the form of a lattice *xyplot()* as we have done before.

We are now in the position to draw the required plot showing the regression line for each subject. Comments for the R code, other than the panel function, are provided for a more complex plot later in the chapter.

```
xyplot(bdi ~ month | subject, bb_long, aspect="xy",
type = c("p"), pch=4, grid= FALSE, cex=.5,
par.strip.text=list(cex=0.5),
coef.list = yslope[,2:3], #provides raw intercept + slope coefficiants
panel = function(x,y,..., coef.list)
    {
    yadd <- 0.64 *y  + coef.list[packet.number(),1]
    xadd <- coef.list[packet.number(),2]
    panel.xyplot(x,y,...)
    panel.abline(a = yadd[1], b= xadd )
    panel.text(5, 40, paste(y , "\n ", round(yadd[1],1)), cex = .5, font = 1)
    # shows bdi_pre
    }
)
```

54.9.2 xyplot and the panel function

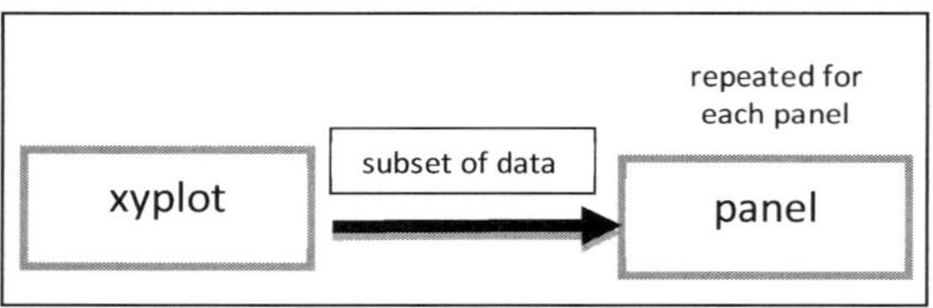

The *xyplot()* function creates many new variables internally; two of which are *x* and *y* based on the x and y values you supply it (the *month* and *bdi* variables from the *bb_long* dataset in this instance). The *xyplot()* function then passes an appropriate subset of these values to a panel function which is called for each panel. If you want to access the first x value in a particular panel you indicate it in the panel function by *x[1]* and similarly for the first y value *y[1]*.

Within the panel function there are several options specified:

- *packet.number()*, provides a simple integer index indicating which panel is currently being drawn
- *panel.xyplot()* draws a scatter plot
- *panel.abline(a, b)* daws a line where a is the intercept and b is the gradient.
- *panel.text()* draws text in the panel, you can concatenate several strings (groups of characters) by enclosing them in the paste() function. To indicate a new line use "\n"
- **...** this strange looking thing indicates that the function should also accept all the additional values that the function would normally receive.

panel.abline(a = yadd[1], b= xadd) requires further explanation. I have used *[1]* to select the first element as this needs to be a single value. When we formed the *yadd* object previously it became a vector which defined 'y' for a particular panel as a set of 5 values (one for each month value). We could have prevented this by editing the relevant line of code to:

```
yadd <- 0.64 *y[1]  + (coef.list[packet.number(),1])
```

and subsequently edited the following line

```
panel.abline(a = yadd, b= xadd )
```

Using either method produces the same result.

Rather than showing the regression line on separate panels, it is also possible to superimpose them on a single panel.

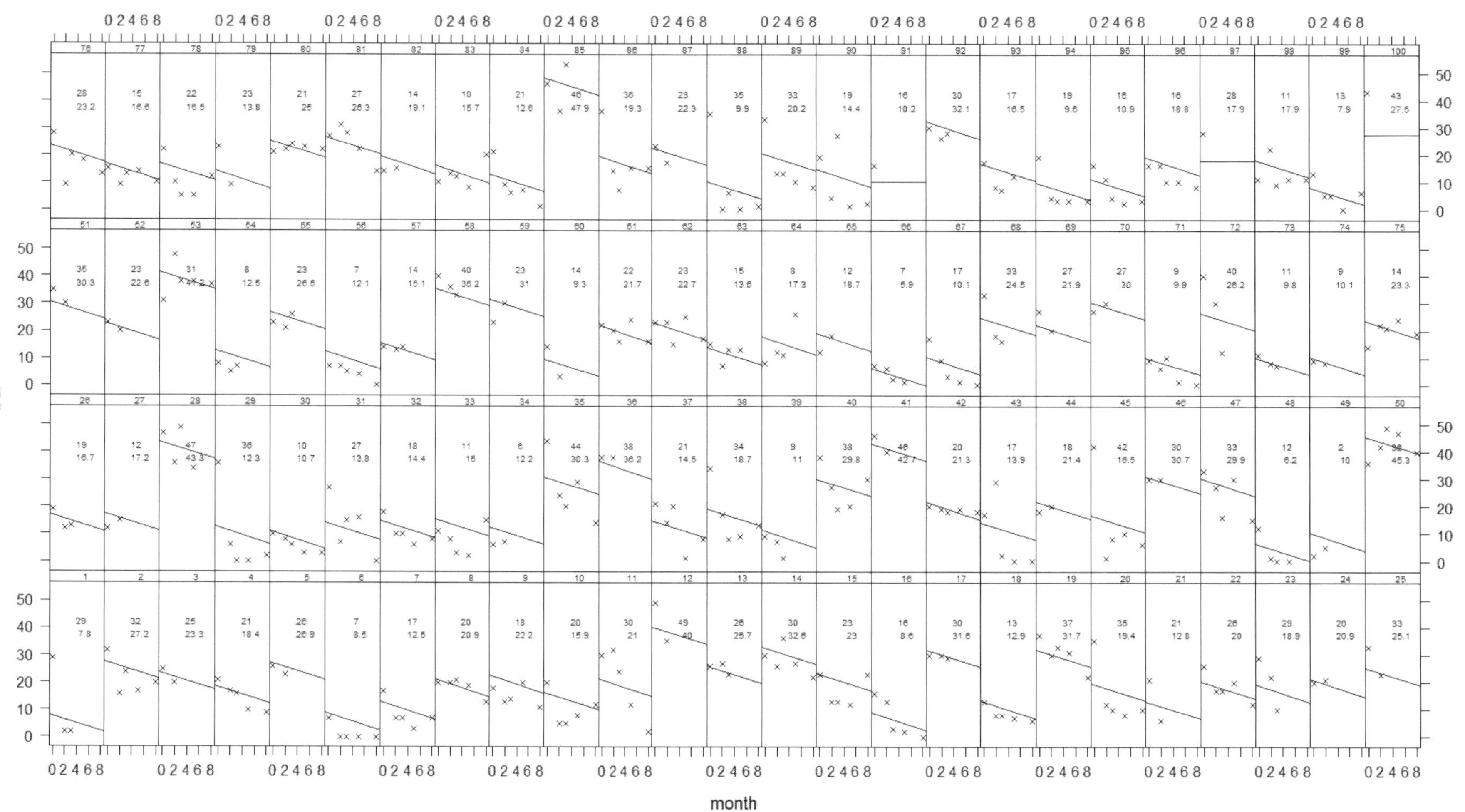

Four points to note:

The horizontal regression lines for the subjects with a single point are irrelevant (we added the zero values to them).

We can see from the above that the regression lines appear to fit the data reasonably well.

The constraint that the regression lines all have the same gradient does not appear to be an obstacle to the adequacy of the fit. However we can check this by creating an equation which allows both the intercept and gradient to vary across subjects.

The two numbers in each panel represent the bdi_pre value and also the result of applying the bdi_pre coefficient to it (0.64). This second value along with the random intercept and fixed intercept value gives the predicted intercept value for the particular subject.

54.10 Drawing multiple regression lines on one panel

To see the distribution of the regression lines across all the subjects on a single graph we need to make use of some other lattice functions within the *xyplot()* function.

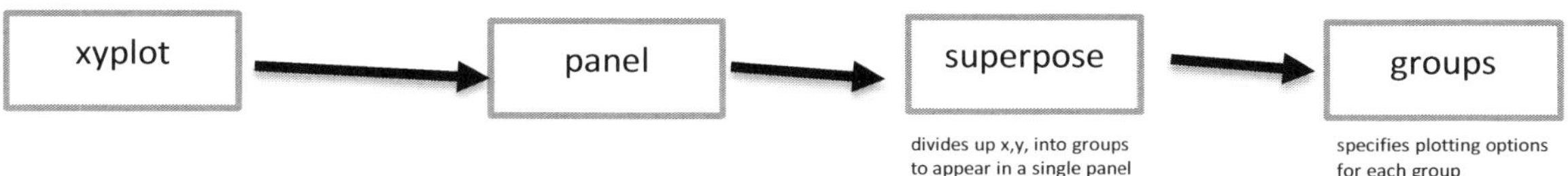
xyplot
panel
superpose
groups
divides up x,y, into groups to appear in a single panel
specifies plotting options for each group

The code below demonstrates how we first draw the raw data once as a scatterplot (xyplot) then specify that we want to superimpose (superpose in lattice language) our regression equations for each subject (i.e. group) as stated in the first line of code.

The panel groups function provides a special value called *group.number* which is the index of the group currently being drawn in the panel. As we know that there are 100 regression equations including the three null value ones, this equates to the required row number for the particular coefficients we need to use.

```
xyplot(bdi ~ month, bb_long, aspect="fill", groups=subject,
    grid= FALSE, cex=.5, ylim=c(-20, 60),
    coef.list = yslope[,2:3], #provides raw intercept + slope coefficiants
        panel = function(x,y,coef.list,    ...) {
        panel.xyplot(x,y)
            panel.superpose(x, y, coef.list, ...,
            panel.groups = function(x,y, coef.list, group.number,...) {
                yadd <- 0.64 *y[1] + coef.list[group.number,1]
                xadd <- coef.list[group.number,2]
                alpha =0.6
                # panel.text(5,300 - 3*group.number, group.number, cex=.5)
                panel.abline(a = yadd, b= xadd )
                } #end panel groups function
                ) # end superpose function
            } # end panel function
    ) # end xyplot
```

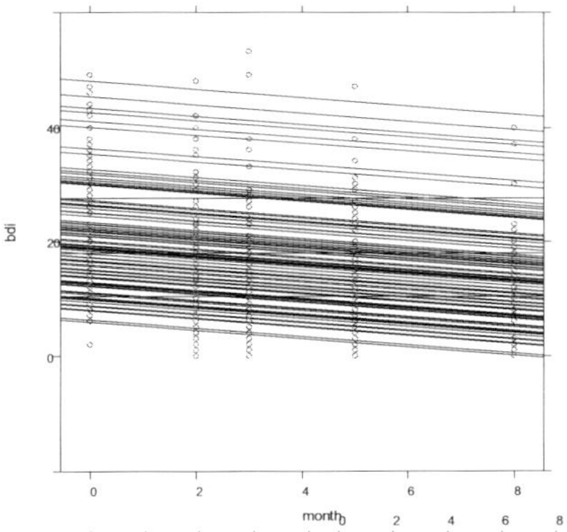

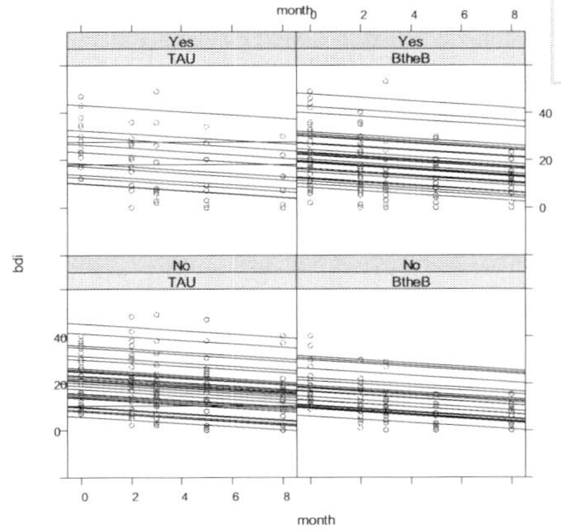

The result of the above code is shown opposite top left. By changing the first line of the above code to that shown below we can extend this approach to show subsets of regression lines. Now we are able to show those in each treatment group along with their drug status.

```
xyplot(bdi ~ month | treat + drug , bb_long, aspect="fill", groups=subject,

grid= FALSE, cex=.5, ylim=c(-20, 60), . . . . . . .
```

54.11 Random Intercept/slope models

Now we will not only allow each subject's regression line to have a different intercept value but also a different slope value as well.

```
> bb_rintercept_slope <-
+ lmer(bdi ~ bdi_pre + month +treat + drug + length + (month|subject),
+ data=bb_long[bb_long$month != 0,],
+ REML = FALSE, na.action= na.omit)
> bb_rintercept_slope
Linear mixed model fit by maximum likelihood  ['merModLmerTest']
Formula: bdi ~ bdi_pre + month + treat + drug + length + (month | subject)
   Data: bb_long[bb_long$month != 0, ]
     AIC       BIC    logLik  deviance  df.resid
1891.0378 1927.3857 -935.5189 1871.0378       270
Random effects:
 Groups   Name        Std.Dev. Corr
 subject  (Intercept) 7.1227
          month       0.4294   -0.18
 Residual             4.8984
Number of obs: 280, groups: subject, 97
Fixed Effects:
 (Intercept)      bdi_pre         month    treatBtheB       drugYes  length   >6m
      5.6139       0.6418       -0.7021       -2.3760       -2.8694        0.1404
> |
```

This time we replace the *1|subject* expression with *(month|subject)*.

```
bb_rintercept_slope <-
lmer(bdi ~ bdi_pre + month +treat + drug +
length + (month|subject),
data=bb_long[bb_long$month != 0,],
REML = FALSE, na.action= na.omit)
bb_rintercept_slope
```

There are now two random effects, one for the intercept and one for the slope. Additionally they have been allowed to correlate with a value of -0.18.

It is also possible to create a model where the two random effects are not allowed to correlate by specifying each of the random effects in separate brackets.

```
> bb_rintercept_slope_ind <- lmer(bdi ~ bdi_pre + month +treat + drug + length+
+ (1|subject) + (0 +month|subject),
+ data=bb_long[bb_long$month != 0,],
+ REML = FALSE, na.action= na.omit)
> bb_rintercept_slope_ind
Linear mixed model fit by maximum likelihood  ['merModLmerTest']
Formula: bdi ~ bdi_pre + month + treat + drug + length + (1 | subject) +     (0 + month | subject)
   Data: bb_long[bb_long$month != 0, ]
     AIC       BIC    logLik  deviance  df.resid
1889.1565 1921.8697 -935.5783 1871.1565       271
Random effects:
 Groups    Name        Std.Dev.
 subject   (Intercept) 6.8974
 subject.1 month       0.3384
 Residual              4.9412
Number of obs: 280, groups: subject, 97
Fixed Effects:
 (Intercept)      bdi_pre         month    treatBtheB       drugYes  length   >6m
     5.58972      0.64499      -0.69453      -2.40674      -2.91381       0.07866
> |
```

(1|subject) + (0 +month|subject)

```
bb_rintercept_slope_ind <- lmer(bdi ~ bdi_pre +
month +treat + drug + length+

(1|subject) + (0 +month|subject),

data=bb_long[bb_long$month != 0,],
REML = FALSE, na.action= na.omit)

bb_rintercept_slope_ind
```

```
> Anova(bb_rintercept)
Analysis of Deviance Table (Type II Wald chisquare tests)

Response: bdi
          Chisq Df Pr(>Chisq)
bdi_pre 67.4438  1  < 2.2e-16 ***
month   23.1770  1  1.478e-06 ***
treat    1.9442  1     0.1632
drug     2.6762  1     0.1019
length   0.0145  1     0.9043
---
Signif. codes:  0 '***' 0.001 '**' 0.01 '*' 0.05 '.' 0.1 ' ' 1
```

Both models produce almost identical parameter values. You may have noticed that in the above outputs there are no p-values or standard errors – this is deliberate as mentioned earlier.

If these are required there are several methods of obtaining p-values and I find the easiest is to use the *Anova()* (note capital 'A') function in the car package.

```
install.packages("car") #only needed first time
require(car)
                              Anova(bb_rintercept)
```

upper case A

Alternatively you can use the *anova()* function in the *lmerTest* package.

```
> anova(bb_rintercept)
Analysis of Variance Table of type 3  with  Satterthwaite
approximation for degrees of freedom
        Sum Sq Mean Sq NumDF   DenDF F.value    Pr(>F)
bdi_pre 1819.50 1819.50     1 104.076  67.444 6.287e-13 ***
month    583.09  583.09     1 199.311  23.177 2.916e-06 ***
treat    106.39  106.39     1  97.116   1.944    0.1664
drug      70.59   70.59     1  98.188   2.676    0.1051
length     0.36    0.36     1 100.255   0.014    0.9045
---
Signif. codes:  0 '***' 0.001 '**' 0.01 '*' 0.05 '.' 0.1 ' ' 1
```

```
install.packages("lmerTest")
#only need above code first time
library(lmerTest)
bb_rintercept <- lmer(bdi ~ bdi_pre + month +treat
+ drug + length +
(1|subject),data=bb_long[bb_long$month != 0,],
REML = FALSE, na.action= na.omit)
anova(bb_rintercept)
```

lower case A

```
> Anova(bb_rintercept_slope)
Analysis of Deviance Table (Type II Wald chisquare tests)

Response: bdi
          Chisq Df Pr(>Chisq)
bdi_pre 67.9865  1  < 2.2e-16 ***
month   20.7690  1  5.182e-06 ***
treat    2.0211  1    0.15513
drug     2.7585  1    0.09674 .
length   0.0073  1    0.93172
---
Signif. codes:  0 '***' 0.001 '**' 0.01 '*' 0.05 '.' 0.1 ' ' 1
```

```
#back to using the car Anova
Anova(bb_rintercept_slope)
```

upper case A

Only month and baseline *bdi* score are now statistically significant, regardless of including a random slope or not. The difference in treatments has disappeared. So the relevant question is which is the most parsimonious model?

54.12 Comparing mixed models

```
> anova(bb_rintercept, bb_rintercept_slope, test="Chisq")
Data: [
Data: bb_long
Data: bb_long$month != 0
Data:
Models:
bb_rintercept: bdi ~ bdi_pre + month + treat + drug + length + (1 | subject)
bb_rintercept_slope: bdi ~ bdi_pre + month + treat + drug + length + (month | subject)
                    Df    AIC    BIC  logLik  Chisq Chi Df Pr(>Chisq)
bb_rintercept        8 1887.5 1916.6 -935.75
bb_rintercept_slope 10 1891.0 1927.4 -935.52 0.4542      2     0.7969
```

There are two standard approaches for comparing mixed models. The first compares the mis-fit measures (BIC and AIC) where smaller values indicate better fit. The second method uses the likelihood ratio test comparing the:

- complex model *to* simple model
- using more df where model *to* less df model
- =intercept + slope model *to* intercept only model

The above idea is represented in R code by: *anova(bb_rintercept, bb_rintercept_slope)*

If you want to know how the above *anova()* function works it is equivalent to the following

```
> pchisq(2 * (logLik(bb_rintercept_slope) - logLik(bb_rintercept)), df = 2, lower.tail = FALSE)
'log Lik.' 0.7968552 (df=10)
```

In the above we have compared the two models and in effect assessed the possible advantage of adding a second correlated random effect. Given that the result is statistically insignificant at the 0.05 level we can say that there is no advantage in adding the additional random effect.

model	AIC	BIC	logLik	deviance
Bb_rintercept	**1887**	**1917**	-935.7	1871
bb_rintercept_slope	1891	1927	-935.5	1871
bb_rintercept_slope_ind	1889	1922	-935.6	

We could also have inspected the model mis-fit measures. Both the AIC and BIC are smallest for the random intercept only model.

```
> bb_rintercept
Linear mixed model fit by maximum likelihood  ['merModLmerTest']
Formula: bdi ~ bdi_pre + month + treat + drug + length + (1 | subject)
   Data: bb_long[bb_long$month != 0, ]
      AIC       BIC    logLik deviance df.resid
1887.492 1916.570 -935.746 1871.492       272
Random effects:
 Groups   Name        Std.Dev.
 subject  (Intercept) 6.984
 Residual             5.014
Number of obs: 280, groups: subject, 97
```

The above analysis has demonstrated the importance of taking into account clustering that might occur within data, in this instance concerning the repeated measurements within each individual. One very useful measure is the intra-class correlation coefficient, which describes how strongly the repeated measures resemble each other within a subject. This is calculated by using the various estimated standard deviations:

$$(\text{subject\_intercept sd})^2 / ((\text{residual sd})^2 + (\text{subject\_intercept sd})^2) = 6.9842^2 / (6.9842^2 + 5.0140^2)$$

```
> 6.9842^2 / (6.9842^2  + 5.0140^2)
[1] 0.6598965
```

In R: 6.9842^2 / (6.9842^2 + 5.0140^2)

Giving a value of 0.6599 = 0.66. Quoting Bland & Altman, 1990. *The intraclass correlation coefficient can be used, for example, as an index of correlation between repeated measures by the same method, i.e. as an index of repeatability, because in that case there is no ordering f the repeated measures.* We can therefore say that the estimated average correlation between the repeated measures within a subject is .66.

54.13 Graphing multiple regression lines

Often we want to see graphically how two or more regression equations fit the data. Taking our example, we will create two new models, one a random intercept only and the other a random intercept and slope (correlated). However, this time they will only include the predictors we have found to be statistically significant (*month* and *bdi_pre*).

If you are starting the chapter here you can download the dataset from:

www.robin-beaumont.co.uk/virtualclassroom/book2data/bb_long.rdata

Save the file locally. Take a note of where you put it then type the following code into the R Console window to load the file. load(file=file.choose())

It is always good to know what one is aiming for and to produce these two regression lines on a per subject basis we need the estimated coefficients from the models along with the raw data. Concentrating on the model coefficients we need to end up with a dataframe, which I have called *allcoef2*, consisting of columns each representing one of the coefficients. Looking below there are 6 columns besides the subject one, the first three are the estimates from the random intercept model and the last three are from the random slope model. Notice that whereas the random intercept model has a fixed value for the month coefficient (*ri_month*) that for the random slope model (*rs_month*) varies across subjects.

```
> allcoefs2
  subject ri_intercept ri_bdi_pre   ri_month rs_intercept rs_bdi_pre   rs_month
1       1  -11.0384907  0.6262672 -0.6986236  -11.4788720  0.6241178 -0.5725298
2       2    2.4763912  0.6262672 -0.6986236    2.0742297  0.6241178 -0.5978890
3       3    5.0437624  0.6262672 -0.6986236    5.1789411  0.6241178 -0.7206300
4       4    3.0490132  0.6262672 -0.6986236    3.5350133  0.6241178 -0.7952603
5       5    6.6659887  0.6262672 -0.6986236    6.8763988  0.6241178 -0.7376569
6       6   -0.7439651  0.6262672 -0.6986236   -1.1776543  0.6241178 -0.6045872
7       7   -0.9793739  0.6262672 -0.6986236   -1.3418666  0.6241178 -0.6160923
```

The code below produces the two models discussed above:

```
bb_rintercept2 <- lmer(bdi ~ bdi_pre + month +(1|subject),

data=bb_long[bb_long$month != 0,], REML = FALSE, na.action= na.omit)

summary(bb_rintercept2)

bb_rintercept_slope2 <- lmer(bdi ~ bdi_pre + month  + (month|subject),

data=bb_long[bb_long$month != 0,], REML = FALSE, na.action= na.omit)

summary(bb_rintercept_slope2)
```

Now we have the models along with the coefficients we need to do some work to extract these values and force them into the format shown in the above *allcoef2* object.

Deepayan Sarkar (Sarker, 2008) has not only developed the package and written the standard book on Lattice, but also provides an excellent online tutorial on graphing regression lines and curves using the package. I have used his approach when developing the previous lattice plot showing the regression line and will now use it again to show the random slope regression lines.

As in the previous model the first stage is to extract the sets of coefficient values, but this time from both the models and place then in separate dataframes which we eventually combine. Focusing on the first model, we use the *coef()* function which unfortunately returns more than we require so to specify that we only want the first item from the list we use [[1]]. We then convert this to a dataframe and place it in a object called *temp1*

```
temp1 <- as.data.frame(coef(bb_rintercept2)[[1]])
```

```
> row.names(temp1) # works returns subject no
 [1] "     1" "     2" "     3" "     4" "     5"
 [6] "     6" "     7" "     8" "     9" "    10"
[11] "    11" "    12" "    13" "    14" "    15"
```

The dataframe *temp1* contains a hidden column which happens to contain the subject id, this can be seen by applying the *row.names()* function to the *temp1* dataframe.

We add this column containing the subject identifier to the *temp1* dataframe giving it the name *subject*.

temp1

(Intercept)
bdi_pre
month
subject

coefficients from random intercept model

```
row.names(temp1) # returns subject no

temp1 <- cbind(temp1, subject = as.numeric(row.names(temp1)))

temp1

str(temp1)
```

```
> str(temp1)
'data.frame':   97 obs. of  4 variables:
 $ (Intercept): num  -11.04 2.48 5.04 3.05 6.67 ...
 $ bdi_pre    : num  0.626 0.626 0.626 0.626 0.626 ...
 $ month      : num  -0.699 -0.699 -0.699 -0.699 -0.699 ...
 $ subject    : num  1 2 3 4 5 6 7 8 9 10 ...
```

The *str()* function provides summary information concerning an object, in the above the structure of the *temp1* dataframe. It shows we have 4 columns, all of which are of the R (num)eric type.

temp2

(Intercept)
bdi_pre
month
subject

coefficients from random slope model

We repeat the above process for the second model

```
temp2 <- as.data.frame(coef(bb_rintercept_slope2)[[1]])

row.names(temp2) # returns subject no

temp2 <- cbind(temp2, subject = as.numeric(row.names(temp2)))

temp2

str(temp2)
```

We now have two temporary dataframes, each providing the coefficients for a particular model. If you wanted to graph more regression lines from other models you would repeat the above process for each model.

We will now combine the two sets of coefficients together using the *merge()* function matching on the subject column and placing the result in another dataframe called *allcoefs*.

A word of warning; this only works if the matching variable, subject here, is the same data type in all the dataframes you wish to combine, hence my checking in the previous stages using the *str()* function.

```
allcoefs <- merge(temp1, temp2, by="subject", all = TRUE)

allcoefs
```

Mainly for aesthetic purposes we will now give the resultant columns some sensible names, firstly inspecting what they are currently called using the *names()* function.

```
> names(allcoefs)
[1] "subject"        "(Intercept).x" "bdi_pre.x"
[4] "month.x"        "(Intercept).y" "bdi_pre.y"
[7] "month.y"
```

```
names(allcoefs)
```

Renaming these columns will be helpful when looking at the results. I use the *names()* function again but this time assign new values to the columns. I have added the prefix *ri* to indicate that they come from the random intercept model and *rs* to indicate that they come from the random slope model.

```
names(allcoefs) <- c("subject", "ri_intercept", "ri_bdi_pre",   "ri_month",

                              "rs_intercept", "rs_bdi_pre",   "rs_month")

str(allcoefs)
```

```
> str(allcoefs) # checking subject is a number in allcoefs
'data.frame':   97 obs. of  7 variables:
 $ subject     : num  1 2 3 4 5 6 7 8 9 10 ...
 $ ri_intercept: num  -11.04 2.48 5.04 3.05 6.67 ...
 $ ri_bdi_pre  : num  0.626 0.626 0.626 0.626 0.626 ...
 $ ri_month    : num  -0.699 -0.699 -0.699 -0.699 -0.699 ...
 $ rs_intercept: num  -11.48 2.07 5.18 3.54 6.88 ...
 $ rs_bdi_pre  : num  0.624 0.624 0.624 0.624 0.624 ...
 $ rs_month    : num  -0.573 -0.598 -0.721 -0.795 -0.738 ...
```

As in the previous example where we developed a lattice xyplot I'm now going to create a column with values of 1 to 100 for the calculated coefficients and null values to match.

```
subjectsall <- data.frame(subject = 1:100)
```

str(subjectsall)

If you have discovered that the subject variable has changed to a factor you can convert it to a numeric variable with the following code (this should not be necessary):

```
allcoefs$subject <- as.numeric(as.character(allcoefs$subject))
```

We now merge the *subjectsall* dataframe with the dataframe containing all the coefficients from the two models, ensuring that those subjects which only had a single bdi measurement are not removed. We place the result in a dataframe called *allcoefs2*.

```
allcoefs2 <- merge(subjectsall, allcoefs, by="subject", all = TRUE)
```

We now have the data in the necessary format for plotting both the subject level regression lines using the lattice *xyplot()* function.

Because we are now plotting two regression lines as well as the raw data for each subject there is more complexity involved but the basic process remains the same: we first plot the raw data as points then plot the first regression line and then the second one.

For the first regression line we use columns two to four of *allcoefs2*, and for the second regression line we use columns five to seven multiplying the suitable coefficients by the appropriate observed x value.

```
xyplot(bdi ~ month | subject, data= bb_long, layout=c(25,4),
  aspect="fill", type = c("p"), pch=4, grid= FALSE, cex=.6,
  par.strip.text=list(cex=0.6),
  coef.list = allcoefs2, # provides raw intercept + slope x2
  panel = function(x,y,..., coef.list, subscripts )
        {
        yadd1 <- y * (coef.list[packet.number(),3]) + (coef.list[packet.number(),2])
        xadd1<- coef.list[packet.number(),4]
        panel.xyplot(x,y,...)
        # abline only works if you specify the first element of x,
        # i.e. the first of the 5 repeated bdi measures for each panel = yadd[1]
        panel.abline(a=yadd1[1], b= xadd1 )
        yadd2 <- y * (coef.list[packet.number(),6]) + (coef.list[packet.number(),5])
        xadd2<- coef.list[packet.number(),7]
        panel.abline(a=yadd2[1], b= xadd2 ) # below adds slope ce
        panel.text(5, 40, paste("r slope", "\n ", round(xadd2[1],3)), cex = .5, font = 1)
        }
)
```

Some comments:

par.strip.text=list(cex=0.6) This refers to the small area above each panel, *cex* is a size value and setting it to 0.6 is in effect saying make the text at the top of each panel 60% of the standard size. On a previous line the xyplot is set 60% of the standard size as well.

panel.text() works the same way as in the previous xyplot but this time I have added some enhancements:

paste() adds together three pieces of text here, the "r slope" text followed by a new line character "*\n* " then the value of the slope parameter for the particular subject which is rounded to 3 decimal places: *round(xadd2[1],3)*.

The size of the text within the panel is also set to be 60% of the standard size. *font=1* indicates plain text (the default), alternatives are *2* for bold face, *3* for italic and *4* for bold italic.

To change the font use the option *family* = one of the following "serif", "sans", "mono", or "symbol". In windows, mono is mapped to "TT Courier New", serif is mapped to"TT Times New Roman", sans is mapped to "TT Arial", mono is mapped to "TT Courier New", and symbol is mapped to "TT Symbol" (TT=True Type) (Kabacoff, 2011 p.54).

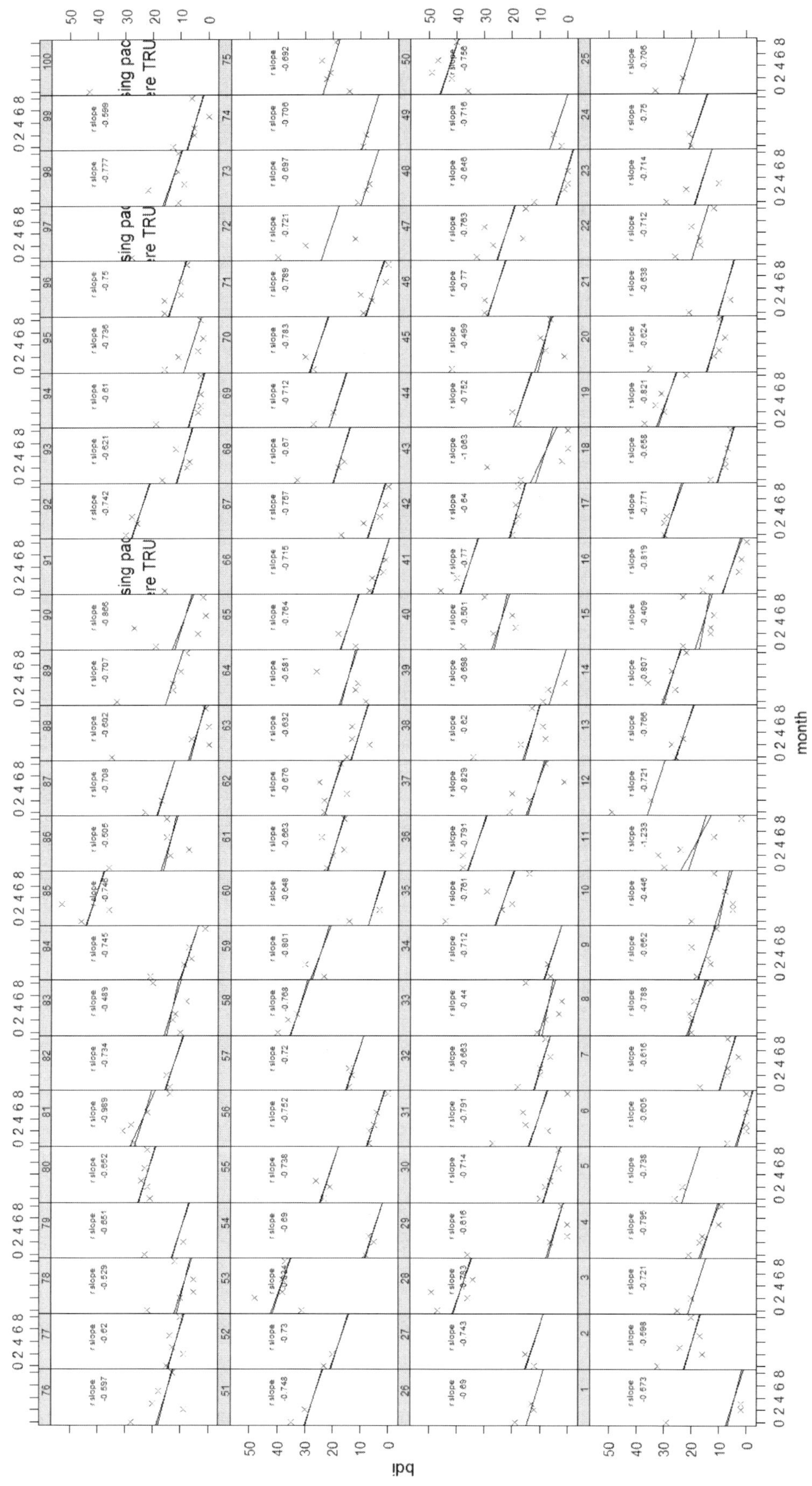
bdi
month

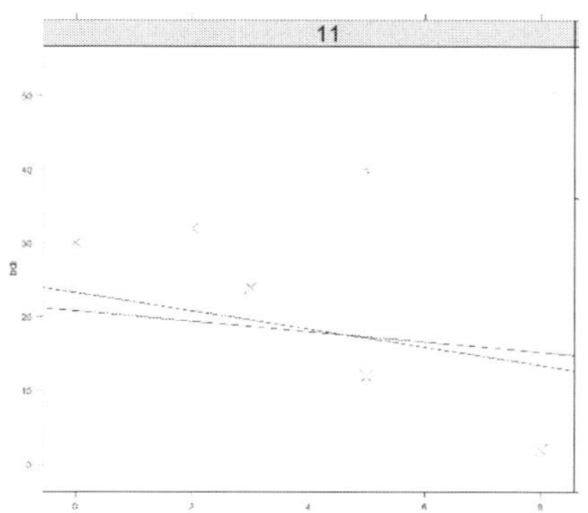

The addition of the random slope makes very little difference to the fit of the line to each panel's data. To highlight this I have enlarged the panel opposite that appears to have the largest discrepancy between the two regression lines. The plot provides a clear visual image of why the addition of the random slope was statistically insignificant.

We could have superimposed all the regression lines for each model on a single chart as we did for the random intercept line but I feel in this instance, where we have very little discrepancy, this is not worthwhile.

54.14 Working correlation / model covariance matrices

In the above examples we have either allowed the random effects to be correlated or forced them to not be (technically called orthogonal). We can also specify how we would like the measures within a unit/cluster to be modelled. In this example a cluster is a set of repeated measures within an individual.

For this chapter we will only consider how to model the relationship between the repeated measures and ignore the possibility of modelling correlation structures for the errors. Correlation structures for errors can be specified in the R *nlme* library and SPSS.

See the summary section to this chapter for details of how these relate to the example we have been working with.

Correlation structures (also called covariance patterns) can take many forms including those which can be defined by the user. Common correlation structures are the identity, where the repeated measures are assumed to not be correlated with each having the same variance, to unstructured where both the correlation and variance for each repeated measure is allowed to vary and be estimated from the data.

The table below lists the most common types along with the commands to specify them in both the R *geepack*, and *nlme* packages. For this chapter we will only consider the *geepack* library for developing GEE models.

Correlation structure name	**Number of parameters**	**Variance at each time point**	**Correlation (covariance) between measurement times**	**R geepack**	R nlme
Scaled identity	0	constant	No correlation (i.e. independent)	Corstr="independence"	NULL
Compound symmetry	1	constant	constant	Corstr="exchangeable"	corCompSymm
Unstructured	Where r = number of repeated measures number of parameters = r(r-1)2	different at each time	Different for each time pairing	Corstr="unstructured"	Default?
AR(1) =Autoregressive	1 = ρ	constant	Correlation gets less as time points get further apart (i.e. $t_1, t_2 = \rho$ but $t_1, t_3 = \rho^2$)	Corstr="ar1"	corAR1
Error (R_i matrix)					Weights=
Diagonal	1	different for each	No correlation (i.e. independent)		

When using the scaled identity correlation matrix you are performing the equivalent to a standard regression analysis without considering the possibility of the repeated measures within an individual being correlated.

We will now investigate how to use these correlation structures with a *GEE* model and our data.

54.15 GEE – Generalised estimating equations

Generalised estimating equations (GEEs) focus on the possible group differences and try to draw our attention away from the intricacies of the repeated measures in assessing the group differences.

GEE models allow us to assess an outcome that has several repeated values ideally across many individuals. For the GEE to work best the number of subjects should be large and the number of repeated measures small. Because GEE models do not contain random effects these models have limited modelling capabilities but at the same time are much simpler to specify in R as you can only have a single grouping variable specified by the *id* option.

To produce a GEE model I will use the *geepack* R package which you install and load with the following commands:

```
install.packages("geepack")

library(geepack)
```

We then can produce several models considering the different working correlation matrices for the repeated measures by specifying the *corstr* option:

```
bb_gee_indeptest <- geeglm(bdi ~
bdi_pre + month + treat + drug +
length,
data=bb_long[bb_long$month != 0,],
id=subject, family=gaussian,
corstr = "independence")
summary(bb_gee_indep)
```

```
>
> bb_gee_indeptest <- geeglm(bdi ~ bdi_pre + month + treat + drug + length,
+ data=bb_long[bb_long$month != 0,], id=subject, family=gaussian,
+ corstr = "independence")
> summary(bb_gee_indeptest)

Call:
geeglm(formula = bdi ~ bdi_pre + month + treat + drug + length,
    family = gaussian, data = bb_long[bb_long$month != 0, ],
    id = subject, corstr = "independence")

 Coefficients:
              Estimate Std.err  Wald Pr(>|W|)
(Intercept)     7.3229  2.1716 11.37  0.00075 ***
bdi_pre         0.5740  0.0887 41.84  9.9e-11 ***
month          -0.9378  0.1801 27.12  1.9e-07 ***
treatBtheB     -3.3225  1.7197  3.73  0.05335 .
drugYes        -3.5687  1.7362  4.22  0.03983 *
length    >6m   1.7107  1.4283  1.43  0.23102
---
Signif. codes:  0 '***' 0.001 '**' 0.01 '*' 0.05 '.' 0.1 ' ' 1

Estimated Scale Parameters:
            Estimate Std.err
(Intercept)     73.6    10.4

Correlation: Structure = independenceNumber of clusters:   97   Maximum cluster size: 4
```

Note that these are the same values as produced by the naive wrong analysis earlier. The default estimates produced by the *geeglm()* function are what are called robust or sandwich estimates.

```
bb_gee_exch <- geeglm(bdi ~ bdi_pre +
month + treat + drug + length,
data=bb_long[bb_long$month != 0,],
id=subject, family=gaussian,
corstr = "exchangable")
summary(bb_gee_exch)
```

```
> summary(bb_gee_exch)

Call:
geeglm(formula = bdi ~ bdi_pre + month + treat + drug + length,
    family = gaussian, data = bb_long[bb_long$month != 0, ],
    id = subject, corstr = "exchangable")

 Coefficients:
              Estimate Std.err  Wald Pr(>|W|)
(Intercept)     5.5415  2.0919  7.02   0.0081 **
bdi_pre         0.6415  0.0798 64.70  8.9e-16 ***
month          -0.6961  0.1542 20.38  6.3e-06 ***
treatBtheB     -2.2907  1.6716  1.88   0.1706
drugYes        -2.8004  1.6604  2.84   0.0917 .
length    >6m   0.1477  1.4937  0.01   0.9212
---
Signif. codes:  0 '***' 0.001 '**' 0.01 '*' 0.05 '.' 0.1 ' ' 1

Estimated Scale Parameters:
            Estimate Std.err
(Intercept)     75.9    10.6

Correlation: Structure = exchangeable ar1 unstructured userdefined fixed  Link = identity

Estimated Correlation Parameters:
      Estimate Std.err
alpha    0.695   0.111
```

The *corstr=exchangable* option fits the repeated measures to a compound symmetry structure where the correlation between each repeated measure is the same as is each variance. Note that the 0.69 is very similar to the value of the intra-class correlation calculated earlier.

```
> summary(bb_gee_ar1)

Call:
geeglm(formula = bdi ~ bdi_pre + month + treat + drug + length,
    family = gaussian, data = bb_long[bb_long$month != 0, ],
    id = subject, corstr = "ar1")

 Coefficients:
             Estimate Std.err  Wald Pr(>|W|)
(Intercept)    6.3446  2.1431  8.76   0.0031 **
bdi_pre        0.5962  0.0813 53.84  2.2e-13 ***
month         -0.7060  0.1611 19.22  1.2e-05 ***
treatBtheB    -2.4779  1.6511  2.25   0.1334
drugYes       -2.3695  1.6321  2.11   0.1466
length   >6m   0.6810  1.5238  0.20   0.6549
---
Signif. codes:  0 '***' 0.001 '**' 0.01 '*' 0.05 '.' 0.1 ' ' 1

Estimated Scale Parameters:
            Estimate Std.err
(Intercept)     75.6    10.6

Correlation: Structure = ar1  Link = identity

Estimated Correlation Parameters:
      Estimate Std.err
alpha    0.801  0.0804
Number of clusters:   97   Maximum cluster size: 4
```

```
bb_gee_ar1 <- geeglm(bdi ~ bdi_pre + month
+ treat + drug + length,
data=bb_long[bb_long$month != 0,],
id=subject, family=gaussian,
corstr = "ar1")
summary(bb_gee_ar1)
```

Here the *corstr=ar1* option fits the repeated measures to a compound symmetry structure where the correlation between each repeated measure gets less as the time between each gets further away. In contrast, the variance for each repeated measure is assumed to be the same.

```
> summary(bb_gee_unstruct)

Call:
geeglm(formula = bdi ~ bdi_pre + month + treat + drug + length,
    family = gaussian, data = bb_long[bb_long$month != 0, ],
    id = subject, corstr = "unstructured")

 Coefficients:
             Estimate Std.err  Wald Pr(>|W|)
(Intercept)    5.8188  2.1027  7.66   0.0057 **
bdi_pre        0.6295  0.0802 61.53  4.3e-15 ***
month         -0.6979  0.1547 20.35  6.4e-06 ***
treatBtheB    -2.3517  1.6578  2.01   0.1560
drugYes       -2.7180  1.6449  2.73   0.0985 .
length   >6m   0.1921  1.4859  0.02   0.8971
---
Signif. codes:  0 '***' 0.001 '**' 0.01 '*' 0.05 '.' 0.1 ' ' 1

Estimated Scale Parameters:
            Estimate Std.err
(Intercept)     75.8    10.6

Correlation: Structure = unstructured  Link = identity

Estimated Correlation Parameters:
          Estimate Std.err
alpha.1:2    0.676   0.112
alpha.1:3    0.702   0.134
alpha.1:4    0.639   0.151
alpha.2:3    0.754   0.147
alpha.2:4    0.638   0.135
alpha.3:4    0.749   0.125
Number of clusters:   97   Maximum cluster size: 4
```

```
bb_gee_unstruct <- geeglm(bdi ~ bdi_pre +
month + treat + drug + length,
data=bb_long[bb_long$month != 0,],
id=subject, family=gaussian,
corstr = "unstructured")
summary(bb_gee_unstruct)
```

The *corstr=unstructured* option fits the repeated measures to an unstructured structure where there are separate estimates for each repeated measure variance and also for each pair of repeated measurements.

The estimated correlations between each of the repeated measures (i.e. the working correlation estimates for the unstructured matrix) is produced at the end of the output shown opposite, you can also extract the values from the *geeglm* object:

```
bb_gee_unstruct$geese$alpha
```

```
> bb_gee_unstruct$geese$alpha
alpha.1:2 alpha.1:3 alpha.1:4 alpha.2:3 alpha.2:4 alpha.3:4
    0.676     0.702     0.639     0.754     0.638     0.749
```

We see that they vary between 0.638 and 0.754

From the above analyses, all the working correlation matrices used (except the inappropriate independence one), resulted in only three significant predictors, the *intercept*, *baseline dbi* and the *month*. We therefore come to the same conclusions as those from the mixed models analysis, but with possibly much less effort.

54.16 Comparing GEE models

The traditional idea about GEE is that the working model correlation matrix is just a side issue and the important aspect is parameter estimation for the various inputs. In addition, because GEE models do not use Maximum likelihood estimation techniques we can't use the techniques discussed previously, specifically when comparing models.

Two situations exist when comparing models: in the first, each model has a different number of fixed parameters; in the second, the only thing that has changed is the specified working correlation structure for the repeated measures.

If you are comparing models with different configurations of fixed parameters then you can use the *Anova()* function in the *car* package.

Where two or more models have the same fixed parameter structure, but different working correlation matrices, you can inspect the differences between traditional standard errors and those produced using the robust method. A small difference indicates a good fit (Everitt & Dunn, 2001 p.244). Alternatively, the *QIC()* function in the *MESS* package can be considered (Hin & Wang, 2009).

The *QIC()* function returns 4 measures of fit; the smaller values indicating a closer fit, so they can be thought of as dissimilarly measures. However, quoting Hin & Wang, 2009, "CIC does not account for penalty in terms of the number of correlation parameters estimated, and therefore direct comparison of correlation structures with different numbers of correlation parameters using CIC may not be advisable".

```
> QIC(bb_gee_indeptest, bb_gee_exch, bb_gee_ar1, bb_gee_unstruct)
                   QIC  QICu Quasi Lik   CIC params  QICC
bb_gee_indeptest 20640 20625    -10306 13.28      6 20640
bb_gee_exch      21261 21261    -10624  5.99      6 21261
bb_gee_ar1       21192 21193    -10591  5.67      6 21193
bb_gee_unstruct  21236 21236    -10612  5.79      6 21237
```

Looking at the results from the various GEE models we have the smallest QIC value for the *ar1* correlation structure. Also referring back to the section showing the correlations for the observed repeated measures we can see that this seems a logical decision, as the correlations appear to get less as we move away from each time point.

```
> summary(bb_gee_ar1)

Call:
geeglm(formula = bdi ~ bdi_pre + month + treat + drug + length,
    family = gaussian, data = bb_long[bb_long$month != 0, ],
    id = subject, corstr = "ar1")

 Coefficients:
             Estimate Std.err  Wald Pr(>|W|)
(Intercept)    6.3446  2.1431  8.76   0.0031 **
bdi_pre        0.5962  0.0813 53.84  2.2e-13 ***
month         -0.7060  0.1611 19.22  1.2e-05 ***
treatBtheB    -2.4779  1.6511  2.25   0.1334
drugYes       -2.3695  1.6321  2.11   0.1466
length   >6m   0.6810  1.5238  0.20   0.6549
---
Signif. codes:  0 '***' 0.001 '**' 0.01 '*' 0.05 '.' 0.1 ' ' 1

Estimated Scale Parameters:
            Estimate Std.err
(Intercept)     75.6    10.6

Correlation: Structure = ar1  Link = identity

Estimated Correlation Parameters:
      Estimate Std.err
alpha    0.801  0.0804
Number of clusters:   97   Maximum cluster size: 4
```

How does the above GEE analysis compare with the Random intercept only model we selected from the mixed models approach? Well they both provide pretty much identical results.

```
> bb_rintercept <- lmer(bdi ~ bdi_pre + month +treat + drug + length + (1|subject),
+ data=bb_long[bb_long$month != 0,], REML = FALSE, na.action= na.omit)
> bb_rintercept
Linear mixed model fit by maximum likelihood
Formula: bdi ~ bdi_pre + month + treat + drug + length + (1 | subject)
   Data: bb_long[bb_long$month != 0, ]
  AIC  BIC logLik deviance REMLdev
 1887 1917 -935.7     1871    1867
Random effects:
 Groups   Name        Variance Std.Dev.
 subject  (Intercept) 48.777   6.9841
 Residual             25.140   5.0140
Number of obs: 280, groups: subject, 97

Fixed effects:
             Estimate Std. Error t value
(Intercept)   5.59244    2.24232   2.494
bdi_pre       0.63967    0.07789   8.213
month        -0.70477    0.14639  -4.814
treatBtheB   -2.32912    1.67026  -1.394
drugYes      -2.82497    1.72674  -1.636
length   >6m  0.19712    1.63823   0.120
```

```
> Anova(bb_rintercept)
Analysis of Deviance Table (Type II

Response: bdi
           Chisq Df Pr(>Chisq)
bdi_pre  67.4508  1  < 2.2e-16 ***
month    23.1776  1  1.477e-06 ***
treat     1.9445  1     0.1632
drug      2.6766  1     0.1018
length    0.0145  1     0.9042
---
Signif. codes:  0 '***' 0.001 '**'
```

The GEE analysis results in p-values which are slightly higher because GEE uses robust standard errors which are therefore also slightly wider than those produced from the mixed model approach.

54.16.1 Standard error estimation in the geepack and gee packages

Quoting the *geeglm()* function help page:

By default, the geeglm() function returns the sandwich estimates. Specifying std.err="fij" the fully iterated jackknife estimate is returned. Additionally, the computationally less demanding approximate jackknife estimate (std.err="jack") or a one-step jackknife estimate (std.err="j1s") can be obtained.

```
> library(gee)
> bb_gee2_indeptest <- gee(bdi ~ bdi_pre + month + treat + drug + length,
+ data=bb_long[bb_long$month != 0,], id=subject, family=gaussian,
+ corstr = "independence")
Beginning Cgee S-function, @(#) geeformula.q 4.13 98/01/27
running glm to get initial regression estimate
 (Intercept)      bdi_pre        month   treatBtheB        drugYes length    >6m
   7.3229026    0.5739634   -0.9378426   -3.3225357   -3.5686627     1.7106746
> summary(bb_gee2_indeptest)

 GEE:  GENERALIZED LINEAR MODELS FOR DEPENDENT DATA
 gee S-function, version 4.13 modified 98/01/27 (1998)

Model:
 Link:                      Identity
 Variance to Mean Relation: Gaussian
 Correlation Structure:     Independent

Call:
gee(formula = bdi ~ bdi_pre + month + treat + drug + length,
    id = subject, data = bb_long[bb_long$month != 0, ], family = gaussian,
    corstr = "independence")

Summary of Residuals:
         Min          1Q       Median           3Q          Max
-23.67794830 -5.41769675   0.01510632   5.32678825  27.26883264
```

So several types of estimation technique are available for the analysis, however *geepack* does not offer a standard non-robust standard error estimate. To obtain them we need to use the older *gee* library.

```
install.packages("gee")
library(gee)
bb_gee2_indeptest <- gee(bdi ~ bdi_pre + month + treat + drug + length,
data=bb_long[bb_long$month != 0,], id=subject, family=gaussian,
corstr = "independence")
summary(bb_gee2_indeptest)
bb_gee2_exch2 <- gee(bdi ~ bdi_pre + month + treat + drug + length,
data=bb_long[bb_long$month != 0,], id=subject, family=gaussian,
corstr = "exchangeable")
summary(bb_gee2_exch2)
```

```
 Correlation Structure:     Independent

Call:
gee(formula = bdi ~ bdi_pre + month + treat + drug + length,
    id = subject, data = bb_long[bb_long$month != 0, ], family = gaussian,
    corstr = "independence")

Summary of Residuals:
         Min          1Q       Median           3Q          Max
-23.67794830 -5.41769675   0.01510632   5.32678825  27.26883264


Coefficients:
               Estimate Naive S.E.   Naive z Robust S.E.  Robust z
(Intercept)   7.3229026 1.72764947  4.238651  2.17156173  3.372183
bdi_pre       0.5739634 0.05497471 10.440500  0.08873535  6.468261
month        -0.9378426 0.23650222 -3.965471  0.18010037 -5.207333
treatBtheB   -3.3225357 1.10068864 -3.018597  1.71966056 -1.932088
drugYes      -3.5686627 1.14716808 -3.110846  1.73618346 -2.055464
length   >6m  1.7106746 1.11056457  1.540365  1.42827169  1.197724

Estimated Scale Parameter:  75.22992
Number of Iterations:  1

Working Correlation
     [,1] [,2] [,3] [,4]
[1,]    1    0    0    0
[2,]    0    1    0    0
[3,]    0    0    1    0
[4,]    0    0    0    1
>
```

We can see that the robust standard errors compared to the traditional standard error are greater for the *intercept*, *bdi_pre*, *treatBtheB*, *drugs* and *length* but less for *month*. Generally robust standard errors compared to the naive ones are:

- reduced for time-dependent predictors (i.e. month)
- increased for time-independent predictors.

```
 Variance to Mean Relation: Gaussian
 Correlation Structure:     Exchangeable

Call:
gee(formula = bdi ~ bdi_pre + month + treat + drug + length,
    id = subject, data = bb_long[bb_long$month != 0, ], family = gaussian,
    corstr = "exchangeable")

Summary of Residuals:
     Min       1Q   Median       3Q      Max
-25.1446  -6.3843  -0.9311   4.2592  24.9868


Coefficients:
             Estimate Naive S.E.  Naive z Robust S.E. Robust z
(Intercept)    5.5458    2.31574  2.39484     2.09214   2.6508
bdi_pre        0.6414    0.08054  7.96316     0.07978   8.0394
month         -0.6968    0.14302 -4.87206     0.15416  -4.5202
treatBtheB    -2.2940    1.73063 -1.32554     1.67164  -1.3723
drugYes       -2.8025    1.78857 -1.56691     1.66049  -1.6878
length   >6m   0.1519    1.69586  0.08957     1.49342   0.1017

Estimated Scale Parameter:  77.54
Number of Iterations:  5

Working Correlation
       [,1]   [,2]   [,3]   [,4]
[1,] 1.0000 0.6918 0.6918 0.6918
[2,] 0.6918 1.0000 0.6918 0.6918
[3,] 0.6918 0.6918 1.0000 0.6918
[4,] 0.6918 0.6918 0.6918 1.0000
```

If we repeat the analysis with the exchangeable working correlation structure we see that the naive and robust standard errors are much closer, indicating that the working correlation structure is a much better fit.

54.17 Summary

In this chapter we have used four R packages to develop the various models,

- *lme4* (Linear mixed effects) was developed by Douglas Bates (Bates, 2006)
- *nlme* (Linear and non linear mixed effects) was developed by José Pinheiro (Pinheiro & Bates, 2000)
- *geepack* was developed by Yan in 2002 (Halekoh, Højsgaard & Yan, 2006)
- *gee* which is now superseded by *geepack*.

Each has its own advantages and disadvantages. *nlme* is the only package that allows the specification of a pattern covariance for the errors, and if necessary a separate one for each random effect. *lme4* can be thought of providing a easier way of performing mixed models but does have limitations, for example you can only imply the correlation pattern from the model you specify as there is no *cor=* or *Weights=* options. Accordingly, we have the following implied pattern covariances in *lme4*:

Covariance pattern	Lme4 Model specification	Estimates produced
Diagonal (no correlation, variances different)	1|group + (0+x1|group) + (0+x2|group)	grouping name variance corr group intercept v group X1 v group X2 v
Unstructured (correlations and variances different)	(1+x1 + x2|group)	grouping name variance corr group intercept v group X1 v v group X2 v v
Partially diagonal	(1+x1|group) + (0+x2|group)	grouping name variance corr group intercept v group X1 v v group X2 v

There is also the possibility of interactions, which I personally find difficult to interpret other than for the first situation below:

result|subject	random slope and intercept for the effect of result for each subject
result|subject:context	random slope and intercept for the effect of result for each subject by context combination
result:context|subject	random slope and intercept for *the interaction effect* of result by context for each subject.

In contrast to many chapters, in this one we have not considered investigating any possible violations in the model such as non-normal error distributions. Bates, 2006 as well as the various online articles concerning *lme4* discuss this in detail.

What I have attempted in the chapter is to provide an example of how to approach the analysis of repeated measures by combining both a thorough graphical investigation before and alongside any model building. These models can become extremely complex and there is a danger of obtaining results that bear little resemblance to the actual data so it is essential that you get to know the data very well using these graphical techniques. The results from any model should be in agreement with your informal graphical analysis.

54.18 Tricks and tips

This is a fast-developing research area with a great deal of information on the web.

Jack Weiss, Adjunct Assistant Professor at the University of North Carolina, runs a course on *gee*, and *lme4* with some excellent examples and R code.

http://www.unc.edu/courses/2010spring/ecol/562/001/docs/lectures/lecture14.htm

UCLA provide a large number of pages about mixed models and R including a very readable introduction to the mathematical theory of mixed models.

http://www.ats.ucla.edu/stat/dae/

The IT department at the University of North Texas runs a short course on mixed models focusing on *SPSS* and *nlme*. The page below contains the material along with an excellent set of links to pdfs.

http://www.unt.edu/rss/class/Jon/SPSS_SC/Module9/M9_LMM/SPSS_M9_LMM.htm

The book, *Applied multivariate data analysis* (Everitt & Dunn, 2001) contains a section entitled *Random effects models for longitudinal data* (chapter 10, Models for multivariate response variables) with an example concerning Alzheimer's disease and cognitive functioning. The dataset if available from:

http://mnstats.morris.umn.edu/multivariatestatistics/data.html

Another excellent book is *Linear Mixed Models: A Practical Guide Using Statistical Software*, 2nd ed. (West, Welch, & Galecki, 2014). The accompanying R package is called, *WWGbook*.

There is an active discussion forum concerning linear mixed models at:

http://glmm.wikidot.com/faq

55 Sample size requirements

Both for grant applications and reporting of findings it is commonplace to provide a rationale for the sample sizes proposed or obtained. In the past the process of calculating the sample size requirements was akin to a mystical experience which usually involved visiting a statistician. However in the past two decades, software has become available to ease the pain, especially for the simpler research designs.

To calculate the required sample size it is necessary to have several pieces of information:

```
Effect size conventions
d = .20 - small
d = .50 - medium
d = .80 - large
```

1. A clear understanding of the proposed research design, considering the following aspects: prospective vs. retrospective, single site vs. multicentre, crossover vs. parallel and clustered vs. non-clustered. Random sampling and allocation is assumed with most sample size calculations.
2. A clearly defined primary outcome measure.
3. An estimate of the effect size, using published results from comparable studies or a pilot study. As a last resort you can use the three levels suggested by Cohen, 1988 (as reported by the *Gpower* software) given opposite for the *d* type effect size measure.
4. A statement giving the level of error you are prepared to accept for falsely coming to the conclusion that you have an effect at the specified effect size (type two error= β [beta]). Often instead of specifying β the value 1- β (called **power**) is reported which is the probability of accepting the specified effect size when it's true. A typical power value is .8 or above indicating that we are prepared to accept that four out of five times (or more) we will make the correct decision that the specified effect size exists in the population.
5. A statement giving the level of error you are prepared to accept for falsely coming to the conclusion that the trial demonstrated zero effect (type one error = α [alpha]). A typical value is 0.05 (this value, which we set, is called the critical value) or less indicating that we are prepared to incorrectly come to the conclusion that there is no effect, when actually one exists at the specified level.

We can think of the above a being two sets of possible values of a statistic from a trial. One set of values for when there is no effect and one for the specified effect size. Our single result (i.e. *t* value or other statistical value) will be in ONE of these distributions but we don't know which. This is shown diagrammatically for the two sample *t* text, produced from the *Gpower* software.

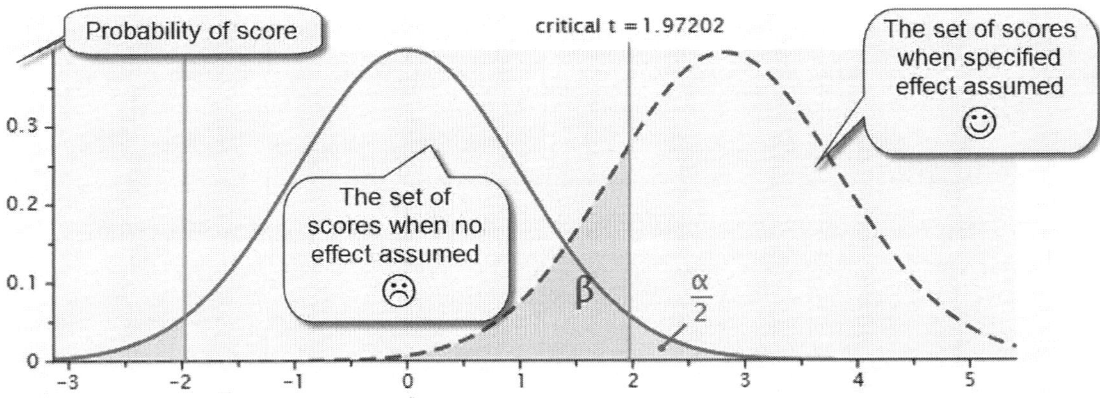

> **Key point:**
>
> One of the fundamental aims of research design is to produce a plan which ensures a high probability of detecting an effect at a specific level if one exists (power), and at the same time minimising the probability of falsely detecting an effect when one does not exist (type one error).

An example should help to highlight some of the above issues which I will demonstrate by using the excellent free standalone application called *Gpower* (Erdfelder, Buchner & Lang, 2009) available to download from http://www.psycho.uni-duesseldorf.de/abteilungen/aap/gpower3/

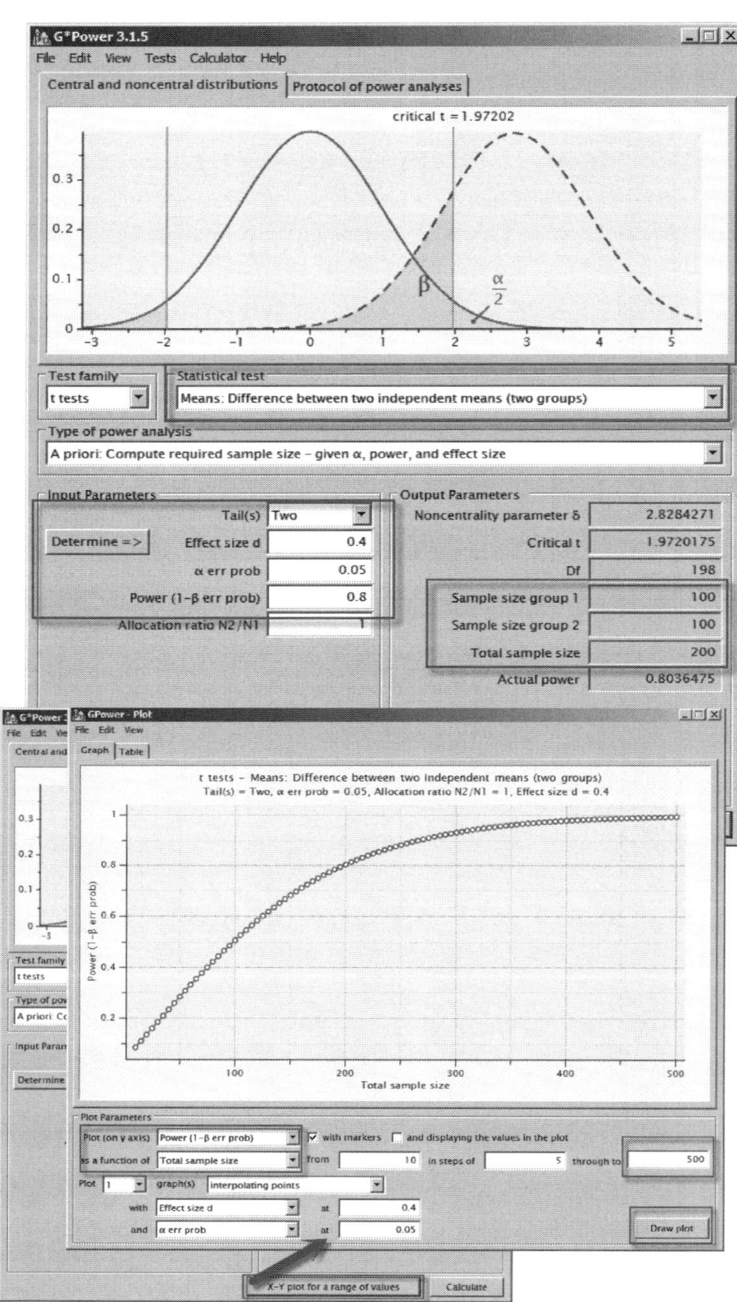

A researcher is proposing a two independent groups design, and she is planning on using a 2 tailed test (*t* test). She believes from previous research that an effect size of d (Cohen's d) =0.4 is expected (calculated from $\mu_1=1$; $\mu_2=2$; sd=2.5), setting alpha=0.05, and power = 0.8. Using *Gpower* and setting these options produced the plot shown opposite of possible t values for the two situations.

We can see in the right hand window opposite, labelled *Sample size group 1*, that she we will need 100 participants to provide data **in each group** to attain the required power.

It is important to note here that we are talking about participants that provide data and not ones that are initially enrolled in the trial. Again context and a literature review will inform the expected attrition rate which invariably mean that she will need to increase the number further.

She also inspects a plot showing power (y axis) against total sample size. From the above plot we can see that it would require another 50 in each group to gain a power level of .85. However reducing the number in each group by the same amount would mean a power of just .5, indicating that she would then have the same chance of gaining a significant result - given that the population had the specified effect size as gaining heads when tossing a coin!

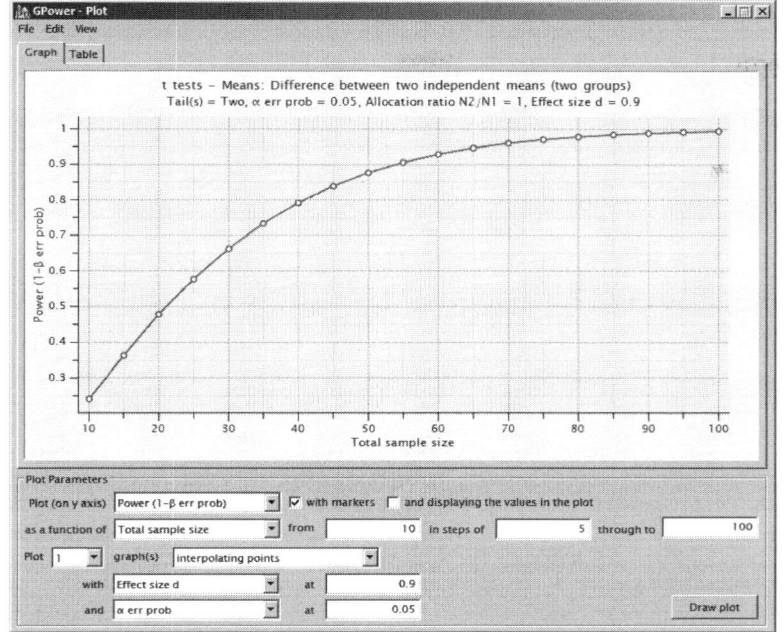

She then carries out a pilot trial and realises that the previous published research had grossly underestimated the effect size measure. So she now reruns the above analysis with a proposed effect size measure of d=0.9 and with her updated value produces a revised graph shown opposite.

In contrast, she now only needs a total sample size of 40 to get the required power of 0.8.

She then discovers that the patients she has used for the pilot were not typical but had an unusually sensitive effect to the drug. Consequently she reduces the effect size down to 0.4 again. However, a colleague of hers suggests that she uses a paired design, so she decides to setup a Matched pairs cohort study and proposes to use the paired sample *t* test to analyse the data. She runs a *Gpower* analysis using this new value and changes the statistical test to a paired *t* test, the results of which are shown opposite and below.

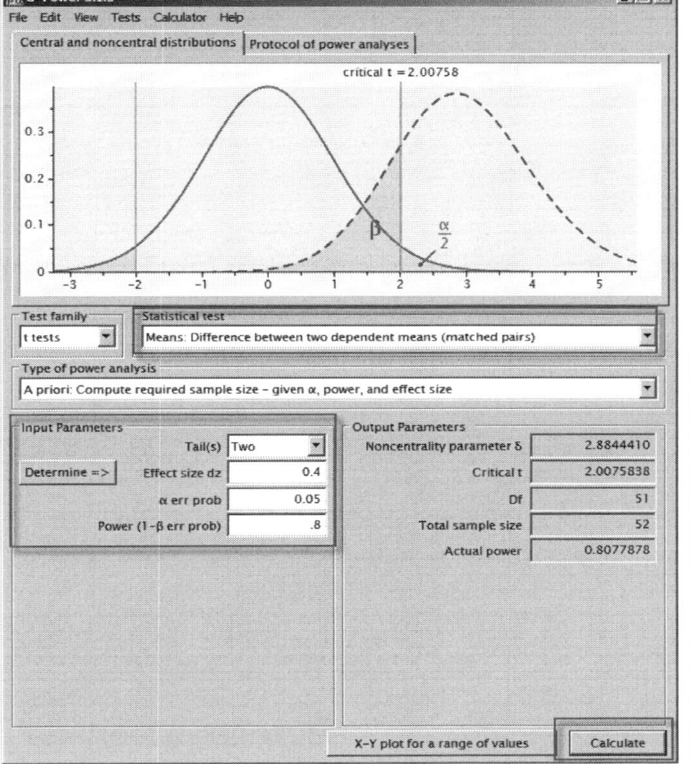

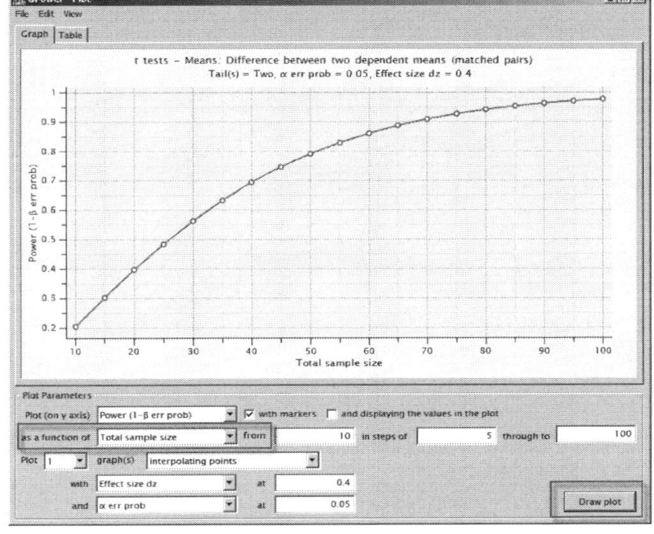

So she now only needs 52 subjects each providing data (two observations) to gain the required power of 0.8.

The above example shows clearly how context, experience and guesswork (read here expert opinion) affect the required sample size, and this is true whatever design you adopt.

GPower provides required sample size calculations for most of the simple designs, however for more complex designs it is necessary to consult a statistician.

Certain factors help reduce the required sample size substantially and should always be borne in mind:

- If the outcome measure can be measured in several different ways use the most accurate and reliable method.
- Use repeated measures designs and consider increasing the number of measures for each subject as this tends to reduce sample size.
- Crossover and cluster trials usually need expert statistical advice. Starting points are Machin & Campbell, 2005 for crossover trials. For cluster trials both Hayes & Moulton, 2009 or West, Welch & Galecki, 2014 provide good starting points.

55.1 Tips and Tricks

There are many sample size calculators available for specific studies such as the excellent sample size calculator for randomised cluster trials developed at Aberdeen University available from:
http://www.abdn.ac.uk/hsru/research/delivery/behaviour/methodological-research/

Commercial packages are also available such as *PASS* ($250 academic license, with a 7 day free trial) and *StudySize* ($125 license, with a 14 day free trial).

In several of the above screen dumps you will notice a term called the noncentrality parameter (δ= delta or capital delta =Δ), this value indicates how far away the distribution with the effect size specified is from the no effect (i.e. null) distribution. More detail is provided in the following chapter *Confidence intervals for effect sizes: Noncentral distributions*.

56 Confidence intervals for effect sizes: noncentral distributions

Much of the focus in this book has been on the null distribution which often prompts people to ask why concentrate on the distribution of no effect or zero correlation (☹). Surely it would be more favourable to have a more optimistic viewpoint and focus on the distribution of the observed effect or correlation etc. where this is modelled by the noncentral distribution (☺). The reasons for this pessimistic approach are many; ranging from the purely statistical to the psychological and philosophical. I tend to think of the two distributions as being the opposite sides of a coin - the optimistic versus the pessimistic viewpoint.

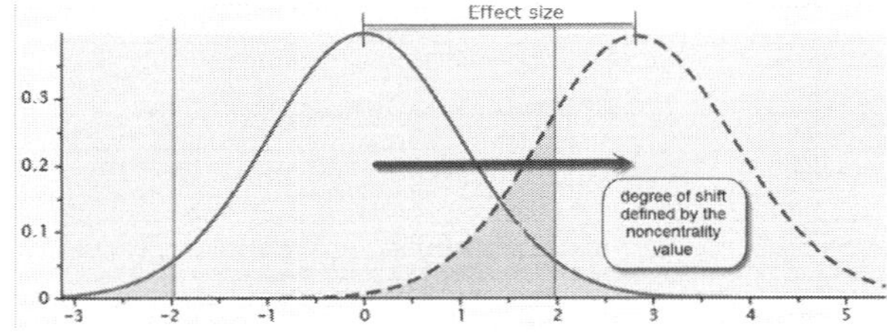

The noncentral distribution, defined by the noncentrality parameter, was introduced in the previous chapter concerning sample size calculations and this chapter provides a more mathematical explanation of these concepts. Here we will consider, in turn, some of the most important statistical distributions, technically called Probability Density Functions (PDF), the *t*, the F, *r* and the chi-squared, by seeing what happens to them when the value no longer centres on zero or some other default value.

56.1 The noncentral *t* distribution

Previously we did not mention what the specific difference was between the null *t* distribution and the *t* distribution of the specified effect size - called the noncentral *t* distribution. For illustration purposes take the first example in the sample size chapter, reproduced above, where we had shifted the distribution to the right. You can see this from the diagram; as the distribution moves, the shape subtly changes. My YouTube video described in the *Tricks and tips* section demonstrates this for a variety of curves.

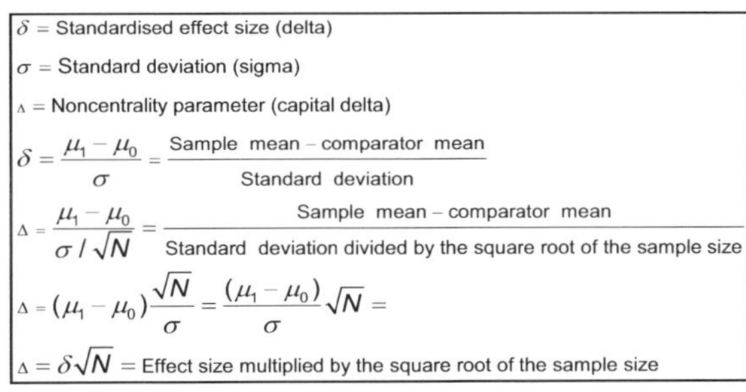

δ = Standardised effect size (delta)

σ = Standard deviation (sigma)

Δ = Noncentrality parameter (capital delta)

$$\delta = \frac{\mu_1 - \mu_0}{\sigma} = \frac{\text{Sample mean} - \text{comparator mean}}{\text{Standard deviation}}$$

$$\Delta = \frac{\mu_1 - \mu_0}{\sigma / \sqrt{N}} = \frac{\text{Sample mean} - \text{comparator mean}}{\text{Standard deviation divided by the square root of the sample size}}$$

$$\Delta = (\mu_1 - \mu_0)\frac{\sqrt{N}}{\sigma} = \frac{(\mu_1 - \mu_0)}{\sigma}\sqrt{N} =$$

$\Delta = \delta\sqrt{N}$ = Effect size multiplied by the square root of the sample size

As with all PDFs we can measure the degree of shift by the appropriate effect size measure (which we have discussed in the previous chapters) or by specifying a related value known as the noncentrality parameter (sometimes both are the same value). Taking the *t* distribution as an example, we can show that the *noncentrality parameter (ncp)* for the *t* distribution is equal to the standardised effect size measure (Cohen's *d*) multiplied by the square root of the sample size (√N), or a slightly more complex value, the simpler situation being shown in the box opposite.

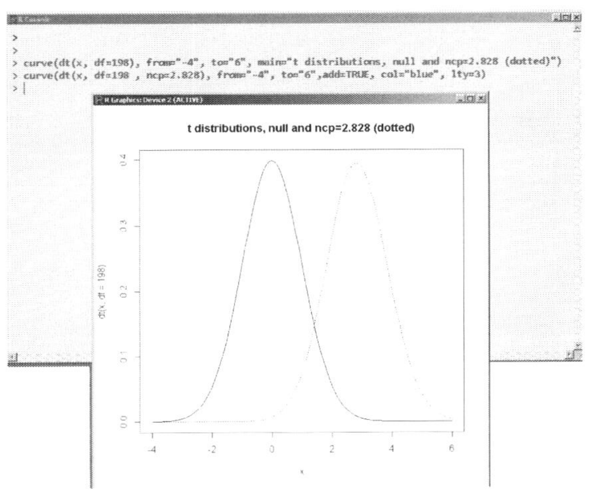

We can also reproduce the two curves obtained in *GPower* (shown above) in R by making use of two specific R functions; *dt()* and *curve()* along with the *ncp* and degrees of freedom (*df*) values.

draw the curve of this function

t pdf for df=198

x axis range from -4 to 6

```
curve(dt(x, df=198), from="-4", to="6",
main="t distributions, null and ncp=2.828 (dotted)")
```

give the plot a title

another *t* density function for df=198 but this time the ncp=2.828

```
curve(dt(x, df=198 , ncp=2.828),
from="-4", to="6",add=TRUE, col="blue", lty=3)
```

add a new plot on top of the old one

change the line colour and make it dotted. See the glossary entries line type and line colour for details.

R function	Name	Takes	Returns	Details
dt()	*t* Density function	*t* value and degrees of freedom (df) +ncp	Density (y value at given *t* value)	Useful for creating PDF plots
pt(t)	*t* cumulative probability function	*t* value and degrees of freedom (df) +ncp	Area under t curve to left of *t* value	Useful for returning a p- value for a *t* value using p_value<-2*pt(negative *t* value, df)
qt()	*t* inverse cumulative probability distribution = *t* quantile function	Area under t curve to left of *t* value +ncp	Returns t value	Useful for discovering t critical values at a specific CV value
rt()	*t* random variable function	Sample size, degrees of freedom (df)) +ncp	A random variable from the specified distribution	Useful for creating simulated datsets

The above R code may look rather strange with an *x* which has not been defined anywhere, however this is a special value that the R function *curve()* uses, in effect setting it to a range of values to draw the curve.

R provides four related functions for the *t* PDF, listed opposite.

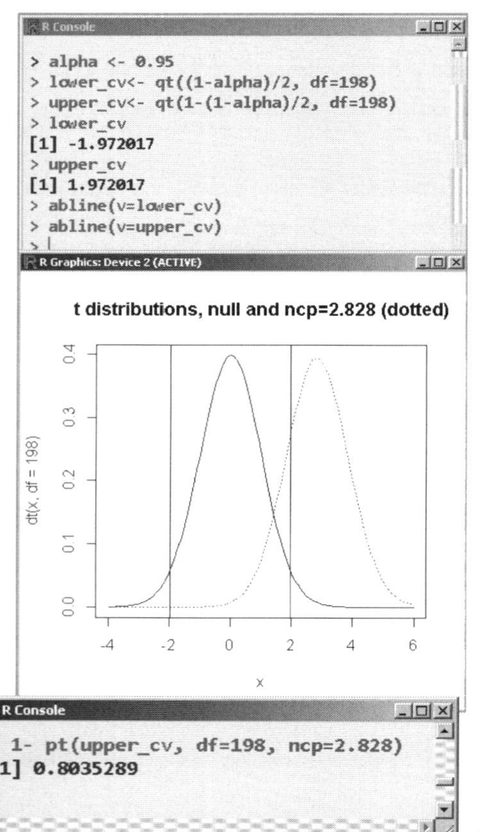

You can also use the *qt()* function to find the *t* critical values at a specific alpha level and then use them to draw the lines on the plot as shown opposite.

```
alpha <- 0.95

lower_cv<- qt((1-alpha)/2, df=198)

upper_cv<- qt(1-(1-alpha)/2, df=198)

lower_cv          negative t value

upper_cv          positive t value

abline(v=lower_cv)

abline(v=upper_cv)
```

We can use the *pt()* function to return the power given a specific *t* value, *df*, and *ncp*. The value returned by the *pt()* function is the area to the left of the value, but we want the area to the right. By making use of the fact that the total area is known (i.e. 1) we can subtract this value from the observed value to obtain the value we want.

```
1 -pt(upper_cv, df=198, ncp=2.828)
```

You can also add the shading to the plot indicating the area of the noncentral distribution that equates to power. To do this I have used three clever lines of code written by Fernando Hosa at http://tinyurl.com/m8r3qwa.

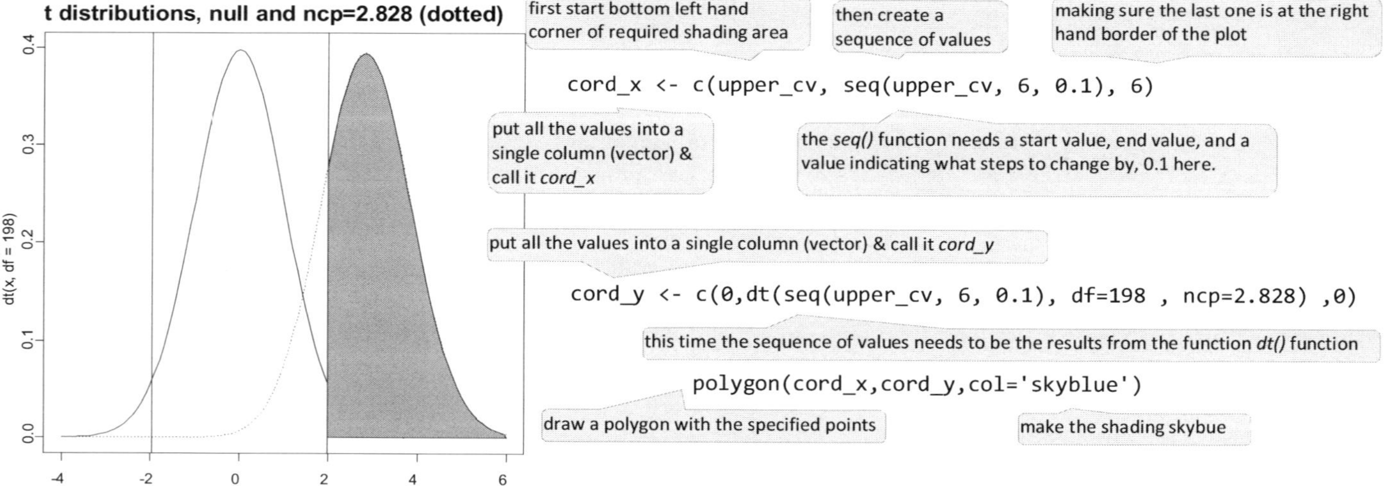

t distributions, null and ncp=2.828 (dotted)
dt(x, df = 198)
x
first start bottom left hand corner of required shading area
then create a sequence of values
making sure the last one is at the right hand border of the plot
put all the values into a single column (vector) & call it cord_x
the seq() function needs a start value, end value, and a value indicating what steps to change by, 0.1 here.
put all the values into a single column (vector) & call it cord_y
this time the sequence of values needs to be the results from the function dt() function
draw a polygon with the specified points
make the shading skybue


```
cord_x <- c(upper_cv, seq(upper_cv, 6, 0.1), 6)

cord_y <- c(0,dt(seq(upper_cv, 6, 0.1), df=198 , ncp=2.828) ,0)

polygon(cord_x,cord_y,col='skyblue')
```

You may have realised that I should have also included the miniscule area in the noncentral *t* distribution to the left of the *t* value (-1.972) as well. We can use the same approach to calculate it, `pt(lower_cv, df=198, ncp=2.828)` providing a value of .0000000901 (see the glossary entry **E notation**). Adding this truly miniscule amount to the .8035 we already have from the right hand tail is nothing to worry about in this situation. However sometimes the left hand tail may provide a larger proportion to the overall power value so it is worthwhile to remember that the power is defined by both tail areas. Clearly we could have combined both values in a single line of code:

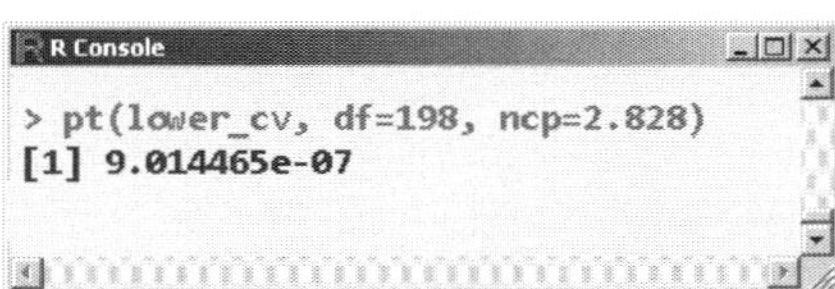


```
power <- 1 -pt(upper_cv, df=198, ncp=2.828) + pt(lower_cv, df=198, ncp=2.828)

power
```

56.1.1 Obtaining power values from R's build-in function

There is also a function in R *power.t.test()* which you can use. Because it calculates the *ncp* within the function, all you need to provide is the actual difference in mean values. Unfortunately, here this value is also called delta, but it is different from the one described on the previous page. You also need to specify the standard deviation of the entire sample - regardless of group. To obtain the power, including both the left and right tails, you need to specify *strict=TRUE*. From the example discussed above the standardised effect value d=0.4 was calculated from $\mu_1=1$; $\mu_2=2$ and sd=2.5 giving a delta value here of 2-1 = 1.

can also be one.sample or paired

required power

required type one error level

```
power.t.test(type="two.sample", power=0.8, sd = 2.5, sig.level = 0.05, delta= 1, strict=TRUE)
```

difference between the two means

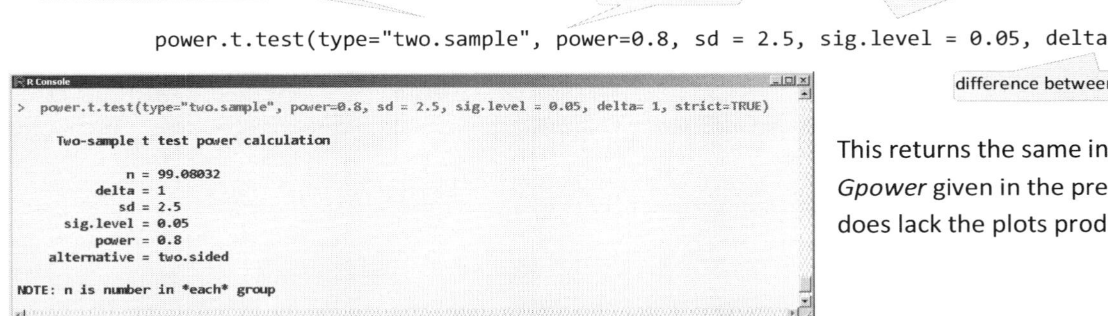

R Console
> power.t.test(type="two.sample", power=0.8, sd = 2.5, sig.level = 0.05, delta= 1, strict=TRUE)
Two-sample t test power calculation
n = 99.08032
delta = 1
sd = 2.5
sig.level = 0.05
power = 0.8
alternative = two.sided
NOTE: n is number in *each* group

This returns the same information as that from *Gpower* given in the previous chapter, but does lack the plots produced with that output.

56.1.2 Confidence intervals for the effect size measure

From the above you should realise now that effect size, noncentrality parameters and power are all tightly bound together. Because the noncentral *t* distribution is not symmetrical there is a problem calculating a confidence interval for it. Luckily R comes to the rescue with a package called *MBESS* (Methods that are especially applicable to researchers working within the Behavioral, Educational, and Social sciences).

Taking the two independent groups example above, we have a Cohen's *d* value of 0.4, along with one hundred participants in each group. To obtain a 95% confidence interval for this effect size (Cohen's *d*):

```
install.packages("MBESS", dependencies=TRUE)
```

install and load the MBESS package

```
library(MBESS)
```

provide a confidence interval for the standardised mean difference (i.e. Cohen's d)

```
ci.smd(smd=.4, n.1=100, n.2=100, conf.level=.95)
```

Cohen's d

size in each group

desired confidence level

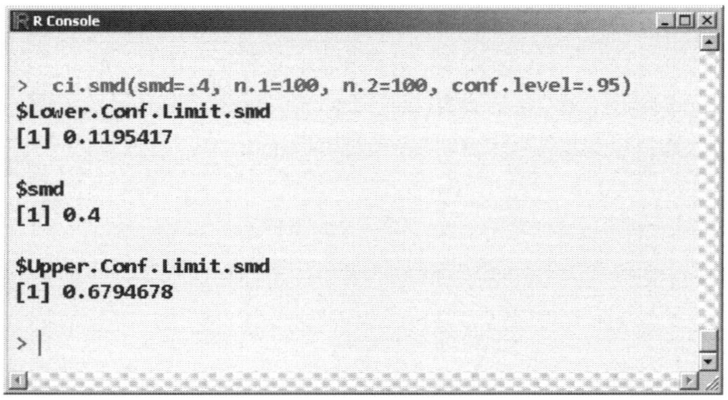

```
R Console
>  ci.smd(smd=.4, n.1=100, n.2=100, conf.level=.95)
$Lower.Conf.Limit.smd
[1] 0.1195417

$smd
[1] 0.4

$Upper.Conf.Limit.smd
[1] 0.6794678

>|
```

This gives us a 95% confidence interval for Cohen's d of 0.1195 to 0.6794.

You can also obtain Cohen's *d* by specifying the noncentrality parameter (called lambda in the *MBESS* package):

```
lambda2delta(lambda=2.8284, n.1=100, n.2=100)
```

Also *MBESS* provides a function, *delta2lamdba()*, which takes Cohen's d and returns the ncp value.

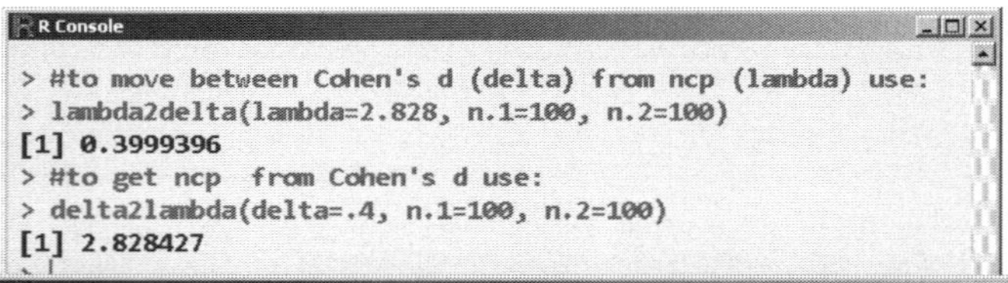

```
R Console
> #to move between Cohen's d (delta) from ncp (lambda) use:
> lambda2delta(lambda=2.828, n.1=100, n.2=100)
[1] 0.3999396
> #to get ncp  from Cohen's d use:
> delta2lambda(delta=.4, n.1=100, n.2=100)
[1] 2.828427
```

```
R Console
> ci.sm(Mean= 4.5, SD=7.24, N=15, conf.level=0.95)
[1] "The 0.95 confidence limits for the standardized mean are given as:"
$Lower.Conf.Limit.Standardized.Mean
[1] 0.05731527

$Standardized.Mean
[1] 0.621547

$Upper.Conf.Limit.Standardized.Mean
[1] 1.167475
```

We can also use the *MBESS* package to obtain a confidence interval for the standardised mean, in contrast to the *mean difference* above. Referring back to the example presented in the single sample *t* test chapter, we had an observed mean of 95.53 with a population mean of 100, giving a difference of 4.5, sd=7.24, sample size of 15 (df=14). This gave a Cohen's *d* value of 4.5/7.24 = .62 Using the same *MBESS* *ci.sm()* function, but this time specifying the *mean* and *sd* values:

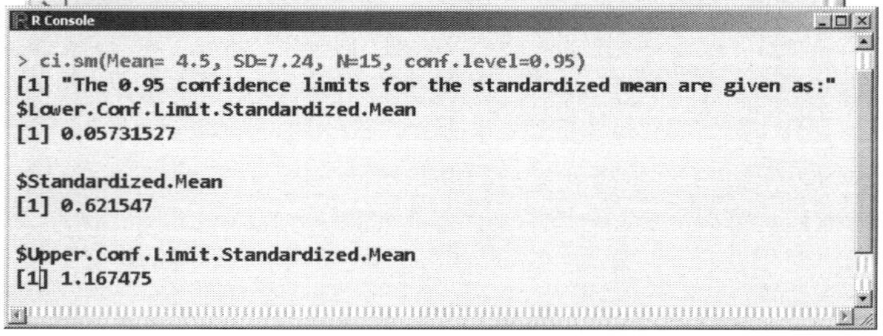

```
library(MBESS)
 ci.sm(Mean= 4.5, SD=7.24, N=15, conf.level=0.95)
```

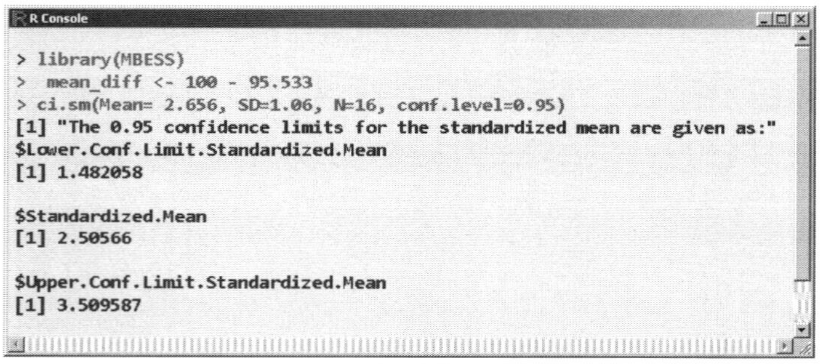

```
R Console
> library(MBESS)
>  mean_diff <- 100 - 95.533
> ci.sm(Mean= 2.656, SD=1.06, N=16, conf.level=0.95)
[1] "The 0.95 confidence limits for the standardized mean are given as:"
$Lower.Conf.Limit.Standardized.Mean
[1] 1.482058

$Standardized.Mean
[1] 2.50566

$Upper.Conf.Limit.Standardized.Mean
[1] 3.509587
```

Finally considering the paired *t* test, using the example which introduced it we had a mean difference of 2.656, sd=1.06 with 16 subjects providing pre-post values (df=15) and a *t* value of -2.65.

```
library(MBESS)
 mean_diff <- 100 - 95.533

ci.sm(Mean= 2.656, SD=1.06, N=16, conf.level=0.95)
```

Besides working out a confidence interval for the effect size it is also possible to calculate one for the noncentrality parameter value.

56.1.3 Confidence intervals for the *t* distribution noncentrality parameter

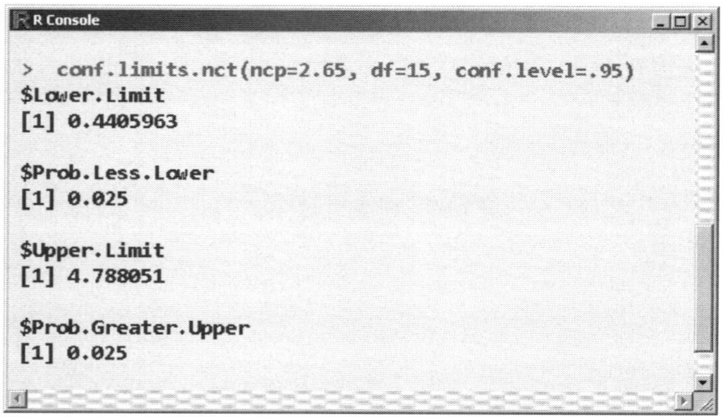

```
R Console
>  conf.limits.nct(ncp=2.65, df=15, conf.level=.95)
$Lower.Limit
[1] 0.4405963

$Prob.Less.Lower
[1] 0.025

$Upper.Limit
[1] 4.788051

$Prob.Greater.Upper
[1] 0.025
```

To demonstrate the construction of a confidence interval for the *t* distribution noncentrality parameter we will use the last example above where we have an absolute *t* value of 2.65 and df=15. For a 95% confidence interval for the actual ncp we use the following *MBESS* command:

```
library(MBESS)

conf.limits.nct(ncp=2.65, df=15, conf.level=.95)
```

56.2 The noncentral *F* distribution

group	mean	sd	n
CBT	3.0	7.3	29
con	-.4	7.9	26
FT	7.1	7.1	17

To demonstrate this distribution I will use the anorexia dataset described in the *Introductory tutorial* chapter, where we obtained an F value of 5.422 df(2, 69) with an associated p-value of 0.0065. The group means and standard deviations are shown opposite.

k = number of groups

N = Total number; n_i = number in group i

$i = 1,\ldots,k$ =specific group

σ = standard deviation (sigma)

f = effect size

$f = \frac{\sigma_\mu}{\sigma}$ = Standard deviation of standardised means

$\sigma_\mu^2 = \sigma^2 - \sigma_{total}^2$ or you can use:

$$\sigma_\mu = \sqrt{\sum_{i=1}^{k} \frac{n_i}{N} (\mu_i - \mu)^2}$$

λ = noncentrality parameter (lamda)

$\lambda = f^2 N$ = Effect size squared multiplied by total sample size

$df_1 = K - 1$

$df_2 = N - k$

In this instance the effect size suggested by Cohen, 1988, is a value called f and is calculated from the standard deviation of the effect divided by the pooled standard deviation.

We can find the pooled standard deviation using the R Commander, menu option **Statistics->summaries->numerical summaries**:

```
+   maxwidth=80, maxheight=30)

> summary(anorexia)
  Treat       Prewt            Postwt          weightgain
 CBT :29   Min.   :70.00   Min.   : 71.30   Min.   :-12.200
 Cont:26   1st Qu.:79.60   1st Qu.: 79.33   1st Qu.: -2.225
 FT  :17   Median :82.30   Median : 84.05   Median :  1.650
           Mean   :82.41   Mean   : 85.17   Mean   :  2.764
           3rd Qu.:86.00   3rd Qu.: 91.55   3rd Qu.:  9.100
           Max.   :94.90   Max.   :103.60   Max.   : 21.500

> library(abind, pos=4)

> library(e1071, pos=4)

> numSummary(anorexia[,"weightgain"], statistics=c("mean", "sd", "quantiles"),
+    quantiles=c(0,.25,.5,.75,1))
    mean       sd    0%    25%  50% 75% 100%  n
 2.763889 7.983598 -12.2 -2.225 1.65 9.1 21.5 72
```

```
RGui (64-bit) - [R Console]
File Edit View Misc Packages Windows Help
>  sdsquared_effect<- (29/72*(3 - 2.763)^2) + (26/72*(-.4 - 2.763)^2) + (17/72*(7.2  - 2.763)^2)
>   sdsquared_effect
[1] 8.283697
>   #needs its square root
>   sd_effect<- sqrt (sdsquared_effect)
>   sd_effect
[1] 2.878141
> |
```

You can then convert the expression σ_μ into an R expression, shown opposite, to calculate the standard deviation of the effect.

R can also be used to calculate the effect size measure *f* along with the noncentrality parameter.

```
RGui (64-bit) - [R Console]
File Edit View Misc Packages Windows Help
> # know sd for whole group is 7.9835
>     variance_within <-7.9835^2 -  sdsquared_effect
>   variance_within
[1] 55.45258
>   f_effect_size <-  sd_effect/sqrt( variance_within)
>    f_effect_size # gpower gives 0.388938
[1] 0.3865015
>   # ncp
>   cnp_mine <-  f_effect_size^2 * 72
>   cnp_mine
```

As you can see from the above, it is rather a long process to get to the noncentrality parameter and one short cut is to make use of a value produced by many ANOVA results, eta (η^2) squared. This value can be used to easily obtain the effect size measure *f* as shown opposite. In fact according to Cohen (1988, p.284)

```
RGui (64-bit) - [R Console]
File Edit View Misc Packages Windows Help
>
>  eta_square<- 615/(615+3911)
>   eta_square
[1] 0.1358816
>   f_effect_size <- sqrt (eta_square/(1-eta_square))
>    f_effect_size
[1] 0.3965461
>
```

η^2 = eta squared = $\frac{SS_{between}}{SS_{total}} = \frac{SS_{treatment}}{SS_{total}} = \frac{SSB}{SST}$

f = effect size

$f = \sqrt{\frac{\eta^2}{1-\eta^2}}$; $\eta^2 = \frac{f^2}{1+f^2}$

"*Effect sizes in behavioural science are generally small, and, in terms of f, will generally be found in the 0.00 – 0.40 range. With f small, f^2 is smaller, and $1+f^2$, the denominator of eta squared (η^2) is only slightly greater than one. The result is that for small values of f such as are typically encountered, η is approximately equal to f, being only slightly smaller and therefore η^2 is similarly only slightly smaller than f^2. . . . For very large effect sizes, say f >0.40, f and η diverge too much for this rough and ready approximation, and f^2 and η^2 even more so.*"

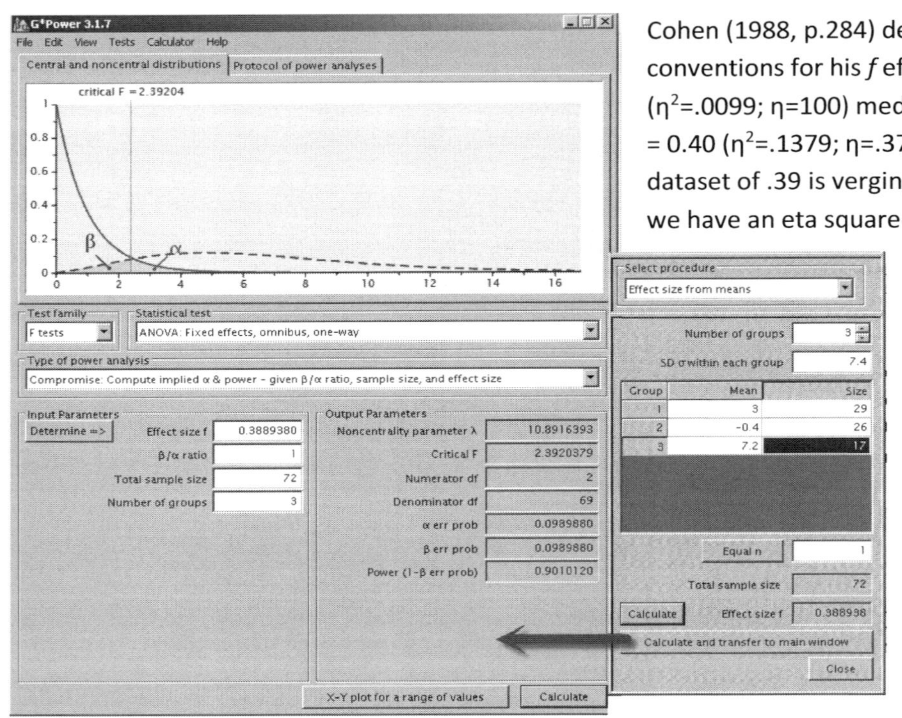

Cohen (1988, p.284) defined the following effect size conventions for his *f* effect size measure: small $f^2 = 0.10$ (η^2=.0099; η=100) medium $f = 0.25$ (η^2=.0588; η=.243) , large $f = 0.40$ (η^2=.1379; η=.371). So the value obtained from our dataset of .39 is verging on the large effect size. Also note that we have an eta squared value of .135, when considering its square root of 0.3674235 is similar to .39.

You can avoid all the above calculations by using *Gpower*, with the setup shown opposite. The graph produced is of both the null and noncentral *F* PDF with a ncp=10.89, very similar to our R computed value of 10.75. *Gpower* also reports a Cohen *f* effect size of .388, similar to our value of .39 in R.

We can produce a similar graph using the R curve commands described earlier:

```
curve(df(x, df1=2, df2=69), from="0", to="17",
main="F distributions, null and ncp=10.755 (dotted)")

curve(df(x, df1=2, df2=69, ncp=10.755),
from="0", to="17",add=TRUE, col="blue", lty=3)
```

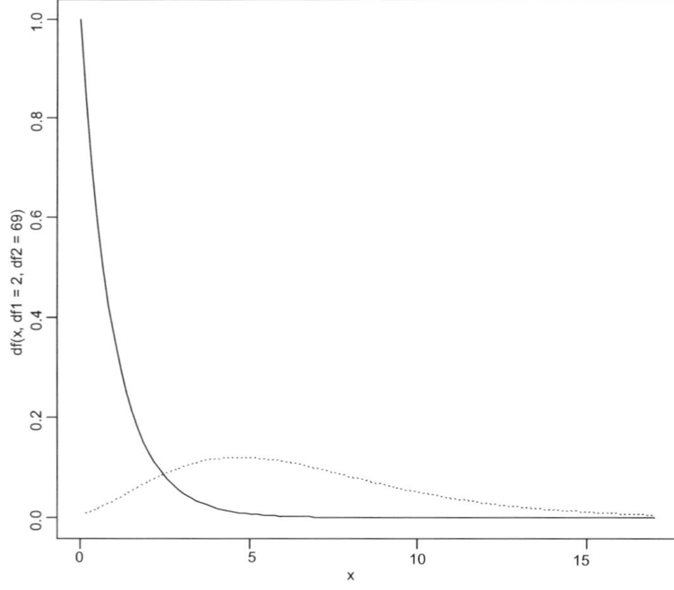

This time, because we want the density values for the *F* distribution rather than the *t*, we have replaced the *dt()* command with the *df()* one.

We could also use the same techniques as described previously to shade the area to indicate power. That is the probability of rejecting the null hypothesis when the specific alternative hypothesis is true.

56.2.1 Confidence interval for the *f* effect size

Obtaining a confidence interval for the *f* effect size measure is easy using the *MBESS* package, where the equivalent value is called the Square Root of the Signal-To-Noise Ratio. Taking the results from the ANOVA (given above) and using the *ci.srsnr()* function produces a confidence interval of 0.1127 to 0.6126:

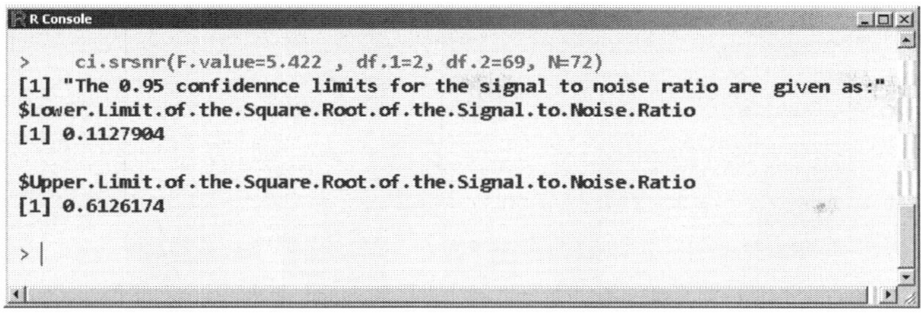

```
R Console
>    ci.srsnr(F.value=5.422 , df.1=2, df.2=69, N=72)
[1] "The 0.95 confidennce limits for the signal to noise ratio are given as:"
$Lower.Limit.of.the.Square.Root.of.the.Signal.to.Noise.Ratio
[1] 0.1127904

$Upper.Limit.of.the.Square.Root.of.the.Signal.to.Noise.Ratio
[1] 0.6126174

> |
```

```
library(MBESS)

ci.srsnr(F.value=5.422 , df.1=2, df.2=69,
N=72)
```

56.2.2 Confidence interval for the *F* distribution noncentrality parameter

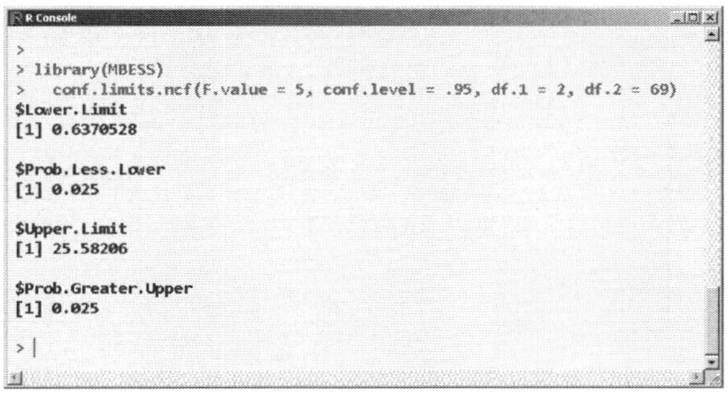

```
R Console
>
> library(MBESS)
>    conf.limits.ncf(F.value = 5, conf.level = .95, df.1 = 2, df.2 = 69)
$Lower.Limit
[1] 0.6370528

$Prob.Less.Lower
[1] 0.025

$Upper.Limit
[1] 25.58206

$Prob.Greater.Upper
[1] 0.025

> |
```

To obtain a confidence interval for the *F* distribution noncentrality parameter using *MBESS*, means you need to provide the *F* value, degrees of freedom values and the required confidence level to the *conf.limits.ncf()* function:

```
library(MBESS)

conf.limits.ncf(F.value = 5, conf.level = .95,
                df.1 = 2, df.2 = 69)
```

56.3 The Correlation = (r) distribution when rho ≠ 0

This was introduced in the chapter *Measuring the degree to which two variables co-vary*. The code used to produce the graph in the chapter is given below. The *r* distribution is not part of R's standard installation and you need to install the *SuppDists* library to access it.

In the code below I have used a loop to draw each curve with the *add=TRUE* option to stop it deleting the previous curve. I have also added a legend.

```
install.packages("SuppDists")
require(SuppDists)
n <- 20 # set the sample size to 20
i <- 0 # set the loop counter to zero
mylinetype <- c( 1,2, 3, 4, 5)
param_vals <- c(.6, .7, .8, .9, .95)
mylinethinkness <- c( 1,2, 3, 2, 1)
yrange<- c(0,22)
xrange<- c( -1, 1)
plot(xrange,yrange, type="n", xlab = "Pearson's r", ylab = "Density", lty=0,
main = paste("Showing change in distribution for different rho values", "n=" , n))
for (rho_val in c(.6, .7, .8, .9, .95))
{ i= i+1
curve(dPearson(x,n , rho=rho_val), -1, 1, xlab = "Pearson's r",
 ylab = "Density", lty=mylinetype[i],  lwd=mylinethinkness[i],  add=TRUE)
}
legend(.3,20,param_vals,  lty=mylinetype,  lwd=mylinethinkness, title= "Rho values")
```

You can adapt the above code, editing the following three lines specifically, to produce different colored lines instead of the dotted and dashed changes specified above:

```
mycolor <- c(  "darkorchid3","red", "green", "blue", "black")

curve(dPearson(x,n , rho=rho_val), -1, 1, xlab = "Pearson's r", ylab = "Density",
col=mycolor[i], add=TRUE)

legend(.3,20,param_vals, fill=mycolor, title= "Rho values")
```

56.4 Noncentral chi-squared (χ^2) distribution

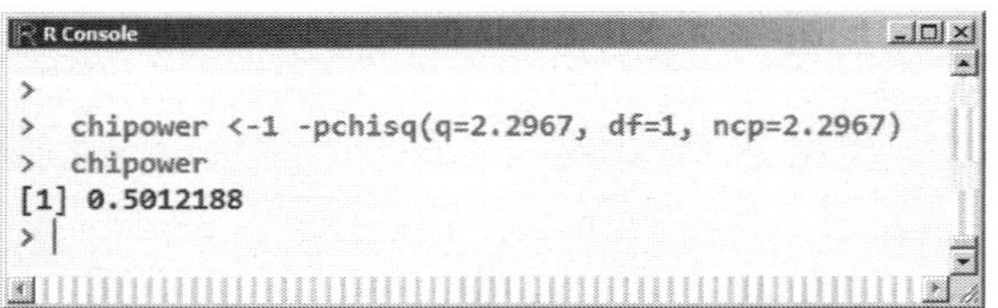

The noncentral chi-squared distribution is part of R's standard installation and just takes an additional value compared to the usual chi-squared distribution, the noncentrality parameter, Lambda =λ. We can demonstrate this by using the chi-squared result from the *Comparing several independent categories: Contingency tables* chapter of X^2=2.2967 (df=1, p-value=0.1297)

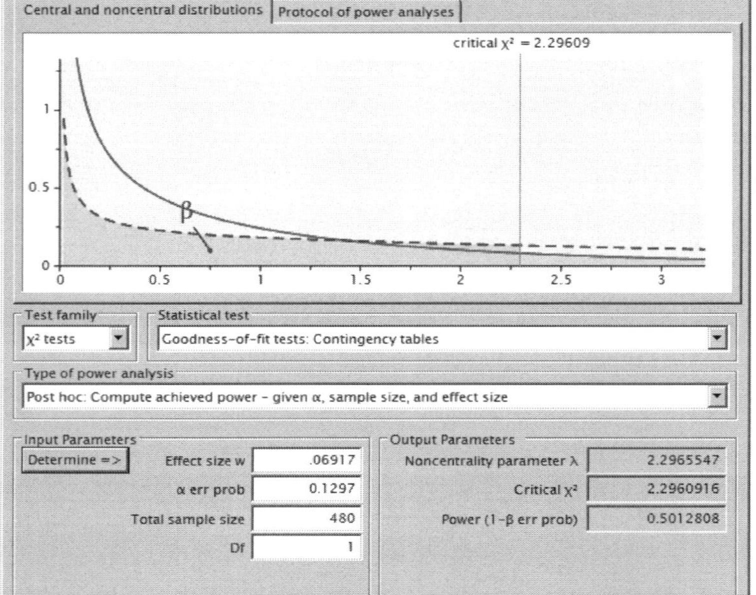

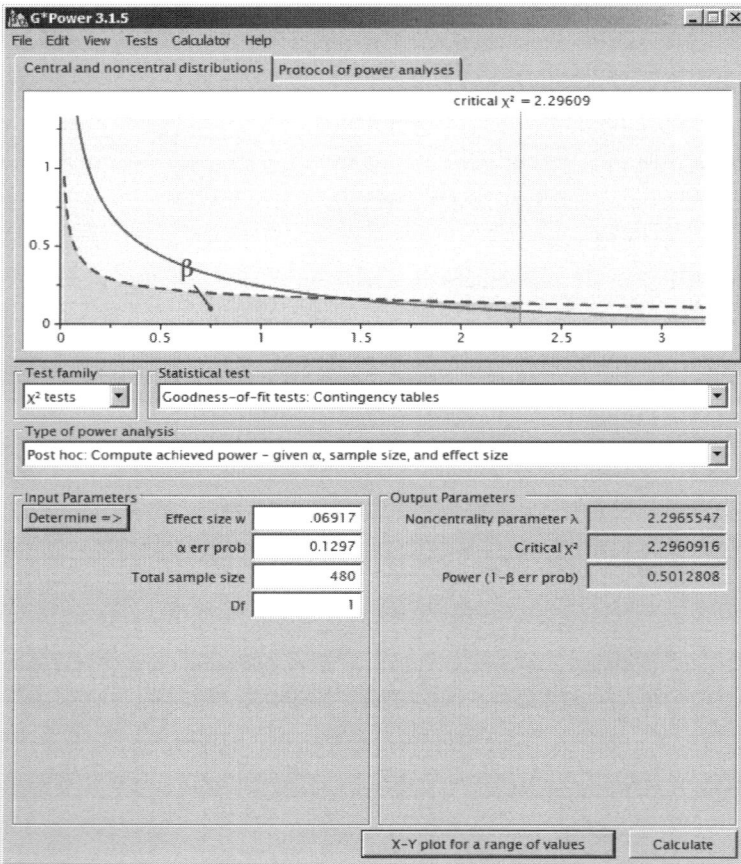

```
chipower <-1 -pchisq(q=2.2967, df=1,
ncp=2.2967)

 chipower
```

While the same value can be obtained using *Gpower*, it requires Cohen's *w* value to be entered. Details of its calculation being provided in the next section:

56.4.1 Effect size measure – Cohen's w

$$w = \sqrt{\frac{X}{N}} \Rightarrow x = w^2 N$$

x = chi squared value

$$w = \sqrt{\frac{2.2967}{480}} = 0.06917$$

Cohen's effect size measure *w* is calculated using the simple relationship between the observed chi-squared value and the total count (N), shown above. Using the above chi-squared result of X^2=2.2967 (df=1, p-value=0.1297) produces a value of 0.06917 for Cohen's *w*. As usual, Cohen defined three levels of *w* representing small medium and large effects:

Cohen's effect size measures				
situation	index	Effect size		
		small	medium	large
Chi-square	*w*	.1	.3	.5

56.4.2 Confidence interval for the chi-squared distribution noncentrality parameter

Before constructing a confidence interval for the noncentral chi-squared distribution, it is useful to plot the null and noncentral distributions:

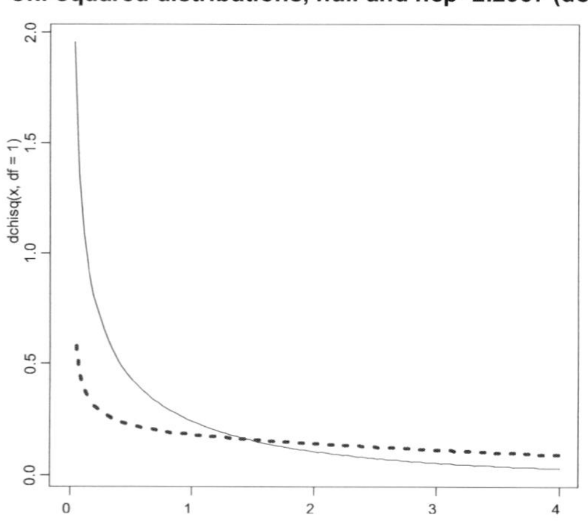


```
curve(dchisq(x, df=1), from="0", to="4",

main="chi-squared distributions, null and ncp=2.2967 (dotted)")

curve(dchisq(x, df=1, ncp= 2.2967), from="0", to="4",add=TRUE,
col="grey10", lwd=4, lty=3)
```

We can see that both curves, which represent df=1, appear to go off to infinity at the x=0 value, technically they are said to be asymptotic. Consequently, any confidence interval approaching zero becomes infinitely large and will crash your computer.

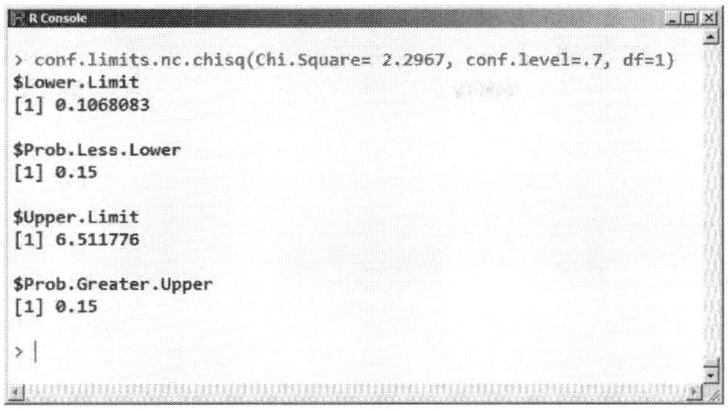

```
R Console
> conf.limits.nc.chisq(Chi.Square= 2.2967, conf.level=.7, df=1)
$Lower.Limit
[1] 0.1068083

$Prob.Less.Lower
[1] 0.15

$Upper.Limit
[1] 6.511776

$Prob.Greater.Upper
[1] 0.15

>
```

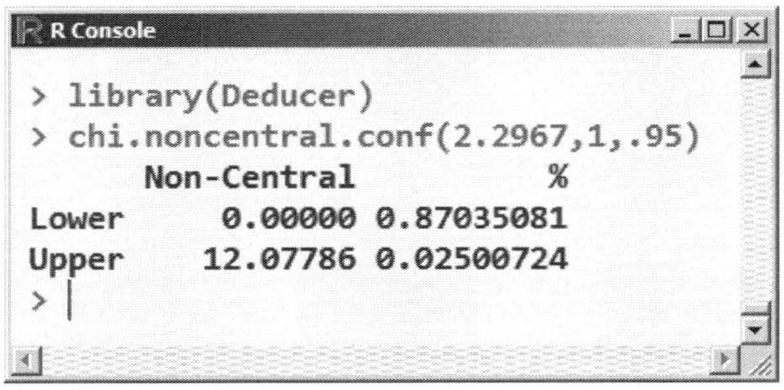

Given the above proviso, it would appear easy to obtain a confidence interval for an observed chi-squared value noncentrality parameter using the *MBESS* package. However, the first line of code below produces an error because the required interval extends to this asymptotic region. By trial and error we can however obtain a 75% confidence interval.

```
conf.limits.nc.chisq(Chi.Square= 2.2967, conf.level=.75, df=1) # produces and error
```

```
conf.limits.nc.chisq(Chi.Square= 2.2967, conf.level=.75, df=1) # works
```

```
R Console
> library(Deducer)
> chi.noncentral.conf(2.2967,1,.95)
      Non-Central          %
Lower     0.00000 0.87035081
Upper    12.07786 0.02500724
>
```

Alternatively, we can use the *Deducer* package which produces a failsafe zero lower limit:

```
install.packages("Deducer", dependencies=TRUE)

library(Deducer)

chi.noncentral.conf(2.2967,1,.95)
```

56.4.3 Confidence interval for the chi-squared distribution effect size measure

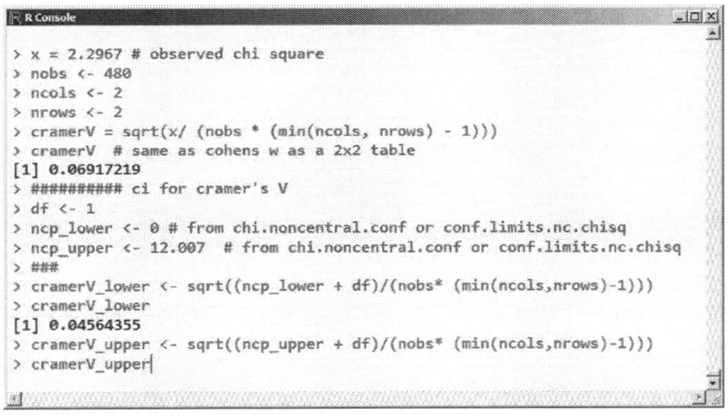

```
R Console
> x = 2.2967 # observed chi square
> nobs <- 480
> ncols <- 2
> nrows <- 2
> cramerV = sqrt(x/ (nobs * (min(ncols, nrows) - 1)))
> cramerV  # same as cohens w as a 2x2 table
[1] 0.06917219
> ########## ci for cramer's V
> df <- 1
> ncp_lower <- 0 # from chi.noncentral.conf or conf.limits.nc.chisq
> ncp_upper <- 12.007  # from chi.noncentral.conf or conf.limits.nc.chisq
> ###
> cramerV_lower <- sqrt((ncp_lower + df)/(nobs* (min(ncols,nrows)-1)))
> cramerV_lower
[1] 0.04564355
> cramerV_upper <- sqrt((ncp_upper + df)/(nobs* (min(ncols,nrows)-1)))
> cramerV_upper
```

Once again there is a slight problem here, this time it relates to the number of rows/ columns in the table. If we have a 2 by 2 table:

Cohen's *w* = phi coefficient (ϕ) = Cramer's phi (V = ϕ_c)

So if we have a 2x2 table and we know an R library that produces a confidence interval for any of the above we can use it. Unfortunately I haven't managed to find one, so will do it from scratch:

```
x = 2.2967 # observed chi-squared
nobs <- 480
ncols <- 2
nrows <- 2
cramerV = sqrt(x/ (nobs * (min(ncols, nrows) - 1)))
cramerV  # same as cohens w as a 2x2 table
########## ci for cramer's V
df <- 1 # same as (ncols-1)*(nrows -1)
ncp_lower <- 0 # from chi.noncentral.conf or conf.limits.nc.chisq
ncp_upper <- 12.007  # from chi.noncentral.conf or conf.limits.nc.chisq
###
cramerV_lower <- sqrt((ncp_lower + df)/(nobs* (min(ncols,nrows)-1)))
cramerV_lower
cramerV_upper <- sqrt((ncp_upper + df)/(nobs* (min(ncols,nrows)-1)))
cramerV_upper
```

This results in a Cramer estimate of .06917 with a 95% confidence interval of 0.04564 to 0.1646144 and because this is a 2x2 table this is equivalent to a confidence interval for Cohen's w.

Cohen's $w = Cramer'sV\sqrt{\min(R,C)-1}$
min(R,C) =gives the smaller of the number of rows or columns
R code equivalent:
cohens_w <- cramerV*sqrt(min(ncols, nrows) - 1)

There is a simple relationship between Cohen's *w* and Cramer's *V* for all table sizes shown opposite.

Taking the blood group and illness prevalence examples from the *Comparing several independent categories: Contingency tables* chapter, I have now edited the values to produce a significant p-value to prevent the problems with confidence interval estimation mentioned earlier.

```
d<- matrix ( c(11, 7, 29, 28, 31, 8, 22, 44, 476,
90, 211, 543), ncol=4, byrow=TRUE)
col_names <- c("O", "AB", "B", "A")
row_names<- c( "severe", "mild", "absent")
dimnames(d) <- list( condition = row_names,
blood_group = col_names)
result1<- chisq.test(d, simulate.p.value = TRUE)
result1
result2<- chisq.test(d)
result2
```

```
R Console
> result<- chisq.test(d, simulate.p.value = TRUE)
> result

        Pearson's Chi-squared test with
        simulated p-value (based on
        2000 replicates)

data:  d
X-squared = 32.8908, df = NA,
p-value = 0.0004998

> result2<- chisq.test(d)
> result2

        Pearson's Chi-squared test

data:  d
X-squared = 32.8908, df = 6, p-value = 1.101e-05
```

The result is a chi-squared of 32.89 (df=6) and a total sample size of 1500, obtained by using the R command *sum(d)*.

Using the MBESS *conf.limits.nc.chisq()* function we obtain a 95% confidence interval of 10.217 to 53.406. If we plug these values into the code we previously developed, we can now calculate the confidence intervals for Cramer's *V* and also use the above code to convert Cramer's *V* into Cohen's w.

```
R Console
> conf.limits.nc.chisq(Chi.Square= 32.89, conf.level=.95, df=6)
$Lower.Limit
[1] 10.21705

$Prob.Less.Lower
[1] 0.025

$Upper.Limit
[1] 53.40678

$Prob.Greater.Upper
[1] 0.025
```

So we have a Cohen's *w* = 0.148 (classed as a small effect) and Cramer's *V* = 0.1047 with a 95% confidence interval of 0.073 to 0.1407.

```
R Console
> x = 32.89 # observed chi square
> nobs <- 1500
> ncols <- 4
> nrows <- 3
> cramerV <- sqrt(x/ (nobs * (min(ncols, nrows) - 1)))
> cramerV  # different to cohens w as NOT a 2x2 table
[1] 0.1047059
> cohens_w <- cramerV*sqrt(min(ncols, nrows) - 1)
> cohens_w
[1] 0.1480766
> ########## ci for cramer's V
> df <- 6 # same as (ncols-1)*(nrows -1)
> ncp_lower <- 10.217 # from conf.limits.nc.chisq (MBESS)
> ncp_upper <- 53.406 # from conf.limits.nc.chisq (MBESS)
> ###
> cramerV_lower <- sqrt((ncp_lower + df)/(nobs* (min(ncols,nrows)-1)))
> cramerV_lower
[1] 0.07352324
> cramerV_upper <- sqrt((ncp_upper + df)/(nobs* (min(ncols,nrows)-1)))
> cramerV_upper
[1] 0.1407196
> check_cohenW<- sqrt(32.89/1500)
> check_cohenW
[1] 0.1480766
> |
```

We can also use the Cramer *V* to Cohen's *w* formula to convert the above Cramer's *V* confidence interval values to Cohen's *w* giving a *w* 95% confidence interval of 0.103 to 0.199.

```
R Console
> cohens_w_lower <- cramerV_lower*sqrt(min(ncols, nrows) - 1)
> cohens_w_lower
[1] 0.1039776
> cohens_w_upper <- cramerV_upper*sqrt(min(ncols, nrows) - 1)
> cohens_w_upper
[1] 0.1990075
> |
```

I have taken my time in explaining how to obtain the confidence interval for Cohen's *w* and the above steps can be shortened but I felt it is useful to demonstrate the process in full.

A useful fact to remember for the noncentral chi-squared distribution is that the observed chi-squared statistic (or likelihood-ratio) can be used as an estimate of the noncentrality parameter for the noncentral chi-squared distribution, just as we did at the start of this section.

56.5 Summary

The distribution of the specified alternative hypothesis is a topic that is rarely presented in introductory statistics books (Cumming, 2012) but I feel is important to include it. This chapter presented several of the most common theoretical distributions including the t, F, chi-squared and correlation. It also presented these in relationship to various effect size measures. Confidence interval calculations for both the noncentrality parameter and the specific effect size have been demonstrated. While methods to calculate these values has been available for some time in R the *MBESS* library makes producing them much easier.

56.6 Tips & Tricks

Remember that *Gpower* provides the easiest approach to calculating some of the values produced in this chapter and only if you can't get what you want from it, would I turn to the *MBESS* library.

Deducer can be difficult to install, specifically on 64 bit Windows. One problem is that it relies upon many other packages including the *rJava* package which it then has problems finding. This is sometimes solved by simply closing and restarting R. If this fails try typing the following into the *R Console* window.

```
Sys.setenv(JAVA_HOME="C:\\Program Files\\Java\\jre7\\")
```

For further details see: http://stackoverflow.com/questions/7019912/using-the-rjava-package-on-win7-64-bit-with-r .

57 Publication quality graphics

The majority of the plots and tables presented in this book would not be suitable for submission to academic journals. To help rectify this I now provide an introduction to the *ggplot2* package, specifically the *ggplots()* functon within it, which does allow you to produce appropriate plots. You can either interact with the *ggplots()* function directly or through a variety of point and click interfaces. We will start by considering direct interaction with *ggplot2*.

57.1 ggplot2 – interacting directly

The R package *ggplot2* provides a large number of graphical elements which are of publication quality. This chapter will take several typical plots produced in other chapters of this book and reproduce them here using the *ggplot2* package.

```
install.packages("ggplot2",dependencies=TRUE)

library(ggplot2)
```

In *ggplot2* you to create plots in one of two ways. Firstly by a quick and dirty method using the *qplot()* function in which you specify a *qplot()* similarly to the way you have developed plots using the *plot()* function in other chapters, under the bonnet *qplot()* creates a *ggplot()* object. The more complex way is to create a *ggplot()* object directly which offers you much more control and opportunities for greater complexity. Unfortunately along with this added freedom comes a more complex way of developing the plots, now you need to think of plots as a series of *layers* where you can add as many as you want. Each layer has two main aspects; geometrics and aesthetics, all wrapped in a single *ggplot()* object. To understand this we need to consider each of these in turn.

57.1.1 Geometrics – *geoms()*

The geometric aspect of each layer specifies a particular plot aspect, such as a series of points, a boxplot or a text box. Each of these geometric aspects also has a number of aesthetic aspects.

geom=	aes=	alpha	colour	size	shape	linetype	weight	fill	Other
		Required/optional							
bar		✓	✓	✓		✓	✓	✓	
point		✓	✓	✓	✓			✓	
line	Connect observations, ordered by x value	✓	✓	✓		✓			
smooth		✓	✓	✓		✓	✓	✓	
histogram		✓	✓	✓		✓	✓	✓	
boxplot		✓	✓	✓			✓	✓	
text		✓	✓	✓					angle, hjust, vjust
density		✓	✓	✓		✓	✓	✓	
errorbar		✓	✓	✓		✓			width
hline		✓	✓	✓		✓			
vline		✓	✓	✓		✓			
density2d		✓	✓	✓		✓			
path	Connect observations in original order	✓	✓	✓		✓			
errorbar		✓	✓	✓		✓			width
dotplot		✓	✓					✓	

57.1.2 Aesthetics - *aes()*

Aesthetics are concerned with appearances, such as the colour, size, weight and degree of transparency (alpha). Some aspects only relate to certain geometrics such as fill. The table opposite provides the aesthetics associated with some of the common geometrics.

I will now demonstrate these aspects with a variety of graphs.

57.1.3 Boxplots

To demonstrate boxplots in the *ggplot2* package the code below reproduces the one created earlier the *Introductory tutorial* chapter using the R Commander menu.

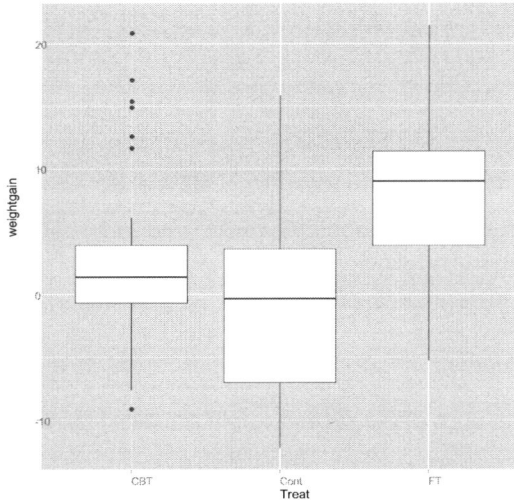

```
library(MASS) #load the MASS library that has the data

data(anorexia, package="MASS") #get the data

qplot(Treat, weightgain,  geom="boxplot", data=anorexia) #or

ggplot(anorexia, aes(x=Treat, y=weightgain)) + geom_boxplot()
```

We can also use the *notch=TRUE, varwidth=TRUE* options as described in the boxplots chapter.

The above boxplots are for a single variable divided by another which is a factor. We can also plot several columns of data, by melting the columns into one long one using the *melt()* function in the *reshape2* package. Taking as an example the boxplots presented in the *Comparing pre-post test median differences* chapter where three columns were plotted side by side on a boxplot.

contains 2 columns, called pre and post

```
mydataframe <- read.delim("http://www.robin-
beaumont.co.uk/virtualclassroom/book2data/wilcoxon_paired_ex1.dat")
```

add a third column, showing the difference

```
mydataframe$difference <
                        - mydataframe$post - mydataframe$pre

mydataframe$difference

library(reshape2)
```

melt all three columns into one, where the melt function automatically renames the two new columns, *variable* and *value*

```
longdf <- melt(mydataframe)

library(ggplot2)

ggplot(data=longdf) +
 geom_boxplot(aes(x=variable,y=value))
```

```
> longdf
     variable value
1        post     7
2        post     6
3        post     8
4        post     7
5        post     6
6        post     9
7         pre     1
8         pre     4
9         pre     3
10        pre     5
11        pre     4
12        pre     5
13 difference     6
14 difference     2
15 difference     5
16 difference     2
17 difference     2
```

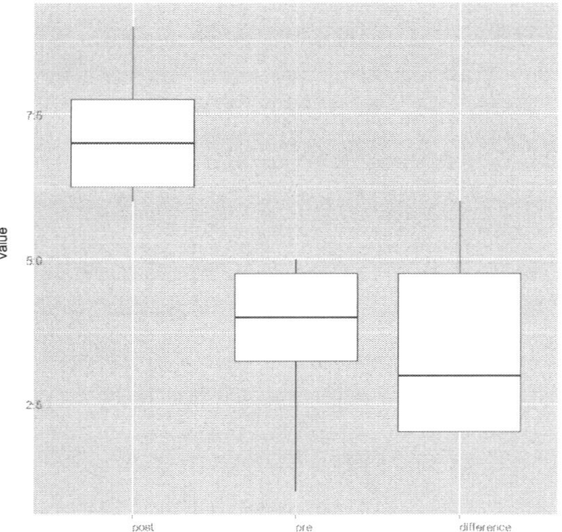

or

```
qplot(variable, value,  geom="boxplot",
data=longdf)
```

To show the individual points as well:

```
qplot(variable, value, geom=c("boxplot", "jitter"),
data=longdf)
```

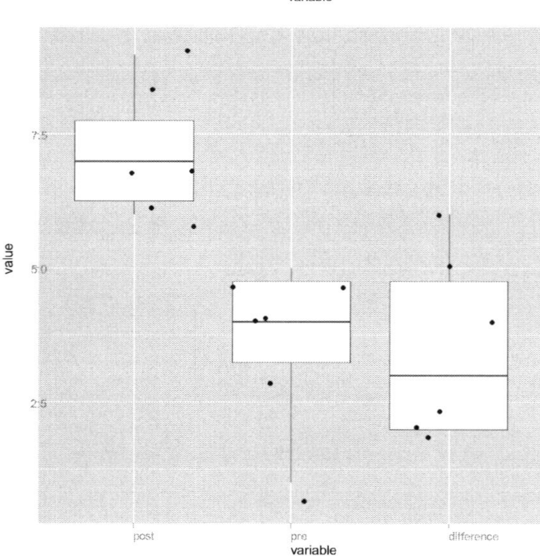

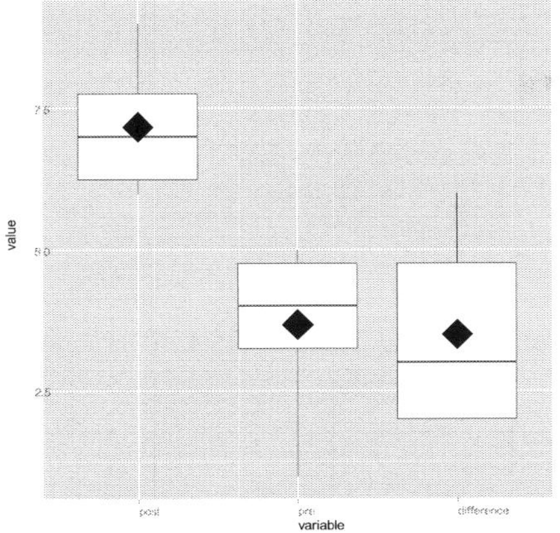

You can also add the mean for each group. This is a useful addition as comparing this value to the median (the centre line in the box) and the relative lengths of each whisker gives an indication as to the skewness of the data.

```
ggplot(data=longdf, aes(x=variable, y=value)) +

geom_boxplot() +

stat_summary(fun.y="mean", geom="point", shape=23, size=10, fill="black")
```

57.1.4 Scatterplots – simple

This example is taken from the *examples and interpretation* section of the correlation chapter.

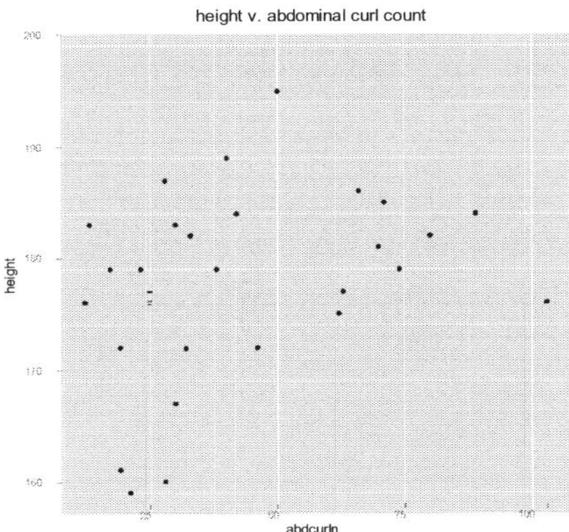


```
library(ggplot2)
mydataframe <- read.delim("http://www.robin-
beaumont.co.uk/virtualclassroom/book2data/correlation_vo2max.dat")
names(mydataframe)

qplot(abdcurln, height, data = mydataframe,
main="height v. abdominal curl count")    #  or

ggplot(data=mydataframe, aes(x=abdcurln, y=height)) +
geom_point() +
labs(title = "height v. abdominal curl count")
```

```
> library(ggplot2)
>    qplot(abdcurln, height, data = mydataframe, main="height v. abdominal curl count")
Warning: Removed 11 rows containing missing values (geom_point).
```

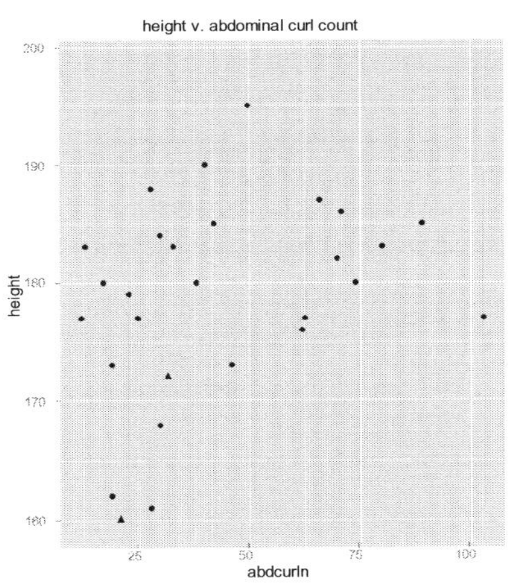

Notice that R gives a warning that 11 rows were removed from the analysis because there were incomplete. You can also produce plots that differentiate various subgroups, in this dataset there is a variable called sex indicating gender.

```
qplot(abdcurln, height, data = mydataframe,
   main="height v. abdominal curl count",
     shape = factor(sex))   #  or
    ggplot(data=mydataframe, aes(x=abdcurln, y=height)) +
   geom_point() +
   labs(title = "height v. abdominal curl count") +
   aes(shape = factor(sex))
```

The symbols are too small in the above scatterplots. To increase their size add *size=I(5)* in the *qplot()* function or *aes(size=5)* in the *ggplot()* function as shown below, which also specifies hollow symbols.

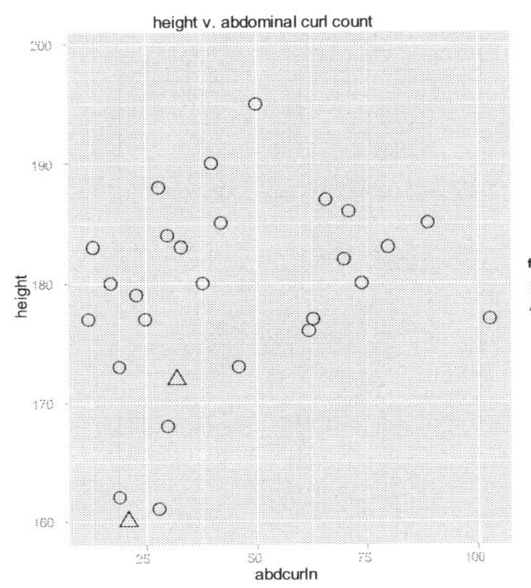


```
ggplot(data=mydataframe, aes(x=abdcurln, y=height)) +
geom_point(size=5) +
labs(title = "height v. abdominal curl count") +
aes(shape = factor(sex)) +
scale_shape(solid = FALSE)
```

You can replace the *shape=* option with *colour=* option if you have a color printer or plan to just present the results on screen. If you have a black and white printer just use the "grey10" color option.

57.1.5 Linear Regression lines

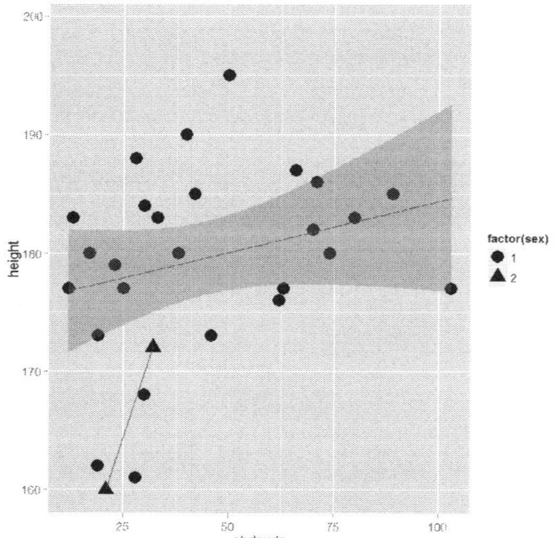

Carrying on with the abdominal curls dataset. To add linear regression lines to each factor just involves a very small edit of the code.

```
qplot(abdcurln, height, data = mydataframe, method="lm",

shape = factor(sex), geom = c("point", "smooth") )

# or

ggplot(data=mydataframe, aes(x=abdcurln, y=height)) +

aes(shape = factor(sex)) +

geom_point(size=5) +

stat_smooth( method="lm")
```

The screen shot opposite failed to produce a confidence interval curve for the females because of insufficient numbers (2). The R code above also increased the size of the symbols. You can easily request an overall regression line by removing the *shape=* expression.

For more complex regression lines see Chang, 2013 R Graphics Cookbook.

57.1.6 Scatterplot matrix

ggplot() does not produce scatterplot matrices and the help file recommends you use the *ggpairs()* function in the *GGally* package. The latest version of the package allows you to use incomplete datasets, reporting how many cases it drops for each comparison. The standard R function *pairs()* also copes happily with incomplete cases so does the *pairs.panels()* function in the *psych* package, which has been demonstrated in the correlation chapter. The *scatterplotMatrix()* function in the car package produces a similar scatterplot matrix and was used to produce the one in the *examples and interpretation* section of the correlation chapter. The code is given below.

```
library(car)

 scatterplotMatrix(~abdcurln+height+mftvomax+weight, reg.line=lm, smooth=FALSE, spread=FALSE,
                         span=0.5, id.n=0, diagonal = 'histogram', data=mydataframe)
```

If you use the *ggpairs()* function you need first to select only those columns you want in the scatterplot matrix and place them in a new dataframe. The latest version of *ggpairs()*, in contrast to previous versions, does not require the removal of all those rows containing empty values (i.e. NAs). I will demonstrate how letting *ggpairs()* remove the empty values (NAs) pairwise is better than simply removing all those rows that contain one or more empty values.

```
> reduced <- mydataframe[, c("abdcurln","height","mftvomax","weight")]
> mydataframe_complete = reduced[complete.cases(reduced),]
> str(mydataframe_complete)
'data.frame':	19 obs. of  4 variables:
 $ abdcurln: num  71 33 62 70 19 30 28 46 40 19 ...
 $ height  : num  186 183 176 182 162 168 188 173 190 173 ...
 $ mftvomax: num  45.2 51 43 50.2 55.4 47.4 47.4 63.8 45.8 50 ...
 $ weight  : num  86 83 78 84 75 73 82.5 74.5 70 73 ...
> length(mydataframe)
[1] 39
```

The code below copies the relevant columns into a temporary dataframe called *reduced* than applies the *complete.cases()* function to it. This removes any rows with one or more empty values, placing the result into another dataframes called *mydataframe_complete*. As you can see, we now have 19 cases left. The *complete.cases()* function is a rather crude approach as it does not carry out the deletion on a pairwise basis. We will now compare what happens when you use these two datasets with the *ggpairs()* function in the *GGally* package.

```
reduced <- mydataframe[, c("abdcurln","height","mftvomax","weight")]
mydataframe_complete = reduced[complete.cases(reduced),]
str(mydataframe_complete); str(reduced)
```

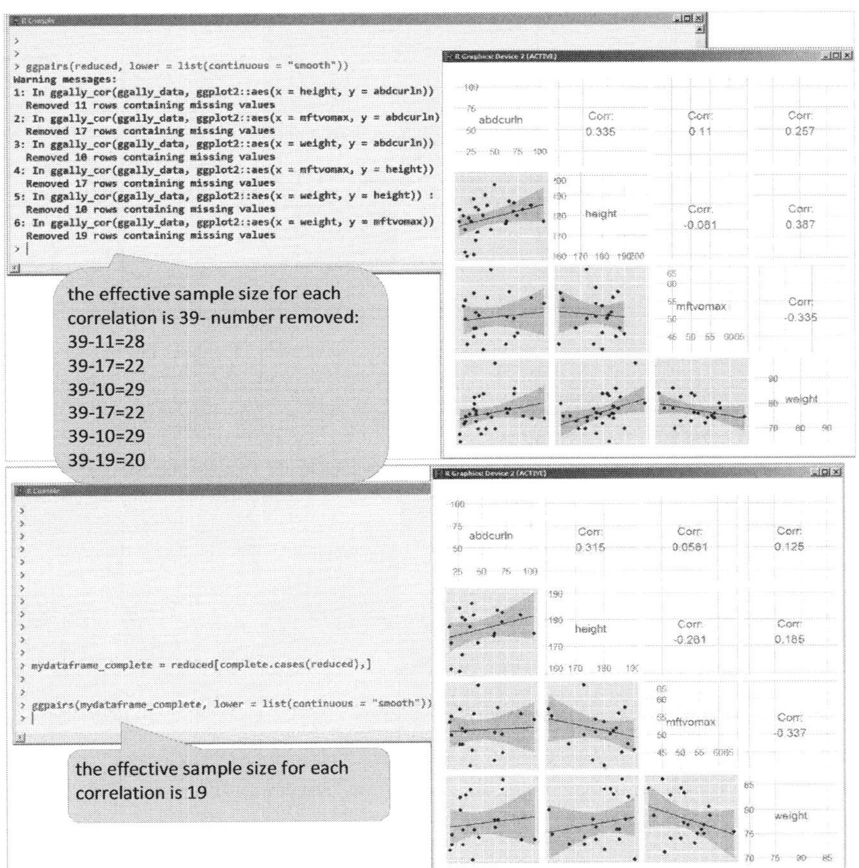

the effective sample size for each correlation is 39- number removed:
39-11=28
39-17=22
39-10=29
39-17=22
39-10=29
39-19=20
the effective sample size for each correlation is 19

The result opposite is from using the *reduced* dataframe, having first installed and loaded the *GGally* package:

```
install.packages("GGally")

library(GGally)

ggpairs(reduced,
lower = list(continuous = "smooth"))
```

You can see the 6 listed warnings in the screenshot on the far left indicate the number of rows that were removed for each pairwise comparison. In effect, each correlation starts with the complete number of rows and then only removes those pairs that are incomplete.

```
ggpairs(mydataframe_complete,
lower = list(continuous = "smooth"))
```

In contrast, using the above code with the *mydataframe_complete* dataframe we can see fewer points with several outliers in the abdcurln/weight and height/weight scatterplots are missing. This affects the reported correlations. We can see what is going on if we request the correlations using the *cor()* function with the *use=pairewise.complete.obs* option. These results mimic the two above by creating a dataframe with no missing values. We have effectively reduced the sample size for all the correlations.

```
> round(cor(reduced,use="pairwise.complete.obs"), 4)
         abdcurln  height mftvomax  weight
abdcurln   1.0000  0.3350   0.1103  0.2566
height     0.3350  1.0000  -0.0810  0.3867
mftvomax   0.1103 -0.0810   1.0000 -0.3346
weight     0.2566  0.3867  -0.3346  1.0000
> round(cor(mydataframe_complete), 4)
         abdcurln  height mftvomax  weight
abdcurln   1.0000  0.3155   0.0561  0.1255
height     0.3155  1.0000  -0.2611  0.1851
mftvomax   0.0561 -0.2611   1.0000 -0.3373
weight     0.1255  0.1851  -0.3373  1.0000
```

With small datasets I recommend that whenever you request correlation matrices you always set the *use* option to pairwise, that is *use="pairwise.complete.obs"*.

The quick-R website has many examples of scatterplots, including 3-d ones at

57.1.7 Lines for repeated measures

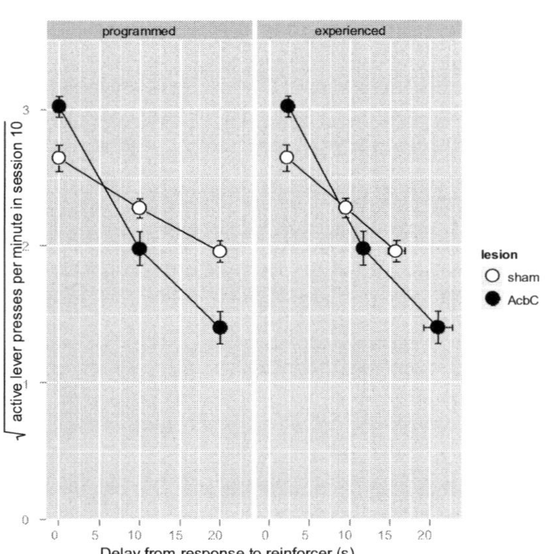

Creating publication quality graphics of repeated measures is a topic that could become several chapters. While generally graphs of repeated measures are often of low quality, Rudolf Cardinal's site at Cambridge University presents excellent, but admittedly complex examples. The one shown opposite is for an experiment, concerning key presses for two groups of subjects (programmed/experienced) under two conditions (sham/AcbC). His work can be found at:

`http://egret.psychol.cam.ac.uk/statistics/R/graphs2.html`

The example opposite makes use of the *facet_grid()* function within the *ggplot2* package producing the equivalent to a panel plot described in the repeated measures chapter where the *lattice* package was used.

57.1.7.1 Data preparation

To create the above graph you need two datasets; *mdfig30a*, *mdfig30b*. Both of these datasets can be found at the above website or on my site. Typing the following two lines of code will download and import them into R:

```
fig30A<- read.csv(("http://www.robin-beaumont.co.uk/virtualclassroom/book2data/MD_fig30A.csv"))
fig30A
fig30B<- read.csv(("http://www.robin-beaumont.co.uk/virtualclassroom/book2data/MD_fig30B.csv"))
fig30B
```

```
> fig30A<- read.csv(("http://www.robin-beaumont.co.uk/virtualclassroom/book2data/MD_fig30A.csv"))
> fig30A
  delay     sham    SEMsham     AcbC    SEMAcbC
1     0 2.640708 0.09574035 3.015727 0.07431359
2    10 2.276907 0.07223806 1.977225 0.12287066
3    20 1.956931 0.07676898 1.400463 0.11313513
> fig30B<- read.csv(("http://www.robin-beaumont.co.uk/virtualclassroom/book2data/MD_fig30B.csv"))
> fig30B
  shamdelay shamresponses SEMshamdelay SEMshamresponses AcbCdelay AcbCresponses SEMAcbCdelay SEMAcbCresponses
1  2.127342      2.640708    0.3954407       0.09574035  2.307076      3.015727    0.3887686       0.07431359
2  9.483373      2.276907    0.4826215       0.07223806 11.680260      1.977225    0.5989249       0.12287066
3 15.812267      1.956931    1.0514280       0.07676898 20.936766      1.400463    1.8152600       0.11313513
```

Both the above files contain 3 records and both are incomplete. The first file provides the delay value for the x axis while the second file contains all the other values required.

```
> mdfig30_long
          type     delay lesion sqrtpresses       ysem      xsem
1   programmed  0.000000   sham    2.640708 0.09574035 0.0000000
2   programmed 10.000000   sham    2.276907 0.07223806 0.0000000
3   programmed 20.000000   sham    1.956931 0.07676898 0.0000000
4   programmed  0.000000   AcbC    3.015727 0.07431359 0.0000000
5   programmed 10.000000   AcbC    1.977225 0.12287066 0.0000000
6   programmed 20.000000   AcbC    1.400463 0.11313513 0.0000000
7  experienced  2.127342   sham    2.640708 0.09574035 0.3954407
8  experienced  9.483373   sham    2.276907 0.07223806 0.4826215
9  experienced 15.812267   sham    1.956931 0.07676898 1.0514280
10 experienced  2.307076   AcbC    3.015727 0.07431359 0.3887686
11 experienced 11.680260   AcbC    1.977225 0.12287066 0.5989249
12 experienced 20.936766   AcbC    1.400463 0.11313513 1.8152600
```

The screenshot opposite shows what we want to finish up with indicating that we need to both combine columns from the files and also stack various column. The following code achieves this creating a new dataframe called *mdfig30_long* by stacking specific rows from each of the datafiles on top of each other using the *rbind()* function:

first column is called *type*, value is "programmed"

second column is called *delay*, values are from the *delay* column of the *fig30a*

third column is called *lesion*, value is "sham"

```
mdfig30_long = rbind( data.frame(type="programmed",  delay=fig30A$delay, lesion="sham",
                      sqrtpresses = fig30A$sham, ysem = fig30A$SEMsham, xsem = 0),
```

create first 3 rows

fourth column is called *sqrtpresses*, values are from the *sham* column of the *fig30a* dataframe

fifth column is called *ysem*, values are from the *SEMsham* column of the *fig30a* dataframe

sixth column is called *xsem*, value is 0

```
data.frame(type="programmed",  delay=fig30A$delay, lesion="AcbC", sqrtpresses = fig30A$AcbC,
                      ysem = fig30A$SEMAcbC, xsem = 0 ),
```

create next 3 rows

```
data.frame(type="experienced", delay=fig30B$shamdelay, lesion="sham", sqrtpresses = fig30B$shamresponses,
                      ysem = fig30B$SEMshamresponses,
                      xsem = fig30B$SEMshamdelay ),
```

create next 3 rows

```
data.frame(type="experienced", delay=fig30B$AcbCdelay, lesion="AcbC", sqrtpresses = fig30B$AcbCresponses,
                      ysem = fig30B$SEMAcbCresponses,
                      xsem = fig30B$SEMAcbCdelay ) )
```

the *data.frame()* function makes each column the same length

end of the *rbind()* function

create last 3 rows

57.1.7.2 Developing the plot

We can now use the *mdfig30_long* dataframe to create the required plot. I have annotated the data below to indicate how each column is used in the plot.

The plot produced on Rudolf Cardinal's site contains more annotations than those I have given, including the use of what is called a theme. A theme is a piece of R code that sets certain graphical parameters such as font size and margin size which can then be used by calling up the theme.

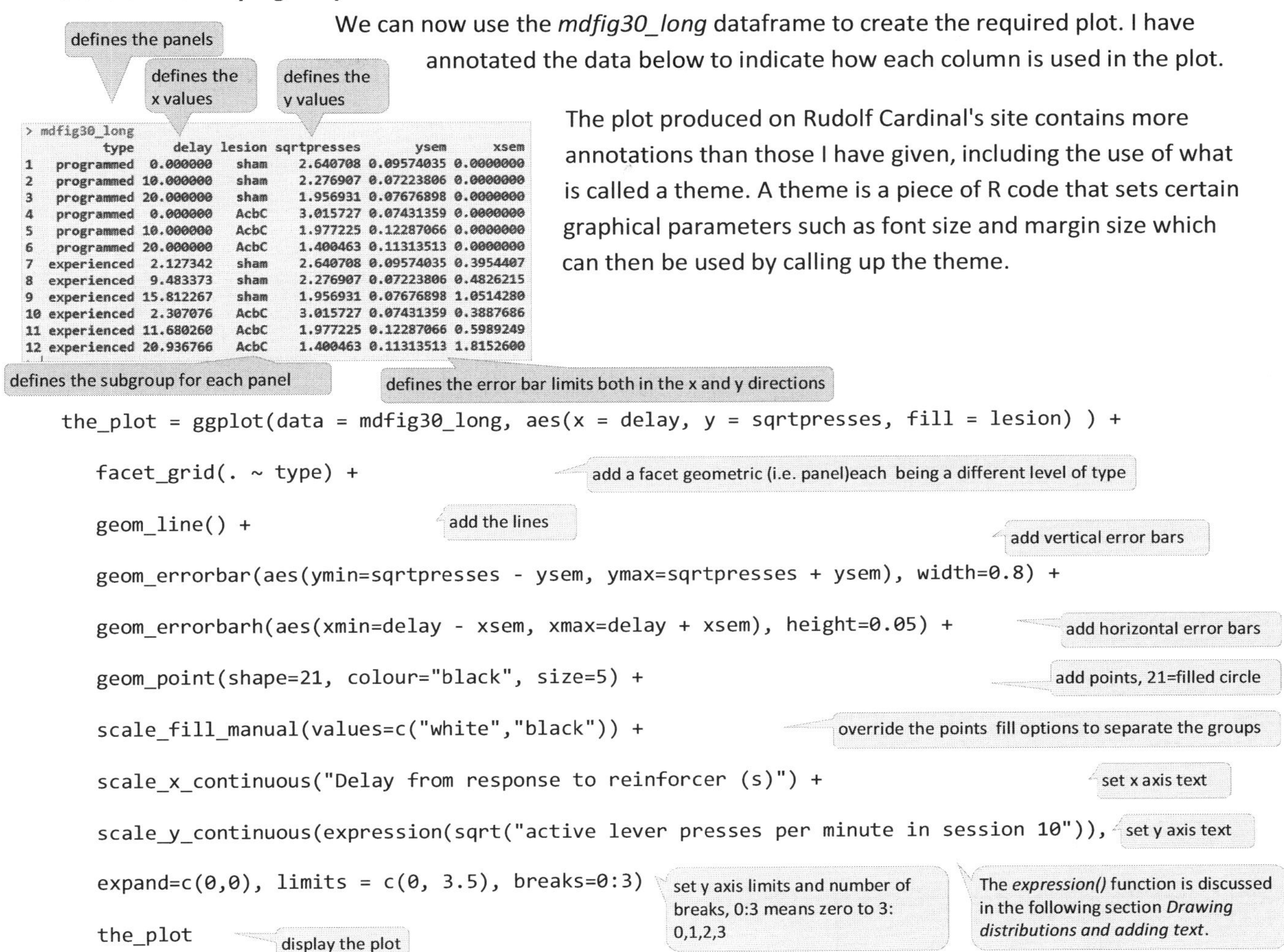

```
> mdfig30_long
          type     delay lesion sqrtpresses       ysem      xsem
1   programmed  0.000000   sham    2.640708 0.09574035 0.0000000
2   programmed 10.000000   sham    2.276907 0.07223806 0.0000000
3   programmed 20.000000   sham    1.956931 0.07676898 0.0000000
4   programmed  0.000000   AcbC    3.015727 0.07431359 0.0000000
5   programmed 10.000000   AcbC    1.977225 0.12287066 0.0000000
6   programmed 20.000000   AcbC    1.400463 0.11313513 0.0000000
7  experienced  2.127342   sham    2.640708 0.09574035 0.3954407
8  experienced  9.483373   sham    2.276907 0.07223806 0.4826215
9  experienced 15.812267   sham    1.956931 0.07676898 1.0514280
10 experienced  2.307076   AcbC    3.015727 0.07431359 0.3887686
11 experienced 11.680260   AcbC    1.977225 0.12287066 0.5989249
12 experienced 20.936766   AcbC    1.400463 0.11313513 1.8152600
```

```
the_plot = ggplot(data = mdfig30_long, aes(x = delay, y = sqrtpresses, fill = lesion) ) +
  facet_grid(. ~ type) +
  geom_line() +
  geom_errorbar(aes(ymin=sqrtpresses - ysem, ymax=sqrtpresses + ysem), width=0.8) +
  geom_errorbarh(aes(xmin=delay - xsem, xmax=delay + xsem), height=0.05) +
  geom_point(shape=21, colour="black", size=5) +
  scale_fill_manual(values=c("white","black")) +
  scale_x_continuous("Delay from response to reinforcer (s)") +
  scale_y_continuous(expression(sqrt("active lever presses per minute in session 10")),
  expand=c(0,0), limits = c(0, 3.5), breaks=0:3)
  the_plot
```

defines the panels
defines the x values
defines the y values
defines the subgroup for each panel
defines the error bar limits both in the x and y directions
add a facet geometric (i.e. panel)each being a different level of type
add the lines
add vertical error bars
add horizontal error bars
add points, 21=filled circle
override the points fill options to separate the groups
set x axis text
set y axis text
set y axis limits and number of breaks, 0:3 means zero to 3: 0,1,2,3
The *expression()* function is discussed in the following section *Drawing distributions and adding text*.
display the plot

As in the *Repeated Measures* chapter there is a fair degree of data manipulation and also development of sophisticated R code to specify the plot. Producing such plots can take a significant amount of time requiring careful planning; with paper based sketches, and data structure details. Another useful technique is to use a very small dummy dataset to check that the plot produced by the R code is actually valid.

57.1.8 Kaplan-Meier plots

ggplot2: Tutorial by Ramon Saccilotto at the Basel Institute for Clinical Epidemiology and Biostatistics.

You can produce Kaplan-Meier plots in *ggplot2* directly, but it involves several dozen lines of R code. Details at:

http://www.ceb-institute.org/bbs/wp-content/uploads/2011/09/handout_ggplot2.pdf

Matt Cooper is a Biostatistician working at the Telethon Institute for Child Health Research (ICHR) Victoria University New Zealand.

Alternatively, Matt Cooper's blog provides an approach that includes a risk table at the bottom of the plot:

`http://mcfromnz.wordpress.com/2012/05/05/kaplan-meier-survival-plot-with-at-risk-table-by-sub-groups/`

As a starting point, you can produce basic ones using the *RcmdrPlugin.KMggplot2* R Commander plugin described later.

57.1.9 Drawing distributions and adding text

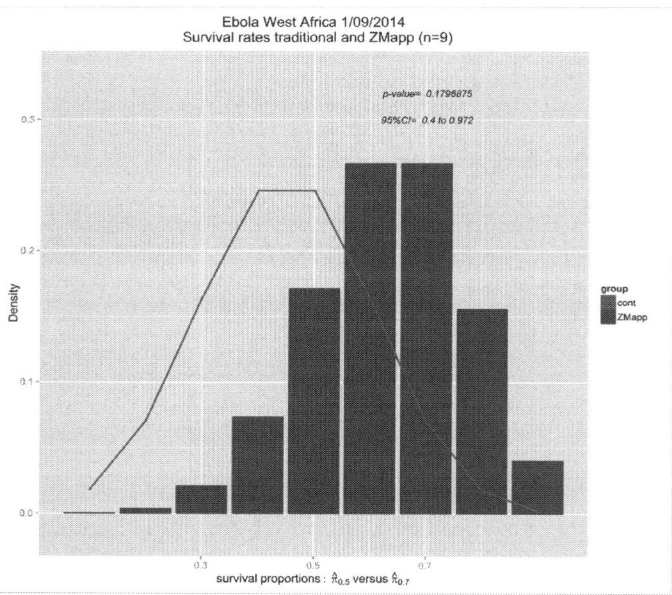

You can also draw the various distributions as presented in the *Confidence intervals for effect sizes: Noncentral distributions* chapter. I will demonstrate this using information about the *Ebola* outbreaks given on the WHO website. At present the overall mortality rate is around 50%. In comparison I will assume that seven out of nine people who have received the experimental drug ZMapp have survived (at the time of writing it was 5 out of 7 with 2 more undergoing treatment). Further information about the Ebola dataset can be found in the *Meta-analysis* chapter.

Before we can create the actual plot we need to; create the distribution of the values from two binomial distributions and join them together, and secondly, create and format specific additional text for the plot regarding the binomial test (p-value and 95%CI) and the x-axis text with subscripts.

57.1.9.1 Creating the distributions and formatted text

```
> totaldata
   group x           y
1   cont 1 0.017578125
2   cont 2 0.070312500
3   cont 3 0.164062500
4   cont 4 0.246093750
5   cont 5 0.246093750
6   cont 6 0.164062500
7   cont 7 0.070312500
8   cont 8 0.017578125
9   cont 9 0.001953125
10 ZMapp 1 0.000413343
11 ZMapp 2 0.003857868
12 ZMapp 3 0.021003948
13 ZMapp 4 0.073513818
14 ZMapp 5 0.171532242
15 ZMapp 6 0.266827932
16 ZMapp 7 0.266827932
17 ZMapp 8 0.155649627
18 ZMapp 9 0.040353607
```

We first create the two distributions and combine them.

```
xval  <- 1:9

dataf <- data.frame(group="cont", x = xval, y = dbinom(xval, 9, 0.5))

dataf2 <- data.frame(group="ZMapp", x = xval, y = dbinom(xval, 9, 0.7))

totaldata <- rbind(dataf, dataf2)

totaldata
```

We now obtain the binomial test results we want to show on the plot:

```
result_ZMapp <- binom.test(7, 9, alternative='two.sided', p=.5, conf.level=.95)

mytext <- paste ("p-value= ", result_ZMapp$p.value)
```

```
> result_ZMapp <- binom.test(7, 9, alternative='two.sided', p=.5, conf.level=.95)
> mytext <- paste ("p-value= ", result_ZMapp$p.value)
> mytext2 <- paste ("95%CI= ", round(result_ZMapp$conf.int[1],3), "to" ,
+ round(result_ZMapp$conf.int[2],3 ))
> mytext
[1] "p-value=  0.1796875"
> mytext2
[1] "95%CI=  0.4 to 0.972"
```

```
mytext2 <- paste ("95%CI= ",
round(result_ZMapp$conf.int[1],3), "to",
round(result_ZMapp$conf.int[2],3 ))

mytext; mytext2
```

expression()
[3] = makes 3 a subscript
2^3 = makes 3 a superscript
bar(xy) = xy with bar
chi = χ
hat(x) = adds a hat to x
infinity = infinity symbol
lambda = λ
nu = μ
over(x, y) = x over y
pi = π
x ~~ y = adds extra space
x ~y = keeps a space

For examples type into the R Console: *demo(plotmath)*

We also wish to add some text with subscripts and the greek letter pi (=π):

```
text3 <- expression(survival~proportions:~hat(pi)[0.5]~versus~hat(pi)[0.7])
```

The above expression uses a form of text description known as *plotmath* more details of which can be found at:
http://stat.ethz.ch/R-manual/R-devel/library/grDevices/html/plotmath.html

We are now ready to create two versions of the plot; colored, and greyscale. First the color version.

57.1.9.2 Plotting the distributions and formatted text

```
library(ggplot2)
library(plyr)
plot <- ggplot(totaldata,
aes(x,y, colour=group, fill=group)) +
geom_bar(subset=.(group=="ZMapp"),stat="identity")+
geom_line(subset=.(group=="cont"),size = 1) +
scale_y_continuous(expand = c(0.1, 0)) +
scale_x_continuous(breaks=c(3,5,7), labels=c("0.3", "0.5", "0.7")) +
xlab(text3) +
ylab("Density") +
labs(title = "Ebola West Africa 1/09/2014 \n Survival rates traditional and ZMapp (n=9)") +
theme_grey(base_size = 14, base_family = "sans") +
theme(plot.title = element_text(size = rel(1.2), vjust = 1.5)) +
annotate("text",x=7,y=.32,label=mytext, fontface = 3, size=4) +
annotate("text",x=7,y=.3,label=mytext2, fontface = 3, size=4)   +
scale_fill_manual(values=c("red", "blue")) +
scale_colour_manual(values=c("red", "blue"))
print(plot)
```

- library(plyr): this additional library is needed for the .() function which is used by the subset command below
- plot <- ggplot(totaldata,: use the whole dataset
- aes(...): set the color of the line and fill of the bars by the group factor
- geom_bar: draw a bar chart with the ZMapp subset of data
- geom_line: draw a line chart with the cont subset of data
- scale_y_continuous: sets the gaps between the data and ends of the y-axis
- scale_x_continuous: sets position and text of the labels for the y-axis
- xlab: sets x-axis title
- ylab: sets y-axis title
- labs: sets plot title the *\n* forces a line break
- theme_grey: sets overall look of the plot to the grey_theme
- theme: adjust the title text
- annotate (mytext): place *mytext* on the plot
- annotate (mytext2): place *mytext2* on the plot; *x* and *y* are the top right coordinates. fontface 3 is italic.
- scale_fill_manual / scale_colour_manual: these options set the colors to map the line and bar data series to – also used to specify the legend.
- print(plot): show the plot in the *R Graphics Device* window

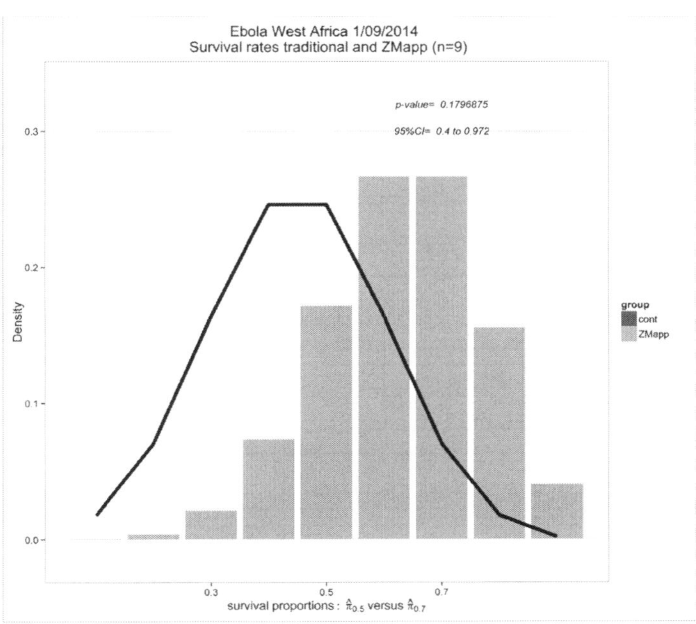

To produce a greyscale version of the above requires the editing of the *scale_fill_manual()* and *scale_colour_manual()* lines, show below:

```
plot <- ggplot(totaldata,
 . . . . .
 . . . . . .
scale_fill_manual(values=c("gray40", "gray75")) +
scale_colour_manual(values=c("gray40", "gray75"))
print(plot)
```

To change the labels in the legend you can edit the *scale_fill_maual()* function, adding a *labels= option*:

```
scale_fill_manual(values=c("gray40", "gray75"), labels=c("trad care","experimental drug"))
```

57.2 Using point and click front ends that generate ggplot2 plots

You can use three different environments that take the point and click approach to plot development to produce *ggplot2* graphics along with the R code:

- A R Commander plugin called *RcmdrPlugin.KMggplot2*
- The R library Deducer
- A interactive web version of *ggplot2* has been developed by Jeroen Ooms who is a Phd student at ucla. Details at http://www.stat.ucla.edu/~jeroen/live.html

I'll demonstrate each of these.

57.2.1 The R Commander plugin for ggplot2 plots - RcmdrPlugin.KMggplot2

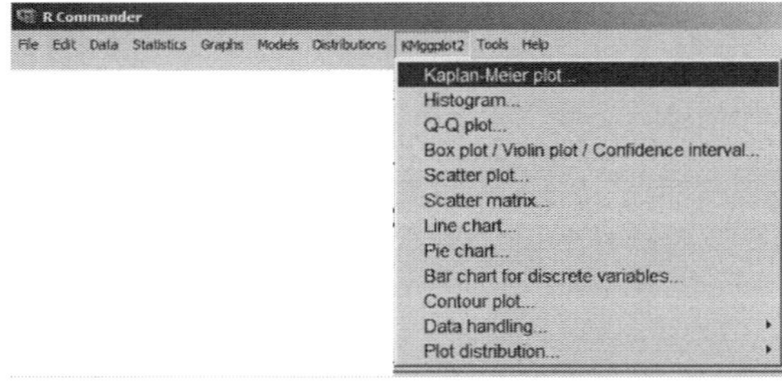

The *RcmdrPlugin.KMggplot2* R Commander plugin adds a menu option to R Commander (opposite). It is installed in the usual manner with the following R command. I recommend that before you install it you update all you packages you have installed and check to see which version of R you have. At present the plugin only seems to work for R windows 32bit.

```
install.packages("tcltk2") # specifically you need the latest version of this

install.packages("RcmdrPlugin.KMggplot2")

library(Rcmdr)
```

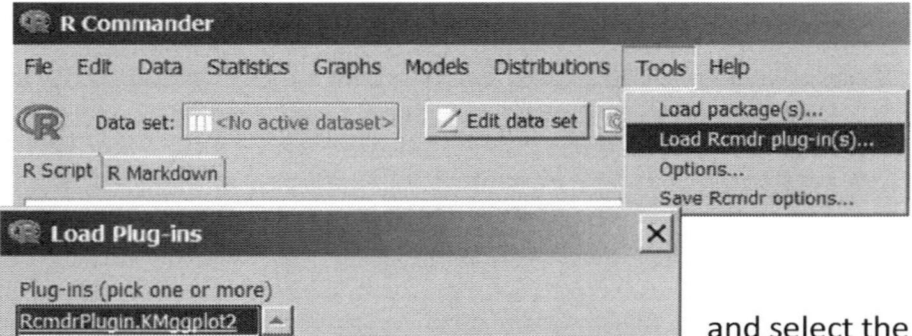

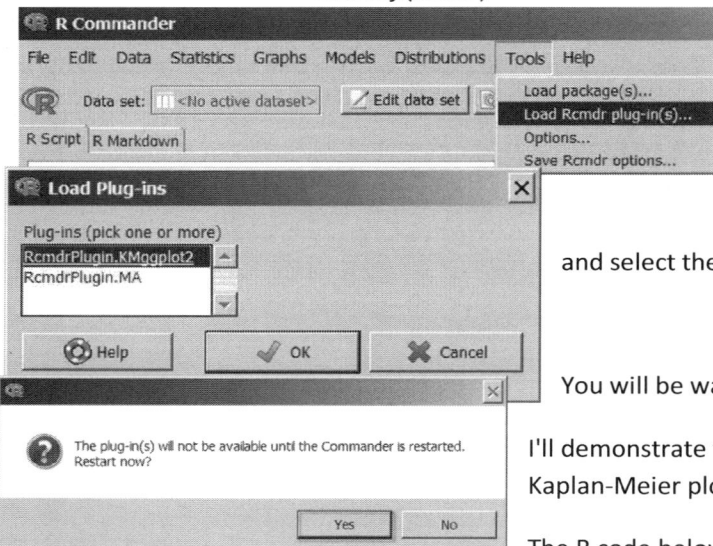

The last line of the above R code loads R Commander. Once in R Commander select the menu option

Tools->load Rcmdr plug-ins

and select the *RcmdrPlugin.KMggplot2* plugin.

You will be warned that R Commander will close and restart, click **YES.**

I'll demonstrate the use of this plugin by reproducing the final Kaplan-Meier plot produced in the *Plotting survival over time* chapter.

The R code below loads the data, displays the column names and displays the data.

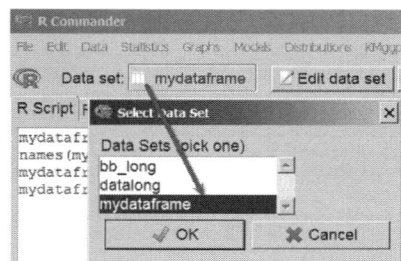

```
mydataframe <- read.delim("http://www.robin-beaumont.co.uk/virtualclassroom/book2data/survival_collett_p290.dat", header=TRUE)

names(mydataframe); mydataframe
```

You then need to make the mydataframe dataframe the active one by clicking on the area to the left of the *Data set: text* and selecting the *mydataframe* option from the list. Click on **OK**.

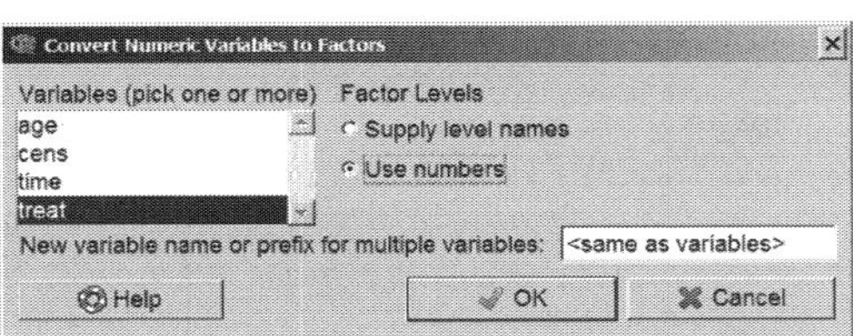

The treat variable also needs to be converted to a factor by either selecting the R Commander menu option:

Data->manage variables in active dataset->convert numeric variables to factors

Complete the dialog box as shown opposite, or simply type the following in the R Commander *Script* window then selecting and clicking on the **submit** button:

```
mydataframe$treat <- as.factor(mydataframe$treat)
```

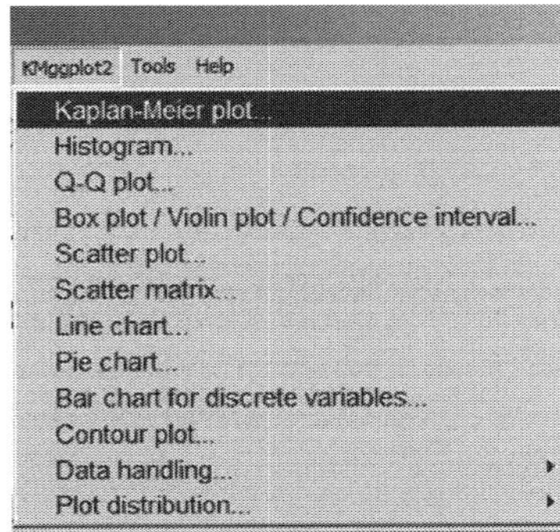

From the R Commander **KMggplot2** menu select the **Kaplen-Meier plot** option.

To produce the graph below right complete the dialog box as shown below left. The **Title** text box allows you to type text and scroll to add the whole of the line *Ovarian Cancer survival (n=26).*

You also can specify a **No at risk** table below the plot as discussed in the *Plotting survival over time: K M (Kaplan-Meier) plots* chapter. You can indicate the number of columns you want for the table by editing the **Tick count** value.

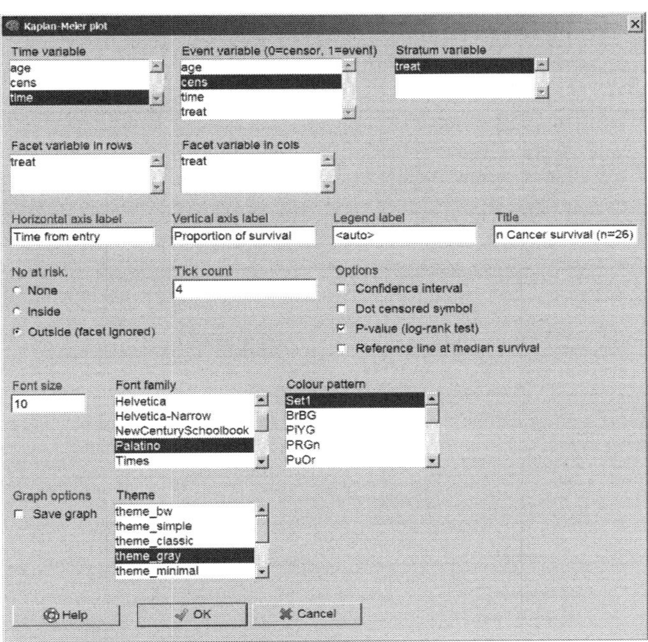

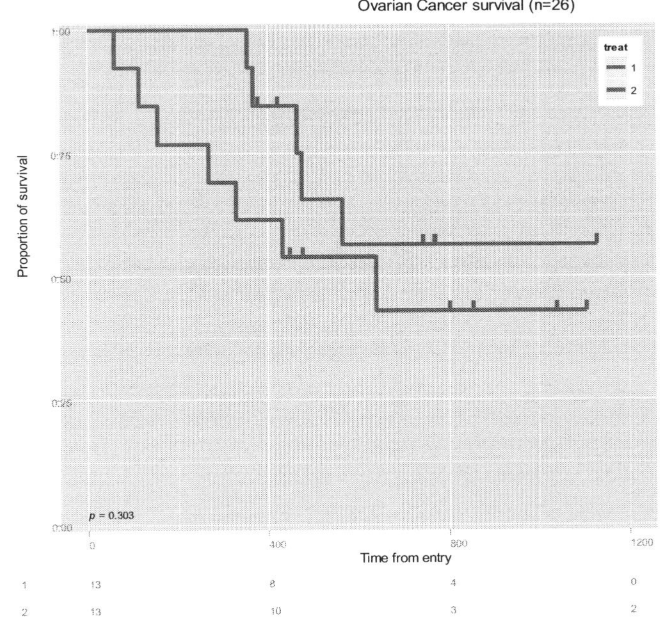

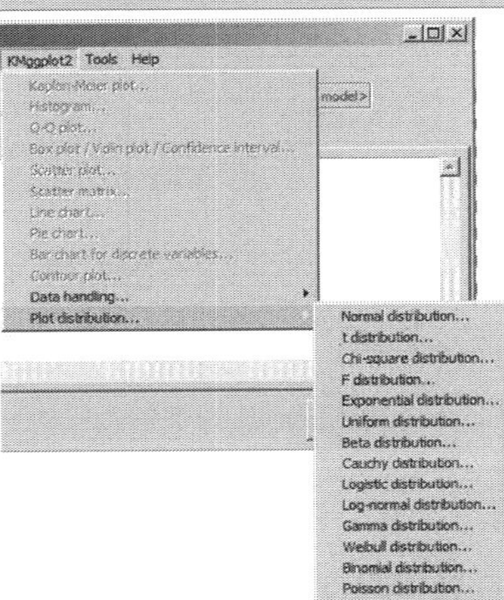

The *RcmdrPlugin.KMggplot2* R Commander plugin also allows you to produce many more *ggplots* including equivalents to the distribution graphs produced by R Commander, the menu is shown opposite.

57.2.2 Deducer

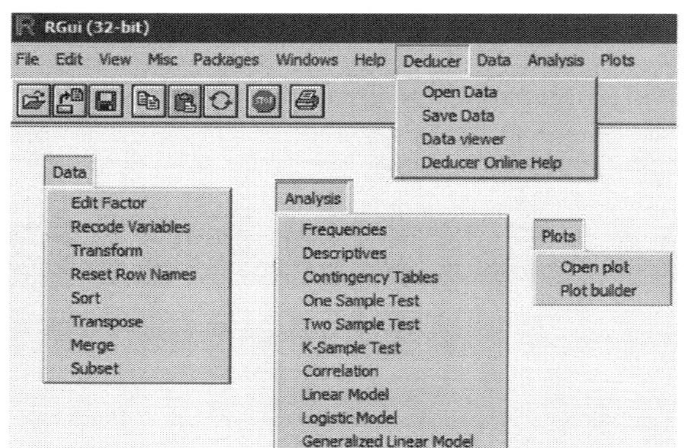

Deducer is an R extension similar to R Commander. You can install and load it as you would any R package:

```
install.packages("Deducer", dependencies=TRUE)

library(Deducer)
```

Deducer can be tricky to install, see the *Tips & tricks* section of the *Confidence intervals for effect sizes: Non-central distributions* chapter for details.

After loading *Deducer* the main R menu is modified adding four menu options as shown opposite. We will not consider the Analysis options, instead concentrating on the plot builder and data menus.

To demonstrate Deducer's plotting capabilities, we will reproduce the plot produced in the previous linear regression lines section. Download the following datafile and save it on your device.

`http://www.robin-beaumont.co.uk/virtualclassroom/book2data/correlation_vo2max.dat`

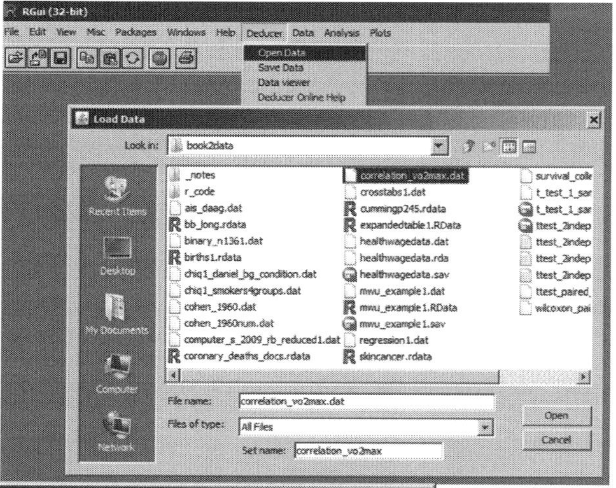

Select the menu option

Deducer->OpenData

A standard select file dialog box appears, select the *correlation_vo2max.dat* file.

The **Set name** option allows you to change the dataframe names associated with this data, While I have kept it as the default, being the same as the filename, I could have also called it *mydataframe* or anything else.

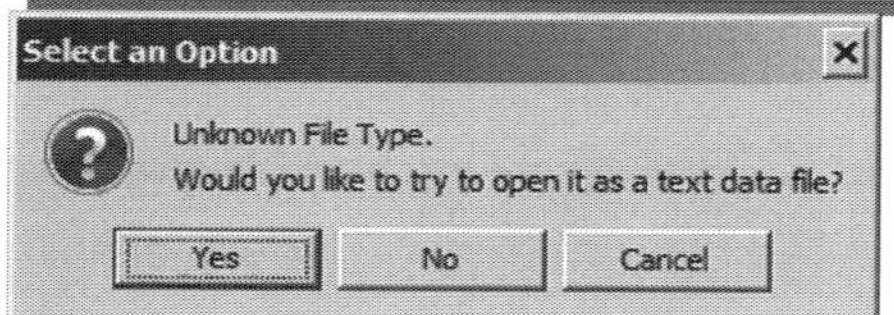

A dialog box appears querying the file type, click on the *Yes* button.

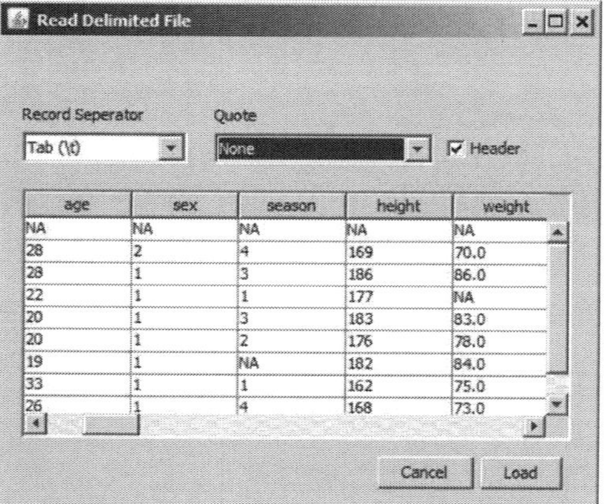

A dialog appears showing the data allowing you to indicate if the file has a header (it does) and what character is the column separator, being a *tab* in this instance.

Click the **Load** button.

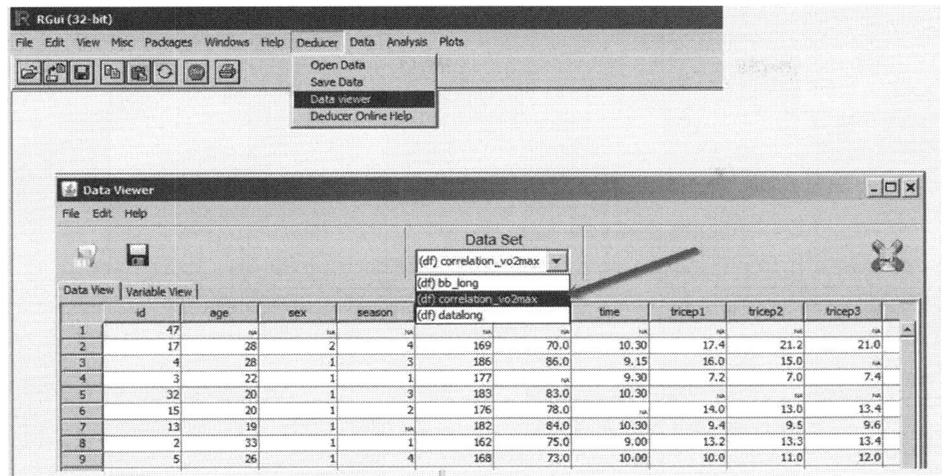

You can see the data with the menu option:

Deducer->Data viewer

When the data viewer window appears, you can select any of the dataframes you have loaded. Make sure the dataframe you just opened is the dataset in view. To close the window click the **X** at the far top right of the window.

We are now in a position to start building the plot now.

57.2.2.1 Plot building with the plot builder

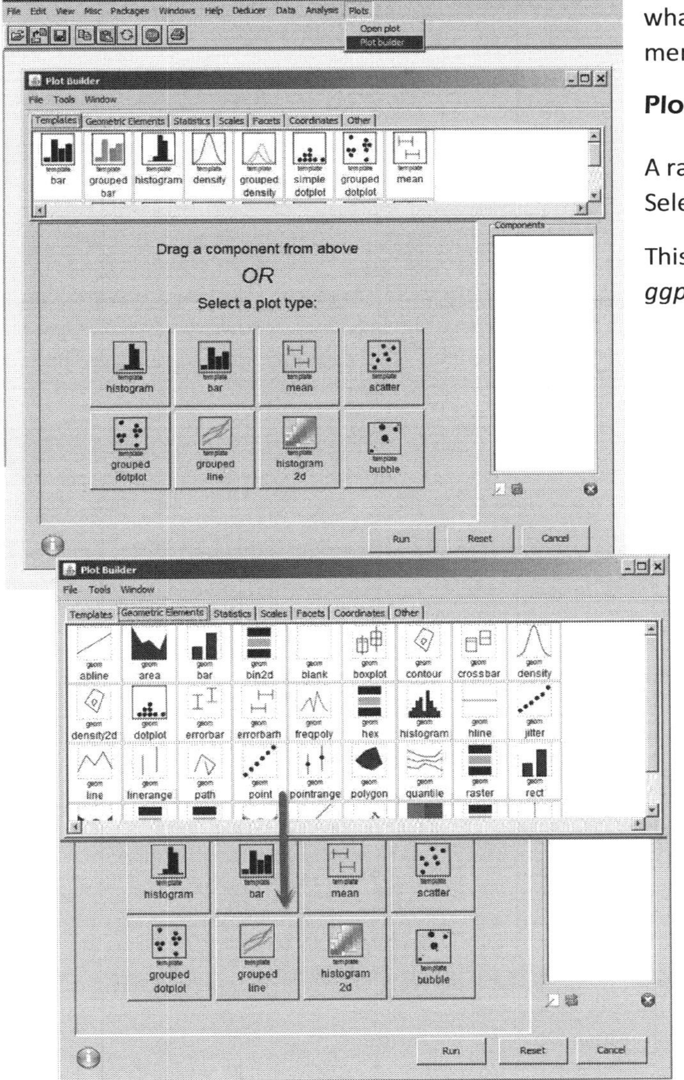

The plot builder is similar to the SPSS plot builder but follows what I feel is a more logical approach. To call it up select the menu option:

Plots->Plot builder

A rather busy window appears most of which you can ignore. Select the **Geometric Elements** tab.

This tab equates with the Geometrics – *geoms()* aspect of the *ggplot2* package.

The top panel expands to show all the geometric elements available.

We will start by creating a scatterplot of the points differentiating between males and females. Select the point icon and drag it down into the central panel.

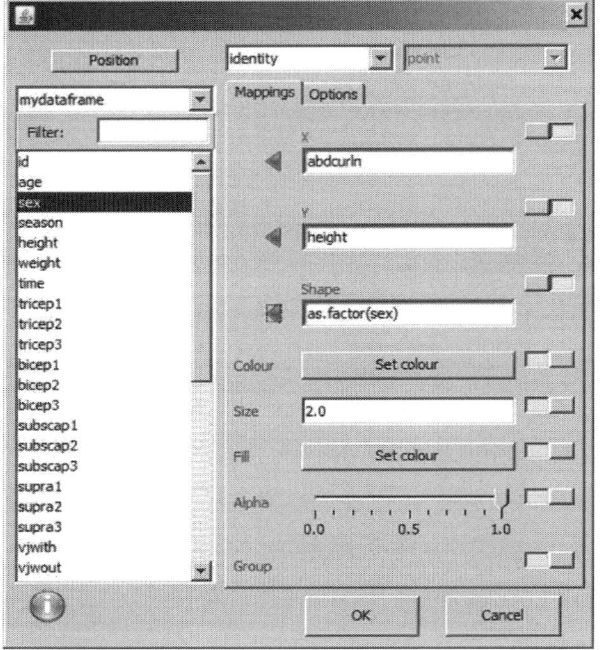

A window appears in which you specify the various characteristics of this *geom()*, set the following:

x = abdcurln

y=height

To indicate that you want the shape of the point to depend upon the level of a particular factor click on the little **slider** to the right of the shape box. Doing this moves it from the right to the left indicating that the shape is now defined by a variable.

Move the **sex** variable across. Not that the plot builder has also converted the variable to a factor.

Finally click on the **OK** button.

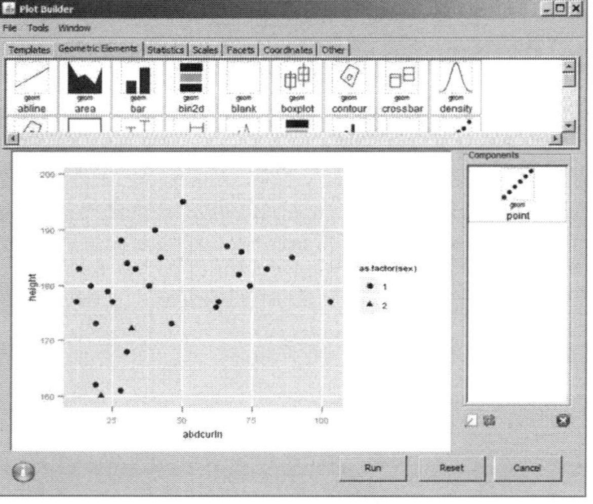

The central panel displays the result, shown opposite.

We now need to add the regression lines.

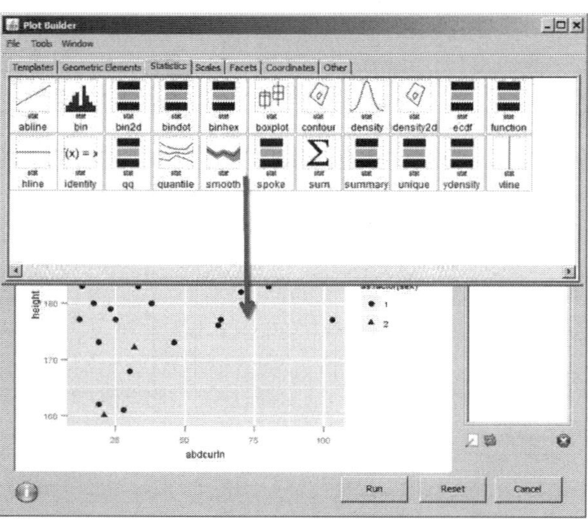

Select the **Statistics** tab and then draw the **smooth** icon into the central panel.

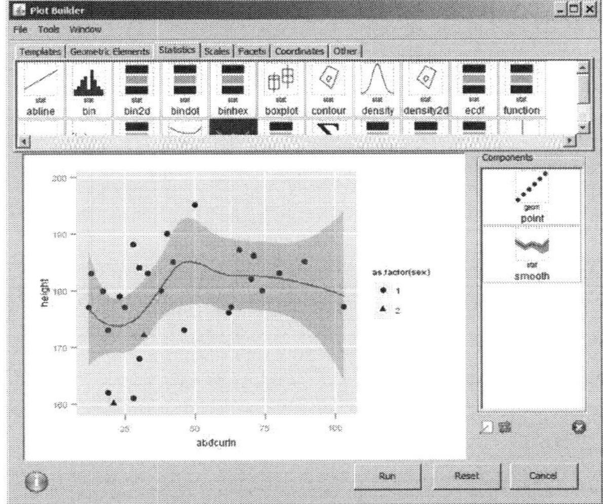

The result is not quite what we want because:

- line is curvy, not straight
- only one line shown, not a separate one for each gender.

To change these characteristics we need to open the aesthetics panel for the smooth geom object. That sounds rather complex but all it means is that you double click on the **smooth** icon in the components panel to the right of the central one, as shown opposite.

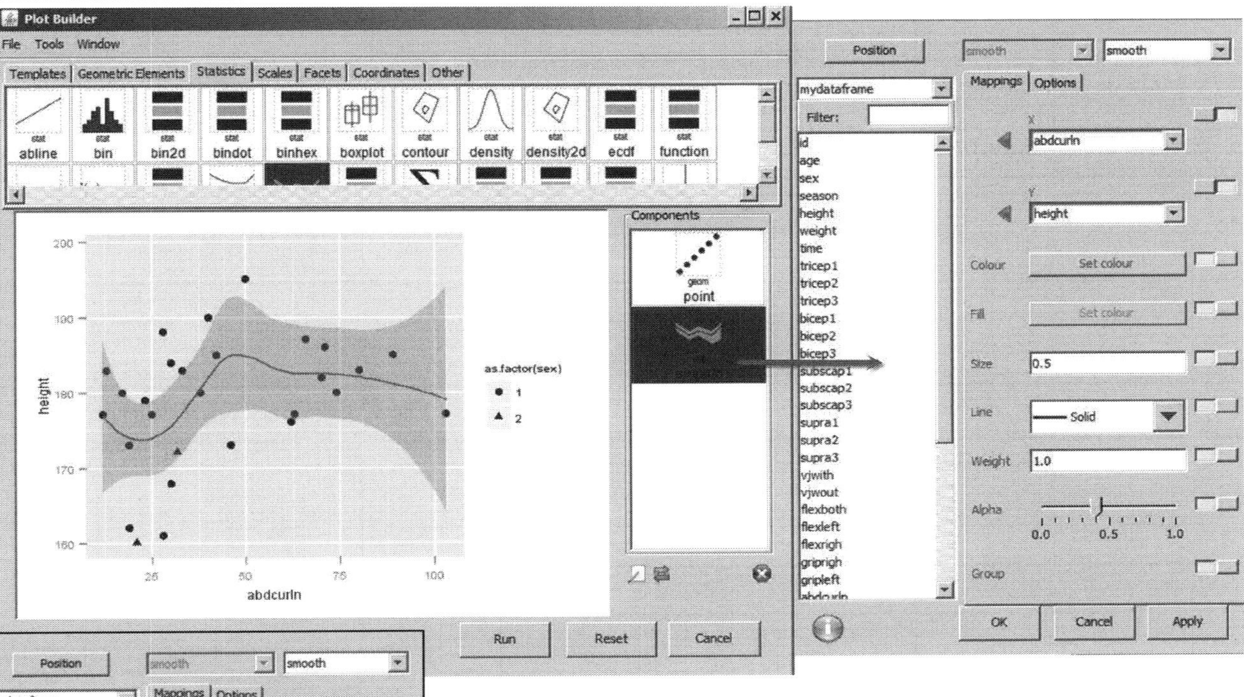

To set the line type to vary for each gender:

- select the slider to the right of the line box, moving it from right to left (shown above).
- move the *sex* variable across to the **Line** box (shown opposite).

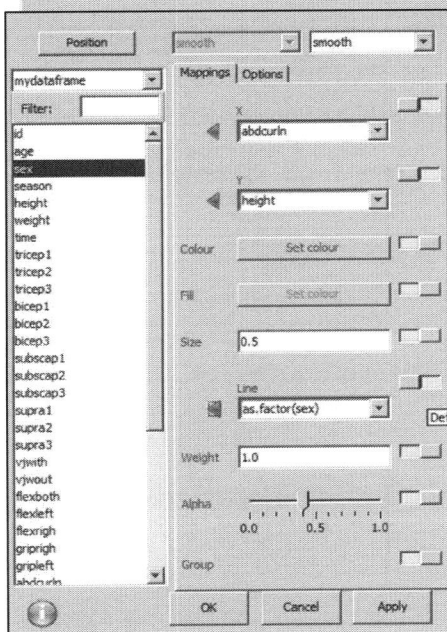

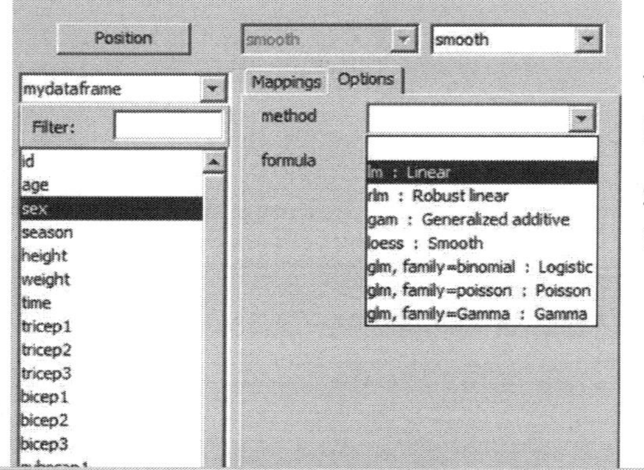

The only edit left to do is to specify that we require linear regression lines rather than the default **LOESS** lines.

Select the **Options** tab and select from the drop down box *Lm:Linear*. Finally click on the **OK** button.

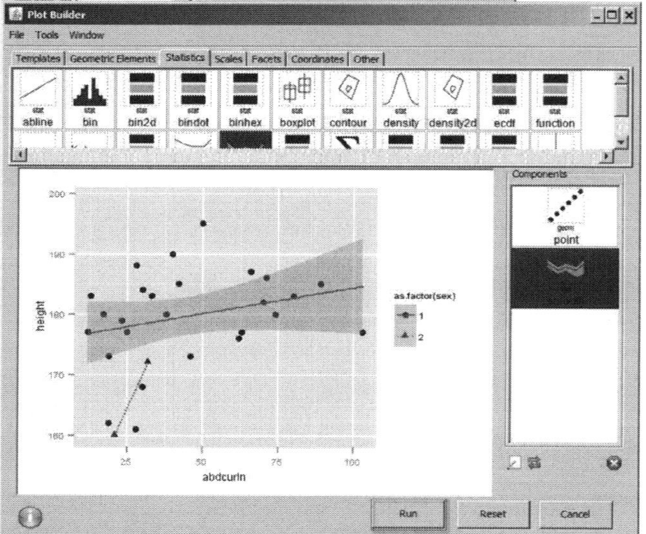

The result is now very similar to that produced using the *ggplot()* functions described previously.

Clicking on the **Run** button produces the plot in a separate window.

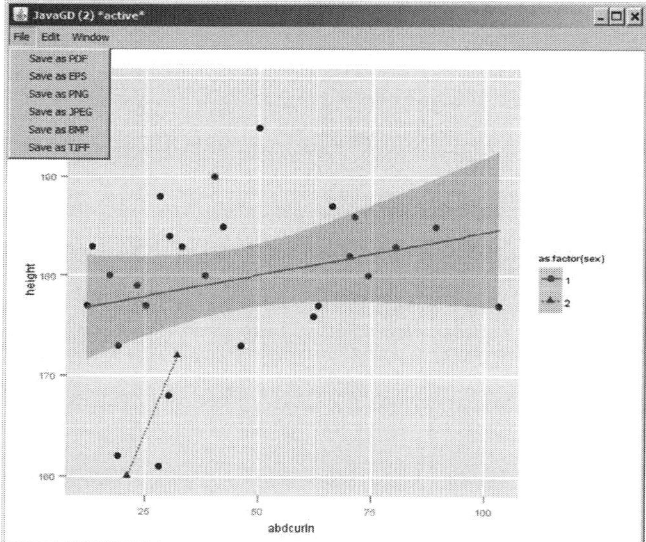

The first menu option **File** provides you with several save as options.

In the *R Console* window we can see the R code produced by the plot builder.

The R code produced is very similar to that we produced previously.

I had previously renamed the dataframe to *mydataframe* which I did not describe above so the code in your window will look slightly different to that below.

```
-- Deducer Command --
dev.new()
ggplot() +
        geom_point(aes(x = abdcurln,y = height,shape = as.factor(sex)),data=mydataframe) +
        stat_smooth(aes(x = abdcurln,y = height,linetype = as.factor(sex)),data=mydataframe,method = lm)
-- End Command   --
There were 13 warnings (use warnings() to see them)
```

57.2.3 ggplot2 builder on the web - Jeroen Ooms

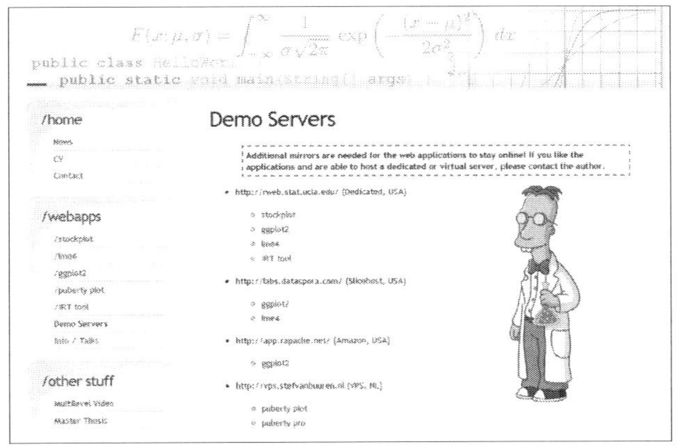

While Deducer is an extension to R that you download, Jeroen Ooms provides a online interactive version of *ggplot2* available at:

`www.stat.ucla.edu/~jeroen/live.html`

Select any of the *ggplot2* sites.

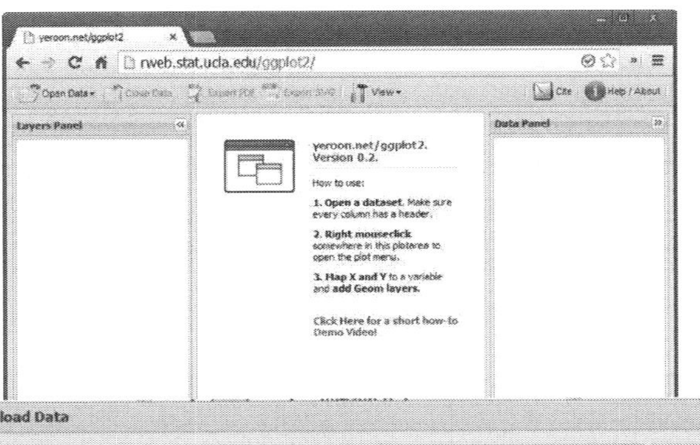

The sparse window has only one option available **Open Data**.

Click on the option and select **Upload File**.

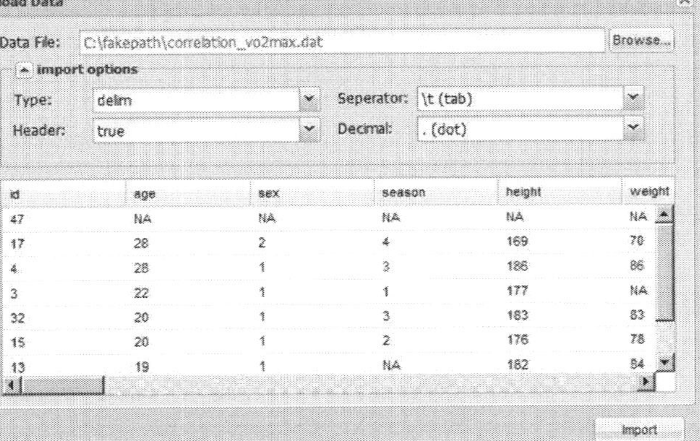

The *Upload Data* dialog box appears. Click on the **Browse** button (far right) and the usual file select dialog box appears select the file downloaded from the previous section:

`correlation_vo2max.dat`

Click on the **Import** button.

The panel on the far right now fills with the data (you may need to expand the categories by clicking on them to see the list)

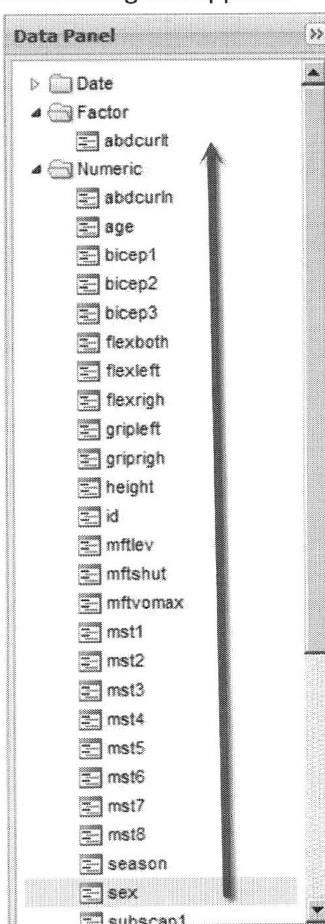

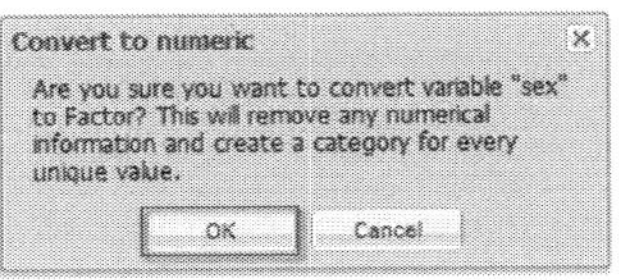

The sex variable needs to be converted to a factor which is achieved by dragging it to the factor category. A warning dialog box appears which you respond by clicking on the **OK** button.

For any graphs you develop you need to specify the x and y coordinates. This is done by right clicking on the main plotting area and then selecting the variable for each axis. For this example set:

$x = abdcurln,\ y = height$

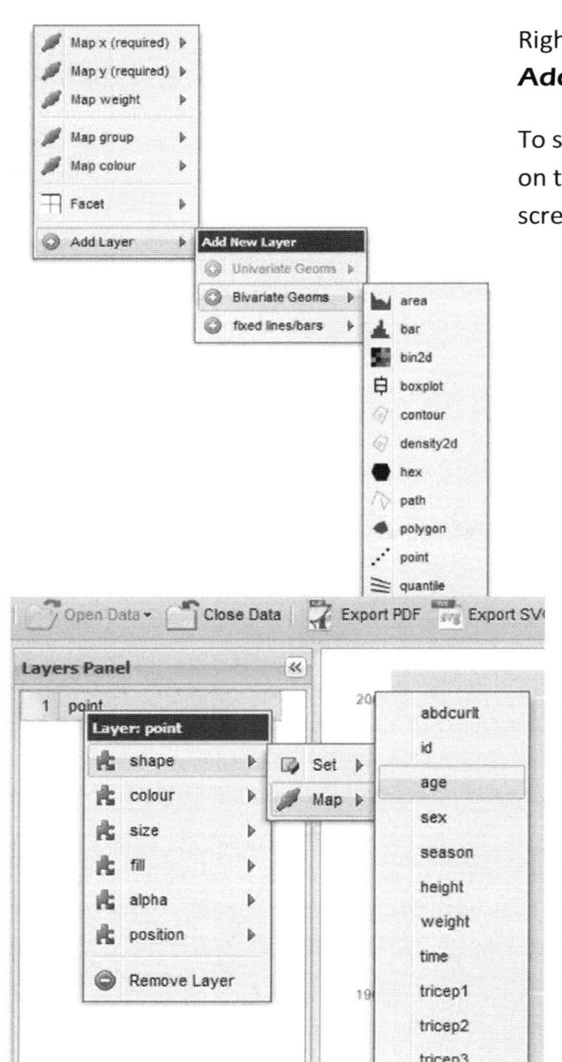

Right clicking a second time on the main panel allows the selection of **Add Layer** where you can then select *Bivariate Geoms* and then point.

To see the effect this has had you need to click on the **Draw plot** button at the bottom of the screen.

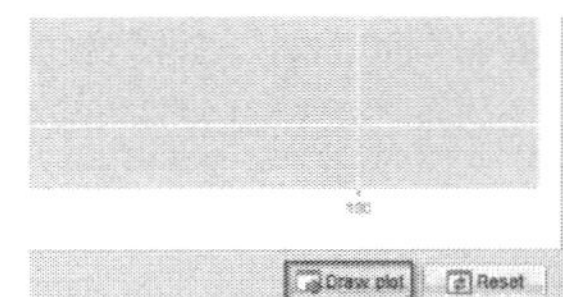

To change the symbol for the two different sexes you need to edit the *point* entry in the **Layers Panel** (far left) as shown opposite. Open up the menu by right clicking on the *point* entry.

Remember to click on the **Draw plot** button at the bottom of the screen to see how the plot has changed.

To add the regression line(s) repeat the stages above but select **Add layer->Bivariate Geoms->path** this time. You would also need to repeat the **map** by *sex* option as well. I also changed the size of the symbols in layer one to 5.

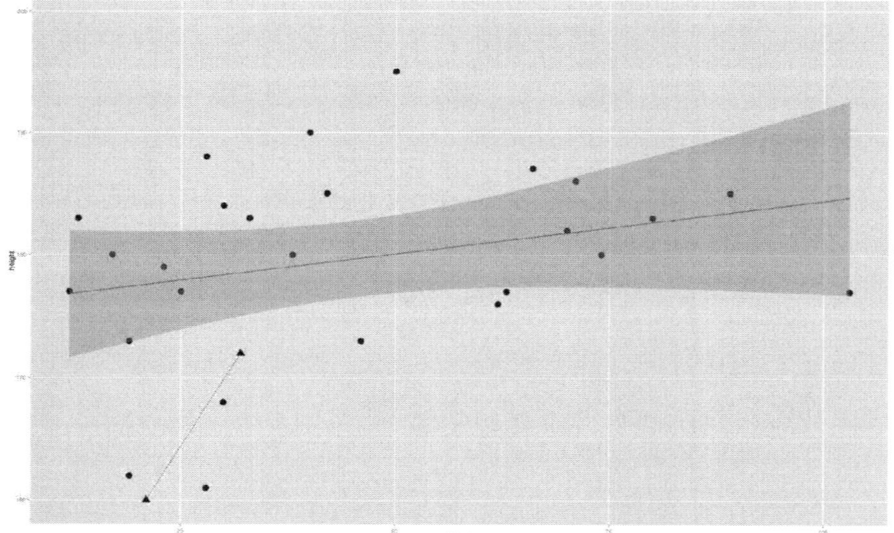

You can see the R code generated by clicking on the view button at the top of the screen, selecting the code panel option.

The result of which is:

```
ggplot(myData, aes(x=abdcurln, y=height)) + geom_point(size=5, aes(shape=sex)) +
geom_smooth(method="lm", aes(linetype=sex))
```

The code produced is very similar to that created by Deducer and originally in R.

57.3 Summary

This chapter has provided a very brief introduction to the many ways you can begin to produce publication quality plots using *ggplot2*. The chapter started by using *ggplot2* directly and finished by demonstrating three point and click interfaces to *ggplot2*. At the time of writing *Deducer* appears the most mature and complete implementation of a *ggplot2* interface.

57.4 Tips and Tricks

The current reference for *ggplot2*, along with examples for each of the functions can be found at: http://docs.ggplot2.org/current/

There are many snippets of code containing calls to the *ggplot2* functions on the Internet. A good place to start is http://stackoverflow.com/ . The search box is at the very top of the screen.

A mixed approach to developing *ggplot2* plots can be used. For example, firstly by using one of the point and click interfaces to get a basic plot then manually editing the R code produced to fine turn the plot.

The current reference for *Deducer*, along with examples and details of extensions can be found at:

http://www.deducer.org/

If you wish to develop charts that look like those on google in R *googleVis,* is a R package providing an interface between R and Google Chart Tools allowing you to visualise data with Google Chart Tools without uploading the data to Google.

The definitive article concerning mathematical notation is Murrell & Ihaka, 2000. Some nice examples in R can be found in R help by typing *?plotmath()* into the R Console window, alternatively you can see some spectacular examples at:

http://blog.snap.uaf.edu/2013/03/25/mathematical-notation-in-r-plots/

There are many guidelines produced concerning formats for particular journals, however the APA (American Psychological Association) provide the most detailed and wide ranging recommendations, Adelheid & Pexman, 2010 provide full details.

58 More regression techniques

The desire to demonstrate the relationship between one or more inputs to an output has resulted in regression being developed in very many ways. This section investigates some of these including; multiple inputs and where they might be all nominal and possibly not independent, outputs that are binary or counts and situations where you can consider ways of identifying clustering of inputs. Identifying such clustering allowing you to either combine or effectively drop duplicate inputs. This section ends with a technique, Structural Equation Modelling (SEM) which combines most of these techniques allowing you to simultaneously analyse models that consist of more than one regression analysis. To demonstrate SEM I have made use of the free software application Ωnyx from http://onyx.brandmaier.de/what-is-onyx/. Ωnyx not only provides a graphical approach to developing SEM models but also produces the relevant R code.

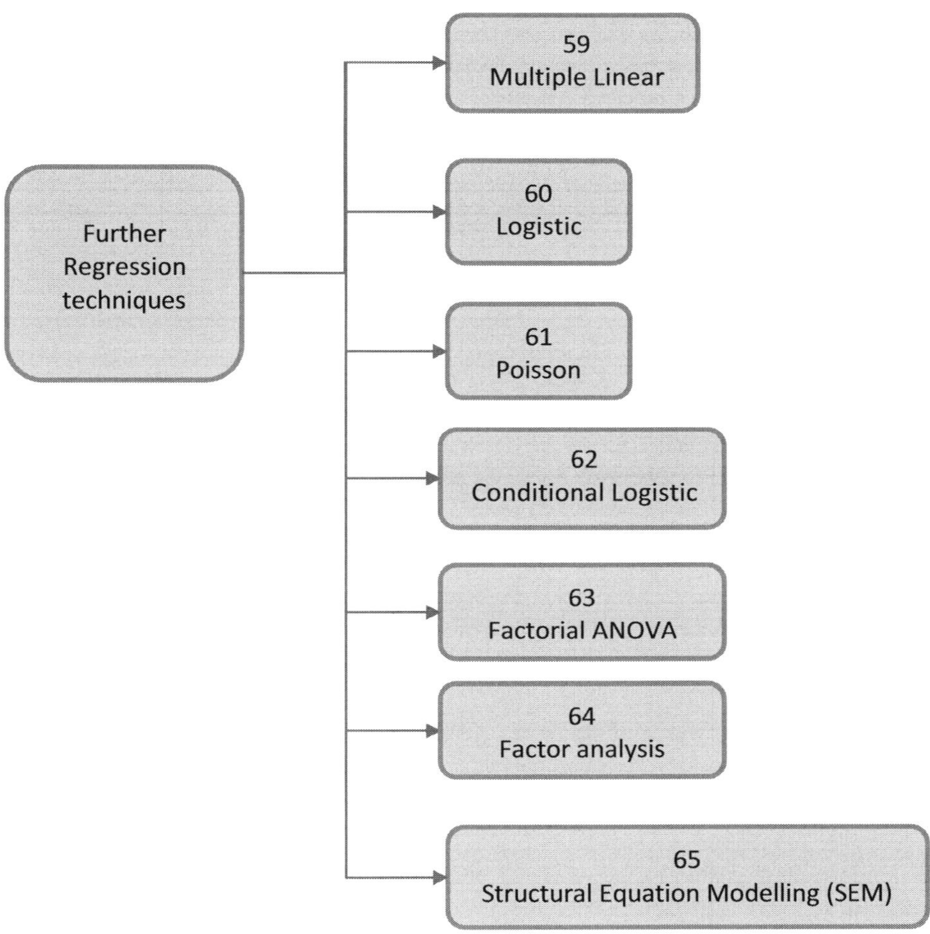

59 Multiple linear regression: measuring the influence of several variables on a continuous variable

The chapter assumes you have read the introductory chapter on simple linear regression, *Measuring the influence of one variable on another.*

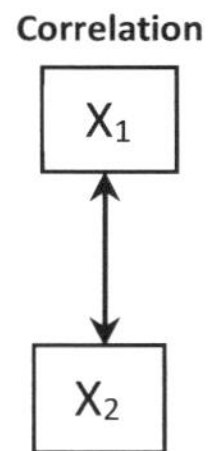

Multiple linear regression is used when you are assessing the influence several variables may have upon a single interval/ratio variable, called the dependent variable or outcome. Under certain circumstances, a dependent variable that has an ordinal **level of measurement,** can be treated in the same way (Rhemtulla, Brosseau-Liard & Savalei, 2012).

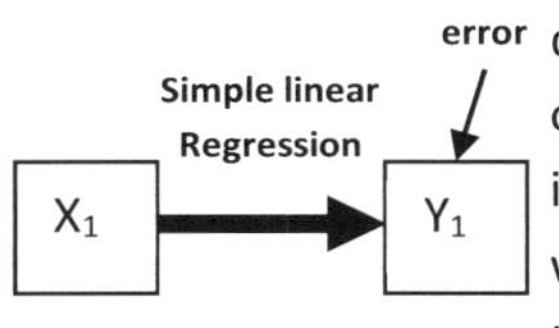

A good way of thinking about multiple linear regression is as an extension of both correlation and simple linear regression, both topics we have already covered. In correlation, the direction of the influence is in both directions, as indicated by the double-headed arrows opposite. However, in simple linear regression there is only one input and one output, both of interval/ratio level of measurement. In addition there is an error, also called the residual, which is the discrepancy between the predicted values from the model and those observed in the dataset.

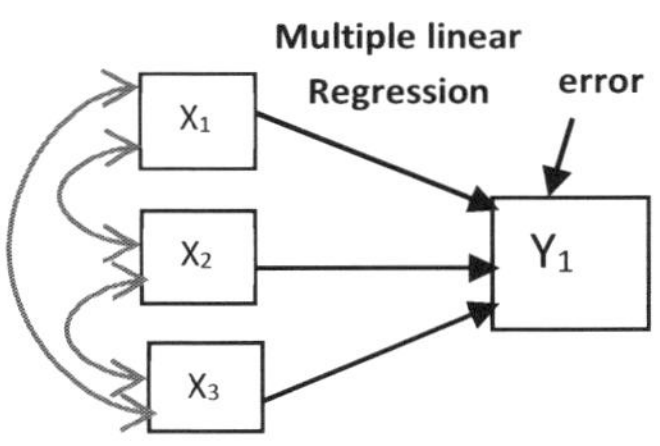

In contrast, in multiple linear regression, we have several input variables which can be of any level of measurement along with a single outcome which does still have an interval/ratio level of measurement.

Other things to note about the multiple linear regression model include the following, all of which will be discussed in detail later:

- the model takes into account the possible correlation between the input variables, but when they are highly correlated problems occur with the model produced. This is because the model estimates are only accurate when the input variables show a low level of correlation between themselves. In contrast, we want a high correlation between the inputs and the output.
- the predicted value $\hat{y}_i$ is not considered to be a perfect measure but is assumed to follow a normal distribution and therefore possesses a standard deviation. This is the same as in simple linear regression.
- there can be as many input variables as you want – but then you need large sample sizes. A rather old fashioned rule of thumb is that you should have at least 10 observations for each input. Therefore, for a model with 4 inputs you would expect a sample of at least 40 complete observations. However, the required sample size is also dependent upon the degree of correlation between the inputs, as well as other factors, the higher the correlation the larger the sample required (Brooks & Barcikowski, 2012 p.9).

The above multiple linear regression model can be expressed mathematically:

$\hat{y}_i=a+b_1x_i+b_2x_i+b_3x_i$ In fact we can extend the model to show an unlimited number of inputs: $\hat{y}_i=a+b_1x_i+b_2x_i+b_3x_i+b_4x_i+b_5x_i \ldots +b_nx_i$

The analysis of multiple linear regression focuses on three inter-related aspects:

- overall fit - Multiple R squared (R^2) assessment. Where R is the correlation between the observed and predicted values and its squared value representing the proportion of variability accounted for by the model.
- parameter estimation / evaluation of the regression coefficients (b's) and their standardised versions betas (β) etc. the modeller needs to decide which inputs to keep and which should be removed from the model?
- assessment of statistical validity of the model – LINE criteria, extrapolation, subgroups, causal implications etc.

The LINE criteria were discussed in the chapter, *measuring the influence of one variable on another: regression.*

59.1 The theory behind Multiple Linear Regression

The following sections provide details of the theory behind multiple linear regression. Please feel free to skip them if you just need the practical aspects.

59.1.1 Partial correlation and b's (regression coefficients)

While we have discussed the aspects listed above in the chapter on simple linear regression the first two issues become more complex with more than one independent (i.e. input) variable. To interpret the results from such an analysis it is necessary to understand two additional types of correlation, the **partial** and **part** (also called **semipartial**) correlations.

Scores from which another variable's influence has been removed are called partial scores or residual scores.

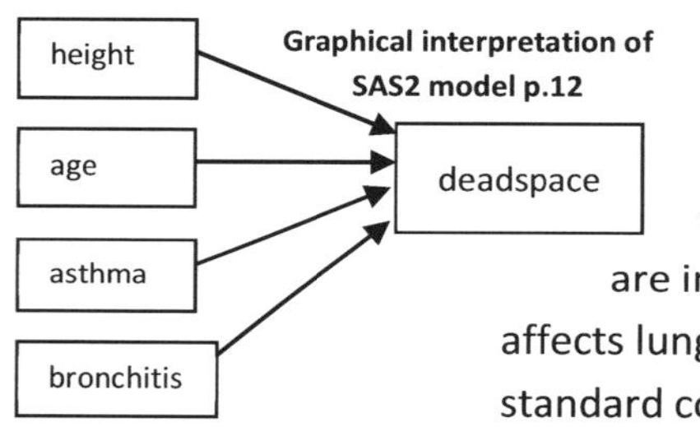

Graphical interpretation of SAS2 model p.12

To explain the concept of a partial correlation I will use the dataset presented in the excellent Multiple Linear Regression chapter in **SAS2**. The data is from patients who have had their lung function assessed by measuring their deadspace. Here we are interested to find out if such things as height, age, asthma and bronchitis status affects lung deadspace, and to what extent. Because partial and part correlations are based on standard correlations (also called zero order correlations) we will start by looking at them.

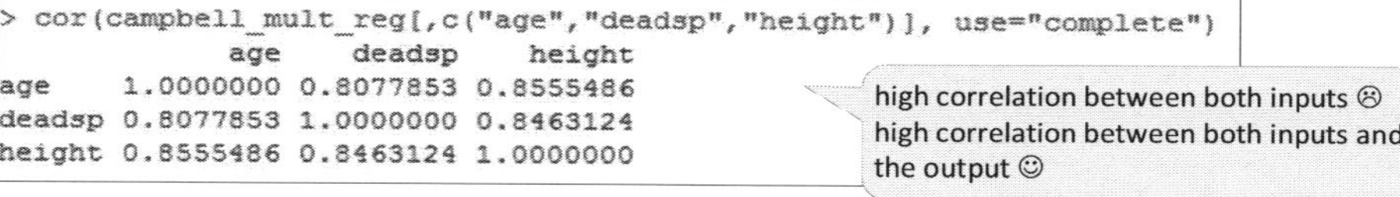

```
> cor(campbell_mult_reg[,c("age","deadsp","height")], use="complete")
             age     deadsp    height
age    1.0000000 0.8077853 0.8555486
deadsp 0.8077853 1.0000000 0.8463124
height 0.8555486 0.8463124 1.0000000
```

high correlation between both inputs ☹ high correlation between both inputs and the output ☺

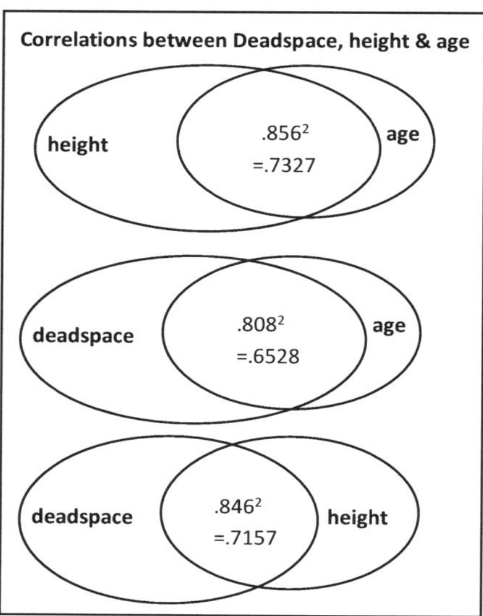

Taking the squared correlations between deadspace and height, we can express this graphically in the form of an informal Venn diagram. This value, the coefficient of determination (R^2) represents the overlapping proportion of the two circles. All the variables show a high level of overlap. What would be useful to know is how much of a correlation exists once we have removed a particular variable (or variables) effect on both variables taking part in the correlation. This is what a partial correlation does; it removes the effect of other variable(s) upon BOTH the variables being correlated.

Partial correlation takes into account effect upon BOTH variables

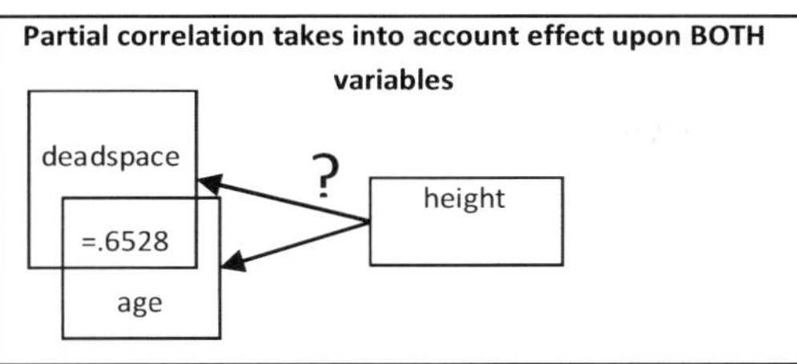

Partial correlation for deadspace & age controlling for height

deadspace
d
a
b
c
age
ć
height

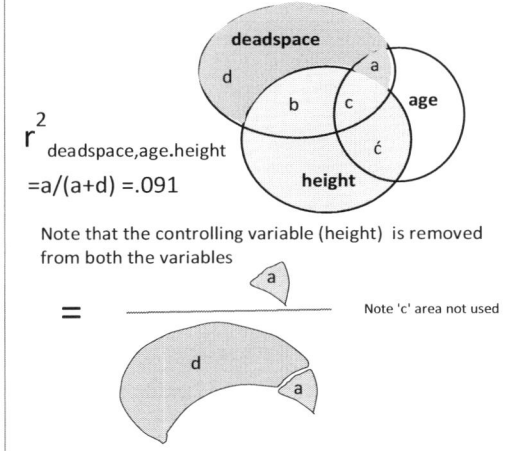

$r^2_{deadspace,age.height}$ $=a/(a+d)=.091$

Note that the controlling variable (height) is removed from both the variables

= a / (d + a)

Note 'c' area not used

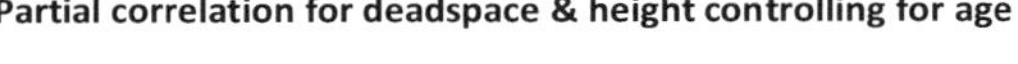

Partial correlation for deadspace & height controlling for age

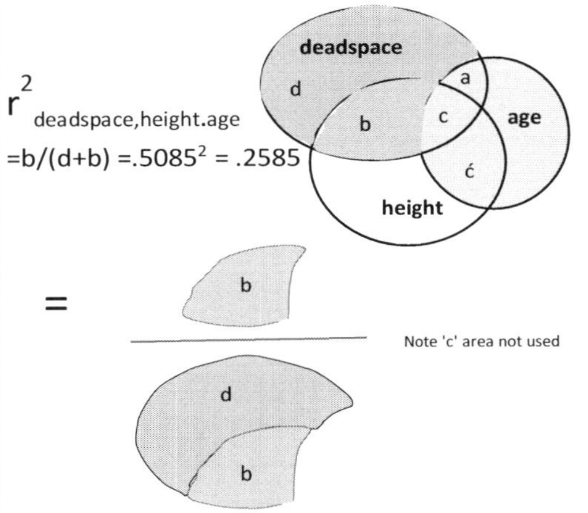

$r^2_{deadspace,height.age}$ $=b/(d+b)=.5085^2=.2585$

Conceptualise this as a regression with two inputs age & height

This is the way standard linear regression works (type III sums of squares). Ordering of the inputs is NOT relevant because:

- d is the same size for each calculation
- unique contribution of each input (i.e a & b) after all the others have been taken into account.

Take for example the correlation between deadspace and age, what is left after we have removed any effect height has upon this correlation? The partial correlation provides the answer.

How to write Partial correlations

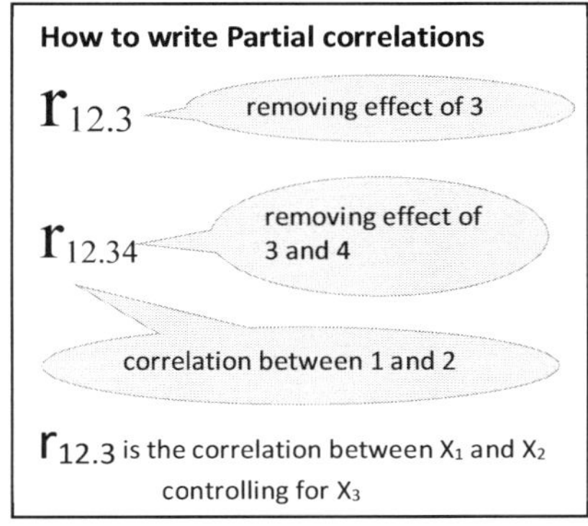

$r_{12.3}$ is the correlation between X_1 and X_2 controlling for X_3

The result (explained overleaf) is 0.303 ($r^2_{deadspace,age.height}$=.091). Comparing this to the original correlation of .808 (r^2=.6528), there is a drop of over 50% when we take into account the effect height has upon both variables. Udny Yule, a Scottish statistician who died in 1951, developed the various notations presented here in 1907 in the Proceedings of the Royal Society, volume 79.

A specific type of Venn diagram (called a Ballentine) shown opposite provides a graphical explanation. Each circle represents the variance for a particular variable and the overlapped areas the shared variance (=covariance). Therefore taking the covariance between deadspace and height (b+c), the correlation between them is this value divided by the product of their square roots of their respective variances (i.e. standard deviations) which is (b+ c)/ (√deadspace*√height) where here I mean deadspace= a+b+c+d and height=height+b+c+ć.

Partial correlations are represented as ratios, the first element being the variance shared by the two variables having already removed the influence of any other specified variables (area a or b in the example opposite). The other element in the ratio is this value plus that variance unique to the first variable having removed the influence of the other specified variables.

Partial correlations are very useful for seeing how much a correlation is affected by other variables, in a sense you can think what it is doing is creating some type of 'pure' correlation between the two variables.

Partial correlations relate to the b's in simple and multiple regression in different ways. In simple linear regression we had $b=cov(x,y)/var(x)$ which is equal to $r_{xy}(sd(y)/sd(x))$, that is the correlation between the input and output variables multiplied by an expression consisting of the standard deviations (sd) of the output divided by that of the input. Consequently, the standardised b (β=beta) is simply r_{xy}, hence, only in simple linear regression is this value correctly interpreted as a correlation.

In contrast this interpretation is not possible in multiple regression because here the the b's (technically called the regression coefficients) are modified to take into account the other input variables by being formed from both a partial correlation and a partial standard deviation. I have not described partial standard deviations but they are formed in much the same way as partial correlations. In fact the general multiple linear regression equation can be rewritten in the form of partial correlations and partial standard deviations:

$$\hat{y} = a + r_{12.34...n}(s_{1.23..n}/s_{2.13...n})X_1 + r_{13.24...n}(s_{1.3..n}/s_{3.12...n})X_2 + r_{14.23...n}(s_{1.4....n}/s_{4.12...n})X_3 + r_{15.23...n}(s_{1.5..n}/s_{5.12...n})X_3 \ldots$$
$$+ r_{1n.23...(n-1)}(s_{1.n}/s_{n.1...n})X_n$$ (Yule & Kendall, 1953 p.287)

Ignoring the many subscripts in this rather austere equation highlights the fact that underneath it all we are still only dealing with correlations and standard deviations, admittedly partial ones. Probably the best way to think of the regression coefficients in multiple regression is as gradients between a specific input and the output variable.

59.1.2 Part (semi-partial) correlations

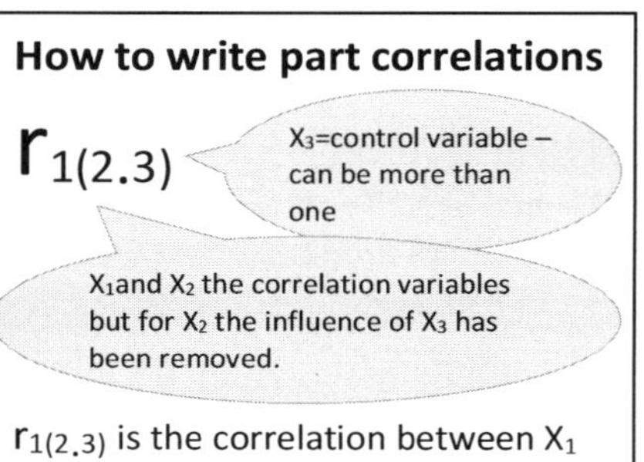

How to write part correlations
$r_{1(2.3)}$
X3=control variable – can be more than one
X1and X2 the correlation variables but for X2 the influence of X3 has been removed.
$r_{1(2.3)}$ is the correlation between X_1 and X_2, controlling for X_3 in X_2

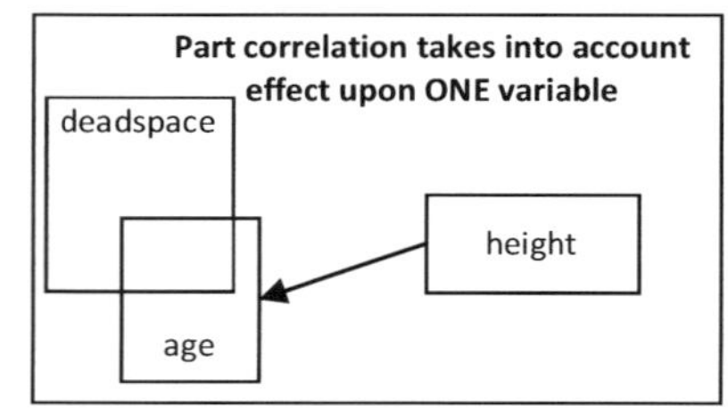

Another type of correlation, the part correlation (also called semi-partial or *sp* for short), also plays a very important part in multiple linear regression. The part correlation indicates roughly the relative weight each input variable provides to the model. Again, you can demonstrate this relationship in the form of an equation that links the part correlations for each of the input variables with a measure of overall model fit, that is the Multiple R squared value (Howell 2007, p.512):

$$R^2_{0.123..p} = r^2_{01} + r^2_{0(2.1)} + r^2_{0(3.12)} + \ldots + r^2_{0(p.123..p-1)}$$

Multiple R squared	Correlation between outcome + first input	Part correlation between outcome & second input, controlling for other inputs	Part correlation between outcome & third input, controlling for other inputs	Part correlation between outcome & last input, controlling for other inputs

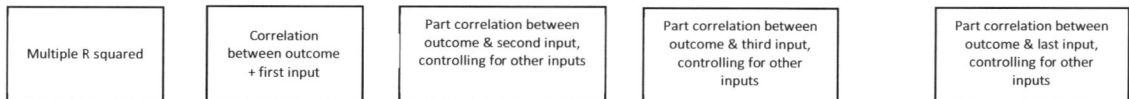

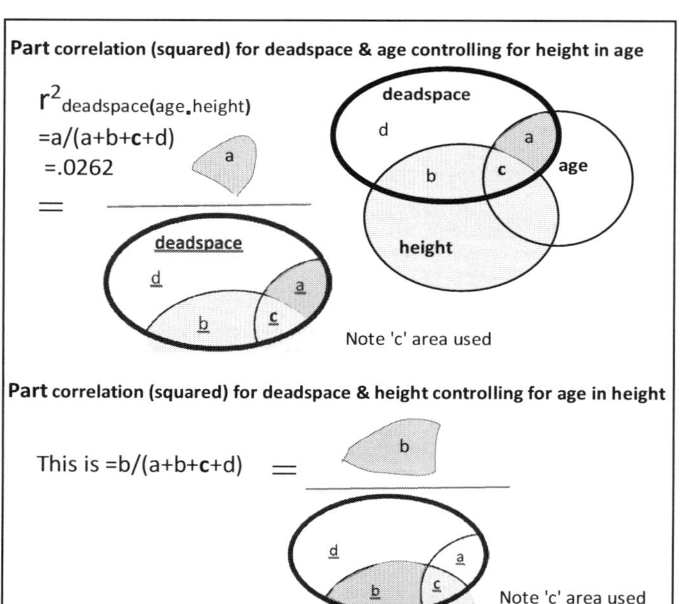

In contrast to partial correlations **Part** correlations take into account the effect variable(s) may have on only one of the variables being correlated. In multiple linear regression this is for each input variable where the effect of all the other inputs on it are removed.

What we are talking about here is statistical control, which is different and nowhere near as good, as experimental control – it is a second best approach!

Part correlations allow the calculation of the unique effect upon overall fit (i.e. multiple R^2) each new predictor brings to the model. Put another way it measures how much R^2 will drop if the variable is omitted from the model.

For example the part correlation between deadspace and age controlling for height in the age variable is 0.162 ($r^2_{deadspace(age.height)} = .0262$) (details later on as to how this value was obtained) indicating that the R^2 value would drop by 0.0262 if the variable were dropped from the model. This suggests that it plays a very small part in the model. As with the partial correlation calculations on the previous page, the ordering of the variables makes no difference here. In contrast, we will now consider a regression approach where order does matter.

59.1.3 Sequential / hierarchical regression (type I sums of squares)

In the above sections we have assumed that each input variable was well behaved, that is each was given equal importance. However, often a researcher wishes to indicate that variables are of decreasing importance or some other type of hierarchy. We will now consider a scenario where the researcher feels height is more important than age in predicting lung deadspace. Here we are considering a **sequential linear regression** . Thinking of the relationship this way should immediately make it clear that the order in which the inputs are entered into the equation matters in this case. However, this only matters when the input variables are correlated, as they usually are. This is demonstrated opposite where the size of both the partial and part correlations for the first input is increased due to it grabbing the shared area 'c', but 'c' only exists if the inputs are correlated.

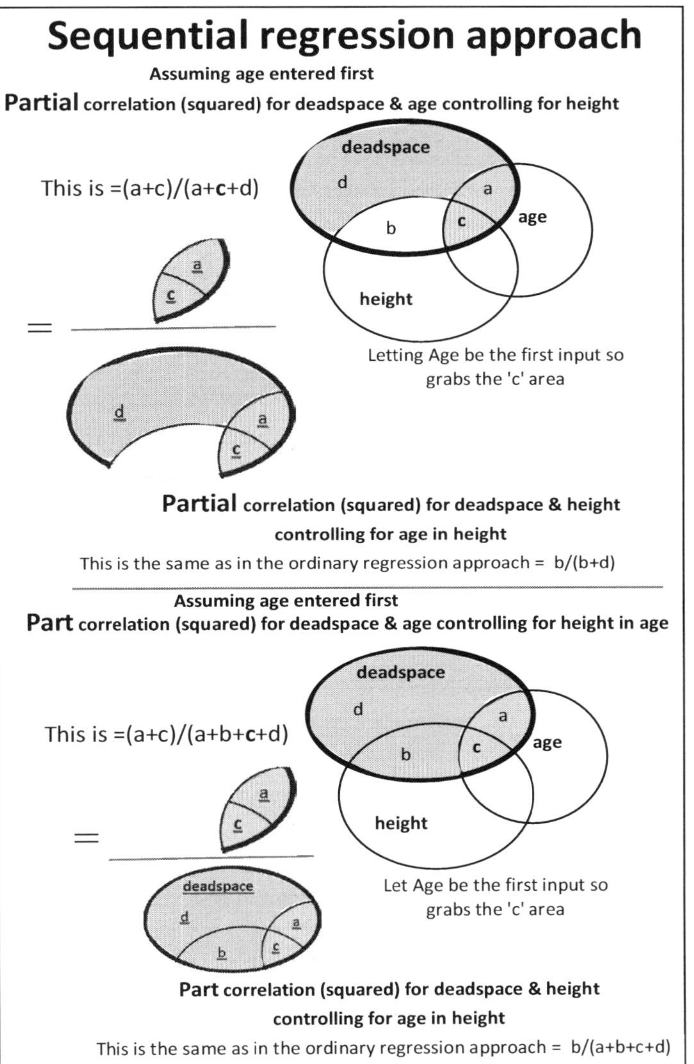

Sequential regression approach
Assuming age entered first
Partial correlation (squared) for deadspace & age controlling for height
This is =(a+c)/(a+c+d)
deadspace
d
a
b
c
age
height
Letting Age be the first input so grabs the 'c' area
Partial correlation (squared) for deadspace & height controlling for age in height
This is the same as in the ordinary regression approach = b/(b+d)
Assuming age entered first
Part correlation (squared) for deadspace & age controlling for height in age
This is =(a+c)/(a+b+c+d)
Let Age be the first input so grabs the 'c' area
Part correlation (squared) for deadspace & height controlling for age in height
This is the same as in the ordinary regression approach = b/(a+b+c+d)

This sequential way of adding the input variables in regression can be automated, and various computerised linear regression procedures, called *Stepwise Regression Procedures* or something similar, use the part correlation values to assess inclusion or exclusion in the formulation of the models.

The take home message here is to remember that with these types of sequential regression models the order in which the inputs are entered generally affects the results. Because of this and many other concerns, statisticians often have serious concerns about these automated procedures. I advise you to usually stick to the traditional block regression approach described earlier and as discussed later you can always compare specific models using R Commander.

deadspace d a b c age f height

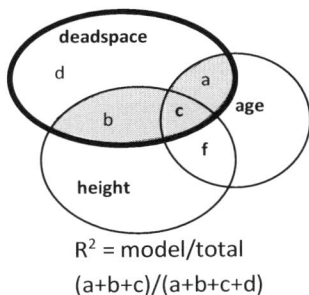

R^2 = model/total (a+b+c)/(a+b+c+d)

59.1.4 Venn diagram representation of Multiple R squared (R^2)

The multiple R value represents the correlation between the actual and predicted output values. Its squared value is interpreted as the proportion of the output variance (a+b+c+d) which co-varies with that of all the combined input variables (a+b+c). In the diagram, d represents the variability in $\hat{Y}$ not accounted for by the model (residual or error) while the combined areas a+b+c represents that accounted for by the model.

It is therefore interpreted in much the same way as in simple linear regression, and as in that situation, the less biased, adjusted value is usually reported.

59.1.5 Venn diagram representation of Regression slopes (b) and beta (β)

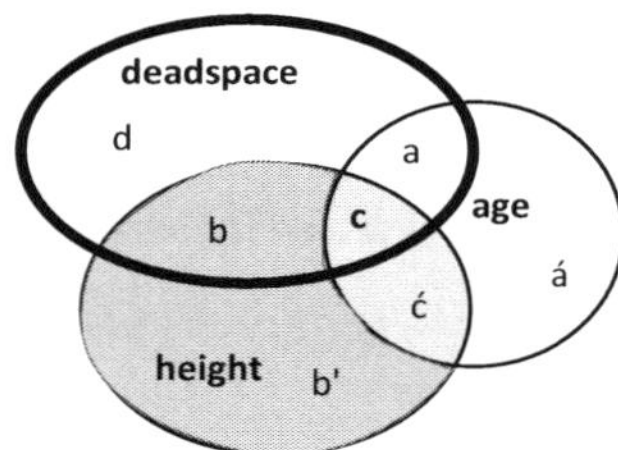

The Venn diagram can also help explain the regression slopes (i.e. the b's) in multiple regression:

Conceptually b: partial correlation / partial sd

$b_{age} = a/(a+á)$; $b_{height} = b/(b+b')$

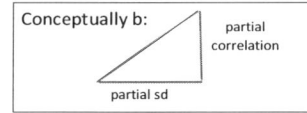

We already know that a and b are the partial correlations for each input variable and not surprisingly the (a+á) and (b+b') can be seen as the squared partial standard deviations mentioned earlier. Beta is simply b_i*sd(y)/sd(x_i). It should not be interpreted as a simple correlation or any other type of correlation (see the previous regression formula) as was the case in simple linear regression. Now it is simply a slope that is an effect size. For reasons given below when multicollinearity is present *b* and beta become inflated, often ending up with betas greater than one.

59.1.6 Multicollinearity and Variance Inflation Factors (VIFs)

Multiple R statistically significant

But

No individual inputs statistically significant

Possible Multicollinearity problem

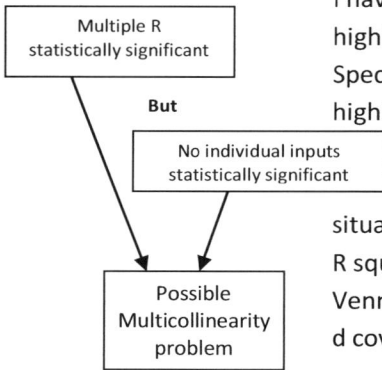

I have mentioned before that there is a problem if two or more of the input variables are highly correlated, a situation known as multicollinearity or simply often called, collinearity. Specifically it is not only a two input variable problem but where one independent variable is highly correlated with one or more of the other independent variables. While this does not affect the overall model fit, that is the multiple R squared value and its associated p-value are still valid, the actual estimates of the b's are affected. You can end up with a situation where you have a highly statistically significant p-value associated with the Multiple R squared value but none of the individual b's appear statistically significant. In terms of the Venn diagram this can be conceptualised as the situation where there is a large proportion of d covered by all the inputs (a+b+c) yet the unique individual areas might be very small.

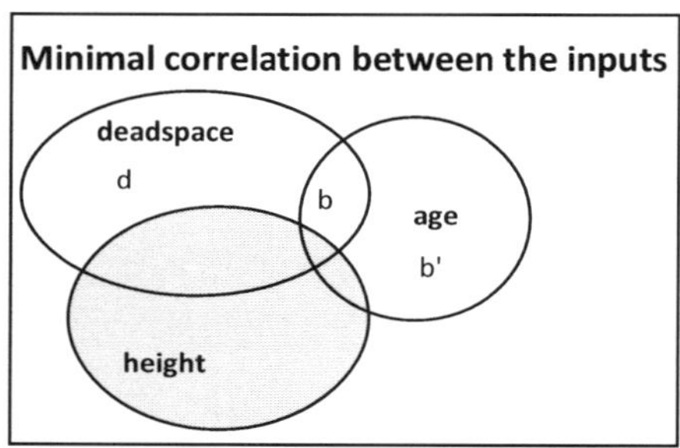

The situation, opposite left (top) is where there is little correlation between the two inputs. In contrast, the situation below it is where they are highly correlated representing multicollinearity.

High degree of correlation between the inputs

deadspace d b b' age height

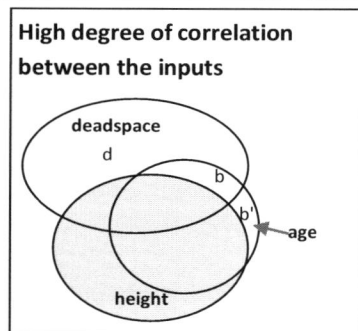

In the previous diagram we saw how the regression slopes (b's) calculation is based upon the two areas of the variable. In the situation with multicollinearity these areas become very small in contrast to the residual (d). The variance (a measure of its accuracy) of b is proportional to d/(b+b') this value gets inflated as there is just not enough information to produce a valid value! While this explanation is at a rather abstract level a more mathematical one is given by considering the equation used to estimate the variance of the *b's*.

$$\text{var}(b_i) = \frac{\sigma^2}{(n-1)s_i^2} \times \frac{1}{1-R^2_{i|others}}$$

σ^2 (sigma squared) is the variance of the error term (residuals) and s_i^2 is the variance for input *i*. The equation consists of two components; the first is the standard error and the second consists of an expression known as the Variance Inflation Factor (VIF), described in more detail later. Often the whole expression is referred to as the standard error. When there is no correlation between the inputs the VIF value equals 1 but as the correlations increase so does this value. In fact, when collinearity is high, VIF tends to infinity as the expression tends to 1/0. Further mathematical detail regarding multicollinearity can be found in Friendly & Kwan, 2009.

While identifying multicollinearity is relatively easy, dealing with it is not. The simplest solution is to either omit one or more of the offending highly correlated inputs or combine them into a single input. Centering or standardising variables can help (see later). If this fails or can not be done you can use more complex regression techniques such as factor analysis, ridge regression (*MASS* package) or structural equation modelling.

59.1.7 Sums of squares – the traditional Venn diagram

It is usual to introduce multiple linear regression with details of various sums of squares, as I did in the simple linear regression chapter, but unfortunately, this gets very messy with multiple regression. In contrast, by using the various Venn diagram depictions we have already considered the various sums of squares can also be explained when we carry out the analyses latter.

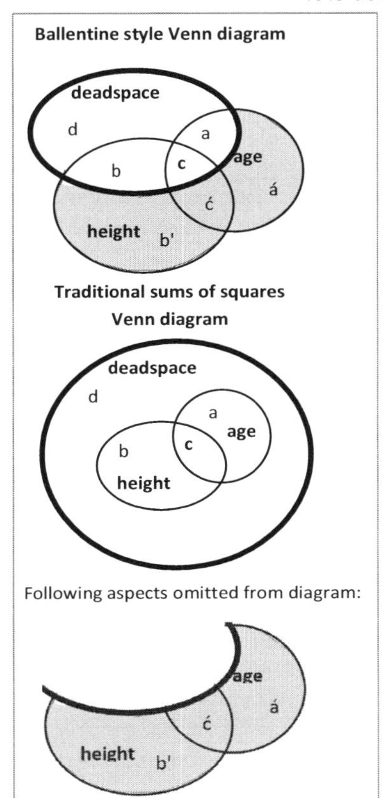

Because the particular format of the type of Venn diagram used above (the Ballentine style) gets untenable when you have more than two inputs I will use a more traditional style Venn diagram following (Norman and Streiner, 2008 p. 146; Ip, 2001; Kennedy, 2002; Cohen, Cohen & West et al. 2003). Now we leave out the variance in the inputs that is not made use of by the model while the residual for the output (marked d) becomes an all encompassing circle. Looking back at the previous Ballentine diagrams you will see that these omitted aspects do not play a part in the description of the various partial and part correlations so we are not dropping anything too important for descriptive purposes.

Consider an example with 3 inputs, called A, B and C. Using the traditional style Venn diagram along with the traditional block regression approach here the ratio of the total area of the inner circles (=the regression sums of squares) to the overall total is equivalent to R^2.

Notice that as the complexity increases the model gets divided into more and more parts. The shaded areas represent the part correlations and are the unique parts for each variable. Another important thing to note is that the unique areas (i.e. part correlations) for each input would possibly change if we either added or removed an input. If we had as many variables as datapoints we would end up with the explained variance filling the whole Venn diagram – there being no residual. Unfortunately such a model is useless as we would just end up with as many values as we started with whereas the idea is to achieve a final model that is as simple as possible (i.e. as few parameters as possible) that adequately models our dataset.

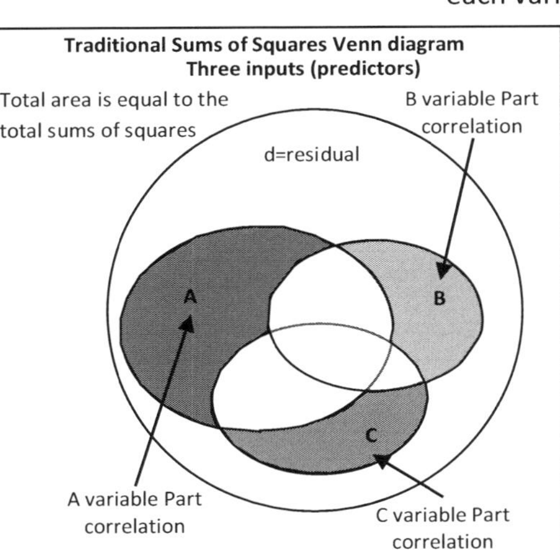

Regrettably, both varieties of Venn diagram fall apart when you are trying to explain more than 4 inputs but hopefully this approach has helped you gain some understanding. There are also problems with using this approach when the overlapping areas might represent a negative value (Ip, 2001).

59.1.8 ANOVA, ANCOVA, Regression

You may have noticed that at the beginning of the chapter I presented a diagram of the multiple regression analysis used in **SAS2**, reproduced opposite.

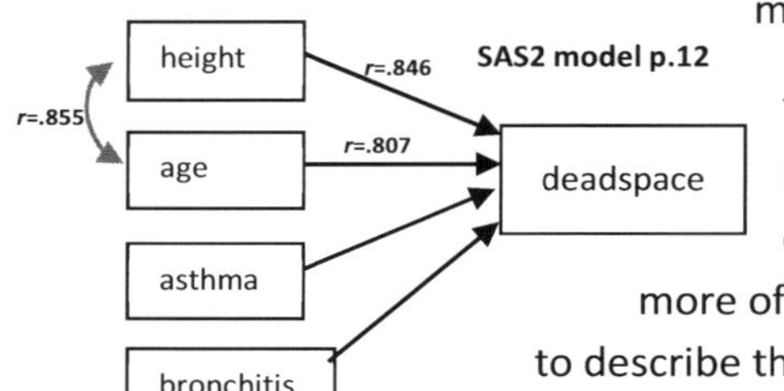

Traditionally the term multiple linear regression was only used when you had input variables that possessed an interval/ ratio **level of measurement** and this criteria is only fulfilled with the height and age inputs. Traditionally when one or more of the input variables possess other levels of measurement other terms are used to describe the model, as shown in the diagram below.

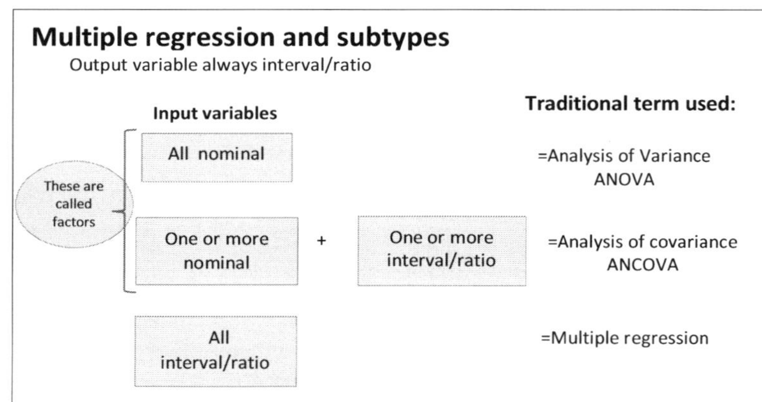

In a modern regression analysis all these variables are treated basically the same regardless of the measurement level. However some statistical software, notably SPSS, provide ANOVA (Analysis of Variance) and ANCOVA (Analysis of covariance) specific analyses, as does R, but the regression analysis approach does provide more, specifically parameter estimates (b's and β's) along with confidence intervals. In contrast, ANOVA techniques do have some advantages particularly when it comes to comparing the relationship between the different levels of the input variables, which are factors here.

59.1.9 Dummy coding - contrasts

The multiple regression equation and diagrams shown in the chapter so far are only really correct for interval/ ratio **level of measurement** inputs and factors that only possess two levels. To demonstrate how the regression equation changes when you have one or more factors that possess more than two levels we will now assume that bronchitis and asthma are mutually exclusive. In other words you can only be; ashmatic, bronchitic or healthy, which is not the case in reality for some unfortunate individuals. The *campbell_mult_reg* dataset is set up to reflect this situation and contains a variable called *condition* which indicates this with three levels (i.e. where k = number of levels; and here $k=3$). To model this variable in the regression equation we make use of a technique known as **dummy** or indicator coding. For an overview see the **contrasts** entry in the glossary section.

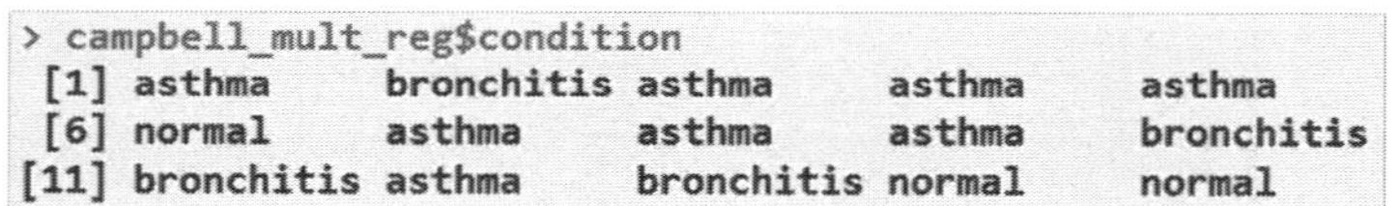

```
> campbell_mult_reg$condition
 [1] asthma     bronchitis asthma     asthma     asthma
 [6] normal     asthma     asthma     asthma     bronchitis
[11] bronchitis asthma     bronchitis normal     normal
```

To represent every factor included in a regression equation, each of which has k levels, we need to include k-1 indicator/dummy variables, so we need two dummy variables to represent our *condition* factor. Luckily when you define a variable to be a factor in R, as has been done with *condition* (see later), R automatically creates the necessary number of dummy variables. However, it is important to have some understanding of what is going on under the bonnet and the next few pages provide some background information.

```
Coefficients:
                     Estimate Std. Error t value Pr(>|t|)
(Intercept)          -48.0963    25.6204  -1.877  0.08993 .
age                    0.7204     2.2611   0.319  0.75658
height                 0.6996     0.2660   2.630  0.02515 *
conditionbronchitis   11.3990     6.1781   1.845  0.09481 .
conditionnormal       26.0513     7.9775   3.266  0.00849 **
```

Consider a multiple linear regression model consisting of three inputs, two of which are continuous and one our factor with three levels: *deadsp ~ age + height + condition*. R automatically creates a four input model as can be seen from the output opposite. Do not worry about what the detailed output means at this stage.

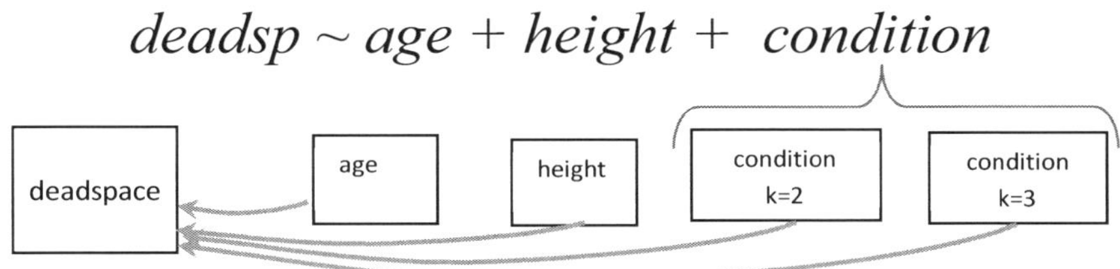

I'll not bore you with the reason why the number of dummy variables created is equal to the number of factor levels minus one. What worries most people is the disappearance of one of the levels? Actually, it has not disappeared but has been subsumed in the intercept parameter estimate.

use the *contrasts()* function to inspect a factor's contrasts

```
> contrasts(campbell_mult_reg$condition)
           bronchitis normal
asthma              0      0
bronchitis          1      0
normal              0      1
```

The 'missing level' is usually called the *reference group* as we usually compare all the other levels of the factor against this value. The problem with the above analysis is that by leaving R to its own devices it has picked the asthmatics level as the reference group (as it occurs first in the alphabet) when it would be most appropriate to select the normal (i.e. control) group for this purpose. We can check out how R has coded the dummy variables by using the *contrasts()* function which shows us that the reference group, which is the group coded with all zeros, is the asthma group. There are several ways to change which factor level is used as the reference group. One way is to use the *contr.treatment()* function which produces a specific type of contrast matrix (see the **contrasts** glossary entry for details of other types). All the function requires is the number of levels you have in the factor along with an indication of which level you want to use as the reference group. The next step is to replace this contrast matrix with the one associated with the factor variable, *condition* in this instance, by using the *contrasts()* function again :

here the *contrasts()* function is used to access the factor's contrasts to allow them to be edited

create a contrast matrix, using dummy coding, for a factor with 3 levels and set the reference group to the third level

```
contrasts (campbell_mult_reg$condition) <- contr.treatment(3, base = 3)
```

run the regression model

```
reg_factor3levelsb <- lm(deadsp ~ age+ height +  condition, data=campbell_mult_reg)
```

display a summary of the results

```
summary(reg_factor3levelsb)
```

```
> contrasts (campbell_mult_reg$condition) <- contr.treatment(3, base = 3)
> reg_factor3levelsb <- lm(deadsp~age+height +  condition, data=campbell_mult_reg)
>  summary(reg_factor3levelsb)
Coefficients:
            Estimate Std. Error t value Pr(>|t|)
(Intercept) -22.0450    27.5591  -0.800  0.44235
age           0.7204     2.2611   0.319  0.75658
height        0.6996     0.2660   2.630  0.02515 *
condition1  -26.0513     7.9775  -3.266  0.00849 **
condition2  -14.6523     8.2939  -1.767  0.10774
```

```
> contrasts(campbell_mult_reg$condition)
           1 2
asthma     1 0
bronchitis 0 1
normal     0 0
```

Now we have the correct reference group labelled *condition1* and *condition2* which is not very useful, although we can find out what they mean by once again using the *contrasts()* function, telling us that 1=asthma; 2= bronchitis.

A way to specify the column names for the contrast matrix is by using the *colnames()* function.

access the column names of the contrasts matrix

make the column names equal

```
colnames(contrasts(campbell_mult_reg$condition))<- c("asthma_v_normal", "Bronchitis_v_normal")
contrasts(campbell_mult_reg$condition) # show the contrast matrix again
```

```
> colnames(contrasts(campbell_mult_reg$condition))<- c("asthma_v_normal", "Bronchitis_v_normal")
> contrasts(campbell_mult_reg$condition)
           asthma_v_normal Bronchitis_v_normal
asthma                   1                   0
bronchitis               0                   1
normal                   0                   0
```

```
                              Estimate Std. Error t value Pr(>|t|)
(Intercept)                   -22.0450    27.5591  -0.800  0.44235
age                             0.7204     2.2611   0.319  0.75658
height                          0.6996     0.2660   2.630  0.02515 *
conditionasthma_v_normal      -26.0513     7.9775  -3.266  0.00849 **
conditionBronchitis_v_normal  -14.6523     8.2939  -1.767  0.10774
```

Now when we re-run the model we get sensible names for the contrasts. To gain even more clarity you might want to add a colon before each label e.g. ":Asthma_v_normal", ":Bronchitis_v_normal", which will separate the variable name 'condition' from the contrast label.

We will consider the detailed interpretation of the estimates for the various contrasts later. For now, it is worth noting that, for dummy contrasts, when you form a linear regression with the factor being the only input, the intercept becomes the mean value for the reference group and each contrast becomes the difference in the mean value for the particular subgroup. To demonstrate this I have created the simple regression model below and added the summary table of observed means for comparison.

```
> with(campbell_mult_reg, mean(deadsp))
[1] 66.93333
```

```
> tapply(campbell_mult_reg$deadsp, campbell_mult_reg$condition, mean)
   asthma bronchitis     normal
 52.87500   72.25000   97.33333
```

```
reg_factor3levels_only <- lm(deadsp ~ condition, data=campbell_mult_reg)

summary(reg_factor3levels_only)
```

```
                             Estimate Std. Error t value Pr(>|t|)
(Intercept)                    97.333      9.664  10.072 3.32e-07 ***
conditionasthma_v_normal      -44.458     11.332  -3.923  0.00202 **
conditionBronchitis_v_normal  -25.083     12.785  -1.962  0.07338 .
```

Note that 97-44.45 = 52.55 and 97-25=72 are the same as the observed means.

Comparing this output with the previous outputs at the top of this page and the previous one you can see these values are attenuated, where the disparity between the normals and asthmatics is reduced from 44 to 26 but is still statistically significant.

Going one step further it is possible to produce custom contrast matrices, details of which are provided in both Field, 2013 and Crawley, 2005 and the online tutorial cited in the glossary entry, **contrasts**. This short section on dummy coding and contrasts is woefully inadequate, but does provide some insight into the process.

It is now time to move onto some actual data analysis. Using the dataset described at the start of this chapter (from **SAS2**) we will first produce a multiple regression with two interval/ratio (i.e. continuous) input variables and then move to the situation of a regression with both a nominal (factor) and an interval/ratio input. I'll discuss the situation where all the inputs are nominal in the ANOVA chapter.

59.2 Two continuous variables

We will start by obtaining the dataset which is an SPSS file (*.sav) from:

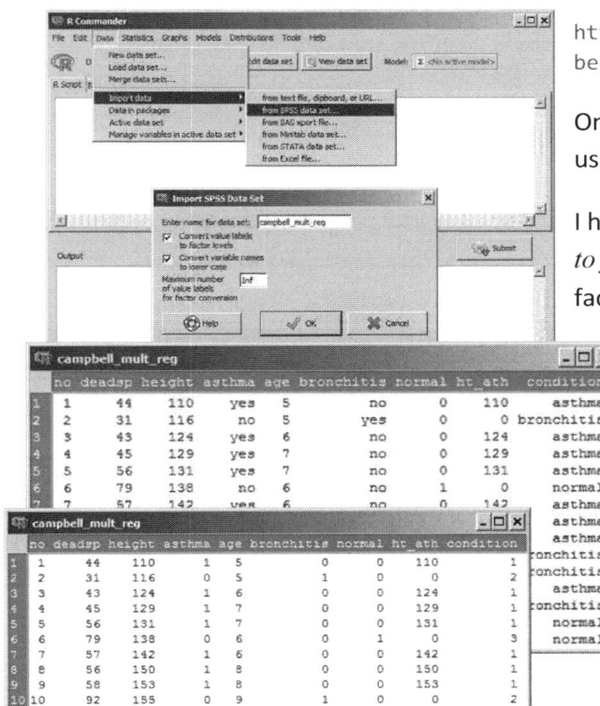

no	deadsp	height	asthma	age	bronchitis	normal	ht_ath	condition
1	44	110	1	5	0	0	110	1
2	31	116	0	5	1	0	0	2
3	43	124	1	6	0	0	124	1
4	45	129	1	7	0	0	129	1
5	56	131	1	7	0	0	131	1
6	79	138	0	6	0	1	0	3
7	57	142	1	6	0	0	142	1
8	56	150	1	8	0	0	150	1
9	58	153	1	8	0	0	153	1
10	92	155	0	9	1	0	0	2
11	78	156	0	7	1	0	0	2
12	64	159	1	8	0	0	159	1
13	88	164	0	10	1	0	0	2
4	112	168	0	11	0	1	0	3
15	101	174	0	14	0	1	0	3

```
http://www.robin-
beaumont.co.uk/virtualclassroom/book2data/campbell_mlr_p12.sav
```

Once you have stored the file locally import the file into R Commander using the menu option shown on the left.

I have called the dataset *campbell_mult_reg*. The *convert value labels to factor levels* option is selected allowing the values to be imported as factor levels.

If you had un-selected the option you would have imported the actual text values shown below left.

If you didn't want to us the R Commander menu you could have typed the following R code:

```
library(foreign)
campbell_mult_reg <-
  read.spss("http://www.robin-
beaumont.co.uk/virtualclassroom/book2data/campbell_mlr_p12.sav",
   use.value.labels=TRUE,
max.value.labels=Inf,
to.data.frame=TRUE)
names(campbell_mult_reg)
```

59.2.1 Preliminaries

Before carrying out the actual regression analysis it is sensible to carry out the following checks.

59.2.1.1 Scatterplots and correlation matrices

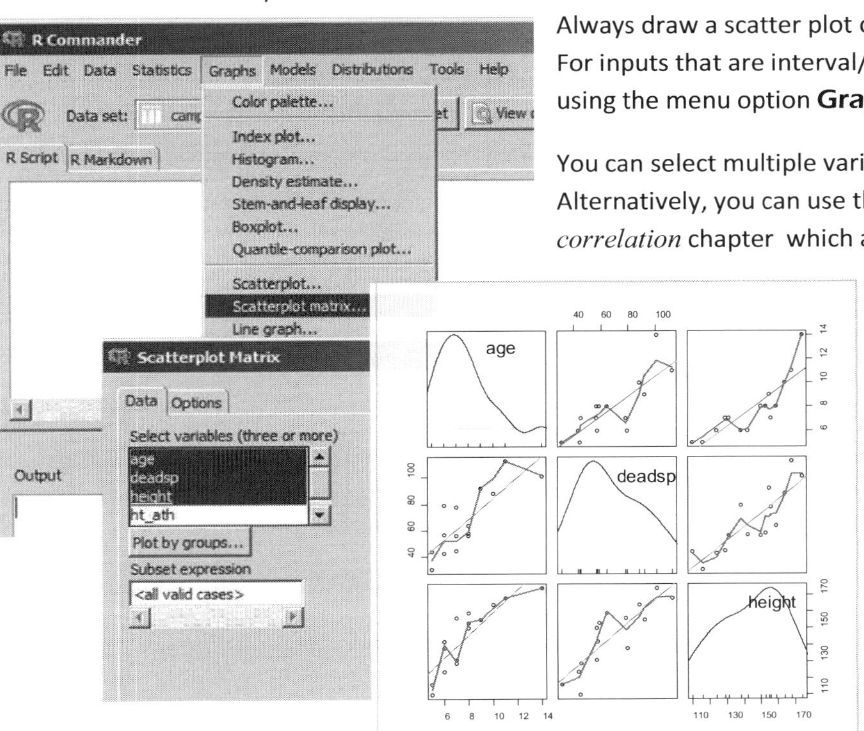

Always draw a scatter plot of each of the inputs against the output. For inputs that are interval/ratio data this is easy in R Commander using the menu option **Graphs -> Scatterplot matrix**

You can select multiple variables by holding down the *Ctrl* key. Alternatively, you can use the psych package as described in the *correlation* chapter which also produces the correlations across the diagonal. Unfortunately, such an option is not available in R Commander. To produce the results given at the beginning of this chapter I used the following R Commander menu option:

Statistics->summaries-> correlation matrix

59.2.1.2 Partial and semi-partial correlations

In R Commander in the *Correlation Matrix* dialog box, shown opposite, you can obtain partial correlations but there is no option for part correlations.

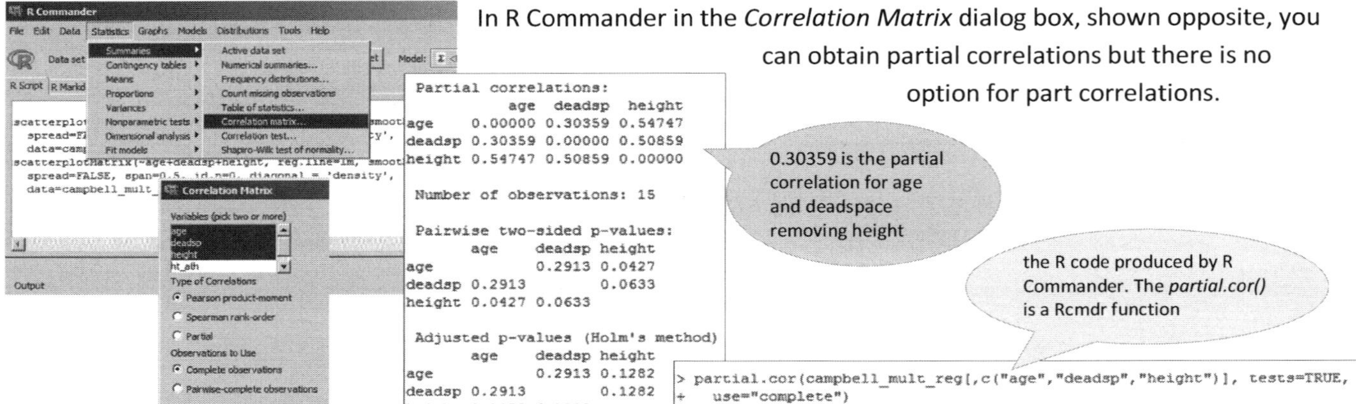
Partial correlations:
age deadsp height
age 0.00000 0.30359 0.54747
deadsp 0.30359 0.00000 0.50859
height 0.54747 0.50859 0.00000
Number of observations: 15
Pairwise two-sided p-values:
age deadsp height
age 0.2913 0.0427
deadsp 0.2913 0.0633
height 0.0427 0.0633
Adjusted p-values (Holm's method)
age deadsp height
age 0.2913 0.1282
deadsp 0.2913 0.1282
height 0.1282 0.1282
0.30359 is the partial correlation for age and deadspace removing height
the R code produced by R Commander. The partial.cor() is a Rcmdr function
> partial.cor(campbell_mult_reg[,c("age","deadsp","height")], tests=TRUE,
+ use="complete")
Correlation Matrix
Pearson product-moment
Spearman rank-order
Partial
Complete observations
Pairwise-complete observations
Pairwise p-values

To obtain part correlations I'll demonstrate the *ppcor* package which you can also use to produce partial correlations. The *ppcor* package contains the *pcor.test()* function which provides a single partial correlation along with a p-value where *spcor.test()* does a similar thing for the part correlation. The following code installs the *ppcor* package, loads it and produces the equivalent partial correlations.

```
> partial_Rsqr <- pcor.test(campbell_mult_reg$deadsp, campbell_mult_reg$age, campbell_mult_reg$height)
> partial_Rsqr
   estimate   p.value statistic  n gp  Method
1 0.3035852 0.2697052  1.103742 15  1 pearson
```

partial correlation for age and deadspace removing height

```
install.packages("ppcor", dependencies=TRUE); library(ppcor)
partial_Rsqr <- pcor.test(campbell_mult_reg$deadsp,
campbell_mult_reg$age, campbell_mult_reg$height)
partial_Rsqr
```

semi-partial correlation between deadspace & age removing height from age

```
> spcor.test(campbell_mult_reg$deadsp, campbell_mult_reg$age, campbell_mult_reg$height)
   estimate   p.value statistic  n gp  Method
1 0.1617159 0.5702576 0.5676722 15  1 pearson
```

```
spcor.test(campbell_mult_reg$deadsp, campbell_mult_reg$age, campbell_mult_reg$height)
```

We can also use two related functions, *spcor()*, for the part (semi-partial) and *pcor()*, for the partials. These functions require rows from a dataframe which consist only of numeric columns and produce a matrix of partial or part correlations. For example using the *pcor()* function and columns 2,3 and 5 from the campbell_mult_reg dataframe produces the partial correlation matrix opposite.

```
> pcor(campbell_mult_reg[c(2,3,5)])
$estimate
          deadsp    height       age
deadsp 1.0000000 0.5085853 0.3035852
height 0.5085853 1.0000000 0.5474722
age    0.3035852 0.5474722 1.0000000
```

partial correlation for age and deadspace removing height

0.299 semi-partial correlation of deadspace & height removing age from height

```
pcor(campbell_mult_reg[,c(2,3,5)])
```

and for the semi-partial correlations:

```
> spcor(campbell_mult_reg[,c(2,3,5)])
$estimate
          deadsp    height       age
deadsp 1.0000000 0.2997993 0.1617159
height 0.2633061 1.0000000 0.2916313
age    0.1571729 0.3227222 1.0000000

$p.value
          deadsp    height       age
deadsp 0.0000000 0.2763265 0.5702576
height 0.3444124 0.0000000 0.2908996
age    0.5814177 0.2375468 0.0000000

$statistic
          deadsp   height       age
deadsp 0.0000000 1.088609 0.5676722
height 0.9454830 0.000000 1.0561503
age    0.5513152 1.181141 0.0000000
```

0.1617 semi-partial correlation deadspace & age removing height from age

these values represent the square root change in R^2 of adding each variable to the regression model

```
spcor(campbell_mult_reg[,c(2,3,5)])
```

Before we carry out any modelling it would also be nice to be able to check the variance inflation factors given that we have a high correlation between the two input variables, however in R Commander we need to first create the model to do this so it will have to wait.

59.2.2 Creating and assessing the model in R Commander

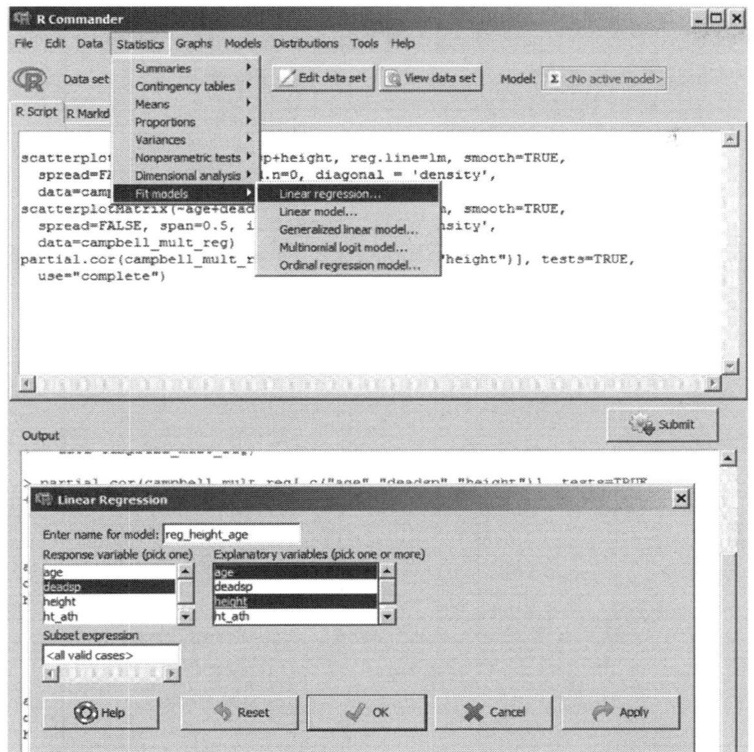

To perform a multiple linear regression in R Commander select the menu option:

Statistics-> Fit models-> Linear regression

Enter a name for your model in the **Enter name for model** box, I have called it *reg_height_age*.

Select the outcome (response) variable which is *deadsp*.

Select the input (explanatory) variables which are *age* and *height*. Then click on the **OK** button. To select multiple options where they are not in a single block:

Hold down the *Ctrl* key then *right mouse click* on each.

If you want a continuous selection just hold the *shift* key instead. The output of the regression appears in the output window.

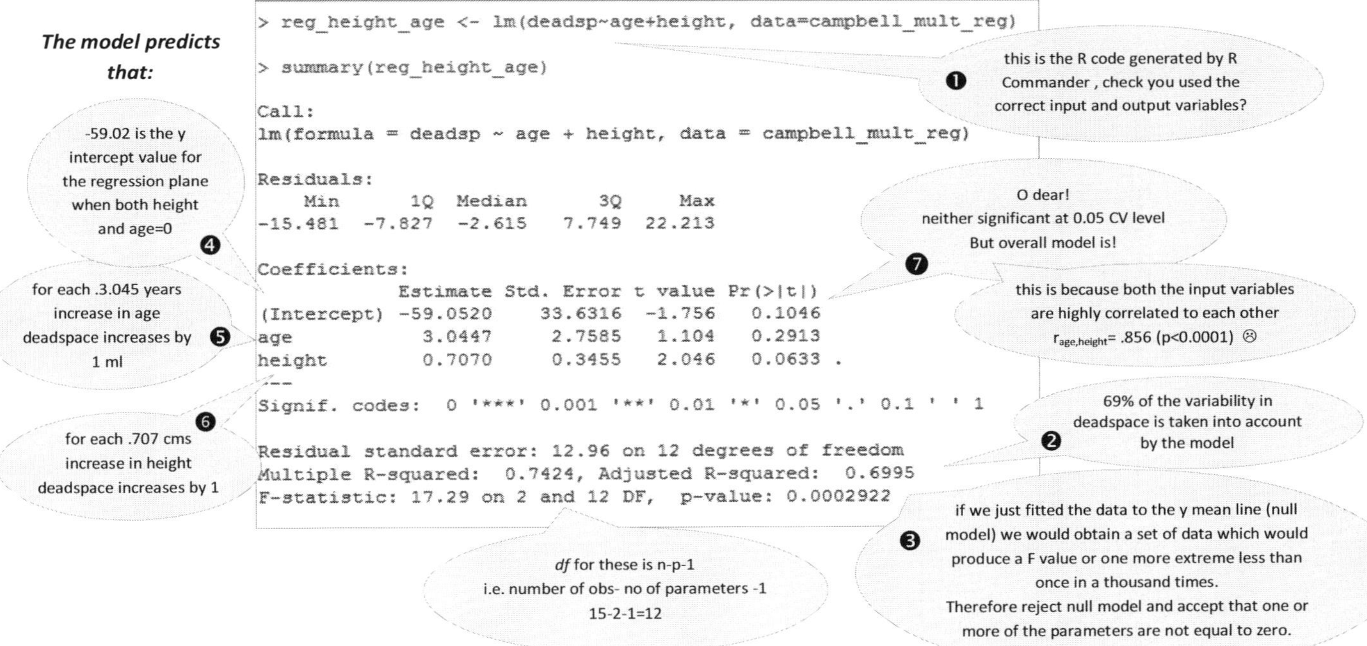

```
> reg_height_age <- lm(deadsp~age+height, data=campbell_mult_reg)

> summary(reg_height_age)

Call:
lm(formula = deadsp ~ age + height, data = campbell_mult_reg)

Residuals:
    Min      1Q  Median      3Q     Max
-15.481  -7.827  -2.615   7.749  22.213

Coefficients:
            Estimate Std. Error t value Pr(>|t|)
(Intercept) -59.0520   33.6316  -1.756   0.1046
age           3.0447    2.7585   1.104   0.2913
height        0.7070    0.3455   2.046   0.0633 .
---
Signif. codes:  0 '***' 0.001 '**' 0.01 '*' 0.05 '.' 0.1 ' ' 1

Residual standard error: 12.96 on 12 degrees of freedom
Multiple R-squared:  0.7424, Adjusted R-squared:  0.6995
F-statistic: 17.29 on 2 and 12 DF,  p-value: 0.0002922
```

The model predicts that:
❶ this is the R code generated by R Commander , check you used the correct input and output variables?
❹ -59.02 is the y intercept value for the regression plane when both height and age=0
❼ O dear! neither significant at 0.05 CV level But overall model is!
this is because both the input variables are highly correlated to each other $r_{age,height}$= .856 (p<0.0001) ☹
❺ for each .3.045 years increase in age deadspace increases by 1 ml
❻ for each .707 cms increase in height deadspace increases by 1
❷ 69% of the variability in deadspace is taken into account by the model
df for these is n-p-1
i.e. number of obs- no of parameters -1
15-2-1=12
❸ if we just fitted the data to the y mean line (null model) we would obtain a set of data which would produce a F value or one more extreme less than once in a thousand times.
Therefore reject null model and accept that one or more of the parameters are not equal to zero.

I have added numbers to each of the above comments boxes to indicate the order in which I read them. Always check you have specified the model you thought you had defined in the dialog box. To do this you need to understand what the following expression means.

```
reg_height_age <- lm(deadsp~age+height, data=campbell_mult_reg)
```

- give the object the name reg_height_age
- the object gets:
- a linear regression model
- output variable= deadsp
- input variables= age & height
- the variables come from the cambell_mult_reg dataframe

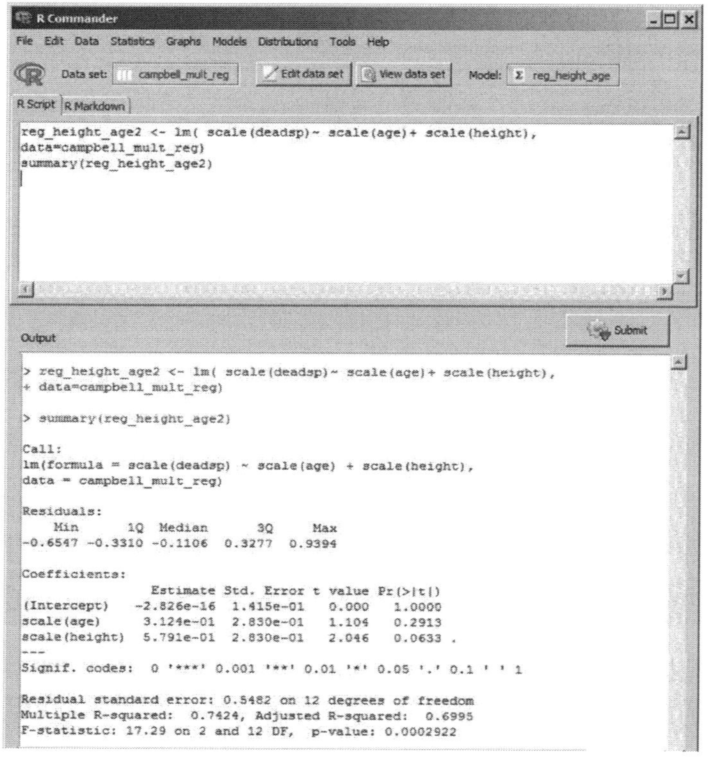
R Commander
File Edit Data Statistics Graphs Models Distributions Tools Help
Data set: campbell_mult_reg
Edit data set
View data set
Model: reg_height_age
R Script
R Markdown
reg_height_age2 <- lm(scale(deadsp)~ scale(age)+ scale(height),
data=campbell_mult_reg)
summary(reg_height_age2)
Output
Submit
> reg_height_age2 <- lm(scale(deadsp)~ scale(age)+ scale(height),
+ data=campbell_mult_reg)
> summary(reg_height_age2)
Call:
lm(formula = scale(deadsp) ~ scale(age) + scale(height),
data = campbell_mult_reg)
Residuals:
Min 1Q Median 3Q Max
-0.6547 -0.3310 -0.1106 0.3277 0.9394
Coefficients:
Estimate Std. Error t value Pr(>|t|)
(Intercept) -2.826e-16 1.415e-01 0.000 1.0000
scale(age) 3.124e-01 2.830e-01 1.104 0.2913
scale(height) 5.791e-01 2.830e-01 2.046 0.0633 .

Signif. codes: 0 '***' 0.001 '**' 0.01 '*' 0.05 '.' 0.1 ' ' 1
Residual standard error: 0.5482 on 12 degrees of freedom
Multiple R-squared: 0.7424, Adjusted R-squared: 0.6995
F-statistic: 17.29 on 2 and 12 DF, p-value: 0.0002922

We can find the standardised b's (betas) by re-running the regression on standardised scores making use of the *scale()* function which standardises variables. The simplest way to do this is to copy the R code generated from the above analysis and then edit it. Do this in the *R Script* window. Then select the edited text and click on the **Submit** button to run it.

```
reg_height_age <- lm(deadsp~age+height,
data=campbell_mult_reg)

summary(reg_height_age)
```

is edited to:

```
reg_height_age2 <- lm( scale(deadsp)~ scale(age)
+ scale(height), data=campbell_mult_reg)

summary(reg_height_age2)
```

We now see that the Intercept has disappeared, except for rounding error, and the betas are age=0.3124 and height=0.5791. Standardising the variables makes no difference to the associated F, R^2 or p-values but can help interpretation across different models. We interpret the betas the same way as the b's but this time it is in standard deviation units.

59.2.2.1 VIFs

The next thing I will do is to quantity the multicollinearity problem by using the VIFs (variance inflation factors) which provides such a measure. We have already identified that we have a multicollinearity problem from two findings; firstly the statistically significant overall model fit (F value) along with the fact that none of the inputs reached statistical significance. Secondly, all the inputs are highly correlated being as high as the correlation between any of the inputs and the output.

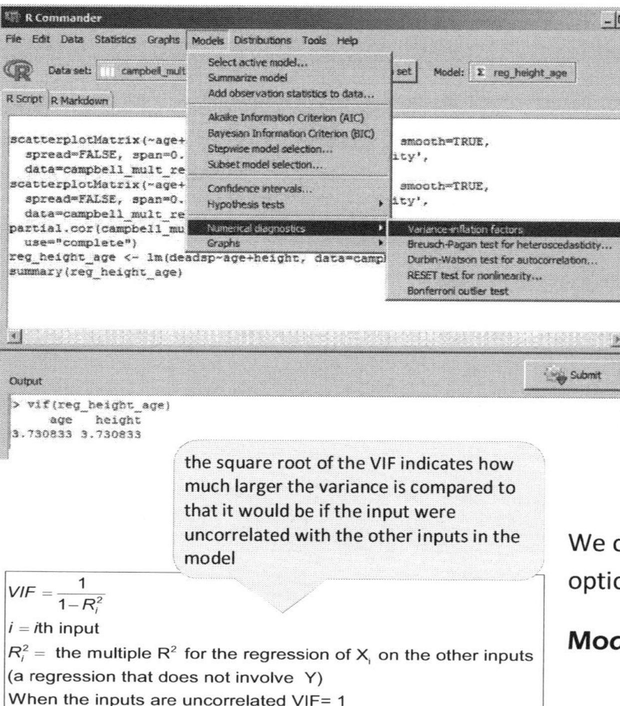
R Commander
File Edit Data Statistics Graphs Models Distributions Tools Help
Select active model...
Summarize model
Add observation statistics to data...
Akaike Information Criterion (AIC)
Bayesian Information Criterion (BIC)
Stepwise model selection...
Subset model selection...
Confidence intervals...
Hypothesis tests
Numerical diagnostics
Graphs
Variance-inflation factors
Breusch-Pagan test for heteroscedasticity...
Durbin-Watson test for autocorrelation...
RESET test for nonlinearity...
Bonferroni outlier test
Submit
> vif(reg_height_age)
age height
3.730833 3.730833
the square root of the VIF indicates how much larger the variance is compared to that it would be if the input were uncorrelated with the other inputs in the model
$VIF = \frac{1}{1-R_i^2}$
i = ith input
R_i^2 = the multiple R^2 for the regression of X_i on the other inputs (a regression that does not involve Y)
When the inputs are uncorrelated VIF= 1

We obtain the VIF for each input using the R Commander menu option:

Models-> Numerical diagnostics->Variance inflation factors

```
> sqrt(3.7308)
[1] 1.931528
```

Alternatively you can type in the R Commander script window *vif(model name).* Both inputs have a VIF of 3.7308. What does this mean? From the equation for the VIF given opposite we can see that the standard error (variance) for our inputs have basically doubled (√3.7308 = 1.93) obtained by using the square root function, *sqrt()* in R.

Different writers suggest different VIF values that indicate a multicollinearity problem. Field, 2013 states that VIF values above 10 indicate multicollinearity whereas others give the value to be 5 and even 2.5! I would suggest that you interpret this value with a great deal of circumspection.

59.2.2.2 *Confidence intervals*

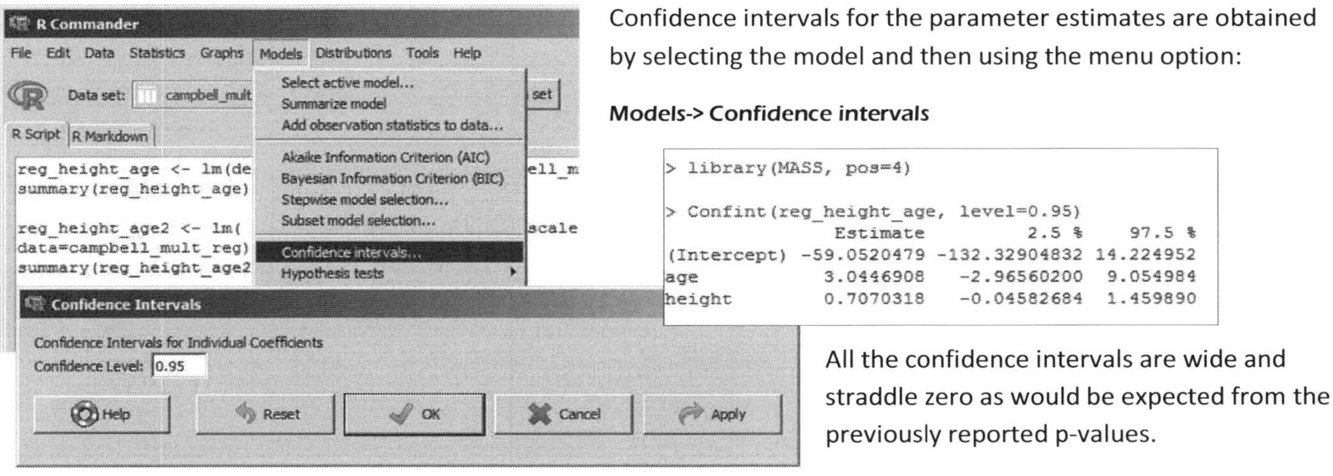

Confidence intervals for the parameter estimates are obtained by selecting the model and then using the menu option:

Models-> Confidence intervals

```
> library(MASS, pos=4)

> Confint(reg_height_age, level=0.95)
               Estimate         2.5 %     97.5 %
(Intercept) -59.0520479 -132.32904832 14.224952
age           3.0446908   -2.96560200  9.054984
height        0.7070318   -0.04582684  1.459890
```

All the confidence intervals are wide and straddle zero as would be expected from the previously reported p-values.

59.2.2.3 *Sums of Squares and the ANOVA table*

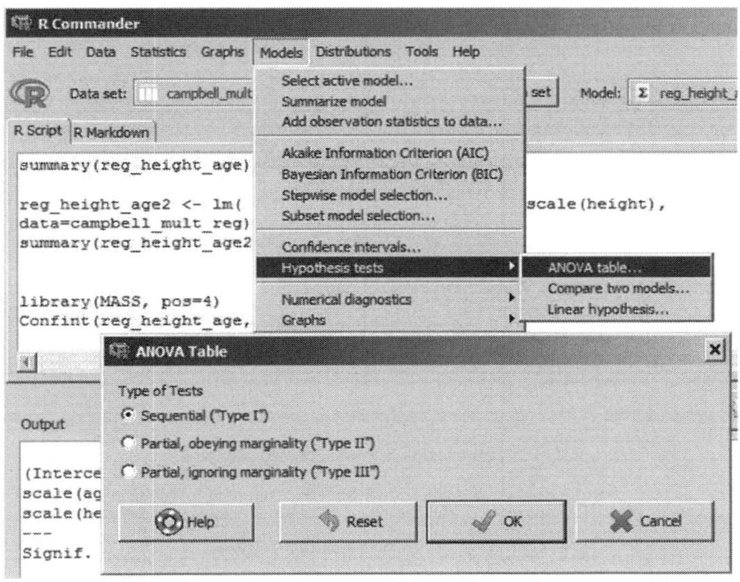

Many statistical packages also produce a table of the various sums of squares known as the ANOVA table which you can obtain with the following menu option:

Models-> Hypothesis tests-> ANOVA tables

You are then asked what type of sums of squares you require, there is nothing to be scared of here, as we have already looked at them using the Venn diagram approach. I have run the option twice to produce each variety shown below.

note these p-values are identical to those produced by the regression analysis

upper case i; alternatively you can use 2 or 3

age first variable so grabs all the shared variance

name	Sequential ' type I'	Partial 'type II' [same as type III for this model]
R code	anova(reg_height_age)	Anova(reg_height_age, type="II")
result	Response: deadsp Df Sum Sq Mean Sq F value Pr(>F) age 1 5108.5 5108.5 30.3964 0.0001334 *** height 1 703.7 703.7 4.1869 0.0632962 . Residuals 12 2016.8 168.1	Response: deadsp Sum Sq Df F value Pr(>F) age 204.74 1 1.2182 0.2913 height 703.66 1 4.1869 0.0633 . Residuals 2016.76 12

in contrast here age only takes its own unique variance

therefore the overlap for age is 5108.5-204.74=4903.76

We can use the above sums of squares to produce our Venn diagrams (right) below:

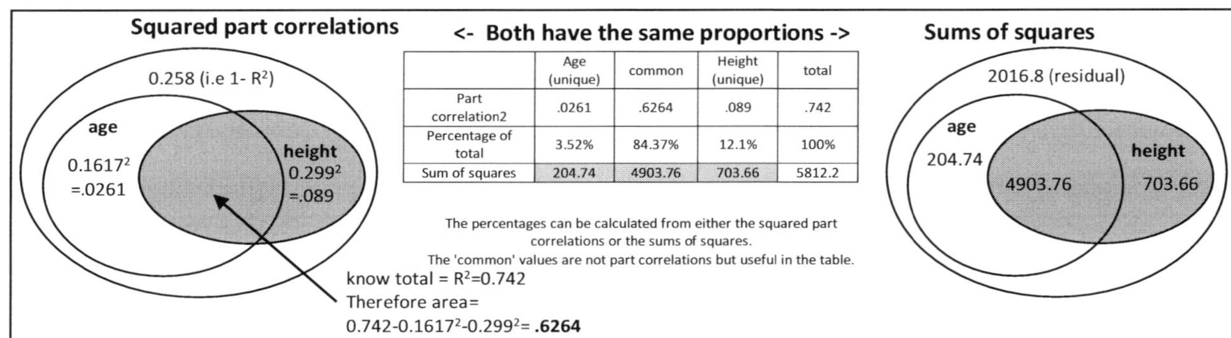

	Age (unique)	common	Height (unique)	total
Part correlation2	.0261	.6264	.089	.742
Percentage of total	3.52%	84.37%	12.1%	100%
Sum of squares	204.74	4903.76	703.66	5812.2

The above diagram demonstrates very nicely the link between the sums of squares and part correlations. You can complete the table in the middle of the diagram many ways. For example if you know the various sums of squares and the R^2 value you can calculate the part correlations. Similarly, if you know the part correlations and just one sum of squares you can calculate the others using the percentages column.

From the above it is clear that age and height share a large amount of the variability (84% worth ☹), my Venn diagram should have a very little area for *age* or *height*, which was not overlapping but rather difficult to draw! By removing either of them would make little difference to the fit of the model that is the R^2 value; and we know exactly how much by looking at the squared part correlations, 0.02 (*age*) and .089's (*height*) worth of .742.

59.2.2.4 Comparing models

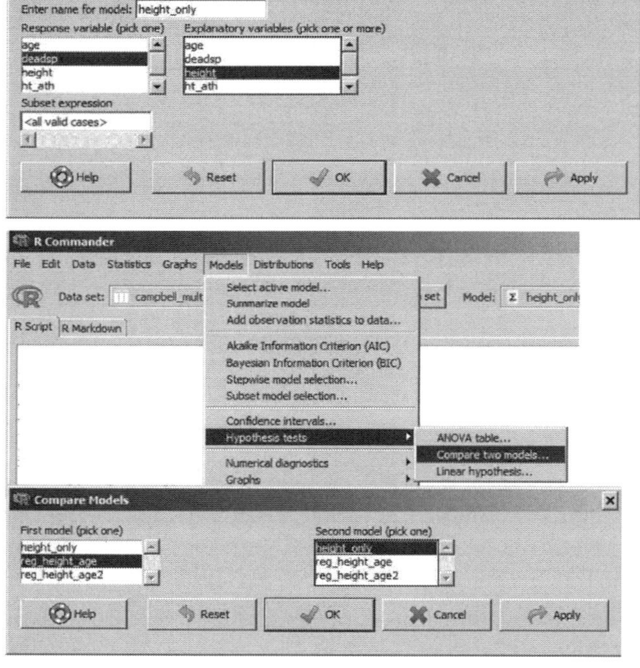

To compare two models you first need to have defined two models with the same output variable. I have used the R Commander menu option **statistics->fit models** to create another regression model which only includes *height* as the input and called it *height_only*.

We can compare the two models formally by using the menu option:

Models-> Hypothesis tests-> compare two models

The results, shown below, indicate that there is no difference between the predictive power of either model (p-value=0.2913).

```
> anova(reg_height_age, height_only)
Analysis of Variance Table

Model 1: deadsp ~ age + height
Model 2: deadsp ~ height
  Res.Df    RSS Df Sum of Sq      F Pr(>F)
1     12 2016.8
2     13 2221.5 -1   -204.74 1.2182 0.2913
```

AIC and BIC
You can find more out about these in the, *Logistic regression* chapter, section: *Model development and evaluation.*

An alternative way of comparing models is to use the AIC or BIC statistics which take into account the complexity of the models (i.e. number of parameters estimated). A smaller value indicating a better fit. To select the particular model you want the AIC value for click on the *Model* box, this will bring up a pick box of all the models you have defined.

To obtain the AIC or BIC value use the R Commander menu option:

Models-> Akaike Information Criterion

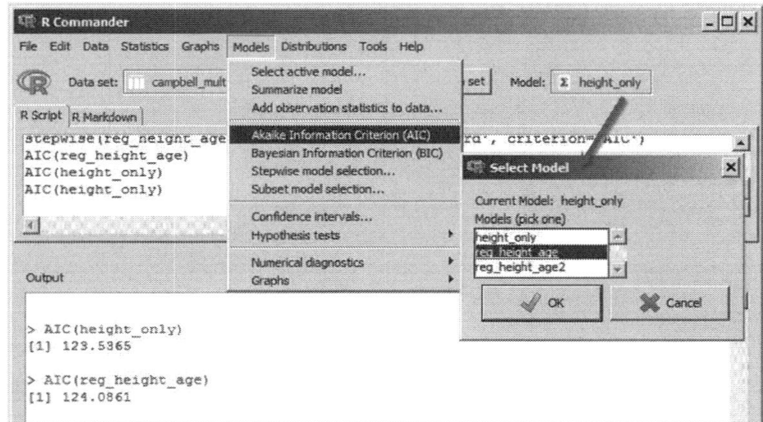

From the output we see that both models produce very similar AIC values with the *height* only model being 123.5 compared to the two input model of 124. This result, along with the ANOVA comparison result and the problem with multicollinearity provides a compelling case for removing age from the model. However, there may be reasons for keeping it, for example looking at the scatterplot matrix we see that the age range is 5 to 14 years, and this may have been a subset of data which originally included a much wider age range? Variable inclusion and exclusion although guided by the maths should always be informed as well by clinical judgement.

59.3 Two predictors, one continuous, one nominal

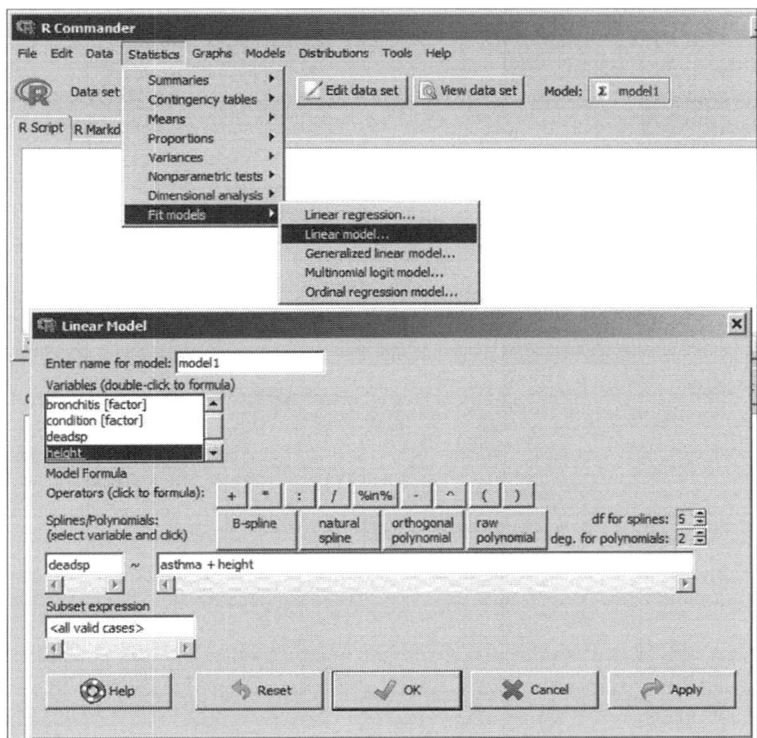

Here we will consider asthma status (nominal) and height as the inputs and deadspace as the output.

To create a linear regression when one or more of the inputs are factors (nominal) you need to use the R Commander menu option:

statistics-> Fit models->linear model

Those variables that are recognised by R Commander to be factors appear in the *variables* pick list with *[factor]* after them. If you had selected the *Linear regression* menu option by mistake they would not have appeared.

I will explain the output on the next page.

Instead of using the menu option you could have typed into the *R Script* window the following R code which R Commander creates for you:

```
model1 <- lm(deadsp ~ asthma +height, data=campbell_mult_reg)

summary(model1)
```

To get some idea of the relationship between height, dead space and asthma status you can draw a 3D graph in R Commander using the Graph menu option:

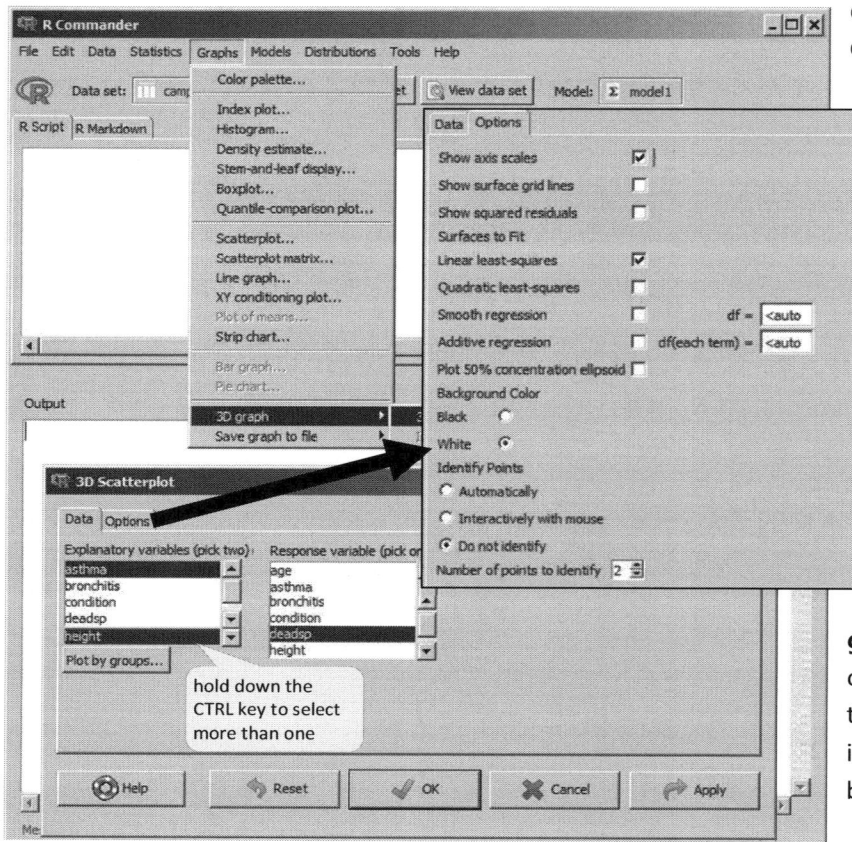

Graphs-> 3D graph-> 3D scatterplot

hold down the CTRL key to select more than one

The resultant graph is a little difficult to interpret; but by dynamically turning it in R Commander you get a much clearer picture. Notice that the regression equation is now a *flat surface*. It appears that those with a lower deadspace have a greater preponderance to suffer from asthma?

To save a 3D graph you need to use the R Commander menu option, **Graphs -> 3d graph -> save graph to file**. The quality of the picture you save is dependent upon the size of the graph window, so it is a good idea to maximise the graph window first before saving it.

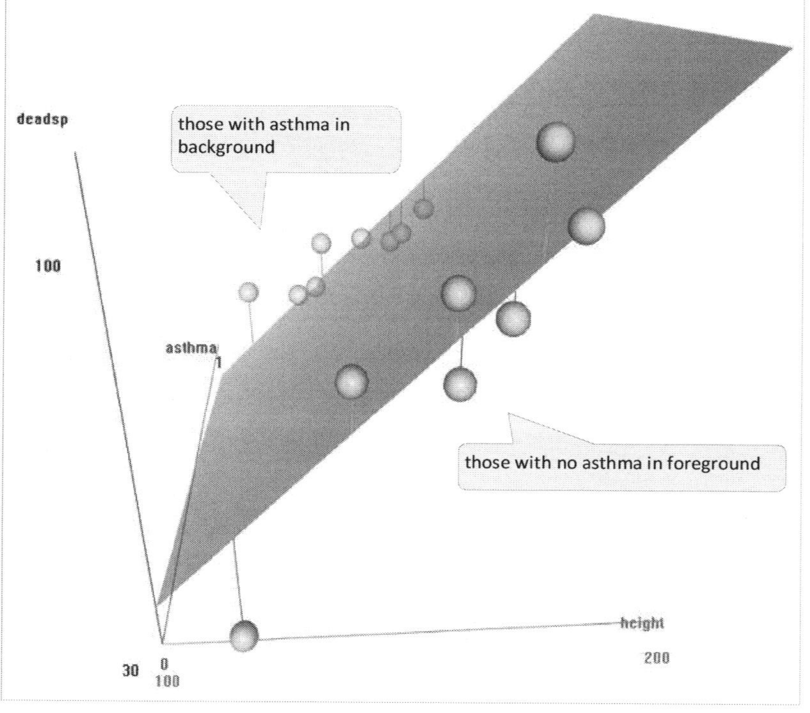

Unfortunately, the regression plane in the 3D diagram is not what we are modelling at the moment. The current model is specifying two parallel regression lines based upon the value of the asthma variable. To understand this situation we need to examine the model output.

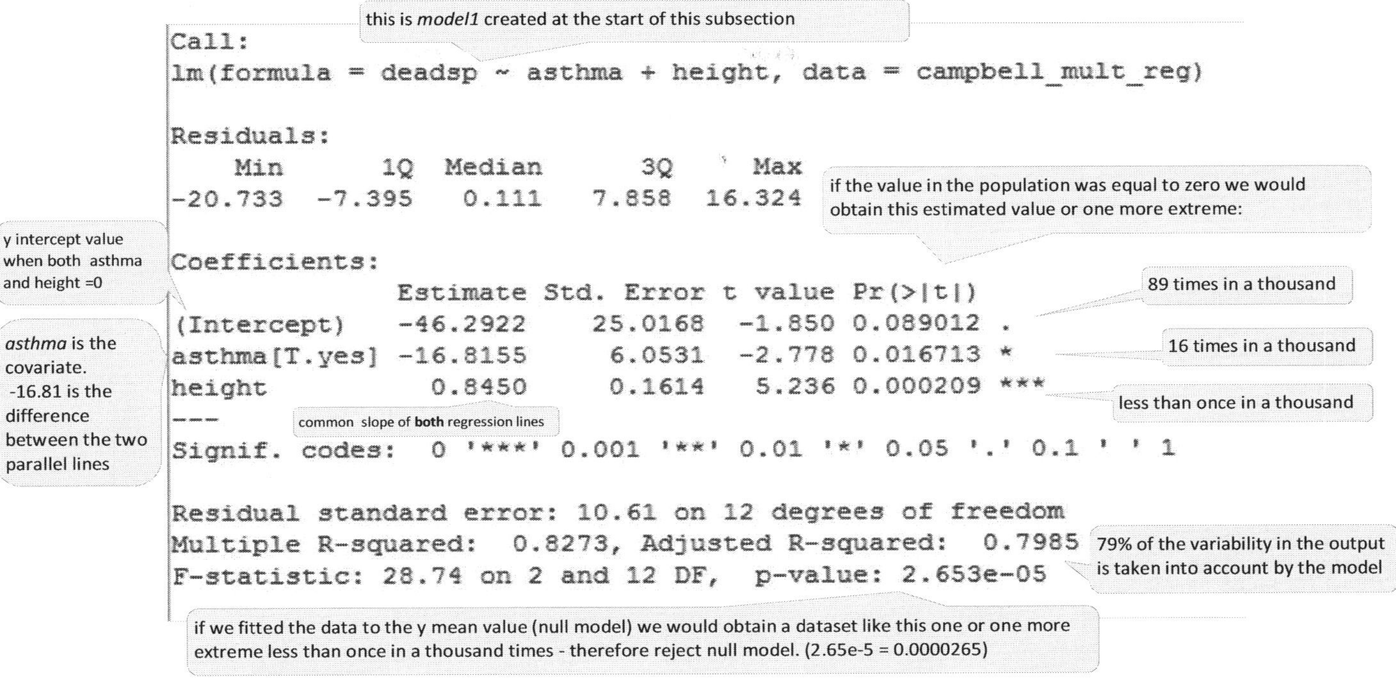
this is *model1* created at the start of this subsection
Call:
lm(formula = deadsp ~ asthma + height, data = campbell_mult_reg)
Residuals:
Min 1Q Median 3Q Max
-20.733 -7.395 0.111 7.858 16.324
if the value in the population was equal to zero we would obtain this estimated value or one more extreme:
y intercept value when both asthma and height =0
Coefficients:
Estimate Std. Error t value Pr(>|t|)
(Intercept) -46.2922 25.0168 -1.850 0.089012 .
asthma[T.yes] -16.8155 6.0531 -2.778 0.016713 *
height 0.8450 0.1614 5.236 0.000209 ***
89 times in a thousand
16 times in a thousand
less than once in a thousand
asthma is the covariate. -16.81 is the difference between the two parallel lines
common slope of **both** regression lines
Signif. codes: 0 '***' 0.001 '**' 0.01 '*' 0.05 '.' 0.1 ' ' 1
Residual standard error: 10.61 on 12 degrees of freedom
Multiple R-squared: 0.8273, Adjusted R-squared: 0.7985
F-statistic: 28.74 on 2 and 12 DF, p-value: 2.653e-05
79% of the variability in the output is taken into account by the model
if we fitted the data to the y mean value (null model) we would obtain a dataset like this one or one more extreme less than once in a thousand times - therefore reject null model. (2.65e-5 = 0.0000265)

From the above output we have a common slope for both regression lines of 0.845, a intercept value of -46.29 for the non-asthmatics and – (46.29 + 16.81)= -63.1 for the asthmatics.

We can graph this exact situation by first using R Commander to plot the points, shown opposite. Making sure that the options *tab* has all the *plot options* cleared.

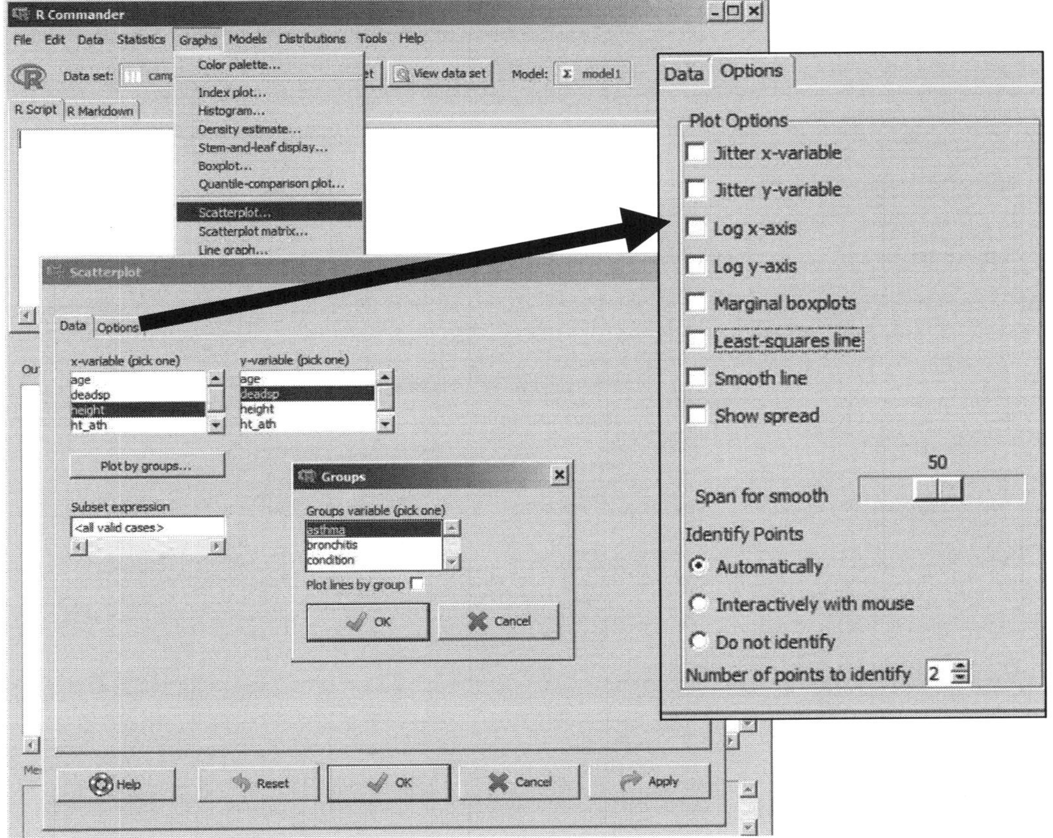
R Commander
File Edit Data Statistics Graphs Models Distributions Tools Help
Color palette...
Index plot...
Histogram...
Density estimate...
Stem-and-leaf display...
Boxplot...
Quantile-comparison plot...
Scatterplot...
Scatterplot matrix...
Line graph...
View data set
Model: model1
R Script R Markdown
Scatterplot
Data Options
x-variable (pick one)
age
deadsp
height
ht_ath
y-variable (pick one)
age
deadsp
height
ht_ath
Plot by groups...
Subset expression
<all valid cases>
Groups
Groups variable (pick one)
asthma
bronchitis
condition
Plot lines by group
OK
Cancel
Help
Reset
OK
Cancel
Apply
Data Options
Plot Options
Jitter x-variable
Jitter y-variable
Log x-axis
Log y-axis
Marginal boxplots
Least-squares line
Smooth line
Show spread
50
Span for smooth
Identify Points
Automatically
Interactively with mouse
Do not identify
Number of points to identify 2

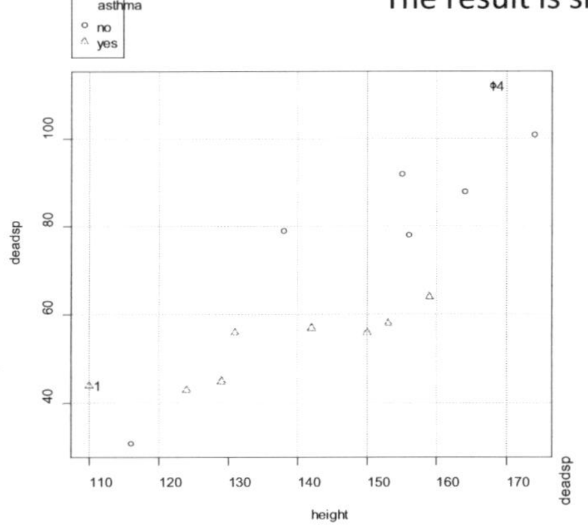

The result is shown opposite and below.

Typing in the following two lines into the R script window, highlighting and then clicking on the submit button produces the necessary regression lines.

```
abline(a=-46.29, b=0.845, col = "blue", lty = 1)
abline(a=-63.1, b=0.845, col = "red", lty = 3)
```

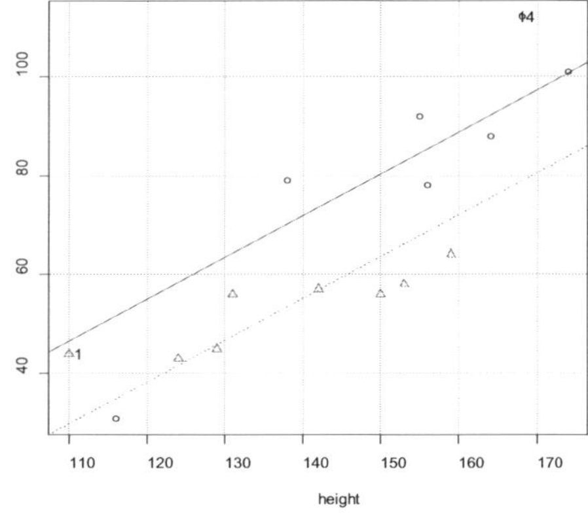


```
> Confint(model1, level=0.95)
                 Estimate        2.5 %     97.5 %
(Intercept)   -46.2921628 -100.7990584  8.214733
asthma[T.yes] -16.8155127  -30.0041448 -3.626880
height          0.8450468    0.4934035  1.196690
```

We can also obtain confidence intervals for the estimates where these values reflect the p-values. Possibly the most interesting aspect is the relatively small width for the height variable suggesting the closeness to the line for each observed and predicted height.

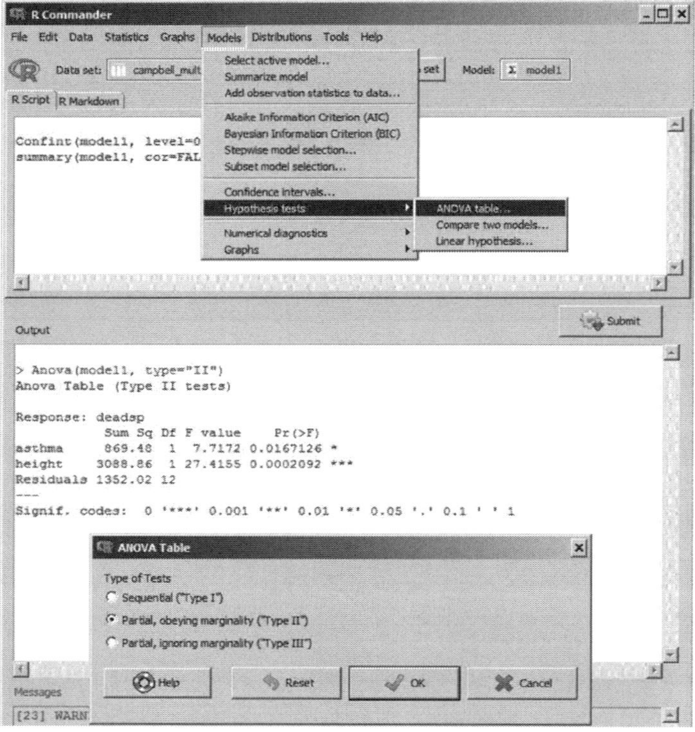

As in the previous example we can obtain an ANOVA table:

Models-> Hypothesis tests-> ANOVA table

As before the type II/type III sums of squares and the Part correlation share the same proportions. That is they both represent a proportion accounted for by the model. We note that R^2=0.827 this is the total amount of variability that is accounted for by the model and we can see how this is divided between the sums of squares from which we can calculate the part correlations.

	asthma	common	height	total
Sum of squares	869		3089	
Percentage of total				100%
Part correlation^2				.827

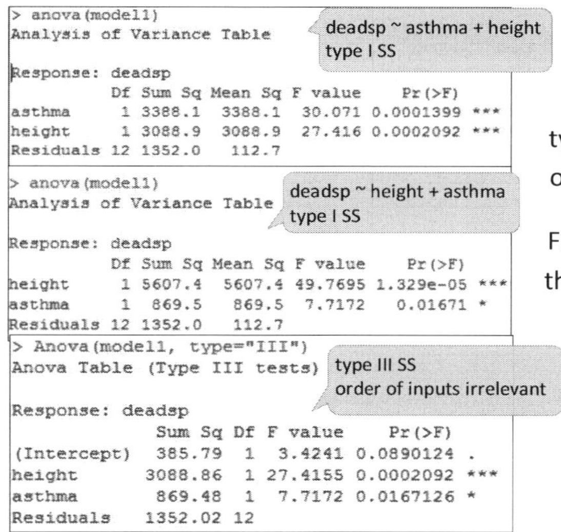

```
> anova(model1)
Analysis of Variance Table

Response: deadsp
          Df Sum Sq Mean Sq F value    Pr(>F)
asthma     1 3388.1  3388.1  30.071 0.0001399 ***
height     1 3088.9  3088.9  27.416 0.0002092 ***
Residuals 12 1352.0   112.7

> anova(model1)
Analysis of Variance Table

Response: deadsp
          Df Sum Sq Mean Sq F value    Pr(>F)
height     1 5607.4  5607.4 49.7695 1.329e-05 ***
asthma     1  869.5   869.5  7.7172   0.01671 *
Residuals 12 1352.0   112.7

> Anova(model1, type="III")
Anova Table (Type III tests)

Response: deadsp
             Sum Sq Df F value    Pr(>F)
(Intercept)  385.79  1  3.4241 0.0890124 .
height      3088.86  1 27.4155 0.0002092 ***
asthma       869.48  1  7.7172 0.0167126 *
Residuals   1352.02 12
```


To demonstrate that order is important in type I sums of squares when the inputs are correlated, I have repeated the analysis with the two inputs entered in different orders opposite. In contrast the type III sums of squares (the traditional block regression approach) only measures the unique contribution of each.

From the results opposite we see that asthma has a total of 3388 so the shared amount is 3388 - 869 = 2519. We can also see that the total sums of squares for the regression is 3388 + 3089 = 6477 and the residual is 1352, so the total sum of squares is equal to 7829. While these numbers seem abstract, if we convert them to percentages it makes them easier to understand.

	asthma	common	height	total
Sum of squares	869	2519	3089	7829
Percentage of total	13.39	38.89	47.72	100%
Part correlation²	.1108	.5671	.3947	.827 (R squared)
Part correlation	.333	.1035	.6282	
Part correlations² calculated from the percentage of total x R squared value				

Representing these results in the form of a Venn diagram adds more clarity.

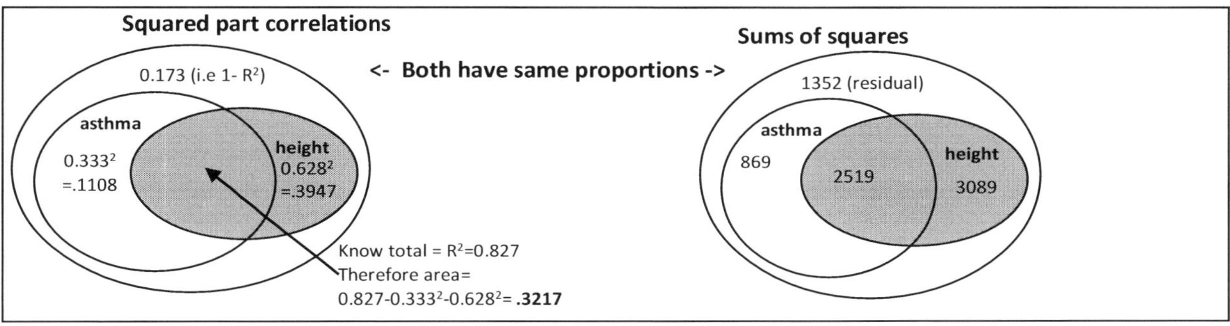

Formally the part correlation is defined as the *value of the signed square root of the increase in R^2 for that particular variable*, therefore remember the larger this value is the more weight the variable has in the model. From the table and above it is clear that height has a much larger influence on the model than asthma status.

If you are confused by the term null hypothesis or statistically significant see the glossary entries; **effect size, p-value, inference, sampling distributions and standard errors, (se)** and **clinical significance**.

For the above model to be valid it is necessary that the actual subsets of data for the covariate (asthma) levels have parallel regression lines. To assess the validity of this we create an interaction term, if the interaction term is shown to be statistically significant (that is we reject the null hypothesis that its value in the population is equal to zero) we cannot use this model. We will create the interaction term in the next section. But before we do the mathematics we will draw these lines of best fit through the two subsets of data and see if they do, and how far they deviate from the parallel constrained ones.

59.3.1 Drawing non-parallel regression lines

To produce the possible non-parallel regression lines we change the options in the Scatterplot dialog box slightly by clicking the *Plot lines by group* option, as shown opposite. Also in the **Options** tab you need to select *Least-squares line* in the **Plot Options** group.

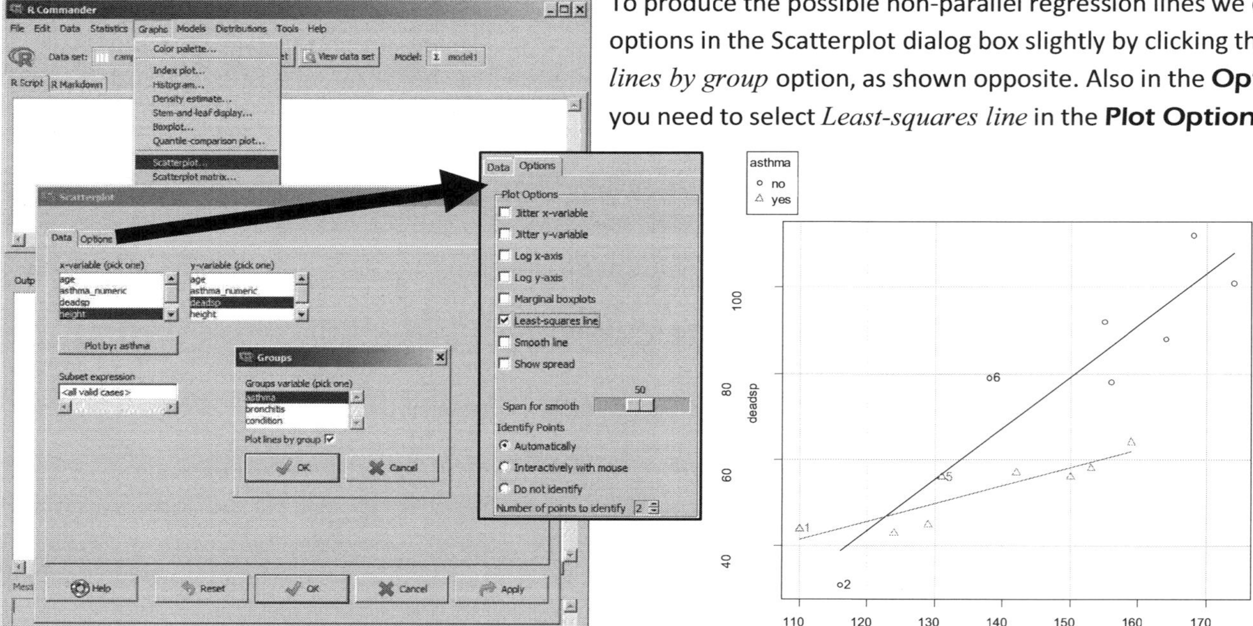

The graph above shows separate regression lines for each level of the asthma factor. The gradients of the two lines look quite different (i.e. not parallel). But to assess this difference between the gradients formally an interaction term is added to the regression equation. If we find that the p-value associated with its parameter estimate is significant (that is it is sufficiently unlikely to be zero in the population) we accept that the gradients are different rather than just due to random sampling. To draw the confidence intervals for each level see the *publication quality graphics* chapter.

59.3.2 Including and testing an interaction term

```
> as.numeric(asthma)
 [1] 2 1 2 2 2 1 2 2 2 1 1 2 1 1 1
```

An interaction term is created from input variables already in the regression equation by multiplying their values. For example, an interaction term between asthma and height could be calculated directly in R by typing *as.numeric(asthma) * height.* The *as.numeric(asthma)* part of the expression makes sure that R uses the actual values rather than the text labels of the factor levels.

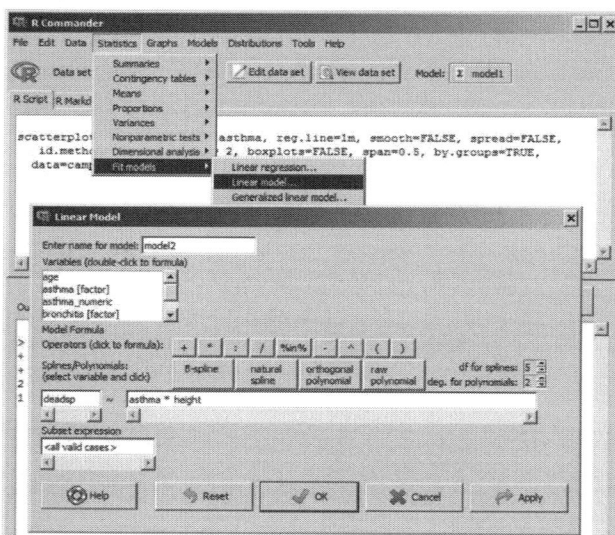

The regression coefficient (b) associated with this new variable represents the change in the slope for different group memberships. A larger sum of squares associated with this value indicates a greater difference between slopes and a more significant (i.e. smaller) p-value. To produce the model this time we use the star sign (*) indicating that we want a model with an interaction term as well as the other inputs, which are now called main effects.

```
deadsp ~ asthma * height
```

Remember here that because the star sign is used with the model sign ~ it does not simply mean multiplication.

If you can't be bothered to use the dialog box you can copy the following code to the script window, highlight it and then click on the **Submit** button to run it.

```
model2<- lm(deadsp ~ asthma * height, data=campbell_mult_reg)
summary(model2)
```

The results and comments are given below.

asthma variable:
1=asthmatic
0=non-asthmatic

this is the R code generated by R Commander , check you used the correct input and output variables?

```
lm(formula = deadsp ~ asthma * height, data = campbell_mult_reg)

Residuals:
   Min      1Q Median
-8.578 -5.750 -1.401  4.

Coefficients:
                     Estimate Std. Error t value Pr(>|t|)
(Intercept)          -99.4624    25.2079  -3.946  0.00229 **
asthma[T.yes]         95.4726    35.6106   2.681  0.02137 *
height                 1.1926     0.1636   7.291 1.56e-05 ***
asthma[T.yes]:height  -0.7782     0.2448  -3.179  0.00877 **
---
Signif. codes:  0 '***' 0.001 '**' 0.01 '*' 0.05 '.' 0.1 ' ' 1

Residual standard error: 8.003 on 11 degrees of freedom
Multiple R-squared:   0.91, Adjusted R-squared:  0.8855
F-statistic: 37.08 on 3 and 11 DF,  p-value: 4.803e-06
```

difference in intercept value for asthmatics i.e. -99.462+95.473=-3.99 compared to non asthmatics=-99.4

-99.46 The y intercept value for the regression line when all the inputs=0,

1.19 x height +0 x asthma Slope of regression line for non-asthmatics (asthma=0)

(1.193-778) x height =0.415 x height Slope of regression line for asthmatics

interaction term - allows testing of ANCOVA model

because interaction significant keep both asthma and height, parameters even if not significant

setting c.v. to 0.05 means that we accept that the interaction term is not equal to zero, therefore we keep it.

88% of variability is taken into account by the model (compared to 79% in the previous model)

-.0.778 The difference in slopes for factor levels of asthma

df for these is n-p-1 i.e. number of obs- no of parameters -1 15-3-1=11

if we fitted the data to the y mean line (null model) we would obtain an F value equal to or one more extreme less than once in a thousand times. Therefore, reject null model and accept that one or more of the parameters above are beneficial in modelling the data

Because the interaction term is statistically significant we keep it AND both the variables that formed it in the equation. This is regardless of the p-value of the two variables that made the interaction. We now know that the slope is different and steeper for asthmatics (0.415) compared to non-asthmatics (1.9) and looking back at the scatterplot with the regression lines they crossover at around the 122cm height value. The regression equation is now:

$$\text{predicted\_deadspace} = a + (b_{height} \times \text{height}) + (b_{asthma} \times \text{asthma}) + (b_{height\_asthma} \times \text{height} \times \text{asthma})$$

$$\text{predicted\_deadspace} = -99.46 + (1.19 \times \text{height}) + (95.47 \times \text{asthma}) - (0.778 \times \text{height} \times \text{asthma})$$

Remembering that the coding for the asthma input is 1=asthmatic; 0=non-asthmatic. The above equation can be considered for both asthmatics and non-asthmatics separately.

For asthmatics:

```
predicted_deadspace = -99.46 + (1.19 x height)  + (95.47 x 1) - (0.778 x height x 1)
predicted_deadspace = -99.46 + 1.19 x height  + (95.47) - (0.778 x height)
predicted_deadspace = 95.47-99.46 + 1.19 x height       - (0.778 x height)
predicted_deadspace = -3.99       + 1.19 x height       - (0.778 x height)
```

For non-asthmatics:

```
predicted_deadspace = -99.46 + (1.19 x height)  + (95.47 x 0) - (0.778 x height x 0)
predicted_deadspace = -99.46 + (1.19 x height)
```

```
> vif(model2)
     asthma      height asthma:height
  73.915773    2.193651     67.593634
```

It is pertinent here to demonstrate the effect centering and standardisation has on VIFs concerning interaction terms. The output opposite is for the un-transformed variables showing very high VIF values indicating multicollinearity for this model.

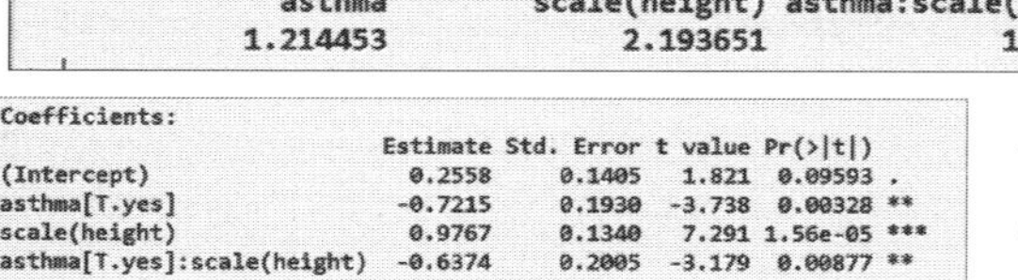

```
> vif(model2_st)
          asthma      scale(height) asthma:scale(height)
        1.214453           2.193651             1.995473
```

```
Coefficients:
                            Estimate Std. Error t value Pr(>|t|)
(Intercept)                   0.2558     0.1405   1.821  0.09593 .
asthma[T.yes]                -0.7215     0.1930  -3.738  0.00328 **
scale(height)                 0.9767     0.1340   7.291 1.56e-05 ***
asthma[T.yes]:scale(height) -0.6374     0.2005  -3.179  0.00877 **
```

Robinson & Schumacker, 2009 demonstrate very clearly how these values can be reduced by centering the inputs. The same effect can be achieved by standardising them using the *scale()* function. Re-running the analysis on the standardised height and deadspace variables (shown opposite) reduces the VIF to very low values while it does not affect the parameter estimation for height and the interaction.

```
Call:
lm(formula = (deadsp - mean(deadsp)) ~ asthma * scale(height -
    mean(height)), data = campbell_mult_reg)

Residuals:
   Min     1Q Median     3Q    Max
-8.578 -5.750 -1.401  4.065 13.889

Coefficients:
                                          Estimate Std. Error t value Pr(>|t|)
(Intercept)                                  6.049      3.322   1.821  0.09593 .
asthma[T.yes]                              -17.062      4.565  -3.738  0.00328 **
scale(height - mean(height))                23.097      3.168   7.291 1.56e-05 ***
asthma[T.yes]:scale(height - mean(height)) -15.073      4.741  -3.179  0.00877 **
---
```

To check the equivalence between centered and standardised variables I have also re-run the analysis using only centered variables –obtaining identical parameter estimates etc.

59.4 Relative importance of inputs

```
> average_partial_estimates <- calc.relimp(model2, type = "lmg")
> average_partial_estimates
Response variable: deadsp
Total response variance: 559.2095
Analysis based on 15 observations

3 Regressors:
asthma height asthma:height
Proportion of variance explained by model: 91%
Metrics are not normalized (rela=FALSE).

Relative importance metrics:

                    lmg
asthma        0.2719108
height        0.5553942
asthma:height 0.0827023

Average coefficients for different model sizes:

                      1X          2Xs         3Xs
asthma        -30.125000 -16.8155127 95.4726312
height          1.033323   0.8450468  1.1925647
asthma:height        NaN         NaN -0.7782494
```

relative importance metrics

same as the regression coefficients

Genizi, 1993 amongst other commentators have suggested that rather than simply using the part correlations to assess the relative important of each input their average value over all possible ordering of inputs should be considered. The *lmg()* function in the *relaimpo* (**rela**tive **impo**rtance of regressors in linear models) package (Groemping, 2006) provides this value. The R code below installs and loads the *relaimpo* package and then runs the *lmg()* function on *modal2* we produced on the previous page.

```
install.packages("relaimpo")
library(relaimpo)
average_partial_estimates <-
calc.relimp(model2, type = "lmg")
```

lower case L

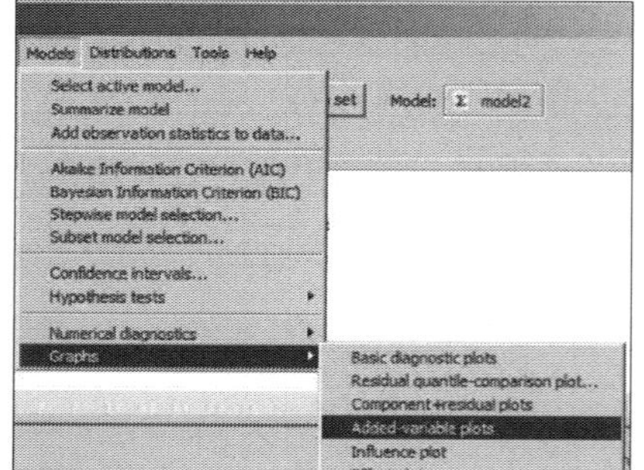

59.4.1 Added variable plots

Plots allowing you to see the impact a particular input has upon the regression equation are called partial regression or individual coefficient plots. In R Commander they are called *added variable plots,* which you obtain using the following menu option, having first selected a model.

Models-> Graphs-> Added-variable plots

The added variable plot has several useful characteristics:

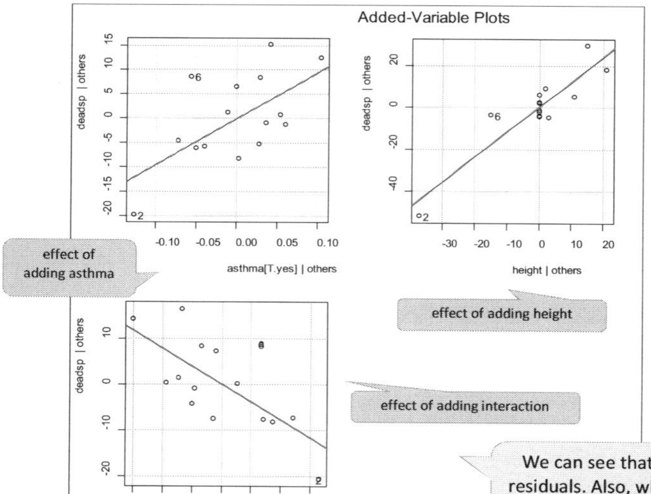

- The gradient of the slope for each input (i) is equivalent to the particular regression coefficient (b_i). Therefore the steeper the gradient (either positive or negative) the bigger the effect size. Also the similarity in the gradient between plots provides an indication of the degree of multicollinearity between the inputs.
- The residuals from the regression line in these plots are identical to the residuals for the original model (Y against all the independent variables including X_i) so we can use them to assess the distribution of the residuals and also identify cases that have excessive lack of fit.

We can see that cases 2 and 6 appear to produce the largest residuals. Also, while the regression coefficients for both height and asthma are positive that for the interaction is negative.

59.4.2 ANOVA table

The ANOVA table can be produced using the same R Commander menu option as described in the previous example. Alternatively, you can enter the following R code directly.

upper case i; alternatively you can use 2 or 3

grabs all sequentially

```
anova(model2)

Anova(model2, type="II")

Anova(model2, type="III")
```

each term after all others, ignoring interactions first (same as SAS 'anova model' not SAS type II)

same as standard regression

SPSS produces this by default

each input 'last in'

Type I – Sequential

```
Response: deadsp
               Df Sum Sq Mean Sq F value    Pr(>F)
asthma          1 3388.1  3388.1  52.897 1.597e-05 ***
height          1 3088.9  3088.9  48.226 2.441e-05 ***
asthma:height   1  647.5   647.5  10.109   0.00877 **
Residuals      11  704.5    64.0
```

704.55 (residual)
asthma
height
3088.9
3388.1
A
B
C
D
704.5
interaction

type="II"

```
Anova Table (Type II tests)

Response: deadsp
               Sum Sq Df F value    Pr(>F)
asthma         869.48  1  13.575 0.003598 **
height        3088.86  1  48.226 2.441e-05 ***
asthma:height  647.47  1  10.109 0.008770 **
Residuals      704.55 11
```

704.55 (residual)
asthma
height
3088.86
869.48
A
B
C
D
704.5
interaction

type="III"

```
Anova Table (Type III tests)

Response: deadsp
              Sum Sq Df F value    Pr(>F)
(Intercept)    997.1  1 15.5684   0.00229 **
asthma         460.4  1  7.1879   0.02137 *
height        3404.8  1 53.1583 1.561e-05 ***
asthma:height  647.5  1 10.1089   0.00877 **
Residuals      704.5 11
```

704.55 (residual)
asthma
height
3404.8
460.4
A
B
C
D
704.5
interaction

To demonstrate that order matters for the type one (I) sums of squares I have re-run the analysis with the inputs reversed:

```
model2_temp<- lm(deadsp ~ height * asthma, data=campbell_mult_reg)
summary(model2_temp)
```

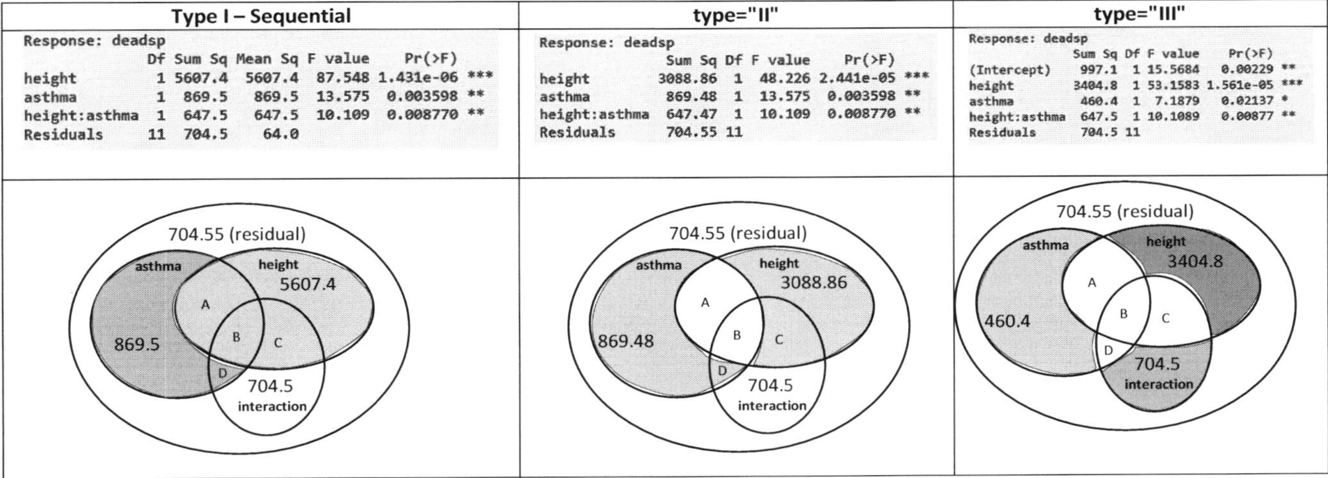

Type I – Sequential

```
Response: deadsp
               Df Sum Sq Mean Sq F value    Pr(>F)
height          1 5607.4  5607.4  87.548 1.431e-06 ***
asthma          1  869.5   869.5  13.575 0.003598 **
height:asthma   1  647.5   647.5  10.109 0.008770 **
Residuals      11  704.5    64.0
```

type="II"

```
Response: deadsp
               Sum Sq Df F value    Pr(>F)
height        3088.86  1  48.226 2.441e-05 ***
asthma         869.48  1  13.575 0.003598 **
height:asthma  647.47  1  10.109 0.008770 **
Residuals      704.55 11
```

type="III"

```
Response: deadsp
              Sum Sq Df F value    Pr(>F)
(Intercept)    997.1  1 15.5684   0.00229 **
height        3404.8  1 53.1583 1.561e-05 ***
asthma         460.4  1  7.1879   0.02137 *
height:asthma  647.5  1 10.1089   0.00877 **
Residuals      704.5 11
```


	asthma	common	height	interaction	total
Percentage of total	11.7%	32.69%	46%	9.5%	100%
Sum of squares	869.5	A+B= (869.5+5607.4)-(869.48+3088.86)= 2418.56	3404.8	704.5	7397.36
Part correlation^2	.106	(.297)	.418	0.086	.91
Part correlation	.3255	(.544)	.646	.293	
Relative importance metric (see next page)	.2719		.555	.082	.91

The situation with the various sums of squares is now a little more complex. The important aspect to note is that the standard regression approach (type III sums of squares) ignores ALL the overlapping areas so order does not matter.

In contrast with the type I sequential approach, all the areas are used but is highly dependent upon the order the inputs are entered. However having said that, regardless of the approach taken in this instance, we would have ended up with the same decision to keep all three inputs because the interaction term is statistically significant in all three approaches. With more complex models this is not likely to be the case.

59.5 Regression diagnostics

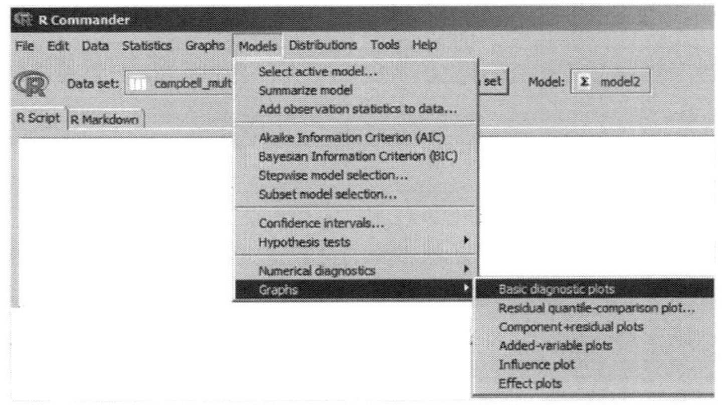

In the introductory chapter on linear regression the large section *influence measures* investigated this aspect and with multiple regression the same issues and approach to investigating them applies.

R Commander provides several menu options for regression diagnostics, with the first option providing a general overview:

Models->Graphs->Basic diagnostic plots

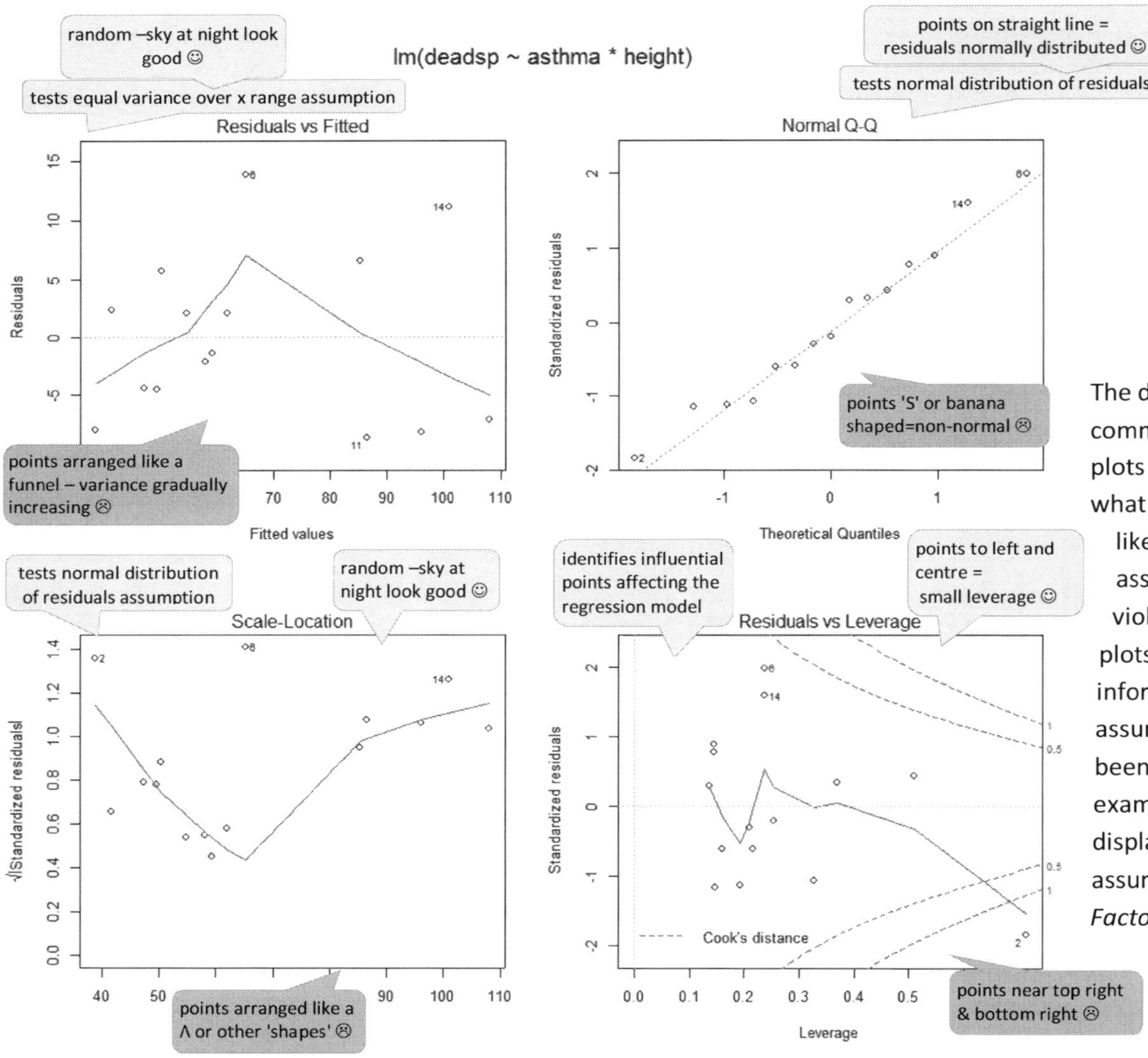

lm(deadsp ~ asthma * height)

random –sky at night look good ☺

tests equal variance over x range assumption

Residuals vs Fitted

points arranged like a funnel – variance gradually increasing ☹

points on straight line = residuals normally distributed ☺

tests normal distribution of residuals assumption

Normal Q-Q

points 'S' or banana shaped=non-normal ☹

tests normal distribution of residuals assumption

random –sky at night look good ☺

Scale-Location

points arranged like a Λ or other 'shapes' ☹

identifies influential points affecting the regression model

points to left and centre = small leverage ☺

Residuals vs Leverage

points near top right & bottom right ☹

The darker shaded comment boxes in the plots opposite indicate what the plots would look like when the assumptions are violated. From the above plots we can see that informally the regression assumptions seem to have been adhered to. For an example of a set of plots displaying violated assumptions see the *Factorial ANOVA* chapter.

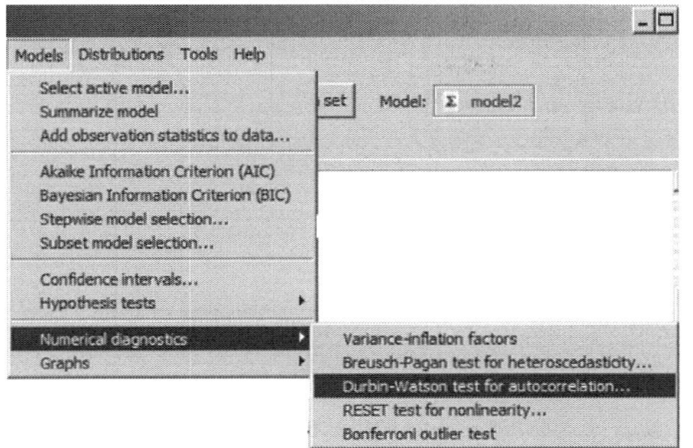

If you don't like the graphical approach there are numerical equivalents, available from the R Commander menu, shown opposite.

A recent paper (Amphanthong, 2012) suggests that two measures derived from the residuals provide the greatest reliability for identifying outliers in multiple regression; the PRESS (predicted residual sum of squares) residuals and the Mahalanobis distance measure, which comes in two varieties.

59.5.1 Identifying outliers: PRESS and Mahalanobis distances

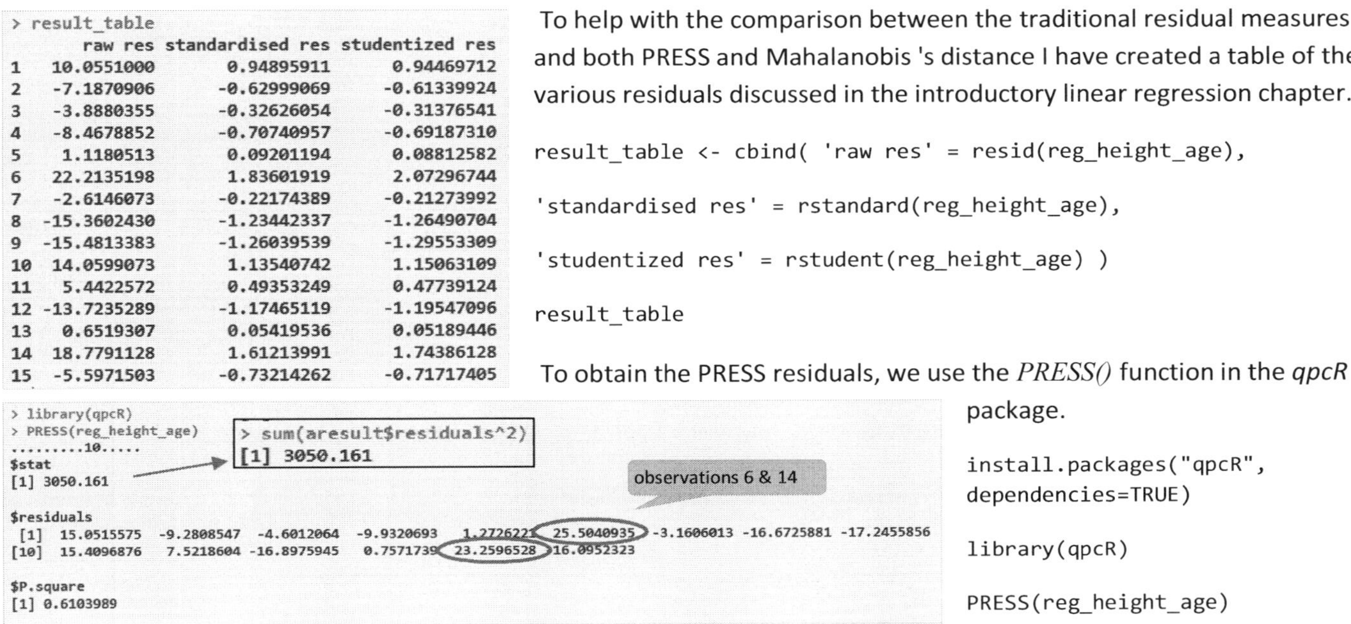

```
> result_table
      raw res standardised res studentized res
1  10.0551000       0.94895911      0.94469712
2  -7.1870906      -0.62999069     -0.61339924
3  -3.8880355      -0.32626054     -0.31376541
4  -8.4678852      -0.70740957     -0.69187310
5   1.1180513       0.09201194      0.08812582
6  22.2135198       1.83601919      2.07296744
7  -2.6146073      -0.22174389     -0.21273992
8 -15.3602430      -1.23442337     -1.26490704
9 -15.4813383      -1.26039539     -1.29553309
10 14.0599073       1.13540742      1.15063109
11  5.4422572       0.49353249      0.47739124
12 -13.7235289      -1.17465119     -1.19547096
13  0.6519307       0.05419536      0.05189446
14 18.7791128       1.61213991      1.74386128
15 -5.5971503      -0.73214262     -0.71717405
```

```
> library(qpcR)
> PRESS(reg_height_age)
.........10.....
$stat
[1] 3050.161

$residuals
 [1]  15.0515575  -9.2808547  -4.6012064  -9.9320693   1.2726221  25.5040935  -3.1606013 -16.6725881 -17.2455856
[10]  15.4096876   7.5218604 -16.8975945   0.7571739  23.2596528  16.0952323

$P.square
[1] 0.6103989
```

```
> sum(aresult$residuals^2)
[1] 3050.161
```


To help with the comparison between the traditional residual measures and both PRESS and Mahalanobis 's distance I have created a table of the various residuals discussed in the introductory linear regression chapter.

```
result_table <- cbind( 'raw res' = resid(reg_height_age),

'standardised res' = rstandard(reg_height_age),

'studentized res' = rstudent(reg_height_age) )

result_table
```

To obtain the PRESS residuals, we use the *PRESS()* function in the *qpcR* package.

```
install.packages("qpcR",
dependencies=TRUE)

library(qpcR)

PRESS(reg_height_age)
```

The PRESS residual is calculated by removing each value in turn from the dataset then recalculating the residual for that particular value. The PRESS statistic (*$stat*) is the squared sum of these values (shown opposite) and can be used to compare models in the same way as BIC and AIC. The P-square value (*$P.square*), which has nothing to do with the p-value, can be thought of as the PRESS equivalent to R squared, its value is similar to the adjusted R squared value of 0.6995 given in the earlier *model2* output.

Focusing on the PRESS residuals we come to the same conclusions as before, cases 6 and 14 have the highest values.

When considering multiple linear regression the Mahalanobis distance is computed only for the x-variables (Varmuza & Filzmoser, 2009 p.147), and is thus equivalent to outlier detection for the x-variables using the Mahalanobis distances.

To obtain the Mahalanobis distances we create a temporary dataframe consisting of only the input (i.e. x) variables:

```
subset_height_age_only <- data.frame(age, height)

subset_height_age_only
```

```
> subset_height_age_only <- data.frame(age, height)
> subset_height_age_only
  age height
1   5    110
2   5    116
```

The traditional Mahalanobis distance is based on the difference between the mean of the input variables and the observed values. It can also be shown that the squared Mahalanobis distance approximately follows a chi-squared distribution χ^2_m with *m* degrees of freedom where *m* is the number of x-variables (*ibid.* p.61). We can then set a cutoff value for the Mahalanobis distance, based on the level of certainty we want (usually 95%) along with the degrees of freedom. For our example, we set the usual 95% level along with 2 degrees of freedom; $\chi^2_{2,0.975}$ and use the standard R functions *qchisq()* and *sqrt()*. The square root function is needed because we are using Mahalanobis distances rather than their squared values.

```
> cutoff <- sqrt(qchisq(0.975,2))
> cutoff
[1] 2.716203
```

```
cutoff <- sqrt(qchisq(0.975,2));   cutoff
```

The cutoff value of 2.7 provides a comparator for our dataset. The robust Mahalanobis distance uses a different measure for the centre of the distribution known as the MCD (minimum covariance determinant) but keeps the same cutoff value.

```
> library(chemometrics)
> trad_m2 <- Moutlier(subset_height_age_only,quantile=0.975,plot=FALSE)$md
> trad_m2
 [1] 1.9271879 1.4916674 1.1120331 1.0632707 0.8758346 0.9343266 1.2186745 0.4106641
 [9] 0.7063187 0.5412654 1.7138584 1.3024763 1.0062775 1.3279708 2.8632390
> summary(trad_m2)
   Min. 1st Qu.  Median    Mean 3rd Qu.    Max.
 0.4107  0.9051  1.1120  1.2330  1.4100  2.8630
> # $rd = robust
> robust_m2 <- Moutlier(subset_height_age_only,quantile=0.975,plot=TRUE)$rd
> robust_m2
 [1] 2.1808552 1.7413913 1.3713224 1.2690385 1.1036429 0.7533447 0.8678059 0.1787208
 [9] 0.4328034 0.2806078 1.2526815 0.9584361 0.7226988 0.9683566 2.0493566
> summary(robust_m2)
   Min. 1st Qu.  Median    Mean 3rd Qu.    Max.
 0.1787  0.7380  0.9684  1.0750  1.3200  2.1810
```

To obtain Mahalanobis distances, we use the *Moutlier()* function in the *chemometrics* package which was developed to support Varmuza & Filzmoser, 2009.

```
install.packages("chemometrics", dependencies=TRUE)

library(chemometrics)

trad_m2 <- Moutlier(subset_height_age_only, quantile=0.975,plot=FALSE)$md

trad_m2

summary(trad_m2)

robust_m2 <- Moutlier(subset_height_age_only,quantile=0.975,plot=TRUE)$rd

robust_m2

summary(robust_m2)
```

traditional Mahalanobis distances

robust Mahalanobis distances

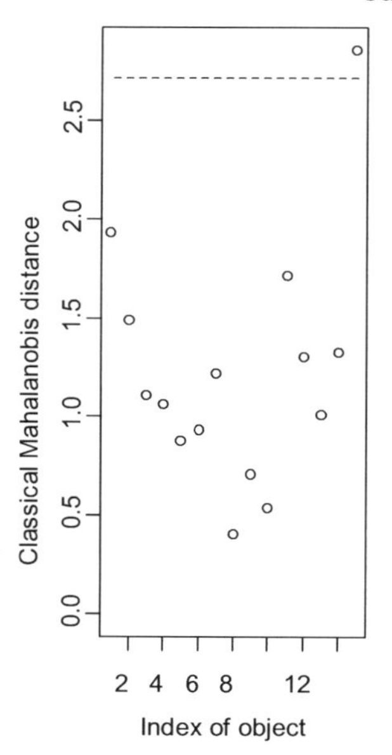

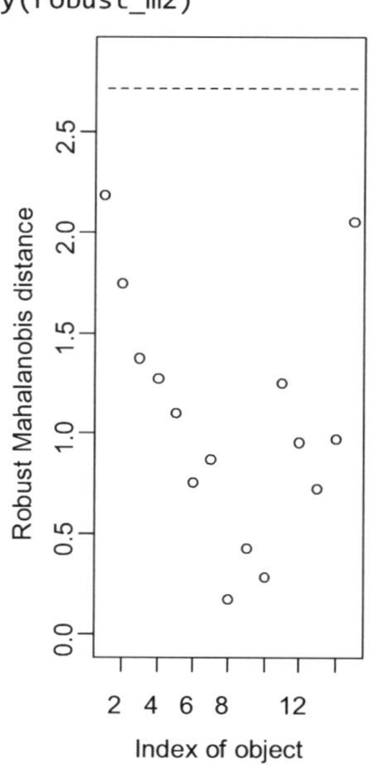

The option *plot=True* produces two plots showing classical and robust Mahalanobis distances versus the observation numbers. The automatically generated horizontal dotted line is the cutoff value.

We see that for traditional Mahalanobis distances one observation is identified as a possible outlier, but when the robust version is used this disappears.

59.5.2 Correlated errors

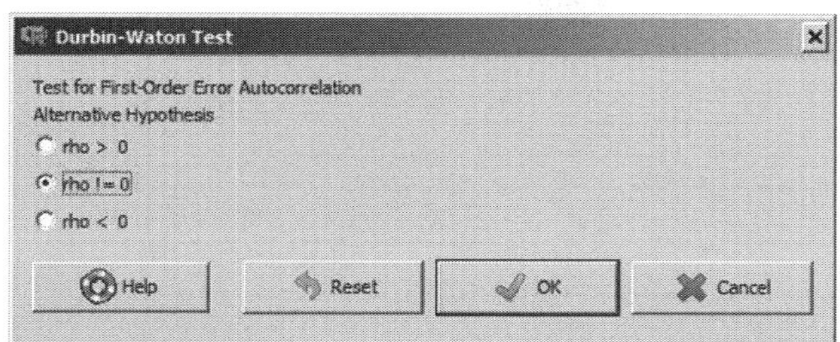

One important assumption of multiple regression is that the cases are independent. This is equivalent to saying that the residuals (errors) are not correlated. The simplest way to check this is to assess how the data were collected. The Durbin-Watson test for autocorrelation provides a quantitative way of assessing possible non-independence between cases. The Durbin-Watson test uses a statistic called DW which is approximately equal to 2(1 – r), where r is the sample autocorrelation of the residuals, so a DW value of 2 indicates no observed autocorrelation.

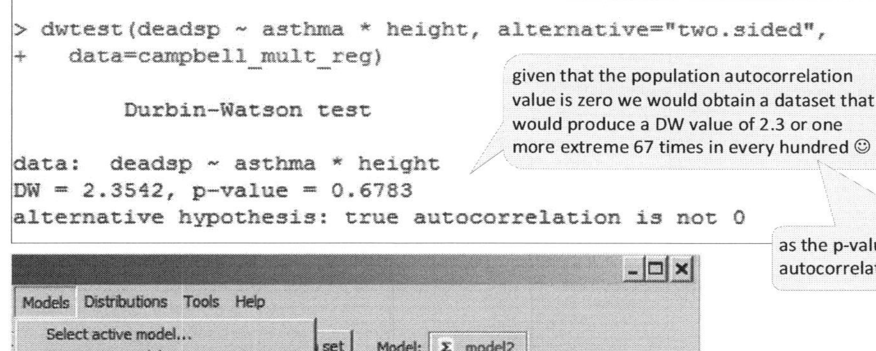

```
> dwtest(deadsp ~ asthma * height, alternative="two.sided",
+   data=campbell_mult_reg)

        Durbin-Watson test

data:  deadsp ~ asthma * height
DW = 2.3542, p-value = 0.6783
alternative hypothesis: true autocorrelation is not 0
```

given that the population autocorrelation value is zero we would obtain a dataset that would produce a DW value of 2.3 or one more extreme 67 times in every hundred ☺
as the p-value is >0.05 we accept that that autocorrelation value is equal to zero ☺

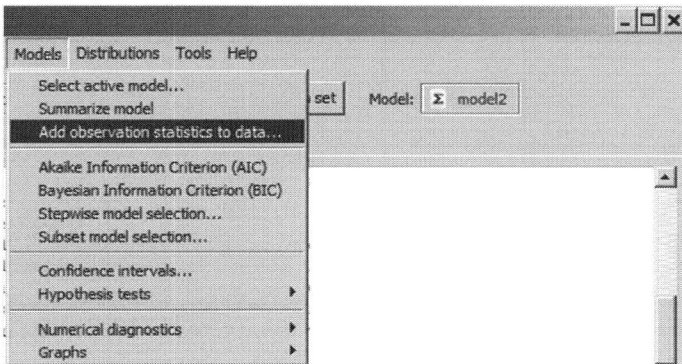

You can think of autocorrelation as the correlation between cases. The Durbin-Watson test assumes that the autocorrelation value is zero in the population so the p-value is interpreted in the usual way.

59.5.3 Predicted values & others

It is often useful to obtain the actual values produced from the model for each case as well as other values such as the leverage influence statistics and various types of residuals described in the introductory linear regression chapter. These can be obtained using the R Commander menu option:

Models-> Add observation statistics to data

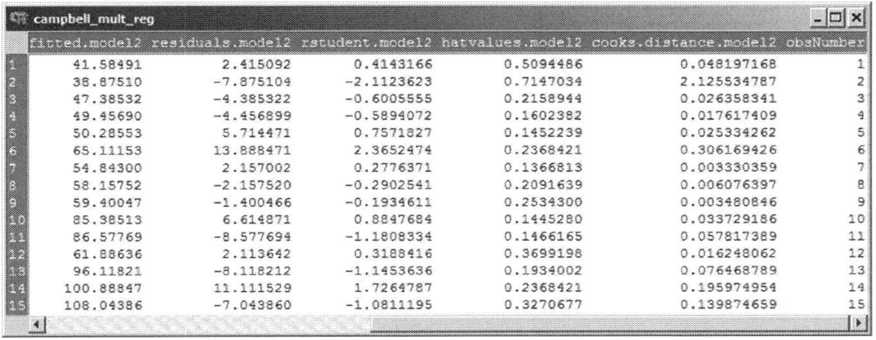

campbell_mult_reg

	fitted.model2	residuals.model2	rstudent.model2	hatvalues.model2	cooks.distance.model2	obsNumber
1	41.58491	2.415092	0.4143166	0.5094486	0.048197168	1
2	38.87510	-7.875104	-2.1123623	0.7147034	2.125534787	2
3	47.38532	-4.385322	-0.6005555	0.2158944	0.026358341	3
4	49.45690	-4.456899	-0.5894072	0.1602382	0.017617409	4
5	50.28553	5.714471	0.7571827	0.1452239	0.025334262	5
6	65.11153	13.888471	2.3652474	0.2368421	0.306169426	6
7	54.84300	2.157002	0.2776371	0.1366813	0.003330359	7
8	58.15752	-2.157520	-0.2902541	0.2091639	0.006076397	8
9	59.40047	-1.400466	-0.1934611	0.2534300	0.003480846	9
10	85.38513	6.614871	0.8847684	0.1445280	0.033729186	10
11	86.57769	-8.577694	-1.1808334	0.1466165	0.057817389	11
12	61.88636	2.113642	0.3188416	0.3699198	0.016248062	12
13	96.11821	-8.118212	-1.1453636	0.1934002	0.076468789	13
14	100.88847	11.111529	1.7264787	0.2368421	0.195974954	14
15	108.04386	-7.043860	-1.0811195	0.3270677	0.139874659	15

The values are added to the current dataframe as shown below. To save these values you need to save the dataframe.

Many of these values have been used to produce the basic diagnostic plots described on the previous page. The R code this option generates is given below for reference:

```
campbell_mult_reg$fitted.model2 <- fitted(model2)
campbell_mult_reg$residuals.model2 <- residuals(model2)
campbell_mult_reg$rstudent.model2 <- rstudent(model2)
campbell_mult_reg$hatvalues.model2 <- hatvalues(model2)
campbell_mult_reg$cooks.distance.model2 <- cooks.distance(model2)
campbell_mult_reg$obsNumber <- 1:nrow(campbell_mult_reg)
```

In the above R code the *fitted()* function was used to return the predicted values (i.e. the y hat values) from those specific input values provided by the dataset but if you want to predict for a new set of values then you need to use the *predict()* function. For example, say we want to use *model2* and find what the predicted deadspace is for an asthmatic with a height of 120cm. The values for the input variables you provide to the *predict()* function must be in the form of a dataframe so I wrap them in the *data.frame()* function.

```
> predict(model2, data.frame(asthma="yes", height=120))
       1
45.72806
```

```
predict(model2, data.frame(asthma="yes", height=120))
```

The function returns the value 45.728. You can also request confidence or prediction intervals for it with the following code:

```
predict(model2, data.frame(asthma="yes", height=120), interval="pred")
```

```
predict(model2, data.frame(asthma="yes", height=120), interval="conf", level=.95)
```

You can also create a dataframe with several values and put that in the *predict()* function, however the columns must have identical names to the input variables you wish to provide values for.

59.6 Strategy

Most regression analyses consider regression diagnostics to be backroom activity, and are often not reported. This is because the primary aim of most of these analyses is to formulate and report on the predictive power of a model rather than an assessment of the underlying assumptions that helped form the model. Readers of the paper assume that the model must have passed these tests for the model to be presented!

The diagram below provides an overview of the process specifying the diagnostics we have used.

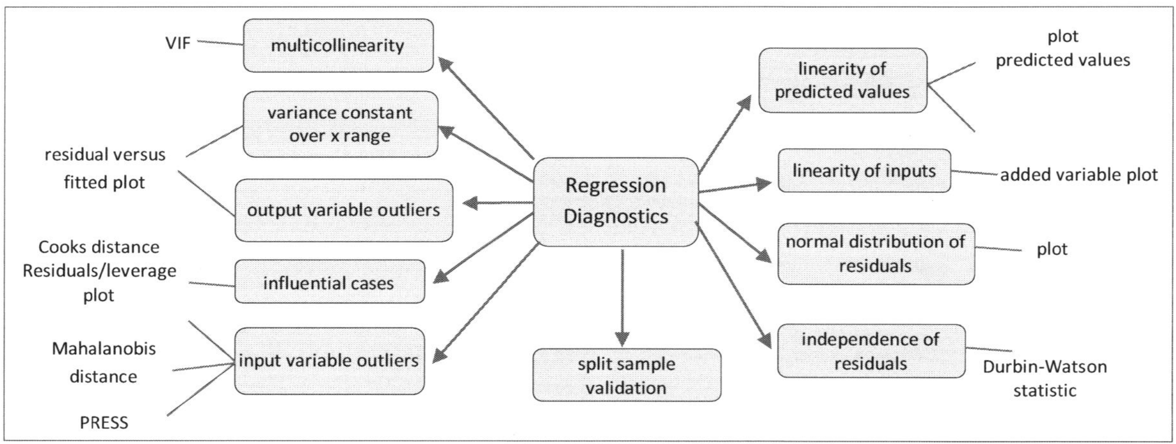

59.7 Writing up the results

15 subjects provided lung deadspace (ml), height (cm), asthma status with the aim to produce a regression model predicting deadspace from the other inputs, an interaction term was included. The overall model was statistically significant (F=37.08; df(3,11), p-value <.001) with an adjusted R squared value of .8855 indicating that 88% of the variability in deadspace could be accounted for by the model *(I would include the confidence interval here as well)*. Because the interaction term was statistically significant (see table) all three inputs were retained in the model. *I would then produce a table containing results with the following heading:*

	R^2 (adjusted 95% CI)	B (95% CI)	SE B	P-value
Constant (intercept)				
height				
Asthma status				
Asthma x deadspace				

I would also include a table showing the correlation matrix or possibly a scatterplot matrix with the correlations across the diagonal.

Multicollinearity was investigated and all VIF values were below 5 (centered variables).

I would include the scatter plot showing both non-parallel regression lines and if I were producing this for a journal, I would follow the example in the professional graphics chapter.

59.8 Summary

This chapter introduced a technique which can be used to assess the influence several inputs might have upon a single interval/ratio variable.

The concepts of partial and semi-partial correlation (part correlation) helped explain some of the complexities of this approach by way of various Venn diagrams. The Venn diagram was also used to explain multicollinearity and why the variance of the *b's* become inflated in that situation.

We saw that while the mechanics of creating such a model was easy by using R Commander the subsequent interpretation, along with the decision making process was less so. Two aspects were highlighted; that of model development (assessing which inputs should be included or excluded and the effect of interaction terms) and the various regression diagnostics.

The development of regression diagnostics is an active research area and a recent addition is the *tableplot()* (Friendly & Kwan, 2009) which has been implemented in the *perturb* and *tableplot* R packages, however at the present time this is only available for numeric inputs rather than factors.

I introduced the techniques of standardising and centering variables, both of which change the interpretation of the partial regression coefficients (*b's*). Considering raw variables, that is those without any transformation, the *b's* represent the effect of each variable when all other variables are zero. With centering, the *b's* represent the effect of each variable when the others are at their mean values, as does standardising (which also converts the scale from the original units to standard deviations).

There are many types of Regression besides Multiple Linear Regression and the following chapters in this book look at several of them, such as Logistic regression, Poisson regression and repeated measures using multilevel modelling.

60 Logistic regression: a binary outcome

In the simple linear regression chapter we discussed in detail the possibility of modelling a linear relationship between two ordinal/interval ratio variables; the incidence of lung cancer per million and cigarette consumption per person per year for a range of countries. We considered lung cancer incidence to be our output variable, that is for a given cigarette consumption we estimated the lung cancer incidence. We then considered situations with several possible inputs in the *Multiple linear regression* chapter. However, up until now, we have only considered an outcome measure that has an interval/ratio **Level of measurement**.

In contrast, consider the situation where we want to calculate (i.e. predict) an individual's **odds** (see glossary entry) of getting lung cancer. The logical way to go about this would be to carry out some type of prospective cohort design and find out who developed lung cancer and who did not. Lung cancer status would technically be called our *primary outcome variable*, so now we have a binary (nominal/dichotomous) variable recording disease status. For each subject we would simply classify them as 0=healthly; 1=diseased, or any arbitrary number we fancied giving them, after all it is just nominal data. Now unfortunately, we can't use our simple regression solution, as we are using a binary outcome measure and also want to produce odds rather than actual predicted values.

While, theoretically, the numbers can be arbitrary, for most computer programs it is best to use the values given here to help with interpretation.

Luckily, a technique called logistic regression comes to our rescue.

Logistic Regression

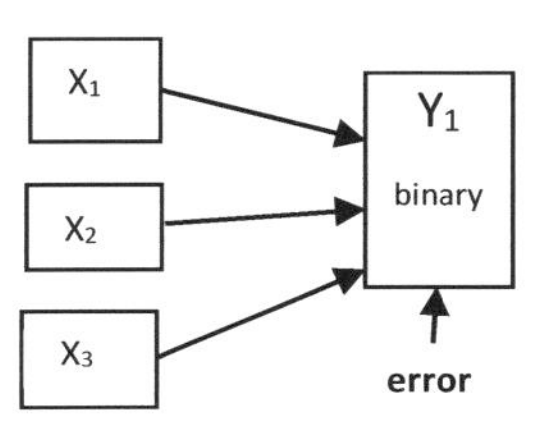
X1
X2
X3
Y1
binary
error

Logistic regression is used when we have a binary outcome (dependent) variable and wish to carry out some type of prediction (regression).

To demonstrate the technique we will use the results from a treadmill stress test along with the probability of suffering from Coronary Heart Disease - CHD (taken from http://www.stat.tamu.edu/spss/LogisiticReg.html). The dataset consists of 17 treadmill stress tests along with an associated diagnosis of CHD. Note that the shortest time was 594 seconds and the longest 1110 seconds. We also have a fairly equal number of healthy and CHD patients.

Seconds on treadmill	Presence of coronary Heart Disease (CHD) 0=health, 1= diseased
1014	0
684	0
810	0
990	0
840	0
978	0
1002	0
1110	0
864	1
636	1
638	1
708	1
786	1
600	1
750	1
594	1
750	1

Binary outcome

How do we analyse this dataset? Well if the presence of CHD were a real number, we could use the standard simple regression technique, but there are three problems with that:

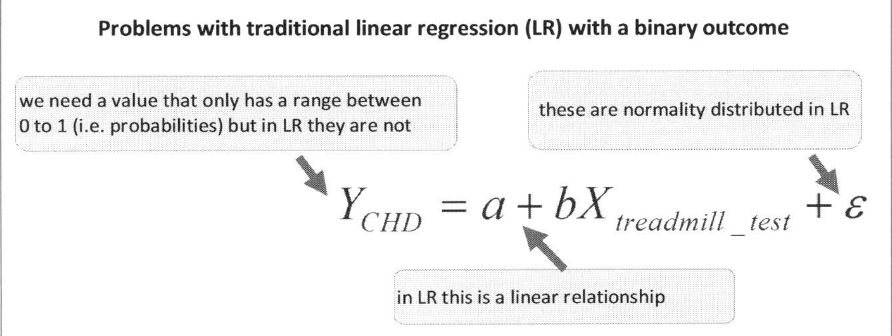
Problems with traditional linear regression (LR) with a binary outcome
we need a value that only has a range between 0 to 1 (i.e. probabilities) but in LR they are not
these are normality distributed in LR
$Y_{CHD} = a + bX_{treadmill_test} + \varepsilon$
in LR this is a linear relationship

As we are now dealing with a binary outcome the errors are no longer normally distributed but assumed to follow a binominal distribution and modelled using what is called a *link function*. The following subsections provide more detail, which can be skipped over if you are only interested in the practical aspects.

60.1 The theory behind logistic regression

The logistic function provides us with a range of values between zero and one. The y value varies between only zero and one, achieved by using the following formula:

$$y = \frac{1}{1 = e^{-z}}$$
$$y(1+e^{-z}) = 1$$
$$y + ye^{-z} = 1$$
$$ye^{-z} = 1 - y$$
$$e^{-z} = \frac{1-y}{y}$$
$$\log_e(e^{-z}) = \log_e(\tfrac{1-y}{y})$$
$$-z\log_e(e) = \log_e(\tfrac{1-y}{y})$$
$$z(1) = -\log_e(\tfrac{1-y}{y})$$
$$z = \log_e(\tfrac{1-y}{y})^{-1}$$
$$z = \log_e(\tfrac{y}{1-y})$$

$$y = \frac{1}{1+e^{-z}}$$

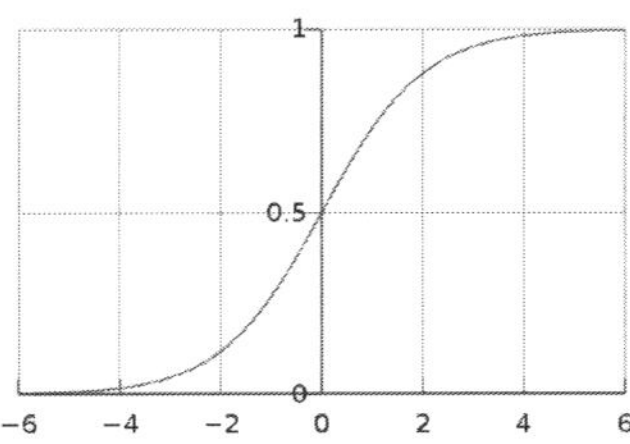

z is what is called the exponent and e is just a special number always equal to 2.718, in other words it is a constant, and said to represent 'organic growth'. e can be represented as either e or *exp*. So in our equation we have 2.718 raised to the power of whatever value $-z$ takes in the equation all under 1, technically called a reciprocal. Not a very pretty thing for those of you who hate maths! You may remember that any number raised to the power of zero is equal to 1 so when z = 0 then e^{-0} = 1 and the logistic function is equal to 1/1+1 = ½ as we can see in our diagram, when z=0 Y=.5.

By making the z in the above logistic function equal the right hand side of our regression equation minus the error term, and by using some simple algebra (shown opposite) we now have $\frac{1-y}{y} = e^{-(a+bx)}$. The next thing is to try to stop our regression equation being an exponent. The expression in parentheses on the right hand side of above equation is called an estimated *Linear Predictor (LP)*. By remembering that we can convert a natural logarithm back to the original number by using exponentiation, technically we say one is the inverse function of the other, we can easily move between the two. If it helps think of squaring a number and then square rooting it to get back to the original value. The square root function is just the inverse of the square function. The diagram opposite provides some examples. Applying the logarithmic function to the above equation (Norman & Streiner, 2008 p.161) we end up with: $\log(\frac{y}{1-y}) = a + bx$

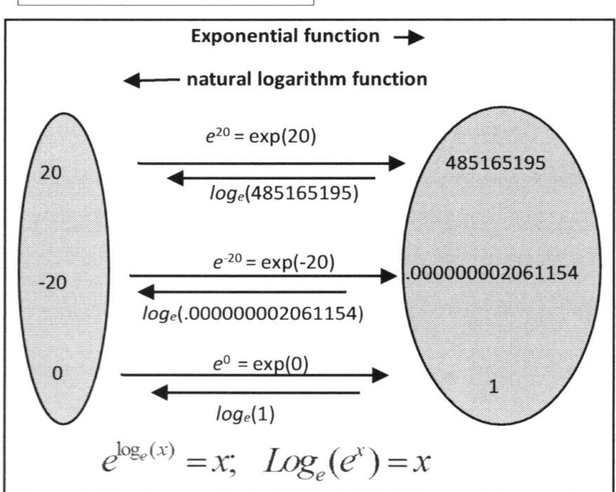

Iteration	-2Log likelihood =deviance
1	13.05511
2	12.58274
3	12.55047
4	12.55022
5	12.55022
6	12.550

Example R code:

```
glm(chd_status ~ time, family=binomial(), trace = TRUE, epsilon = 1e-14)
```

The left hand side of this equation is called the **logit function** of y, representing the log odds value, presenting the log odds of success. This value can be converted into either an odds or probability value (see glossary entry **odds**). The right hand side is something more familiar to you, it looks like the simple regression equation. Unfortunately calculating the b's here can't make use of the method of least squares, because for one thing it is not a linear relationship, so instead the computer uses an iterative method to produce a *Maximum Likelihood* (see glossary entry **Likelihood**) to estimate the *a* and *b* values. In most statistical applications, you can request to see how the computer arrived at the final estimate. For example, the table opposite, is from a set of results in R using the *trace=TRUE* option to demonstrate this.

Unfortunately, sometimes the computer fails to find a solution and displays the error message *failure to converge* but, luckily there are alternative "solution hunting" computer algorithms than can help with this problem (see Marschner, 2011 for how to do this in R). The computer stops searching when the difference in deviances between two successive estimates is smaller than a set value, known as the convergence tolerance or epsilon (ε). You can set this tolerance in R using the *epsilon=value* option i.e. *epsilon=1e-14*.

To sum up, the logistic regression equation produces a Log **odds** value, which is also called the *logit* of *y*, while the coefficient values (*a* and *b*) are *log odds ratios* ($\log_e$(OR)).

$=\log_e$(odds) of y

$$\text{logit(y)} = \log\left(\frac{y}{1-y}\right) = a + bx$$

$=\log_e$(odds **ratios**)

Before we consider actually carrying out a logistic regression analysis reflecting on simple linear regression is useful. In that situation we had; a) a measure of overall fit, b) an estimate of each parameter along with an accompanying confidence interval, so we would be expecting something similar to that in the output to a logistic regression model.

60.1.1 Model fit -2log likelihood and pseudo R squared

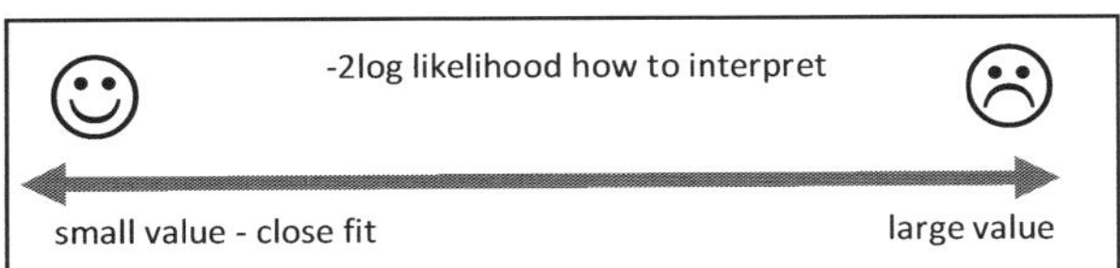

In logistic regression we have a value analogous to the residual/ error sum of squares found in linear regression called the Log likelihood, which when multiplied by -2 is written as -2log(likelihood), -2LL or -2LogL and called the *deviance*. The minus 2log version is used because it has been shown to follow a chi-squared distribution when the value in the population is equal to zero, which means we can associate a p-value with it. The degree of fit of the model to the observed data is reflected in the -2log likelihood value. The larger the value, the greater the discrepancy between the model and the data. In contrast, the minimum value it can attain, when we have perfect fit, is zero (Miles & Shevlin, 2001 p.158), and in contrast the likelihood would equal its maximum of 1.

But what is a small or a big -2log likelihood value? Defining this is a problem, therefore as in other situations we have come across where there is no definitive answer several possible solutions are offered. Firstly the associated p-value with the -2log likelihood only tells us if we can accept or reject the model, not the degree to which the model is a good fit.

model chi-squared
-2(log likelihood of current model – log likelihood of previous model)
=
-2log(likelihood of current model / likelihood of previous model)
=
a ratio and has a chi-squared distribution when both same value

expected value = -2log(1)=0 given the null hypothesis is true

Different statistical applications provide a variety of fit measures. R by default provides none whereas SPSS provides three measures of model fit all based on the -2log likelihood; the model chi-squared and two R square statistics, often called *pseudo R squares* because they are rather pale imitations of the R square in linear regression.

The model chi-squared (sometimes called the *traditional fit measure*) is *-2(log likelihood of current model – log likelihood of previous model)* for our example 12.55- 23.508 = -10.958. It is called a *Likelihood ratio test* because subtracting logarithms is the same as the logarithm of one raw value divided by the other, as shown opposite. If both models are equivalent (i.e. fit equally well) then the expected value in the population is zero and follows a chi-squared distribution.

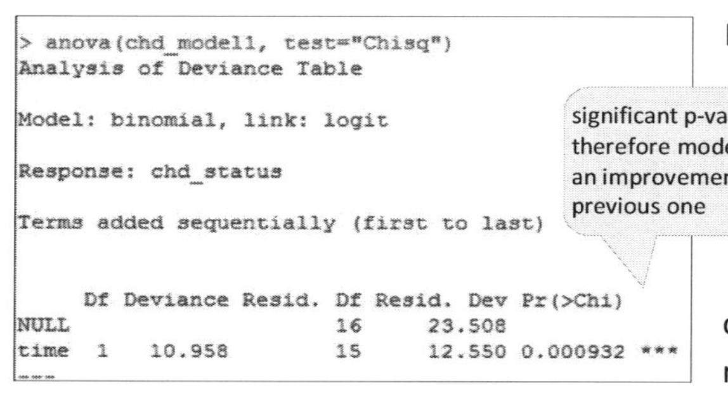

```
> anova(chd_model1, test="Chisq")
Analysis of Deviance Table

Model: binomial, link: logit

Response: chd_status

Terms added sequentially (first to last)


     Df Deviance Resid. Df Resid. Dev Pr(>Chi)
NULL                   16     23.508
time  1   10.958       15     12.550 0.000932 ***
---
```


In R, the model chi-squared value (in SPSS called the *Omnibus Test of model Coefficients*) is obtained using the *anova()* function. The *Pr(>Chi)* column is the p-value, this very small value (i.e. highly significant) indicates that the present model is an improvement over the previous one. The -2log likelihood final estimate is 12.55 compared to the initial value of 23.508 for the null model, containing only a constant term, that is no independent variables, reflecting what we did in the simple regression chapter.

60.1.2 Pseudo R squared values Cox & Snell and Nagelkerke's R

Both the Cox & Snell and Nagelkerke's R squared values can be interpreted in a similar manner to how you would the R squared value in linear regression, they are **effect size** measures representing the percentage of variability in data that is accounted for by the model. As with the R squared value in linear regression a higher value indicates a better model fit whereas, the Cox & Snell value cannot always reach 1, this problem is dealt with by the Nagelkerke R square value. According to Campbell, 2006 p.37 you should 'consider only the rough magnitude' of them.

Cox & Snell R Square	Nagelkerke R Square
.475	.634

We will look at the equations for these values later as it is necessary to convert them to R code to obtain the values.

As an alternative to the traditional chi-squared fit measure you can compute the Hosmer - Lemeshow statistic. The Hosmer - Lemeshow statistic is considered to be a more robust measure of model fit and one variety of it, *C*, is based on grouping cases into deciles (tenths) and then comparing the observed probability with the expected probability within each decile. Because there are several different varieties of the Hosmer -Lemeshow statistic, each of which produces a different value, you should report the software used and any other relevant details (see, Hosmer, Hosmer, le Cessie & Lemeshow, 1997; Hosmer & Lemeshow, 2000 p.150).

Hosmer and Lemeshow C statistic (produced using deciles)		
Chi-square	df	Sig.
9.01	8	.3412

want a insignificant p-value larger the better

Assuming we are using the decile approach to calculate the Hosmer - Lemeshow *C* statistic, we would expect a good fitting model to produce a high p-value (i.e. insignificant value).

Notice that we interpret the associated p-value here in a confirmatory way as we did the p-value associated with Levene's statistic in the *t* test. A large p-value (i.e. one close to 1) indicates that the modelled situation is highly likely to occur, and demonstrates a good fit (☺). A small p-value indicates a poor fit (☹) and suggests that you should look for an alternative way to model the relationship.

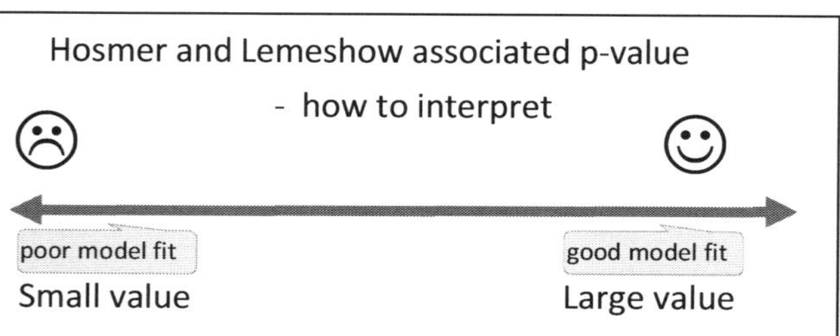

In addition to assessing the model by considering the overall fit, we can also consider each of the parameter estimates individually which we will now do. In the one input model this is a rather redundant procedure but it will provide a framework for interpreting models that have more than one input.

60.1.3 The coefficients - log odds ratios, odds ratios and % change in odds

1=chd

```
Deviance Residuals:
    Min       1Q   Median       3Q      Max
-2.0627  -0.3439   0.2547   0.5998   1.5405

Coefficients:
             Estimate Std. Error z value Pr(>|z|)
(Intercept) 12.727363   5.802709   2.193   0.0283 *
time        -0.015683   0.007255  -2.162   0.0307 *
---
Signif. codes:  0 '***' 0.001 '**' 0.01 '*' 0.05 '.' 0.1 ' ' 1

(Dispersion parameter for binomial family taken to be 1)

    Null deviance: 23.508  on 16  degrees of freedom
Residual deviance: 12.550  on 15  degrees of freedom
AIC: 16.55

Number of Fisher Scoring iterations: 5
```

In contrast to ordinary linear regression the a and b coefficients do not appear to have a t test associated with them but instead a z value. There is nothing here to worry about as the same principle applies to this value, it being the estimate divided by its standard error (i.e. signal/noise). This value can also be used to provide a p-value, assuming that the parameter value in the population is equal to zero. Those of you who do not take my word will notice that 12.72/5.8 is equal to 2.193 and -0.0156/0.00725 is equal to -2.162. The box on the left describes how these values relate to their associated p-values.

Finding the p-values
Given a z score of 2.193 what is the probability of obtaining a value of 2.193 or one more extreme in either the positive or negative direction when the population from which it came has a mean of zero and a standard deviation of one?

```
> pnorm(c(2.193), mean=0, sd=1, lower.tail=FALSE)
[1] 0.01415369
```

but need both tails so multiply by 2 =0.02830738 same as that obtained in the R logistic regression output

Because we have turned all the values in the equation to logs the coefficients are also logs, representing log odds ratios.

The intercept term, misleadingly called the *constant*, is the log odds ratio of a subject who has a *theoretical* treadmill time score of zero to be suffering from CHD. Remember that In this dataset a person who was 'healthy' was coded with a zero, in contrast to those suffering from CHD, who were coded with a '1' on the output variable. It is important here to stress the term *theoretical* because inspecting the dataset indicates the minimum treadmill time anyone achieved was 594 seconds so here this extrapolation is of no practical value.

If the intercept was found to have not been statistically significant then for a logistic model it indicates that the logit (i.e. log odds) is zero, which implies that there is a probability of 0.5 belonging to either of the outcome groups.

```
> tapply(treadmill$time, treadmill$chd_status, length)
0 1
8 9
```

```
> exp(coef(chd_model1))
 (Intercept)         time
3.368398e+05 9.844398e-01
```

The *Estimate* column, provides the natural logs odds ratios and, as these values are rather difficult to interpret, it is advisable to apply the antilog function to them, a process called exponentiation, to gain more clarity. This is achieved in R by applying the *exp()* function.

exp(b) is $e^{-0.015683}$ = .9844 and for the intercept is $e^{12.727}$ = 33683.9. Taking the intercept first; 33683 is the odds of having CHD with a treadmill time of zero. We therefore know that the odds of NOT having CHD with a treadmill time of zero is 1/33683.9 = 2.968e-5=0.0002968. (see glossary entry **odds ratio**). Given the discussion in the previous paragraph it must be stressed these are purely theoretical values.

The estimate of .9844 for time is also an **odds ratio**, and can be interpreted in terms of the change in odds of the outcome for a unit change in the input, *time.* Our value is very slightly less than one so we can say there is a *decrease in the odds* of just over 1% [i.e. (1-.9844) x 100] of becoming diseased for every extra second managed on the treadmill.

Because this is an interval/ratio predictor we can ask how much the odds of becoming diseased will change for more than one second change in treadmill tolerance. For example if you wanted to see how much the odds would change for each additional minute tolerated on the treadmill (i.e. 60 seconds), you take the odds ratio and raise it to the 60th power, that is $.9844^{60} = 0.3893107$. This implies that an increase of one minute on the treatmill test (1-0.3893107) x 100 = 61.06893. This means that the model predicts that you decrease your odds of becoming diseased by 61 percent. This may seem a very large value for a single minute but remember we are talking here odds ratios (range from zero to infinity) not probabilities (range zero to 1). In the next section, *Interpreting Log odds, odds ratios and probabilities for the individual,* we will see how this odds value relates to the probability of disease. Instead of raising the odds ratio to the required power we could have simply multiplied the log odds value and then exponentiated the result - both give the same value.

Exp(B) (Assuming significant i.e. p < 0.05)	**As input increases:**
Value < 1	Outcome odds decrease
Value = 1	Outcome odds stay same
Value >1	Outcome odds increase

In the above paragraph we had an odds ratio value which was less than one. However, often the value is more than one, in which case we just reverse the sentence (see table opposite).

Note that while the actual *exp(b)* value is an odds ratio we can also interpret it as a % change in odds for the outcome for a unit change in the associated input (Field, 2009 p.271).

60.1.4 Classification table, sensitivity, specificity and false negatives/positives

It is difficult for most people to see what exactly even the odds ratios signify. A much easier way to understand the output from a logistic regression is to look at a table showing those cases which were correctly and incorrectly classified, such a table is called a *classification table*.

classification table			
Observed	**Predicted**		
	healthy	diseased	Percentage Correct
healthy	6 (=tn)	2 (=fp)	75.0
diseased	1 (=fn)	8 (=tp)	88.9
Overall Percentage	7	10	82.4
a. The cut value is .5; tn= true negative; fp = false positive etc.			

This table provides a quick and easy way of seeing how well the model performs with the observed data. Taking our Treadmill test and CHD we can see from the classification table opposite that the model correctly predicts 75% of the healthy cases and nearly 90% of the diseased ones.

<table>
<tr><th colspan="6">Comparing the predicted and observed counts</th></tr>
<tr><td rowspan="4">Considering the rows</td><td>Sensitivity of prediction = True positive (tp) rate</td><td>tp/(tp+fn)</td><td>8/(8+1)=88.9%</td><td>given that they are diseased the probability of being correctly classified is 90%</td><td>=p(pred_diseased|diseased)</td></tr>
<tr><td>Specificity of prediction = True negative (tn) rate</td><td>tn/(tn+fp)</td><td>6/(6+2) = 75%</td><td>given that they are not diseased the probability of being correctly classified is 75%</td><td>= p(pred not diseased |not diseased)</td></tr>
<tr><td>False positive ratio (FPR) for true not diseased</td><td>fp/(tn+fp)</td><td>2/(2+6) =2/8= .25 = 25%</td><td>given that they are not diseased being incorrectly classified as diseased</td><td>=p(pred_diseased|not_diseased)</td></tr>
<tr><td>False negative ratio (FNR) for true diseased</td><td>fn/(fn+tp)</td><td>1/(1+8) = 1/9 = 0.111 = 11%</td><td>given that they are diseased being incorrectly classified as healthy</td><td>= P(pred not diseased |diseased)</td></tr>
<tr><td rowspan="2">Considering the columns</td><td>Positive predictive value</td><td>tp/(tp+fp)</td><td>8/(8+2) = 0.8</td><td>Given that they are classified as diseased the probability of being diseased</td><td>= P(pred diseased |all diseased)</td></tr>
<tr><td>Negative predictive value</td><td>tn/(tn+fn)</td><td>6/(6+1) = 0.8571</td><td>Given that they are classified as NOT diseased the probability of not being diseased</td><td>= P(pred not diseased |all not diseased)</td></tr>
<tr><td></td><td colspan="5">Checks: specificity = 1-FPR -> 1-.25 = .75 qed; FNR = 1- sensitivity -> 1- .8888 = 0.1111 qed</td></tr>
</table>

These values can also be viewed from the perspective of true positive/negative rates etc. See the chapter *Sensitivity, specificity and predicted values* for details.

We can also consider the correct and incorrect classification for individual cases which will be investigated when we analyse the data.

60.1.5 Interpreting Log odds, odds ratios and probabilities for the individual

Log odds:	negative values	0	positive values
Odds:	zero or more	1	> 1
Probability:	zero or more	0.5	one or less

So far we have mentioned how to interpret Log odds and odds from the table of coefficient estimates, in other words b and $exp(b)$. We can also use the actual logistic regression equation and obtain values for a specific x value. For our treadmill/CHD example this is the time (i.e. the input) achieved on the treadmill. Because there is a relationship between odds and probabilities, we can also obtain a value indicating the probability that a case with value X for the input has a particular outcome. In our example this will be the probability that an individual who stays on the treadmill a specific number of seconds is either healthy or suffering from CHD.

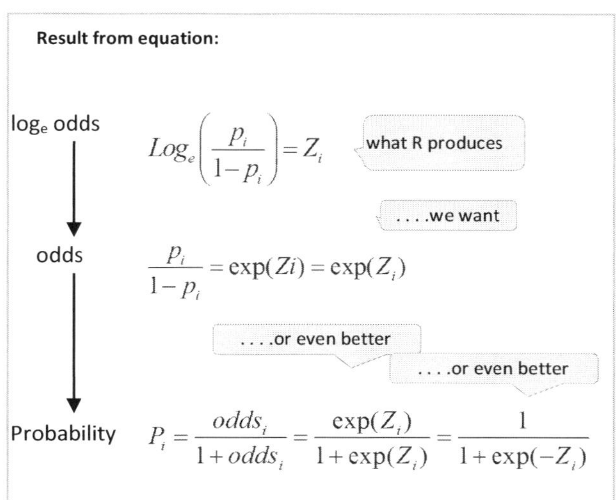

$$Log_e\left(\frac{p_i}{1-p_i}\right) = Z_i$$

$$\frac{p_i}{1-p_i} = \exp(Zi) = \exp(Z_i)$$

$$P_i = \frac{odds_i}{1+odds_i} = \frac{\exp(Z_i)}{1+\exp(Z_i)} = \frac{1}{1+\exp(-Z_i)}$$

Letting Z_i represent the linear predictor described at the beginning of the chapter we can see how the logistic regression equation can be transformed to produce probabilities (opposite).

For our Treadmill/CHD example.

loge(odds ratios)

loge(odds)

$$Logit(p_{chd}) = Log_e\left(\frac{p_{chd}}{1-p_{chd}}\right) = 12.7273 + (1.0157)X$$ therefore

$$\frac{p_{chd}}{1-p_{chd}} = \exp(12.7273 + (1.0157)X)$$

resulting in $$p_{chd} = \frac{1}{1+\exp-(12.7273+(1.0157)X)}$$

```
> 33683.9/(33683.9 +1)
[1] 0.9999703
```

X in the above equation is the time in seconds the person managed on the treadmill. We can also convert odds to probabilities with the above equation. Taking the previous odds of having chd with a treadmill time of zero was 33683.9 so the equivalent probability is 33683.9/(1+33683.9) = 0.9999703.

Predictions for individual with a specific treadmill time

treadmill time (secs)	Logit(y) log odds of health	exp(logit(y)) = odds of health	p of CHD	p of Health	change in p of Health
500	4.877	131.2757	0.9924	0.0076	
550	4.092	59.8775	0.9836	0.0164	0.0089
600	3.307	27.3113	0.9647	0.0353	0.0189
650	2.522	12.4572	0.9257	0.0743	0.0390
700	1.737	5.6820	0.8503	0.1497	0.0753
750	0.952	2.5917	0.7216	0.2784	0.1288
800	0.167	1.1821	0.5417	0.4583	0.1798
850	-0.618	0.5392	0.3503	0.6497	0.1914
900	-1.403	0.2459	0.1974	0.8026	0.1529
950	-2.188	0.1122	0.1009	0.8991	0.0965
1000	-2.973	0.0512	0.0487	0.9513	0.0522
1050	-3.758	0.0233	0.0228	0.9772	0.0259
1100	-4.543	0.0106	0.0105	0.9895	0.0123

We can also consider other cases using R, concentrating on the period between 500 and 1100 seconds in 50 second intervals, this being the observed range of values for our dataset.

Notice how the probability of suffering from CHD changes at a greater rate for the treadmill times of 750 to 900 seconds, and this is the same region of treadmill times where the odds of disease for the individual change from being less than one (i.e. p=0.5) to more than one.

This is rather upsetting for those who are at the extremes of the observed treadmill times. Taking the lower levels, if they improve their times from 500 to 550 seconds, the change in probability of having CHD is only .008, less than 1 in a hundred. Similarly, those who can withstand the treadmill for 950 seconds or greater again vary little in terms of CHD risk.

60.2 Simple logistic regression in R Commander and R

We will now consider several datasets to undertake logistic regression, beginning with a simple one using the treadmill dataset.

60.2.1 Data preparation

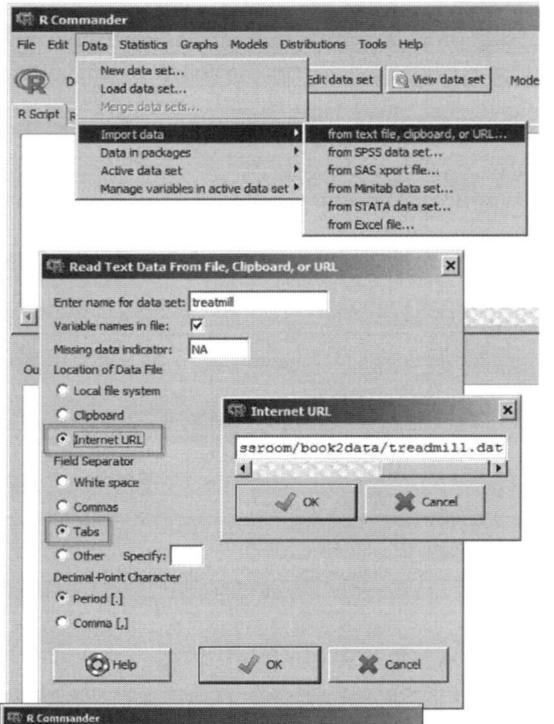

To obtain the treadmill dataset described in the previous section first load R Commander by typing in the following command:

```
library(Rcmdr)
```

To load the sample treadmill dataset, select the R Commander menu option:

Data-> Import data-> from text, the clipboard or URL

I have given the resultant dataframe the name *treadmill*, also indicating that it is from a URL (i.e. the web) and the columns are separated by *tab* characters. Clicking on the **OK** button brings up the internet URL box in which you need to type the following to obtain my sample data:

```
http://www.robin-beaumont.co.uk/virtualclassroom/book2data/treadmill.dat
```

Clicking on the **View data set** or **Edit data set** buttons displays the data. One important aspect to note is that the outcome *chd_status* is a numeric value rather than a factor with two levels, such as "healthy" and "chd". While this does not impinge significantly on the analysis I would advise you to whenever possible use a numeric variable for the output rather than a factor. Obviously, you can change it while you carry out the analysis or create a new variable of the other variety as we subsequently will do.

Data Editor

	time	chd_status	var3
1	1014	0	
2	684	0	
3	810	0	
4	990	0	
5	840	0	
6	978	0	
7	1002	0	
8	1110	0	
9	864	1	
10	636	1	
11	638	1	
12	708	1	
13	786	1	
14	600	1	
15	750	1	
16	594	1	
17	750	1	
18			

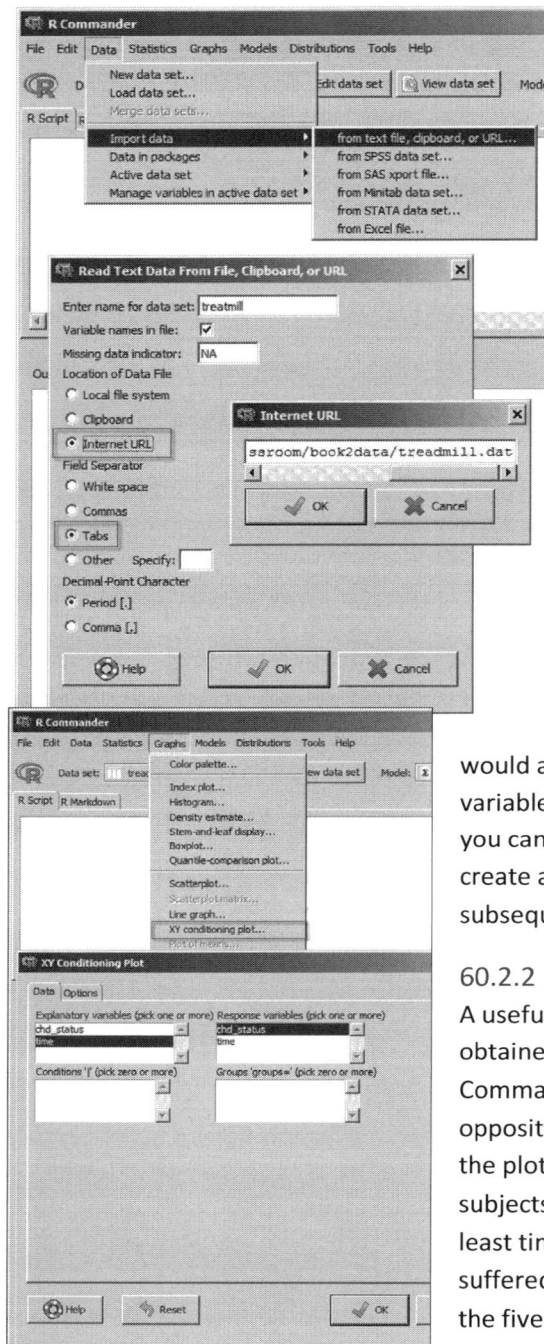

60.2.2 Data description

A useful graphical description is the *XY conditioning plot* obtained from the R Commander menu shown opposite. We can see from the plot that three of the subjects who managed the least time on the treadmill suffered from CHD whereas the five that managed the longest times were all healthy. Also note that case 2 who was diagnosed as healthy has a much lower treadmill time than the other healthy subjects.

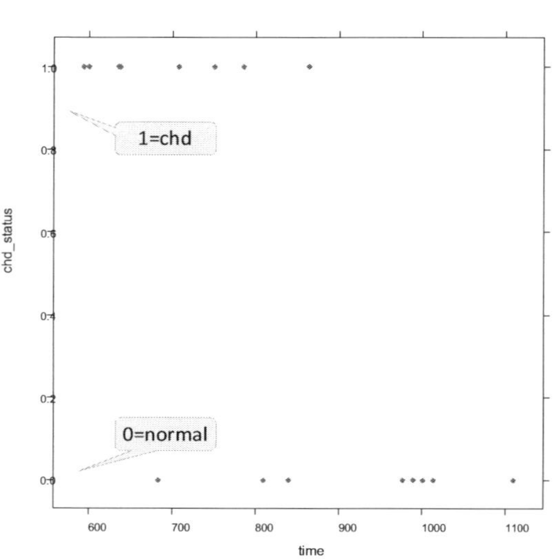

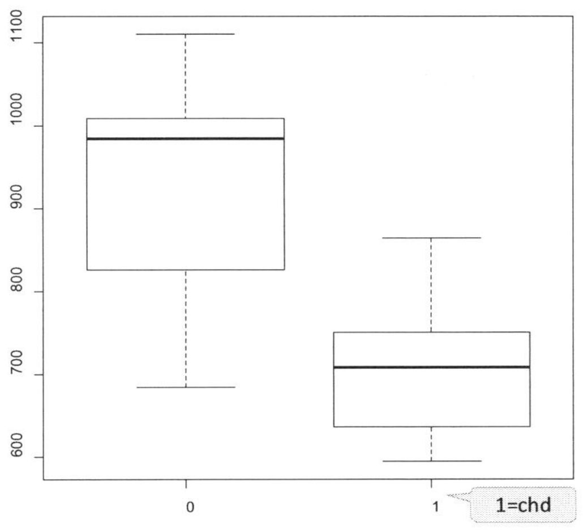

Another useful graphic is the boxplot, however because we have not defined the outcome in R Commander as a factor we are unable to use the R Commander menu to produce the boxplot. Use the following instead:

```
boxplot( time ~ chd_status,  data=treadmill)
```

We can see that those who subsequently were found to have CHD appear to have managed less time on the treadmill.

We can also obtain numerical summaries:

```
summary(treadmill)

tapply(treadmill$time, treadmill$chd_status, mean)
```

```
> summary(treadmill)
      time          chd_status
 Min.   : 594.0   Min.   :0.0000
 1st Qu.: 684.0   1st Qu.:0.0000
 Median : 786.0   Median :1.0000
 Mean   : 809.1   Mean   :0.5294
 3rd Qu.: 978.0   3rd Qu.:1.0000
 Max.   :1110.0   Max.   :1.0000
> tapply(treadmill$time, treadmill$chd_status, mean)
       0        1
928.5000 702.8889
```

We can see that on average the healthy group managed an extra 226 seconds (928.5 - 702.8) on the treadmill.

```
> ## Calculate number per time category
> with(temp_data, table(chd_status, time_category))
          time_category
chd_status <mean >mean
         0     1     7
         1     8     1
```

It would be interesting to see how many of the subjects are either healthy or diseased when we consider the mean treadmill time (809 sec) as the boundary. This requires us to create a new factor which classifies each subject, based upon them achieving the mean treadmill time or not. The result is shown opposite; only one out of eight healthy subjects obtain a treadmill time score of less than the average whereas one of the nine who have CHD achieved a treadmill score above the average. It looks like the mean is a good way of classifying the subjects into those who are healthy or unhealthy.

first need to a specify break point and call it *themean*

```
themean <- mean(treadmill$time)
```

create a new temporary dataframe, call it *temp_data*, containing the old dataframe plus the new category column called time_category

```
temp_data <- cbind(treadmill, time_category=
```

cut() needs:

```
cut(treadmill$time,
```

name of the variable to use to create the factor

one more break than levels you wish to create, as we want two levels we need three breaks

```
breaks= c(-Inf, themean, +Inf ), include.lowest = TRUE))
```

-Inf and + Inf are universal constants in R meaning negative and positive infinity

```
temp_data
```

show your work so far

now give the levels sensible names

```
levels(temp_data$time_category) <- c("<mean", ">mean")

temp_data
```

show your work so far

calculate the count per time category

```
with(temp_data, table(chd_status, time_category))
```

a comment line

```
# or do it the old fashioned way:

table(temp_data$chd_status, temp_data$time_category)
```

table() does not have a data= option so we need to use the *with* command instead

We could have also shown this result graphically by adding a vertical line to a simplified XY conditioning plot using the code *plot(chd_status ~ time,data=treadmill); abline(v=809)*.

60.2.3 Model development and evaluation

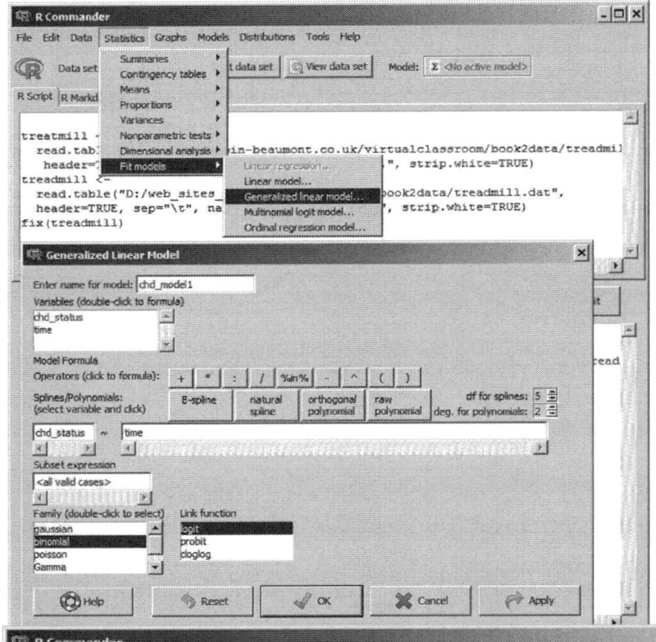

Using the R Commander menu option:

Statistics-> Fit models ->
Generalised linear model

Then enter an appropriate name for the model, I have called it *chd_model1*. Place the outcome variable, *chd_status* in the box to the left of the model sign (~). Add the *time* input to the box on the right of the sign.

The family and link boxes should be by default set to *family=binomial* and *link=logit.* Finally,

Click **OK**.

The result then appears in the R Commander output window.

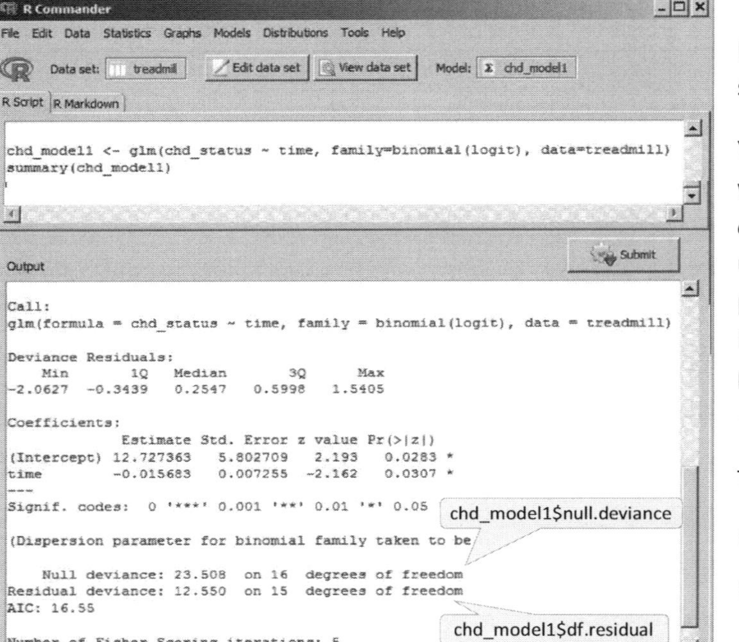

Most of the output has been explained in the previous section.

You will notice that in the R output there is a number with AIC beside it meaning *Akaike information criterion*, it is another measure of model fit based upon the -2log likelihood value and the number of parameter estimates, smaller the value better the fit. It is usually used to compare models (Crawley, 2005 p.208).

$AIC = 2(\textit{number} \text{ of parameters in model}) + -2\log(\textit{likelihood})$

The equivalent R code being:

```
(2 * 2) + chd_model1$deviance

(2 x 2) + 12.550 = 16.55
```

Instead of using the R Commander menu to produce the logistic regression model you could have typed the following R code, being the same as that generated by R Commander.

```
chd_model1 <- glm(chd_status ~ time, family=binomial(logit), data=treadmill)

summary(chd_model1)
```

60.2.3.1 Drawing the observed and expected values

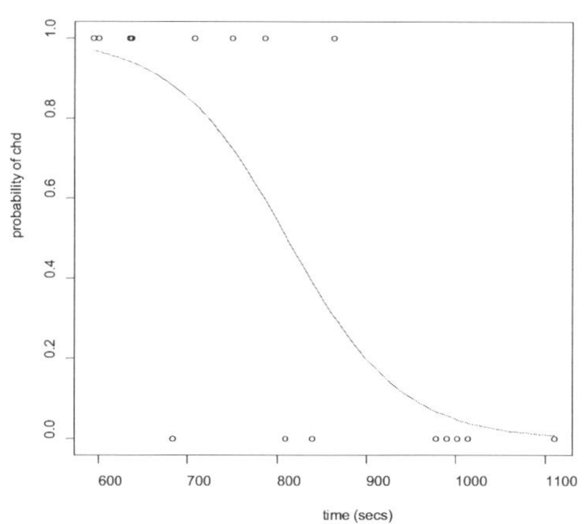

It is extremely useful to draw a graph of observed versus expected (model) values and the two lines below of R code achieve this. The first line draws the observed values at the top and bottom of the plot as we did in the preliminary analysis. The second line is more complex and makes use of the *curve()* and *predict()* functions, both discussed in detail later.

```
plot (treadmill$time, treadmill$chd_status, xlab= "time (secs)",
ylab= "probability of chd")

curve(predict(chd_model1, data.frame(time = x), type="response"),
add=TRUE, col="red")
```

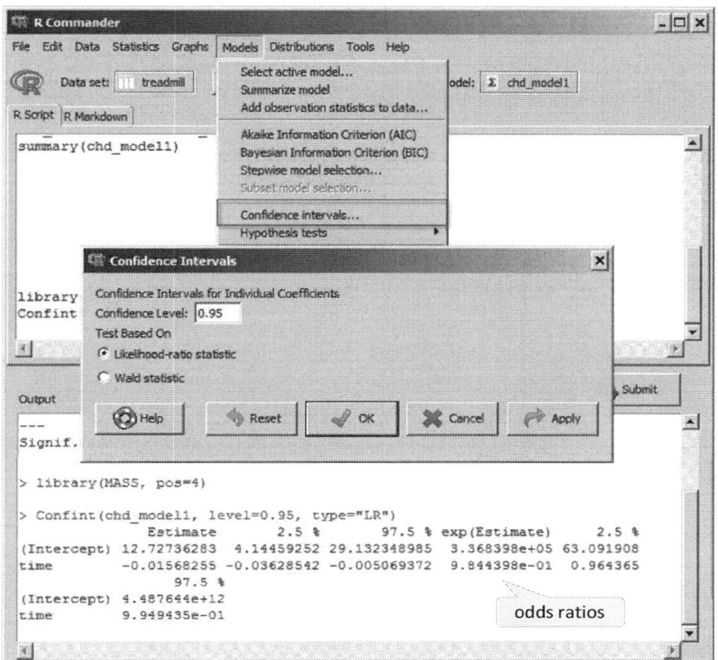

60.2.3.2 Confidence intervals for the estimates along with odds ratios

You can easily obtain confidence intervals for the model estimates by selecting the R Commander menu option:

Models-> Confidence intervals

The 2.5% and 97.5% values represent the lower and upper limits of the confidence intervals.

The output also automatically produces the exponential values representing the odds ratios.

Instead of using the menu options you could have typed the following R code into the *R Script* window, then selected it and clicked on the **Submit** button.

```
confint(chd_model1)
```

Notice also the confidence interval for *exp(x)* does not include within its bounds 1 (equal odds) which is to be expected given the significant p-value associated with it (i.e. 0.03). Also note the very wide confidence interval for the *Intercept* due to the few observations along with their range being far from zero.

60.2.3.3 The Hosmer-Lemeshow statistic

There appear to be several different varieties of the Hosmer - Lemeshow statistic (Hosmer & Lemeshow, 2000 p.169) and the R code for this chapter provides a demonstration of them all. For now I will use the *HLgof.test()* function, which stands for the Hosmer-Lemeshow goodness of fit tests, in the *MKmisc* R package.

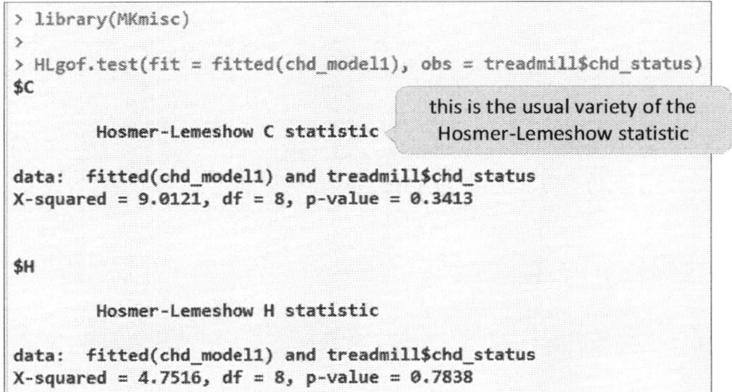

```
> library(MKmisc)
>
> HLgof.test(fit = fitted(chd_model1), obs = treadmill$chd_status)
$C

        Hosmer-Lemeshow C statistic

data:  fitted(chd_model1) and treadmill$chd_status
X-squared = 9.0121, df = 8, p-value = 0.3413


$H

        Hosmer-Lemeshow H statistic

data:  fitted(chd_model1) and treadmill$chd_status
X-squared = 4.7516, df = 8, p-value = 0.7838
```



```
install.packages("MKmisc")

library(MKmisc)

HLgof.test(fit = fitted(chd_model1),
obs = treadmill$chd_status)
```

install the package

load the package

calculate two varieties of the Hosmer - Lemeshow statistic

60.2.3.4 Pseudo R squared values

The formula's for the Cox & Snell and Nagelkerke equations are given below.

$$R^2_{Cox\_snell} = 1 - \exp\left[-\left\{\frac{2(LL_{\text{model}} - LL_{null})}{n}\right\}\right]$$

For an explanation of the Nagelkerke equation see Thompson, 2006 p.101. *LL* means log likelihoods in the equations, which can also be manipulated to -2log likelihoods (see the R code below).

$$R^2_{\text{Nagelkerke}} = \frac{1 - \exp\left[-\left\{\frac{2(LL_{\text{model}} - LL_{null})}{n}\right\}\right]}{1 - \exp\left\{\frac{2LL_{\text{null}}}{N}\right\}} = \frac{R^2_{Cox\_snell}}{1 - \exp\left\{\frac{2LL_{\text{null}}}{N}\right\}}$$

Looking at the R output we have a -2log(likelihood) but we need the log likelihoods so we just need to divide our -2log(likelihood) values by 2 and slap a minus sign in front of them, at last a simple calculation you may say. The n or N is for the total number of observations, in this case 17. In R you can use the length function to find the number of items in a particular dataset i.e. *length(time)*.

R term	R code	SPSS term
null deviance	chd_model1$null.deviance	initial -2log likelihood
deviance	chd_model1$deviance	final -2log likelihood
residual deviance	chd_model1$df.residual	

Every statistical application seems to use different terms when it comes to logistic regression as shown in the table opposite!

Another way to obtain the log likelihood in R is to use the *logLik()*, function notice the capital 'L', to obtain the log likelihood, therefore for our data *logLik(themodel)* produces -6.275108 (df=2) from the 12.55 in the output on the previous page which was the -2log likelihood value. The R code below implements the above equations.

need the log likelihoods rather than the -2logLL's two ways of obtaining them

```
logLik(chd_model1) # produces the final log likelihood value

Likelihood_final = -chd_model1$deviance/2

Likelihood_init = -chd_model1$null.deviance/2

cox_snell<- 1- exp(-(2*(Likelihood_final - Likelihood_init)/length(treadmill$time)))

cox_snell2 <- 1-exp((chd_model1$deviance -
chd_model1$null.deviance)/length(treadmill$time))

nagelkerke <-(1 - exp((chd_model1$deviance -
chd_model1$null.deviance)/length(treadmill$time)))/(1 - exp( -
chd_model1$null.deviance/length(treadmill$time)))

nagelkerke1 <- cox_snell/(1-exp(2*(Likelihood_init)/length(treadmill$time)))

cat("Likelihood_final:   ", Likelihood_final, "\n")

cat("Likelihood_init:    ", Likelihood_init, "\n")

cat("cox_snell:          ", cox_snell, "\n")

cat("cox_snell2:         ", cox_snell2, "\n")

cat("nagelkerke:         ", nagelkerke, "\n")

cat("nagelkerke1:        ", nagelkerke1, "\n")
```

Margin notes:

- first method: use the logLik() function
- second method: divide the -2logLL's by 2 and add a minus sign
- cox_snell method 1
- cox_snell method 2 quick and easy version
- nagelkerke's method 1 Using the formula by companion to the Arigosti book for R by Laura A. Thompson
- nagelkerke's method 2 Using the formula in Andy Field's book p.269
- the cat() function 'cancatenate' allows you to put more than one thing on a line of output
- line feed
- both methods produce identical result. You would only ever report one
- both methods produce identical result. You would only ever report one

```
Likelihood_final:    -6.275108
Likelihood_init:     -11.75407
cox_snell:           0.4751192
cox_snell2:          0.4751192
nagelkerke:          0.6342255
nagelkerke1:         0.6342255
```

60.2.3.5 *Obtaining predicted values*

You can use either R Commander or the *predict()*function to obtain the predicted values for each of the cases in the dataset. Considering first the predicted values from those input values in the dataset:

```
> predict_rate <- predict(chd_model1, type="response")
> predict_rate
         1          2          3          4          5          6 
0.04012715 0.88084908 0.50612307 0.05741252 0.39031623 0.06848639 
         7          8          9         10         11         12 
0.04803689 0.00919122 0.30526198 0.94009709 0.93830620 0.83536173 
        13         14         15         16         17 
0.59889782 0.96503544 0.72421101 0.96807516 0.72421101 
```

```
predict_rate <- predict(chd_model1, type="response")

predict_rate
```

produces probabilities of having CHD. If you want the original scale of log odds specify type="link"

The multiple logistic regression example, given later in this chapter, describes how to obtain predicted values using the R Commander menu option.

Now let's consider obtaining predicted values for some new data, specifically those time values shown in the table previously. The *predict()* function is a bit inflexible and has several requirements for it to work properly:

- any data you pass to it, which you wish it to apply the model, must be in the form of a dataframe. Therefore in the code below I have first put the new time values in a vector with the *c()* function then converted it to a dataframe using the *data.frame()* function.
- the dataframe must be given column names that exactly match the names of the input variables in the model. Therefore, I have given the column consisting of *newtimes* the name *time* in the dataframe.

```
newtime <- c(500, 550, 600, 650, 700, 750, 800, 850, 900, 950, 1000, 1050, 1100)
newdata <- data.frame(time= newtime)
new_time_predicts <- predict(chd_model1, newdata, type="response")
new_time_predicts
```

```
> newtime <- c(500, 550, 600, 650, 700, 750, 800, 850, 900, 950, 1000, 1050, 1100)
> newdata <- data.frame(time= newtime)
> predict(chd_model1, newdata, type="response")
         1          2          3          4          5          6          7          8          9         10 
0.99250567 0.98372887 0.96503544 0.92647096 0.85189913 0.72421101 0.54520598 0.35370137 0.19989729 0.10237920 
        11         12         13 
0.04949170 0.02321836 0.01073506 
```

There are three enhancements we can make to this:

- instead of typing in each individual *newtime* value use the *seq()* function to produce a sequence of values.
- make the output more attractive by putting it in a dataframe and
- adding the *newtime* values beside each result.

use the *seq()* function to produce a sequence of values from 500 to 1,100 in steps of 50

make it look better by putting it in a dataframe and adding the column of original values beside the results

```
newtime <- seq(from=500, to=1100, by=50)

predictions <- data.frame(time=newtime,
"prob of chd"= new_time_predicts)
predictions
```

the probability is always for the outcome coded as 1

```
> predictions
   time prob.of.chd
1   500  0.99250567
2   550  0.98372887
3   600  0.96503544
4   650  0.92647096
5   700  0.85189913
6   750  0.72421101
7   800  0.54520598
8   850  0.35370137
9   900  0.19989729
10  950  0.10237920
11 1000  0.04949170
12 1050  0.02321836
13 1100  0.01073506
```

You can also obtain the result for one specific time on the treadmill by using the code below.

```
predict(chd_model1, data.frame(time= 1125), type="response")
```

60.2.3.6 *Classification table, sensitivity, specificity and false negatives/positives*

The introductory section of this chapter displayed a classification table along with other associated measures, the code for which is reproduced below

add overall + individual row/column lables

```
table(predict_rate>.5, treadmill$chd_status)
class_table <- table(predict_rate>.5, treadmill$chd_status)
dimnames(class_table) = list(predicted=c("healthy", "CHD"),
observed=c("healthy", "CHD"))
class_table
##  stats for classification table
# Sensitivity=True positive (tp) rate =  tp/(tp+fn) = 8/(8+1)=88.9%
# =p(pred_diseased|diseased)
100 * ( class_table[2,2] / (class_table[2,2] + class_table[1,2])  )
# Specificity=True negative (tn) rate = tn/(tn+fp) = 6/(6+2) = 75%
# =  p(pred not diseased |not diseased)
100 * ( class_table[1,1] / (class_table[1,1] + class_table[2,1])  )
# False positive ratio (FPR)  = fp/(tn+fp) = 2/(2+6) =2/8= .25 = 25%
# =p(pred_diseased|not_diseased)
100 * ( class_table[2,1] / (class_table[2,1] + class_table[1,1])  )
# False negative ratio (FNR) = fn/(fn+tp) = 1/(1+8) = 1/9 = 0.111 = 11%
# = P(pred not diseased |diseased)
100 * ( class_table[1,2] / (class_table[1,2] + class_table[2,2])  )
# Checks: specificity = 1-FPR -> 1-.25 = .75
# FNR = 1- sensitivity -> 1- .8888 = 0.1111 qed
#Positive predictive value = tp/(tp+fp) = 8/(8+2) = 80%
100 * ( class_table[2,2] / (class_table[2,2] + class_table[2,1])  )
#Negative predictive value = tn/(tn+fn) = 6/(6+1) = 0.8571 = 85.71%
100 * ( class_table[1,1] / (class_table[1,1] + class_table[1,2])  )
```

60.2.3.7 *Obtaining and plotting confidence intervals for predicted values for a interval/ratio input*

In multiple linear regression we were able to add an additional option to the *predict()* function to obtain either a confidence or prediction interval but unfortunately this option is not available in logistic regression *glm* models. The excellent stakeoverflow newsgroup has several explanations of how to obtain these values and below I provide an edited version of one posting.

call *predict()* with *type = "link"*, and

the first produces predictions on the scale of the linear predictor, the second returns the standard errors of the predictions.

call *predict()* with *se.fit = TRUE.*

```
new_time_predicts <- predict(chd_model1, newdata, type = "link", se.fit = TRUE)

critval <- 1.96
```

setting the confidence interval range to the usual value of 95%

```
upr <- new_time_predicts$fit + (critval * new_time_predicts$se.fit)
lwr <- new_time_predicts$fit - (critval * new_time_predicts$se.fit)
fit <- new_time_predicts$fit
upr; lwr; fit
fit2 <- chd_model1$family$linkinv(fit)
upr2 <- chd_model1$family$linkinv(upr)
lwr2 <- chd_model1$family$linkinv(lwr)
lwr2;  fit2;  upr2
```

now for fit, upr and lwr we need to convert them back. The *glm* object has a functon within it called *linkinv()* which will do that

Adapted from
http://stackoverflow.com/questions/14423325/confidence-intervals-for-predictions-from-logistic-regression

Now you can add these confidence intervals to the plot. In the code below the first two lines repeat the previous plot before adding the confidence intervals.

```
plot (treadmill$time, treadmill$chd_status, xlab= "time (secs)", ylab= "probability of chd")

curve(predict(chd_model1, data.frame(time = x),
        type="response"), add=TRUE, col="red")

lines(newtime, upr2, col="blue",lty=2, lwd=1)

lines(newtime, lwr2, col="red", lty=3, lwd=1)
```

The confidence intervals are very wide because of the very small sample size.

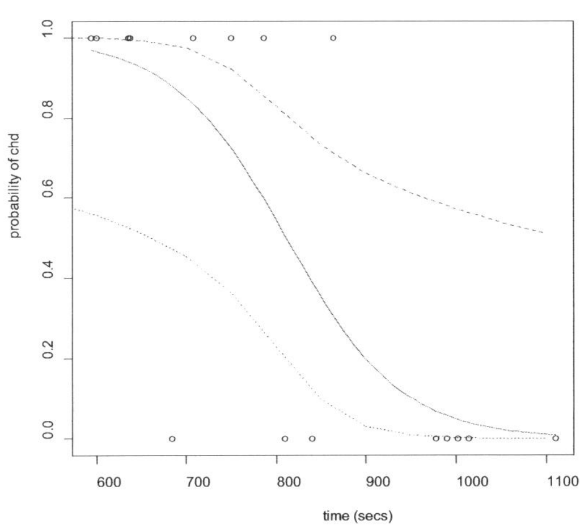

60.2.4 Diagnostics

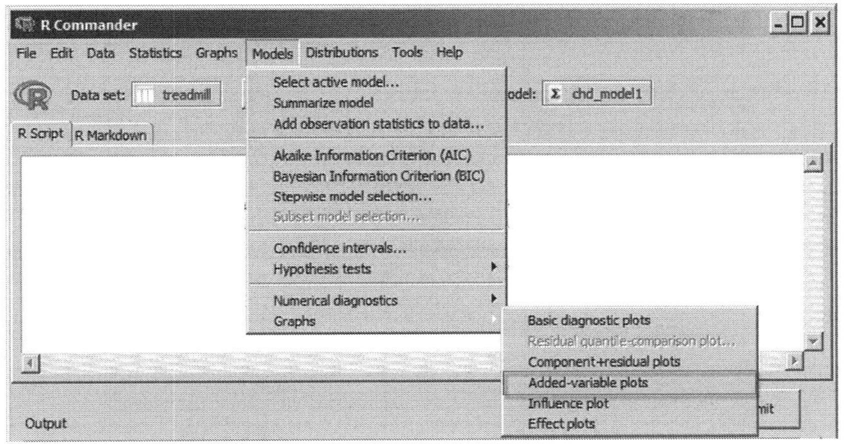

Because the predictor is a binary variable there is no point is looking at the residuals as one would do in linear regression. Maindonald and Braun (2010, p.255) recommend added-variable plots which can easily be obtained in R Commander.

Techie point
Everitt & Hothorn, 2010 recommend also using bubble plots. These require the model to have two or more interval/ratio inputs, so are not discussed any further here.

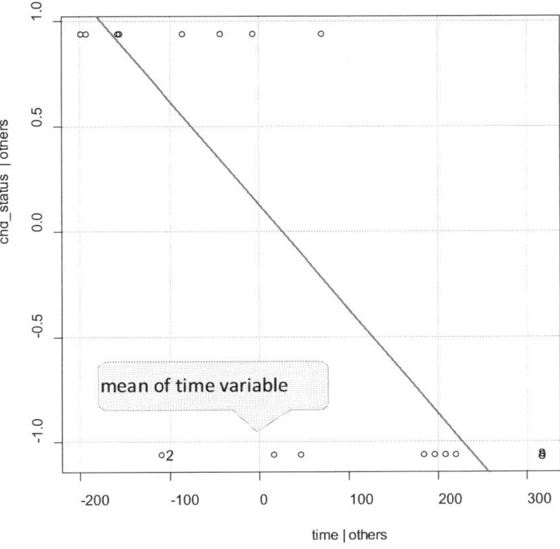

Possible outliers are displayed along with their case (row) number. Case 2 being identified again!

Another related aspect is the power of the model which is discussed next.

60.3 Sample size estimation and power

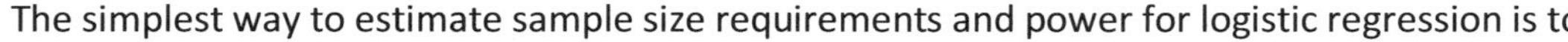

The simplest way to estimate sample size requirements and power for logistic regression is to download and install the free program *GPower*. Most of the information below has been taken from http://www.mormonsandscience.com/gpower-guide.html

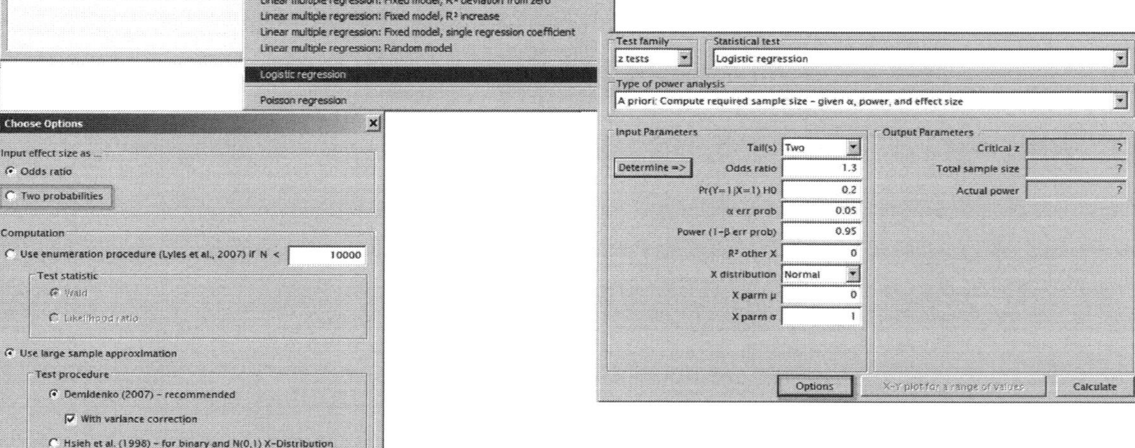

Select the menu option:

Tests->correlation and regression->Logistic regression

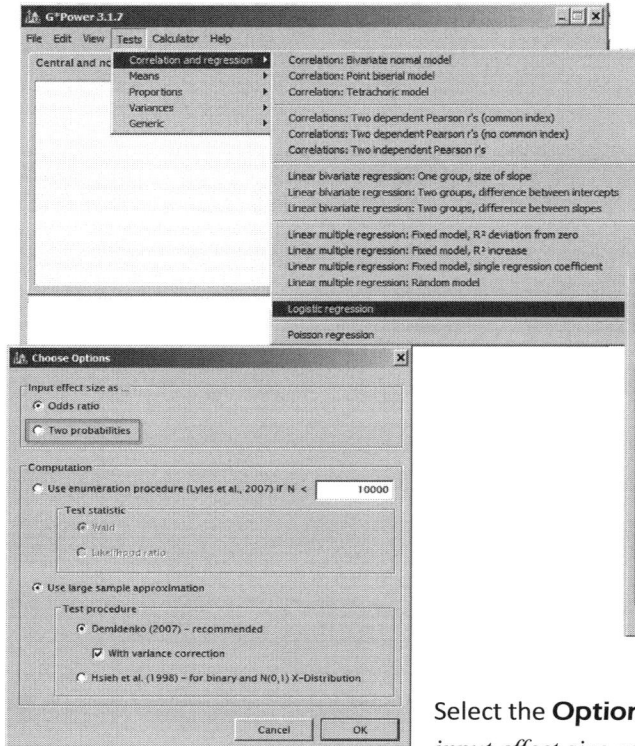

Select the **Options** button at the bottom of the screen and then change the *input effect size as* option in the *Choose Options* dialog box to say *input effect size as . . .* **Two probabilities**.

```
> 809 - sd(treadmill$time)
[1] 647.4495
> 809 + sd(treadmill$time)
[1] 970.5505
```

As with any required sample size estimation procedure we now have to use 'informed' knowledge to decide upon certain values.

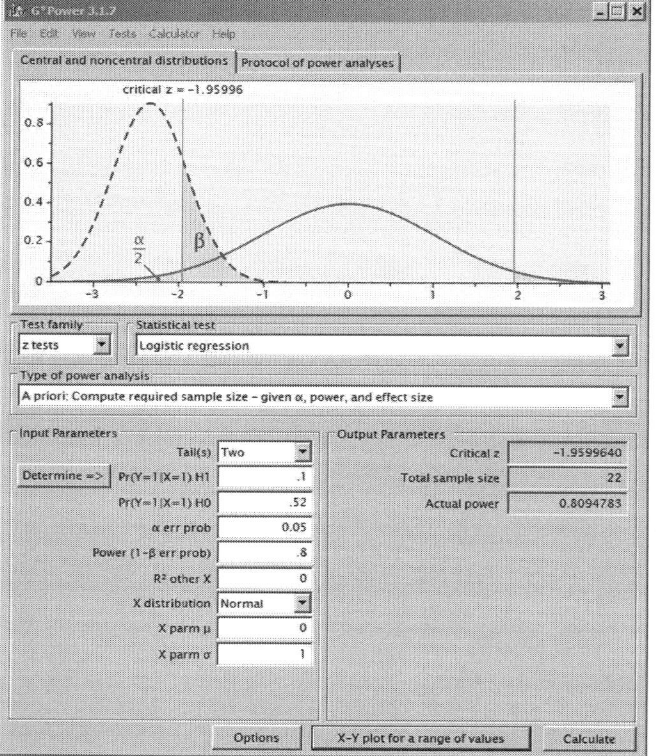

Pr(Y=1 | X=1) H1 represents the probability of the outcome = 1 (i.e. having CHD) when the main predictor is one SD unit (i.e. one z score) above its mean, and all the other inputs, if there are any, set at their mean values. In our example which we will assume was a pilot study, we can find the value of the standard deviation for the time input using *sd(treadmill$time)* providing a value of 161.55. We already now the mean is 809 therefore a time value of one zscore from the mean is 647 or 970. Looking at the table *Predictions for individual with a specific treadmill time* we can see that the probability of chd for minus one sd is .92 compared to .54 for the mean or 0.07, say 0.1, for plus one sd of the treadmill time.

Pr(Y=1 | X=1) H0 represents the probability of the outcome = 1 (i.e. having CHD) when all the inputs are set to their mean value. We know this value is .52.

R-squared other X = Here you indicate the expected squared multiple correlation coefficient (R^2) between the input you are considering at the moment and all the other inputs, if there are any. As there are none, enter 0. If you have more than one input you can find R^2 using GPower's calculation tool in linear multiple regression (it is under Exact or Multivariate tests).

You usually set the Alpha level to 0.05, power to 0.8 and make it a 2 tailed test.

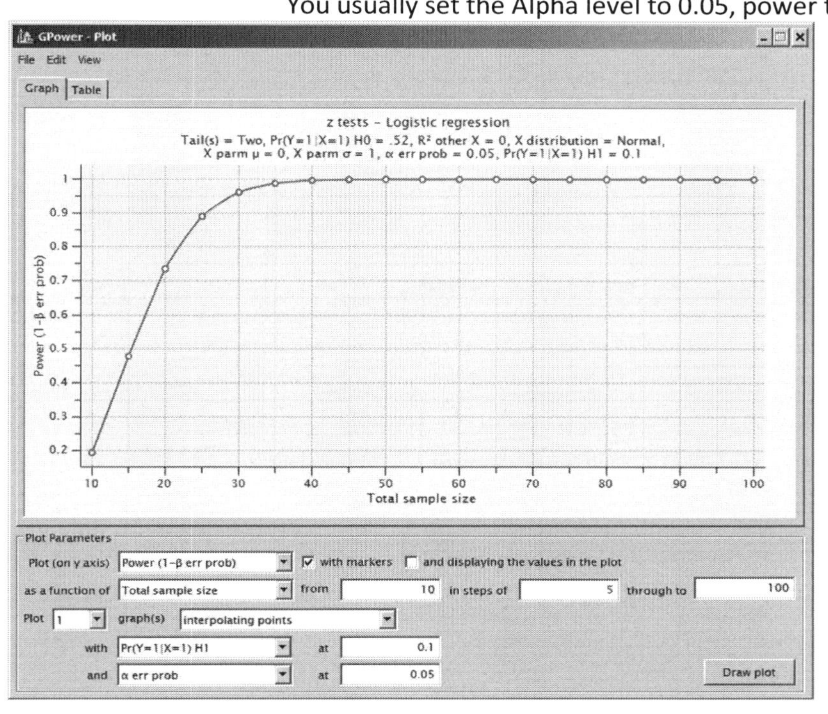

Click on the **Calculate** button which will update the output parameters. Then click on the **x-y plot for a range of values** button.

Select *Power* as the y axis and *Total sample size* for the x axis, finally click on the **Draw plot** button.

We see that if our pilot was typical (I think the results are too good!) we would only need 22 subjects to obtain a power of .8 and we only need around 25 to achieve a power of .9!

60.4 Reporting a simple logistic regression

Taking our treadmill example:

17 male subjects all aged 30 to 33 years with a BMI of 23 to 26 undertook treadmill stress tests, all used the same machine, and instructor. Concurrently all subjects were assessed for the presence of Coronary Heart Disease (CHD) using a standardised protocol. A logistic regression analysis of the data was carried out using R (3.1.1). The presence of CHD was the binary outcome measure and time on treadmill (sec) the continuous predictor. The logistic model was found to be an appropriate model (chi-square 10.95; df=1; p-value <.001, Hosmer Lemeshow goodness of fit chi-squared 6.77; p-value =.342) with the model predicting correctly 75% (n=8) and 90% (n=9) of healthy and diseased subjects respectively. A log odds ratio of – 0.0157 (se=.0072; Wald=4.6714; df=1; p-value< .0307) and odds ratio (OR) of .9844 (95% CI .9705 to .9985) indicated that for every extra minute managed on the treadmill test the odds of disease decreased by 61%. A treadmill test of 10 minutes suggested a probability of CHD of .9647 (96%) in contrast to a 15 minute result of .1974 (19%) and one of just over 18 minutes of .0105 (1%).

I might also include the R square measures (particularly the Nagelkerke value) and would also include the table of the parameter estimates along with the graph.

60.5 Multiple logistic regression

To demonstrate multiple logistic regression I will use a small dataset concerned with prostate cancer, where the outcome is if the disease has spread to other areas (the *nodes* variable where 0= no; 1=yes). The data is from Stevens, 2002 p.149.

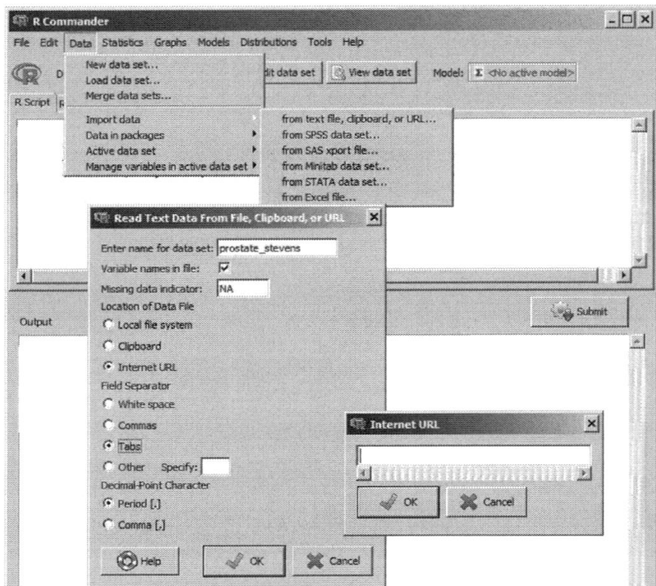

To obtain the data, first load R Commander by typing in the following command:

```
library(Rcmdr)
```

Then load the *prostate_brown_1980* dataset, by selecting the R Commander menu option:

Data-> from text, the clipboard or URL

I have given the resultant dataframe the name *prostate_stevens*, also indicating that it is from a URL (i.e. the web) and the columns are separated by *tab* characters. Clicking on the **OK** button brings up the *Internet URL* box in which you need to type the following:

```
http://www.robin-beaumont.co.uk/virtualclassroom/book2data/prostate_brown_1980.dat
```

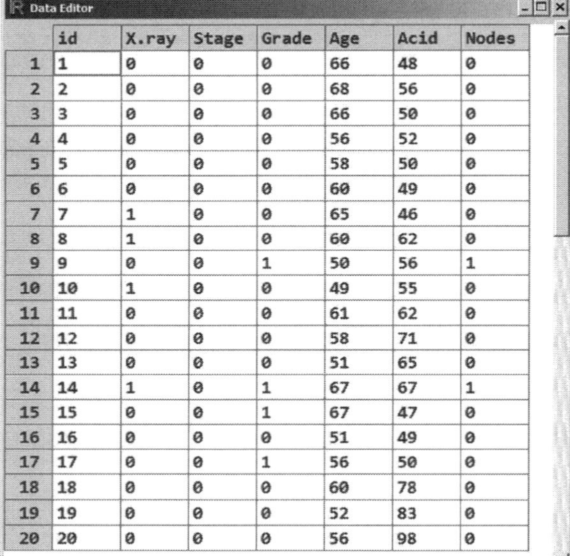

	id	X.ray	Stage	Grade	Age	Acid	Nodes
1	1	0	0	0	66	48	0
2	2	0	0	0	68	56	0
3	3	0	0	0	66	50	0
4	4	0	0	0	56	52	0
5	5	0	0	0	58	50	0
6	6	0	0	0	60	49	0
7	7	1	0	0	65	46	0
8	8	1	0	0	60	62	0
9	9	0	0	1	50	56	1
10	10	1	0	0	49	55	0
11	11	0	0	0	61	62	0
12	12	0	0	0	58	71	0
13	13	0	0	0	51	65	0
14	14	1	0	1	67	67	1
15	15	0	0	1	67	47	0
16	16	0	0	0	51	49	0
17	17	0	0	1	56	50	0
18	18	0	0	0	60	78	0
19	19	0	0	0	52	83	0
20	20	0	0	0	56	98	0

Clicking on the **View data set** or **Edit data set** buttons allows you to inspect the data, shown opposite.

name	description	type
id	Unique case number	
X.ray	results of an X-ray examination (0=negative, 1=positive)	input
Stage	A measurement of the size and position of the tumour observed by palpitation with the fingers via the rectum. (0=small, 1=large)	input
Grade	Another indicator of the seriousness of the cancer, this one is determined by a pathology reading of a biopsy taken by needle before surgery. (0=less serious, 1=more serious)	input
Age	age of patient at diagnosis (years)	input
Acid	level of serum acid phosphatase (in King-Armstrong units)	input
Nodes	nodal involvement (0=no, 1=yes)	output

I will assume that you have carried out all the preliminary analysis stages described for the previous simple regression analysis and move straight to model development.

60.5.1 Model development

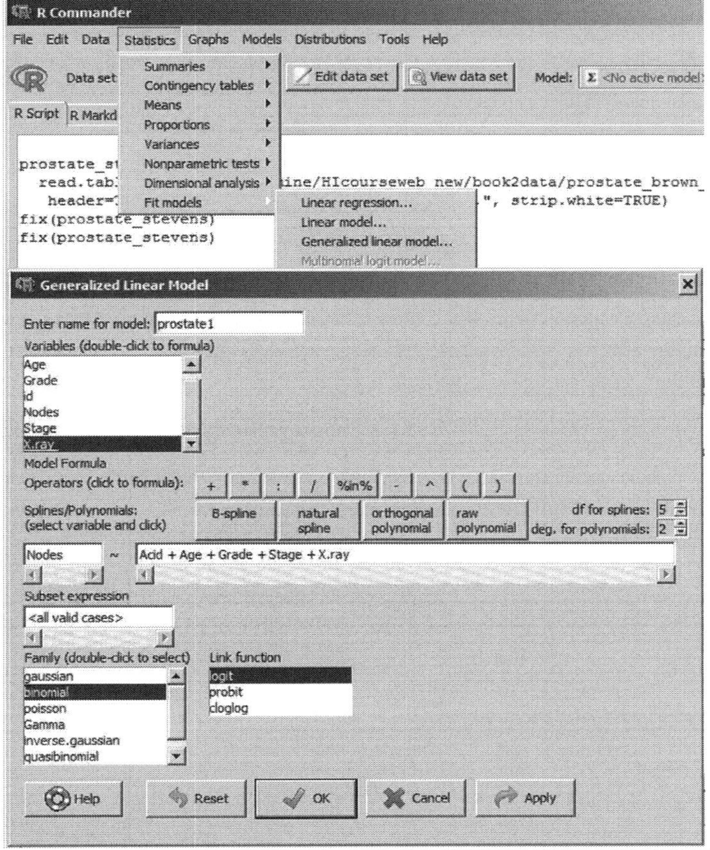

Using the R Commander menu option:

Statistics-> Fit models ->
Generalised linear model

Then enter an appropriate name for the model, I have called it *prostate1*. Place the outcome variable, *Nodes* in the box to the left of the model sign (~). Then add the inputs: *Acid, Age, Grade, Stage, X.ray* to the box on the right of the sign.

The family and link boxes should be set by default to *family=binomial* and *link=logit.* Finally,

Click **OK**.

R Commander produced the following code which you could have entered directly.

```
prostate1 <- glm(Nodes ~ Acid + Age +
Grade + Stage + X.ray,
family=binomial(logit), data=prostate_stevens)

summary(prostate1)
```

The results are shown below.
As before the exponentiated values (=odds ratios) are easier to interpret. These are obtained by using the *confint(model)* function either directly or using the R Commander menu.

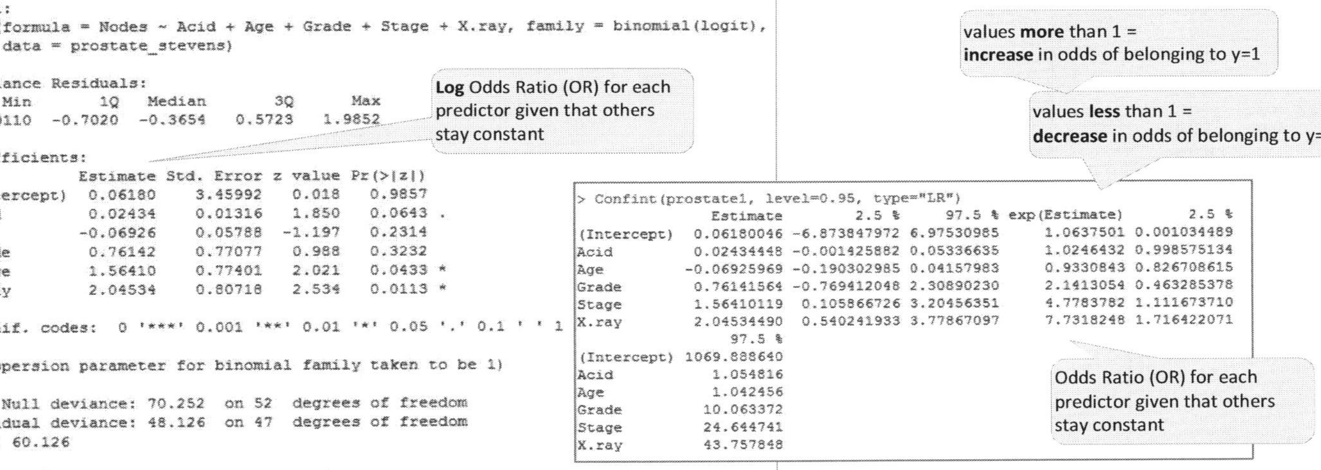

```
> prostate1 <- glm(Nodes ~ Acid + Age + Grade + Stage + X.ray,
+   family=binomial(logit), data=prostate_stevens)

> summary(prostate1)

Call:
glm(formula = Nodes ~ Acid + Age + Grade + Stage + X.ray, family = binomial(logit),
    data = prostate_stevens)

Deviance Residuals:
    Min       1Q   Median       3Q      Max
-2.0110  -0.7020  -0.3654   0.5723   1.9852

Coefficients:
            Estimate Std. Error z value Pr(>|z|)
(Intercept)  0.06180    3.45992   0.018   0.9857
Acid         0.02434    0.01316   1.850   0.0643 .
Age         -0.06926    0.05788  -1.197   0.2314
Grade        0.76142    0.77077   0.988   0.3232
Stage        1.56410    0.77401   2.021   0.0433 *
X.ray        2.04534    0.80718   2.534   0.0113 *
---
Signif. codes:  0 '***' 0.001 '**' 0.01 '*' 0.05 '.' 0.1 ' ' 1

(Dispersion parameter for binomial family taken to be 1)

    Null deviance: 70.252  on 52  degrees of freedom
Residual deviance: 48.126  on 47  degrees of freedom
AIC: 60.126

Number of Fisher Scoring iterations: 5
```

```
> Confint(prostate1, level=0.95, type="LR")
               Estimate        2.5 %     97.5 % exp(Estimate)       2.5 %
(Intercept)  0.06180046 -6.873847972 6.97530985     1.0637501 0.001034489
Acid         0.02434448 -0.001425882 0.05336635     1.0246432 0.998575134
Age         -0.06925969 -0.190302985 0.04157983     0.9330843 0.826708615
Grade        0.76141564 -0.769412048 2.30890230     2.1413054 0.463285378
Stage        1.56410119  0.105866726 3.20456351     4.7783782 1.111673710
X.ray        2.04534490  0.540241933 3.77867097     7.7318248 1.716422071
                 97.5 %
(Intercept) 1069.888640
Acid           1.054816
Age            1.042456
Grade         10.063372
Stage         24.644741
X.ray         43.757848
```


We have two inputs which are continuous variables (*Acid* and *Age)* so the odds ratios can be interpreted for these in terms of the change in odds of the outcome for a unit change in the inputs. For *Acid* we can say the model predicts a change in the odds of nodal involvement of 1.06 for a one unit increase in *Acid*. This can be converted to a percentage, (1.06-1)x100 = 6%, the model predicts a 6% odds **increase** for nodal involvement with a one unit increase in *Acid*.

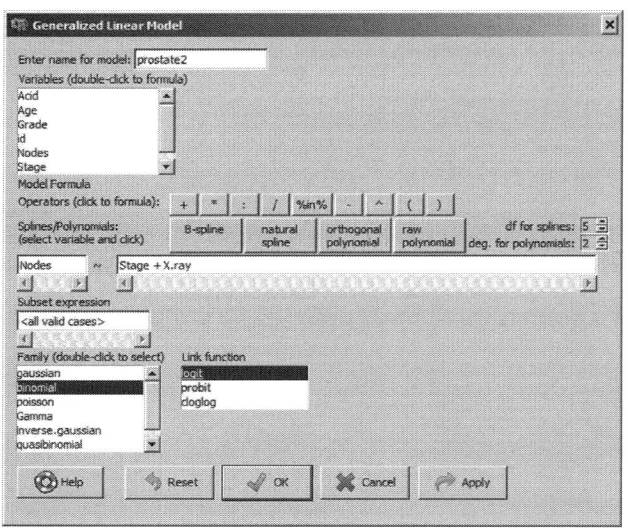

For *Age* we can say the model predicts a change in odds of nodal involvement of 0.933 for a one unit increase in *Age*. Again this can be converted to a percentage, (1-0.933)x100 = 6.7%, the model predicts a 6.7% odds *decrease* for nodal involvement with a one unit increase in *Age*.

For the nominal variables we need to remember the category we gave the zero value to for each. The *Stage* input predicts that those in stage category 1 compared to those in the zero category have 4.7 times the odds of exhibiting nodal involvement. Finally, the *X.ray* input predicts that those who have a positive X-ray result have 7.7 times the odds of those with a negative X-ray of displaying nodal involvement.

```
glm(formula = Nodes ~ Stage + X.ray, family = binomial(logit),
    data = prostate_stevens)

Deviance Residuals:
    Min       1Q   Median       3Q      Max
-1.9166  -0.9907  -0.4934   0.5892   2.0815

Coefficients:
            Estimate Std. Error z value Pr(>|z|)
(Intercept)  -2.0446     0.6100  -3.352 0.000802 ***
Stage         1.5883     0.7000   2.269 0.023274 *
X.ray         2.1194     0.7468   2.838 0.004541 **
---
Signif. codes:  0 '***' 0.001 '**' 0.01 '*' 0.05 '.' 0.1 ' ' 1

(Dispersion parameter for binomial family taken to be 1)

    Null deviance: 70.252  on 52  degrees of freedom
Residual deviance: 53.353  on 50  degrees of freedom
AIC: 59.353

Number of Fisher Scoring iterations: 4
```

The only statistically significant inputs are *Stage* and *X.ray* (taking the cut-off value to be 0.05) indicating that the parameter values for the other inputs are actually equal to zero in the population. We can now consider a new model including only the statistically significant inputs.

Using the R Commander menu option:

Statistics-> Fit models ->
Generalised linear model

I have called the new model *prostate2*, shown opposite. Alternatively, we could have entered the following R code:

```
prostate2 <- glm(Nodes ~ Stage + X.ray,
family=binomial(logit), data=prostate_stevens)

summary(prostate2)
```

The results show that both inputs remain statistically significant and we could use these parameter values to predict any individual value, as demonstrated in the previous example. We can compare the two models, *prostate1* and *prostate2* a number of ways, the simplest being inspection of the AIC values. Unfortunately, 60.12 versus 59.35 suggests equivalence. Another way is to compare the two models using a chi-squared test (also called a likelihood ratio test here).

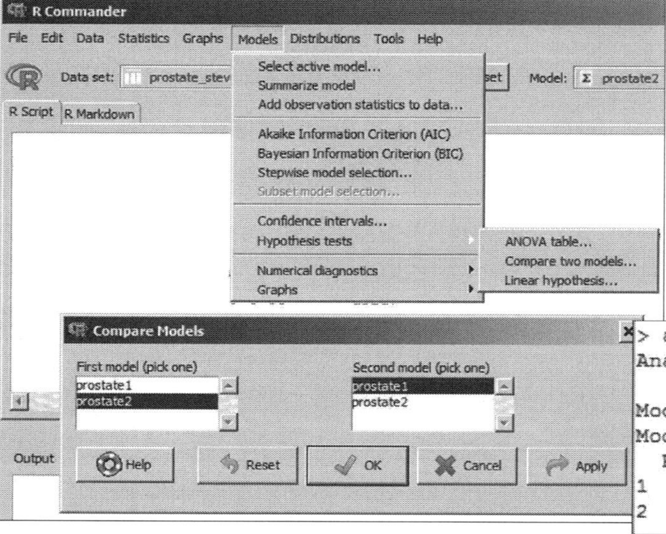


```
> anova(prostate2, prostate1, test="Chisq")
Analysis of Deviance Table

Model 1: Nodes ~ Stage + X.ray
Model 2: Nodes ~ Acid + Age + Grade + Stage + X.ray
  Resid. Df Resid. Dev Df Deviance Pr(>Chi)
1        50     53.353
2        47     48.126  3   5.2276   0.1559
```

The p-value is greater than the usual critical value of 0.05 indicating that both models offer the same level of predictive power; therefore, we select the simpler model (i.e. *model2*) as the preferred one.

60.5.2 Tables of counts for binary inputs

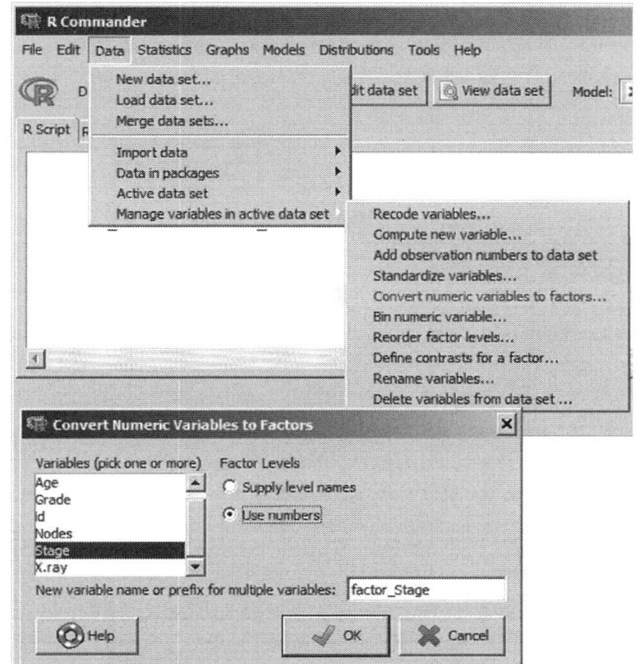

Both the *X.Ray* and *Stage* inputs in the model are binary variables and it would be nice to see how the values of each of these relate to the outcome, *Nodes*.

To do this in R Commander we need to either convert or create three new variables from the factor versions of *Stage, Nodes and X.ray*. This is achieved by using the menu option opposite for each of the variables. Alternatively, using the *xtabs()* function, shown below, we do not need to do this as the function converts the variables on the fly:

```
count_table1 <- xtabs(~Nodes+ Stage, data=
prostate_stevens)
count_table1
count_table2 <- xtabs(~Nodes+ X.ray, data=
prostate_stevens)
count_table2
```

```
> count_table1 <- xtabs(~Nodes+ Stage, data= prostate_stevens)
> count_table1
     Stage
Nodes  0  1
    0 21 12
    1  5 15
> count_table2 <- xtabs(~Nodes+ X.ray, data= prostate_stevens)
> count_table2
     X.ray
Nodes  0  1
    0 29  4
    1  9 11
```

The diagonal counts (↘) represent those counts where both the input (*X.ray* or *Stage*) and outcome (*Nodes*) are in agreement, in an ideal world we would like to see all the counts in these cells. That is both diagnostic tests have 100% sensitivity and specificity, but in reality this will never be the case.

```
> chi_test1 <- chisq.test(count_table1)
> chi_test1

        Pearson's Chi-squared test with Yates' continuity correction

data:  count_table1
X-squared = 5.9727, df = 1, p-value = 0.01453

> chi_test2 <- chisq.test(count_table2)
> chi_test2

        Pearson's Chi-squared test with Yates' continuity correction

data:  count_table2
X-squared = 9.269, df = 1, p-value = 0.002331
```

We can also consider the possible differences in proportions by carrying out a chi-squared test.

```
chi_test1 <- chisq.test(count_table1)
chi_test1
chi_test2 <- chisq.test(count_table2)
chi_test2
```

Comparing the two p-values produced here to those produced in the logistic regression model we can see they are quite similar, and in this instance the analysis is quite similar – however if we had retained an interval/ratio input such as *Age* or *Acid* the logistic regression approach offers more.

The values above are the observed values but what would be better would be to compare them with those predicted by the chosen model which we will do now.

60.5.3 Predicted and observed values using R Commander - the classification table

In the previous, simple logistic regression example, we used the R *predict()* function to obtain the predicted values from the dataset. Alternatively we can use the R Commander menu option shown opposite.

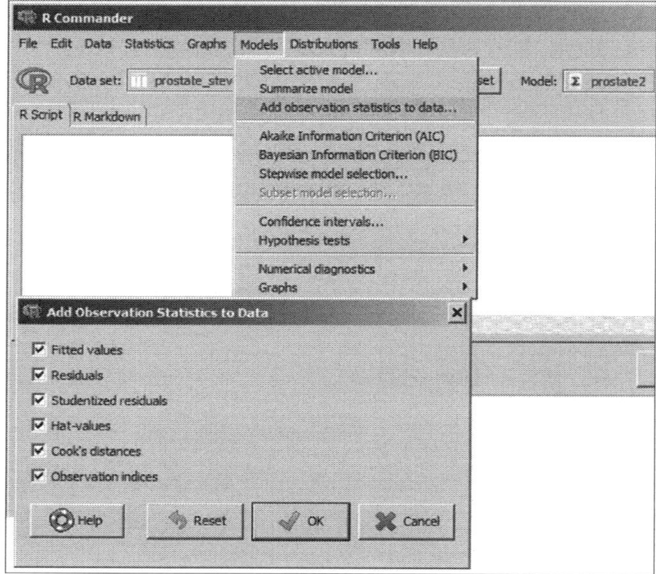

This option adds various columns to the dataframe by generating the following R code.

```
prostate_stevens$fitted.prostate2 <-
fitted(prostate2)
prostate_stevens$residuals.prostate2 <-
residuals(prostate2)
prostate_stevens$rstudent.prostate2 <-
rstudent(prostate2)
prostate_stevens$hatvalues.prostate2 <-
hatvalues(prostate2)
prostate_stevens$cooks.distance.prostate2 <-
cooks.distance(prostate2)
prostate_stevens$obsNumber <-
1:nrow(prostate_stevens)
```

fitted.prostate2	residuals.prostate2	rstudent.prostate2
0.1145964	-0.4933795	-0.4984988
0.1145964	-0.4933795	-0.4984988
0.1145964	-0.4933795	-0.4984988
0.1145964	-0.4933795	-0.4984988
0.1145964	-0.4933795	-0.4984988
0.1145964	-0.4933795	-0.4984988
0.5186952	-1.209342	-1.273765
0.5186952	-1.209342	-1.273765
0.1145964	2.081509	2.15309
0.5186952	-1.209342	-1.273765

We can then use these values to create a classification table with the following R code.

```
class_table <- table(prostate_stevens$fitted.prostate2>.5, prostate_stevens$Nodes)

dimnames(class_table) = list(predicted=c("no nodal", "nodal"), observed=c("no nodal",
"nodal"))

class_table
```

```
> class_table <- table(prostate_stevens$fitted.prostate2>.5, prostate_stevens$Nodes)
> dimnames(class_table) = list(predicted=c("no nodal", "nodal"), observed=c("no nodal", "nodal"))
> class_table
          observed
predicted  no nodal nodal
  no nodal       29     9
  nodal           4    11
```

The results show that of those who had no nodal involvement (observed column 1) 29 of the 33 (i.e. 88%) were correctly classified by the model. In contrast the model performs more poorly for those with observed nodel involvement where we have only 11 of the 20 (i.e. 55%) being correctly classified. In other words, you might as well flip a coin to decide which group they will belong to! Given that it is probably more important to mis-classify patients as having nodel involvement when they actually do not, rather than the other way round, this model is of little use. In other words it is missing a potentially fatal complication in 9 out of 20 patients (45%). One should not be seduced by the statistically significant p-values.

60.6 Analysing data in different formats

The two detailed examples given in this chapter have used data which is in a specific format, that is each row of data represents a single case. This is the way most such datasets appear when we are dealing with raw data. However often an analyst wishes to analyse a dataset from a secondary data source where it might be in one of many formats, two of which I will discuss below.

60.6.1 Summary counts or proportions

chd	normal	total	time
80	10	90	450
77	18	95	500
76	20	96	550
70	18	88	600
82	52	134	650
40	10	50	700
32	8	40	750
34	26	60	800
30	54	84	850
21	75	96	900
13	108	121	950
2	196	198	1000

In this format the output variable (the dependent variable) is provided as a table, often representing the number of observations equal to a particular value of the outcome. I have adapted the example given on Koji Yatani's Statistics for HCI Research website (http://yatani.jp) to illustrate this.

Opposite we have a table of counts representing the number of those found subsequently to be suffering from CHD divided by how long they lasted on the treadmill test. I have used 50 second time bands, along with the total number tested for each band. If we had the data in the format used at the start of this chapter we would have 1152 rows representing this data, instead now we only have 12. We could have also been given the proportions for each time band that were subsequently found to be suffering from CHD. Either form of the dataset is treated the same by R. Sticking with the actual counts we can carry out a logistic regression using the following R code:

```
 time <- c(450, 500, 550, 600, 650, 700, 750, 800, 850, 900, 950, 1000)
 chd <- c(80,  77, 76, 70, 82, 40, 32,  34, 30, 21, 13, 2)
total <- c(90, 95, 96, 88, 134, 50, 40, 60, 84, 96, 121, 198)
normal <- (total - chd)
thedata<- cbind( chd, normal, total,  time)
thedata<- data.frame(thedata)
str(thedata)
testresult <- cbind(chd, normal)
result <- glm(testresult ~ time, family = binomial)
summary(result)
```

```
Call:
glm(formula = testresult ~ time, family = binomial)

Deviance Residuals:
    Min       1Q   Median       3Q      Max
-5.1940  -1.3056  -0.3593   1.9061   4.1905

Coefficients:
              Estimate Std. Error z value Pr(>|z|)
(Intercept)  6.3478873  0.3561199   17.82   <2e-16 ***
time        -0.0085786  0.0004668  -18.38   <2e-16 ***
---
Signif. codes:  0 '***' 0.001 '**' 0.01 '*' 0.05 '.' 0.1 ' ' 1

(Dispersion parameter for binomial family taken to be 1)

    Null deviance: 586.87  on 11  degrees of freedom
Residual deviance:  78.67  on 10  degrees of freedom
AIC: 134.2
```

We can also bind the two columns representing each of the counts for the output on the fly within the *glm()* function, the results are identical as shown opposite:

```
result2 <- glm( cbind(chd, total - chd) ~ time,
family = binomial, data= thedata)
summary(result2)
```

```
glm(formula = cbind(chd, total - chd) ~ time, family = binomial,
    data = thedata)

Deviance Residuals:
    Min       1Q   Median       3Q      Max
-5.1940  -1.3056  -0.3593   1.9061   4.1905

Coefficients:
              Estimate Std. Error z value Pr(>|z|)
(Intercept)  6.3478873  0.3561199   17.82   <2e-16 ***
time        -0.0085786  0.0004668  -18.38   <2e-16 ***
---
Signif. codes:  0 '***' 0.001 '**' 0.01 '*' 0.05 '.' 0.1 ' ' 1

(Dispersion parameter for binomial family taken to be 1)

    Null deviance: 586.87  on 11  degrees of freedom
Residual deviance:  78.67  on 10  degrees of freedom
AIC: 134.2

Number of Fisher Scoring iterations: 5
```

With this type of data we can easily produce plots of the observed proportions along with the logistic curve.

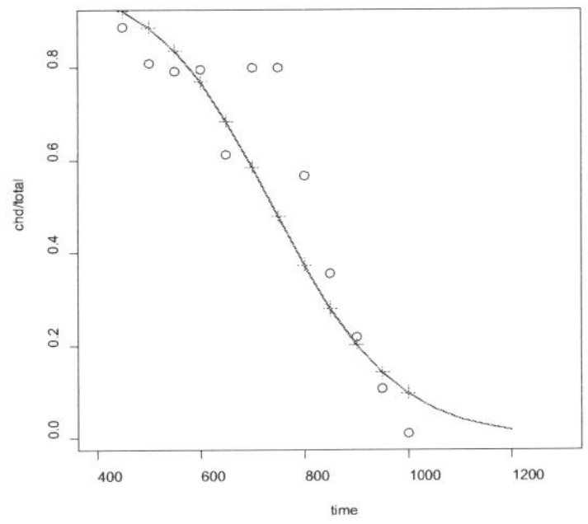

```
predict_rate <- predict(result, type="response")
predict_rate
plot(time, chd/total, cex=1.5, xlim = c(400, 1300) )
points(time, predict_rate, pch=3, col="red", cex=1.5)
x <- seq(400, 1200, 50)
y <- 1 - (1 / (1 +    exp(result$coef[1] + result$coef[2]*x)))
lines(x, y, col="blue", lwd=2)
```

draw observed proportions of those suffering from CHD from dataset

add the proportions predicted by the model

create a vector of values 400 to 1200 at intervals of 50

now to draw the smoothed line of predictions

60.6.2 Summary tables

Gender	Dept	Admitted	Rejected
Male	A	512	313
	B	353	207
	C	120	205
	D	138	279
	E	53	138
	F	22	351
Female	A	89	19
	B	17	8
	C	202	391
	D	131	244
	E	94	299
	F	24	317

This example demonstrates how you can use summary data from a table to carry out a logistic regression as if it were raw data. This example has been taken from William King's R tutorials at http://ww2.coastal.edu/kingw/statistics/R-tutorials/logistic.html . The data are from 1973 showing admissions by gender to the top six graduate programs at the University of California, Berkeley. From the table, there appears to be a bias for admitting men. The datafile is called *UCBAdmissions* and forms part of the R core installation. Assume that the outcome of interest is the probability of being admitted.

Once again we see the outcome variable is being represented by two columns of data *Admitted* and *Rejected*, which we can class as a binary variable with two levels *yes* and *no*.

To get the data into a suitable structure to allow a logistic regression analysis requires several steps each described below.

```
gender=rep(c("Male","Female"), c(6,6))
dept=rep(LETTERS[1:6],2)
yes=c(512,353,120,138,53,22,89,17,202,131,94,24)
no=c(313,207,205,279,138,351,19,8,391,244,299,317)
ucb_dataframe = data.frame(gender, dept, yes, no)
ucb_model1 = glm(cbind(yes,no) ~ gender*dept, family=binomial(logit), data=ucb_dataframe)
summary(ucb_model1)
```

create the gender factor, two levels each having six values

create the department factor, six levels repeat the level sequence twice

add the counts of admissions

add the counts of rejects

put them in a dataframe

run the model

and get a summary of the results

60.7 A strategy for logistic regression analysis

We can make use of the recommendations from Daniel, 1991 p.411, as discussed in the simple linear regression chapter:

1. Identify the model – have you the appropriate variables for the analysis, if it is an observational study decide which are the input variables and which is the output. Are you happy with only modelling association not causation?

2. Review logistic regression assumptions – the validity of the conclusions from the analysis depends upon the appropriateness of the model, appropriate data types, and study design ensuring **Independent measures** (to maintain the independence of errors assumption). Whereas in linear regression we assumed a linear relationship between the predictor and outcome variables, now it is between the logit of the outcome variable and the predictor. In others words statistical validity needs to be considered.

3. Always graph the observed values of one or two key inputs against the logistic curve.

4. Evaluate the logistic regression equation, Hosmer & Lemeshow fit measure, R^2 (Nagelkerke most important), and b (log odd ratio, se, p-value), $exp(b)$ (odds ratio) along with their associated p-values and if possible confidence intervals. Also consider the probability of the outcome for specific levels of the independent variable (may help drawing up a table showing the range of odds ratio values against probabilities).

5. Logistic Regression diagnostics to identify possible values that need further investigation or removing from analysis (data cleaning).

6. Consider sensible minimum and maximum input variable ranges for which the model provides useful predictions.

There are more technical aspects that we could consider, such as *overdispersion* discussed in the chapter on *Poisson regression*. Also see (Field, 2009 p.276, p.309; Campbell, 2006 p.42).

Campbell, 2006 provides a number of very useful checklists from which I have created the following pointers:

- Are you sure that the outcome variable is a simple binary variable?
- Is there the concept of survival associated with the outcome, if so then survival analysis might be better?
- Does the dataset represent independent observations for both the predictor and outcome variable. Repeated measures or matched designs need a different type of logistic regression. For further details see the chapter *Conditional logistic regression* or Campbell, 2006.
- The model provides an odds ratio which can be recalculated to be probabilities, only reinterpret again to 'relative risk' if you must, then only when you are sure the data is from a prospective trial and also has a low incidence (i.e. <10 %) of the disease.
- Does the relationship look loglinear when graphed.

60.8 Summary

In this chapter we have looked in detail at modelling the possible association between a binary outcome, representing probability/odds of an event happening, and both continuous and nominal inputs. In doing so we considered the transformation needed to create a suitable model using the logit function. We also discovered that, in contrast to the linear regression model, with logistic regression we needed to use an iterative process to pinpoint the parameter estimates using the Maximum Likelihood Estimator (MLE) which is itself based on likelihood. We also discovered that the natural logarithm of the likelihood value multiplied by -2 had a chi-squared distribution allowing the assessment of overall model fit as it behaved analogously to the residual/error sum of squares in linear regression.

Various pseudo R square values were described along with warnings concerning their interpretation, the Hosmer & Lemeshow statistic (also with a chi-squared distribution) being offered as a weak alternative. The rather unfriendly raw parameter estimates (i.e. B =log odd ratios) and their more user friendly exponentiated (i.e. antilog) versions were discussed along with p-values and confidence intervals. The conversion to probabilities was also made to demonstrate the non-linear relationship. The user friendly classification table was described and in the 2 by 2 case how this related to sensitivity, specificity and false negative/positives.

To summarise, the following values should be inspected in any set of results:

Global fit measures:

- Model chi-squared (Likelihood ratio test)
- Pseudo R squared values; Cox & Snell, Nagelkerke, Hosmer & Lemeshow
- Logistic curve graphs with observed values and/or classification tables depending upon the type of inputs.

Individual parameter measures:

- Wald statistic / z value + p-value
- Effect size (odds ratio)
- Confidence interval

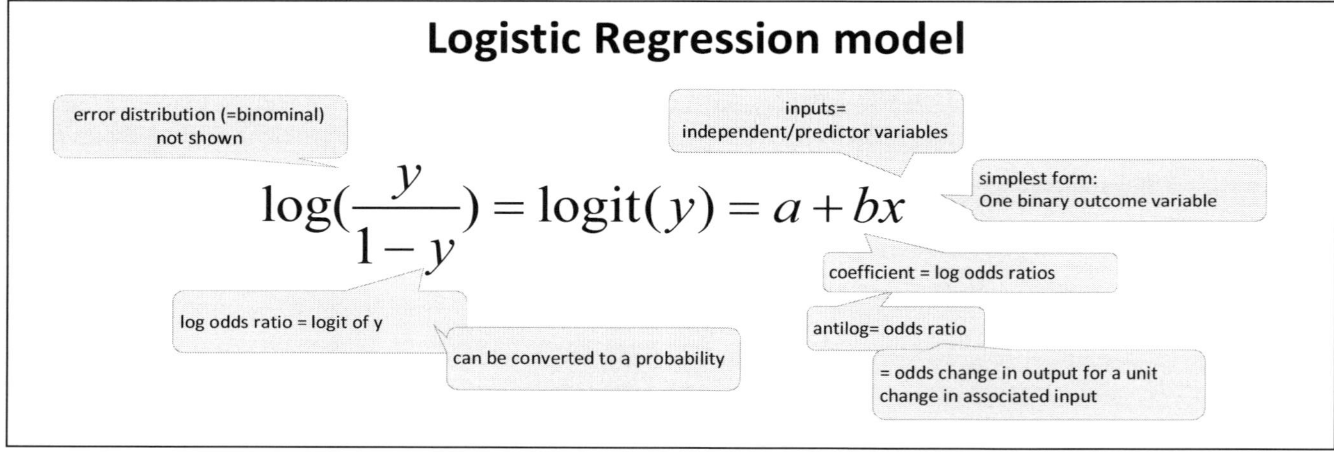

60.9 Tips and Tricks

In the last twenty years logistic regression has become much more popular and there are several good books describing it from basic introductions (Campbell, 2006; Field, 2009; Agresti & Finlay, 2008; Norman & Streiner, 2008) to complex monographs (Agresti, 2002; Fleiss, Levin & Paik, 2003). Also there are several websites providing tutorials and let's not forget YouTube.

Paper based tutorials on the web:

A three part tutorial written by Richard Williams, Sociology Department, University of Notre Dame:

Logistic Regression I: Problems with the Linear Probability Model (LPM):
http://www.nd.edu/~rwilliam/stats2/l81.pdf

Logistic Regression II: The Logistic Regression Model (LRM):
http://www.nd.edu/~rwilliam/stats2/l82.pdf

Logistic Regression III: Hypothesis Testing, Comparisons with OLS;
http://www.nd.edu/~rwilliam/stats2/l83.pdf

Chan Yiong Huak, Head of the Biostatistics Unit, Yong Loo Lin School of Medicine, National University of Singapore also provides an excellent tutorial for logistic regression using SPSS (Chan, 2004a).

For R specific logistic regression analyses look at the chapter in Cawley, 2005 or alternatively David Hitchcock, assistant professor of statistics at the University of South Carolina provides an R code script for a logistic regression at:
http://www.stat.sc.edu/~hitchcock/diseaseoutbreakRexample704.txt

For details concerning multiple nominal inputs, see:

http://courses.ncssm.edu/math/Stat_Inst/PDFS/Categorical%20Data%20Analysis.pdf

The Institute for Digital Research and Education (IDRE) at UCLA provide a good tutorial including an example of using the *ggplot2* package (see the *Publication quality graphics* chapter) to draw confidence intervals for logistic models:

http://www.ats.ucla.edu/stat/r/dae/logit.htm

YouTube

An excellent tutorial showing how to import CSV data (medical example) into R using R Commander and also carrying out logistic regression:
http://www.youtube.com/watch?v=oKt52byH83o

Logistic regression (several predictors) using R and the Deducer package:
http://www.youtube.com/watch?v=PE5mgNbq7xQ

61 Poisson (log-linear) regression

Poisson regression allows you to use counts (whole numbers) as the outcome (dependent variable) in regression analysis. Examples of counts include:

1. Number of epileptic seizures each patient had in a two week period
2. Number of bacteria in a fixed volume of growth suspension after 24 hours incubation
3. Number of patients visiting a family doctor (GP) complaining of tiredness over the last year
4. Number of skin cancers reported in a particular town for a set period
5. Number of coronary deaths in male doctors in a particular country for a set period
6. Number of cancers and distance from a nuclear power station in the last decade

Lambda (λ) = mean
In the Poisson distribution the single parameter **λ** is also the mean of the distribution. This can be confusing as here it is always a count (whole number). We do not obtain it by calculating the average for divisions of the time period, but simply add the values up.

In all the above examples within the time period the actual count can vary between zero and infinity but cannot be negative. Consider example 1 above, a patient might have zero or more seizures on a particular day of the two week period but over the two weeks will have experienced the total of all seizures for that period. Such a situation is frequently modelled using a Poisson distribution, which takes a single parameter value called lambda (λ) which represents the count for the specified period. Using such an approach means that the linear regression equation, which was introduced in an earlier chapter, is adapted to take this into account in several ways, most importantly we now have a dependent variable which is the log of its original value. This is why Poisson regression is frequently called log-linear regression.

$\log(y) = \text{Intercept} + b_1(x_1) + b_2(x_2) + b_3(x_3) \ldots$ implies:

$y = \exp(\text{Intercept} + b_1(x_1) + b_2(x_2) + b_3(x_3) \ldots)$

$y = \exp(\text{Intercept}) * \exp(b_1(x_1)) * \exp(b_2(x_2)) * \exp(b_3(x_3)) \ldots)$

The above is an example of what is known as a Generalised Linear Model (GLM, pronounced glim). It consists of four aspects. Firstly the above expression; $Intercept + b1(x1) + b2(x2) + b3(x3)$ …. which is called a linear predictor (*LP*) so we can write the above as: $y = exp(LP)$. The second aspect is the function that affects the linear predictor, in this instance the log transformation. The third aspect to a GLM model is the distribution of the expected value for any linear combination (y_i). In standard linear regression this was taken to be normal but in this instance we assume it is Poisson. There is also the question of variability (i.e. variance) of the predicted values and this is the final aspect that is defined for these models.

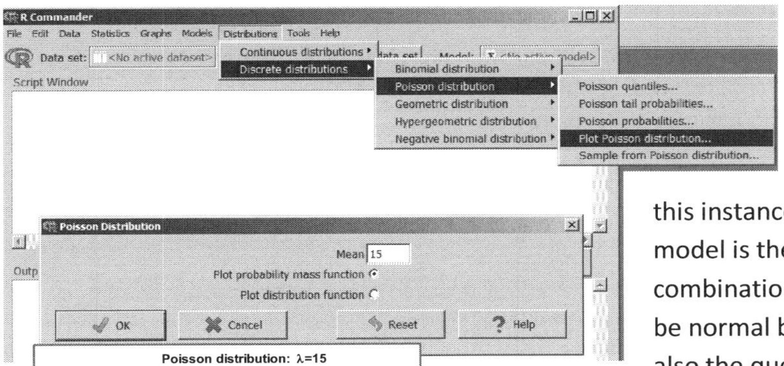

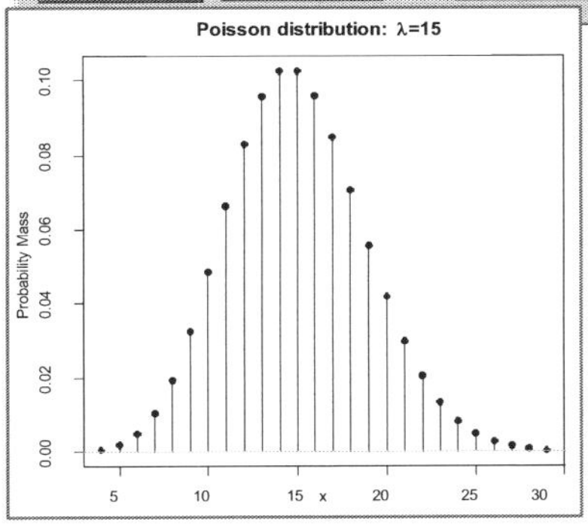

We can see what a particular Poisson distribution looks like using R Commander as shown opposite. Note that to define a Poisson distribution all we need to specify is the mean which must be a positive integer (whole number), I have decided to plot the Poisson distribution with a lambda of 15. From the distribution plot we can see that the values can only be whole positive numbers which makes it eminently suitable for modelling counts. Note that we do NOT define the spread, as we did when defining the normal distribution, here the variance is assumed to be equal to lambda.

Because the Poisson distribution used in the Poisson regression model is defined by a single value (lambda) a certain assumption holds called *equidispersion*, which is often broken when trying to apply the model to real data. Luckily, this problem is easily spotted and the model can be adapted to accommodate such a violation. What exactly is the equidispersion assumption? Well, it is where the variance of the predicted values at each value of the input variable is assumed to equal the mean. When the data poorly fits this model the problem of overdispersion is said to exist and can result in grossly **de**flated standard errors and therefore grossly **in**flated *t* values in the output (Cameron & Trivedi, 2013). In each of the examples, we will consider this aspect.

There are three common reasons why overdispersion might arise:

- The underlying data requires the model to take into account this changing or larger variance.
- An important input variable is missing from the model
- The data points are not independent

The skill of the researcher is required to decide which one or more of the above three reasons has resulted in overdispersion.

Births	Hospital type	Caesareans
	0=private 1=public	
236	0	8
739	1	16
970	1	15
2371	1	23
309	1	5
679	1	13
26	0	4
1272	1	19
3246	1	33
1904	1	19
357	1	10
1080	1	16
1027	1	22
28	0	2
2507	1	22
138	0	2
502	1	18
1501	1	21
2750	1	24
192	1	9

Adjustment of the outcome in Poisson regression is common, particularly in epidemiology. Examples of this include adjusting the output count by the time each subject has been in a particular state (i.e. person years) or other denominator value such as the population size to produce rates. I will demonstrate this in one of the later examples but let's start with a more straightforward one.

61.1 Caesarians, births and hospital type

Here we have the total number of births along with caesareans for a number of private and public hospitals (factor with two levels) and are interested to know if the number of caesareans is related to the total number of births and also if this is influenced by hospital type.

Caesareans is clearly a variable representing a count, as is births.

You can download the data from:

http://www.robin-beaumont.co.uk/virtualclassroom/book2data/births1.rdata

Once you have stored the file locally use the following R command to load it.

```
load(file=file.choose())
```

The code below loads the dataset and creates a variable indicating the rate of caesareans per 1000 births.

```
attach(births1)
        names(births1)
        births1

births1$c_per1000 <- with(births1, (caesareans/ births)*1000)
```

create a new variable = caesareans per 1000 births

61.1.1 Preliminaries

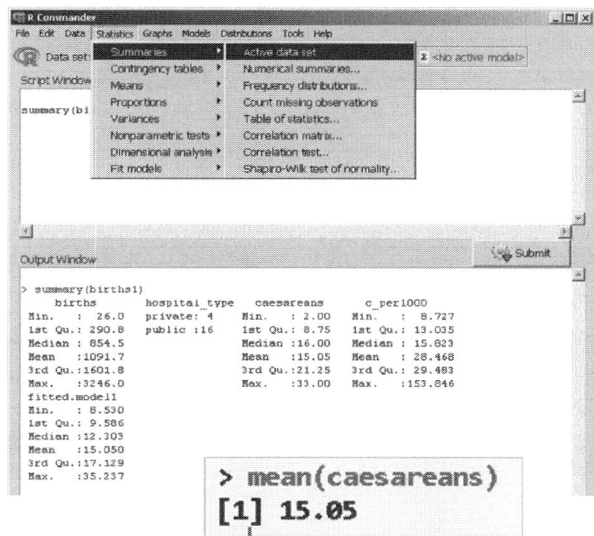

It is essential to get a feel for the data before considering any model. Always start by using the **Statistics->Summaries->Active data set** menu option in R Commander, or the equivalent R code *summary(dataframe),* replacing dataframe with the name of yours.

The most striking thing is that there are only 4 private hospitals but 16 public. Such a small number of private hospitals in comparison to public ones will obviously effect the accuracy of any estimate from the private sector.

We can also see that the mean number of caesareans is 15.05. This is the value that the model will use to create the Poisson distribution and will assume that the variance of the counts is also equal to this. We can also see that the distribution of caesareans is skewed as the 1st and 3rd quartiles are not equidistant from the mean. This will be noticeable with the graphical summaries.

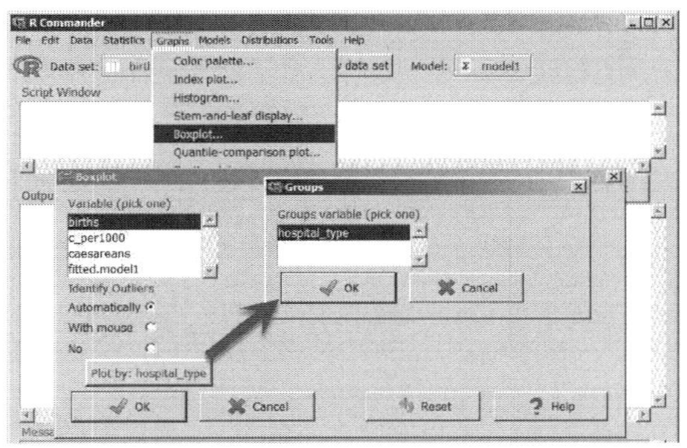

I generally find it useful to produce boxplots when you have data in groups. To produce the boxplots we will consider the raw counts and the incidence of caesareans per 1000 births. You can produce them either using the R Commander menu (opposite) or the R code below.

```
boxplot(caesareans~hospital_type, data=births1,
id.method="y") # raw data boxplot
```

```
boxplot(births1$c_per1000~hospital_type, data=births1, id.method="y")
# boxplot per 1000
```

Raw data

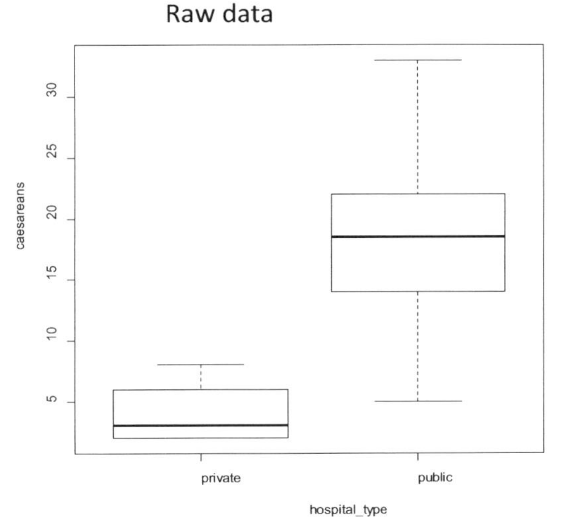

Per 1000 births

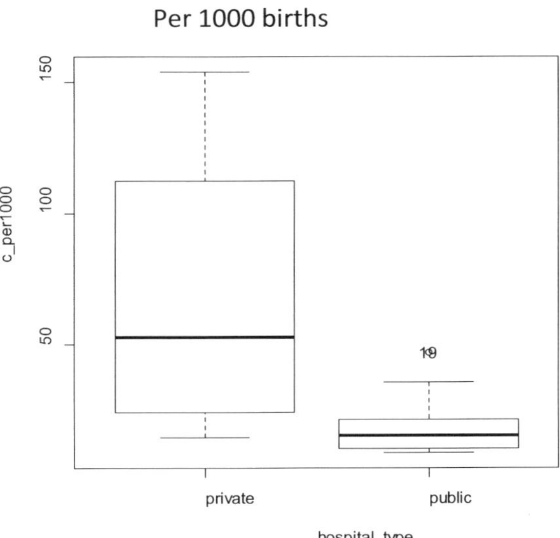

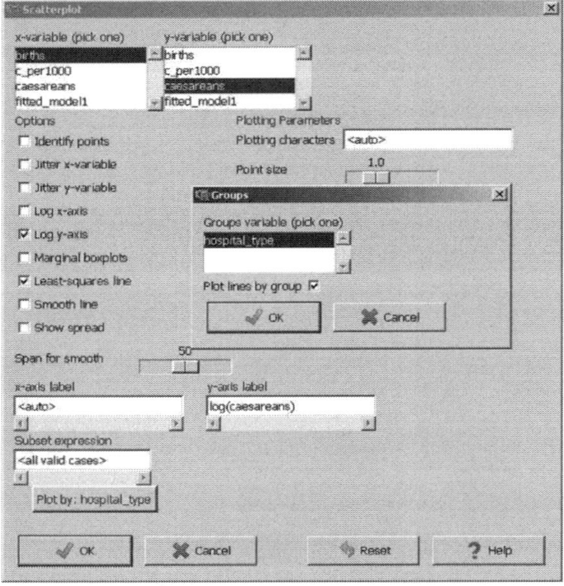

The two above boxplots show the problem that occurs if you just consider counts when they come from different sized samples, an example of what some call *ignoring the denominator*. Looking at the raw data it would appear that public hospitals have far more caesareans than private ones, but considering the caesareans on a per 1000 births it is clearly the other way round. Another useful way of looking at this relationship is to use scatterplots.

Using the R Commander menu option **Graphs->Scatterplot**, I have created a log scatterplot and requested separate regression lines for both the private and public hospitals data. I selected the log option after first producing a scatterplot with the raw scales, which showed an excessively curved relationship.

This graph is very useful in this situation highlighting several important aspects:

- All four private hospitals have far fewer births than the public ones
- There looks like a definite (log) linear relationship between caesareans and number of births for the public hospitals
- It is very unlikely we will be able to make any meaningful conclusions about the private hospitals with so few points.

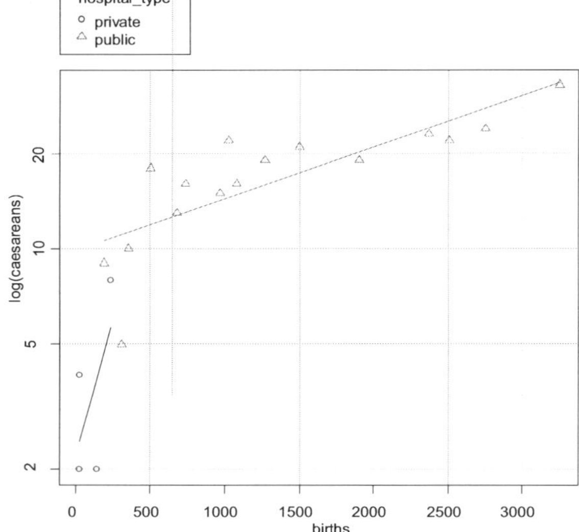

61.1.2 Modelling the count and a continuous input variable

We will now begin to develop a Poisson regression model of our data by first considering the simplest model with one continuous input and then adding more. Here the simple model is the number of caesareans as the output and the number of births as the input:

Mathematically the model looks like this:

```
log(ceasareans) = Intercept + b₁(births)
```

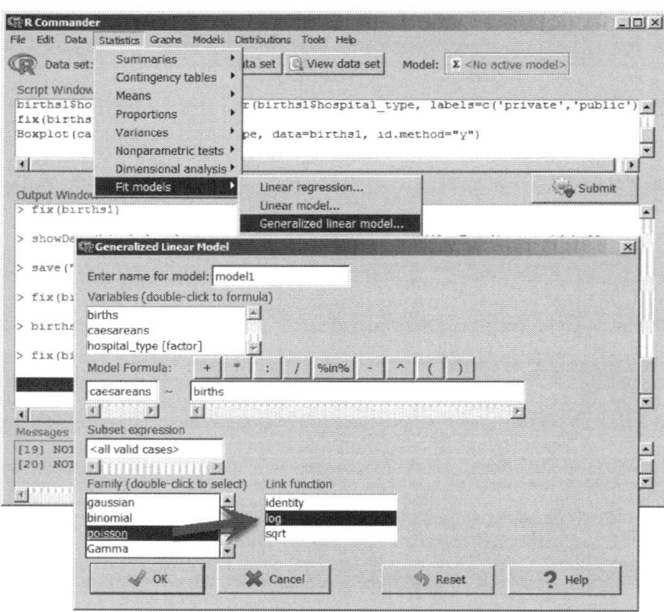

We can create the model using the R Commander menu option:

Statistics->Fit models->Generalised linear model

I have called this model *model1* and specified the **Family** to be *Poisson*, and by double clicking on the *Poisson* term selected the **Link function** to be the *log*.

Alternatively, I could have typed directly into the *Script* window the following equivalent code:

```
model1<- glm(formula = caesareans ~ births,
family=Poisson(log), data=births1) summary(model1)
```

```
glm(formula = caesareans ~ births, family = poisson(log), data = births1)

Deviance Residuals:
     Min        1Q    Median        3Q       Max
-2.81481  -0.73305  -0.08718   0.74444   2.19103

Coefficients:
             Estimate Std. Error z value Pr(>|z|)
(Intercept) 2.132e+00  1.018e-01  20.949  < 2e-16 ***
births      4.406e-04  5.395e-05   8.165 3.21e-16 ***
---
Signif. codes:  0 '***' 0.001 '**' 0.01 '*' 0.05 '.' 0.1 ' ' 1

(Dispersion parameter for poisson family taken to be 1)

    Null deviance: 99.990  on 19  degrees of freedom
Residual deviance: 36.415  on 18  degrees of freedom
AIC: 127.18

Number of Fisher Scoring iterations: 4
```

As is the case with any regression, actual parameter (i.e. coefficient) estimates can be interpreted as effect size measures. For example the estimate for the births coefficient, 4.406e-04 = 0.0004406, indicates that increasing by one birth increases the log count of caesareans by 0.0004406.

```
log(caesareans) = 2.132 + 0.0004406 *
births
```

Ignoring the intercept, we note that the estimate for the births parameter is statistically highly significant by noting the three stars beside it. We can plug these values into the mathematical equation above and by reversing the log effect using the *exp()* R function find the number of caesareans the model predicts for a given number of births. Say we want to find the expected number of caesareans for 1000 births we type:

```
Countcaesareans = exp(2.132)*exp(.0004406 *1000)
Countcaesareans
```

```
> Countcaesareans = exp(2.132)*exp(.0004406 *1000)
> Countcaesareans
[1] 13.09984
```

Providing a value of 13.009 expected caesareans for 1000 births.

Rather than typing in the values, we could have made use of the fact that the object produced from the *glm()* function, *model1* in this instance, contains the coefficient values which we can extract using the dollar ($) sign:

```
exp(model1$coefficients[1]) * exp(model1$coefficients[2]*1000)
```

```
> exp(model1$coefficients[1])*exp(model1$coefficients[2]*1000)
(Intercept)
   13.10035
```

It is important to remember that because we are working with a log linear scale the increase in predicted caesareans is not constant. For example for 1000 and 1001 births we have an increase of 0.0057 caesareans but for 10 to 11 births we have an increase in 0.0037 approximately predicted caesareans.

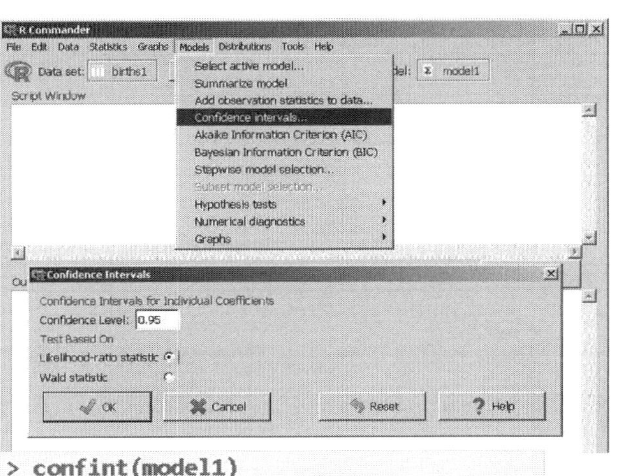

We can also obtain confidence intervals for the coefficients either by using the R Commander menu option shown opposite or by typing into the *Script* window:

```
confint(model1)
```

In a subsequent example we will make these values more interpretable by applying the *exp()* function to them.

```
> confint(model1)
Waiting for profiling to be done...
                   2.5 %       97.5 %
(Intercept) 1.9278030427 2.3269703811
births      0.0003344146 0.0005460732
```

It is possible to access if overdispersion is a problem by comparing the Residual deviance value to its degrees of freedom. From the above we can see that the residual deviance is 36 and its df=18 giving a ratio of 36/18=2. There appears to be no consensus as to the exact value that indicates if overdispersion is a problem, what we know is that a value near 1 suggests there isn't one. If we decide that a value of 2 is problematic there are four methods available to take into account this overdispersion problem:

- Use the quasi-likelihood approach, which means changing the variance from 1 to a value estimated from the data called the 'dispersion parameter' by using *family=quasiPoisson()* in R.
- Use sandwich estimators for the standard errors by using the *sandwich* package in R.
- Use the negative binomial distribution instead (using the *glm.nb()* function in the R package *MASS*) .
- Use a resampling technique (covered in the *Resampling* chapter)

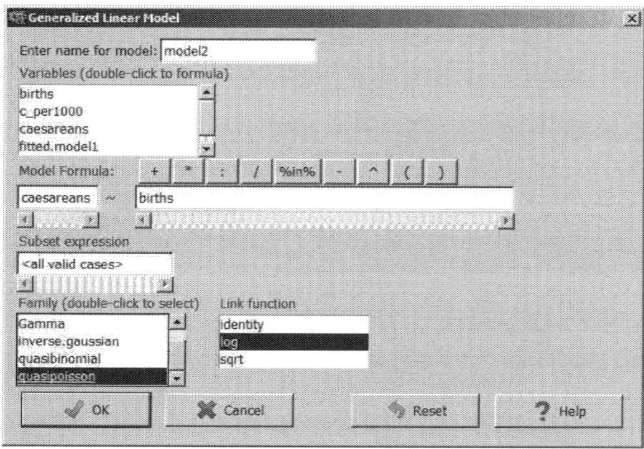

It is important to note that we could also have a large Residual deviance compared to its degrees of freedom if the model is mis-specified, such as having left out important input variables. If this is the case, none of the above methods will reduce overdispersion.

We will now use the first three techniques in our example and compare the results.

61.1.3 Quasi-likelihood approach

The screen shot opposite shows how to repeat the analysis using the quasi –likelihood approach in R Commander, the equivalent R code is:

```
model2 <- glm(caesareans ~ births, family=quasiPoisson(log), data=births1)
summary(model2)
```

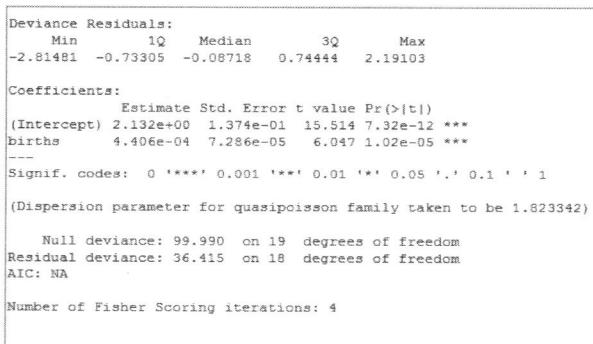

```
Deviance Residuals:
     Min        1Q    Median        3Q       Max
-2.81481  -0.73305  -0.08718   0.74444   2.19103

Coefficients:
             Estimate Std. Error t value Pr(>|t|)
(Intercept) 2.132e+00  1.374e-01  15.514 7.32e-12 ***
births      4.406e-04  7.286e-05   6.047 1.02e-05 ***
---
Signif. codes:  0 '***' 0.001 '**' 0.01 '*' 0.05 '.' 0.1 ' ' 1

(Dispersion parameter for quasipoisson family taken to be 1.823342)

    Null deviance: 99.990  on 19  degrees of freedom
Residual deviance: 36.415  on 18  degrees of freedom
AIC: NA

Number of Fisher Scoring iterations: 4
```

In *Model2* the standard errors of the estimates have increased as the variance has now increased by a factor of 1.8. However, this change is too small to affect the associated p-values which still are less than 0.001. The *t* values are simply the estimates divided by their standard errors (i.e. 2.132/01374 = 15.514).

61.1.4 Sandwich estimation of standard errors

Sandwich estimators are used to take the 'noise' from the data into account. If you want to know more about them, I recommend you read the excellent introduction in SAS2. In R you need to use the appropriately named 'sandwich' package to produce them as unfortunately R Commander does not provide the facility.

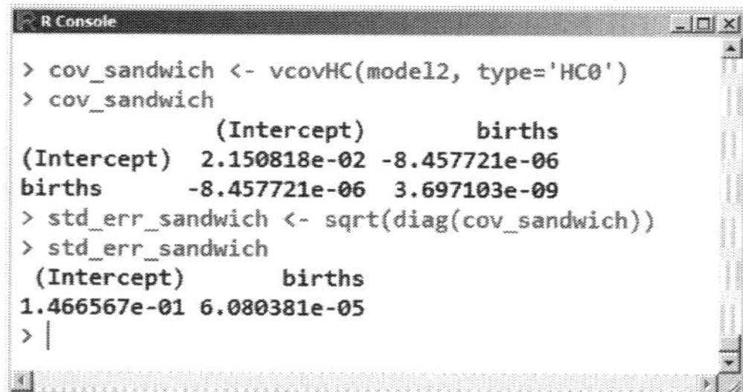

The code below, including the commented first line, allows you to install the sandwich package and produce sandwich estimates of the standard errors.

```
# install.packages("sandwich")
library(sandwich)
cov_sandwich <- vcovHC(model2, type='HC0')
cov_sandwich
std_err_sandwich <- sqrt(diag(cov_sandwich))
std_err_sandwich
```

install package if necessary by removing the #

load it

the standard errors are the square root of the diagonals of the cov_sandwich object

```
> 4.406e-4
[1] 0.0004406
> 4.406e-4 / 7.286e-5
[1] 6.047214
> 4.406e-4 / 6.0803e-5
[1] 7.246353
```

The above code produces robust sandwich estimates of the intercept SE of 0.1466567 and a births SE estimate of 0.0000608381. We can see that the new estimates change the *t* value from 6.04 to 7.2, making the associated p-value more statistically significant. At the end of this section we will compare these values with the others produced.

We can enhance the R code to produce a nicely formatted table of the results.

```
result_table <- cbind( 'Estimate' = coef(model2),
        'Robust SE' = std_err_sandwich,
        'Pr(>|z|)' = 2* pnorm(abs(coef(model2)/std_err_sandwich), lower.tail=FALSE),
        LL = coef(model2) - 1.96 * std_err_sandwich,
        UL = coef(model2) + 1.96 * std_err_sandwich
                    )
result_table
```

```
> result_table <- cbind( 'Estimate' = coef(model2),
+ 'Robust SE' = std_err_sandwich,
+ 'Pr(>|z|)' = 2* pnorm(abs(coef(model2)/std_err_sandwich), lower.tail=FALSE),
+ LL = coef(model2) - 1.96 * std_err_sandwich,
+ UL = coef(model2) + 1.96 * std_err_sandwich
+ )
> result_table
               Estimate    Robust SE     Pr(>|z|)           LL           UL
(Intercept) 2.1320930114 1.466567e-01 6.960943e-48 1.8446459091 2.4195401137
births      0.0004405458 6.080381e-05 4.312754e-13 0.0003213703 0.0005597213
> |
```

You can also use the *round(x, digits = 0)* function to reduce the number of decimal places reported. Reported opposite are the results to both three and five decimal places.

```
> round(result_table, 3)
            Estimate Robust SE Pr(>|z|)    LL    UL
(Intercept)    2.132     0.147        0 1.845 2.420
births         0.000     0.000        0 0.000 0.001
> round(result_table, 5)
            Estimate Robust SE Pr(>|z|)      LL      UL
(Intercept)  2.13209   0.14666        0 1.84465 2.41954
births       0.00044   0.00006        0 0.00032 0.00056
> |
```

61.1.5 Negative binomial distribution

The third method of managing overdispersion involves using the negative binomial (NB) distribution in place of the Poisson. This is achieved in R by using the *glm.nb()* function in the *MASS* package.

The negative binomial (NB) distribution is defined by two values (parameters) instead of the one for the Poisson distribution. Besides the mean it requires a clumping parameter (θ theta) which defines the degree to which the values cluster together. This value is usually calculated from the sample data as is the mean (nu=μ).

```
R Console
> model_nb1<- glm.nb(caesareans ~ births, data=births1)
> summary(model_nb1)

Call:
glm.nb(formula = caesareans ~ births, data = births1, init.theta = 24.87058596,
    link = log)

Deviance Residuals:
     Min        1Q    Median        3Q       Max
-2.46104  -0.63940  -0.05668   0.60280   1.76987

Coefficients:
             Estimate Std. Error z value Pr(>|z|)
(Intercept) 2.096e+00  1.245e-01  16.842  < 2e-16 ***
births      4.683e-04  7.247e-05   6.463 1.03e-10 ***
---
Signif. codes:  0 '***' 0.001 '**' 0.01 '*' 0.05 '.' 0.1 ' ' 1

(Dispersion parameter for Negative Binomial(24.8706) family taken to be 1)

    Null deviance: 66.522  on 19  degrees of freedom
Residual deviance: 26.172  on 18  degrees of freedom
AIC: 128.01

Number of Fisher Scoring iterations: 1


              Theta:  24.9
          Std. Err.:  28.0

 2 x log-likelihood:  -122.014
```

```
install.packages("MASS")

library(MASS)

model_nb1<- glm.nb(caesareans ~ births,
data=births1)

summary(model_nb1)
```

How theta and its standard error (*Std.Err*) are calculated is described at the end of this chapter.

On the next page we compare the results from the various methods used to account for overdispersion along with the original model.

Distribution	Model name	Intercept Estimate s.e. sig level	Births Estimate s.e. sig level	Hospital Type Estimate s.e. sig level	Births*hospital Estimate s.e. sig level	AIC Smaller value = better fit	BIC Smaller value = better fit	Chi-squared p-value higher value = better fit (max=1)
Poisson	model1	2.132 1.018 -E01 ***	4.406 –E04 5.395 –E05 ***			127	129.16	0.0062
QuasiPoisson	model2	2.132 1.37 -E01 ***	4.406 –E04 7.28 –E05 ***			NA	NA	0.0062
Negative binomial	model_nb1	2.096 .245 -E01 ***	4.683 –E04 7.247 –E05 ***			128.01	131.00	0.095
Sandwich	(used model2)	2.136 1.46 -E01 ***	4.405 –E04 6.08 –E05 ***					-
QuasiPoisson	model3	2.35 2.54 –E01 ***	3.26 –E04 6.14 -E05 ***	1.045 2.78 -E01 **		NA	NA	0.3863
QuasiPoisson	model4	.777 .458 0.109	0.004 0.002 0.092	1.62 0.47 **	-.004 .0027 p-val=0.1142	NA	NA	0.4916
Poisson	model4_Poisson	.777 .4800 0.105	0.004 0.002 0.087	1.62 0.49 **	-.004 .0028 p-val=0.1105	110	114	0.4916

Taking the first four models above the thick line, we can see from the comparison table that there is very little difference in the Standard Errors (SE) produced from any method and the resultant p-values. However, this is not always the case. In addition, the coefficient values are identical for the Poisson, quasiPoisson and sandwich techniques the only one producing a slightly different value being the Negative binomial.

61.1.6 Adding a second Nominal input variable

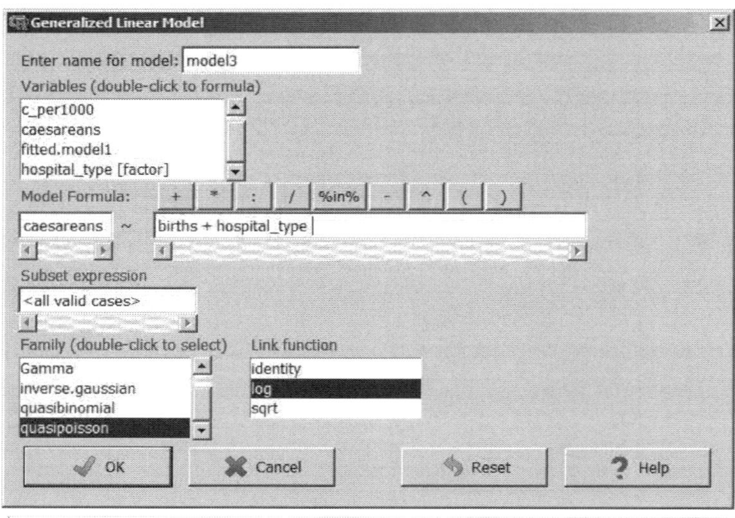

We will now add a second input variable, the hospital type. Given the very small number in the private sector group this is more an exercise of the mechanics of the process rather than expecting any important outcome.

I have created a new model called *model3*, using either the R Commander dialog box, or the equivalent R code:

```
model3 <- glm(caesareans ~ births +
hospital_type, family=quasiPoisson(log),
data=births1)

summary(model3)
```

```
    Min       1Q   Median       3Q      Max
-2.3270  -0.6121  -0.0899   0.5398   1.6626

Coefficients:
                         Estimate Std. Error t value Pr(>|t|)
(Intercept)             1.351e+00  2.548e-01   5.302 5.85e-05 ***
births                  3.261e-04  6.146e-05   5.306 5.81e-05 ***
hospital_type[T.public] 1.045e+00  2.781e-01   3.759  0.00157 **
---
Signif. codes:  0 '***' 0.001 '**' 0.01 '*' 0.05 '.' 0.1 ' ' 1

(Dispersion parameter for quasipoisson family taken to be 1.038096)

    Null deviance: 99.990  on 19  degrees of freedom
Residual deviance: 18.039  on 17  degrees of freedom
AIC: NA

Number of Fisher Scoring iterations: 4
```

Notice now that the overdispersion problem has disappeared with a Residual deviance of 18 (df=17) representing a near perfect ratio of 1.

This would indicate that the overdisperson problem was due to the omission of an import input variable rather than any other reason. We can check this by rerunning the analysis with the Poisson link function.

```
Call:
glm(formula = caesareans ~ births + hospital_type, family = poisson(log),
    data = births1)

Deviance Residuals:
    Min       1Q   Median       3Q      Max
-2.3270  -0.6121  -0.0899   0.5398   1.6626

Coefficients:
                          Estimate Std. Error z value Pr(>|z|)
(Intercept)              1.351e+00  2.501e-01   5.402 6.58e-08 ***
births                   3.261e-04  6.032e-05   5.406 6.45e-08 ***
hospital_type[T.public] 1.045e+00  2.729e-01   3.830 0.000128 ***
---
Signif. codes:  0 '***' 0.001 '**' 0.01 '*' 0.05 '.' 0.1 ' ' 1

(Dispersion parameter for poisson family taken to be 1)

    Null deviance: 99.990  on 19  degrees of freedom
Residual deviance: 18.039  on 17  degrees of freedom
AIC: 110.8
```

In the screen shot opposite, I repeated the analysis using the Poisson distribution. An important finding is that the Residual deviance remains the same. To reiterate, this indicates that the overdispersion problem in this instance was due to mis-specification of the model, that is the omission of an important input rather than the non-Poisson distribution of the caesarean counts.

Also, notice that by using the Poisson distribution the significance level for the hospital variable has changed from two stars to three.

We could use the same techniques described in the previous model to produce confidence intervals. However, I will now turn our attention to the problem of measuring model fit.

61.1.7 Overall fit - Deviance

In standard regression we have various sums of squares, the total, divided into that due to the model (☺) and that left over (☹). In Poisson regression we have a similar situation except this time it is called the deviance which is analogous to the residual sum of squares. We have the null deviance (No predictors=baseline deviance), and the residual deviance (which is the difference between the deviance of the current model and an ideal one where all the predicted values are equal to those observed). For large samples the distribution of any deviance is approximately chi-squared with $n-p$ degrees of freedom, ($n=$ number of observations; $p=$ number of parameters). Thus, the deviance is used to directly test the goodness of fit of the current model to another one.

In R this is achieved by using the residual deviance (*model1$deviance*) along with its degrees of freedom (*model1$df.residual*).

```
model1fit<- pchisq(model1$deviance, model1$df.residual, lower.tail=FALSE)

model1fit
```

```
> model1fit<- pchisq(model1$deviance, model1$df.residual, lower.tail=FALSE)
> model1fit
[1] 0.006240737
```

This very small value indicates a very poor fit.

```
model3fit<- pchisq(model3$deviance, model3$df.residual, lower.tail=FALSE)
model3fit
```

```
> model3fit<- pchisq(model3$deviance, model3$df.residual, lower.tail=FALSE)
> model3fit
[1] 0.3863719
```

A much better fit. Two other measures of overall fit are the AIC and the BIC. However, these are estimated only when the Poisson link function is used. Both measures allow the comparison of models where a smaller value indicates a better fit. Let us now look at this aspect in more detail.

61.1.8 Comparing models

There are two main approaches to model comparison. The first is a qualitative approach, where the reviewer just inspects the values obtained from each model, such as the BIC and AIC measures where smaller indicates a better fit. The second is to make use of the fact that for the same dataset, and models, which had defined but different degrees of freedom, there is a known approximate distribution so we can obtain a p-value indicating the relative fit. According to the R help for *anova.glm()* obtained by typing into the *R Console* window *?anova.glm()*:

"For models with known dispersion (e.g., binomial and Poisson fits) the chi-squared test is most appropriate, and for those with dispersion estimated by moments (e.g. gaussian, quasibinomial and quasiPoisson fits) the F test is most appropriate. Mallows' Cp statistic is the residual deviance plus twice the estimate of σ^2 times the residual degrees of freedom, which is closely related to AIC (and a multiple of it if the dispersion is known). You can also choose "LRT" and "Rao" for likelihood ratio tests and Rao's efficient score test."

I have used this information to produce the table below:

		Anova() options	
Link function	**Model comparison**	**R code**	**BIC / AIC**
Poisson + binomial	chi-squared test	anova(model1, model2, test='Chisq')	√
QuasiPoisson	F test	anova(model, model2, test='F') ?	
Negative binomial	likelihood ratio test	anova(model, model2, test=' Chisq ')	√

To use the *anova.glm()* function you simply type *anova(glm_object1, glm_object2... test="Chisq")*. Because you are passing the *anova()* function a *glm* object it knows how to handle it. You can pass the *anova()* function a single *glm* object to obtain the deviance along with its p-value for each additional parameter.

While the R help extract above mentions the use of the F test rather than the chi-squared for the quasiPoisson link function, most examples I have come across seem to use the chi-squared and Maindonald & Braun, 2010 p.267 also recommend it, particularly for comparing factor levels on a pairwise basis. In the next section concerning interaction terms, we will see that the two methods can produce very different p-values.

To compare the fit of *model1* (Poisson) to *model3* (quasiPoisson) we can use:

```
anova(model1, model3, test="Chisq")  # note capital C
```

```
> anova(model1, model3, test='Chisq')
Analysis of Deviance Table

Model 1: caesareans ~ births
Model 2: caesareans ~ births + hospital_type
  Resid. Df Resid. Dev Df Deviance  Pr(>Chi)
1        18     36.415
2        17     18.039  1   18.376 2.585e-05 ***
---
Signif. codes:  0 '***' 0.001 '**' 0.01 '*' 0.05 '.' 0.1 ' ' 1
```

In contrast to the single model chi-squared assessment where a high p-value indicated a good fit in this situation, we now want a small (significant) p-value to indicate one models superiority of the other. For example, opposite we have a p-value of 0.00002585 indicating that the second model, *model3*, is a better fit than *model1*.

61.1.9 Interaction terms

Adding additional inputs assumes they produce parallel regression lines unless we add an interaction term. The idea is that if the interaction term is significant at say the p-value=0.05 then we accept there is a significant interaction and the lines are not parallel. To indicate an interaction term in R or R Commander we use the asterisk sign *. As with the other models we can either use the R Commander dialog box or type directly into the R Console window.

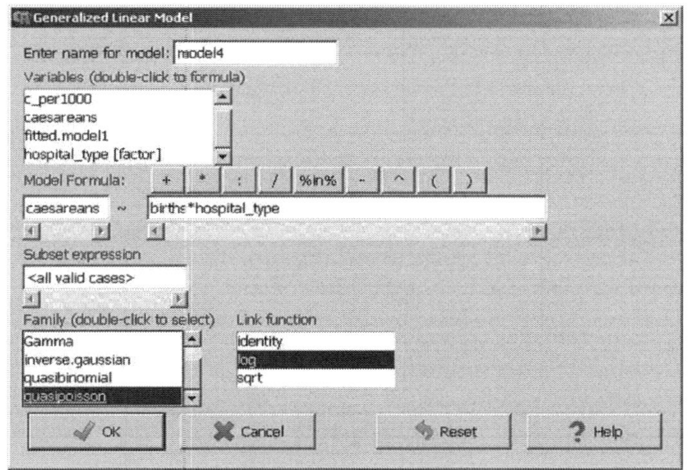


```
model4 <- glm(caesareans ~ births*hospital_type,
family=quasiPoisson(log), data=births1)
```

Note that we do not need to include the two separate inputs as well as the interaction because the asterisk '*' is part of a model specification, indicated by the ~ sign, because of this the * is interpreted differently. The output opposite demonstrates this where we have both the separate effects and the interactive effects of both inputs modelled.

```
Call:
glm(formula = caesareans ~ births * hospital_type, family = quasipoisson(log),
    data = births1)

Deviance Residuals:
     Min        1Q    Median        3Q       Max
-2.33547  -0.56560  -0.08887   0.53369   1.58787

Coefficients:
                              Estimate Std. Error t value Pr(>|t|)
(Intercept)                   0.777965   0.458466   1.697  0.10908
births                        0.004827   0.002695   1.791  0.09217 .
hospital_type[T.public]       1.621577   0.471440   3.440  0.00337 **
births:hospital_type[T.public] -0.004503  0.002695  -1.671  0.11421
---
Signif. codes:  0 '***' 0.001 '**' 0.01 '*' 0.05 '.' 0.1 ' ' 1

(Dispersion parameter for quasipoisson family taken to be 0.9119925)

    Null deviance: 99.990  on 19  degrees of freedom
Residual deviance: 15.454  on 16  degrees of freedom
AIC: NA
```

This is a strange result! We now have only one significant predictor *hospital type* when the interaction between births and hospital type is modelled but is itself insignificant. It is also interesting to note that using the quasiPoisson link function instead of the Poisson produces very similar standard errors now that we have included the interaction.

```
> model4_poisson <- glm(caesareans ~ births*hospital_type, family=poisson(log), data=births1)
> summary(model4_poisson)

Call:
glm(formula = caesareans ~ births * hospital_type, family = poisson(log),
    data = births1)

Deviance Residuals:
     Min        1Q    Median        3Q       Max
-2.33547  -0.56560  -0.08887   0.53369   1.58787

Coefficients:
                              Estimate Std. Error z value Pr(>|z|)
(Intercept)                   0.777965   0.480077   1.620  0.10513
births                        0.004827   0.002822   1.711  0.08713 .
hospital_type[T.public]       1.621577   0.493663   3.285  0.00102 **
births:hospital_type[T.public] -0.004503  0.002822  -1.596  0.11059
---
Signif. codes:  0 '***' 0.001 '**' 0.01 '*' 0.05 '.' 0.1 ' ' 1

(Dispersion parameter for poisson family taken to be 1)

    Null deviance: 99.990  on 19  degrees of freedom
Residual deviance: 15.454  on 16  degrees of freedom
AIC: 110.22

Number of Fisher Scoring iterations: 4

> confint(model4_poisson)
Waiting for profiling to be done...
                                      2.5 %       97.5 %
(Intercept)                    -0.2886297253 1.6189729416
births                         -0.0006682773 0.0105789820
hospital_type[T.public]         0.7472615442 2.7089229335
births:hospital_type[T.public] -0.0102561446 0.0009935276
> |
```

I feel that the non significance of the interaction is due to the small number of values it has, 4. We will return to this on the next page.

We can also produce confidence intervals in the usual way using either the R Commander menu option or the R code:

```
confint(model4_Poisson)
```

```
> anova(model3, model4, test="Chisq")
Analysis of Deviance Table

Model 1: caesareans ~ births + hospital_type
Model 2: caesareans ~ births * hospital_type
  Resid. Df Resid. Dev Df Deviance Pr(>Chi)
1        17     18.039
2        16     15.454  1    2.585  0.09226 .
---
Signif. codes:  0 '***' 0.001 '**' 0.01 '*' 0.05 '.' 0.1 ' ' 1
> anova(model3, model4, test="F")
Analysis of Deviance Table

Model 1: caesareans ~ births + hospital_type
Model 2: caesareans ~ births * hospital_type
  Resid. Df Resid. Dev Df Deviance      F Pr(>F)
1        17     18.039
2        16     15.454  1    2.585 2.8345 0.1117
>  anova(model3, model4, test="LRT") # give chi square
Analysis of Deviance Table

Model 1: caesareans ~ births + hospital_type
Model 2: caesareans ~ births * hospital_type
  Resid. Df Resid. Dev Df Deviance Pr(>Chi)
1        17     18.039
2        16     15.454  1    2.585  0.09226 .
---
```

The question we need to consider is, is the inclusion of this additional complexity using the interaction term worth it? Such a question can be answered by using the *anova()* model comparison function again, this time comparing *model3* and *model4*. I will also use three of the choices available for the test option:

```
anova(model3, model4, test="Chisq")

anova(model3, model4, test="F")

anova(model3, model4, test="LRT") # give chi-squared
```

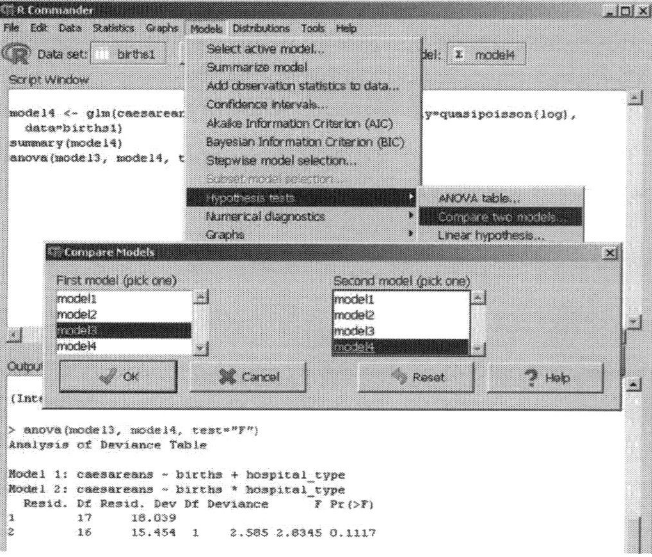

The result shown opposite indicates that both the *LRT* and *Chisq* choices produce a chi-squared test value, in contrast the *F* option carries out an F test option. While both p-values are not highly significant, they are rather different; 0.092 and 0.1117. Personally, I would use the chi-squared test and repeat it with both models based on the Poisson link function in this instance.

Unfortunately while we can do a similar thing with the R Commander menu option **Models-> Hypothesis test ->Compare two models** it only carries out an F test rather than a chi-squared test, so it is not much use.

Considering the negative binomial distribution, passing such a model to the *anova()* function automatically carries out a chi-squared likelihood ratio test regardless of what test we request. The following code produces two new models using the negative binomial distribution and then compares them:

```
model_nb2<- glm.nb(caesareans ~ births + hospital_type,maxit= 2000,  data=births1)
summary(model_nb2)
model_nb3<- glm.nb(caesareans ~ births * hospital_type, data=births1)
summary(model_nb3)
```

```
> anova(model_nb2, model_nb3, test='LRT')
Warning in anova.negbin(model_nb2, model_nb3, test = "LRT") :
  only Chi-squared LR tests are implemented
Likelihood ratio tests of Negative Binomial Models

Response: caesareans
                  Model        theta Resid. df    2 x log-lik.   Test    df LR stat.   Pr(Chi)
1 births + hospital_type 3.350407e+10        17       -104.7981
2 births * hospital_type 2.489804e+05        16       -102.2177 1 vs 2     1 2.580439 0.1081917
> anova(model_nb2, model_nb3, test='Chisq')
Likelihood ratio tests of Negative Binomial Models

Response: caesareans
                  Model        theta Resid. df    2 x log-lik.   Test    df LR stat.   Pr(Chi)
1 births + hospital_type 3.350407e+10        17       -104.7981
2 births * hospital_type 2.489804e+05        16       -102.2177 1 vs 2     1 2.580439 0.1081917
> anova(model_nb2, model_nb3, test='F')
Warning in anova.negbin(model_nb2, model_nb3, test = "F") :
  only Chi-squared LR tests are implemented
Likelihood ratio tests of Negative Binomial Models

Response: caesareans
                  Model        theta Resid. df    2 x log-lik.   Test    df LR stat.   Pr(Chi)
1 births + hospital_type 3.350407e+10        17       -104.7981
2 births * hospital_type 2.489804e+05        16       -102.2177 1 vs 2     1 2.580439 0.1081917
> |
```

```
#compare them
anova(model_nb2, model_nb3,
test='LRT') # ignores and does Chisq

anova(model_nb2, model_nb3,
test='Chisq')

anova(model_nb2, model_nb3, test='F')
# ignores and does Chisq
```

We note here that the p-value is equal to 0.108 which is very close to the F test p-value of 0.1117. We conclude that the addition of the interaction term adds nothing to the model and can be dropped.

61.1.10 Drawing the points using R code

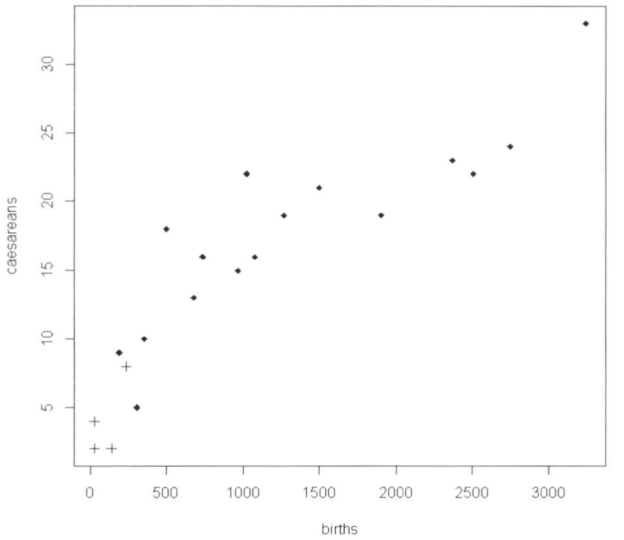

At the start of this example we used R Commander to produce a useful plot of the *log(caesareans)* against births and also hospital type for the observed data. It would also be very useful to see what the predicted values look like from the various models in comparison to the raw data - an exercise that is often a sobering experience! We start the process using the code below to produce a plot similar to the one we created earlier in R Commander.

```
attach(births1)
# stops needing to type dataframe$name format

names(births1)  #check the names of the variables

# [1] "births" "hospital_type" "caesareans" "c_per1000"
# [5] "fitted_model1"

plot(births, caesareans, type='n')

yvals<- split(caesareans, hospital_type)

xvals<- split(births, hospital_type)

plot(births, caesareans, type='n') # first create an empty plot

points(xvals[[1]], yvals[[1]],pch=3)
                         # double[[]] used as referencing elements of a list

points(xvals[[2]], yvals[[2]],pch=18  #points(x, y)
```

There are many things to note in the above code. We start by creating a blank plot by using the *type='n'* option.

```
> levels(hospital_type)
[1] "private" "public"
```

```
> yvals
$private
[1] 4 2 2 8

$public
 [1] 16 15 23  5 13 19 33 19 10 16 22 22 18 21 24  9
```

To plot the predicted values we need to split the data into two reflecting the values for the hospital_type variable which we know has two levels, this is achieved using the *split()* function. We first split the values in the caesareans variable into two lists, the first one consisting of the private hospitals and call this *yvals* and do the same thing for the births variable. Because the *split()* function produces lists rather than dataframes to reference a particular element in each list we need to use double square brackets, for example *[[1]]* refers to the first element and *yvals[[1]]* refers to the first element (called *private*) of the *yvals* list.

Finally we use the *points(x,y)* function along with the y and y co-ordinates we have defined to create the plot.

There are many ways to produce the above effect, and the chapter entitled, *Densityplots for subgroups defined by factor levels* provides a different strategy.

61.1.11 Drawing the regression curves.

We are now in a position to add the predicted curves to the scatterplot. To graph the predicted curves we need to consider both the x and y values. Let us firstly consider the values for the y axis.

Crawley, 2005 p.57, makes the sensible recommendation that you need at least 100 distinct y values to produce a reasonable curve. We know from the summary statistics that the maximum number of births for a hospital is 3246, so we need a series ranging from zero to around 3246, say 3300. Dividing this range into 100 intervals; 3300/100 = 33 which is the approximate increment required for each y step. To summarize, the series needs to start at zero and go to 3300 in steps of around 30. To produce such a series of values in R we use the command *seq(0,3300,30)* placing these values in an object called *xseq*.

To produce the y values, the output from the equation, we use the *predict()* R function. For the *predict()* function we need to specify the model we wish it to use, which in this instance is *model3*. We need also to specify the list of values for it to use with each of the input variables. For the births variable we specify *xseq*, and for the *hospital_type* variable we create an object the same length as *xseq* found by using the *length(xseq)* function. We then repeat the level name of the *hospital_type* we wish to use and place it in an object called *hospital_level*. We recreate the *hospital_level* twice, each time with the different *hospital level,* as we want to create a set of predicted values for both private and public hospitals.

The *predict()* function can return the value in either the original units used by the input variables (i.e. raw counts here) or the log value when we use the Poisson link function. To specify that we need raw counts, to match the y axis on our scatterplot, we use the option *type="response"*.

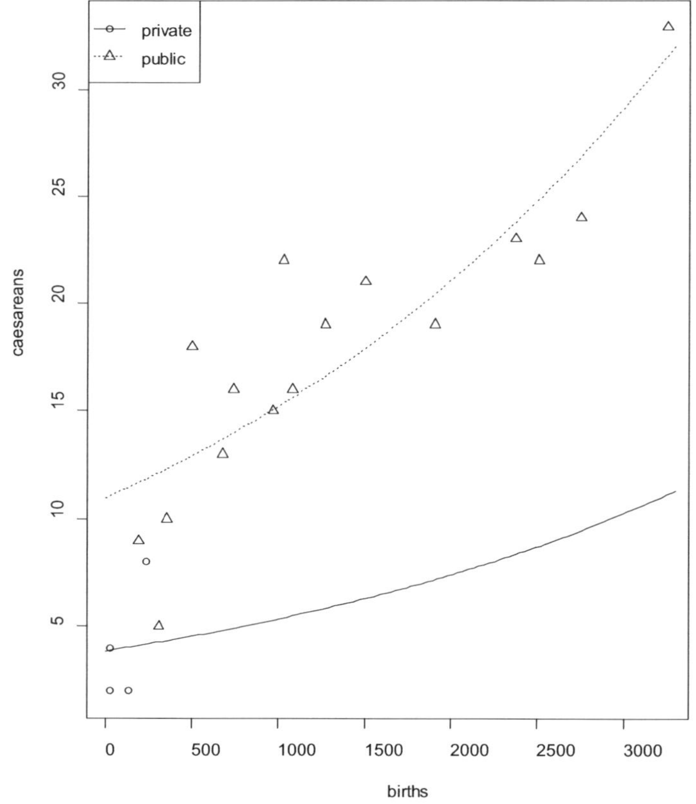

Finally we use the *lines()* function to draw the lines, making the first one solid with the option *lty=1* and the second one dashed with *lty=3*.

```
xseq<- seq(0,3300,30)

length(xseq) # returns  111

levels(hospital_type)
# returns [1] "private" "public" to check name

hospital_level<- factor( rep("private",
length(xseq)) )
  yseq <- predict(model3, data.frame(births =
xseq,
  hospital_type =hospital_level),
type="response")
  lines(xseq, yseq, lty=1)

  hospital_level<- factor( rep("public",
length(xseq)) )
  yseq <- predict(model3, data.frame(births =
xseq, hospital_type =hospital_level),
  type="response")
  lines(xseq, yseq, lty=3)
```

The graph also contains a legend which is achieved with the following code:

```
legend(x= "topleft",                    # places a legend at the appropriate place
 lty=c(1,3),                            # gives the legend appropriate symbols (lines)
 levels(hospital_type),                 # puts text in the legend
 pch= c(1,2),                           # gives the legend appropriate symbols (marker)
 lwd=c(1,1),col=c("black","black"))     # gives the legend lines the correct color and width
```

The above graph demonstrates very clearly the danger of over extrapolation as we only have 4 points for the private hospitals which are also at the very low level of the births axis. Comparing this graph to the one produced by R Commander earlier (where we allowed the gradient of the lines to vary and the lines stopped with the last datapoint) we can see that the regression line for the private hospitals fitted the 4 points much better. However, there is clearly a problem with making any decision with such a small sample.

In conclusion, we can say that this example has allowed us to apply some interesting techniques but the actual results lack any usefulness for the comparison we set out to consider. Probably the best we can do is to omit the private hospital data from our model for the time being until further private hospital data is collected.

61.2 Rates modelled by offsets

Data Editor

	Cases	Town	Age	Population	age_mid
1	1	msp	15-24	172675	20
2	16	msp	25-34	123065	30
3	30	msp	35-44	96216	40
4	71	msp	45-54	92051	50
5	102	msp	55-64	72159	60
6	130	msp	65-74	54722	70
7	133	msp	75-84	32185	80
8	40	msp	85+	8328	90
9	4	dfw	15-24	181343	20
10	38	dfw	25-34	146207	30
11	119	dfw	35-44	121374	40
12	221	dfw	45-54	111353	50
13	259	dfw	55-64	83004	60
14	310	dfw	65-74	55932	70
15	226	dfw	75-84	29007	80
16	65	dfw	85+	7538	90

Often rather than considering the raw counts we consider these counts in relation to some denominator such as population size. I have taken as an example of this the incidence of non-melanoma skin cancers among women in Minneapolis-St Paul, Minnesota, (msp) and Dallas-Fort Worth, Texas (dfw). The data has been reported many times and is available to download from Smyth, 2011 at the Australasian Data and Story Library (OzDASL).

Before looking at the data, one would expect sun exposure to be greater in Dallas than in Minnesota and hence cancer, and to see the incidence rise with age.

Type the following into the *R Console* window to download and load the data into R:

```
load(url("http://www.robin-beaumont.co.uk/virtualclassroom/book2data/melanoma.rdata"))
```

Pasting the code below into the *R Console* window gives the *mydataframe* dataframe focus, displays the names of the columns and displays the data.

```
ls() # list the objects returns [1] "mydataframe"
attach(mydataframe)
names(mydataframe)
mydataframe
```

within the melanoma R object there is a dataframe called *mydataframe*

The first thing we need to do is create the calculated variable that will represent the rate for each population, as usual, we can do this in R Commander or directly using R code.

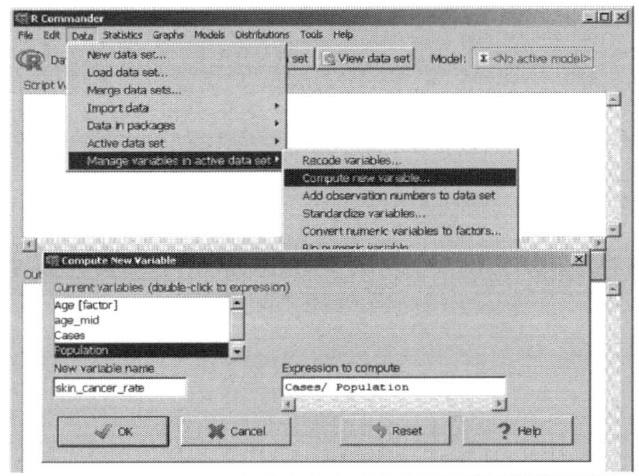

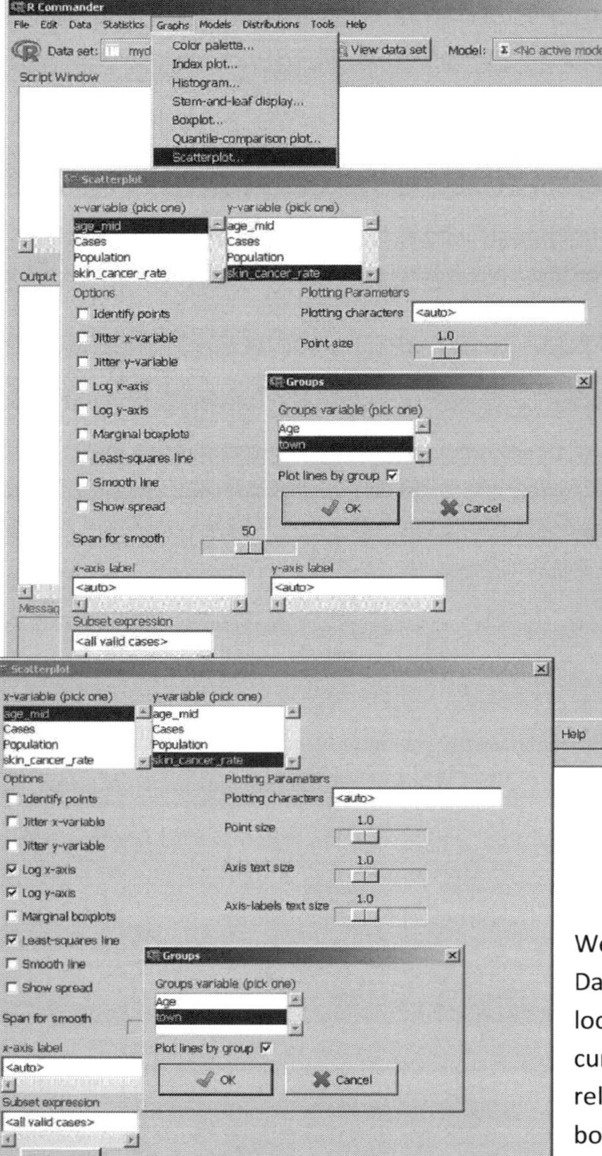


```
mydataframe$skin_cancer_rate <- Cases / Population
```

We should also produce summary statistics for each variable to give us a taste for the data. We can see that the mean number of cases across city populations is 110.3

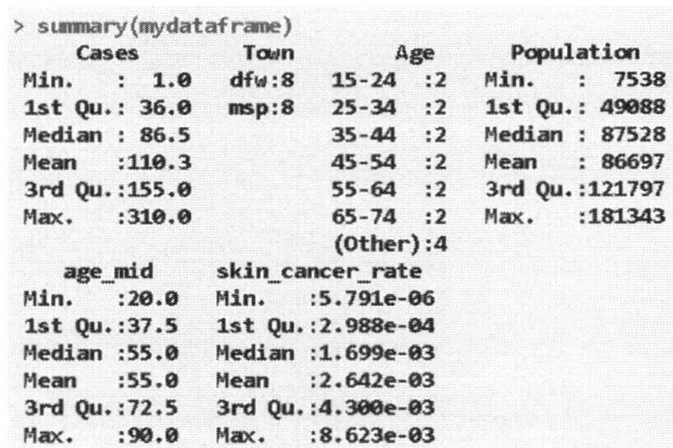

```
> summary(mydataframe)
     Cases          Town         Age       Population
 Min.   :  1.0   dfw:8   15-24  :2   Min.   :  7538
 1st Qu.: 36.0   msp:8   25-34  :2   1st Qu.: 49088
 Median : 86.5           35-44  :2   Median : 87528
 Mean   :110.3           45-54  :2   Mean   : 86697
 3rd Qu.:155.0           55-64  :2   3rd Qu.:121797
 Max.   :310.0           65-74  :2   Max.   :181343
                         (Other):4
    age_mid     skin_cancer_rate
 Min.   :20.0   Min.   :5.791e-06
 1st Qu.:37.5   1st Qu.:2.988e-04
 Median :55.0   Median :1.699e-03
 Mean   :55.0   Mean   :2.642e-03
 3rd Qu.:72.5   3rd Qu.:4.300e-03
 Max.   :90.0   Max.   :8.623e-03
```

The next stage is to get a feel for the relationship between the outcome (number of non-melanoma skin cancers) and input(s). Such a graph is produced in R Commander with the options shown opposite.

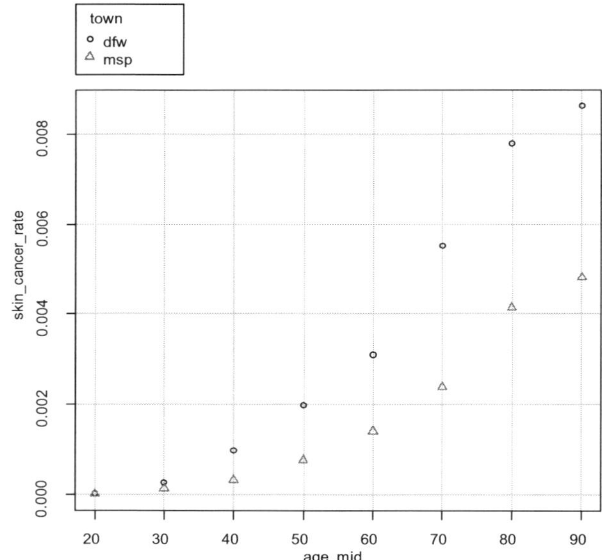

We can see that the rate of non-melanoma skin cancers is higher for Dallas (*dfw*) compared to Minneapolis (*msp*) as we suspected. Also looking at the shape over the age range the rate appears to follow a curved projectory - where such a shape suggests that the relationship can be converted to a straight one by log transforming both the x and y values. We can graph this by selecting the options shown opposite.

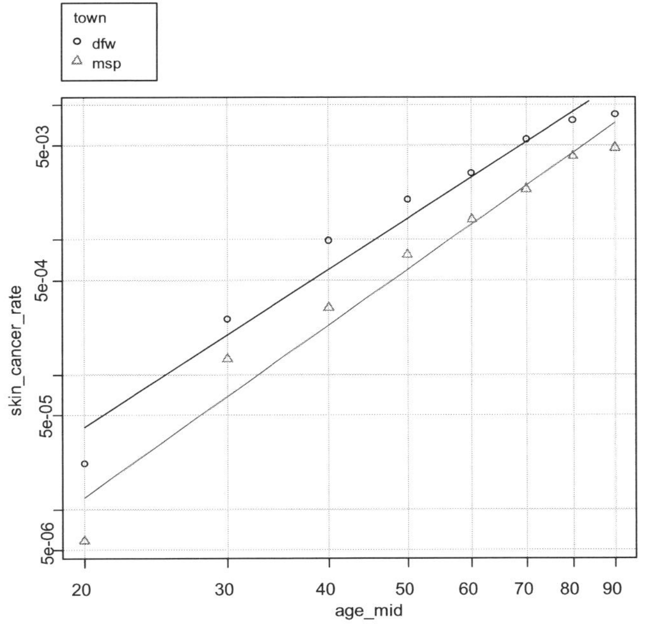

Graphing the log log relationship (as we have transformed both x and y values here), certainly makes it appear increasingly linear (shown opposite).

Mathematically considering a model with both age and city, as dummy input variables looks like this:

```
log(cases/population)        = Intercept + b2(city) +
b1log((age_mid)

log(cases) - log(population) = Intercept + b2(city) +
b1log((age_mid)

log(cases) = Intercept + b2(city) + b1log((age_mid) +
log(population)
```

The term *log(population)* is an adjustment for each outcome (log cases) and might be different for each, as it is for each city in this situation. Such a term is called an *offset* and in R we specify the offset using the term, *offset(variable)* in the model formula.

To create the model in R Commander we set the dialog box up as show opposite. There is no option to specify that a particular input is an offset so we just type it in, *offset(log(Population))*. I have named the model *model1*.

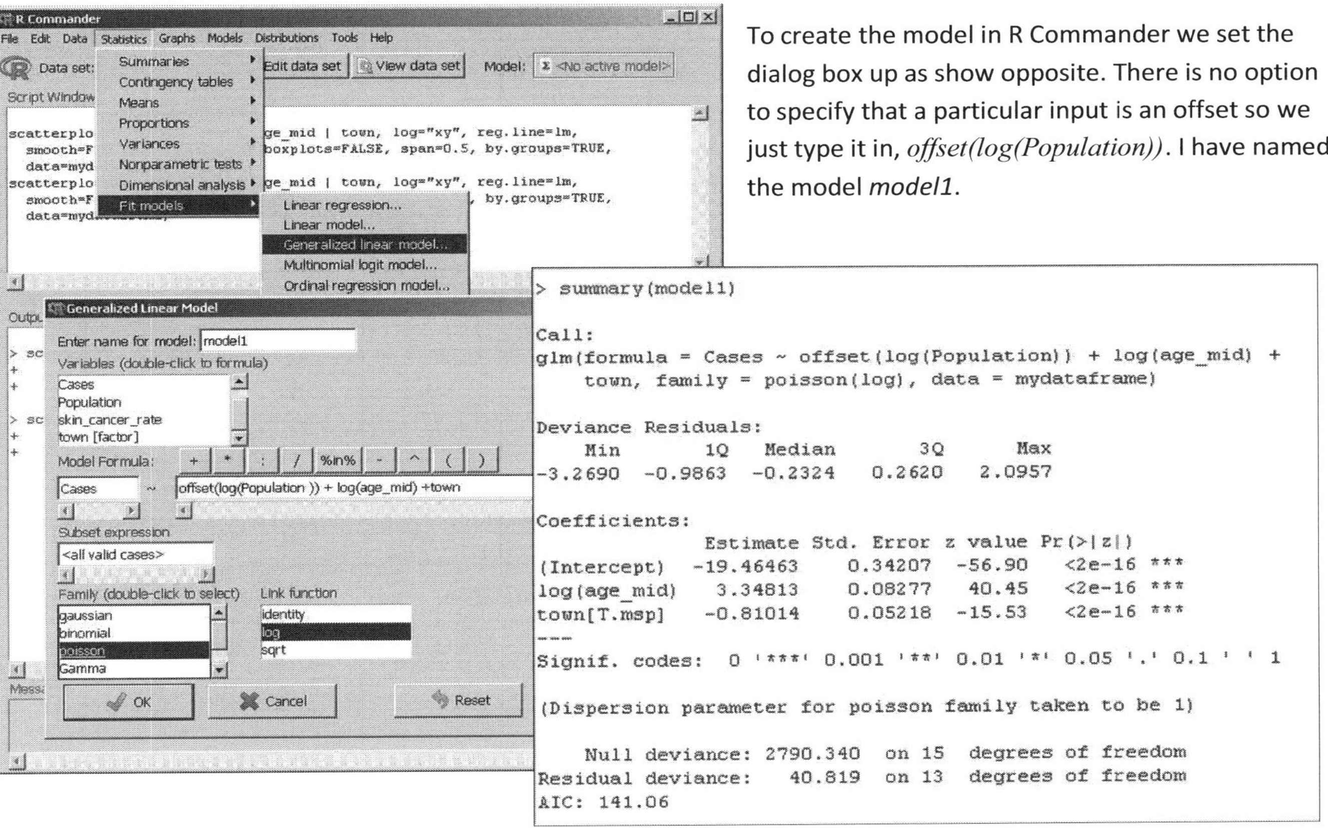

```
> summary(model1)

Call:
glm(formula = Cases ~ offset(log(Population)) + log(age_mid) +
    town, family = poisson(log), data = mydataframe)

Deviance Residuals:
    Min       1Q   Median       3Q      Max
-3.2690  -0.9863  -0.2324   0.2620   2.0957

Coefficients:
             Estimate Std. Error z value Pr(>|z|)
(Intercept) -19.46463    0.34207  -56.90   <2e-16 ***
log(age_mid)   3.34813    0.08277   40.45   <2e-16 ***
town[T.msp]   -0.81014    0.05218  -15.53   <2e-16 ***
---
Signif. codes:  0 '***' 0.001 '**' 0.01 '*' 0.05 '.' 0.1 ' ' 1

(Dispersion parameter for poisson family taken to be 1)

    Null deviance: 2790.340  on 15  degrees of freedom
Residual deviance:   40.819  on 13  degrees of freedom
AIC: 141.06
```

We can see that both inputs, *log(age_mid)* and town are highly statistically significant. However, there is a problem with overdispersion with a residual deviance/df ratio of 3.13. As in the previous examples we can carry out the above analysis directly using R code.

```
model1 <- glm(Cases ~ offset(log(Population)) +log(age_mid) + Town,
family=Poisson(log), data=mydataframe)
summary(model1)
```

We can also obtain a chi-squared goodness of fit statistic:

```
model1fit<- pchisq(model1$deviance, model1$df.residual, lower.tail=FALSE)
model1fit
```

```
> model1fit<- pchisq(model1$deviance, model1$df.residual, lower.tail=FALSE)
> model1fit
[1] 0.0001019375
```

This is a low p-value indicating a poor fit. We can check to see if this poor model is any better than a single parameter (null) model. We can check this with a chi-squared test:

```
# comparing to null
model1chisq <- model1$null.deviance -  model1$deviance
model1chisq
model1df <- model1$df.null   -  model1$df.residual
model1df
model1_chisq <- pchisq(model1chisq, model1df, lower.tail=FALSE)
model1_chisq
```

```
> model1chisq <- model1$null.deviance -  model1$deviance
> model1chisq
[1] 2749.521
> model1df <- model1$df.null   -  model1$df.residual
> model1df
[1] 2
> model1_chisq <- pchisq(model1chisq, model1df, lower.tail=FALSE)
> model1_chisq
[1] 0
```

We obtain a p-value of zero indicating that while is it a poor fit it is better than a single parameter model.

Possibly the fit of the model can be improved by adding an interaction between age and town.

```
Enter name for model: model2
Variables (double-click to formula)
Age [factor]
age_mid
Cases
Population
Model Formula: + * : / %in% - ^ ( )
Cases ~ offset(log(Population)) + log(age_mid) *town
Subset expression
<all valid cases>
Family (double-click to select)    Link function
gaussian                           identity
binomial                           log
poisson                            sqrt
Gamma
OK   Cancel   Reset   Help
```

We can also create an interaction term between *log(age_mid)* and town using the '*' in the model formula, remembering that this use of the asterisk will produce estimates for both the two separately, as well as their interaction.

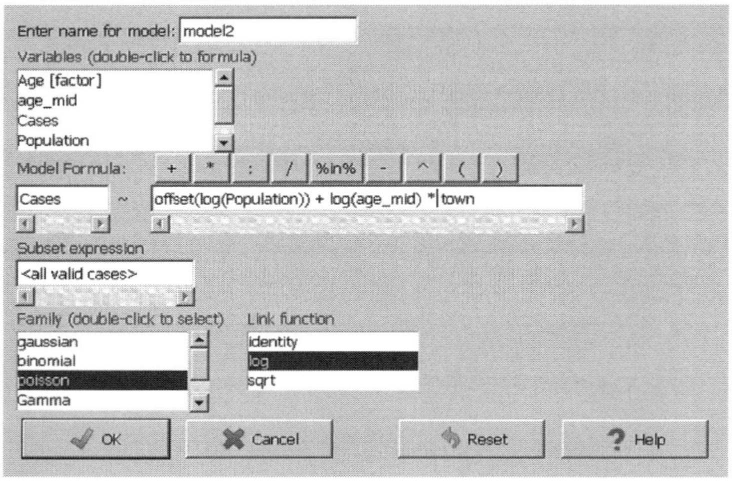

```
Call:
glm(formula = Cases ~ offset(log(Population)) + log(age_mid) *
    town, family = poisson(log), data = mydataframe)

Deviance Residuals:
    Min       1Q   Median       3Q      Max
-3.6256  -1.4621  -0.1760   0.6059   1.8159

Coefficients:
                         Estimate Std. Error z value Pr(>|z|)
(Intercept)             -19.06488    0.39840 -47.854  < 2e-16 ***
log(age_mid)              3.25098    0.09668  33.627  < 2e-16 ***
town[T.msp]              -2.25628    0.78096  -2.889  0.00386 **
log(age_mid):town[T.msp]  0.34894    0.18772   1.859  0.06305 .
---
Signif. codes:  0 '***' 0.001 '**' 0.01 '*' 0.05 '.' 0.1 ' ' 1

(Dispersion parameter for poisson family taken to be 1)

    Null deviance: 2790.340  on 15  degrees of freedom
Residual deviance:   37.291  on 12  degrees of freedom
AIC: 139.54
```

As in the previous example, the Residual deviance is around three times that of its degrees of freedom, indicating that we once again have a problem with overdispersion. Therefore, repeating the two above analyses but using the quasiPoisson approach gives the results shown on the following page.

```
Call:
glm(formula = Cases ~ offset(log(Population)) + log(age_mid) +
    Town, family = quasipoisson(log), data = mydataframe)

Deviance Residuals:
    Min       1Q   Median       3Q      Max
-3.2690  -0.9863  -0.2324   0.2620   2.0957

Coefficients:
             Estimate Std. Error t value Pr(>|t|)
(Intercept) -19.46463    0.55976 -34.773 3.26e-14 ***
log(age_mid)  3.34813    0.13544  24.720 2.58e-12 ***
Town[T.msp]  -0.81014    0.08538  -9.489 3.30e-07 ***
---
Signif. codes:  0 '***' 0.001 '**' 0.01 '*' 0.05 '.' 0.1 ' ' 1

(Dispersion parameter for quasipoisson family taken to be 2.677821)

    Null deviance: 2790.340  on 15  degrees of freedom
Residual deviance:   40.819  on 13  degrees of freedom
AIC: NA
```

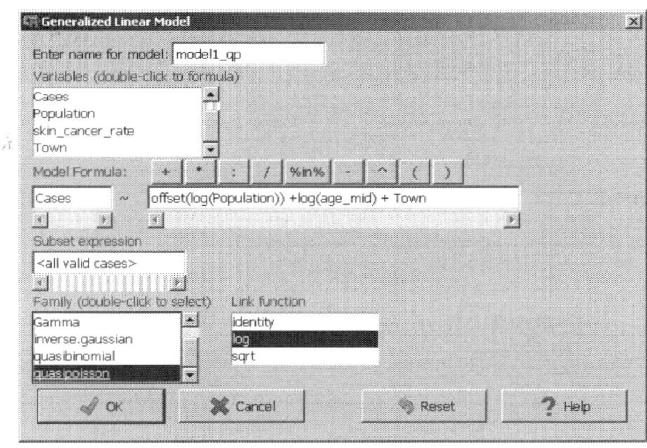


```
> summary(model2_qp)

Call:
glm(formula = Cases ~ offset(log(Population)) + log(age_mid) *
    Town, family = quasipoisson(log), data = mydataframe)

Deviance Residuals:
    Min       1Q   Median       3Q      Max
-3.6256  -1.4621  -0.1760   0.6059   1.8159

Coefficients:
                         Estimate Std. Error t value Pr(>|t|)
(Intercept)              -19.0649     0.6430 -29.650 1.35e-12 ***
log(age_mid)               3.2510     0.1560  20.835 8.65e-11 ***
Town[T.msp]               -2.2563     1.2604  -1.790   0.0987 .
log(age_mid):Town[T.msp]   0.3489     0.3030   1.152   0.2719
---
Signif. codes:  0 '***' 0.001 '**' 0.01 '*' 0.05 '.' 0.1 ' ' 1

(Dispersion parameter for quasipoisson family taken to be 2.604805)

    Null deviance: 2790.340  on 15  degrees of freedom
Residual deviance:   37.291  on 12  degrees of freedom
AIC: NA
```

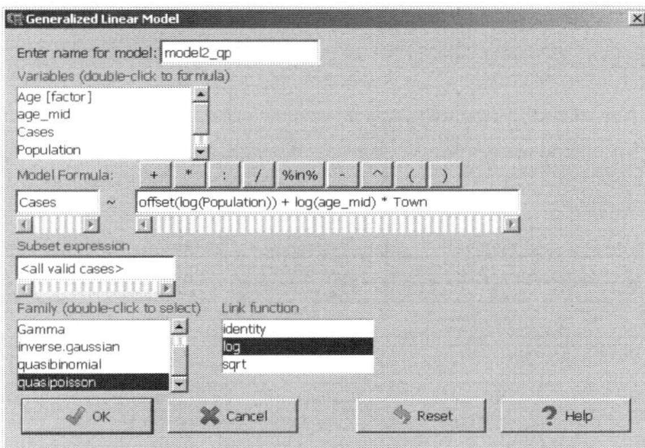

Comparing the Poisson and quasiPoisson approaches, while the significant levels are the same as for *model1*, when we include the *Town*log(age_mid)* interaction it becomes less significant. We can compare the two models using the *anova()* function which produces a p-value of 0.2445 demonstrating that the addition of the interaction does nothing to increase the fit of the model, so we can drop it. We still have a poorly fitting model, so let's try another approach.

61.3 A growth model approach

Given the curved shape of the original scatterplot it might be sensible to add a squared term to the age variable. Such an approach is common in a group of models known as growth models as they present growth patterns found in nature. So now, we want to model mathematically:

$$\log(cases/population)= Intercept + b_2(city) + b_1\log((age\_mid) + b_3[\log((age\_mid)*\log((age\_mid)]$$

```
> anova(model1_qp, model2_qp, test = "Chisq")
Analysis of Deviance Table

Model 1: Cases ~ offset(log(Population)) + log(age_mid) + Town
Model 2: Cases ~ offset(log(Population)) + log(age_mid) * Town
  Resid. Df Resid. Dev Df Deviance Pr(>Chi)
1        13     40.819
2        12     37.291  1   3.5286   0.2445
```

You might think that we would achieve this in R by simply mimicking the above equation but unfortunately, this is not the case. As you may remember '*' was interpreted a special way when it is part of an R model, indicated by the ~ sign. To let R know that we want the '*' to mean just multiplication we wrap the expression in the upper case i function where *I()* means 'as Is'.

In R we write the above model as:

```
model1_growth<- glm
(offset(log(population)) + Town + log(age_mid) + I(log(age_mid) *(log(age_mid)))
```

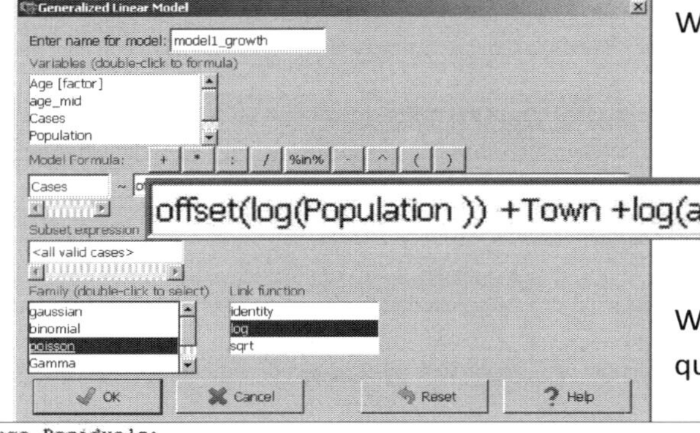

We can also do this in R Commander as shown opposite.

We can repeat the model for both Poisson (left below) and quasiPoisson (right below) link functions.

```
Deviance Residuals:
     Min        1Q    Median        3Q       Max
-1.59295  -0.95258  -0.01212   0.48171   2.03360

Coefficients:
                                Estimate Std. Error z value Pr(>|z|)
(Intercept)                     -35.3123     3.5738  -9.881  < 2e-16 ***
Town[T.msp]                      -0.8027     0.0522 -15.378  < 2e-16 ***
log(age_mid)                     11.3901     1.7948   6.346 2.21e-10 ***
I(log(age_mid) * log(age_mid))   -1.0138     0.2248  -4.510 6.47e-06 ***
---
Signif. codes:  0 '***' 0.001 '**' 0.01 '*' 0.05 '.' 0.1 ' ' 1

(Dispersion parameter for poisson family taken to be 1)

    Null deviance: 2790.340  on 15  degrees of freedom
Residual deviance:   17.584  on 12  degrees of freedom
AIC: 119.83
```

```
Deviance Residuals:
     Min        1Q    Median        3Q       Max
-1.59295  -0.95258  -0.01212   0.48171   2.03360

Coefficients:
                                Estimate Std. Error t value Pr(>|t|)
(Intercept)                    -35.31230    4.29741  -8.217 2.86e-06 ***
Town[T.msp]                     -0.80270    0.06277 -12.788 2.37e-08 ***
log(age_mid)                    11.39008    2.15817   5.278 0.000195 ***
I(log(age_mid) * log(age_mid))  -1.01380    0.27028  -3.751 0.002767 **
---
Signif. codes:  0 '***' 0.001 '**' 0.01 '*' 0.05 '.' 0.1 ' ' 1

(Dispersion parameter for quasipoisson family taken to be 1.44598)

    Null deviance: 2790.340  on 15  degrees of freedom
Residual deviance:   17.584  on 12  degrees of freedom
AIC: NA
```

We see that for both the Poisson and quasiPoisson approaches the *log(age_mid)*log(age_mid)* input is statistically significant.

```
> exp(model1_growth$coefficients)
                   (Intercept)                    Town[T.dfw]
                  2.067534e-16                   2.231562e+00
                  log(age_mid) I(log(age_mid) * log(age_mid))
                  8.843995e+04                   3.628365e-01
```

We can make more sense of the above coefficients by reversing the log transformation to produce *Incident ratios* (IR's also called relative risks), using the *exp(model1_growth$coefficiants)* command. The result shows that town *dfw (= Dallas)* has 2.23 times the skin cancer rate to that of Minneapolis which does not change across age bands.

```
> Confint(model1_growth, level=0.95, type="LR")
                                Estimate       2.5 %      97.5 %
(Intercept)                  -36.1150049 -43.3469524 -29.3332501
Town[T.dfw]                    0.8027018   0.7010775   0.9057421
log(age_mid)                  11.3900790   7.9752818  15.0148351
I(log(age_mid) * log(age_mid)) -1.0138030  -1.4668513  -0.5852444
```

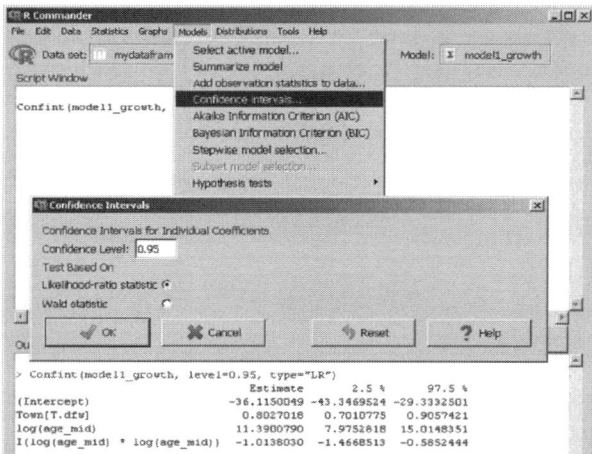

To produce confidence intervals for the model estimates we either use the confidence interval menu option in R Commander or type into the *R Console* window *confint(modelname, level=0.95, type="LR")* alternatively you could use, the *type= "wald"* option which occasionally produces slightly narrower intervals.

```
> exp(confint( model1_growth, level=0.95, type="LR"))
Waiting for profiling to be done...
                                      2.5 %       97.5 %
(Intercept)                    1.495057e-19 1.822768e-13
Town[T.dfw]                    2.015924e+00 2.473767e+00
log(age_mid)                   2.908177e+03 3.317875e+06
I(log(age_mid) * log(age_mid)) 2.306506e-01 5.569697e-01
```

These intervals are for the original estimates not the IRR's. To obtain confidence intervals for the IRR's we apply the *exp()* function on masse to the confidence intervals of the model object producing a 95% confidence interval for *Town IRR* of 2.01 to 2.47.

```
> model1_growth_fit<- pchisq(model1_growth$deviance, model1_growth$df.residual, lower.tail=FALSE)
> model1_growth_fit
[1] 0.128905
```

We can also assess the fit of the *model1_growth*, using the chi-squared statistic. The obtained p-value is 0.1289 representing an improvement over the previous value but still not an indication of a good fit.

The final part of the analysis would be to produce a graph of the observed values and those predicted from the final model as we did in the examples in the previous two chapters.

61.4 Person years as the offset

deaths	personyrs	smoker	age_grp_start
32	52407	1	1
2	18790	0	1
104	43248	1	2
12	10673	0	2
206	28612	1	3
28	5712	0	3
186	12663	1	4
28	2585	0	4
102	5317	1	5
31	1462	0	5

The final example I would like to consider is the example in SAS1 (page 86) which concerns coronary deaths in GPs (General Practitioners = family doctors) and how they are affected by smoking status and age at the start of the study. Here the offset variable is the length of time variable called *person years*.

Type the following into the *R Console* window to download and load the data into R:

within rdata there is a dataframe called coronary_deaths_docs

```
load(url(http://www.robin-beaumont.co.uk/virtualclassroom/book2data/coronary_deaths_docs.rdata))
```

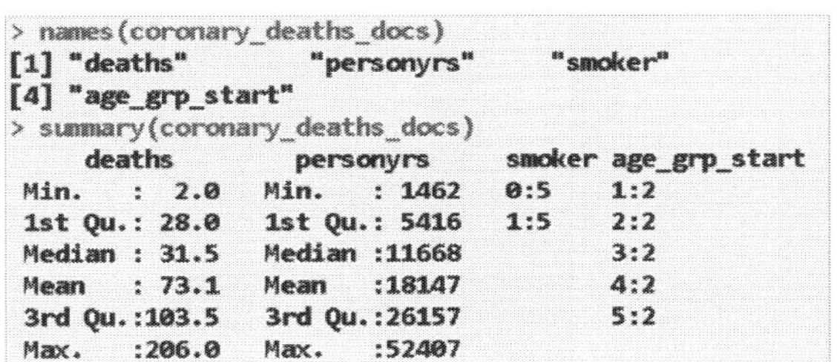

```
> names(coronary_deaths_docs)
[1] "deaths"        "personyrs"     "smoker"
[4] "age_grp_start"
> summary(coronary_deaths_docs)
     deaths          personyrs       smoker age_grp_start
 Min.   :  2.0   Min.   : 1462    0:5    1:2
 1st Qu.: 28.0   1st Qu.: 5416    1:5    2:2
 Median : 31.5   Median :11668           3:2
 Mean   : 73.1   Mean   :18147           4:2
 3rd Qu.:103.5   3rd Qu.:26157           5:2
 Max.   :206.0   Max.   :52407
```

The code below lists the dataframe and the column names, attaches it and produces the summary statistics shown opposite.

```
ls() # list the objects returns [1] coronary_deaths_docs"
attach(coronary_deaths_docs)
names(coronary_deaths_docs)  #  list the field names
summary(coronary_deaths_docs) # give summary statistics
```

```
> death_rate <- deaths / personyrs
> summary(death_rate)
     Min.   1st Qu.    Median      Mean   3rd Qu.      Max.
0.0001064 0.0014440 0.0060510 0.0082260 0.0137200 0.0212000
>
```

You can see that the rate variable (death rate from coronary disease per person year) has a mean of 0.008. This indicates 8 coronary deaths per 1000 person years. We will use this rate variable to produce scatterplots to give us a greater understanding of the data.

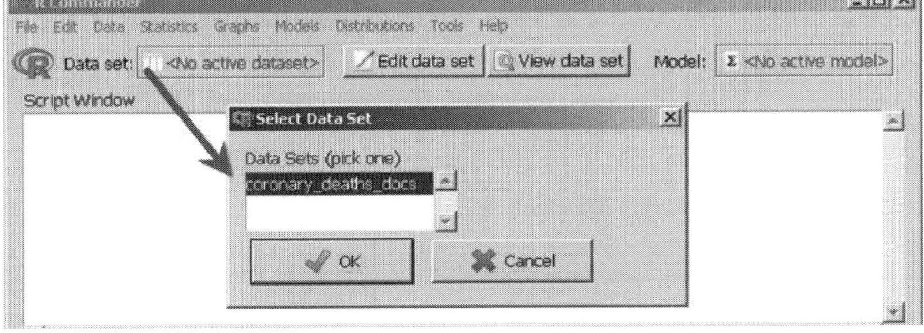

Starting R Commander with the function *library(Rcmdr)* from within the R Console window requires us to inform R Commander to load the dataset. You achieve this by clicking on the dataset box then selecting it, as shown opposite.

We also need to convert the age group variable from a factor to a numeric variable so that we can plot it, achieved with the following code:

```
coronary_deaths_docs$age_grp_start <- as.numeric(coronary_deaths_docs$age_grp_start)
```

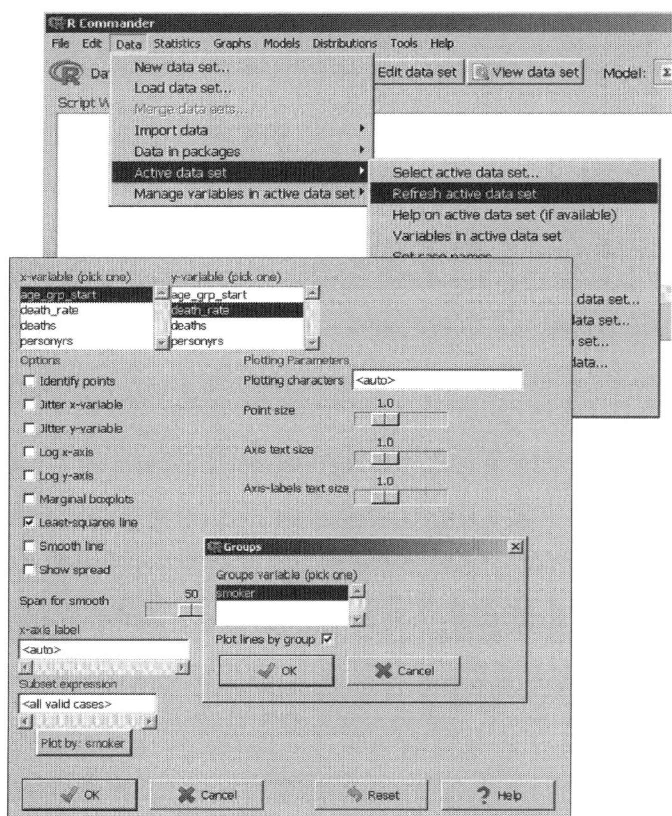

After the conversion, we need to update the dataset, and in R Commander we achieve this with the menu option **Data->active data set->Refresh active dataset**, shown opposite.

As in the previous examples we create a scatterplot in R Commander to show the adjusted count (i.e. rate) on the y axis and the *age_grp_start* on the x axis at the same time producing separate regression lines for smoker and non-smoker.

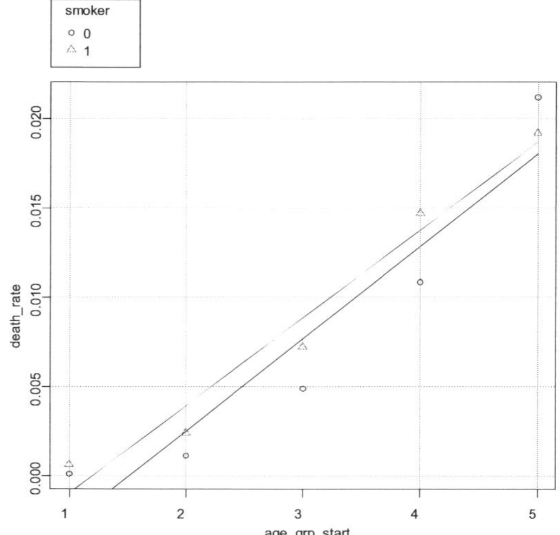

Both the smoker and non-smoker groups have an increased death rate as the start age increases as would be expected, furthermore smokers (coded 1) have an increased risk at each age except the highest.

To carry out the Poisson regression we need to either convert the age group variable back to being a factor or copy it to a new variable that is a factor; I have taken the second approach using the following code:

```
coronary_deaths_docs$age_grp_start_fact <- as.factor(coronary_deaths_docs$age_grp_start)
```

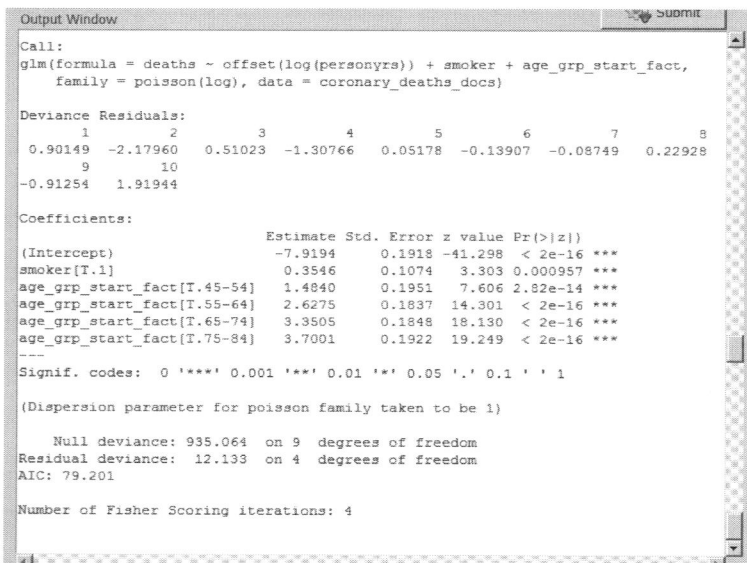

```
Output Window
Call:
glm(formula = deaths ~ offset(log(personyrs)) + smoker + age_grp_start_fact,
    family = poisson(log), data = coronary_deaths_docs)

Deviance Residuals:
       1         2         3         4         5         6         7         8
 0.90149  -2.17960   0.51023  -1.30766   0.05178  -0.13907  -0.08749   0.22928
       9        10
-0.91254   1.91944

Coefficients:
                          Estimate Std. Error z value Pr(>|z|)
(Intercept)                -7.9194     0.1918 -41.298  < 2e-16 ***
smoker[T.1]                 0.3546     0.1074   3.303 0.000957 ***
age_grp_start_fact[T.45-54] 1.4840     0.1951   7.606 2.82e-14 ***
age_grp_start_fact[T.55-64] 2.6275     0.1837  14.301  < 2e-16 ***
age_grp_start_fact[T.65-74] 3.3505     0.1848  18.130  < 2e-16 ***
age_grp_start_fact[T.75-84] 3.7001     0.1922  19.249  < 2e-16 ***
---
Signif. codes:  0 '***' 0.001 '**' 0.01 '*' 0.05 '.' 0.1 ' ' 1

(Dispersion parameter for poisson family taken to be 1)

    Null deviance: 935.064  on 9  degrees of freedom
Residual deviance:  12.133  on 4  degrees of freedom
AIC: 79.201

Number of Fisher Scoring iterations: 4
```

For this analysis, I will carry out the Poisson regression just using R code:

```
model1 <- glm(deaths ~ offset(log(personyrs))
+ smoker + age_grp_start_fact,family=
Poisson(log), data=coronary_deaths_docs)

summary(model1)
```

```
> exp(model1$coefficients)
           (Intercept)              smoker[T.1] age_grp_start_fact[T.2]
          3.636179e-04             1.425664e+00            4.410560e+00
age_grp_start_fact[T.3] age_grp_start_fact[T.4] age_grp_start_fact[T.5]
          1.383849e+01             2.851656e+01            4.045104e+01
> |
```

The above estimates are for log values which we can convert to incident ratios as in the previous example by applying the *exp()* function to the model coefficients. We can also create confidence intervals for the incident ratios as described earlier.

We can easily extract the expected counts from the *glm* object

```
model1$fitted.values
```

```
> model1$fitted.values
        1         2         3         4         5         6         7         8
 27.16762   6.83238  98.88309  17.11691 205.25765  28.74235 187.19579  26.80421
        9        10
111.49585  21.50415
```

```
> model1fit<- pchisq(model1$deviance, model1$df.residual, lower.tail=FALSE)
> model1fit
[1] 0.01638998
```

We can also see how well the model fits the data using the chi-square test to produce a p-value of 0.016, remembering that here we want a high p-value to indicate a good fit, this suggests that our model is a poor fit. This is probably largely the effect of having such a small dataset, that is only one value for each age band for each smoking status.

```
> logLik(model1)
'log Lik.' -33.60041 (df=6)
```

Some programs produce the log likelihood value which is used to create the deviance. And because we used the Poisson link function by using the *logLik(glm_model)* function, we can extract this value from the model. If you attempt to use this function when you use the *quasiPoisson()* link function, you get an error.

Noting that in the above model we have a residual deviance/df ratio of 14/4= 3. If we repeat the model using the quasipossion approach instead, smoker loses its significance as well as the first factor level for age. Again, the lack of multiple values for each factor level *age_mid* is a major limitation.

```
> summary(model1_qp)

Call:
glm(formula = deaths ~ offset(log(personyrs)) + smoker + age_grp_start_fact,
    family = quasipoisson(log), data = coronary_deaths_docs)

Deviance Residuals:
       1         2         3         4         5         6         7         8         9
 0.90149  -2.17960   0.51023  -1.30766   0.05178  -0.13907  -0.08749   0.22928  -0.91254
      10
 1.91944

Coefficients:
                    Estimate Std. Error t value Pr(>|t|)
(Intercept)          -7.9194     0.3203 -24.728 1.59e-05 ***
smoker1               0.3546     0.1793   1.978 0.119121
age_grp_start_fact2   1.4840     0.3258   4.554 0.010383 *
age_grp_start_fact3   2.6275     0.3068   8.563 0.001021 **
age_grp_start_fact4   3.3505     0.3086  10.856 0.000409 ***
age_grp_start_fact5   3.7001     0.3210  11.526 0.000324 ***
---
Signif. codes:  0 '***' 0.001 '**' 0.01 '*' 0.05 '.' 0.1 ' ' 1

(Dispersion parameter for quasipoisson family taken to be 2.789156)

    Null deviance: 935.064  on 9  degrees of freedom
Residual deviance:  12.133  on 4  degrees of freedom
```

61.5 Tips and Tricks

This chapter has presented three different examples of Poisson regression and has left out more than it has included. Maindonald & Braun, 2010 is unusual in providing a chapter on the subject and describes solutions to several common problems; such as the lack of observations for a particular factor level and pairwise comparisons. Other excellent books dedicated to the subject and specifically implementing the models in R are Hilbe, 2014, and Hilbe, 2011.

There are other measures of goodness of fit, which I describe in the chapter on Logistic regression.

Applying a quasiPoisson approach to the overdispersion problem is not a solution in all circumstances as demonstrated in one of the examples where the reason was the omission of an important input variable. Another common reason for overdispersion is that the counts may occur in clumps for different levels of one or more of the input factor levels, which is effectively modelled using the quasiPoisson approach. In fact, the dispersion parameter reported by R can be thought of as a measure of this clumping where 1 is equal to no clumping (more detail below).

An extension to Poisson regression is conditional Poisson regression which allows the evaluation of repeated count measures. The *gnm* package contains the *gnm()* function to fit a conditional Poisson regression model with the option $eliminate = subject$ (where subject is a factor for each individual in the data set) further details can be found at: http://statistics.open.ac.uk/sccs/r.htm

61.6 Techie stuff

In this chapter I have barely touched on the negative binomial distribution yet we have ended up using it for most of the models. Specifically I have not provided any details about theta so to remedy that I have included a posting from Joseph Hilbe, whose two books have been cited above. I find it fascinating to read about the historical development of these techniques and such postings provide a unique insight.

What is theta in a negative binomial regression fitted with R?

http://stats.stackexchange.com/questions/10419/what-is-theta-in-a-negative-binomial-regression-fitted-with-r

In this context, θ is usually interpreted as a measure of overdispersion with respect to the Poisson distribution.
The variance of the negative binomial is $\mu+\mu^2/\theta$, so θ really controls the excess variability compared to Poisson (which would be μ), and not the skew.

. . . There seems to be a lot of misinformation about the negative binomial model, and especially with respect to the dispersion statistic and dispersion parameter.

The dispersion statistic, which gives an indication of count model extra-dispersion, is the Pearson statistic divided by the residual DF. μ is the location parameter. For count models, the scale parameter is set at 1. For R *glm()* and *glm.nb()* θ (theta) is a dispersion parameter, also called the ancillary parameter. . . . I give a complete rationale for the various terms in the NB model in my forthcoming book, Modeling Count Data (Cambridge University Press) which is going to press today.

glm.nb() and *glm()* are unusual in how they define the dispersion parameter. The variance is given as $\mu+(\mu^2/\theta)$ rather than $\mu+\alpha\mu^2$, which is the direct parameterization. Direct parameterization is the way NB is modeled in SAS, Stata, Limdep, SPSS, Matlab, Genstat, Xplore, in fact almost all software. When you compare *glm.nb()* results with other software results, remember this. The author of *glm()* (which came from S-plus) and *glm.nb()* apparently took the indirect relationship from McCullagh & Nelder, but Nelder (who was the co-founder of GLM in 1972) wrote his kk system add-on to Genstat in 1993 in which he argued that the direct relationship is preferred. He and his wife used to visit me and my family about every other year in Arizona starting in early 1993 until the year before he died. We discussed this pretty thoroughly, since I had put a direct relationship into the GLM program I wrote in late 1992 for Stata and Xplore software, and for a SAS macro in 1994.

The *nbinomial()* function in the msme package on CRAN allows the user to employ the direct (default) or indirect, as an option, to duplicate *glm.nb()* parameterization, and provides the Pearson statistic and residuals to output. Output also displays the dispersion statistic, and allow the user to parameterize α (or θ), giving parameter estimates for the dispersion. This allows you to assess which predictors add to the extra-dispersion of the model. This type of model is generally referred to as heterogeneous negative binomial. I'll put the *nbinomial()* function into the COUNT package . . . , plus a number of new functions and scripts for graphics. Joseph Hilbe [edited].

62 Conditional logistic regression

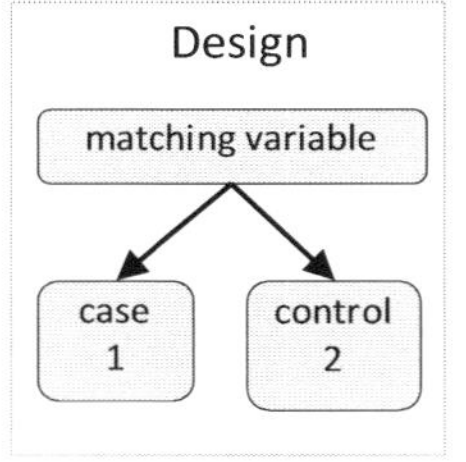

Conditional Logistic Regression

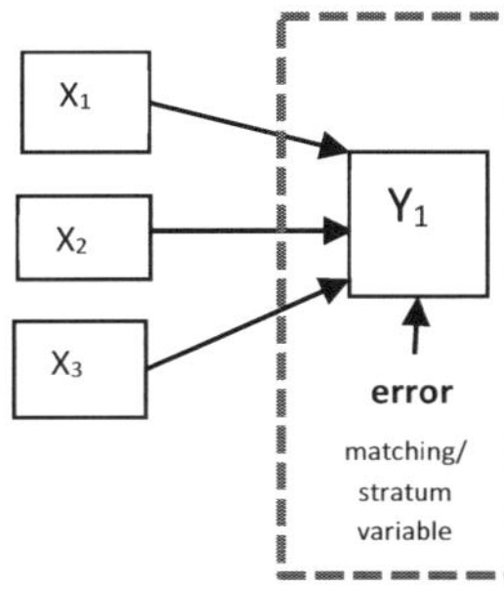

This is an extension to the approach taken in the chapter *Paired nominal data: comparing proportions using McNemar's test*. It can be thought of as the equivalent to a repeated measures design for a binary outcome.

Conditional logistic regression allows us to assess the possible relationship between a binary outcome which is adjusted to some baseline level along with several inputs. Therefore, it is very similar to ordinary multiple logistic regression but now we have the binary outcome adjusted in some way. This is achieved by specifying a matching variable, also called a *stratum variable*. This means that we can't employ the same technique to that used in logistic regression to estimate the parameters in the model, instead using conditional maximum likelihood.

The most common use of conditional logistic regression is in matched case-control studies (SAS2, p.47; Beslow & Day, 1980). This is where you have one subject matched to one or more 'controls'. Campbell, 2006 p.47 provides an example of 35 patients who died (outcome) in hospital from asthma who were individually matched for sex and age with 35 control subjects who had been discharged from the same hospital the previous year. Note that the authors have assumed here that the pairing is equivalent to a repeated measure. The investigators felt that the adequacy of monitoring (=nominal input) of the patients had an effect upon outcome so it was also recorded. Here the stratum is the group of two patients that are matched on sex and age. Other such matching might be an age group, a geographical location (GP practices, hospitals), a specific environment (culture dish, cage or habitat) or some type of genetic similarity (twins or family members).

In conditional logistic regression the matching or stratum variable is not calculated as such – you see no estimate appearing in the output but it is used in the analysis process. Instead the outcome is adjusted to take into account the possible influence this stratum variable might have on it. Technically, we call the levels of the stratum variable "nuisance" parameters to reflect this attitude. Another way of looking at this is to think of the strata as allowing for a different intercept value for each set of observations within each strata level. Because you can't perform conditional logistic regression using the R Commander menus I will only provide an example in R.

62.1 Conditional Logistic regression in R: data preparation

I will continue with the example described above (from Campbell, 2006 p.47). First download the following spss data file from:

`http://www.robin-beaumont.co.uk/virtualclassroom/book2data/condit_log_reg_campbell3_added.sav`

Within the *R Console* window type the following, which will allow you to read the spss file and place it in a dataframe called *hospital_monitoring1*:

```
library(foreign)
hospital_monitoring1 <-
  read.spss(file=file.choose(),
   use.value.labels=TRUE, max.value.labels=Inf, to.data.frame=TRUE)
colnames(hospital_monitoring1) <- tolower(colnames(hospital_monitoring1))
names(hospital_monitoring1)
```

```
> names(hospital_monitoring1)
[1] "pairing"     "outcome"     "monitoring"  "casecontrol" "bmi"
```

You will now see a list of the column names for the *hospital_monitoring1* dataframe.

62.1.1 Data description

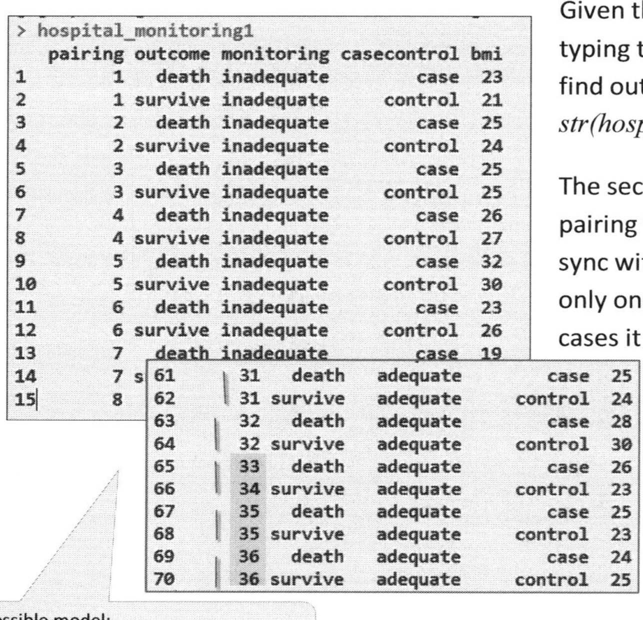

```
> hospital_monitoring1
   pairing outcome monitoring casecontrol bmi
1        1   death inadequate        case  23
2        1 survive inadequate     control  21
3        2   death inadequate        case  25
4        2 survive inadequate     control  24
5        3   death inadequate        case  25
6        3 survive inadequate     control  25
7        4   death inadequate        case  26
8        4 survive inadequate     control  27
9        5   death inadequate        case  32
10       5 survive inadequate     control  30
11       6   death inadequate        case  23
12       6 survive inadequate     control  26
13       7   death inadequate        case  19
14       7 s
15       8
```

```
61    31   death   adequate        case  25
62    31 survive   adequate     control  24
63    32   death   adequate        case  28
64    32 survive   adequate     control  30
65    33   death   adequate        case  26
66    34 survive   adequate     control  23
67    35   death   adequate        case  25
68    35 survive   adequate     control  23
69    36   death   adequate        case  24
70    36 survive   adequate     control  25
```


Given that this is a small dataset, it is sensible to view it all by typing the dataframe name in the *R Console* window. We could find out the size of the dataframe before we did this by typing *str(hospital_monitoring1)* in the R Console.

The second screen shot opposite indicates there is a mistake in the pairing variable in the datafile where the pairing number falls out of sync with the actual observations with pair 33 and 34 possessing only one value. Considering the other variables values for these cases it is reasonable to assume that rows 65 and 66 relate to pair 33 and therefore rows 67 and 68 = pair 34 and rows 69 and 70 = pair 35. Clearly this is a typing error.

You could correct these values using the R data editor obtained by typing: *edit(hospital_monitoring1)*

Alternatively we can rewrite the pairing variable using the *rep()* function which we will do below. It is always good to think what we might be considering in terms of a proposed model and one such model is shown opposite.

Given that we have a binary outcome (=outcome) and a binary input (=monitoring) it seems sensible to obtain a table of counts. Unfortunately because we are dealing with pairs of observations this means some manipulation of the data.

We will first begin by creating some numeric variables from the original SPSS file. The code below also corrects the pairing variable by creating a new variable called *matched* which does the equivalent job.

```
names(hospital_monitoring1)
str(hospital_monitoring1)
attach(hospital_monitoring1)
monitoring_numeric <- as.numeric(monitoring) -1
monitoring_numeric
casecontrol_numeric <- as.numeric(casecontrol) -1
casecontrol_numeric
matched <- rep(1:35, rep(2,35))
matched
```

"pairing" "outcome" "monitoring" "casecontrol" "bmi"

1=inadequate
0=adequate

```
> monitoring_numeric <- as.numeric(monitoring) -1
> monitoring_numeric
 [1] 1 1 1 1 1 1 1 1 1 1 1 1 1 1 1 1 1 1 1 1 1 0 1 0 1 0 1 0 1 0 1 0 1 0 1 0 1 0 1 0 1 0 1 0 1 0 0 1 0 1 0 1 0 0 0 0 0 0
[59] 0 0 0 0 0 0 0 0 0 0 0 0
> casecontrol_numeric <- as.numeric(casecontrol) -1
> casecontrol_numeric
 [1] 0 1 0 1 0 1 0 1 0 1 0 1 0 1 0 1 0 1 0 1 0 1 0 1 0 1 0 1 0 1 0 1 0 1 0 1 0 1 0 1 0 1 0 1 0 1 0 1 0 1 0 1 0 1 0 1 0 1
[59] 0 1 0 1 0 1 0 1 0 1 0 1
> matched <- rep(1:35, rep(2,35))
> matched
 [1]  1  1  2  2  3  3  4  4  5  5  6  6  7  7  8  8  9  9 10 10 11 11 12 12 13 13 14 14 15 15 16 16 17 17 18 18 19 19 20
[40] 20 21 21 22 22 23 23 24 24 25 25 26 26 27 27 28 28 29 29 30 30 31 31 32 32 33 33 34 34 35 35
```

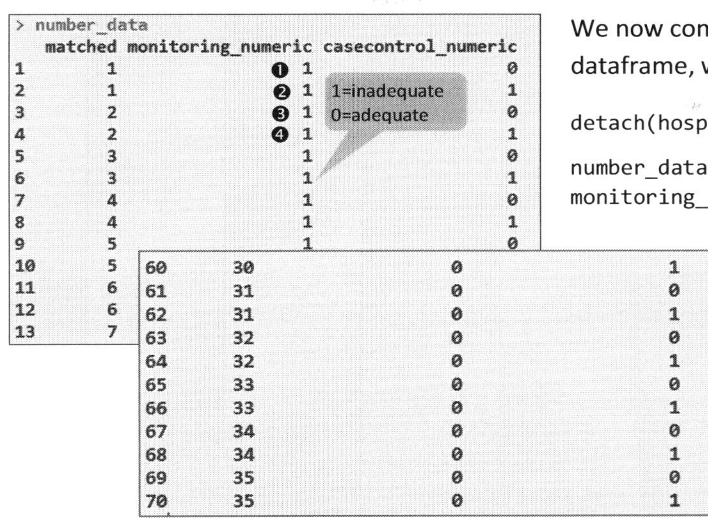

```
> number_data
   matched monitoring_numeric casecontrol_numeric
1        1                ❶ 1                   0
2        1                ❷ 1                   1
3        2                ❸ 1                   0
4        2                ❹ 1                   1
5        3                  1                   0
6        3                  1                   1
7        4                  1                   0
8        4                  1                   1
9        5                  1                   0
10       5
11       6
12       6
13       7
60      30                  0                   1
61      31                  0                   0
62      31                  0                   1
63      32                  0                   0
64      32                  0                   1
65      33                  0                   0
66      33                  0                   1
67      34                  0                   0
68      34                  0                   1
69      35                  0                   0
70      35                  0                   1
```


We now combine the above columns together to form a new dataframe, which I have called *number_data*.

```
detach(hospital_monitoring1) # no need for the old data now

number_data <- data.frame(cbind( matched,
monitoring_numeric, casecontrol_numeric))

number_data
```

The problems with pairs 34 to 35 are now also corrected.

Now we need to change the format from long to wide to be able to produce the table of counts of pairs. The symbols {❶,❷,❸,❹} indicate the position of the first four values demonstrating the data manipulation.

```
> wide_data
   matched monitoring_numeric.0 monitoring_numeric.1
1        1                  ❶ 1                  ❷ 1
3        2                  ❸ 1                  ❹ 1
5        3                    1                    1
7        4                    1                    1
9        5                    1                    1
11       6                    1                    1
13       7                    1                    1
15       8                    1                    1
17       9                    1                    1
19      10                    1                    1
21      11                    1                    0
23      12                    1                    0
25      13                    1                    0
27      14                    1                    0
29      15                    1                    0
31      16                    1                    0
33      17                    1                    0
35      18                    1                    0
37      19                    1                    0
39      20                    1                    0
41      21                    1                    0
43      22                    1                    0
45      23                    1                    0
47      24                    0                    1
49      25                    0                    1
51      26                    0                    1
53      27                    0                    0
55      28                    0                    0
57      29                    0                    0
59      30                    0                    0
61      31                    0                    0
63      32                    0                    0
65      33                    0                    0
67      34                    0                    0
69      35                    0                    0
```

cases | controls | 23 pairs | 10 pairs | 13 pairs | 3 pairs | 9 pairs

```
wide_data <- reshape (number_data,  timevar=
"casecontrol_numeric",

v.names = c("monitoring_numeric"),

idvar = "matched", direction = "wide")

wide_data

attach(wide_data)
```

Finally, we can create the table we require of the pairs.

```
attach(wide_data)

#  for the dnn option = row , column

table(monitoring_numeric.0, monitoring_numeric.1,

dnn = c("case:adeq/inadeq", "control:adeq/inadeq"))
```

It shows that 9 of both the paired controls and cases had adequate monitoring with a further 3 pairs having adequate monitoring for the cases group only.

```
> #  for the dnn option = row , column
> table(monitoring_numeric.0, monitoring_numeric.1,
+ dnn = c("case:adeq/inadeq", "control:adeq/inadeq"))
                  control:adeq/inadeq
case:adeq/inadeq   0  1
                 0  9  3
                 1 13 10
```

Looking at the cases, 23 had inadequate monitoring of which 10 of the matched controls had inadequate monitoring.

There is an alternative approach, which means that you do not need to convert the data from long format to wide format. The approach involves using the *matchTab()* function in the *epicalc* library. The function requires three variables each coded a specific way , detailed opposite.

Variable	Detail
case =	Outcome variables where 0 = control and 1 = case
exposed =	Exposure variable where 0 = non-exposed and 1 = exposed
strata =	Identification number for each matched set

In our dataframe, *number_data* we have the *casecontrol* variable coded the wrong way round with 1=control and 0=case. To reverse the values we make use of the *ifelse()* function. Rather than changing the values in the original variable I create a new one with the suffix '2'.

new variable

old variable

change value to this
[in the new variable] when the condition is true

```
number_data$casecontrol_numeric2 <-ifelse(number_data$casecontrol_numeric ==0, 1, 0)

number_data$casecontrol_numeric2
```

else change the value to this
[in the new variable] when the condition is false

The result is shown below:

```
> number_data$casecontrol_numeric2 <-ifelse(number_data$casecontrol_numeric ==0, 1, 0)
> number_data$casecontrol_numeric2
 [1] 1 0 1 0 1 0 1 0 1 0 1 0 1 0 1 0 1 0 1 0 1 0 1 0 1 0 1 0 1 0 1 0 1 0 1 0 1 0 1 0 1 0 1 0 1 0 1 0 1 0 1 0 1 0 1
[56] 0 1 0 1 0 1 0 1 0 1 0 1 0 1 0
```

We can now use this variable to send to the *matchTab()* function.

```
library(epicalc)

matchTab(number_data$casecontrol_numeric2, number_data$monitoring_numeric,
strata=number_data$matched)

detach(wide_data)
```

```
> library(epicalc)
> matchTab(number_data$casecontrol_numeric2, number_data$monitoring_numeric,  strata=number_data$matched)

Exposure status: $ = 1
 Exposure status: number_data = 1
 Exposure status: monitoring_numeric = 1

Total number of match sets in the tabulation = 35

Number of controls = 1
                        No. of controls exposed
No. of cases exposed  0  1
                     0  9  3
                     1 13 10

Odds ratio by Mantel-Haenszel method = 4.333

Odds ratio by maximum likelihood estimate (MLE) method = 4.333
 95%CI= 1.235 , 15.206
```

Odds ratio of dying in hospital when monitoring inadequate

Using this method we also obtain the odds ratio along with its 95% confidence interval.

Let's move on to the model development now.

62.1.2 Model development/interpretation

To form a conditional logistic regression model the *clogit()* function is used from the survival package. The function takes the following general form, where exposure can be several variables:

```
case.status ~ exposure + strata(matched.set)
```

For our example, using the *hospital_monitoring1* dataframe we started with, this becomes:

```
library(survival)
monitor_model1 <- clogit(as.numeric(outcome) ~ monitoring + strata(pairing),
hospital_monitoring1)
summary(monitor_model1)
confint(monitor_model1)
exp(confint(monitor_model1))
```

```
> monitor_model1 <- clogit(as.numeric(outcome) ~ monitoring + strata(pairing), hospital_monitoring1)
> summary(monitor_model1)
Call:
coxph(formula = Surv(rep(1, 70L), as.numeric(outcome)) ~ monitoring +
    strata(pairing), data = hospital_monitoring1, method = "exact")

  n= 70, number of events= 35

                        coef exp(coef) se(coef)     z Pr(>|z|)
monitoringinadequate 1.4663    4.3333   0.6405 2.289   0.0221 *
---
Signif. codes:  0 '***' 0.001 '**' 0.01 '*' 0.05 '.' 0.1 ' ' 1

                     exp(coef) exp(-coef) lower .95 upper .95
monitoringinadequate     4.333     0.2308     1.235     15.21

Rsquare= 0.092   (max possible= 0.49 )
Likelihood ratio test= 6.74  on 1 df,   p=0.009437
Wald test            = 5.24  on 1 df,   p=0.02206
Score (logrank) test = 6.25  on 1 df,   p=0.01242

> confint(monitor_model1)
                         2.5 %   97.5 %
monitoringinadequate 0.2109554 2.721719
> exp(confint(monitor_model1))
                         2.5 %   97.5 %
monitoringinadequate 1.234857 15.20644
```

4.33 = Odds ratio of dying in hospital when monitoring inadequate

We get the same results here as we did from the *matchTab()* function as discussed previously because we only have one input variable, monitoring.

The dataframe contains the variable *bmi* (= body mass index) which we can also include in a new model:

```
monitor_model2 <- clogit(as.numeric(outcome) ~ monitoring + bmi + strata(pairing),
hospital_monitoring1)
summary(monitor_model2)
```

We interpret these values the same way as in logistic regression.

```
> monitor_model2 <- clogit(as.numeric(outcome) ~ monitoring + bmi + strata(pairing), hospital_monitoring1)
> summary(monitor_model2)
Call:
coxph(formula = Surv(rep(1, 70L), as.numeric(outcome)) ~ monitoring +
    bmi + strata(pairing), data = hospital_monitoring1, method = "exact")

  n= 70, number of events= 35

                         coef exp(coef) se(coef)      z Pr(>|z|)
monitoringinadequate  1.47663   4.37815  0.64411  2.293   0.0219 *
bmi                  -0.02051   0.97970  0.09058 -0.226   0.8208
---
Signif. codes:  0 '***' 0.001 '**' 0.01 '*' 0.05 '.' 0.1 ' ' 1

                     exp(coef) exp(-coef) lower .95 upper .95
monitoringinadequate    4.3782     0.2284    1.2389     15.47
bmi                     0.9797     1.0207    0.8203      1.17

Rsquare= 0.092   (max possible= 0.49 )
Likelihood ratio test= 6.79  on 2 df,   p=0.03353
Wald test            = 5.26  on 2 df,   p=0.07224
Score (logrank) test = 6.29  on 2 df,   p=0.04312
```

The same assumptions apply as those concerning logistic regression. Additionally in conditional logistic regression, we are assuming that the matching is valid, that is the values for the pairing/groupings are NOT independent.

62.2 Tips and Tricks

This chapter has provided a very short introduction to conditional logistic regression. More information is provided in Beslow & Day, 1980, which is freely available online. Unfortunately, there appears little written on this subject and R.

The MRC Cognition and Brain sciences unit has a good health related example at: http://imaging.mrc-cbu.cam.ac.uk/statswiki/FAQ/clogit

Some people dislike using the *ifelse()* function finding it too complex, instead, for the example given earlier, you can easily achieve the same result using the following code.

```
# first create new variable

number_data$casecontrol_numeric3  <- NA

# Then recode the old field into the new one for the specified rows

number_data$casecontrol_numeric3[number_data$casecontrol_numeric==0] <- 1

number_data$casecontrol_numeric3[number_data$casecontrol_numeric==1] <- 0

number_data$casecontrol_numeric3
```

63 Factorial ANOVA

The chapter *A quick tutorial: Analysing data shipped with R*, introduced the process of evaluating several means (the output) divided up by a single nominal variable (i.e. a factor). Now we take this process a step further by considering the mean values for an interval/ratio level variable divided by several nominal variables (factors).

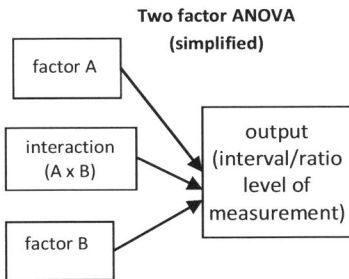

Two factor ANOVA (simplified)

factor A
interaction (A x B)
factor B
output (interval/ratio level of measurement)

> The focus of Factorial ANOVA is the assessment of means between different levels of several factors.

63.1 Headache dataset – a two factor example

To make this more tangible I will use an example given in Stevens, 2007 concerning 98 headache suffers involved in an experiment assessing the effectiveness of different kinds of psychological treatment to noise sensitivity/tolerance.

Each subject went through the following sequence: (1) measurement of initial sensitivity score, (2) relaxation training, (3) treatment, and (4) measurement of final sensitivity scores.

Noise sensitivity (the outcome/output variable) was measured two ways; the level at which the gradually increasingly loud test tone became uncomfortable (*variable=uncomf*) and also definitely unpleasant (*variable=dfunpl*). For the following analysis we will only consider the *uncomf* outcome variable.

All subjects received relaxation training in two stages: They were first asked to listen to the tone at their definitely unpleasant level for up to two minutes (with the option to terminate the exposure if they chose). They were then given instruction on breathing techniques and the use of visual imagery to act as a controlled distraction.

The two input variables (factors) were:

headache_type – headache_type is a factor with two levels representing the two types of headache sufferers in the study; migraine (=1) and tension (=2).

treat_grp – treat_grp is another factor with four levels representing the following four treatment groups. Each subject was randomised to one of four groups:

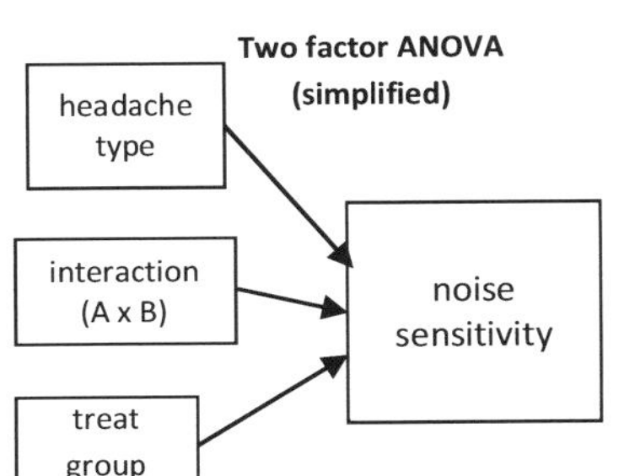

- **1** - subjects in this group listened to the tone again at their definitely unpleasant (DU) level for the length of time that they were able to stand it in stage (1) of the above sequence.
- **2** - as in treatment group 1 but with one extra minute's exposure to the tone.
- **3** - as treatment group 2 but *were explicitly* instructed to use the relaxation technique of breathing and imagery.
- **4** - this was a control group, in that the subject experienced no exposure to the tone between the relaxation training and the final sensitivity measures.

In the above example we have two factors providing the inputs (headache_type and treat_grp). Whilst it is possible to have several input factors, I would recommend not using more than three factors as it becomes increasingly complex to interpret the results.

The major advantage of Factorial ANOVA is the significant reduction of subjects required compared to some other forms of analysis (Stevens, 2007 p.126) for details see the *sample size requirements* chapter.

63.2 Describing ANOVA

Uncomf =outcome variable		**treat_grp factor (4 levels)**				**marginal**
	Level:	1	2	3	4	
headache_type factor (2 levels)	1	11 $\bar{x}$11= 2.22	11 $\bar{x}$12= 3.1	12 $\bar{x}$13= 4.78	11 $\bar{x}$14= 2.19	$\bar{x}$1.=3.11
	2	14 $\bar{x}$21= 3.36	11 $\bar{x}$22= 4.53	16 $\bar{x}$23= 4.69	12 $\bar{x}$24= 2.87	$\bar{x}$2.=3.88
marginal		$\bar{x}$.1=2.86	$\bar{x}$.2=3.81	$\bar{x}$.3=4.71	$\bar{x}$.4=2.54	$\bar{x}$=3.53

The ANOVA technique has a whole language associated with it that needs explaining, so I'll use the headache example to explain some of the terms. The table of means and counts opposite will be the focus.

The experiment is called a **4 x 2 ANOVA**, referred to as a 'four by two design' where the 4 refers to the number of levels for the first factor (*treat_grp*) and 2 for the second one (*headache_type*). You could have more factors if you wanted such as **4 x 2 x 2 ANOVA**, indicating we also have another factor such as gender.

Each square in the table is called a **cell**, representing a specific level for each of the factors.

Subscripts are used to indicate the row and column. For example x_{11}, indicates the cell for those *uncomf* values where both *Treat_grp factor=1* and *headache_type=1*. See the chapter, *Comparing several independent categories: contingency tables*, for more detail.

The number of (independent) observations in each cell are sometimes called **replicates**. For example $\bar{x}_{14}$, represents a cell with 11 replicates. However, other writers use this term to indicate replicated experiments (Winer, 1972 p.391) and prefer to use the term 'observations' for the usual values in each cell.

$\bar{x}$, is referred to as 'x bar' indicating the mean value of a specific cell.

The above observations are **independent** as only one value from each subject is provided in the table. If several values from a subject occurred in the table, such as pre-test scores, it would be a **repeated measures** design.

Marginal values, these represent those values across all levels of a particular factor ignoring the levels, they are important when assessing ANOVA designs as we will see later.

Ways – An ANOVA is often referred to by the number of factors involved. One-way ANOVA refers to a single factor analysis. A two-way ANOVA indicates two factors and a multi-way (or n-way) ANOVA refers to two or more factors, as does the term factorial ANOVA.

Crossing – A design is said to be crossed when there are values for every level of each factor in the design, as in the above example.

Balanced – When there are equal numbers of observations in each cell it is called a balanced design. Such a situation is desirable for several reasons including the fact that it produces the optimal estimation of the various parameters. When there are unequal numbers of observations in the cells the design is said to be un-balanced.

Blocking – A blocking factor is an input variable that is added to the analysis that is believed to affect the outcome but is usually not under the control of the researcher. In the headache example, headache_type could be classed as a blocking factor. Similarly if gender were included in the analysis this could also be classed as a blocking factor. The researcher hypotheses that the results within each level of the blocking factor are homogenous and therefore it is assumed that including the blocking factor reduces sources of variability and thus leads to greater precision. In the analysis process the blocking factor is considered to be just another factor in the analysis.

The term blocking and a related one, **plot** come from the agricultural community as ANOVA was developed by RA Fisher (Fisher, 1936) who initially worked in that area. Doncaster & Davey, 2007, focusing on the Life sciences as well as agriculture, provide a detailed guide including R code for analysing each type of design.

63.3 Main and simple effects

***Uncomf* =outcome variable**		**Treat_grp factor (4 levels)**				**marginal**
	Level:	1	2	3	4	
headache_type factor (2 levels)	1	n=11 $\bar{x}_{11}$= 2.22	n=11 $\bar{x}_{12}$= 3.1	n=12 $\bar{x}_{13}$= 4.78	n=11 $\bar{x}_{14}$= 2.19	$\bar{x}_{1.}$=3.11
	2	n=14 $\bar{x}_{21}$= 3.36	n=11 $\bar{x}_{22}$= 4.53	n=16 $\bar{x}_{23}$= 4.69	n=12 $\bar{x}_{24}$= 2.87	$\bar{x}_{2.}$=3.88
marginal		$\bar{x}_{.1}$=2.86	$\bar{x}_{.2}$=3.81	$\bar{x}_{.3}$=4.71	$\bar{x}_{.4}$=2.54	$\bar{x}$=3.53

Main effects

$\beta_{headache\_type=1}$ = 3.11 − 3.53 = -0.42
$\beta_{headache\_type=2}$ = 3.88 − 3.53 = 0.35
$\beta_{treat\_grp=1}$ = 2.86 − 3.53 = -0.67
$\beta_{treat\_grp=2}$ = 3.81 − 3.53 = 0.28
$\beta_{treat\_grp=3}$ = 4.71 − 3.53 = 1.18
$\beta_{treat\_grp=4}$ = 2.54 − 3.53 = -.99

The *main effect* for a factor level is the difference between the marginal mean for the particular factor level and the grand mean (Winer, 1972 p.317). Here we have ignored all the other factors besides the one in which we are interested. We can calculate the 6 main effects for this dataset shown below where I have used the beta symbol to represent the main effect. These values give us an indication of the effect that a particular factor level (on average) has compared to the overall average. Note that treatment group 4 has the highest noise sensitivity values (☺) regardless of the headache type. This is reflected in the fact that the main effect for treatment group 3 is also the largest. Group 4 being the control group has means below the overall average. Possibly an interesting line of investigation would be to specifically compare this control group against all the others, more about this latter.

A simple effect can relate to main effects or interactions. In the former case they are called *main simple effects* (Winer, 1972 p.347). Main simple effects measure the effect of one level of a factor against a specific level of another factor (Winer, 1972 p.317). I'll demonstrate this later when we look at post-hoc tests.

63.4 The ANOVA analysis process

There are several different strategies available to analyse an ANOVA design. The traditional way is to first carry out what is called an *omnibus test*, to discover if the proposed model fits the data any better than the overall mean. If this is found to be the case then the main effects are investigated. Then if any interaction(s) are found to be statistically significant traditionally the analysis would be abandoned. However the researcher may attempt to pin point where the actual difference in means occur within different factor levels by using post-hoc tests. Another approach is to bypass the omnibus, main effects and interactions and go straight to the post-hoc tests.

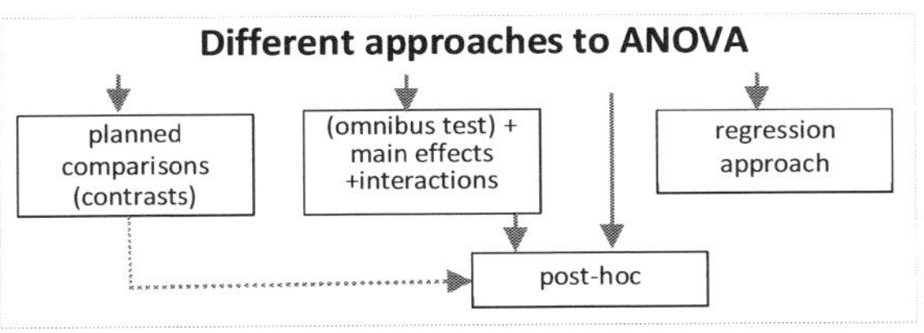

From the above description we can see there is the potential for many comparisons which could result in false positive results (i.e. significant p-values). While most of the approaches attempt to manage this problem the best approach is not to do them, instead the researcher focuses specifically on comparing specific cell means, or groups of them, using a method known as planned comparisons (also called planned contrasts) introduced in the multiple linear regression chapter. The following analysis of the headache data will demonstrate and compare these approaches.

63.5 Analysing the Headache data

63.5.1 Preparing the data

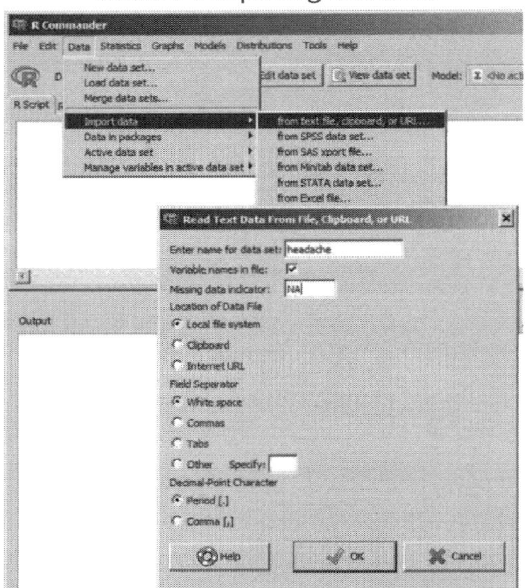

To import the headache data described earlier in the chapter you first need to load R Commander by typing the following command:

```
library(Rcmdr)
```

Now select the R Commander menu option:

Data-> Import data-> from text file, clipboard or URL

Complete the dialog box as shown opposite. I have given the resultant dataframe the name *headache* and indicated that the fields are separated by white space.

The headache datafile is available from the url below. You can either save it locally or use the *Internet URL* option in the above dialog box and load it directly:

```
http://www.robin-beaumont.co.uk/virtualclassroom/book2data/headache_stevens.txt
```

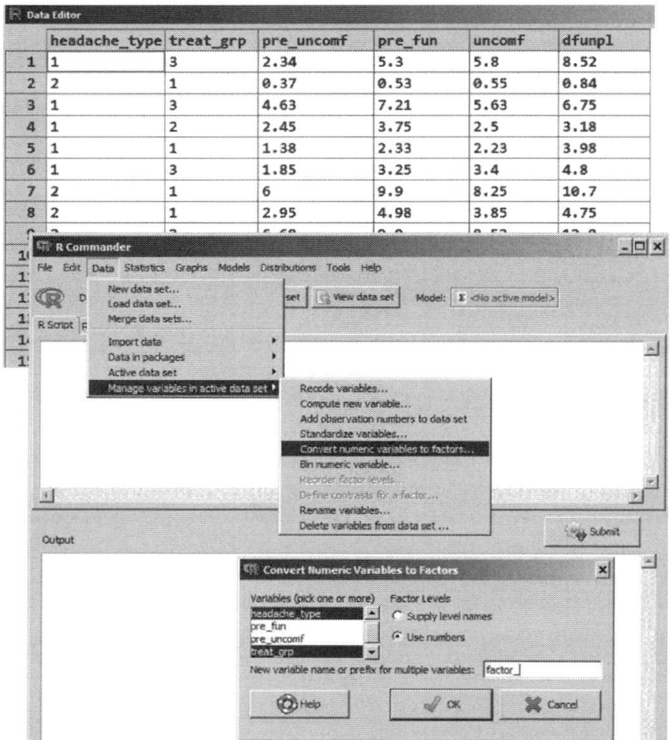

Clicking on the **View data set** button produces a view similar to that shown opposite. While both the *headache_type* and *treat_grp* variables are factors, R is unaware of this at the moment.

We use the following R Commander menu option to convert the two variables to factors:

Data-> Manage variables in active dataset-> Convert numeric variables to factors

Select the two variables in the **Variables** box then select the option *Use numbers* in the **Factor Levels** group. I have also added the following expression in the **New variable name or prefix** box:

```
factor_
```

The two new variables, each with the *factor_ prefix* are shown below. It is always a good idea to look at the summary statistics using the menu option **Statistics->summary->active dataset**

uncomf	dfunpl	factor_headache_type	factor_treat_grp
5.8	8.52	1	3
0.55	0.84	2	1
5.63	6.75	1	3
2.5	3.18	1	2
2.23	3.98	1	1
3.4	4.8	1	3
8.25	10.7	2	1
3.85	4.75	2	1
8.52	12.8	2	2
3.27	7.8	2	3
2.5	3.5	1	1
2.7	4.8	2	3

63.5.2 Plots of cell means – Interaction/profile plots

```
> summary(headache)
 headache_type     treat_grp      pre_uncomf        pre_fun
 Min.   :1.000   Min.   :1.00   Min.   : 0.370   Min.   : 0.530
 1st Qu.:1.000   1st Qu.:1.25   1st Qu.: 1.502   1st Qu.: 2.703
 Median :2.000   Median :3.00   Median : 2.195   Median : 4.400
 Mean   :1.541   Mean   :2.50   Mean   : 3.071   Mean   : 5.583
 3rd Qu.:2.000   3rd Qu.:3.00   3rd Qu.: 3.560   3rd Qu.: 7.277
 Max.   :2.000   Max.   :4.00   Max.   :15.100   Max.   :17.000
     uncomf           dfunpl        factor_headache_type factor_treat_grp
 Min.   : 0.220   Min.   : 0.390   1:45                 1:25
 1st Qu.: 1.377   1st Qu.: 2.590   2:53                 2:22
 Median : 2.330   Median : 4.050                        3:28
 Mean   : 3.532   Mean   : 5.549                        4:23
 3rd Qu.: 4.612   3rd Qu.: 7.463
 Max.   :15.500   Max.   :16.870
```

Because we are dealing with the assessment of mean values across the factor levels it is appropriate to obtain a graphical display of them. Such plots of the means for each of the cells are known as *Interaction* or *profile plots*.

Select the R Commander menu option:

Graphs-> Plot of means

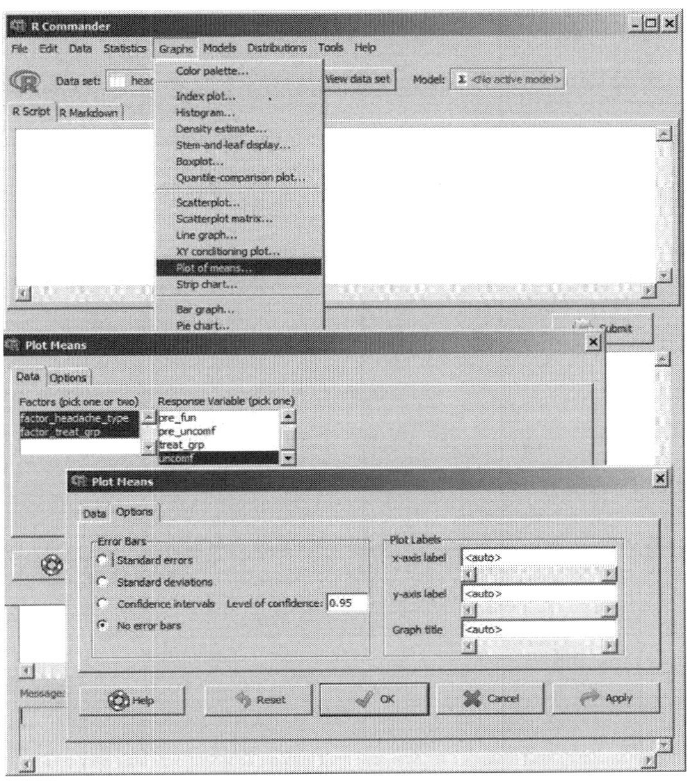

In the **Data** tab window select both the factors by holding down the *CTRL* key. Also select the outcome variable. Finally click on the **Options** tab and select the *No error bars* option. I find the plot much easier to interpret than the table of means displayed earlier. We can see that for both types of headache the means follow a similar profile over the four different treatment groups. We could reasonably create one profile from the other simply by adding a constant value, in other words they appear here to be **additive**. This is an important aspect to take note of from the plot. Informally I would say they show a high level degree of similarity. The less parallel they appear the greater the degree of interactivity is said to exist between them. Therefore, here we have very little interactivity.

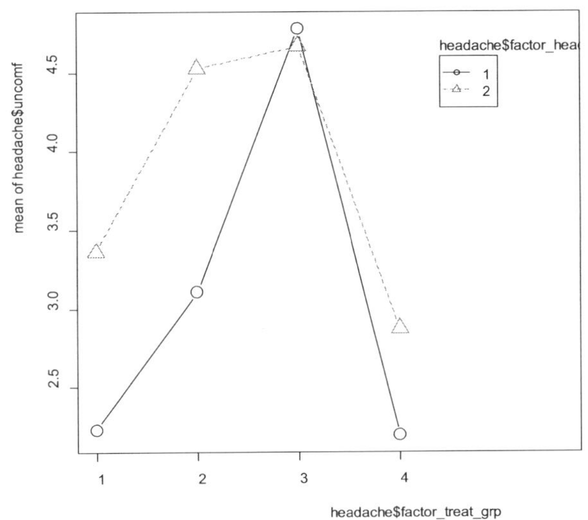

Those in treatment group 3 (the ones that *were explicitly* instructed to use the relaxation technique of breathing and imagery) coped with the loudest noise, while those in the control group (group 4) tolerated the least sound. I have not bothered to produce tables of means etc. as these will be generated automatically when we develop the model.

63.5.3 Homogeneity of Variance assumption

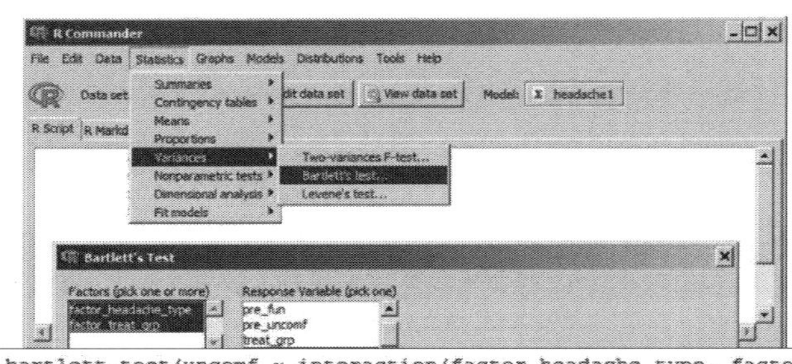

As with one-way ANOVA the variances within the groups (cells here) should be similar to obtain a valid ANOVA result. We can use the following R Commander menu option to obtain a p-value indicating how likely we would have obtained such disparate variances if all the samples had come from a single population:

Statistics-> Variances-> Bartlett's test

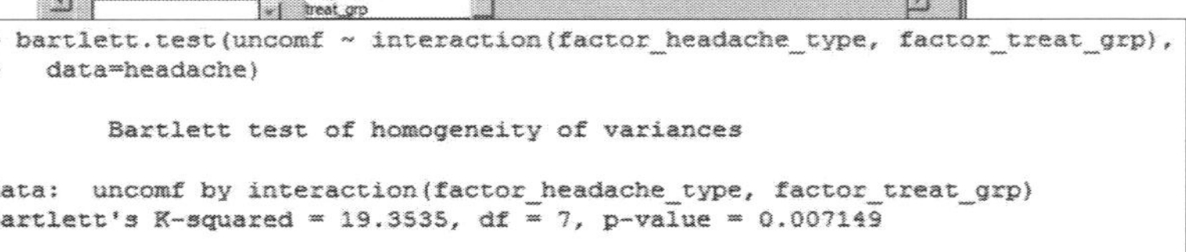

```
> bartlett.test(uncomf ~ interaction(factor_headache_type, factor_treat_grp),
+   data=headache)

        Bartlett test of homogeneity of variances

data:  uncomf by interaction(factor_headache_type, factor_treat_grp)
Bartlett's K-squared = 19.3535, df = 7, p-value = 0.007149
```

The resultant **p-value** indicates that we would have obtained variances such as those observed in the dataset or ones more extreme given that all the samples had been produced from a single population 7 times in a thousand. Because this is less than the usual cut off value of one in twenty (0.05) we come to the conclusion that the variances do not all come from the same population.

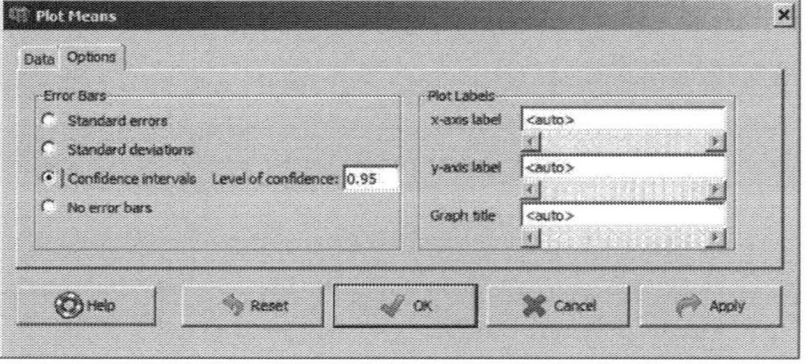

We can see the range of the variances informally by editing the plot means option we produced earlier to produce now a plot of either standard deviations or confidence intervals as shown opposite.

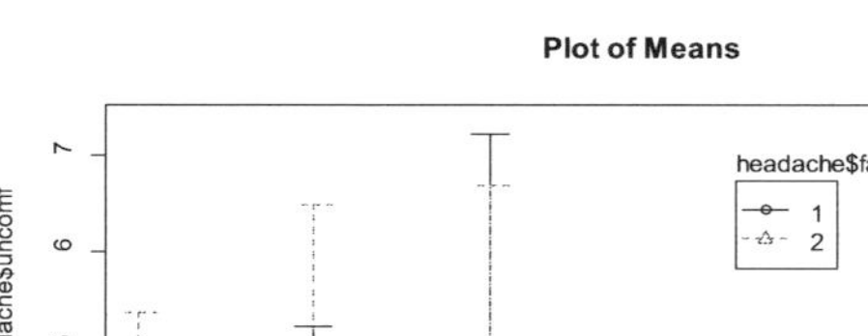

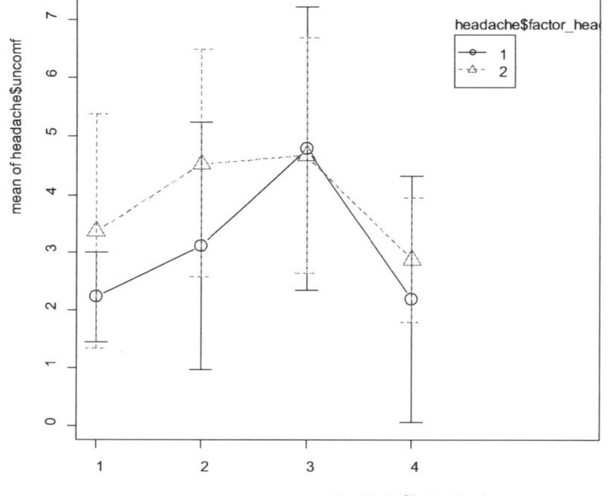

We can see that the CI's for level 1 of the *treat_grp* factor appear much narrower than for the other treatment conditions. Instead of using the menu option we could have obtained the plot using the following R code:

```
interaction.plot(headache$factor_treat_grp,
headache$factor_headache_type, headache$uncomf)
```

The plot of means does not show the individual observations and does not help assess the distribution of the various samples. We can obtain a basic boxplot using the following R code:

```
boxplot(uncomf ~ factor_headache_type *factor_treat_grp,
data=headache)
```

Even this rough and ready boxplot demonstrates its immediate usefulness. For example we can see that for both types of headache for treatment group 3 (shown as 1.3 and 2.3 along the x axis) there are single extreme outliers scoring around 15 compared to the median values that appear similar across all the cells. Looking at the level of overlap between the various boxes possibly 1:2 and 1:4 seem to have lower tolerance levels (☹) to the sound than the other groups.

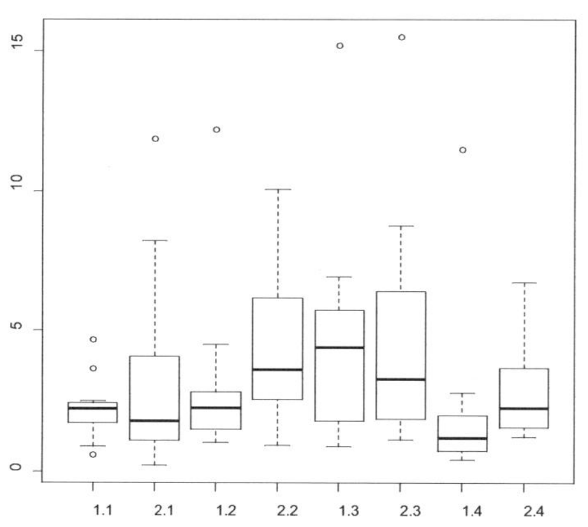

represents headache_type: treatment groups

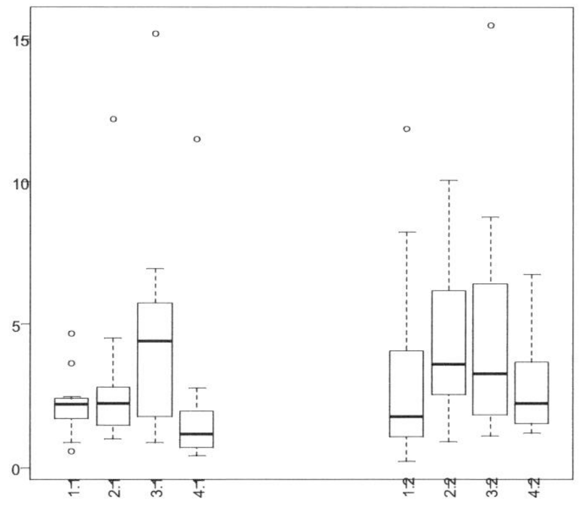

If we look at the height of the boxes, which represent the interquartile range (think equivalent to variance here), they appear to vary wildly with 1:1 (headache type=1; treat_grp=1) being less than a quarter of the height of 2:3, visually confirming the result from Bartlett's test mentioned earlier.

You can improve the look of the boxplot by creating a gap between factor levels.

```
boxplot(uncomf ~ headache$factor_treat_grp +
headache$factor_headache_type,
data=headache, at =c(1,2,3,4, 9,10,11,12), las = 2)
```

specify position along the x axis

specify direction of x axis text

It would be nice to see the actual observations along with the boxplots and this is achieved below.

63.5.4 Beeswarm plots and boxplots

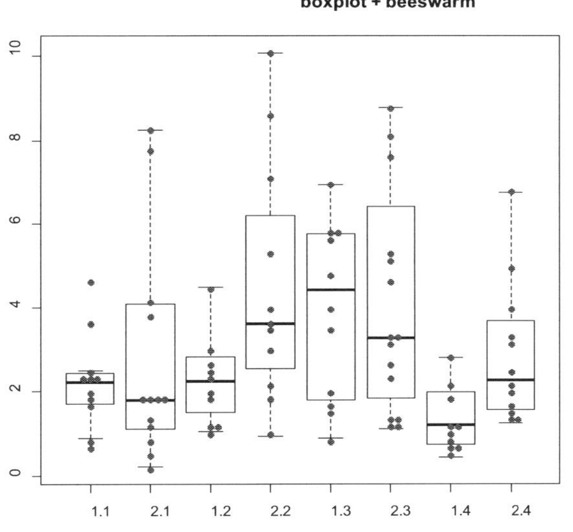

A useful plot that shows both the individual scores and their inferred distribution is the beeswarm plot, which also groups them for each factor level.

To create a beeswarm plot it is necessary to download, install and load the *beeswarm* package. The package contains a function called appropriately *beeswarm()* which works much like the boxplot function. The code below produces the graph shown opposite reinforcing our views obtained from the interaction plots and boxplots.

```
install.packages('beeswarm')

library(beeswarm)

boxplot(uncomf ~ headache_type+ treat_grp, data =
headache, outline = FALSE,

main = 'boxplot + beeswarm')
```

avoids double-plotting outliers, if any

```
beeswarm(uncomf ~ headache_type+ treat_grp, method = "square", data = headache,

col = 4, pch = 16, add = TRUE)
```

set colour

set marker symbol

adds one plot to the other

other options available are: "swarm", "center", "hex"

For an alternative to using the *beeswarm* package see the hints and tips section at the end of this chapter.

Given these findings about the distribution and non-homogeneity of the samples we might question any results obtained from the actual Factorial ANOVA but as a learning experience we will ignore them and press on.

63.5.5 Creating and evaluating a Factorial ANOVA

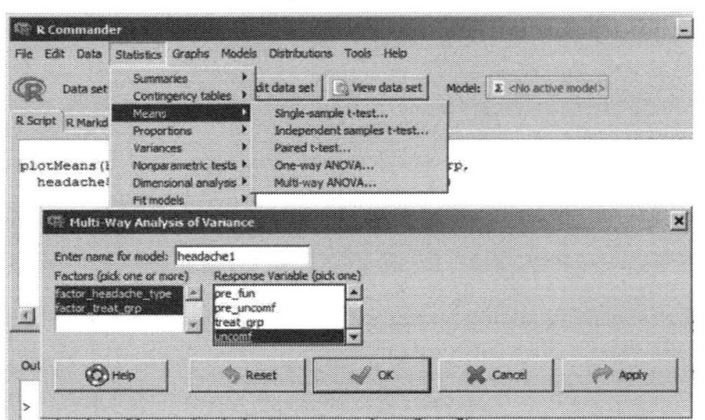

To produce a Factorial ANOVA using the R Commander Menu you can either select **Statistics->Fit models->linear model** or, as shown opposite:

Statistics-> means-> Multi-way ANOVA

I have given the model the name *headache1*. As with the plots of means I have selected the factors and outcome variables as shown opposite.

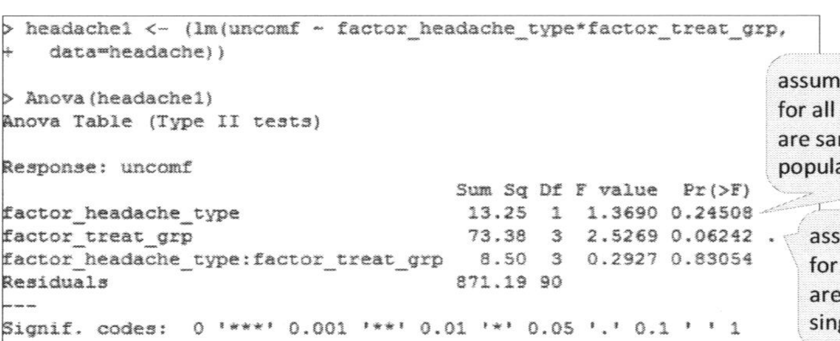

```
> headache1 <- (lm(uncomf ~ factor_headache_type*factor_treat_grp,
+   data=headache))

> Anova(headache1)
Anova Table (Type II tests)

Response: uncomf
                                   Sum Sq Df F value  Pr(>F)
factor_headache_type                13.25  1  1.3690 0.24508
factor_treat_grp                    73.38  3  2.5269 0.06242 .
factor_headache_type:factor_treat_grp   8.50  3  0.2927 0.83054
Residuals                          871.19 90
---
Signif. codes:  0 '***' 0.001 '**' 0.01 '*' 0.05 '.' 0.1 ' ' 1
```

assumes marginal means for all levels of headache are sampled from single population

assumes marginal means for all levels of treat_grp are sampled from a single population

The result indicates there is no statistically significant difference between the means across headache types. And also across treatment groups if we adopt the critical value of 0.05

```
> tapply(headache$uncomf,
+   list(factor_headache_type=headache$factor_headache_type,
+   factor_treat_grp=headache$factor_treat_grp), mean, na.rm=TRUE) # means
                    factor_treat_grp
factor_headache_type        1        2        3        4
                   1 2.229091 3.104545 4.780833 2.191818
                   2 3.360714 4.527273 4.669375 2.870833

> tapply(headache$uncomf,
+   list(factor_headache_type=headache$factor_headache_type,
+   factor_treat_grp=headache$factor_treat_grp), sd, na.rm=TRUE)
+   # std. deviations
                    factor_treat_grp
factor_headache_type        1        2        3        4
                   1 1.148342 3.182095 3.850052 3.172632
                   2 3.500454 2.914207 3.806751 1.690930

> tapply(headache$uncomf,
+   list(factor_headache_type=headache$factor_headache_type,
+   factor_treat_grp=headache$factor_treat_grp), function(x) sum(!is.na(x)))
+   # counts
                    factor_treat_grp
factor_headache_type  1  2  3  4
                   1 11 11 12 11
                   2 14 11 16 12
```

The interaction term indicates how much the lines we discussed earlier interact with one another, that is the less parallel they are the smaller the p-value would be for this parameter and the greater the degree of interaction.

The output also produces the means, standard deviations and counts for each cell but I find their graphical representation more helpful.

Removing the non-significant interaction term and re-running the analysis produces a p-value for factor_treat_grp of 0.0578 (df=3, mean sq=24.46, F=2.586).

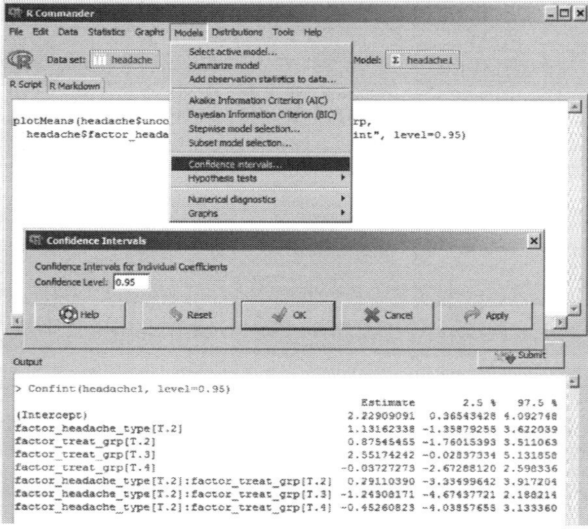

```
> Confint(headache1, level=0.95)
                                               Estimate       2.5 %   97.5 %
(Intercept)                                  2.22909091  0.36543428 4.092748
factor_headache_type[T.2]                    1.13162338 -1.35879255 3.622039
factor_treat_grp[T.2]                        0.87545455 -1.76015393 3.511063
factor_treat_grp[T.3]                        2.55174242 -0.02837334 5.131858
factor_treat_grp[T.4]                       -0.03727273 -2.67288120 2.598336
factor_headache_type[T.2]:factor_treat_grp[T.2]  0.29110390 -3.33499642 3.917204
factor_headache_type[T.2]:factor_treat_grp[T.3] -1.24308171 -4.67437721 2.188214
factor_headache_type[T.2]:factor_treat_grp[T.4] -0.45260823 -4.03857655 3.133360
```

63.5.6 Confidence intervals

Confidence intervals can be produced easily using the menu option:

Models-> Confidence intervals

As is consistent with the above non-statistically significant p-values all the 95% confidence intervals straddle zero.

63.5.7 Investigating the means graphically

To produce the necessary plots you need to download, install and load the *phia* package given in the two lines below. Within the *phia* package the *interactionMeans()* function when wrapped around the *plot()* function provides the necessary graphs.

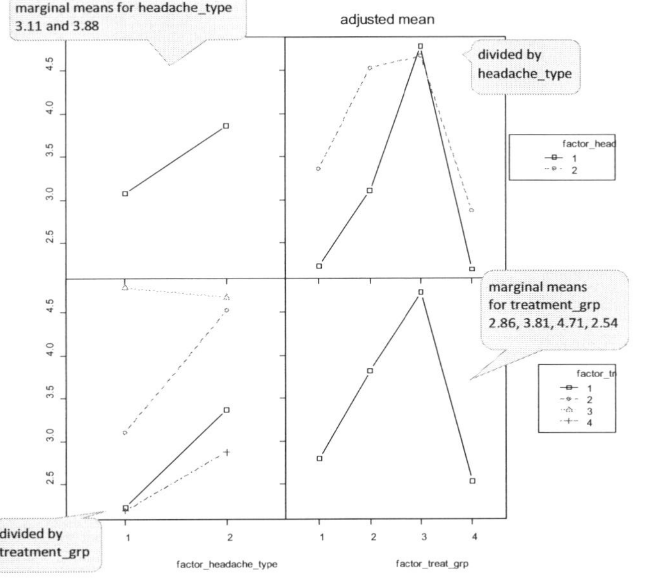


```
install.packages("phia")

library("phia")

adjust_means<- interactionMeans(headache1)

plot(adjust_means)
```

These mean values also give us insight into the fact that Factorial ANOVA is simply linear regression. Typing *summary(headache1)* in the R Commander *Script* or main *R Console* window produces the regression version of the analysis we have carried out, part of which is shown below. The output also produces the R squared value and associated F and p-values (not shown here).

While the output looks very different from the ANOVA output you can see some similarities. We know that the estimates (b's) in linear regression, shown below, are simply slopes and we can see how the slopes are describing those in the various plot of the means.

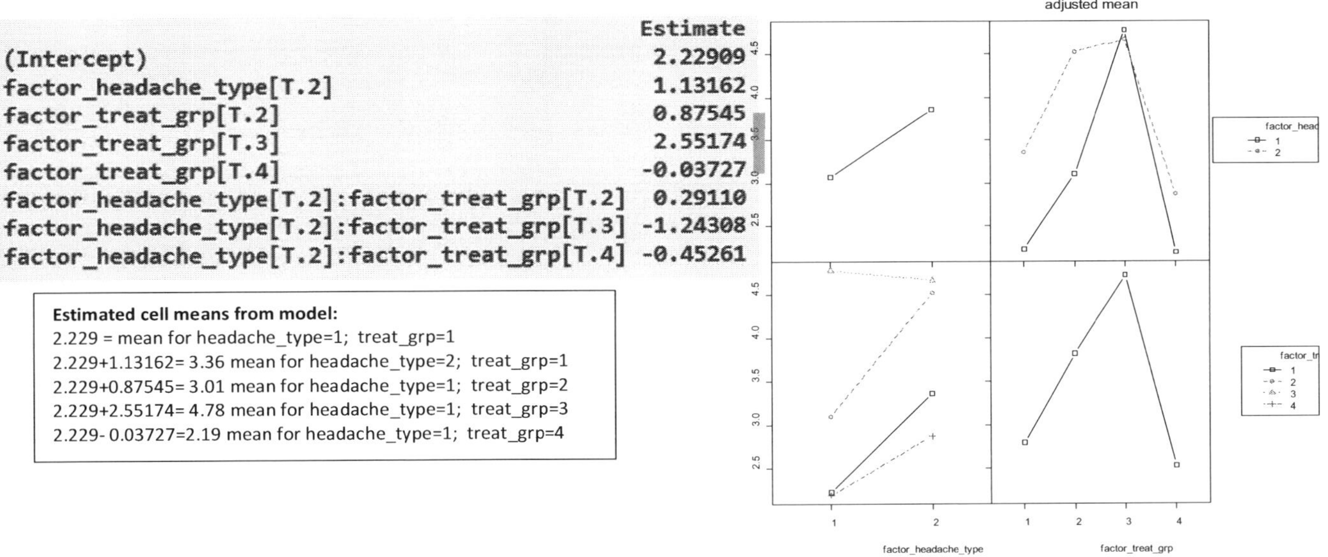

	Estimate
(Intercept)	2.22909
factor_headache_type[T.2]	1.13162
factor_treat_grp[T.2]	0.87545
factor_treat_grp[T.3]	2.55174
factor_treat_grp[T.4]	-0.03727
factor_headache_type[T.2]:factor_treat_grp[T.2]	0.29110
factor_headache_type[T.2]:factor_treat_grp[T.3]	-1.24308
factor_headache_type[T.2]:factor_treat_grp[T.4]	-0.45261

Estimated cell means from model:
2.229 = mean for headache_type=1; treat_grp=1
2.229+1.13162= 3.36 mean for headache_type=2; treat_grp=1
2.229+0.87545= 3.01 mean for headache_type=1; treat_grp=2
2.229+2.55174= 4.78 mean for headache_type=1; treat_grp=3
2.229- 0.03727=2.19 mean for headache_type=1; treat_grp=4

The above estimates can also be thought of as the expected means for various cells from the model. Comparing these values with the observed means (see earlier table) we see they are virtually identical. The estimates for the interaction term are a little more difficult to interpret, they can be thought of as the change in difference in the slopes between the two profile lines for each level of factor *treat_grp* as shown in the top right plot. Rosnow & Rosenthal, 1995 and Pagano, 2011 provide additional information.

63.5.8 Assessing the assumptions

We have already considered the equal variance assumption and because Factorial ANOVA is the same under the bonnet as multiple linear regression we can use the same menu option as we did in the multiple linear regression chapter to get an overview, as shown opposite.

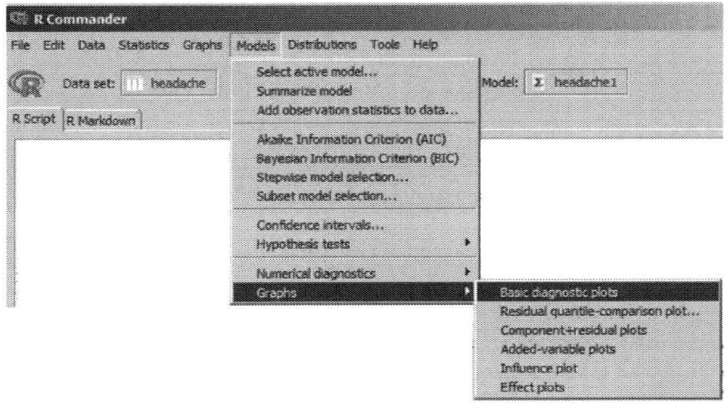

The results, shown below (enlarged to show detail), reinforce our view that the variance of the different cells varies widely, and the values are not normally distributed. We also note that the outliers previously identified in the boxplots are again prominent (cases 95, 36, 49). For an example of a set of data that does comply to the required assumptions see the *regression diagnostics* section in the *Multiple Linear Regression chapter*.

lm(uncomf ~ factor_headache_type * factor_treat_grp)

random –sky at night look good ☺

tests equal variance over factor level range

Residuals vs Fitted

Residuals

Fitted values

spread of points for each combination of levels very different ☹

tests normal distribution of residuals assumption

points on straight line = residuals normally distributed ☺

Normal Q-Q

Standardized residuals

Theoretical Quantiles

points 'S' or banana shaped=non-normal ☹

tests normal distribution of residuals assumption

random –sky at night look good ☺

Scale-Location

√|Standardized residuals|

Fitted values

points arranged like a Λ or other 'shapes' ☹

identifies influential points affecting the regression model

points to left and centre = small leverage ☺

Residuals vs Leverage

Standardized residuals

Cook's distance

Leverage

points near top right & bottom right ☹

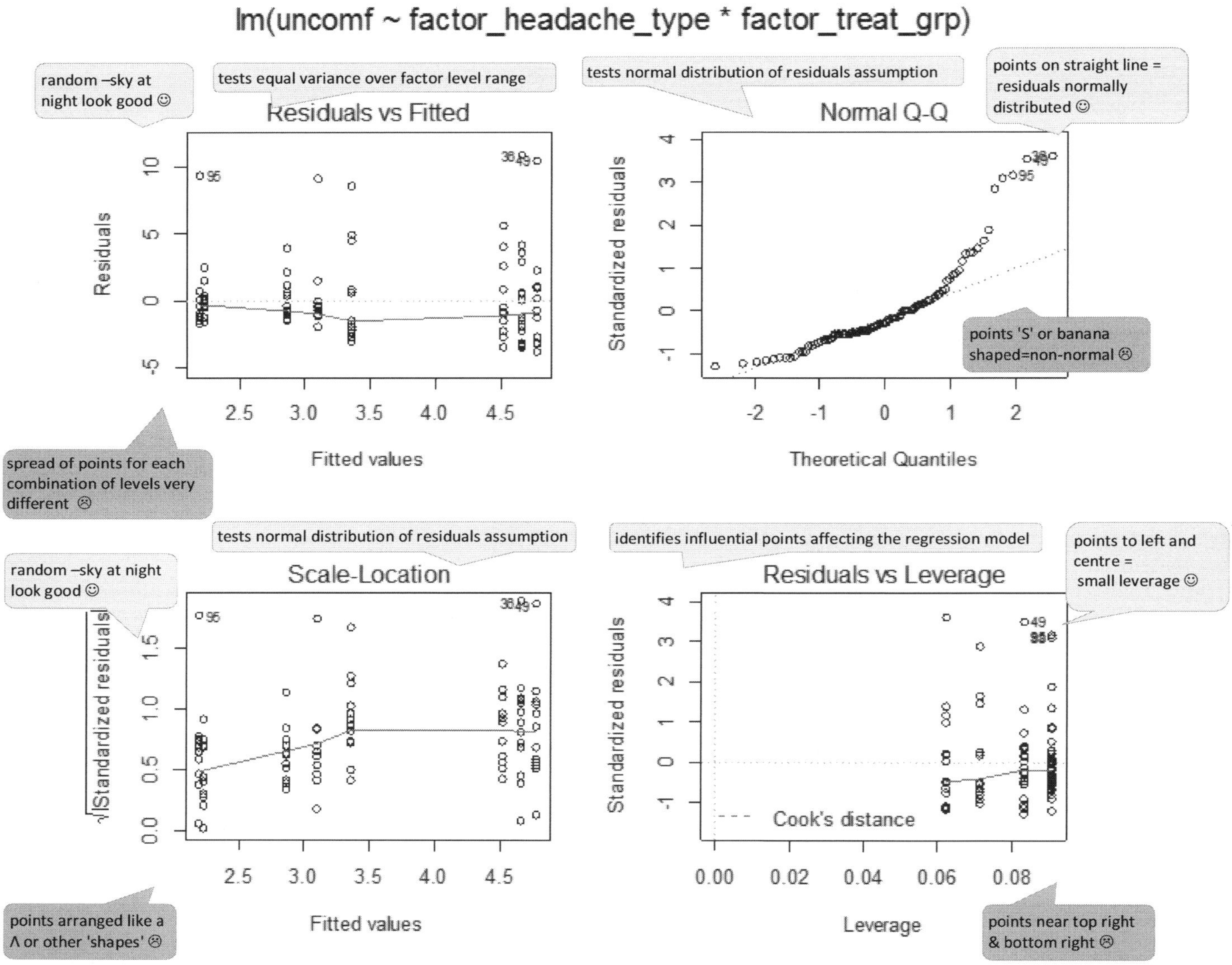

63.6 Post-hoc tests

```
> TukeyHSD(factor_anova)
  Tukey multiple comparisons of means
    95% family-wise confidence level

Fit: aov(formula = uncomf ~ factor_headache_type * factor_treat_grp, data = headache)

$factor_headache_type
         diff        lwr      upr     p adj
2-1 0.7725367 -0.4804057 2.025479 0.2237942

$factor_treat_grp
          diff        lwr       upr     p adj
2-1  0.9994613 -1.3813273 3.3802498 0.6911132
3-1  1.8455139 -0.3954877 4.0865154 0.1437364
4-1 -0.2871551 -2.6402527 2.0659425 0.9886502
3-2  0.8460526 -1.4742674 3.1663726 0.7754509
4-2 -1.2866164 -3.7153744 1.1421416 0.5109251
4-3 -2.1326690 -4.4245676 0.1592296 0.0776312
```

comparision across headache_type factor level means

comparisions across treat_grp factor level means

If we have obtained a statistically significant result of one or more of the parameters we would still not know exactly which means differed from which (see the simple ANOVA tutorial again if you can't see why). To tease out, or as some statisticians say poke the entrails, referring to the fact that we are hunting for something that might not be there and is often dubious, we carry out comparisons between the means of the various factor levels.

Such an approach is known as Post-hoc testing. One of the most common approaches is to use *Tukey's Honest Significant Differences* test. Unfortunately, you can't do this using R Commander menus so therefore type the following into the R Commander script window, highlight it and then click on the **Submit** button to run it.

you need to use the *aov()* function instead of *lm()* for the *TukeyHSD()* function to work

```
factor_anova<- aov(uncomf ~ factor_headache_type*factor_treat_grp, data=headache)

TukeyHSD(factor_anova)
```

```
$`factor_headache_type:factor_treat_grp`
              diff       lwr      upr     p adj
2:1-1:1  1.13162338 -2.758686 5.021933 0.9850933
1:2-1:1  0.87545455 -3.241662 4.992571 0.9977951
2:2-1:1  2.29818182 -1.818935 6.415298 0.6663491
1:3-1:1  2.55174242 -1.478688 6.582173 0.5110748
2:3-1:1  2.44028409 -1.341529 6.222097 0.4860366
1:4-1:1 -0.03727273 -4.154389 4.079844 1.0000000
2:4-1:1  0.64174242 -3.388688 4.672173 0.9996632
1:2-2:1 -0.25616883 -4.146478 3.634140 0.9999992
2:2-2:1  1.16655844 -2.723751 5.056868 0.9822326
1:3-2:1  1.42011905 -2.378332 5.218570 0.9409079
2:3-2:1  1.30866071 -2.224888 4.842210 0.9437066
1:4-2:1 -1.16889610 -5.059205 2.721413 0.9820274
2:4-2:1 -0.48988095 -4.288332 3.308570 0.9999178
2:2-1:2  1.42272727 -2.694389 5.539844 0.9609017
1:3-1:2  1.67628788 -2.354143 5.706718 0.9000205
2:3-1:2  1.56482955 -2.216983 5.346642 0.9024291
1:4-1:2 -0.91272727 -5.029844 3.204389 0.9971310
2:4-1:2 -0.23371212 -4.264143 3.796718 0.9999997
1:3-2:2  0.25356061 -3.776870 4.283991 0.9999994
2:3-2:2  0.14210227 -3.639711 3.923915 1.0000000
1:4-2:2 -2.33545455 -6.452571 1.781662 0.6479537
2:4-2:2 -1.65643939 -5.686870 2.373991 0.9055336
2:3-1:3 -0.11145833 -3.798711 3.575794 1.0000000
1:4-1:3 -2.58901515 -6.619446 1.441415 0.4919918
2:4-1:3 -1.91000000 -5.851839 2.031839 0.8034957
1:4-2:3 -2.47755682 -6.259370 1.304256 0.4659445
2:4-2:3 -1.79854167 -5.485794 1.888711 0.7981109
2:4-1:4  0.67901515 -3.351415 4.709446 0.9995111
```

The first part of the result is shown above which is for the main effects, first headache type and then treatment group.

The next section, shown opposite, is much longer and compares all the separate possible interaction levels. These are basically equivalent to the simple interaction effects I mentioned earlier.

You can also obtain a plot of the confidence intervals, typing the following R commander code:

```
plot(TukeyHSD(factor_anova))
```

When you have many comparisons the labels on the left of the plot disappear. However, you can see more of them by maximising the window in R and the order is identical to that produced in the text output which helps.

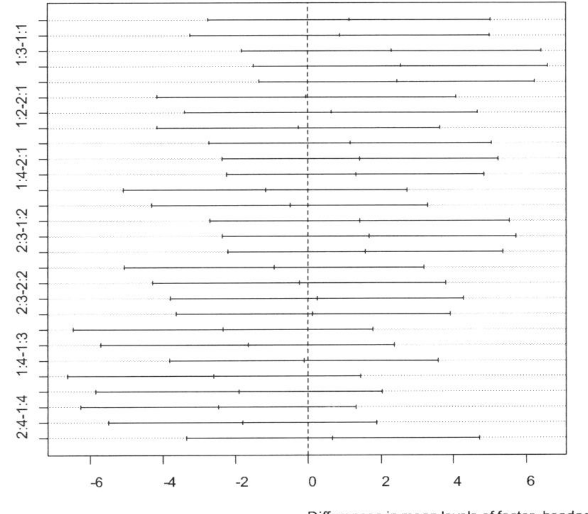

63.6.1 Pairwise comparisons

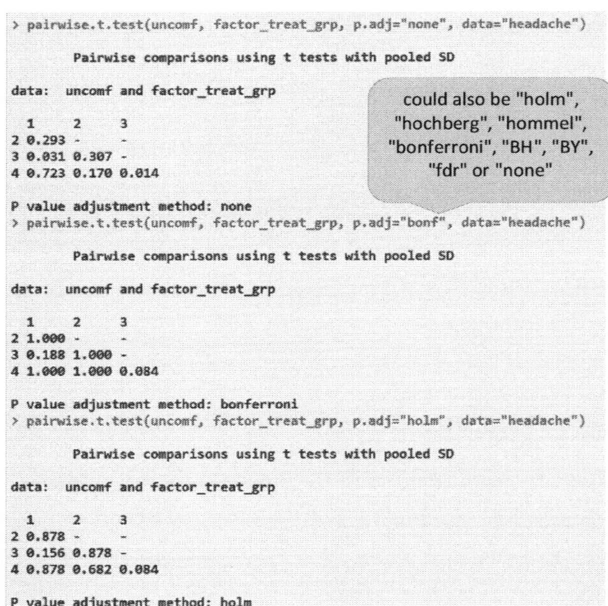

```
> pairwise.t.test(uncomf, factor_treat_grp, p.adj="none", data="headache")

        Pairwise comparisons using t tests with pooled SD

data:  uncomf and factor_treat_grp

  1     2     3
2 0.293 -     -
3 0.031 0.307 -
4 0.723 0.170 0.014

P value adjustment method: none
> pairwise.t.test(uncomf, factor_treat_grp, p.adj="bonf", data="headache")

        Pairwise comparisons using t tests with pooled SD

data:  uncomf and factor_treat_grp

  1     2     3
2 1.000 -     -
3 0.188 1.000 -
4 1.000 1.000 0.084

P value adjustment method: bonferroni
> pairwise.t.test(uncomf, factor_treat_grp, p.adj="holm", data="headache")

        Pairwise comparisons using t tests with pooled SD

data:  uncomf and factor_treat_grp

  1     2     3
2 0.878 -     -
3 0.156 0.878 -
4 0.878 0.682 0.084

P value adjustment method: holm
```

could also be "holm", "hochberg", "hommel", "bonferroni", "BH", "BY", "fdr" or "none"

Many people consider Tukey's Honest Significant Differences test to be too honest, meaning that it requires the differences to be greater than some would expect for them to be statistically significant. Two alternatives are pairwise tests or comparing all groups to a control.

Pairwise tests only consider two variables at a time, one output and a single input. The R function *pairwise.t.test()* offers 8 methods of calculating these comparisons, details of which can be found by typing *?pairwise.t.test()* in the *R Console* window. The code below produces three varieties, unadjusted comparisons (same as doing an ordinary *t* test), Bonferroni corrected comparisons and Holm adjusted ones. All the methods, accept the 'none' option, attempt to avoid the pitfalls of finding statistically significant differences between the means when they do not exit – a constant threat when carrying out multiple tests.

```
pairwise.t.test(uncomf, factor_treat_grp, p.adj="none", data="headache")
pairwise.t.test(uncomf, factor_treat_grp, p.adj="bonf", data="headache")
pairwise.t.test(uncomf, factor_treat_grp, p.adj="holm", data="headache")
```

By default the above function uses the pooled standard deviation, however you can request that the individual factor level standard deviations are used instead by adding *pool.sd = FALSE* to the code.

63.6.2 Comparing multiple groups to a control - Dunnett's test

Dunnett's test allows comparison between a control group and several other groups and works best when interactions are not specified in the model so I'll first create a new model:

```
> test_dunnett=glht(factor_anovab,linfct=mcp(factor_treat_grp="Dunnett"))
> confint(test_dunnett)

         Simultaneous Confidence Intervals

Multiple Comparisons of Means: Dunnett Contrasts


Fit: aov(formula = uncomf ~ factor_headache_type + factor_treat_grp,
    data = headache)

Quantile = 2.3908
95% family-wise confidence level


Linear Hypotheses:
             Estimate lwr     upr
2 - 1 == 0  0.9975  -1.1539  3.1488
3 - 1 == 0  1.8459  -0.1775  3.8693
4 - 1 == 0 -0.2884  -2.4137  1.8369
```

problem uses level 1 as the control

```
factor_anovab<- aov(uncomf ~ factor_headache_type +
factor_treat_grp, data=headache)
```

The necessary function called *glht()*, possibly meaning *General linear hypotheses and multiple comparisons* is within the *multcomp* package which you need to install and load:

```
install.packages("multcomp"); library(multcomp)
```

The *glht()* function is a rather complex object so I will first demonstrate a basic analysis.

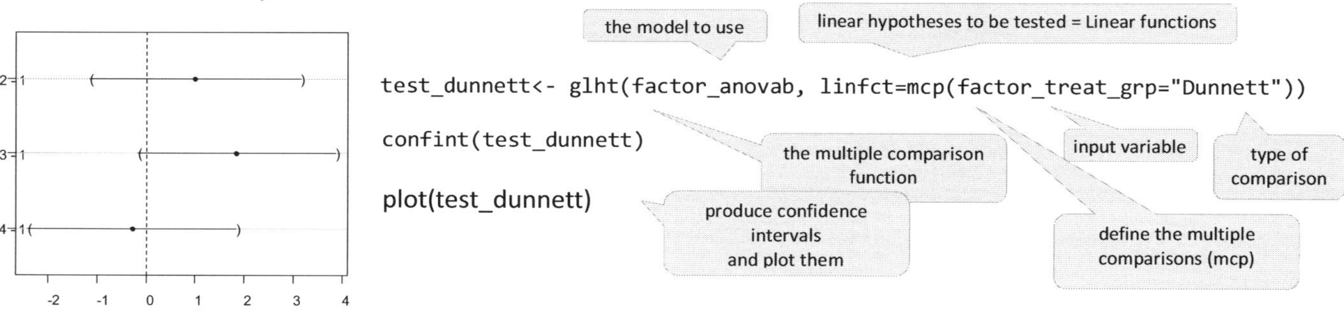


```
test_dunnett<- glht(factor_anovab, linfct=mcp(factor_treat_grp="Dunnett"))
confint(test_dunnett)
plot(test_dunnett)
```

```
> dunnett_base4_noint <- glht(factor_anovab, linfct=mcp(factor_treat_grp=contrasts))
> dunnett_base4_noint

         General Linear Hypotheses

Multiple Comparisons of Means: User-defined Contrasts


Linear Hypotheses:
            Estimate
1 - 4 == 0    0.2884
2 - 4 == 0    1.2859
3 - 4 == 0    2.1343

> confint(dunnett_base4b_noint)

         Simultaneous Confidence Intervals

Multiple Comparisons of Means: User-defined Contrasts


Fit: aov(formula = uncomf ~ factor_headache_type + factor_treat_grp,
    data = headache)

Quantile = 2.3836
95% family-wise confidence level


Linear Hypotheses:
            Estimate lwr      upr
1 - 4 == 0  0.28843 -1.83040  2.40726
2 - 4 == 0  1.28589 -0.90052  3.47230
3 - 4 == 0  2.13433  0.07003  4.19863

> plot(dunnett_base4b_noint)
```

Unfortunately, the above analysis uses level 1 as the control group when the actual control group is level 4. Possibly the simplest way to solve the problem is to recode the factor levels using the R Commander menu option. However I will demonstrate how you can set the *glht()* function to use level '4' as the control group in the code below.

```
factorlevels <-
table(headache$factor_treat_grp)

contrasts <-  contrMat (factorlevels,
type = "Dunnett", base=4)

dunnett_base4_noint <- glht(factor_anovab,
linfct = mcp(factor_treat_grp=contrasts))

dunnett_base4_noint

confint(dunnett_base4_noint)

plot(dunnett_base4_noint)
```

contrMat= contrast matrix

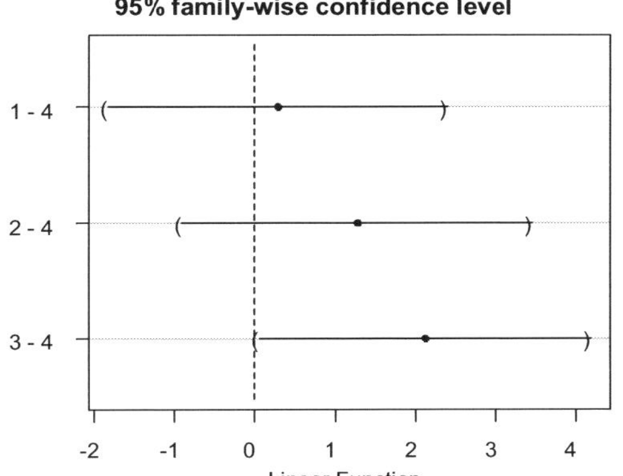


```
summary(dunnett_base4_noint) # to obtain the p-values
```

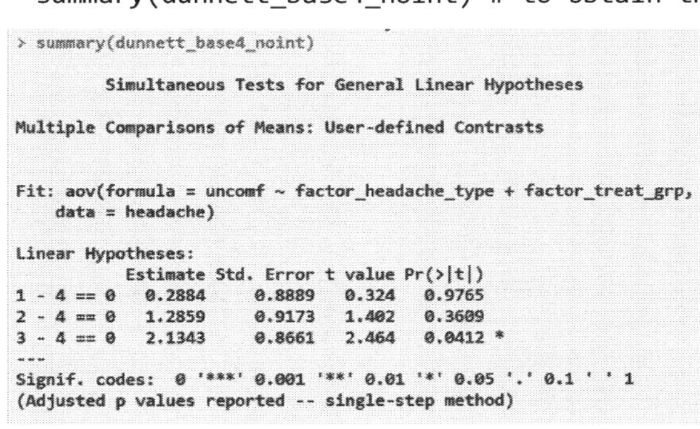

```
> summary(dunnett_base4_noint)

         Simultaneous Tests for General Linear Hypotheses

Multiple Comparisons of Means: User-defined Contrasts


Fit: aov(formula = uncomf ~ factor_headache_type + factor_treat_grp,
    data = headache)

Linear Hypotheses:
            Estimate Std. Error t value Pr(>|t|)
1 - 4 == 0    0.2884     0.8889   0.324   0.9765
2 - 4 == 0    1.2859     0.9173   1.402   0.3609
3 - 4 == 0    2.1343     0.8661   2.464   0.0412 *
---
Signif. codes:  0 '***' 0.001 '**' 0.01 '*' 0.05 '.' 0.1 ' ' 1
(Adjusted p values reported -- single-step method)
```

To instruct the *glht()* function we want to use level 4 as the control group there are several stages we need to go through. We first need to find the number of observations at each level of the factor. These values are saved to an object I have called *factorlevels* as the information is required by the *contrMat()* function. *contrMat()* produces an object known as a contrast matrix. This contrast matrix then instructs the *glht()* function how to do the comparisons.

Defining a contrast matrix in the multcomp R package

Which factor level should be classed as the control group? When the comparison type is Dunnett 1 is the default

contrasts <- contrMat (factorlevels, type = "Dunnett", base=4)

contrMat= contrast matrix

a vector of values indicating the number of observations at each factor level

type of comparison required one of:
"Dunnett",
"Tukey",
"Sequen",
"AVE",
"Changepoint",
"Williams",
"Marcus",
"McDermott",
"UmbrellaWilliams",
"GrandMean"

The definition of various contrasts is a important aspect of focused analysis so I have provided a bit more detail opposite. For further details see the glossary entry **Contrasts** or the wikipeadia article *planned contrasts* or do a google search. Alternatively Crawley, 2005 provides a readable chapter on the subject.

Comparison between treatment groups 3 and 4	P-value
Bonferroni	0.084
Holm	0.084
Tukey HSD	0.078
Dennett	0.041
'None' (same as a 2 sample t test)	0.014

The table opposite lists the results from the above post-hoc analyses, for the difference between treatment groups 3 (treatment) and 4 (control). We see little difference between the p-values varying between .08 and .04 ignoring the inappropriate independent samples *t* test.

The development and evaluation of different approaches to post-hoc and planned comparisons is an active research area and the authors of the *multcomp* R package have produced both a book and several articles concerning it (Bretz, Hothorn and Westfall, 2010; Bretz, Genz & Hothorn, 2001).

63.7 Effect size

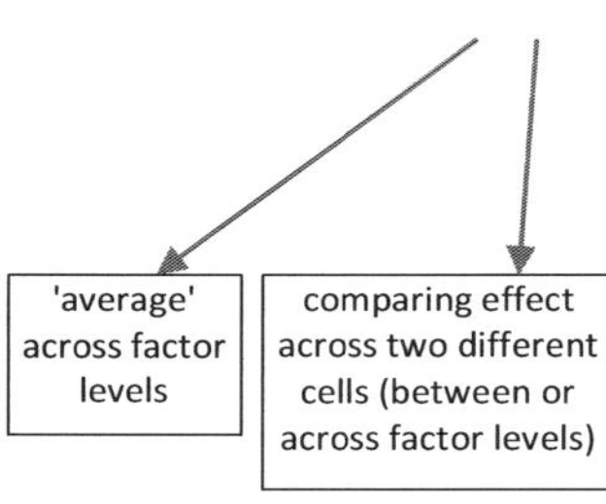

In the introductory tutorial concerning one-way ANOVA an effect size measure was not formally defined, other than being marked on the plot of means indicating that it measures the difference between the means in some way. In the situation with one factor you can develop an average effect size over the various levels of the factor or you can consider the difference between each factor level. An average effect size measure in this situation is eta squared (η^2) or omega squared (ω^2) as described in the chapter entitled, *Confidence intervals for effect sizes*, section, *the non-central F distribution.*

Cohen guidelines		Small	medium	large
η^2 eta squared	one-way anova (regression)	0.01	0.06	0.14
η^2	Anova	0.02	0.13	0.26
ω^2	Anova	0.01	0.06	0.14

When considering more than one factor a partial eta or omega squared value is used instead as this provides a measure of the variance explained in the model for the given factor (it has a upper limit of 1 like a correlation). Unfortunately, R does not have a built in function to produce eta or partial eta squared but the function *etaSquared()* in the *lsr* ("Learning Statistics with R") package does provide both values.

But before using this I will consider the various values produced by the *aov()* function to create partial eta squared using R code. This is a basic adaptation of the code posted by Dr Chris Hummersone (lecturer in sound recording at the University of Surrey) on his blog at http://chris-hummersone.blogspot.co.uk/2011/06/anova-in-r.html .

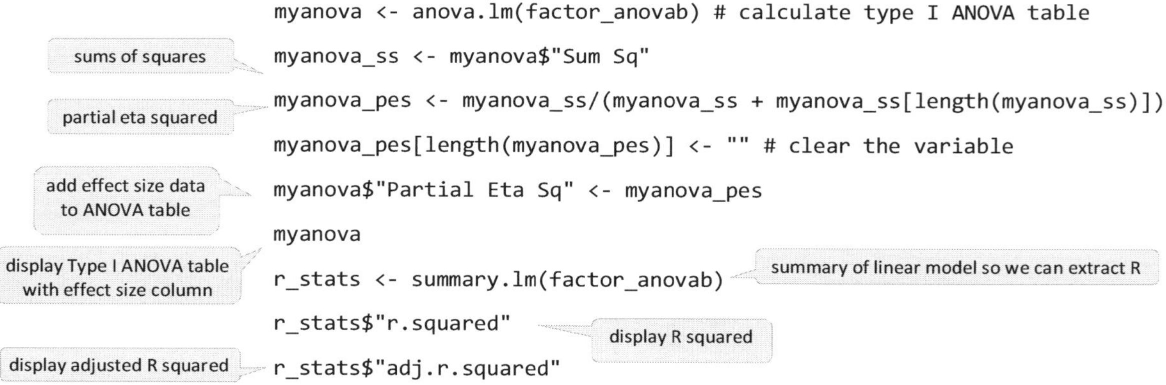


```
myanova <- anova.lm(factor_anovab) # calculate type I ANOVA table
myanova_ss <- myanova$"Sum Sq"
myanova_pes <- myanova_ss/(myanova_ss + myanova_ss[length(myanova_ss)])
myanova_pes[length(myanova_pes)] <- "" # clear the variable
myanova$"Partial Eta Sq" <- myanova_pes
myanova
r_stats <- summary.lm(factor_anovab)
r_stats$"r.squared"
r_stats$"adj.r.squared"
```

```
> myanova <- anova.lm(factor_anovab) # calculate type I ANOVA table
> myanova_ss <- myanova$"Sum Sq" # sums of squares
> myanova_pes <- myanova_ss/(myanova_ss + myanova_ss[length(myanova_ss)]) # partial eta squared
> myanova_pes[length(myanova_pes)] <- "" # clear effect size for residual
> myanova$"Partial Eta Sq" <- myanova_pes # add effect size data to ANOVA table
> myanova # display Type I ANOVA table with effect size column
Analysis of Variance Table

Response: uncomf
                  Df Sum Sq Mean Sq F value   Pr(>F) Partial Eta Sq
factor_headache_type  1  14.52 14.5245  1.5355 0.218406       0.016243
factor_treat_grp      3  73.38 24.4601  2.5859 0.057799       0.076993
Residuals            93 879.69  9.4591
> r_stats <- summary.lm(factor_anovab) # summary of linear model
> r_stats$"r.squared" # display R squared
[1] 0.09084853
> r_stats$"adj.r.squared" # display Adjusted R squared
[1] 0.05174524
```

The result is shown opposite. Referring back to the chapter cited above we see that the first value (0.016) can be classed as less than a small effect size (as value <0.02) and 0.077 can be seen as a small effect size (as value <0.13).

```
> etaSquared(factor_anovab, type=1, anova = TRUE)
                         eta.sq eta.sq.part        SS df        MS        F          p
factor_headache_type 0.01501087  0.01624268  14.52448  1 14.524478 1.535510 0.21840554
factor_treat_grp     0.07583765  0.07699339  73.38029  3 24.460098 2.585892 0.05779867
Residuals            0.90915147          NA 879.69234 93  9.459057       NA         NA
> etaSquared(factor_anovab, type=2, anova = TRUE)
                         eta.sq eta.sq.part        SS df        MS        F          p
factor_headache_type 0.01369518  0.01484015  13.25142  1 13.251420 1.400924 0.23958553
factor_treat_grp     0.07583765  0.07699339  73.38029  3 24.460098 2.585892 0.05779867
Residuals            0.90915147          NA 879.69234 93  9.459057       NA         NA
> etaSquared(factor_anovab, type=3, anova = TRUE)
                         eta.sq eta.sq.part        SS df        MS        F          p
factor_headache_type 0.01369518  0.01484015  13.25142  1 13.251420 1.400924 0.23958553
factor_treat_grp     0.07583765  0.07699339  73.38029  3 24.460098 2.585892 0.05779867
Residuals            0.90915147          NA 879.69234 93  9.459057       NA         NA
```

The screenshot opposite provides comparable values using the *etaSquared()* function in the *lsr* package. Here you can specify the type of sums of squares you wish to use and also specify if you want the standard ANOVA table shown as well.

One problem with both forms of eta squared is that they are sample values rather than population corrected ones (think R and adjusted R here). An equivalent value for the population is a value known as omega (ω^2) squared, a value also known as the coefficient of determination.

Andy Field's R book (Field, 2013) mentions a planned companion package called DSUR which is meant to contain a function to produce partial omega squared values for each estimate of a 2 way ANOVA with an optional interaction but, at the time of writing, the package is not available. Instead you can download the relevant function using the following line of R code.

```
thecode<- source("http://www.robin-beaumont.co.uk/virtualclassroom/book2data/r_code/omega_squared_andy_field.r")
        thecode
```

```
> omega_factorial(11, 2, 4, 14.52, 73.38, 0 ,879.69)
[1] "Omega-Squared A:  0.00360098606548299"
[1] "Omega-Squared B:  0.0412754932530849"
[1] "Omega-Squared AB:  -0.0337102419064035"
```

ignore the AB term here

I have put the code into an object called *thecode* so that I can see it, we could have added the option *echo=TRUE*, but this only shows a selection of the lines, whereas requesting to display *thecode* displays it all. We now have a function called *omega_factorial()*, which takes several values:

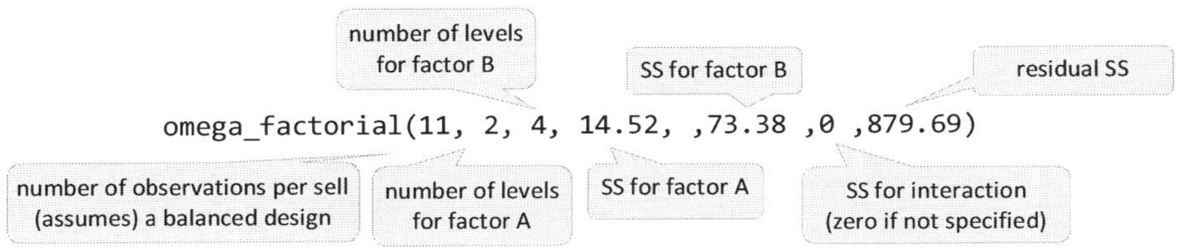

```
omega_factorial(11, 2, 4, 14.52, ,73.38 ,0 ,879.69)
```

number of observations per sell (assumes) a balanced design

number of levels for factor A

number of levels for factor B

SS for factor A

SS for factor B

SS for interaction (zero if not specified)

residual SS

Partial omega squared is a measure of the variance in the output accounted for by the specific input, with the effects of the other inputs removed. This should sound familiar, as it is similar to the part correlation description provided in the *Multiple linear regression* chapter.

There is much ongoing research along with heated debate, concerning the best effect measure for factorial designs. Usually the above partial eta and omega squared statistics are used but Olejnik & Algina, 2003, provide compelling evidence why alternative measures such as the Generalized Eta and Omega Squared Statistics are considered more appropriate by some (Bakeman, 2005). The Generalised Eta Squared value is available in two packages; the *aov.car()* function in the *afex* package, and the *ezANOVA()* function in the *ez* package.

install and load the afex package

```
install.packages("afex", repos="http://R-Forge.R-project.org")

library("afex")

headache$rowid <- rownum(headache)

aov.car (uncomf ~ factor_headache_type + factor_treat_grp + Error(rowid),
return="Anova", data=headache)

aov.car (uncomf ~ factor_headache_type + factor_treat_grp + Error(rowid),
return="nice", data=headache)
```

the aov.car function needs a subject id to work properly so therefore we add one to the headache dataframe

the function also requires you to specify explicitly the error term

this option provides the standard ANOVA output using type III sums of squares.
You can change this by using type=2. Type 1 not available

this option provides the effect size measure for each factor. The default is for the Generalised Eta Squared statistic (ges). Use es="pes" to report the standard eta squared

```
Anova Table (Type III tests)

Response: dv
                      Sum Sq Df  F value Pr(>F)
(Intercept)          1154.69  1 122.0728 <2e-16 ***
factor_headache_type   13.25  1   1.4009 0.2396
factor_treat_grp       73.38  3   2.5859 0.0578 .
Residuals             879.69 93
---
Signif. codes:  0 '***' 0.001 '**' 0.01 '*' 0.05 '.' 0.1 ' ' 1
> aov.car(uncomf ~ factor_headache_type + factor_treat_grp + Error(rowid), return="nice", data=headache)
                Effect    df  MSE      F ges   p
1 factor_headache_type 1, 93 9.46   1.40 .01 .24
2     factor_treat_grp 3, 93 9.46 2.59 + .08 .06
```

The results are shown opposite. 0.01 for headache type and 0.08 for group. Olejnik & Algina, 2003 p.446 indicate these values can be interpreted the same way as traditional eta squared (table previous page); therefore we have a less than small effect for headache type and a small one for group. Comparing these values to those produced from the *etaSquared()* function in the *lsr* package and our home grown R code demonstrates little difference.

63.7.1 Observed versus measured variables

There are two main reasons for the lack of clarity concerning which of the effect size measures is the best; firstly, all the above measures are model dependent, secondly they do not take into account the principle that factors are broadly of two types: whether they can be manipulated by the researcher or not.

For example, factors like hair/eye colour or gender and many demographic values are not able to be manipulated by the researcher. In contrast, such things as which experimental group they are randomly assigned to, or the amount of treatment given is under the direct control of the researcher. These two different types of factors require a different method of assessing their effect size.

$$\eta_G^2 = \frac{ss_{effect}}{\delta(ss_{effect}) + \sum_{Meas} ss_{Meas} + \sum_k ss_k}$$

ss_{effect} = Sum of squares for the effect of interest
δ = 1 f the effect is a manipulated factor, zero if a observed variable
ss_{Meas} = the are the sums of squares for all sources of variance that involve measured factors (rather than manipulated factors) but do not within subjects factors
ss_k = sums of squares for all sources of variance within subjects.

The equation for the generalised eta squared value has a term in the denominator which changes value, dependent upon the factor being an observed or manipulated one. The equation also shows us that it takes into account other things but they do not concern us at present.

The fact that these other issues have not been taken into account in the past when considering effect size measures for Factorial designs is an important oversight , so I will allow Olejnik & Algina, 2003 to speak for themselves:

The importance of dividing factors into observed and manipulated. From Olejnik & Algina, 2003.

The current practice of reporting the partial eta or partial omega squared statistic to estimate an effect size for each factor in multifactor research design is often misleading and inappropriate. Routine use of these measures of effect size does not appropriately consider the design features of a research study when estimating the magnitude of an effect. In the case of measures of association in an ANOVA context, little attention has been given to the design features, which can affect the size of the estimated effects. Textbooks and computer output often imply that the partial eta or partial omega squared statistics are the appropriate effect-size measures to use in all research designs that include more than a single explanatory variable. This is an appropriate recommendation if the researcher manipulates all of the factors in the design.

However, if a research design includes one or more measured factors, partial eta or omega squared statistics often provide an undesirable effectsize measure. Researchers often include one or more measured factors as covariates or as blocking variables [...]. These factors increase the statistical power by reducing the error variance. However, although power of the analysis is increased, the interpretation of the hypothesis test is unchanged. The reduced error variance as a result of blocking or covarying, however, does affect the interpretation of effect-size measures. Reducing the error variance restricts the population for whom the effect is being estimated. Partial eta and omega squared also can provide indices that are inappropriately compared to Cohen's (1988) guidelines for defining the small, medium, and large effects. [in contrast ...]

The generalized eta and omega squared statistics have two major advantages. First, these statistics provide measures of effect size that are comparable across a wide variety of research designs that are popular in education and psychology. Second, these effect-size measures provide indices of effect that are consistent with Cohen's (1988) guidelines for defining the magnitude of the effect. Cohen pointed out three decades ago that design considerations must be used when computing an effect-size measure. Currently, most researchers who choose to report an effect size as the proportion of variance explained have ignored Cohen's caution. Using the procedures outlined in the present article, researchers can correct this omission and provide more comparable effect-size measures.

The *aov.car()* function in the *afex* package allows the specification of these two types of factors. By default all the inputs are assumed to be manipulated variables and you specifically indicate that a particular factor is an observed type factor by adding the *observed=variable_name* option. Unfortunately, in the headache dataset there are no observed type factors. But for demonstration purposes I will assume we have an observed type factor called *gender* indicating the sex of each subject. To include this variable in the analysis we would modify the R code thus:

observed= indicates an observed type variable.

```
aov.car(uncomf ~ factor_headache_type + factor_treat_grp +gender +  Error(rowid),
observed = c("gender"),  return="nice",   data=headache)
```

You can have any number of variables of either type.

63.7.2 Effect size for specific comparisons - minimum efficient scale (mes), Cohen's d

All the above effect size measures have attempted to capture the effect over all levels of the factor. Frequently what is more interesting is to investigate the effect size for two specific cells and appropriate values can be obtained using the *mes()* function from the *compute.es* package. All you need to know is the mean, standard deviation and number of observations for each cell, all values we have from the initial ANOVA output. The following code installs and loads the *compute.es* package and then produces Cohen's *d*, along with a set of many other effect size measures.

Uncomf =outcome variable		**Treat_grp factor (4 levels)**				**marginal**
	Level:	1	2	3	4	
headache_type factor (2 levels)	1	n=11 $\bar{x}_{11}$= 2.22 sd=1.148	n=11 $\bar{x}_{12}$= 3.1 sd=3.18	n=12 $\bar{x}_{13}$= 4.78 sd=3.85	n=11 $\bar{x}_{14}$= 2.19 sd=3.17	$\bar{x}_{1.}$=3.11
	2	n=14 $\bar{x}_{21}$= 3.36 sd=3.50	n=11 $\bar{x}_{22}$= 4.53 sd=2.91	n=16 $\bar{x}_{23}$= 4.69 sd=3.81	n=12 $\bar{x}_{24}$= 2.87 sd=1.69	$\bar{x}_{2.}$=3.88
marginal		$\bar{x}_{.1}$=2.86	$\bar{x}_{.2}$=3.81	$\bar{x}_{.3}$=4.71	$\bar{x}_{.4}$=2.54	$\bar{x}$=3.53

As an example consider the effect size between the means of the treatment groups 3 (= *treatment*) and 4 (=*control*) for migraine suffers.

```
install.packages("compute.es")

library("compute.es")
```

compare control group to treatment group for migraine suffers (=1)

mean group1 — sd group1 — number in cell group1

```
mes(4.78, 2.19, 3.85, 3.17, 12, 11)
```

mean group2 — sd group2 — number in cell group2

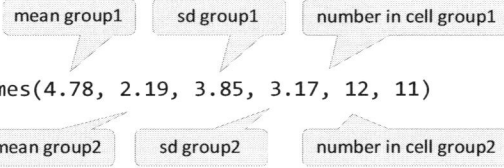

```
> # compare control group to treatment group for migraine suffers (=1):
> mes(4.78, 2.19, 3.85, 3.17, 12, 11)

   EFFECT SIZE CALCULATION (FOR SINGLE INPUT)

Mean Differences ES:

 d [ 95 %CI] = 0.73 [ -0.17 , 1.63 ]
  var(d) = 0.19
  p-value(d) = 0.1
  U3(d) = 76.76 %
  CLES(d) = 69.74 %
  Cliff's Delta = 0.39

 g [ 95 %CI] = 0.7 [ -0.16 , 1.57 ]
  var(g) = 0.17
  p-value(g) = 0.1
  U3(g) = 75.95 %
  CLES(g) = 69.09 %
```

Interpreting Cohen's *d* as described in previous chapters, we have a difference of .73 indicating we have an effect size of three quarters of a standard deviation, representing a medium effect size using Cohen's guidelines. The values in square brackets provide equivalent 95% confidence limits.

63.8 Reporting the results

The reporting of results for a Factorial ANOVA is slightly more complex than other types of analysis because of the different approaches taken to its analysis. However all reports would state that the various assumptions required to produce a valid ANOVA result were investigated and found to be fulfilled (as was not the case in this instance).

A traditional report would describe the main effects and then carry out a post-hoc analysis. A more modern alternative would be to consider specific comparisons (planned contrasts) immediately. Considering the traditional approach:

Main effect headache type:
No statistically significant difference was found between the means for noise tolerance (outcome measure) across type of headache reported, F(1,90)=1.37, p-value=0.245. Generalised Eta Squared= 0.01 (less than a small effect).

Main effect group:
No statistically significant difference was found between the means for noise tolerance (outcome measure) across treatment group (4 groups including a control), F(3,90)=2.53, p-value=0.0624. Generalised Eta Squared= 0.08 (small effect).

Interaction headache type:groups
No statistically significant difference was found between the interactions, F(3,90)=0.293, p-value=0.830.

Post-hoc analysis

Setting the critical value to 0.05; Tukey's Honest Significant Differences test failed to produce a statistically significant value for any main effect or interaction.

Comparing the three treatment groups to the control group using Dunnett's test produced a statistically significant value for the means between group 3 and the control group (estimate=2.13, t=2.464, p-value=0.041).

I would include a table of cell means, standard deviations and counts along with an interaction plot if possible also showing the original data points (see the chapter *publication quality graphics* for examples)

63.9 More complex designs – between and within (nested)

In this chapter I have discussed a two factor design including an interaction term which is the simplest of the designs we could have considered. We could have added another input variable such as gender or the result from an anxiety questionniare.

All the above inputs provide a single measure, that is to say they are *independent* and known as *between* subjects variables. In contrast we might repeatedly measure the noise level they find uncomfortable, ending up with a series of measures from each subject where each measure is considered as a within subject variable.

When you have a design that includes both within and between variables it is called a mixed or split plot design. Mixed effects models are discussed in the repeated measures chapter which takes a more modern approach called multilevel modelling.

There are two additional complexities when working with repeated measures in R. Firstly, R requires datasets to be in long format as described in the *Repeated Measures* chapter, so often a fair amount of work is required manipulating the data before any analysis of such datasets can be undertaken. Secondly, if using the ANOVA approach rather than the multilevel one, the error term(s) used for the analysis have to be specified explicitly and this is often an area of confusion. To give a taster I have included some sample code below:

An example of a single repeated measure design. Here the repeated measure could be a treatment, where each subject receives each of the treatments.

```
1factor_rep <- aov(y ~ treatment + Error(Subject/treatment), data=mydataframe)
summary(1factor_rep)
```

An example with a single repeated measure along with a single between variable. In this example the repeated measure is the treatment (=W), where each subject receives each of the treatments. Additionally each subject is either healthy or suffering from a particular condition (=B), representing the between factor.

```
1factor_rep_1btwn <- aov(y~(W1*B1)+Error(Subject/W1)+(B1), data=mydataframe)
summary(1factor_rep_1btwn)
```

The R personality project website has extended examples of such designs at: https://personality-project.org/r/r.anova.html . Because of the complexity of forming and specifying such datasets for these models an R package has been developed to help called *ez* which contains the *ezANOVA ()* function. Also the *aov.car()* function in the *afex* package which we used earlier in this chapter provides support for these models.

63.10 Summary

This chapter has introduced many new concepts when investigating the effect two factors might have upon an interval/ratio output. Agriculture and Psychology have been the areas where these techniques have been largely developed and reported, whereas other research areas have adopted equivalent techniques such as regression analysis, which we have seen in this chapter are equivalent.

Time was spent plotting the data a number of ways, seeing how the traditional interaction plot, which only displayed the means, can be misleading. The beeswarm plot offered a better alternative.

The recently developed generalised eta squared effect size measure, along with various others, were introduced and compared.

The chapter finished by briefly describing some more complex designs which involved both within and between subject variables.

63.11 Tips and Tricks

Instead of using the *beeswarm* package you can create a simple version using the *stripchart()* and *boxplot()* functions that are in R:

```
stripchart(uncomf ~ factor_headache_type + factor_treat_grp, pch =16,cex = 1, vertical=TRUE,
data=headache)

boxplot(uncomf ~ headache_type+ treat_grp, data = headache,

outline = FALSE,     ## avoid double-plotting outliers, if any

main = 'boxplot + stripchart', add = TRUE)
```

This chapter has introduced the basic type of Factorial ANOVA. For more information consider the following:

- Help with formula development can be found at Will Lowe's web article *Formulae in R: ANOVA and other models, mixed and fixed* at http://conjugateprior.org/2013/01/formulae-in-r-anova/
- The R-stats blog provides a good page with links to many online tutorials for various types of ANOVA at http://www.r-statistics.com/2010/04/repeated-measures-anova-with-r-tutorials/
- An excellent online resource for ANOVA for those used to using SPSS but switching to R is http://egret.psychol.cam.ac.uk/statistics/R/anova.html
- For a traditional paper based book try Field, 2013.

64 Factor analysis

This chapter provides details of a method that can help you to restructure your data by specifically reducing the number of variables. Such an approach is often called a "data reduction" or "dimension reduction" technique. What this means is that we start off with a set of variables, say 20, and then by the end of the process we have a smaller number but which still reflects a large proportion of the information contained in the original dataset. The way that the 'information contained' is measured is by considering the variability within and co-variation across variables, that is the variance and co-variance (i.e. correlation). The reduction might be by discovering that a particular linear combination of our variables accounts for a large percentage of the total variability in the data or by discovering that several of the variables reflect another 'latent variable'.

This process can be used in broadly three ways, firstly to simply discover the linear combinations that reflect the most variation in the data. Secondly, to discover if the original variables are organised in a particular way reflecting another '**latent** variable' (called **Exploratory Factor Analysis – EFA**). Thirdly, we might want to confirm a belief about how the original variables are organised in a particular way (**Confirmatory Factor Analysis – CFA**). It must not be thought that EFA and CFA are mutually exclusive; often what starts as an EFA becomes a CFA.

construct
= latent variable
= factor
(in factor analysis context)

I have used the term factor in the above and we need to consider this term a little more. A factor in this context, is different to a 'factor' in the Analysis of Variance (ANOVA) context. In the factor analysis context a factor is equivalent to a latent variable or construct.

Cope et al (1986), and quoted in Everitt & Dunn 2001 (page 281)

1. My doctor treats me in a friendly manner
2. I have some doubts about the ability of my doctor
3. My doctor seems cold and impersonal
4. My doctor does his/her best to keep me from worrying
5. My doctor examines me as carefully as necessary
6. My doctor should treat me with more respect
7. I have some doubts about the treatment suggested by my doctor
8. My doctor seems very competent and well trained
9. My doctor seems to have a genuine interest in me as a person
10. My doctor leaves me with many unanswered questions about my condition and its treatment
11. My doctor uses words that I do not understand
12. I have a great deal of confidence in my doctor
13. I feel a can tell my doctor about very personal problems
14. I do not feel free to ask my doctor questions

A latent variable is a variable that cannot be measured directly but is measured indirectly through several observable variables (called **manifest** variables). Some examples will help. If we were interested in measuring intelligence (=latent variable) we would measure people on a battery of tests (=observable variables) including short term memory, verbal, writing, reading, motor and comprehension skills etc.

Similarly we might have an idea that patient satisfaction (=latent variable) with a person's GP can be measured by asking questions such as those used by Cope et al, 1986 and quoted in Everitt & Dunn, 2001 p.281. Each question being presented as a five point option from strongly agree to strongly disagree (i.e. Likert scale, scoring 1 to 5).

You might be thinking that you could group some of the above variables (manifest variables) together to represent a particular aspect of patient satisfaction with their GP such as personality, knowledge and treatment. So now we are not just thinking that a set of observed variables relate to one latent variable but that specific subgroups of them relate to a specific latent variable, each of which can relate to other latent variables. Two other things to note; firstly often the observable variables are questions in a questionnaire and can be thought of as items and consequently each subset of items relating to a specific latent variable represents a scale.

Secondly, you will notice in the diagram opposite that, besides the line pointing towards the observed variable X_i from the latent variable (representing its degree of correlation to the latent variable), there is another line pointing towards it labelled error. This error line represents the unique contribution of the variable, that is that portion of the variable that cannot be predicted from the remaining variables. This uniqueness value is equal to $1-R^2$ where R^2 is the standard multiple R squared value. We will look much more at this in the following sections considering a dataset that has been used in many texts concerned with factor analysis. Using a common dataset will allow you to compare this exposition with that presented in other texts.

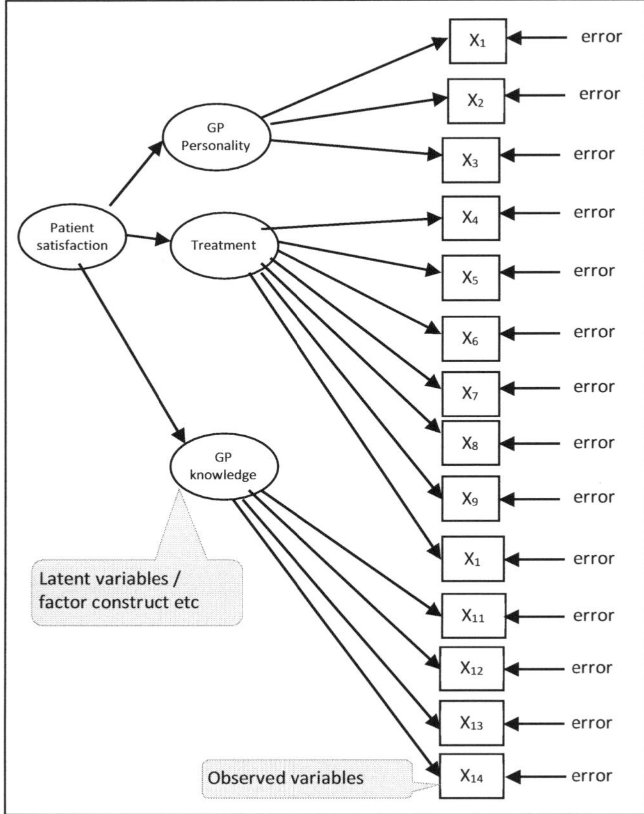

The first example in this chapter will make use of a subset of data from the Holzinger and Swineford, 1939 study where they collected data on 26 psychological tests from seventh – eighth grade children in a suburban school district of Chicago (file called *grnt_fem.dat*). Our subset of data consists of data from 73 girls from the Grant-White School. The six variables represent scores from seven tests of different aspects of educational ability, visual perception, cube and lozenge identification, word meanings, sentence structure and paragraph understanding. We believe that several of these tests measure certain underlying latent variables such as visual perception and reading ability.

There are many varieties of factor analysis involving a multitude of different techniques. However, the common characteristic is that factor analysis is carried out using a computer, although the early researchers in this area were not so lucky, with the first paper introducing factor analysis being published in 1904 by C. Spearman of Spearman's rank correlation coefficient fame, long before the friendly PC was available.

Factor analysis works only on interval/ratio data, and ordinal data at a push. If you want to carry out some type of variable reduction process on nominal data you have to use other techniques or substantially adapt the factor analysis procedure, see Bartholomew, Steele, Moustaki & Galbraith, 2008 for details.

64.1 Hozinger & Swineford 1939 data

Any statistical analysis starts with standard data preparation techniques and factor analysis is no different. Basic descriptive statistics are produced to note any missing/abnormal values and appropriate action taken. Also in addition, two other processes are undertaken:

1. Any **computed variables** (strickly speaking only linear transformations) are **excluded** from the analysis. These are easily identified as they will have a correlation of 1 with the variable from which they were calculated.

2. **All the variables should measure the construct in the same direction**. Considering the GP satisfaction scale, we need all the 14 items to measure satisfaction in the same direction where a score of 1 represents high satisfaction and 5 the least satisfaction or the other way round. The direction does not matter, the important thing is that all the questions score in the same direction. Taking question 1: *My doctor treats me in a friendly manner,* this provides the value 1 when the respondent agrees, representing total satisfaction and 5 when the respondent strongly disagrees and is not satisfied. However, question three is different: *My doctor seems cold and impersonal.* A patient indicating strong agreement to this statement would also provide a value of 1 but this time it indicates a high level of dissatisfaction. The solution is to reverse score all these negatively stated questions.

Unfortunately, it is not possible to carry out a Factor analysis in R Commander, you need to use R directly which we will now do. Factor analysis follows a clearly defined set of stages which we will now follow.

64.1.1 Data preparation

You can obtain the dataset from my website, by typing the following code into the R Console window:

```
hozdata<- read.table("http://www.robin-
beaumont.co.uk/virtualclassroom/book2data/grnt_fem.dat",sep="\t", header=TRUE)

names(hozdata)
```

```
>   names(hozdata)
[1] "visperc"  "cubes"    "lozenges" "paragrap" "sentence"
[6] "wordmean"
```

The 6 variable names are listed opposite.

64.1.2 Data description – assessment of appropriateness for carrying out a factor analysis

The starting point for all factor analysis techniques is the correlation matrix. All factor analysis techniques try to clump subgroups of variables together based upon their correlations and often you can get a feel for what the factors are going to be just by looking at the correlation matrix and spotting clusters of high correlations between groups of variables. Typing the following R code produces the necessary correlation matrix.

```
> hozdatamatrix <- cor(hozdata)
> hozdatamatrix
           visperc     cubes  lozenges  paragrap  sentence  wordmean
visperc  1.0000000 0.4828862 0.4916630 0.3430168 0.3667612 0.2297377
cubes    0.4828862 1.0000000 0.4924264 0.2106977 0.1786094 0.1838991
lozenges 0.4916630 0.4924264 1.0000000 0.3258026 0.3354078 0.3689271
paragrap 0.3430168 0.2106977 0.3258026 1.0000000 0.7243611 0.7430226
sentence 0.3667612 0.1786094 0.3354078 0.7243611 1.0000000 0.6955315
wordmean 0.2297377 0.1838991 0.3689271 0.7430226 0.6955315 1.0000000
```

```
hozdatamatrix <- cor(hozdata)

hozdatamatrix
```

```
> round(hozdatamatrix,4)
         visperc  cubes lozenges paragrap sentence wordmean
visperc   1.0000 0.4829   0.4917   0.3430   0.3668   0.2297
cubes     0.4829 1.0000   0.4924   0.2107   0.1786   0.1839
lozenges  0.4917 0.4924   1.0000   0.3258   0.3354   0.3689
paragrap  0.3430 0.2107   0.3258   1.0000   0.7244   0.7430
sentence  0.3668 0.1786   0.3354   0.7244   1.0000   0.6955
wordmean  0.2297 0.1839   0.3689   0.7430   0.6955   1.0000
```

This can be made more readable by applying the *round()* R function to the above object requesting the results to formatted to 4 significant figures.

```
round(hozdatamatrix,  4)
```

Along with the above results, a basic set of descriptive statistics for each variable is also useful. The easiest way is to use the *describe()* function in the *psych* package with the code below.

```
> describe(hozdata)
          vars  n  mean   sd median trimmed  mad min max range  skew kurtosis   se
visperc      1 73 29.32 6.92     30   29.61 5.93  11  45    34 -0.40    -0.36 0.81
cubes        2 73 24.70 4.53     25   24.61 4.45   9  37    28 -0.13     1.32 0.53
lozenges     3 73 14.84 7.91     13   14.03 7.41   3  36    33  0.82     0.04 0.93
paragrap     4 73 10.59 3.56     10   10.46 2.97   2  19    17  0.37    -0.31 0.42
sentence     5 73 19.30 5.05     20   19.80 4.45   4  28    24 -0.82     0.44 0.59
wordmean     6 73 18.01 8.32     16   17.58 7.41   2  41    39  0.56    -0.29 0.97
```

```
install.packages("psych")

library(psych)

describe(hozdata)
```

We can now see that we have 73 cases and there don't appear to be any missing values or out of range values. Also, the correlation matrix does not contain any '1''s except the expected diagonals.

Looking at the correlation matrix we see that wordmean, sentence and paragraph seem to form one cluster and lozenges, cubes and visperc tests the other cluster. I've highlighted these. Norman and Streiner, 2008 p.197 quote Tabachnick & Fidell, 2001 saying that if there are few correlations above 0.3 it is a waste of time carrying on with the analysis as the variables do not show any level of co-dependency, clearly we do not have that problem.

While eyeballing is a valid method of statistical analysis (!) obviously some type of statistic, preferably with an associated probability density function to produce a p-value, would be useful to help us make a decision about the appropriateness of carrying out a factor analysis. Three such statistics are the Bartlett Test of Sphericity, the Kaiser-Meyer-Olkin Measure of Sampling Adequacy (usually called the MSA) and the determinant.

The Bartlett Test of Sphericity compares the correlation matrix with a matrix of zero correlations (technically called the identity matrix, which consists of all zeros except the 1's along the diagonal). From this test we are looking for a small p-value indicating that it is highly unlikely for us to have obtained the observed correlation matrix from a population with zero correlation. However, there are many problems with the test – a small p-value indicates that you should not continue but a large p value does not guarantee that all is well (Norman & Streiner, 2008 p.198). You can easily obtain the value from the Bartlett Test of Sphericity in R.

```
> cortest.bartlett(hozdata)
R was not square, finding R from data
$chisq
[1] 180.3307

$p.value
[1] 2.038929e-30

$df
[1] 15
```

```
cortest.bartlett(hozdata)
```

The associated p-value is less than <0.001 suggesting that now we can continue and perform a valid factor analysis. The MSA does not produce a p-value but we are aiming for an overall value of 0.8 or more, a value below 0.5 is considered to be useless! You can also obtain MSA values for each variable; here 0.7 is considered the minimum acceptable value. Norman & Streiner, p.198 recommend that you consider removing variables with an MSA below 0.7.

```
> KMO(hozdata)
Kaiser-Meyer-Olkin factor adequacy
Call: KMO(r = hozdata)
Overall MSA =  0.76
MSA for each item =
 visperc    cubes lozenges paragrap sentence wordmean
    0.73     0.73     0.78     0.77     0.80     0.74
```

To carry out the Kaiser-Meyer-Olkin Measure of Sampling Adequacy (MSA), we use the *KMO()* function, recently added to the *psych* package: *KMO(hozdata)*.

If you don't have the latest version of the *psych* package installed you can use the equivalent *kmo()* function (note lower case) which you can download from: http://www.robin-beaumont.co.uk/virtualclassroom/book2data/rcode/kmo_function.r

```
> kmo(hozdata)
$overall
[1] 0.7630899

$report
[1] "The KMO test yields a degree of common variance middling."

$individual
  visperc     cubes  lozenges  paragrap  sentence  wordmean
0.7344951 0.7318282 0.7802956 0.7676797 0.8026988 0.7426252

$AIS
            [,1]         [,2]        [,3]        [,4]        [,5]         [,6]
[1,]  0.61266171 -0.204148611 -0.17744941 -0.06526691 -0.10102096  0.090914525
[2,] -0.20414861  0.676304595 -0.20996453 -0.01667937  0.04246589 -0.007578625
[3,] -0.17744941 -0.209964528  0.61497954  0.02185063 -0.01230462 -0.099997319
[4,] -0.06526691 -0.016679367  0.02185063  0.35435571 -0.14459145 -0.176172792
[5,] -0.10102096  0.042465886 -0.01230462 -0.14459145  0.39902040 -0.132989720
[6,]  0.09091452 -0.007578625 -0.09999732 -0.17617279 -0.13298972  0.370824176

$AIR
           [,1]        [,2]        [,3]        [,4]        [,5]        [,6]
[1,]  0.7344951 -0.31715013 -0.28909053 -0.14007574 -0.20431657  0.19073873
[2,] -0.3171501  0.73182820 -0.32557004 -0.03407130  0.08174693 -0.01513336
[3,] -0.2890905 -0.32557004  0.78029555  0.04680734 -0.02483935 -0.20939871
[4,] -0.1400757 -0.03407130  0.04680734  0.76767966 -0.38452566 -0.48599869
[5,] -0.2043166  0.08174693 -0.02483935 -0.38452566  0.80269878 -0.34572955
[6,]  0.1907387 -0.01513336 -0.20939871 -0.48599869 -0.34572955  0.74262524
```

Simply copy all the code that appears in your web browser when you visit the above link and then paste it into your R Console window. To run the function type:

```
kmo(hozdata)
```

The overall MSA value is near 0.8 and none of the individual MSA's are below 0.7 again giving us the thumbs up for continuing with the factor analysis.

One final aspect to consider is the possibility of multicollinearity. This potential problem was discussed in the chapters investigating multiple regression and in this context, it is assessed by looking at a particular summary value called the determinant. We require the determinant to be greater than 0.00001 (decimal place and four zeros) to indicate that multicollinearity is not a problem. In R you simply use the *det()* function to obtain the value. We obtain a value of 0.073 indicating that multicollinearity is not a problem.

works on the matrix not the raw data

```
> det(hozdatamatrix)
[1] 0.07374135
```

Determinants and multicollinearity
The determinant measures the volume of the uncertainty ellipsoid. The uncertainty ellipsoid is a general extension to the **correlation ellipse** shown in section four of the *Correlation* chapter. Small values indicate smaller volumes implying closer correlations between variables.

64.1.3 Extracting the Factors

There are two common methods to extract the factors (i.e. latent variables), the *Principal components* (PCA) and the *Principal axis* (PA) factoring extraction methods. Strictly speaking the PCA method is not a type of factor analysis but it often gives very similar results. Let us try both and see what we get.

However, there is one other thing we need to decide first. Do we want to specify in advance how many latent variables we want to end up with, or alternatively, do we want the computer to decide for us by using some criteria? If we do not specify either of these than the number of factors produced is equal to the number of variables in the model (as shown opposite). The common method is to let the computer decide for us by simply saying that factors that possess an eigenvalue >1 are retained. However there are several reasons why this is not an altogether good idea setting the bar too low. Norman & Streiner, 2008 discuss a way of setting more appropriate specific levels. For now I'll use the dodgy eigenvalue >1 approach.

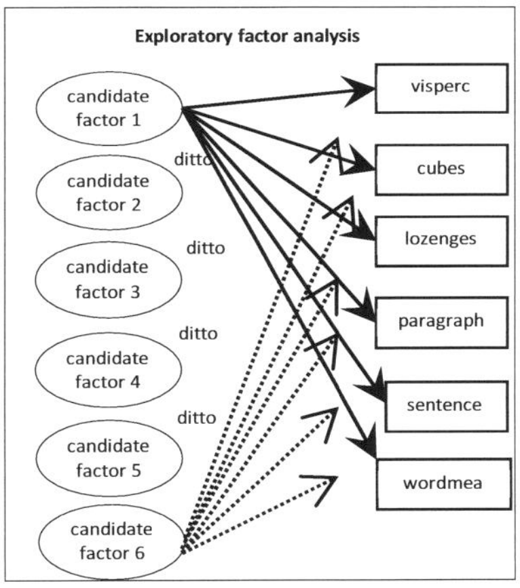

Exploratory factor analysis

candidate factor 1 — visperc, cubes, lozenges, paragraph, sentence, wordmea
candidate factor 2 — ditto
candidate factor 3 — ditto
candidate factor 4 — ditto
candidate factor 5 — ditto
candidate factor 6

In an PCA an eigenvalue represents the variance explained by each principal component, and because there are as many principal components as there are observed variables a value of one indicates that is provides as much variance to the PCA model as the original variable did. You can think it as a measure of practical significance of each principal component. The eigenvector represents a measure of the relationship between each factor and observed variable, called a loading, with 1 being perfectly related and 0 no relationship.

eigenvalues

```
> eigen(hozdatamatrix)
$values
[1] 3.0988862 1.3486867 0.5491883 0.4876152 0.2819058 0.2337179

$vectors
           [,1]       [,2]       [,3]         [,4]        [,5]         [,6]
[1,] -0.3640109 -0.4215835  0.7527461  0.169182179 -0.22542214  0.209043337
[2,] -0.2986254 -0.5681836 -0.2692357 -0.708014222  0.11913152  0.005093146
[3,] -0.3806579 -0.3859478 -0.4824707  0.662484823  0.05945471 -0.175911351
[4,] -0.4670648  0.3339525  0.0661646 -0.170303374 -0.41140500 -0.683879457
[5,] -0.4604434  0.3218774  0.1967648 -0.003670902  0.80348560  0.008239295
[6,] -0.4508446  0.3678991 -0.2915705 -0.046801664 -0.34148863  0.676440051
```

eigenvectors = loadings

For any square matrix in R you can use the standard R function *eigen()* to obtain both eigenvalues and eigenvectors:

```
eigen(hozdatamatrix)
```

We can see that two of the eigenvalues are greater than 1 indicating that these extracted components contribute more to the model than the original variables. Usually these values are obtained as part of the analysis process so we will do it correctly now.

I have run both a Principal Axis and a Principal Component Analysis below. We will start with the PCA extraction method.

The *principal()* function in the *psych* package carries out a PCA analysis and provides the necessary values. We can also apply the *plot()* function to the eigenvalues to plot them, producing what is known as a scree plot.

```
model1<- principal(hozdata, nfactors = 6, rotate = "none")

model1

plot(model1$values, type = "b")
```

PCA analysis

standardised loadings derived from the eigenvectors on the previous page

```
> model1<- principal(hozdata, nfactors = 6, rotate = "none")
> model1
Principal Components Analysis
Call: principal(r = hozdata, nfactors = 6, rotate = "none")
Standardized loadings (pattern matrix) based upon correlation matrix
          PC1   PC2   PC3   PC4   PC5   PC6 h2       u2
visperc  0.64  0.49 -0.56 -0.12 -0.12  0.10  1  0.0e+00
cubes    0.53  0.66  0.20  0.49  0.06  0.00  1  2.2e-16
lozenges 0.67  0.45  0.36 -0.46  0.03 -0.09  1 -4.4e-16
paragrap 0.82 -0.39 -0.05  0.12 -0.22 -0.33  1  4.4e-16
sentence 0.81 -0.37 -0.15  0.00  0.43  0.00  1  1.1e-16
wordmean 0.79 -0.43  0.22  0.03 -0.18  0.33  1  8.9e-16

                       PC1  PC2  PC3  PC4  PC5  PC6
SS loadings           3.10 1.35 0.55 0.49 0.28 0.23
Proportion Var        0.52 0.22 0.09 0.08 0.05 0.04
Cumulative Var        0.52 0.74 0.83 0.91 0.96 1.00
Proportion Explained  0.52 0.22 0.09 0.08 0.05 0.04
Cumulative Proportion 0.52 0.74 0.83 0.91 0.96 1.00

Test of the hypothesis that 6 components are sufficient.
```

eigenvalues

eigenvalues

model1$values

3.0 2.5 2.0 1.5 1.0 0.5

1 2 3 4 5 6

Index

```
> diag(result1$values)
         [,1]    [,2]      [,3]      [,4]      [,5]      [,6]
[1,] 3.099269 0.00000 0.0000000 0.0000000 0.0000000 0.0000000
[2,] 0.000000 1.34828 0.0000000 0.0000000 0.0000000 0.0000000
[3,] 0.000000 0.00000 0.5487278 0.0000000 0.0000000 0.0000000
[4,] 0.000000 0.00000 0.0000000 0.4880944 0.0000000 0.0000000
[5,] 0.000000 0.00000 0.0000000 0.0000000 0.2818006 0.0000000
[6,] 0.000000 0.00000 0.0000000 0.0000000 0.0000000 0.2338282
```

```
> loadings_me <- result1$vectors %*% sqrt(diag(result1$values))
> loadings_me
           [,1]       [,2]       [,3]         [,4]        [,5]          [,6]
[1,] -0.7937945 -0.4274025 -0.2157448 -0.032348119 -0.17731897  0.3289120439
[2,] -0.8105329 -0.3738518  0.1463022 -0.003410413  0.42644484 -0.0003386304
[3,] -0.8221225 -0.3876276  0.0485665 -0.118374188 -0.22252113 -0.3285918010
[4,] -0.6700128  0.4481644 -0.3581413  0.462286979  0.03147945 -0.0852414999
[5,] -0.5258331  0.6594749 -0.1988277 -0.495099411  0.06270906  0.0020240933
[6,] -0.6410089  0.4897229  0.5573341  0.119127163 -0.11859313  0.1019910633
```

standardized loadings well nearly (see text)

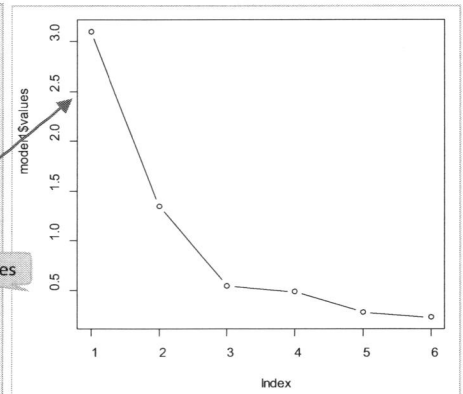

The loadings from a PCA analysis are modified eigenvector values being rescaled by the square root (*sqrt()*) of the eigenvalues. To show this we need to perform matrix multiplication which requires a suitably structured matrix of eigenvalues done by using the *diag()* function to produce what is known as a diagonal matrix. We then perform matrix multiplication using the matrix multiplication operator %*% - yep I know it looks like the keyboard has taken a turn for the worse but this is it!

```
loadings_me <- result1$vectors %*% sqrt(diag(result1$values))

loadings_me
```

The values in the *loadings_me* object are equivalent to the PCA standardised loadings, except the rows and signs are reversed. You can see how these additional steps are achieved in the *principal()* function by typing *psych::principal* into the *R Console* window which brings up the source code for the function. You do not need to carry out the above process when routinely carrying out a PCA analysis but many people find it difficult to see how the loadings from the analysis relate to the eigenvectors so that is why I included it. Now back to the analysis proper.

We can see that two of the extracted factors have an eigenvalue >1 (labelled SS loadings) so now we can re-run the model specifying that we wish to extract only two factors.

```
model2 <- principal(hozdata,
nfactors = 2, rotate = "none")

model2
```

We can find the reproduced correlations and the communalities (the diagonals).

```
factor.model(model2$loadings)
```

And we can also find the differences between the observed and model estimated correlations, where the diagonals represent the uniqueness values (1- R squared).

```
residuals <- factor.residuals(hozdatamatrix,
model2$loadings)

residuals
```

```
> model2 <- principal(hozdata, nfactors = 2, rotate = "none")
> model2
Principal Components Analysis
Call: principal(r = hozdata, nfactors = 2, rotate = "none")
Standardized loadings (pattern matrix) based upon correlation matrix
          PC1   PC2   h2   u2
visperc  0.64  0.49 0.65 0.35
cubes    0.53  0.66 0.71 0.29
lozenges 0.67  0.45 0.65 0.35
paragrap 0.82 -0.39 0.83 0.17
sentence 0.81 -0.37 0.80 0.20
wordmean 0.79 -0.43 0.81 0.19

                       PC1  PC2
SS loadings           3.10 1.35
Proportion Var        0.52 0.22
Cumulative Var        0.52 0.74
Proportion Explained  0.70 0.30
Cumulative Proportion 0.70 1.00

Test of the hypothesis that 2 components are sufficient.

The degrees of freedom for the null model are  15  and the objective function was  2.61
The degrees of freedom for the model are 4  and the objective function was  0.47
The total number of observations was  73  with MLE Chi Square =  31.96  with prob <  1.9e-06
```

```
> factor.model(model2$loadings)
           visperc     cubes  lozenges  paragrap  sentence  wordmean
visperc  0.6503201 0.6599180 0.6488366 0.3369822 0.3363788 0.2993839
cubes    0.6599180 0.7117498 0.6480156 0.1763163 0.1794419 0.1352927
lozenges 0.6488366 0.6480156 0.6499246 0.3771271 0.3756017 0.3403236
paragrap 0.3369822 0.1763163 0.3771271 0.8264319 0.8114095 0.8182446
sentence 0.3363788 0.1794419 0.3756017 0.8114095 0.7967197 0.8030022
wordmean 0.2993839 0.1352927 0.3403236 0.8182446 0.8030022 0.8124267
```

```
> residuals <- factor.residuals(hozdatamatrix, model2$loadings)
> residuals
             visperc         cubes    lozenges    paragrap      sentence    wordmean
visperc   0.34967989 -0.1770317673 -0.15717362  0.00603456  0.0303824116 -0.06964619
cubes    -0.17703177  0.2882501914 -0.15558927  0.03438147 -0.0008325667  0.04860637
lozenges -0.15717362 -0.1555892661  0.35007536 -0.05132449 -0.0401939045  0.02860348
paragrap  0.00603456  0.0343814731 -0.05132449  0.17356811 -0.0870484791 -0.07522195
sentence  0.03038241 -0.0008325667 -0.04019390 -0.08704848  0.2032803113 -0.10747064
wordmean -0.06964619  0.0486063698  0.02860348 -0.07522195 -0.1074706423  0.18757327
```

To carry out a principal axis factor analysis we use the *fa()* function in the *psych* package.

```
model4 <- fa(hozdata, nfactors = 2,fm = "pa", rotate = "none")

model4
```

I'll now provide an explanation of the various aspects of the output along with an insert to allow comparison with the results from the PCA analysis.

PA analysis

```
> model4 <- fa(hozdata, nfactors = 2,fm = "pa", rotate = "none")
> model4
Factor Analysis using method =  pa
Call: fa(r = hozdata, nfactors = 2, rotate = "none", fm = "pa")
Standardized loadings (pattern matrix) based upon correlation matrix
          PA1   PA2   h2   u2 com
visperc  0.56  0.42 0.49 0.51 1.9
cubes    0.45  0.55 0.51 0.49 1.9
lozenges 0.59  0.40 0.50 0.50 1.8
paragrap 0.82 -0.31 0.76 0.24 1.3
sentence 0.78 -0.27 0.69 0.31 1.2
wordmean 0.78 -0.33 0.71 0.29 1.3

                       PA1  PA2
SS loadings           2.75 0.92
Proportion Var        0.46 0.15
Cumulative Var        0.46 0.61
Proportion Explained  0.75 0.25
Cumulative Proportion 0.75 1.00
```

PCA analysis

```
> model2 <- principal(hozdata, nfactors = 2, rotate = "none")
> model2
Principal Components Analysis
Call: principal(r = hozdata, nfactors = 2, rotate = "none")
Standardized loadings (pattern matrix) based upon correlation
          PC1   PC2   h2   u2
visperc  0.64  0.49 0.65 0.35
cubes    0.53  0.66 0.71 0.29
lozenges 0.67  0.45 0.65 0.35
paragrap 0.82 -0.39 0.83 0.17
sentence 0.81 -0.37 0.80 0.20
wordmean 0.79 -0.43 0.81 0.19

                       PC1  PC2
SS loadings           3.10 1.35
Proportion Var        0.52 0.22
Cumulative Var        0.52 0.74
Proportion Explained  0.70 0.30
Cumulative Proportion 0.70 1.00
```

h2 = commonality

u2 = uniqueness (=1-h2)

total variance explained greater with PCA analysis

Comparing the results from both methods shows that the factor loadings (PA1, PA2), or strictly speaking the component loadings for the PCA, differ. Those for the PCA are larger in absolute values as are the communalities (*h2*) and as a consequence the total variance explained is also greater. Here are a few pointers to help you interpret the above:

Factor loadings (PA1, PA2) for the PA = correlation between a specific observed variable and a specific factor. Higher values mean a closer relationship. They are equivalent to standardised regression coefficients (β weights) in multiple regression. **Higher the value the better.**

Communality (h2) for the PA = the total influence on a single observed variable from all the factors associated with it. It is equal to the sum of all the squared factor loadings for all the factors related to the observed variable and this value is the same as R^2 in multiple regression. The value ranges from zero to 1 where 1 indicates that the variable can be fully defined by the factors and has no uniqueness. In contrast a value of 0 indicates that the variable cannot be predicted at all from any of the factors. The communality (h2) can be derived for each variable by taking the sum of the squared factor loadings for each of the factors associated with the variable. So for *visperc* = $0.56^2 + 0.42^2 = 0.49$ and for *cubes* $=0.45^2 + 0.55^2 =0.505$. These values can be interpreted the same way as R squared values in multiple regression, that is they represent the % of variability taken into account by the model. Because we are hoping that the observed dataset is reflected in the model we **want** these values to be as **high** as possible, nearer to one the better.

Uniqueness (u2) for each observed variable it is that portion of the variable that cannot be predicted from the other variables (i.e. the latent variables). Its value is 1-communality. So, for example, for *wordmean* we have 1-0.714 = 0.286. As the communality can be interpreted as the % of the variability that is predicted by the model we can say that uniqueness is the % of variability in a specific observed variable that is NOT predicted by the model. This means that we **want** this value for each observed variable to be as **low** as possible. In the diagram given previously it is represented by the 'error' arrow.

Total variance explained (cumulative var) this indicates how much of the variability in the data has been modelled by the extracted factors. You might think that, given that the PCA analysis models 74% of the variability compared to just 61% for the PA analysis we should go for the PCA results. However this value is higher in the PCA analysis because the initial estimates for the communalities are all set to 1 which is higher than for the PA analysis. In the PA analysis instead an estimate of the R^2 value is used. Whereas the PCA approach makes use of all the variability available in the dataset, in the PA analysis the unique variability for each observed variable is disregarded as we are only really interested in how each variable relates to the latent variable(s). What is an acceptable level of variance explained by the model? Well one would hope for the impossible which would be 100% whereas in reality often analyses are reported which provide 60-70% of the variability explained by the model.

Now we have our factors we need to find a way of interpreting them – to enable this we carry out a process called factor rotation.

64.1.4 Giving the factors meaning - Rotation

```
> model3 <- principal(hozdata, nfactors = 2, rotate = "varimax")
Loading required package: GPArotation
> model3
Principal Components Analysis
Call: principal(r = hozdata, nfactors = 2, rotate = "varimax")
Standardized loadings (pattern matrix) based upon correlation matrix
          RC1  RC2   h2   u2
visperc  0.22 0.78 0.65 0.35
cubes    0.03 0.84 0.71 0.29
lozenges 0.27 0.76 0.65 0.35
paragrap 0.89 0.18 0.83 0.17
sentence 0.87 0.19 0.80 0.20
wordmean 0.89 0.13 0.81 0.19

                       RC1  RC2
SS loadings           2.47 1.98
Proportion Var        0.41 0.33
Cumulative Var        0.41 0.74
Proportion Explained  0.56 0.44
Cumulative Proportion 0.56 1.00
```

PCA analysis

By rotating the factors the factor loadings change with the aim to make interpretation easier. Norman & Streiner provide an excellent discussion explaining how rotation helps interpretation. To specify a rotation method in R you use the *rotate =* option in the *fa()* or *principal()* functions. We will consider two types of rotation; Varimax and Promax. First Varimax.

```
model3 <- principal(hozdata, nfactors = 2,
          rotate = "varimax")

model3
```

Also, an equivalent PCA analysis:

```
model5 <-  fa(hozdata, nfactors = 2,fm = "pa",
rotate = "varimax")

model5
```

```
> model5 <-  fa(hozdata, nfactors = 2,fm = "pa", rotate = "varimax")
> model5
Factor Analysis using method =  pa
Call: fa(r = hozdata, nfactors = 2, rotate = "varimax", fm = "pa")
Standardized loadings (pattern matrix) based upon correlation matrix
          PA1  PA2   h2   u2 com
visperc  0.22 0.66 0.49 0.51 1.2
cubes    0.07 0.71 0.51 0.49 1.0
lozenges 0.26 0.66 0.50 0.50 1.3
paragrap 0.85 0.21 0.76 0.24 1.1
sentence 0.80 0.22 0.69 0.31 1.1
wordmean 0.83 0.16 0.71 0.29 1.1

                       PA1  PA2
SS loadings           2.17 1.50
Proportion Var        0.36 0.25
Cumulative Var        0.36 0.61
Proportion Explained  0.59 0.41
Cumulative Proportion 0.59 1.00
```

PA analysis

com = Hoffman's index of complexity ignore as often unreliable. For details see the end of the chapter.

We can see from both of the above set of results that they are pretty similar. As a result of the rotation now paragraph, sentence and wordmean load heavily on the first factor/component and the other three on the second factor/component.

By selecting the *Varimax* rotation option I have demanded that the factors are uncorrelated (technically orthogonal). However, this might not be the case. The Promax rotation allows for correlated factors, as demonstrated below.

```
> model6 <-  fa(hozdata, nfactors = 2,fm = "pa",rotate="promax")
> model6
Factor Analysis using method =  pa
Call: fa(r = hozdata, nfactors = 2, rotate = "promax", fm = "pa")
Standardized loadings (pattern matrix) based upon correlation matrix
           PA1   PA2   h2   u2 com
visperc   0.06  0.67 0.49 0.51 1.0
cubes    -0.13  0.76 0.51 0.49 1.1
lozenges  0.10  0.66 0.50 0.50 1.0
paragrap  0.87  0.01 0.76 0.24 1.0
sentence  0.82  0.03 0.69 0.31 1.0
wordmean  0.86 -0.04 0.71 0.29 1.0

                       PA1  PA2
SS loadings           2.20 1.47
Proportion Var        0.37 0.25
Cumulative Var        0.37 0.61
Proportion Explained  0.60 0.40
Cumulative Proportion 0.60 1.00

 With factor correlations of
     PA1  PA2
PA1 1.00 0.46
PA2 0.46 1.00
```

```
model6 <-  fa(hozdata, nfactors = 2,fm =
"pa",rotate="promax")

model6
```

So by allowing the two latent variables to correlate we have an estimated correlation of 0.46 while the factor loadings have changed little.

The next thing we do is to disregard those loadings below a certain threshold on each factor often this is something like 0.3 or 0.4 . We'll use this approach for now, but if you want more information about testing for an appropriate threshold, I suggest you read Norman and Streiner, p.205.

```
> print.psych(model6, cut = 0.3 , sort = TRUE)
Factor Analysis using method =  pa
Call: fa(r = hozdata, nfactors = 2, rotate = "promax", fm = "pa")
Standardized loadings (pattern matrix) based upon correlation matrix
         item  PA1   PA2   h2   u2 com
paragrap    4 0.87        0.76 0.24 1.0
wordmean    6 0.86        0.71 0.29 1.1
sentence    5 0.82        0.69 0.31 1.0
cubes       2       0.76 0.51 0.49 1.0
visperc     1       0.67 0.49 0.51 1.0
lozenges    3       0.66 0.50 0.50 1.0
```

Just because the factor scores are not shown does not mean we assume them to be zero!

In R you can use the *print.psych()* function in the *psych* package to specify a cut off value for displaying values as well as a option to indicate you want them sorted.

```
print.psych(model6, cut = 0.3 , sort = TRUE)
```

Possibly, you might be asking yourself why we spent to all this time and effort when we had come to pretty much the same conclusion when we eyeballed the correlation matrix at the start of the procedure, and some people agree. However, factor analysis does often offer more than can be achieved by merely eyeballing a set of correlations and it does offer a level of statistical rigor (although statisticians argue this point).

64.1.5 Making sense of the factors; the Reification problem

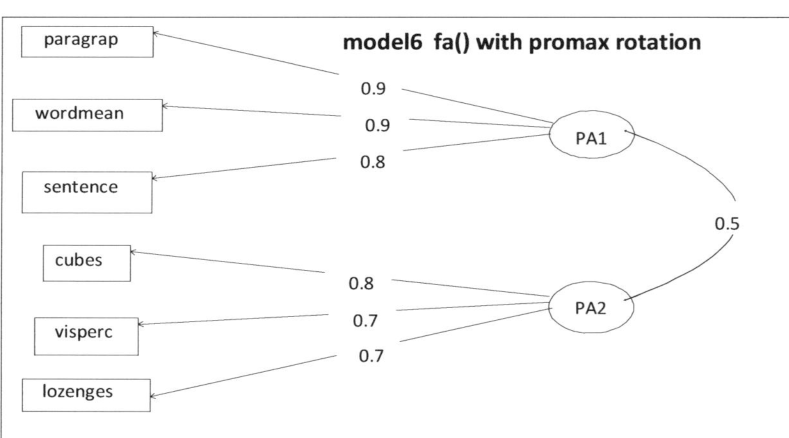

Although the computer presents us with what appears a lovely organised set of variables that make up a factor there is no reason to believe that this subset of variables should equate to something in reality. This is called the *fallacy of misplaced concreteness* or the *reification problem*. Basically it is wrong to assume something exists just because it appears to. Your latent variables may not really exist! We will return to this at the end of this chapter.

However, in this instance, it does seem reasonable to suggest that the first factor could be called reading ability and the second visual perception. The diagram on the left was produced by typing the following command into the R Console window, *fa.diagram(model6)*.

64.1.6 Obtaining factor scores for individuals

We can obtain factor scores for each individual (case) and then compare them across cases. I will demonstrate this using *model2* which has two factor values for each case.

In R, for PCA we just add the option *scores = true*

```
model2a <- principal(hozdata, nfactors = 2, rotate = "none", scores = TRUE)
```

factor scores for each case

```
> head(model2a$scores, 10)
             PC1        PC2
 [1,] -0.4902900  0.6334255
 [2,]  0.3200279  0.1306837
 [3,]  2.5690905  0.4306738
 [4,] -0.5146886  0.1043434
 [5,] -0.2628493  0.1623493
 [6,] -0.6011907 -0.7157889
 [7,] -1.8924652 -0.1504280
 [8,]  0.9820907  1.2751289
 [9,] -0.1901319  0.2211839
[10,]  1.4956947 -0.1289962
```

```
model2a
```

The output from the above model looks no different to of *model2*, so I have not bothered to reproduce it here. However, as part of it, it now has a *scores* aspect which we can see:

```
model2a$scores
```

Alternatively to print out just the top 10 scores, as shown opposite:

```
head(model2a$scores, 10)
```

```
> zscore_hozdata <- scale(hozdata)
> # look at first 10
> head(zscore_hozdata, 10)
         visperc       cubes   lozenges   paragrap   sentence    wordmean
 [1,]  0.53281893 -0.59534845  0.2735921 -0.7267921 -0.4553215 -0.96328413
 [2,]  0.09903698  0.06648561  0.6528115 -0.1653548  0.7317667 -0.00164664
 [3,]  0.96660088  1.83137646  2.6753152  1.7996758  1.1274627  2.76306116
 [4,] -0.19015099  0.06648561 -0.7376597 -0.1653548 -0.2574735 -0.84307945
 [5,]  0.09903698  0.06648561 -0.4848468  0.1153638  0.3360706 -1.20369351
 [6,] -1.34690287  0.06648561 -1.1168792 -0.4460735  0.3360706 -0.24205601
 [7,] -1.78068482 -0.81595981 -1.1168792 -1.5689481 -1.8402576 -0.96328413
 [8,]  0.53281893  1.39015375  1.9168763  0.1153638  0.7317667 -0.00164664
 [9,]  0.09903698 -0.59534845  0.6528115 -0.7267921 -0.4553215  0.23876273
[10,]  0.96660088  0.72831968  0.9056245  0.6768011  0.9296147  2.16203772
```

How are the above factor scores for each case calculated? The answer is that an equation is used where the dependent variable is the predicted factor score and the independent variables are the observed variables. To demonstrate this we first need to obtain two sets of values; z scores of the original values and the factor score coefficient matrix (also called the beta weights). First obtaining the z scores:

```
zscore_hozdata <- scale(hozdata)

head(zscore_hozdata, 10)
```

the beta weights

```
> model2$weights
               PC1        PC2
visperc  0.2067814  0.3630179
cubes    0.1696383  0.4892526
lozenges 0.2162380  0.3323327
paragrap 0.2653226 -0.2875605
sentence 0.2615612 -0.2771628
wordmean 0.2561086 -0.3167913
```

Now to obtain the beta weights which are another part of the *model2a* object:

```
model2a$weights
```

For a Principle Components Analysis (PCA) you can check the individual factor score values produced by R by plugging the standardised variable scores for each individual into the equation below. However, this does not work for the other types of factor extraction; once you have lost some of the variance in the factor extraction process, you can't go back. In these cases the factor scores produced by R are estimates rather than exact values. So returning back to the PCA factors, if we take the equations for the first two factors (i.e. technically principal components here) we have:

the beta weights

FS_1 = (0.207)visperc +(0.170)cubes + (0.216)lozenges + (0.265)paragraph + (0.262)sentence +(0.256)wordmean

FS2 = (0.363)visperc +(0.489)cubes + (0.332)lozenges +(-0.288)paragraph+(-0.277)sentence +(-0.317)wordmean

Now considering the first case that is the first row in the dataset, we can also plug in their standardised scores:

$FS_{1subject1}$ = (0. 207)0.53282 + (0. 170)(-0.59535) + (0. 216)0.27359 + (0. 265)(-0.72679) + (0. 262)(-0.45532) + (0. 256)(-0.96328)

In R:

Answer <- (0.207)*0.53282 + (0.170)*(-0.59535) + (0.216)*0.27359 + (0.265)*(-0.72679) + (0.262)*(-0.45532) + (0.256)*(-0.96328)

= -0.4903132.

Which is the same as the answer produced by R shown on the previous page.

Rather than calculate each row separately we can multiple the two matrices together using the matrix multiplication operator %*%. We want to multiply each set of z scores from each case with the associated beta weights:

```
>    zscore_hozdata %*% model2a$weights
             PC1          PC2
 [1,] -0.4902900  0.633425461
 [2,]  0.3200279  0.130683716
 [3,]  2.5690905  0.430673830
 [4,] -0.5146886  0.104343402
 [5,] -0.2628493  0.162349273
 [6,] -0.6011907 -0.715788943
 [7,] -1.8924652 -0.150427953
 [8,]  0.9820907  1.275128870
 [9,] -0.1901319  0.221183914
[10,]  1.4956947 -0.128996181
```

zscore_hozdata %% model2a$weights*

What do these factor scores tell us about a specific individual? Well, as the first factor is concerned with reading/writing and the second one is concerned with visual comprehension, we can see how the individual has scored on each of these two latent variables.

It is of interest to carry out some basic descriptive statistics on these new variables. Opposite is a simple scatterplot of the factor scores for the PCA. While the degree of correlation is as expected we can see that the values range from around -3 to 3 for both factor scores. Also the mean for each is zero and the standard deviation is 1, in other words they are standardised variables.

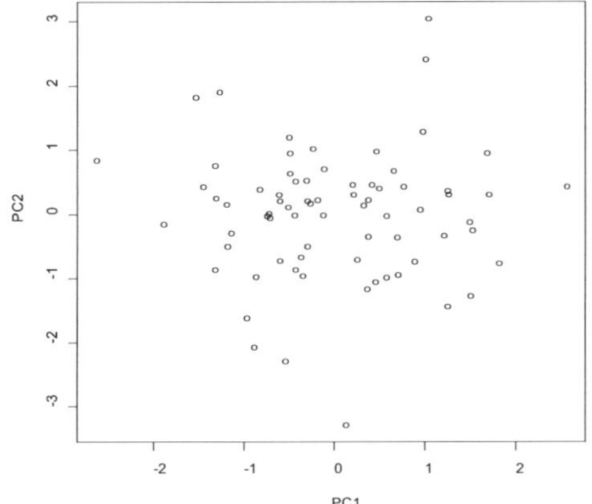


```
plot(model2a$scores)

describe(model2a$scores)
```

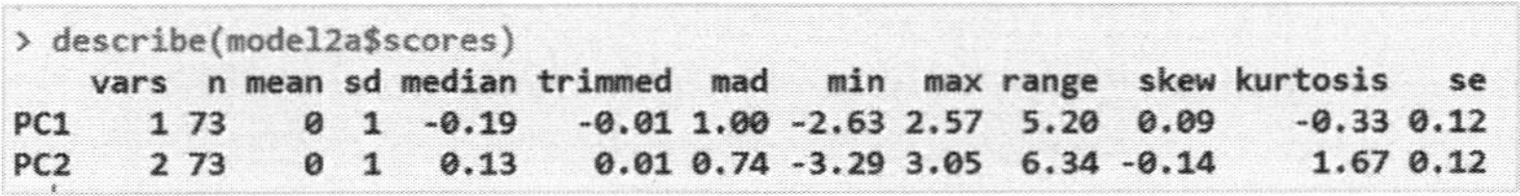

> describe(model2a$scores)

	vars	n	mean	sd	median	trimmed	mad	min	max	range	skew	kurtosis	se
PC1	1	73	0	1	-0.19	-0.01	1.00	-2.63	2.57	5.20	0.09	-0.33	0.12
PC2	2	73	0	1	0.13	0.01	0.74	-3.29	3.05	6.34	-0.14	1.67	0.12

We can also produce a 3d plot. As we only have 2 factors is not really necessary but is useful when you have more.

```
install.packages("scatterplot3d")

library(scatterplot3d)

scatterplot3d(model2a$scores)
```

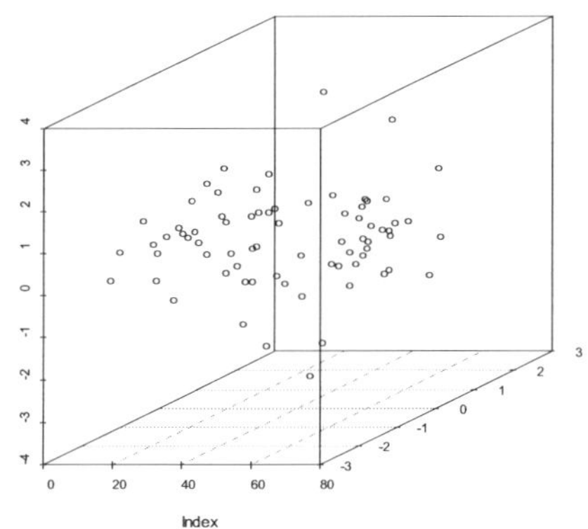

In the above I have demonstrated a typical exploratory factor analysis. In real life much more attention would be paid to possible violations of the assumptions necessary for a valid factor analysis. In particular you would undertake a careful analysis of the residuals mentioned earlier along with a thorough discussion of the possible meaning of the factors in terms of possible validation of specific theories within the research area.

64.2 Using a matrix instead of raw data

You can provide either raw data or a matrix of correlations as input for CPA/factor analysis. The R code below provides an equivalent analysis to that described previously but using a correlation matrix as input.

```
library(psych)
hozdatamatrix <-  matrix(c(
1.000, .696, .743, .369, .184, .230,
.696, 1.000, .724, .335, .179, .367,
.743, .724, 1.000, .326, .211, .343,
.369, .335, .326, 1.000, .492, .492,
.184, .179,.211, .492, 1.000,  .483,
.230, .367, .343, .492, .483, 1.000), ncol = 6, byrow = TRUE)
# now give the columns and rows names
colnames(hozdatamatrix)<- c("wordmean",  "sentence",     "paragrap",     "lozenges",     "cubes", "visperc")
rownames(hozdatamatrix)<- c("wordmean",  "sentence",     "paragrap",     "lozenges",     "cubes", "visperc")
######### you can NOT use the standard cor function i.e. cor(hozdatamatrix)
#  With a correlation matrix as it produces correlations of the correlations.
# you can produce a correlation plot by using the cor.plot function  the darker the shading for the cell the
# higher the correlation
cor.plot(hozdatamatrix)
# According to the psych package manual you should be able to use the
# coor.p() function below to obtain p values for the associated correlation matrix
# but all you get is the original correlations back not the p values,
# a nice email from  the author professor Revelle (package developer)
# said this bug will be fixed in the next version (1.4.3) and sent a bug fix below
##  given below I have called it corr.p_temp
corr.p_temp <-function(r, n, adjust="holm") {
cl <- match.call()
if(missing(n)) stop("The number of subjects must be specified")
t <- (r*sqrt(n-2))/sqrt(1-r^2)
p <- 2*(1 - pt(abs(t),(n-2)))
p[p>1] <- 1
if (adjust !="none") {
if(isSymmetric(p)) {sym <- TRUE
lp <- upper.tri(p)   #the case of a symmetric matrix
   pa <- p[lp]
   pa <- p.adjust(pa,adjust)
   p[upper.tri(p,diag=FALSE)] <- pa
 } else {
p[] <- p.adjust(p ,adjust)   #the case of an asymmetric matrix
 sym <- FALSE}
}
result <- list(r = r,n=n,t=t,p=p,sym=sym, Call=cl)
class(result) <- c("psych", "corr.test")
return(result)
}
aresult1<- corr.p_temp(r = hozdatamatrix, n= 73, adjust="holm");  aresult1
aresult2<-corr.p_temp (r = hozdatamatrix, n= 73);                  aresult2
aresult3<- corr.p_temp (r = hozdatamatrix, n= 73, adjust="none"); aresult3
aresult4<- corr.p_temp (r = hozdatamatrix, n= 73, adjust="holm"); aresult4
det(hozdatamatrix) #  gives same answer as using the raw data 0.07374609
result1 <- eigen(hozdatamatrix) # provides the eigenvalues/vectors
result1
loadings_sd <- result1$vectors %*% sqrt(diag(result1$values))
loadings_sd # the principal function also reorders then and reverses sign.
# # you can see the code for the principal() by typing into the r console window
#  psych::principal
# To carry out a PCA analysis using a correlation matrix need to
#  tell the principal function how many observations formed the
#  basis of the correlations specifying a value for the  the n.obs parameter
model1<- principal(hozdatamatrix, nfactors = 6, n.obs = 73,  rotate = "none")
model1
# to carry out a factor analysis using a correlation matrix
# adapt the fa function in a similar way:
modelb <- fa(hozdatamatrix, nfactors = 2, fm = "pa", n.obs = 73, rotate = "none")
modelb
```

The *table2matrix()* function in the *psych* package can be used to convert an R table to a matrix. Also in the *psych* package are various *read.clipboard()* functions that allow you to copy and paste a matrix of correlations in something like Excel or Word and then paste directly into R (see the *psych* package manual for details).

64.3 Summary - to factor analyse or not

We can use the information from the analysis along with the diagramming technique we introduced earlier to summarize our results in the diagram opposite. Notice that I have left out the lines where the loadings were below 0.69 and I have used the results from the PA extraction with Promax rotation, showing the correlation between the two factors. I could also have put the uniqueness values in but PCA does not take these into account compared to factor analysis. We followed a clearly defined set of stages:

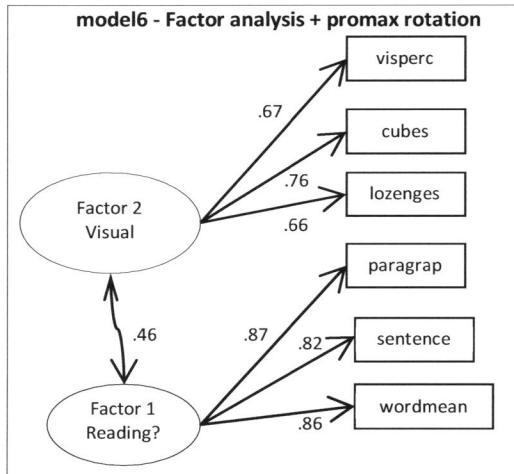

1. Data preparation (most of it had already been done in this example)
2. Observed correlation matrix inspection
3. Statistics to assess suitability of dataset for basis of PCA (KMO, Bartlett's and determinant measures)
4. Factor extraction - PCA
5. Factor rotation – to allow interpretation
6. Factor name attribution
7. Factor score interpretation

We have barely scratched the surface in this short introduction and there has always been a hot debate as to alleged benefits of PCA and factor analysis. The debate focuses on those who consider it an art rather than a scientific approach. This is because at each stage, there are many ways of interpreting the results and also a multitude of ways of proceeding. Not only this but different authorities suggest that the analysis stops at different points in the analysis and also different authors give different interpretations to the various results. Quoting Everitt & Dunn, 2001 p.288:

> Hills (1977) has gone as far as to suggest that factor analysis is not worth the time necessary to understand and carry it out. Similarly Chatfield and Collins (1980) recommend that factor analysis should not be used in most practical situations. Most people feel such criticisms go too far. Factor analysis is simply an additional, and at times very useful, tool for investigating particular features of the structure of multivariate observations. Of course, like many models used in analysing data, the one used in factor analysis is likely to be only a very idealized approximation to the truth in the situations in which it is generally applied. Such an approximation may however prove a valuable starting point for further investigations.
>
> Hills M 1977 Book review. Applied statistics (26) 339-340.
>
> Chatfield C, Collins A J 1980 Introduction to Multivariate Analysis. Chapman & Hall. London.

Loehlin, 2004 p.230-6 provides an excellent in-depth criticism of Latent Variable modelling (of which factor analysis is one example). In contrast to these criticisms a more positive approach can be found in many books about factor analysis, for example chapter 7 entitled factor analysis in Bartholomew, Steele, Moustaki & Galbraith, 2008 as well as the conference proceedings held in 2004 entitled 100 years of factor analysis available at http://www.fa100.info/.

64.4 Tips and Tricks

There is a wealth of information about using R to carry out PCA and factor analysis in R. For PCA there is an excellent youtube video given by Edward Boone who is associate professor at Virginia Commonwealth University at: http://youtu.be/Heh7Nv4qimU you can see his personal page at: www.people.vcu.edu/~elboone2/

There are many ways of carrying out factor analysis in R and the R site, *quick-R* (http://www.statmethods.net/), provides not only general advice about R but also detailed information about carrying out various types of factor analysis with links to sources of additional information all of which can be found at: http://www.statmethods.net/advstats/factor.html

The R *psych* package, used in this chapter, aids factor analysis and the developer of the package maintains an excellent online book including a very detailed chapter on factor analysis (http://personality-project.org/r/book/Chapter6.pdf). Also the factor analysis chapter in Andy Field's *Discovering Statistics using R* (2012) describes non-orthogonal rotation (promax) in much more detail.

R offers many more options than SPSS for both PCA and factor analysis. One very interesting option (in the *psych* package) is the ability to create "Parallel Analysis Scree plots". This is where R produces a random data matrix besides the dataset you are working with and then plots the Eigen values from both on a scree plot allowing you to assess the difference between what your dataset has produced against a random dataset. For more details see the *fa.parallel()* entry in the *psych* package manual.

Graphical innovations to gain greater insight into the various aspects of CPA and factor analysis are provided in the *adegenet* package, for details see: http://cran.at.r-project.org/web/packages/adegenet/vignettes/adegenet-dapc.pdf

Factor analysis forms the basis of an even more complex technique called Structural Equation Modelling (SEM). What we have done in this chapter as well as much more, can be achieved using SEM.

SEM provides much more sophistication than the traditional exploratory factor analysis, although a traditional EFA often is the first step to a full SEM analysis. In SEM we can compare models and also analyse the overall fit of a model both of which are discussed in the next chapter re-analysing this data using a SEM framework.

There is an article describing how to link output from the *pych* package to a sem analysis in R at: http://cran.r-project.org/web/packages/psych/vignettes/psych_for_sem.pdf .

64.5 Techie notes

64.5.1 Kaiser normalisation

```
> model7 <- kaiser(model4, rotate="promax")
> model7

Call: NULL
Standardized loadings (pattern matrix) based upon correlation matrix
          PA1   PA2   h2   u2
visperc   0.06  0.67 0.45 0.55
cubes    -0.12  0.76 0.59 0.41
lozenges  0.10  0.66 0.44 0.56
paragrap  0.87  0.00 0.76 0.24
sentence  0.82  0.03 0.67 0.33
wordmean  0.86 -0.04 0.75 0.25

                       PA1  PA2
SS loadings           2.20 1.46
Proportion Var        0.37 0.24
Cumulative Var        0.37 0.61
Proportion Explained  0.60 0.40
Cumulative Proportion 0.60 1.00
     PA1  PA2
PA1 1.00 0.18
PA2 0.18 1.00
```

Sometimes you wish to compare the results from R to those produced by another statistical application such as SPSS. Unfortunately, the factor loadings in the above are sometimes not the same as those produced in SPSS because SPSS scales the values using something called Kaiser normalisation. Luckily, the *psych* package provides a function to carry out this conversion. The *kaiser()* function works best if its input is the non-rotated form of the analysis (information taken from the *psych* package help file). As an example, to obtain the Kaiser normalised factor loadings from *model4* type the following:

```
model7 <- kaiser(model4, rotate="promax")

model7
```

64.5.2 Hofman's complexity index

Hofman's complexity index is reported by the *fa()* function when you perform an oblique rotation. Pettersson & Turkheimer, 2010 describe it thus:

"Hofman's complexity index represents the average number of latent variables needed to account for the [observed] manifest variables. Whereas a perfect simple structure solution has a complexity of one where each item would only load on one factor. In contrast a solution with evenly distributed items has a complexity greater than one. For each item, the sum of the squared loadings is squared, and then divided by the sum of the fourth powers of the loadings. A mean is then computed across the items. c_i represents the complexity of the *i*th variable in any loading matrix *a*, where *r* represents the number of rows [factors]".

$$c_i = \frac{\left(\sum_{j=1}^{r} a^2 ij\right)^2}{\sum_{j=1}^{r} a^4 ij}$$

The original, difficult to obtain, Hofman paper is; Hofmann, 1978, Complexity and simplicity as objective indices descriptive of factor solutions. Multivariate Behavioral Research, vol.13 (2), p.247-250.

65 Structural Equation Modelling (SEM)

Self test questions
Whereas relatively few people actually carry out SEM analyses a much large number read books and articles reporting such analyses. Therefore, for those of you who only want to learn about SEM rather than carry out a SEM analysis I have included several self test questions in this chapter.

Structural Equation modelling, SEM for short, allows you to develop and test models that consist of regressions, correlations and differences in means between groups. SEM is a statistical technique that has developed from the concepts of covariance and correlation, therefore all the facts you know about correlation, including its limitations and pitfalls apply to SEM. Correlations themselves form the basis of path analysis (PA) and confirmatory factor analysis (CFA), which are forms of SEM. This will be explained diagrammatically on the next few pages.

65.1 The background to SEM

The statistical technique SEM attempts to minimise the difference between two sets of values (matrices of covariances=unstandardized correlations). One set is calculated from the sample data while the other is generated from a hypothesised theoretical model defined by the researcher. This idea can be expressed succinctly in the following equation.

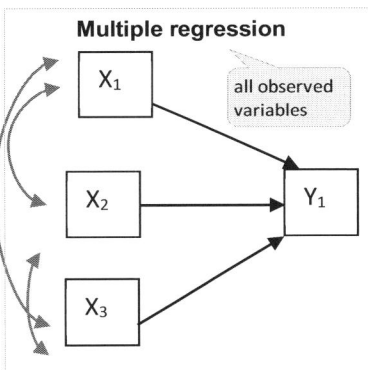

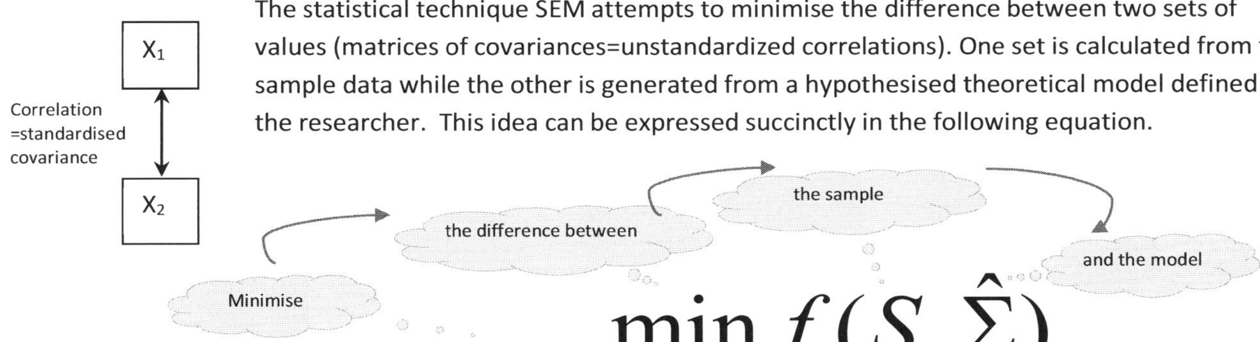

$$\min f(S, \hat{\Sigma})$$

Multiple regression

X1 · X2 · X3 · Y1 · all observed variables

Schumacher and Lomax's excellent book, *A Beginner's guide to Structural Equation Modeling*, 2010, provides details of what SEM is, from which I have compiled the following list:

1. Based on Correlations (Covariance)
2. Complex mathematical approach only made widely available with the use of suitable software
3. Allows the definition of complex relationships using models (mathematically using covariance matrices which can be partially represented by diagrams)
4. Extends regression (Path models)
5. Extends Confirmatory Factor Analysis (CFA)
6. Combines the two to form very complex models = Structural Equation Models (SEM)
7. Allows assessment of the degree to which a proposed model fits the sample data
8. Allows comparison of models

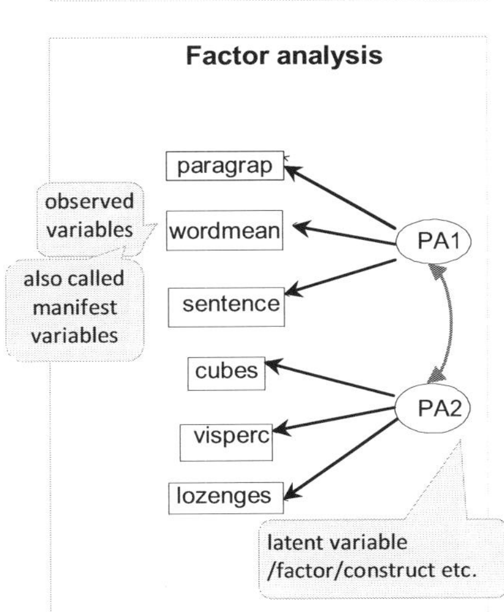

There has been a long-ongoing debate as to what degree SEM models can help establish causal rather than correlational relations (Bollen & Pearl, 2013). Possibly the best approach is to consider the model as correlational and then re-interpret it in terms of context to decide which are causal associations.

Let's take a look at the two building builds of SEM, Path modelling and (Confirmatory) factor analysis (CFA) in a little more detail.

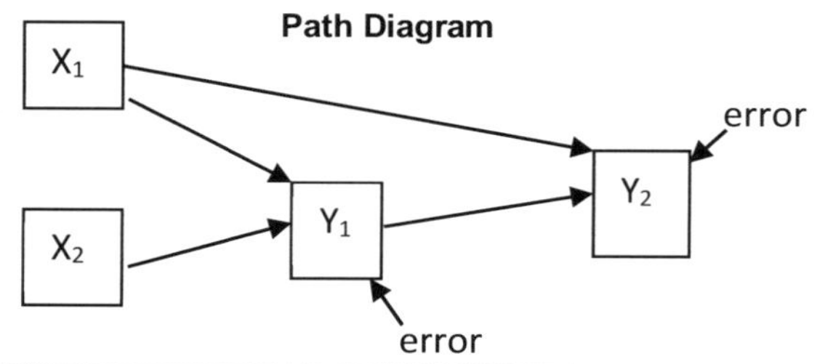

In SEM speak when the diagrams only contain observed variables they are called **path diagrams.** As shown opposite, we can extend the regression idea to consider the relationship between more complex configurations of observed variables. Measured variables are also called manifest variables. We no longer have simply inputs (independent variables) along with a single output (dependent variable), here there is additionally a whole set of intervening variables as well.

In contrast, models that contain both observed and latent variables are called SEM models.

The single arrowed lines are called **paths** and the squares **observed variables**.

The double arrowed lines are **covariances** or **correlations** depending upon which set of values are drawn on the diagram. This is demonstrated in the multiple regression design shown on the previous page where the correlations between the inputs (predictors) are shown as the double headed arrows (also see Kline, 2005 p.32-33).

Unfortunately, there are several different ways of drawing SEM diagrams and each SEM software developer shows or ignores different aspects. A common aspect that is omitted is the estimate of variance of each observed/latent variable. Because a variable's estimated variance is equivalent to the estimate of its covariance with itself, i.e. $cov(x,x) = var(x)$ we can use the double-headed arrows again to represent this; they simply turn back on themselves.

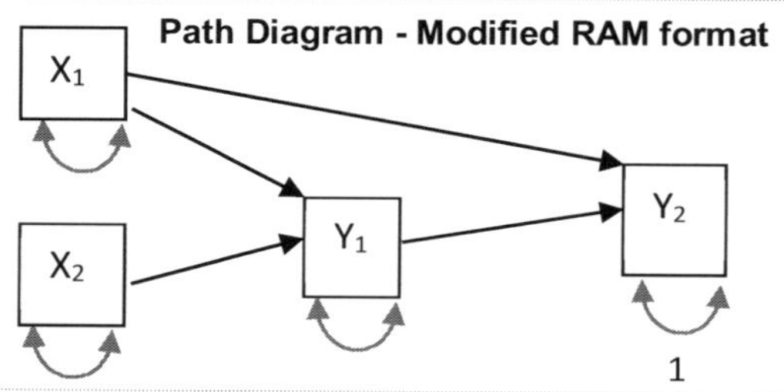

The Modified RAM-format path diagram does this, (Fox, 2006 p.476; Kline, 2011 p.96) as shown opposite. The error 'variables' now simply becomes another covariance arrow. Furthermore, including this visual aspect aids interpretation of the model output, specifically the parameter estimates. You can now calculate exactly the number of parameters that are going to be estimated by adding up all the single and double headed arrows omitting those that have an identification constraint, distinguished by having a value attached to them (e.g. y_2 has a constraint of 1).

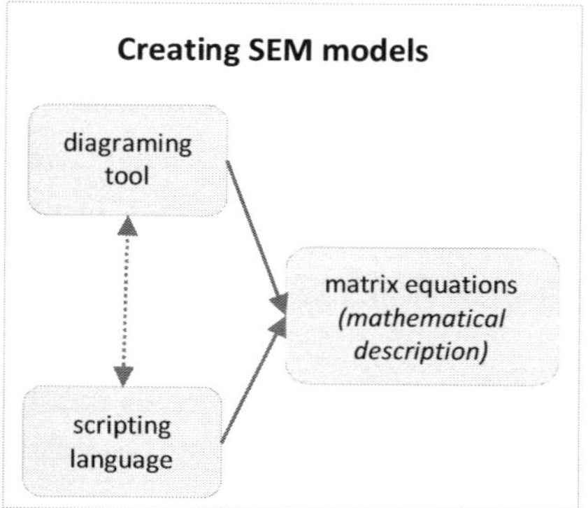

Eventually, whichever style of diagram you use, it will ultimately be converted into a set of matrices, which we will now briefly discuss. Alternatively, you can also specify a SEM model using various scripting languages which we will look at in the practical section in this chapter.

65.1.1 The model equations

There are two main ways of expressing the SEM model as a set of matrices. Firstly those developed by Joreskog & Van Thillo, 1972 culminating in the development of the LISREL software which is still one of the most popular commercial SEM software packages. Secondly, in 1980, McArdle suggested an alternative formulation, simplifying the 8 matrices needed by LISREL to just 4 to create the Reticular action model (RAM). Schumacker & Lomax, 2010 provide a chapter describing the LISREL matrix approach while Fox, 2006 provides details of the RAM approach along with how this is implemented in the R *SEM* package.

If you consider only those variables which have one or more single arrowed path pointing towards them, each group can be considered to be a separate equation (actually a regression equation). In the diagram below, we have one such group.

Let's look at the 'predictors of mortality' regression example opposite produced using the free Ωnyx program. When you right click on the diagram you can view either the RAM or LISREL matrices (menu option below). I have shown the RAM option below.

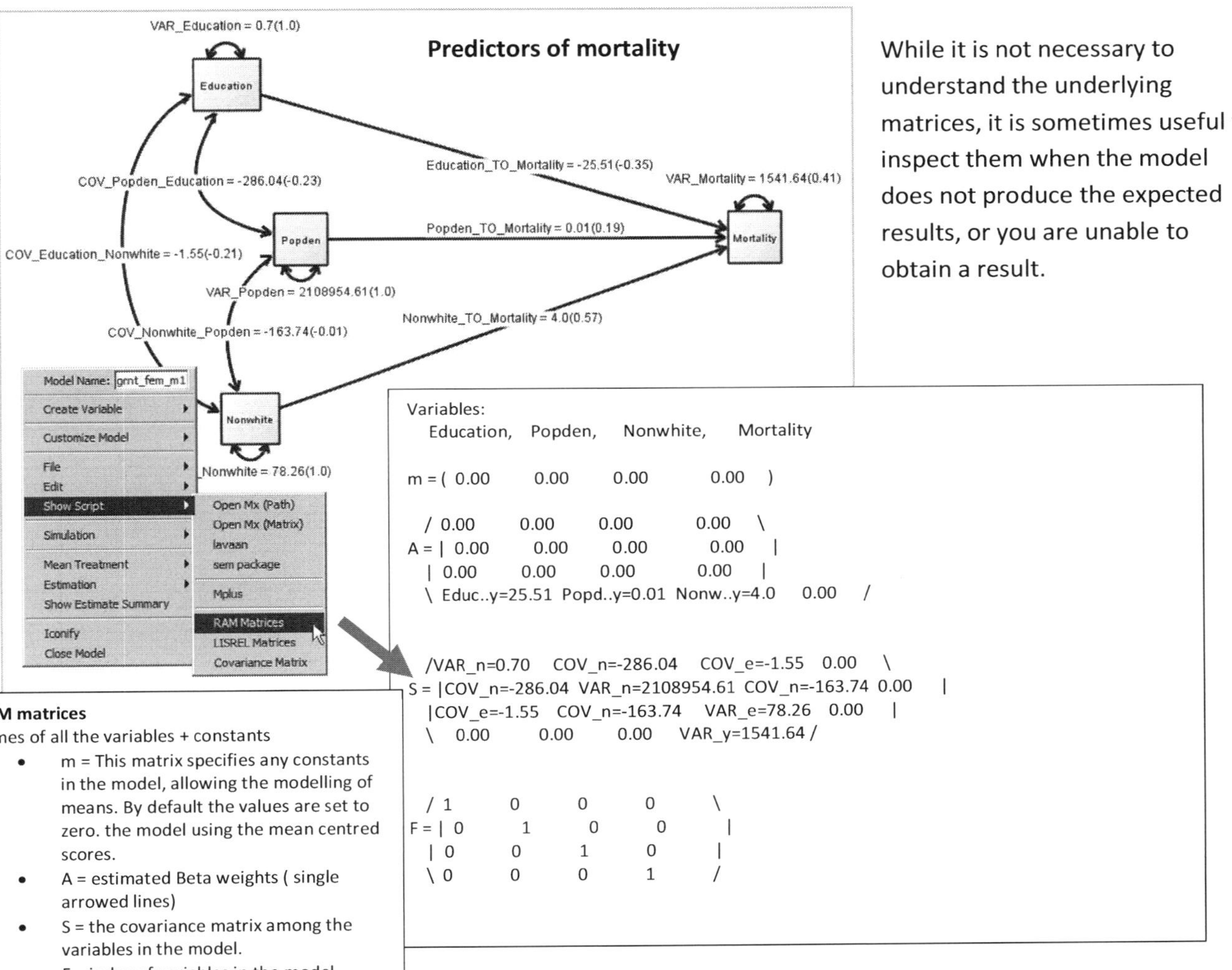

While it is not necessary to understand the underlying matrices, it is sometimes useful to inspect them when the model does not produce the expected results, or you are unable to obtain a result.

RAM matrices

names of all the variables + constants

- m = This matrix specifies any constants in the model, allowing the modelling of means. By default the values are set to zero. the model using the mean centred scores.
- A = estimated Beta weights (single arrowed lines)
- S = the covariance matrix among the variables in the model.
- F = index of variables in the model, observed =1; constants and latent's=0. There are only observed variables in this model. (this matrix is also called J by Fox, 2006)

The above matrices are manipulated to produce the various results shown in a SEM output. Specifically the **S** matrix of OBSERVED values are compared against those PREDICTED by the model stored in a matrix called **C.** which is calculated from all the above matrices.

The computer program then attempts to minimise the difference between the two.

This description leads to two questions:

1. **What is meant by difference between?**
There are various ways of measuring the difference between the two matrices and the computed values are often called fit or discrepancy functions. Such functions go by the name of ordinary least squares (OLS), asymptotically distribution free criterion (ADF), generalised least squares (GLS) and maximum Likelihood (ML) where ML is the standard technique in SEM. In ordinary least squares the result is the sum of the squared difference between the two corresponding items in each matrix. In all these criteria the minimum value is zero when both matrices are the same.

2. **Is there any strategy to finding an improved fit between the two matrices each time around?**
By making use of one of the above computed discrepancy function values we can compare these for difference guesses of our parameters and hopefully gradually home in until we get a sufficiently small value. There are special search algorithms build into SEM computer programs to help it gradually 'home in' but because the search program may go widely astray we can usually also set a maximum number of times (i.e. iterations) it can have a shot at finding a sufficiently close answer.

We will now look at a typical SEM diagram showing a set of model path estimates.

65.1.2 Interpreting a Standardised Path model (i.e. no circles)

The diagram below contains more information than the previous ones so I will discuss each aspect separately:

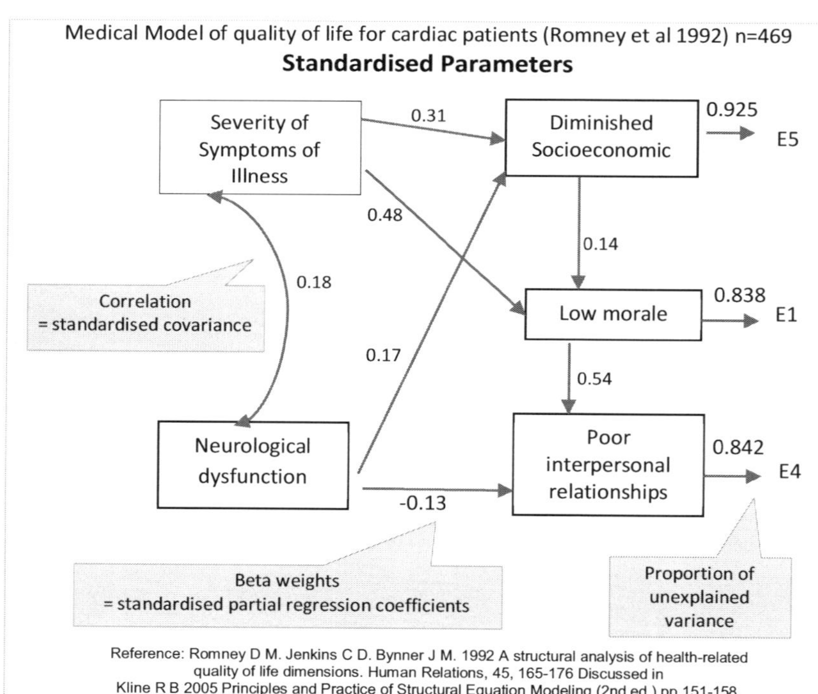

The results are taken from a correlational study concerning 469 cardiac patients where the investigators were keen to find out the associations between various measures.

SEM is usually carried out on large samples and 469 is small compared to many studies. Also usually, the study is of a retrospective/correlational design although often there are attempts in the discussion of such papers to interpret the relationships as causal. We will return to this issue later.

Title of Standardised Parameters - this indicates that the means are set to zero and the standard deviations to one for each variable (i.e. z scores). This allows the various parameters in the diagram to be compared and interpreted in a specific way described below.

Double arrow lines - This indicates a correlation where the modelling process estimates the value. Variables that do not have lines between them specify that the correlation between them is set to zero in the model (i.e. they are independent).

Single arrow lines - These are beta weights (standardised partial regression coefficients) which are identical to those in multiple regression. With this knowledge, it is possible to interpret them as you would do in a regression equation. For example, a level of *Severity of illness* one full standard deviation above the mean predicts an increase in low morale by 0.48 standard deviations above the mean controlling for Neurological dysfunction.

Error terms (E1, E4, E5)– This is the unexplained variance, that which is either measurement error (random) or not explained by the model. This diagram uses the 'e' style to show them rather than the self directed double arrows.

Proportion of unexplained variance values – these are also called the *standardised residual variances* **(SRV).** One minus these values gives the R^2 value (proportion of variance explained). So the SRV x100 gives the % of the variance that is not predicted by the specific inputs that have arrows pointing to that variable. In some diagrams and software (e.g. EQS) the square root value is given instead.

Direction of the single arrow - this does matter as it defines which variable will have the error term (the variable that the arrow points to) and also how to interpret the beta weight (i.e. path coefficient).

To check your understanding I have included some self-test questions opposite.

Self-test:

1. Please complete this sentence. A level of Neurological dysfunction one full standard deviation above the mean predicts an increase in low morale by ____ Standard _____ above the mean controlling for Symptoms of illness.
2. What does the value 0.54 represent between 'Low morale' and 'Poor relationships'?
3. What do the values from E5, E1 and E4 suggest about the model? How do they relate to R^2.
4. What does the double pointed arrow (<-->) indicate.
5. There is no path between 'Neurological dysfunction' and 'Diminished Socioeconomic Status' in the model. What does this say about the model specification?
6. Does the above model more closely represent a Regression or Factor analysis? Give reasons for your answer.

65.1.3 Interpreting a Standardised SEM model (i.e. circles and squares)

Interpreting an SEM model containing both observed and latent variables is the same as for the above path model. Where the model is representing a factor analysis the only difference is one of terminology, demonstrated in the practical section to this chapter.

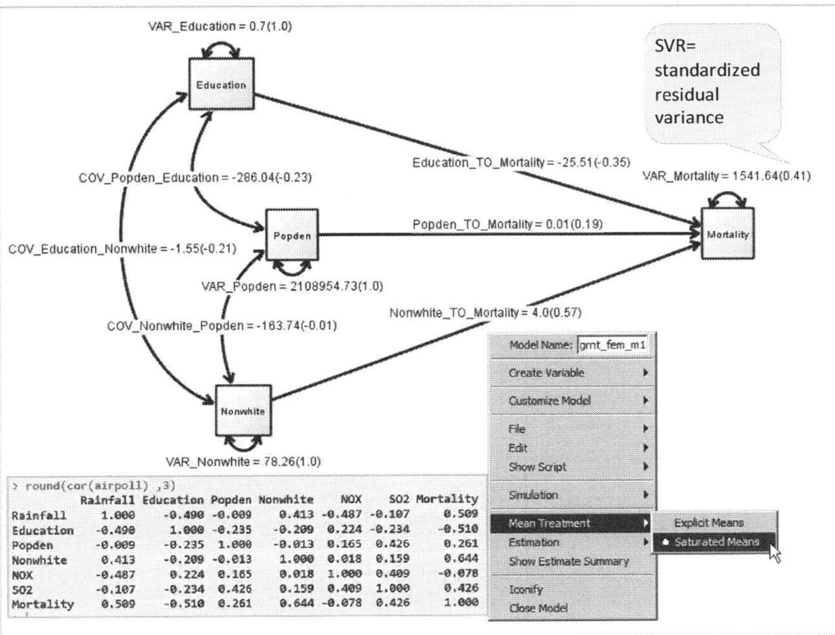


```
> round(cor(airpoll) ,3)
```

	Rainfall	Education	Popden	Nonwhite	NOX	SO2	Mortality
Rainfall	1.000	-0.490	-0.009	0.413	-0.487	-0.107	0.509
Education	-0.490	1.000	-0.235	-0.209	0.224	-0.234	-0.510
Popden	-0.009	-0.235	1.000	-0.013	0.165	0.426	0.261
Nonwhite	0.413	-0.209	-0.013	1.000	0.018	0.159	0.644
NOX	-0.487	0.224	0.165	0.018	1.000	0.409	-0.078
SO2	-0.107	-0.234	0.426	0.159	0.409	1.000	0.426
Mortality	0.509	-0.510	0.261	0.644	-0.078	0.426	1.000

65.1.4 Computing R squared from a regression using SEM

Utilising the path coefficients in the SEM diagram opposite along with the simple correlations (below) we can obtain R^2 by using the equation discussed in the multiple linear regression chapter.

$$
\begin{aligned}
R^2 &= \beta_1 r_{YX_1} + \beta_2 r_{YX_2} + \beta_3 r_{YX_3} \\
&= (-.35)(-.510) + (.19)(.261) + (.57)(.644) \\
&= .1770741 + .0488818 + .3694264 \\
&= .595\ (3\,decimal\ places)
\end{aligned}
$$

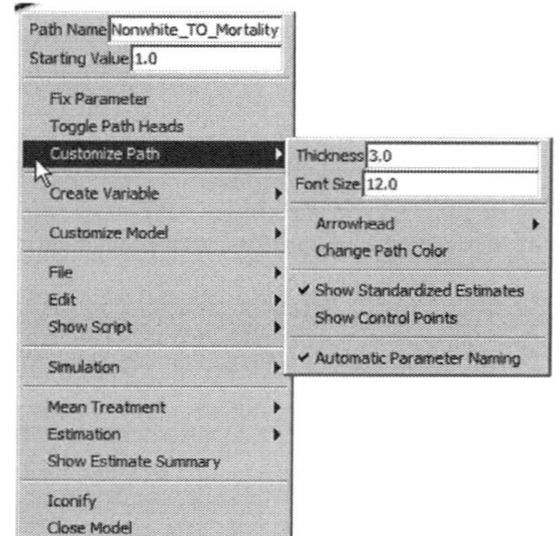

Alternatively, we could have simply calculated:

`(1 - the standardised residual variance (SRV) of Mortality) = 1-.41 = 0.59.`

I hope that this strengthens your belief that the SRV is a measure of what remains unexplained after the regression analysis, as well as the idea that the proportion of variance explained can be calculated from the beta weights and the simple correlations between each input and the output.

To display the standardised path estimates alongside the unstandardised ones in Ωnyx you need to select each path and then select the menu option shown opposite.

65.1.5 Calculating Direct, Indirect and Total Effects

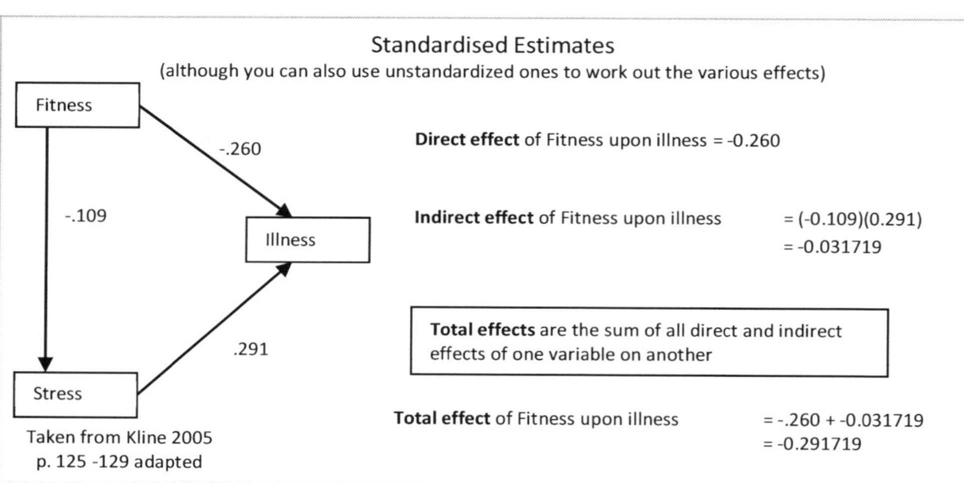

In a structural equation model it is possible to calculate the various effects one or more variables have upon another in the model via other variables as well as directly.

From the above example it may appear easy to calculate such effects, however this is not the case as with complex models the indirect path may follow tortuous routes. To help simplify this problem various rules have been developed. Remember you must **always follow a path that is in the direction of the arrows**. Remember that the SEM model you have created will undoubtedly not contain all the possible paths between each variable so that there will also be **unanalysed effects.** Similarly, there may also be **spurious effects** in the model where you create a path that goes against the flow of the arrows at some point.

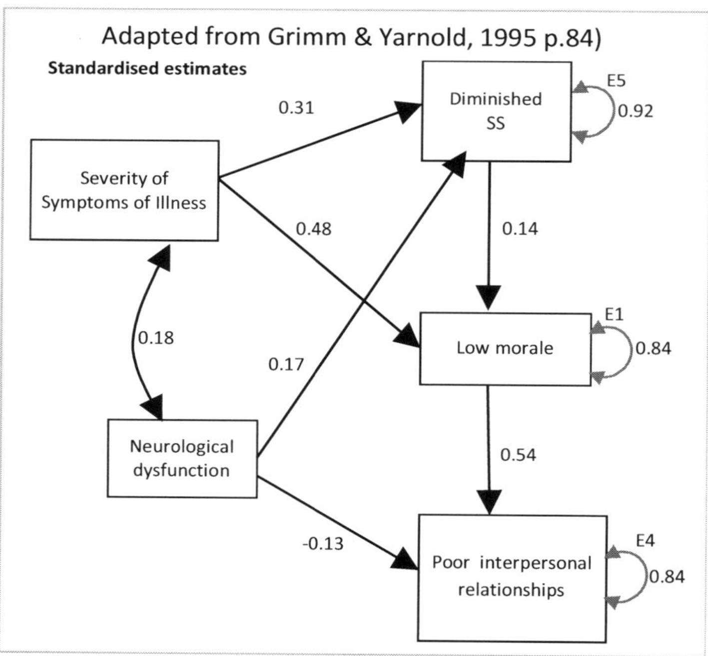

Once again, because you are more likely to need to interpret and calculate these measures from a diagram rather than carrying out a whole analysis yourself I have included an exercise below.

> Self test:
>
> 1. What is the total effect of Severity of Symptoms of Illness on Poor Interpersonal relationships?
> 2. What is the total effect of Neurological dysfunction upon Diminished Socioeconomic Status?
> 3. What is the direct effect of Diminished Socioeconomic Status on Poor interpersonal relationships?
>
> Hint: in reality often pathways which have very small coefficients (i.e. less than 0.1) are ignored.

65.1.6 Covariances rather than raw data are used in model development

From the above discussion it should now be clear that calculation of the various path values (coefficients) only makes use of summary statistics (i.e. covariances or correlations) along with an indication of the original sample size. You may remember in the chapter on factor analysis we saw how a matrix of correlations along with a value specifying the sample size could be used instead of the raw data; the same goes for SEM analysis.

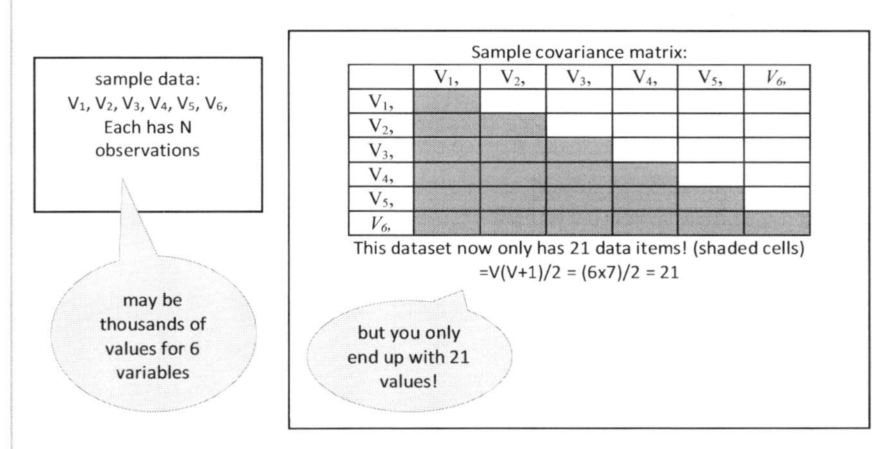
sample data:
V1, V2, V3, V4, V5, V6,
Each has N
observations
may be thousands of values for 6 variables
Sample covariance matrix:
V1, V2, V3, V4, V5, V6,
This dataset now only has 21 data items! (shaded cells)
=V(V+1)/2 = (6x7)/2 = 21
but you only end up with 21 values!

Related to this is the fact that authors are encouraged to publish the observed covariance matrix, to enable others to repeat at least part of the analysis when presenting a SEM analysis.

Just because the 'raw data' is not used in the calculation of the path values, it should not be assumed that sample size is not important, in contrast, it is extremely important. This is because we always need to assess if the various estimated values are any different from zero. This means it is necessary to either produce p-values or confidence intervals, both of which rely upon the calculation of standard errors (unless we use bootstrapping techniques), which itself is directly dependent upon the raw sample size. Furthermore, because SEM models often have a substantial number of variables and paths, a large sample size (as a rule of thumb, of at least 200) is needed to begin to make any sense of the data.

65.1.7 Why and When to use Structural Equation Modelling

The following is a list of the common reasons given for using SEM techniques:

1. Confirm Relationships / hypotheses
2. Test complexity (multivariate)
3. Longitudinal / Multiple groups (panel designs)
4. Test Specific relationships
5. Multiple observed variables
6. Take measurement error into account

To use Structural Equation Modelling successfully often requires time to dedicate oneself to the SEM literature and also specific SEM software with the result that frequently those researchers that become competent tend to apply it to almost all problems. Because of the paradox between the complexity of the mathematics and interpretation of the results compared to the relative ease of being able to draw the diagrams for a structural equation model it is often inappropriately used. One common problem often encountered with developing SEM models is that of identification, which is discussed next.

65.1.8 The identification problem and sample size

The identification aspect is basically asking, do we have enough data to allow us to develop a unique solution for our model. The following discussion, adapted from Schumacker & Lomax, 2010 p.63, provides an explanation.

Suppose our model is X + Y = 10 (in effect we have two free parameters and one data item)

If this is the case, X and Y can take on any number of values (e.g. x=1;y=9; x=2.5 y=7.5 etc.).

Imagine that each time the computer runs a program to find a solution to this problem it just randomly assigns a value between 0.1 and 9.9 to X and therefore provides a different solution for Y each time. When it is impossible to define a unique value to each of our parameters in our model the model is said to be **under-identified** and is useless. When we talk about identification, we are talking about the degrees of freedom for the model (df_m) - in other words the amount of data that can move freely after accounting for the parameter estimates. A under-identified model possesses negative degrees of freedom and no unique solution can be found.

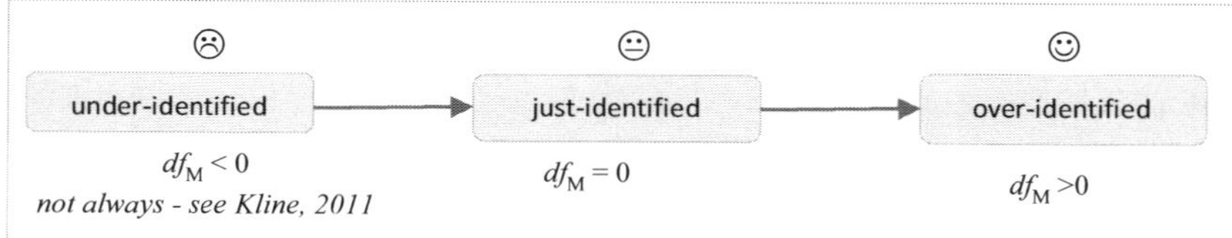

In contrast, the desirable situation is to have a model which is preferably over-identified or, failing that, just identified. Ideally we want a model that estimates the smallest number of parameters but at the same time provides the best fit, thus maximising the accuracy of the estimates of our parameters (or more traditionally we say maximising the power of the study).

This ratio of the of sample size to number of parameters to be estimated in an SEM is important. Quoting Bagozzi & Yi, 2012 p.29: *As Bentler and Chou (1987, p. 91) state that the ratio "may be able to go as low as 5:1 under normal and elliptical theory, especially when there are many indicators of latent variables and the associated factor loadings are large". But they also believe that "a ratio of at least 10:1 may be more appropriate for arbitrary distributions."*. I will revisit this issue when discussing model fit indices.

One solution to the identification problem is to set one or more of the parameters to a specific value, which is then known as a **fixed parameter**. An example of fixed parameters are all the hidden paths in the model indicating a correlation fixed to zero. Looking back at the model matrices section you can see these hidden values are represented by the zeros. It is also possible to constrain more than one parameter estimate to a specific value. Such a value is known as an **identification constraint**. We saw this in the factor models chapter where one of the loadings was set to 1 for each of the factors.

Along with the identification problem comes the problem of assessing how well overall the model fits the observed data.

65.1.9 Model fit

There are many problems with taking an often complex model and obtaining a single value from it to represent the idea of fit. As there is no acceptable solution to this problem every SEM program provides over two dozen fit measures, and most research papers offer a selection! Because single fit measures provide an overall averaged value they unfortunately stop researchers looking at each individual parameter residual. There is always the possibility that the model may fit very well for the majority of the parameters but then have a large residual for one. If this is the case the researcher would then need to consider if the particular variable was correctly specified in the model and also how important is was in comparison to the other variables. I describe this as the *does my bum look big in this dress aspect.* Because of this I recommend that you always start by looking at the standardised residual matrix (Loehlin, 2004 p.69) which will highlight particular hotspots.

Schumacker & Lomax then suggest a three stage strategy (2010 p. 74):

1. Inspect two overall fit measures:

    - Chi-squared - you are aiming for a insignificant result i.e. high p-value (usually above .95) resulting from a small actual value. This value is problematic both when you have df=0 or a large initial sample (number of subjects, not the number of covariances). Various versions of this statistic such as GFI and AGFI have been developed to take these problems into account.

    - RMSEA (Root Mean Square Error of Approximation) - is at present the most popular measure, you can use it to produce confidence intervals or an associated p-value.

2. Inspect significance of individual parameters. The critical ratio (CR) for each parameter estimate is computed by dividing the estimate by its standard error. In most SEM programs it is compared to a z or t value of 1.96 at the .05 level of significance. Some software (i.e. EQS) indicate parameters that are statistically significant by placing a asterisk (*) beside them. Often in the model development process insignificant paths are removed from the model.

3. Inspect magnitude and direction of parameter estimates - does the value make sense in the proposed theoretical model? Has the computer provided impossible values such as negative variances (called Heywood cases) or correlations >1.

```
Observed Statistics                 : 14.0
estimated Parameters                : 10
Restricted Degrees of Freedom       : 4.0
Minus Two Log Likelihood            : 620.5323387476835
Number of Observations              : 60.0
χ²                                  : 1.1368683772161603E-13
AIC                                 : 640.5323387476835
AICc                                : 644.5323387476835
BIC                                 : 661.4757843699045
RMSEA                               : 0.0
SRMR (covariances only)             : 3.769947920989143E-9
CFI (to independent model)          : 1.079116807697788

TLI (to independent model)          : 0.9999999999999953
```

A typical set of model fitting statistics is shown opposite for the morality model obtained from the free Ωnyx software.

The chi-squared actual value is very small (E-13) meaning that we would have a p-value approaching 1, Indicating a good fit. The AIC and BIC measures are used to compare models where smaller values indicate a better fit.

Baggozzi, 2012 quoting various writers recommends the following standards for assessing adequate fitness of SEM models: CFI≥0.95, and SRMR≤.08, also see Schumacker & Lomax, 2010. Again both the values reported opposite indicate a good model fit. I'll discuss the RMSEA next.

65.1.9.1 RMSEA Root Mean Square Error of Approximation

$$RMSEA = \sqrt{\frac{d}{df}}$$

where

$$d = \frac{\chi^2 - df}{N - 1}$$

The RMSEA is the golden boy of fit indices having many attractive attributes. It uses a value *d*, based the non-centrality parameter (Lambda =λ) of the chi-squared (χ^2) distribution which is simply the obtained chi-squared value of the model (i.e. Lambda =λ = obtained chi-squared).

Taking the mortality example shown earlier we have a χ^2 = approximately zero. With 4 df in a sample of 60 cases, producing a *d* value of (0 - 4)/59 = -.066, and a RMSEA equal to √(.066/59), or -.001186. Because this value is less than zero we set the RMSEA value to zero. The maximum value RMSEA can obtain is 1.

The RMSEA measure was developed by Steiger (1998) who considers values below .10 "good" and below .05 indicative of "very good" fit. It is also possible to obtain a confidence limit for RMSEA using the noncentral χ^2, distribution, described in the confidence intervals for effect sizes chapter. The development of the RMSEA stems from Steiger's (Steiger, 1989) belief that the appropriate question for statistical analysis in SEM is not assessing how perfect the fit is but rather, the three following questions.

1. How well does my model fit my statistical population?

2. How precisely have we determined population fit from our sample data?

3. Does the fit still appear good when we take into account the complexity of the model and its number of free parameters?

By developing a confidence interval for the RMSEA we can take into account the three above issues testing a null hypothesis of poor, instead of perfect, fit. If the upper limit of the 90% confidence interval lies above the desired cutoff, for example .10, we can then reject the hypothesis that the fit of the model in the population is worse than our desired level. Similarly, if the CI does not extend beyond this cut-off we can conclude our model is of adequate fit. The width of the confidence interval will also provide information about the accuracy of the estimate which is always useful. Because the RMSEA is based upon the non-central chi-squared distribution it can also form the basis of a power analysis discussed next.

65.1.10 Power and sample size determination

The previous section described the general rule of the ratio of the sample size to the number of parameters. However this rather questionably naïve approach has been improved upon by several writters.

The easiest approach is to consider the RMSEA for two models from which you can readily calculate the power. Schoemann, Preacher & Coffman, 2010 provide an online calculator along with several other comparable measures, along with R code. Similarly the R package SEM Tools contains a function *plotRMSEApower()* which allows the plotting of power across a range of sample sizes much like Gpower. The investigation of power during the planning stages of a study that intends to use SEM techniques can be a long and arduous process as it requires not least the investigators to define their proposed models at the start of the process and also make a guess (quite literally) of what the RMSEA value might be.

Therefore detailed power analysis is not for the faint-hearted and anyone planning to undertake this I would recommend they start by reviewing the literature. I would suggest MacCallum, Browne & Sugawara, 1996; Hancock & Freeman, 2001; Lee, Cai & MacCallum, 2012 and Miles, 2003. Professor David Kaplan, University of Wisconsin Madison, provides a brief review of these at: http://www2.gsu.edu/~mkteer/power.html

Rex Kline who has written a popular introduction to SEM (Kline, 2011) provides a set of notes concerning power estimation for a variety of SEM models, including some R code, at

http://psychology.concordia.ca/fac/kline/SEM/qicss2/qicss2setA.pdf

65.1.11 Stages of Structural Equation Modelling

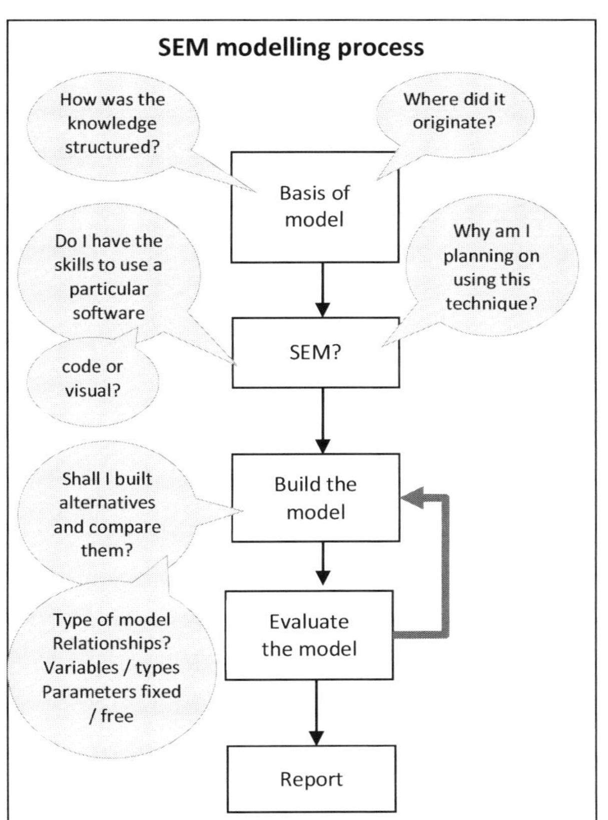

After having discussed identification issues, model fit indices and power analysis it is clear that the development of SEM models is far more complex than just drawing a diagram. Complex models also require the analyst to know the research area in depth and be able to make sensible informed decisions concerning prospective models.

Once you have a working model that the program is able to estimate then you often need to change it to perform comparisons with various other models. Often only after you have done this several times do you finally come to a decision deciding which is the 'best' possible one.

I'll now demonstrate how you can create a simple SEM model by repeating the analysis described in the factor analysis chapter. I'll also demonstrate a more complex model.

65.2 A basic SEM model – the Hozinger & Swineford 1939 data analysed using Onyx (Ωnyx)

Details of the Hozinger & Swineford 1939 dataset were discussed in the factor analysis chapter. In this section we will use a separate program that produces R code called Ωnyx (Ωnyx) to carry out the SEM analysis. This allows us to specify the model using a set of graphical symbols which you may recognise from the factor analysis chapter.

65.2.1 Preliminaries

This time you need to download the dataset (tab-delimited format) from the link below and save it locally:

http://www.robin-beaumont.co.uk/virtualclassroom/book2data/grnt_fem.dat

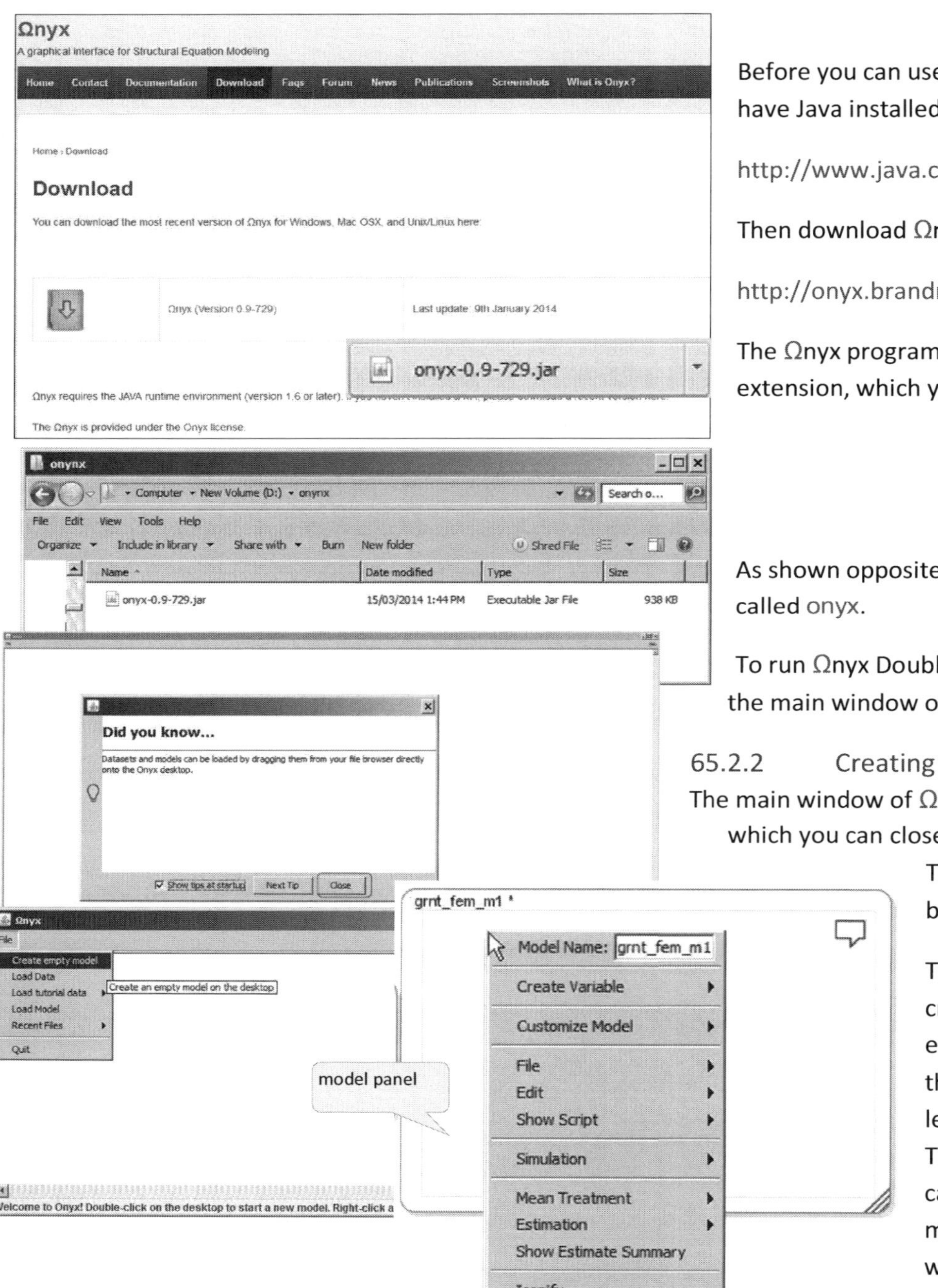

Before you can use the Ωnyx program you need to have Java installed which you can find at:

http://www.java.com/en/

Then download Ωnyx from:

http://onyx.brandmaier.de/download/

The Ωnyx program, being a java file, has a jar extension, which you save to a local folder.

As shown opposite, I created a folder on my d: drive called onyx.

To run Ωnyx Double click on the jar file. This brings up the main window of Ωnyx.

65.2.2 Creating an SEM model in Ωnyx

The main window of Ωnyx contains a tip of the day box which you can close by clicking on the **close** button. The tip also appears on the bottom bar of the Ωnyx window.

The first thing you need to do is create a new model. You can do this either by selecting the option from the dropdown menu or by double left mouse clicking in the desktop. The model panel now appears. You can give the model a name by left mouse clicking anywhere in the window. I have called it *grnt_fem_m1*.

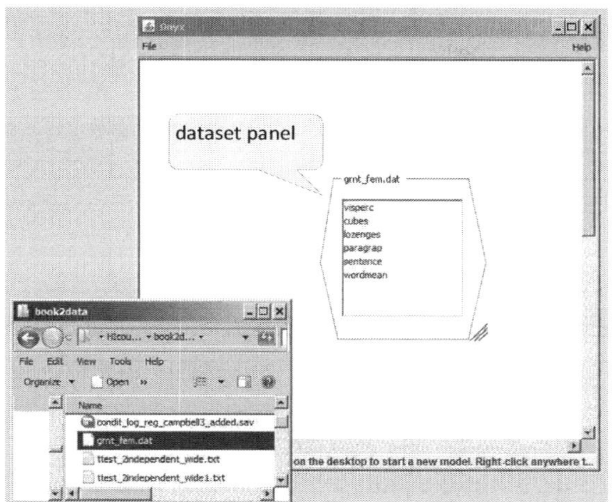

We can now proceed to specify the model by either drawing the symbols and then linking them to the fields in our dataset or, the technique we are going to use now, loading the dataset into Ωnyx and drag the fields across to the model panel.

To load the dataset into Ωnyx, either select the following menu option or simply drag the data file from the file browser window.

File->load data

A hexagonal window (called the dataset panel) appears which represents your dataset.

Ωnyx has context-sensitive popup menus, which means that if you hold the mouse cursor over various elements you get information about them

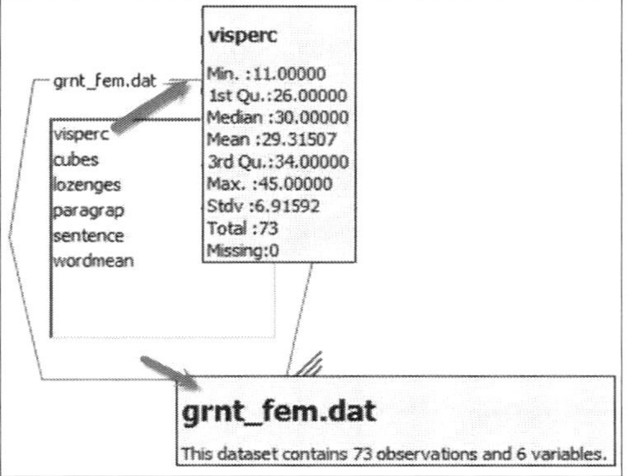

To add one or more variables to your model window simply drag them from the dataset panel to the model panel. You can select more than one using either the *shift* (continuous blocks) or *CTRL* keys (discontinuous blocks).

Ωnyx gives each graphical representation of the variable the same name as the variable in the dataset.

You can move and reshape the elements by selecting and then clicking and dragging.

Similarly, you can move the model panel by dragging its top or bottom border (a hand appears) or resize it by dragging the right hand lower corner.

We are going to create a type of SEM model known as a Confirmatory Factor Analysis (CFA) and to interpret the findings of such an analysis is it best to have the variables standardised. Luckily, you don't need to convert the variables as Ωnyx can do it for you.

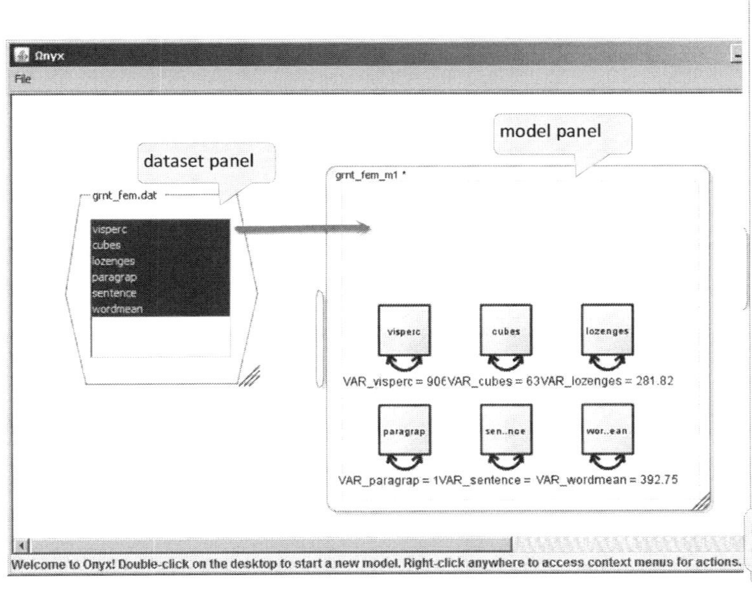

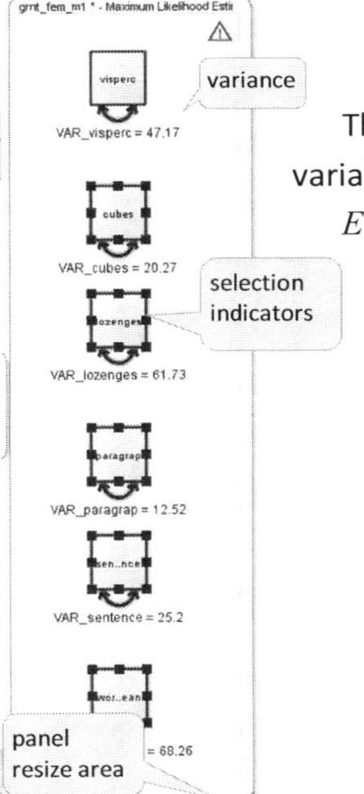

The error variance for each variable is often simply labelled *E1* to *En* where *n* is the number of variables in the model requiring error variances.

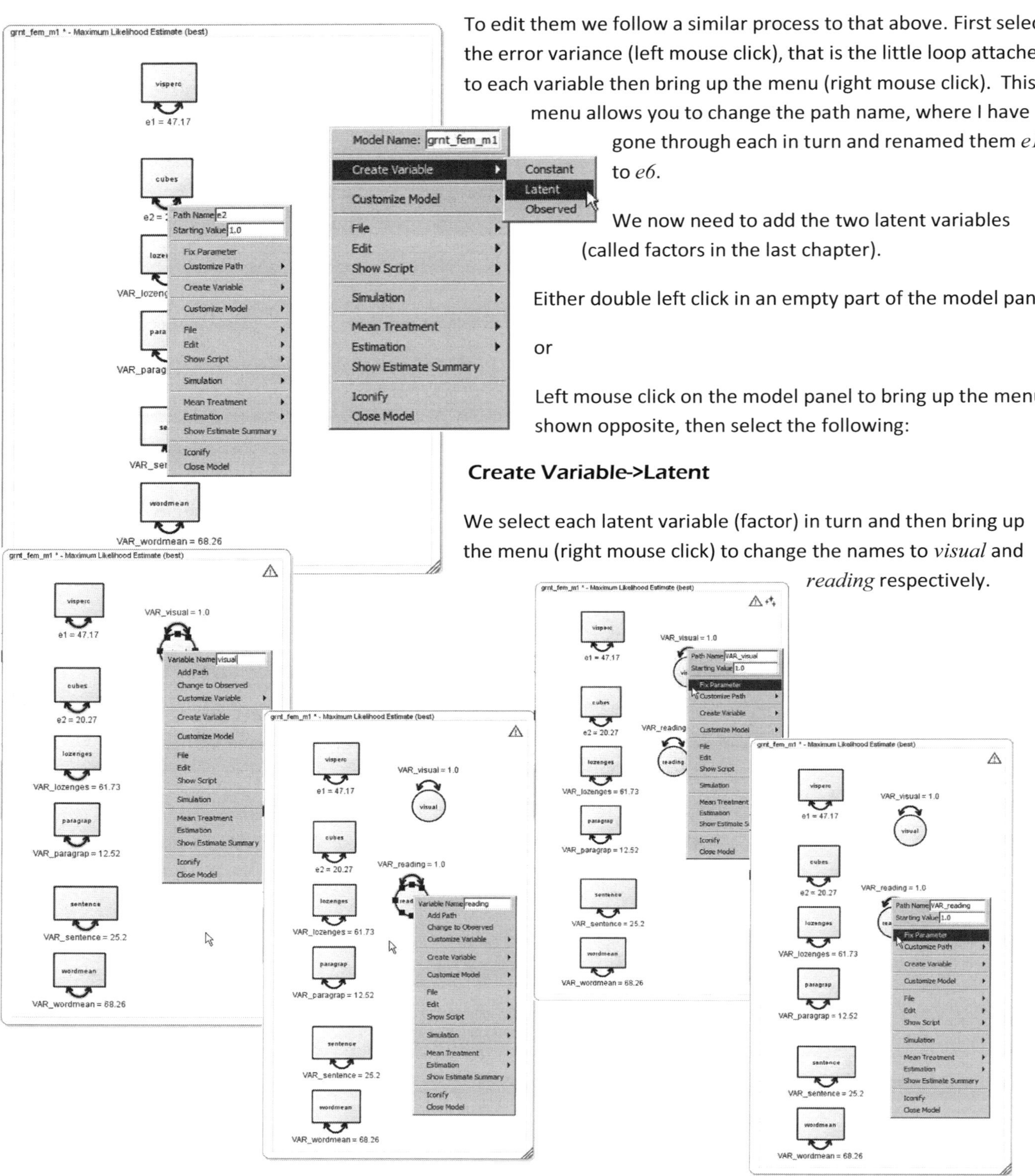

To edit them we follow a similar process to that above. First select the error variance (left mouse click), that is the little loop attached to each variable then bring up the menu (right mouse click). This menu allows you to change the path name, where I have gone through each in turn and renamed them *e1* to *e6*.

We now need to add the two latent variables (called factors in the last chapter).

Either double left click in an empty part of the model panel

or

Left mouse click on the model panel to bring up the menu shown opposite, then select the following:

Create Variable->Latent

We select each latent variable (factor) in turn and then bring up the menu (right mouse click) to change the names to *visual* and *reading* respectively.

We now need to fix both the error variances associated with these latent variables . Select the variance paths (left mouse click) and then bring up the menu (right mouse click). Each time select the **Fix Parameter** option from the menu.

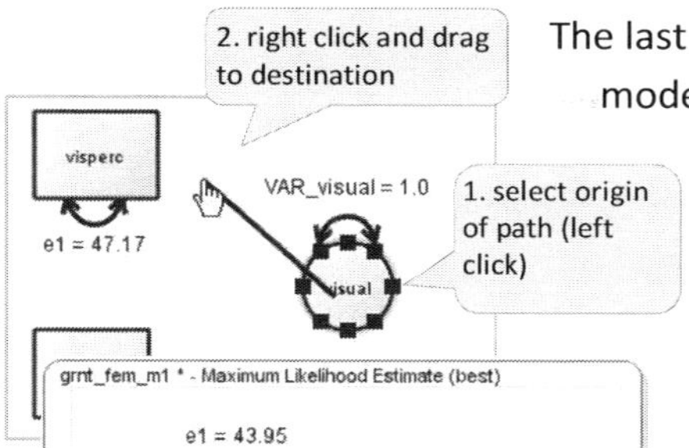

The last thing to do is to specify the relationships between the variables in the model. First the paths between the observed variables and the latent variables.

To draw a path select the variable for the paths origin (left mouse click) then:

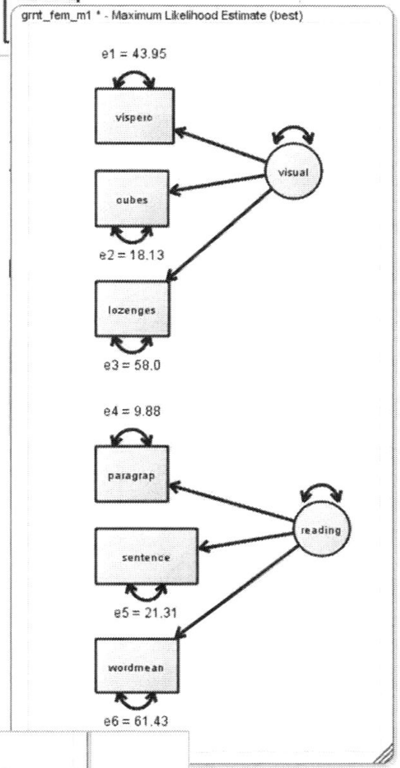

Right mouse click on it and drag to target variable.

I repeat this several times to create the model shown on the left.

Important: note that the arrows go from the latent variables to the observed ones.

The style described above is the traditional way of defining a factor model using what is known as reflective measurement. An alternative, more contentious approach is to have the arrows pointing in the other direction producing a formative measurement model. However there are questions concerning the validity of this latter approach and I suggest that anyone who believes otherwise consult various articles in Psychological Methods, 2007, 12(2), specifically Howell, Breivik, & Wilcox, 2007.

The only other path to add is that for the correlation (i.e. covariance) between the two latent variables.

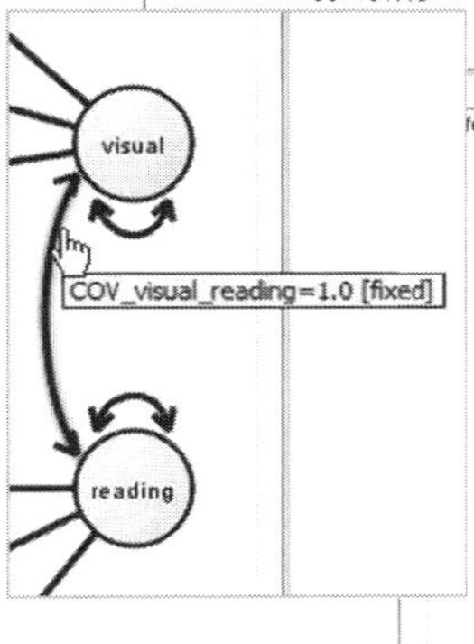

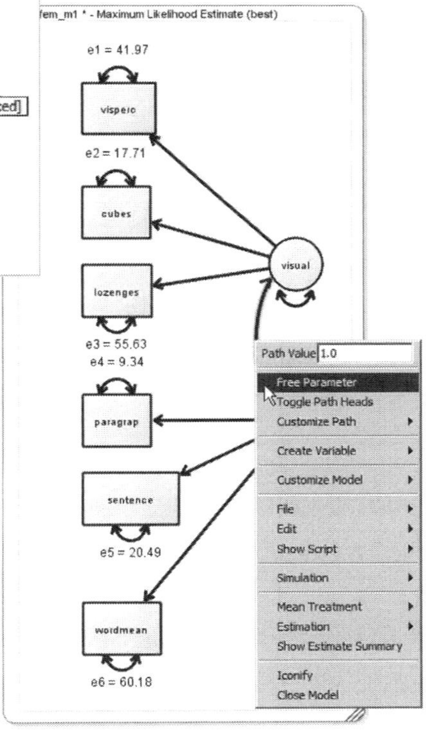

Select one of the latent variables by left mouse clicking on it and hold down the *shift key + right mouse* click then drag to target variable.

The far left screenshot shows that by hovering the mouse cursor over this line we discover it is a fixed parameter estimate whereas we want the model to estimate its values. To change it:

Select the path (left mouse click) then right mouse click on it to bring up the menu and select the following menu option:

Free Parameter

You need to carry out this process for all the paths directed from the observed variables to the latent ones as well. Once again, you can check to see which ones are fixed or allowed to be estimated by hovering over each path.

65.2.3 Model estimation

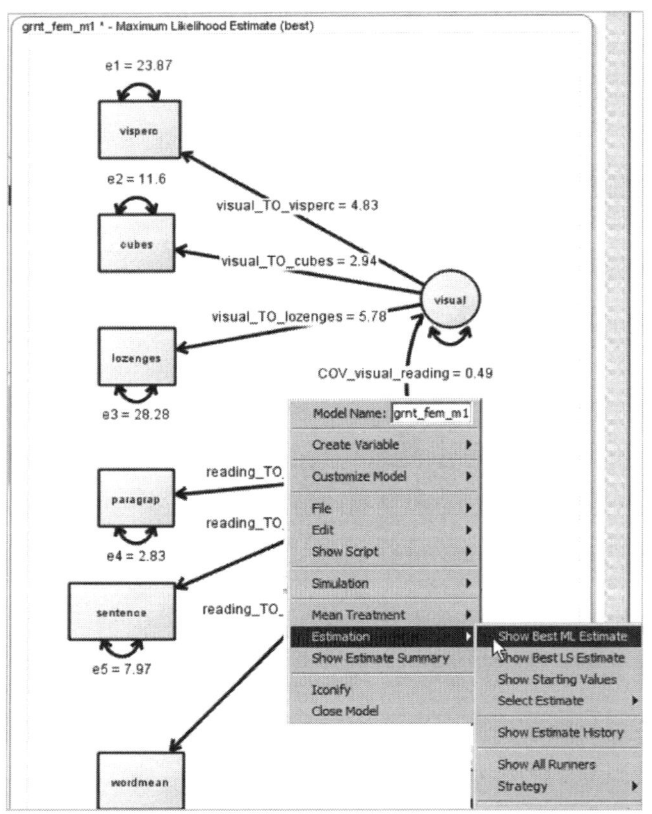

While we have been specifying the model, the Ωnyx program has been calculating the parameter estimates along with several model fit values.

You can update the diagram to show the 'best' parameter estimates by right mouse clicking on an empty part of the model panel and then selecting

Estimation-> Show Best ML Estimate

65.2.3.1 Displaying standardised values

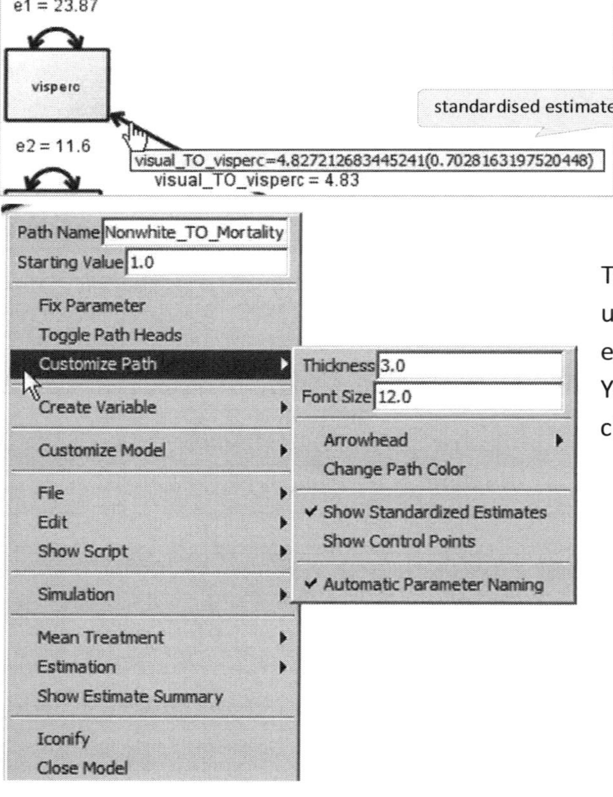

It is also useful to see the standardised estimates. These appear in parentheses when you hover over a path.

To display the standardised path estimates alongside the unstandardized ones in an Ωnyx diagram you need to select each path and then select the menu option shown opposite. You can select multiple paths by using the *shift+ right* mouse click combination.

65.2.4 Obtaining the results

In Ωnyx the results are obtained by right mouse clicking on an empty part of the model panel and then selecting **Estimate summary**.

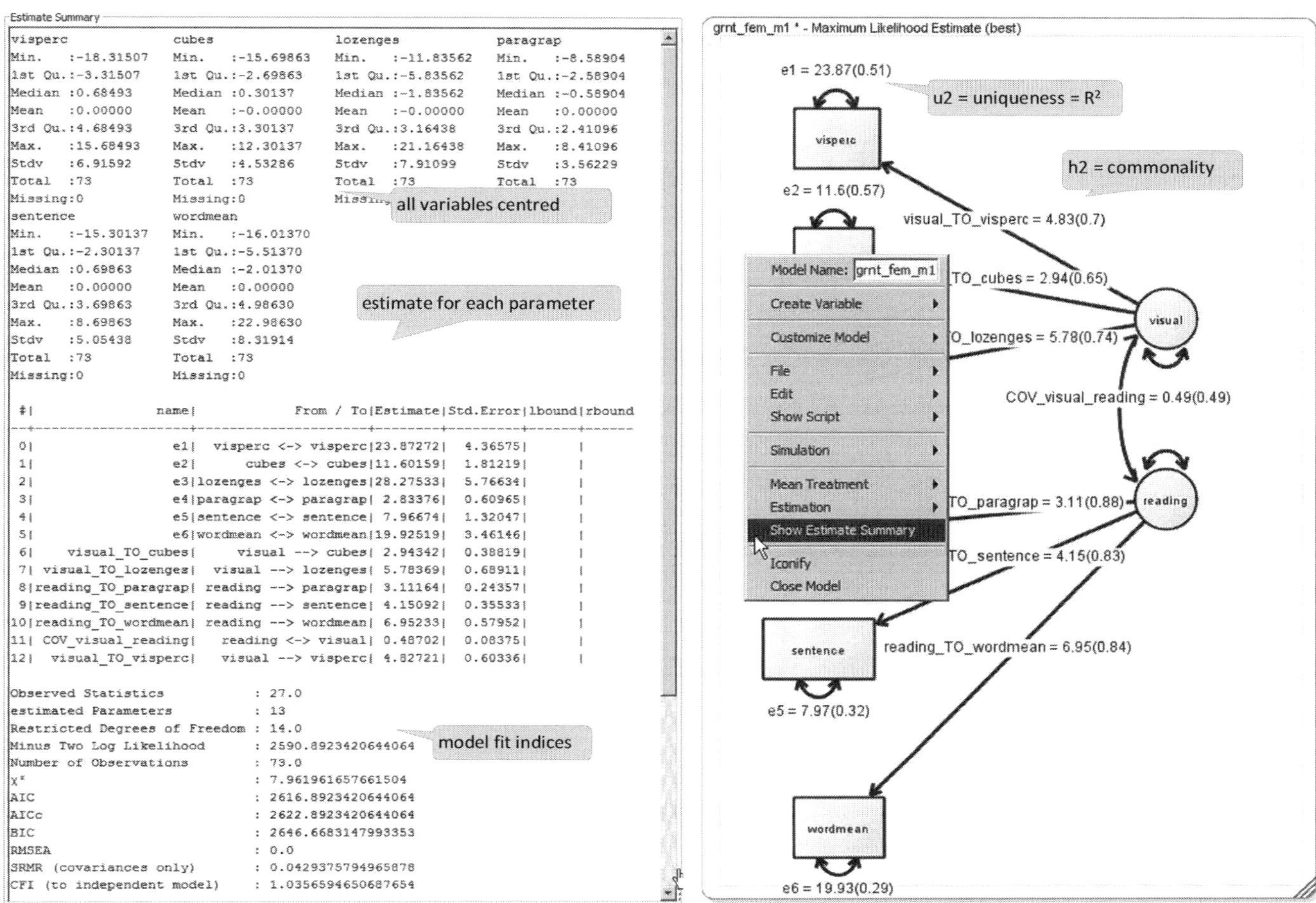

The results are shown above. It is important to realise that Ωnyx calculates the results 200 times and the final part of the estimate summary provides details of the results from which we can see that luckily the results stabilised.

```
This estimate is a local optimum, better estimates exist.
This estimate is reliably converged.
There are 20 local maxium likelihood optima found so far, 14 of them reliable.
This estimate has been found with 7 of 200 starting value sets converged in total
The overall estimation situation has stabilized.
```

Because the above model is very similar to that presented in the factor analysis chapter, mimicking most closely the Promax rotation result, we can interpret the paths between the latent variables and the observed variables as commonality values and the error terms as being the errors. The slight differences between these values and those in the factor analysis chapter, where u2i + h2 i =1, is because here we have constrained several of the paths to zero (i.e. there is no path between lozenges and verbal). This is known as a Confirmatory Factor Analysis (CFA) approach in contrast to the previous chapter where we followed what is known as an Exploratory Factor (EFA) approach.

65.3 Adding intercepts - Mean structures

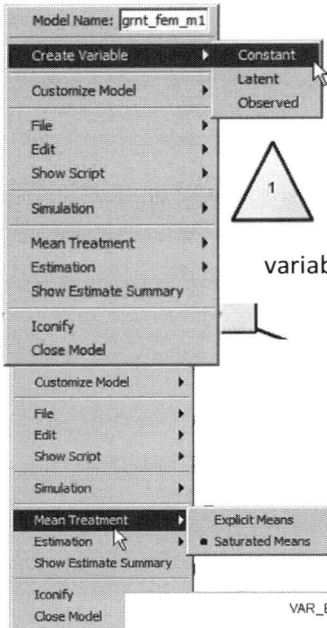

For more complex SEM models, specifically where we wish to investigate possible differences between means in several groups, it is necessary to add this information to the model. This is because so far we have been working with centered variables (i.e. mean =zero), the so-called saturated means model.

Graphically to explicitly specify the means in a SEM model we add a special graphical symbol, a triangle, to the model and then draw lines from it, pointing to those variables for which you want the means to be included in the model. Alternatively, you can select the menu option:

Treatment-> Explicit Means.

Because factor analysis style SEM models nearly always use standardised variables, I will demonstrate the approach with the predictors of mortality regression example described at the start of this chapter. You can download the dataset from

http://www.robin-beaumont.co.uk/virtualclassroom/book2data/airpoll.dat

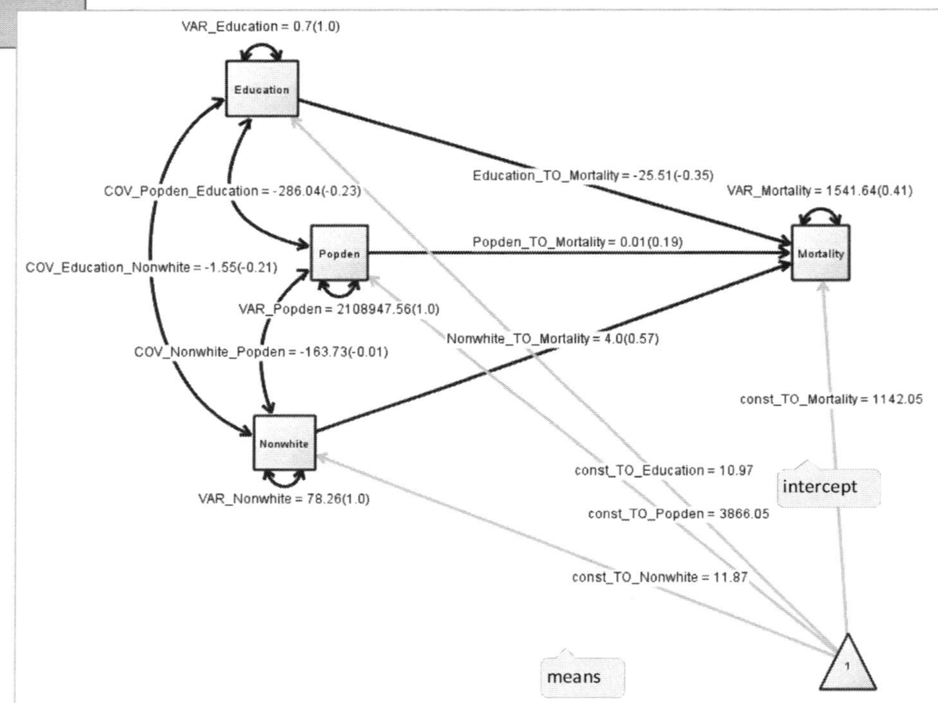

To add means (also called intercepts) to the model we follow three simple steps:

1. add a 'constant' element to the diagram, this automatically changes the model to an explicit means model.
2. add paths from the constant element to each of the observed variables where you wish a mean to also be included in the model.
3. select each path in turn and make it a free parameter rather than fixed.

The results shown above indicate that each value on the new paths from the 'constant' represents an estimate of the mean. Specifically the path to the output (dependent variable) represents the mean intercept value.

65.3.1 Reloading a saved model

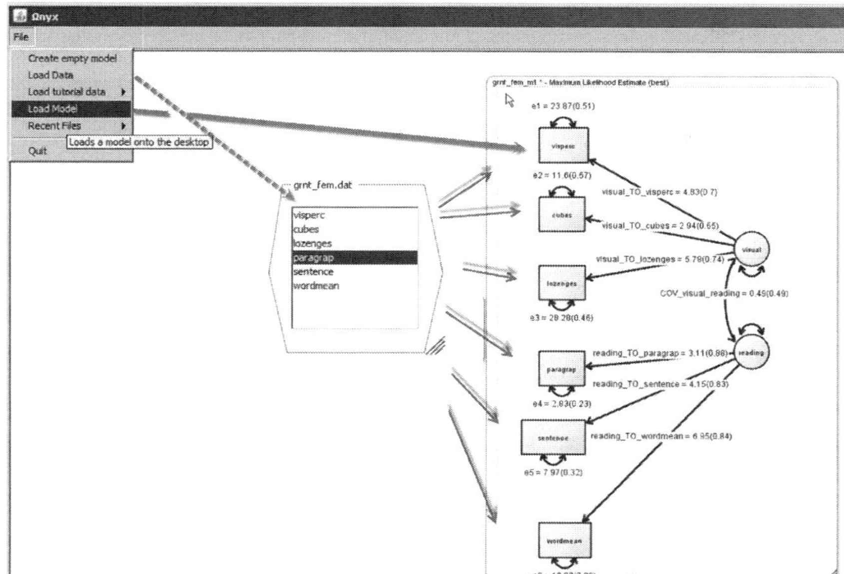

Once you have saved a model in Ωnyx when you reload it all the connections to the data will have disappeared. To reform these connections you simply open the data file again, then drag the appropriate field name into the correct object in the model panel. Repeat this process for each field.

65.4 Linking Ωnyx models to R

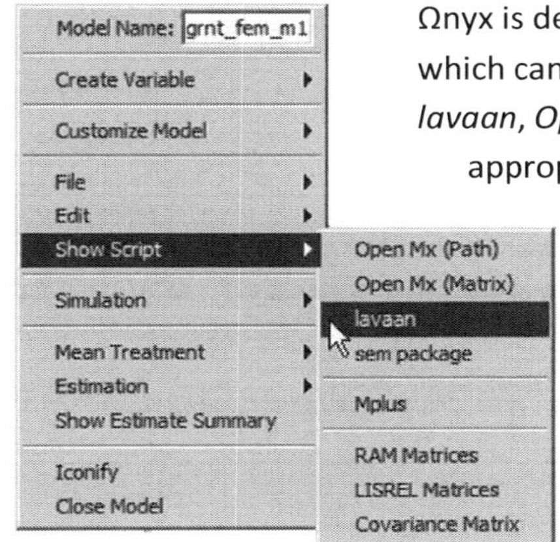

Ωnyx is designed primarily as a pedagogical tool allowing you to specify models graphically which can then be run using actual R packages. Specifically it supports three R packages *lavaan*, *OpenMx* and *SEM*. The way it supports these R packages is by producing the appropriate R code for each. We will begin by looking at the *Lavaan* code.

65.4.1 Lavaan

I will demonstrate the use of Lavaan using the code generated by Ωnyx for the Confirmatory Factor Analysis (CFA) described above. You first need to obtain and load the data to go with the model into R. Copy the following code into the R Console window.

```
hozdata<- read.table("http://www.robin-beaumont.co.uk/virtualclassroom/book2data/grnt_fem.dat",
sep="\t", header=TRUE);  names(hozdata)
```

Within R you first need to install and load the lavaan package:

```
install.packages("lavaan", dependencies = TRUE); library(lavaan)
```

The *Lavaan* package makes use of what it calls lavaan model syntax which describes a latent variable model, and luckily Ωnyx produces this code for us. You can see this syntax by selecting the **lavaan** option in the menu shown above, this appears when you *right mouse click* in an empty area of the model pane. This is shown on the next page.

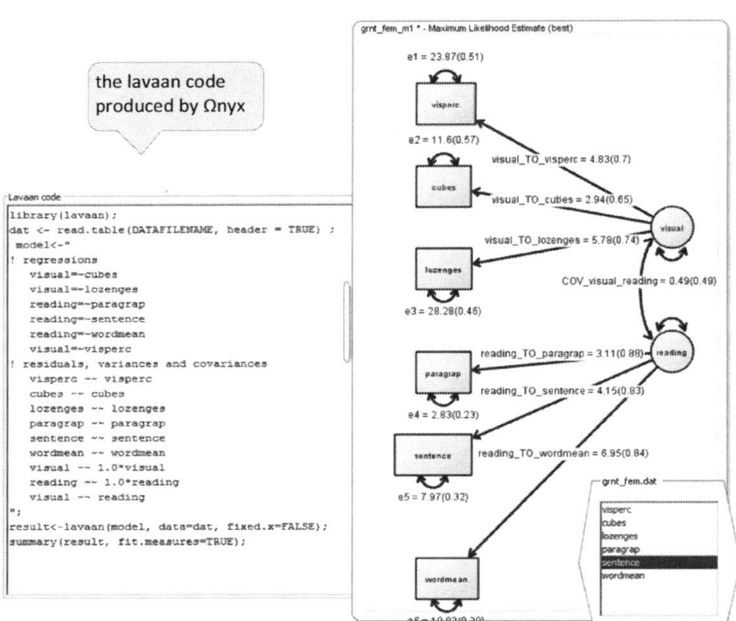

The window that appears shows the lavaan code.

As with R Commander this auto-generated code is more complex than need be, these automated tools lack the aesthetics and creative originality of a human! But it does provide a very useful starting point.

The code is punctuated by two comments lines (!) where each subsection defines a particular aspect of the model. I have reproduced the code below with detailed comments. To obtain the equivalent analysis to Ωnyx in R using the *lavaan* package requires slight modification of the code which we will now do.

```
library(lavaan);
dat <- read.table(DATAFILENAME, header =
TRUE) ;
 model<-"
! regressions
   visual=~cubes
   visual=~lozenges
   reading=~paragrap
   reading=~sentence
   reading=~wordmean
   visual=~visperc
! residuals, variances and covariances
   visperc ~~ visperc
   cubes ~~ cubes
   lozenges ~~ lozenges
   paragrap ~~ paragrap
   sentence ~~ sentence
   wordmean ~~ wordmean
   visual ~~ 1.0*visual
   reading ~~ 1.0*reading
   visual ~~ reading
";
result<-lavaan(model, data=dat,
fixed.x=FALSE);
summary(result, fit.measures=TRUE);
```

ither specify the data filename (text file) you wish to use for the model –by editing this value or remove this line. I usually remove it. See below

load the lavaan package, we have already done that, so you can remove this line

" or '
we now begin to specify the lavaan model, basically we have model_name <-"
note the double quote that is important

a comment line, you could use the # symbol instead

~
Latent variable (continuous variables) definitions. The name of the latent variable is on the left of the "=~" operator, while the terms on the right, separated by "+" operators, are the indicators of the latent variable. Alternatively, as here each is on a separate line. ead "=~" as "is measured by"

~~
Variances and covariances. The "~~" ('double tilde') operator specifies (residual) variances of an observed or latent variable, or a set of covariances between one variable, and several other variables. Several variables, separated by "+" operators can appear on the right. This way, several pairwise (co)variances involving the same left-hand variable can be expressed in a single expression. The distinction between variances and residual variances is made automatically.

variances

"*" are referred to as modifiers. 1 * sets the variance to one, becoming a fixed parameter

covariances - only one in this model

you need to specify the dataframe here. Change data=data to data = hozdata

" or '
To indicate we have finished specifying the lavaan model. Followed either by a SEMi colon (;) or a blank line

now send the model definition to the lavaan function

produce a set of summary results for our model including fit indices

~ (not used in above model)
Regression definitions. The output (dependent) variable is on the left of a "~" operator and the input (independent) variables, separated by "+" operators, are on the right.

Taking into account the annotations described about we now have:

```
lavaan (0.5-16) converged normally after  51 iterations

  Number of observations                            73

  Estimator                                         ML
  Minimum Function Test Statistic                7.962
  Degrees of freedom                                 8
  P-value (Chi-square)                           0.437

Model test baseline model:

  Minimum Function Test Statistic              190.325
  Degrees of freedom                                15
  P-value                                        0.000

User model versus baseline model:

  Comparative Fit Index (CFI)                    1.000
  Tucker-Lewis Index (TLI)                       1.000

Loglikelihood and Information Criteria:

  Loglikelihood user model (H0)              -1295.446
  Loglikelihood unrestricted model (H1)      -1291.465

  Number of free parameters                         13
  Akaike (AIC)                                2616.892
  Bayesian (BIC)                              2646.668
  Sample-size adjusted Bayesian (BIC)         2605.705

Root Mean Square Error of Approximation:

  RMSEA                                          0.000
  90 Percent Confidence Interval          0.000  0.137
  P-value RMSEA <= 0.05                          0.569
```

```
library(lavaan);
dat <- hozdata
 model<-"
! regressions
  visual=~cubes
  visual=~lozenges
  visual=~visperc
  reading=~paragrap
  reading=~sentence
  reading=~wordmean
! residuals, variances and covariances
  visperc ~~ visperc
  cubes ~~ cubes
  lozenges ~~ lozenges
  paragrap ~~ paragrap
  sentence ~~ sentence
  wordmean ~~ wordmean
  visual ~~ 1.0*visual
  reading ~~ 1.0*reading
  visual ~~ reading
";
result<-lavaan(model, data=dat, fixed.x=FALSE);
summary(result, fit.measures=TRUE);
```

The above summary function produces the output on the left showing a range of model fit indices. To gain R squared and standardised values you can edit the summary function to:

```
R-Square:

    visperc          0.494
    cubes            0.428
    lozenges         0.542
    paragrap         0.774
    sentence         0.684
    wordmean         0.708
```

```
summary(result, fit.measures=TRUE, standardized=TRUE, rsq=TRUE)
```

To gain further information about the parameter estimates we use the *parameterEstimates()* function which also provides confidence intervals, given below. I have shaded out two irrelevant columns.

```
parameterEstimates(result, standardized = TRUE)
```

```
> parameterEstimates(result, standardized = TRUE)
        lhs op      rhs    est    se     z pvalue ci.lower ci.upper std.lv std.all std.nox
1    visual =~    cubes  2.943 0.549 5.357  0.000    1.867    4.020  2.943   0.654   0.654
2    visual =~ lozenges  5.784 0.955 6.057  0.000    3.912    7.655  5.784   0.736   0.736
3   reading =~ paragrap  3.112 0.344 9.033  0.000    2.436    3.787  3.112   0.880   0.880
4   reading =~ sentence  4.151 0.502 8.270  0.000    3.167    5.135  4.151   0.827   0.827
5   reading =~ wordmean  6.952 0.820 8.476  0.000    5.345    8.560  6.952   0.841   0.841
6    visual =~  visperc  4.827 0.836 5.777  0.000    3.189    6.465  4.827   0.703   0.703
7   visperc ~~  visperc 23.873 5.945 4.016  0.000   12.221   35.525 23.873   0.506   0.506
8     cubes ~~    cubes 11.602 2.566 4.521  0.000    6.572   16.631 11.602   0.572   0.572
9  lozenges ~~ lozenges 28.275 7.837 3.608  0.000   12.914   43.636 28.275   0.458   0.458
10 paragrap ~~ paragrap  2.834 0.863 3.286  0.001    1.143    4.524  2.834   0.226   0.226
11 sentence ~~ sentence  7.967 1.856 4.292  0.000    4.329   11.605  7.967   0.316   0.316
12 wordmean ~~ wordmean 19.925 4.917 4.052  0.000   10.288   29.563 19.925   0.292   0.292
13   visual ~~   visual  1.000 0.000    NA     NA    1.000    1.000  1.000   1.000   1.000
14  reading ~~  reading  1.000 0.000    NA     NA    1.000    1.000  1.000   1.000   1.000
15   visual ~~  reading  0.487 0.117 4.145  0.000    0.257    0.717  0.487   0.487   0.487
```

indicating regression weight or variance/covariance

estimated value

standard error

p-value given that it equals zero in the population

95% ci

standardized estimate

65.4.1.1 *Modification indices (MI)*

There are been several attempts to automate both model creation and model modification. Unfortunately both approaches can result in nonsense models where the optimum fitting model (usually several) does not reflect a sensible theory in the particular research area. As with factor analysis, SEM modelling is a mixture of both art and science.

Modification measures come in two main varieties:

1. Adding paths - Langrange multiplier tests (LM) provide information about how much better the model would fit (i.e. $\chi 2$ statistic would be reduced) if a particular parameter were added to the model. That is if the parameter were allowed to be estimated rather than being constrained/fixed.
2. Removing paths - Wald tests (backward search) indicate which parameters can be dropped without affecting model fit.

The *lavaan* package takes the first approach with the *modificationIndices()* function, using the output from our model *modificationIndices(result)* the output of which is shown below.

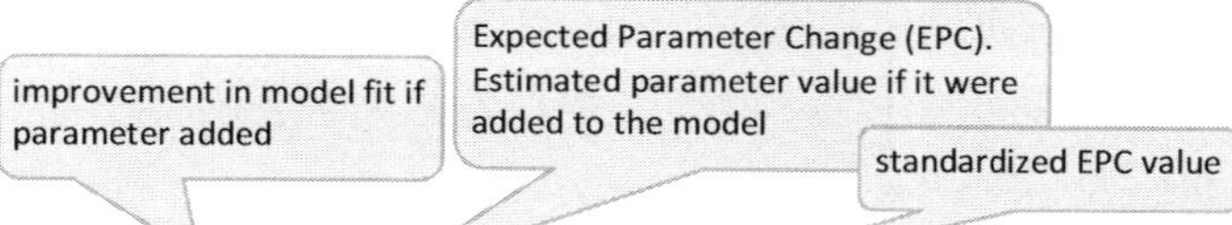
improvement in model fit if parameter added
Expected Parameter Change (EPC). Estimated parameter value if it were added to the model
standardized EPC value

```
> modificationIndices(result)
        lhs op      rhs    mi     epc sepc.lv sepc.all sepc.nox
1    visual =~    cubes 0.000   0.000   0.000    0.000    0.000
2    visual =~ lozenges 0.000   0.000   0.000    0.000    0.000
3    visual =~ paragrap 0.001  -0.013  -0.013   -0.004   -0.004
4    visual =~ sentence 0.244   0.273   0.273    0.054    0.054
5    visual =~ wordmean 0.197  -0.398  -0.398   -0.048   -0.048
6    visual =~  visperc 0.000   0.000   0.000    0.000    0.000
7   reading =~    cubes 1.984  -0.890  -0.890   -0.198   -0.198
8   reading =~ lozenges 0.724   0.989   0.989    0.126    0.126
9   reading =~ paragrap 0.000   0.000   0.000    0.000    0.000
10  reading =~ sentence 0.000   0.000   0.000    0.000    0.000
11  reading =~ wordmean 0.000   0.000   0.000    0.000    0.000
12  reading =~  visperc 0.228   0.472   0.472    0.069    0.069
13    cubes ~~    cubes 0.000   0.000   0.000    0.000    0.000
14    cubes ~~ lozenges 0.228   2.609   2.609    0.074    0.074
15    cubes ~~ paragrap 0.000   0.016   0.016    0.001    0.001
16    cubes ~~ sentence 0.681  -1.179  -1.179   -0.052   -0.052
17    cubes ~~ wordmean 0.037  -0.446  -0.446   -0.012   -0.012
18    cubes ~~  visperc 0.724   3.804   3.804    0.123    0.123
19 lozenges ~~ lozenges 0.000   0.000   0.000    0.000    0.000
20 lozenges ~~ paragrap 0.689  -1.324  -1.324   -0.048   -0.048
21 lozenges ~~ sentence 0.026  -0.386  -0.386   -0.010   -0.010
22 lozenges ~~ wordmean 2.541   6.180   6.180    0.095    0.095
23 lozenges ~~  visperc 1.984 -13.219 -13.219   -0.245   -0.245
24 paragrap ~~ paragrap 0.000   0.000   0.000    0.000    0.000
25 paragrap ~~ sentence 0.197  -1.159  -1.159   -0.065   -0.065
26 paragrap ~~ wordmean 0.244   2.226   2.226    0.076    0.076
27 paragrap ~~  visperc 0.690   1.173   1.173    0.048    0.048
28 sentence ~~ sentence 0.000   0.000   0.000    0.000    0.000
29 sentence ~~ wordmean 0.001  -0.185  -0.185   -0.004   -0.004
30 sentence ~~  visperc 2.067   3.059   3.059    0.089    0.089
31 wordmean ~~ wordmean 0.000   0.000   0.000    0.000    0.000
32 wordmean ~~  visperc 3.916  -6.799  -6.799   -0.120   -0.120
33  visperc ~~  visperc 0.000   0.000   0.000    0.000    0.000
34   visual ~~   visual    NA      NA      NA       NA       NA
35   visual ~~  reading 0.000   0.000   0.000    0.000    0.000
36  reading ~~  reading    NA      NA      NA       NA       NA
```

While blindly looking through the indices and purely deciding on 'improving the model' by adding one or two paths with the highest *mi* values might be tempting this is not really the best strategy as the modelling process should also be informed by knowledge of the research area.

There are no standard rules concerning a specific cut off value of the *mi* to indicate necessary inclusion in the model. Therefore it is often more sensible to only display those values where the change is greater than a specific value, e.g. 10, which can be achieved with the following code.

```
change_results <-modificationIndices(result)

subset(change_results, mi > 10)
```

Looking at the results opposite, we see that the greatest improvement in model fit would be achieved by adding a covariance between *visperc* and *wordmean*. Whether this makes sense in terms of the theory of verbal and spatial intelligence would need knowledge of the research area.

65.4.1.2 Residual analysis – analysing local fit

```
> resid(result, type="raw")
$cov
         cubes  lozngs pargrp sentnc wordmn visprc
cubes     0.000
lozenges  0.392  0.000
paragrap -1.105  0.291  0.000
sentence -1.914  1.536 -0.053  0.000
wordmean -3.126  4.364  0.085 -0.013  0.000
visperc   0.722 -1.388  1.020  2.886 -3.308  0.000

$mean
    cubes lozenges paragrap sentence wordmean  visperc
        0        0        0        0        0        0

> resid(result, type="normalized")
$cov
         cubes  lozngs pargrp sentnc wordmn visprc
cubes     0.000
lozenges  0.085  0.000
paragrap -0.580  0.085  0.000
sentence -0.713  0.315 -0.020  0.000
wordmean -0.706  0.539  0.020 -0.002  0.000
visperc   0.180 -0.197  0.339  0.672 -0.485  0.000

$mean
    cubes lozenges paragrap sentence wordmean  visperc
        0        0        0        0        0        0

> resid(result, type="standardized") # like z scores
$cov
         cubes  lozngs pargrp sentnc wordmn visprc
cubes     0.000
lozenges  0.423     NA
paragrap -1.055  0.179  0.000
sentence -1.184  0.572     NA  0.003
wordmean -1.203  0.986  0.300 -0.035     NA
visperc   0.719     NA  0.642  1.122 -0.905  0.003
```

As mentioned earlier it is all well and good obtaining overall (i.e. global) model fit measures but, as is the case with the simple contingency table chi-squared introduced many chapters previously, often the desire is to see where the actual model fits best and least well. As we did in that chapter, a residual analysis of the values, this time the covariance residual matrix, provides such information.

In effect, each value in the residual matrix reflects the difference between the observed and modal covariance matrix. You can obtain three sets of values, raw difference, standardised scores (which are similar to z scores), and normalised scores which are more difficult to interpret (see Bollen, 1989 p.259).

```
resid(result, type="raw")
resid(result, type="normalized")
resid(result, type="standardized")
# like z scores
```

If we look at the standardized residuals, we can see that the largest is for the covariance between cubes and sentence which misfits by around three quarters of a standard deviation. Given that none of them is greater than 1.96 reinforces the belief that this is not only a good model fit at the global level (from the RMSEA value) but also a good model fit from the individual parameter level as well.

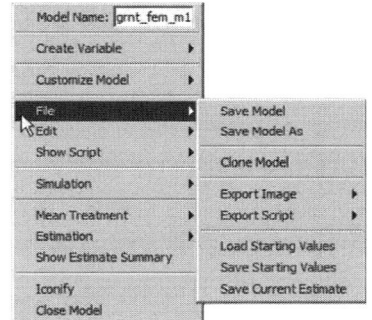

Model Name: grnt_fem_m1
Create Variable
Customize Model
File
Edit
Show Script
Simulation
Mean Treatment
Estimation
Show Estimate Summary
Iconify
Close Model
Save Model
Save Model As
Clone Model
Export Image
Export Script
Load Starting Values
Save Starting Values
Save Current Estimate

65.4.1.3 Comparing models

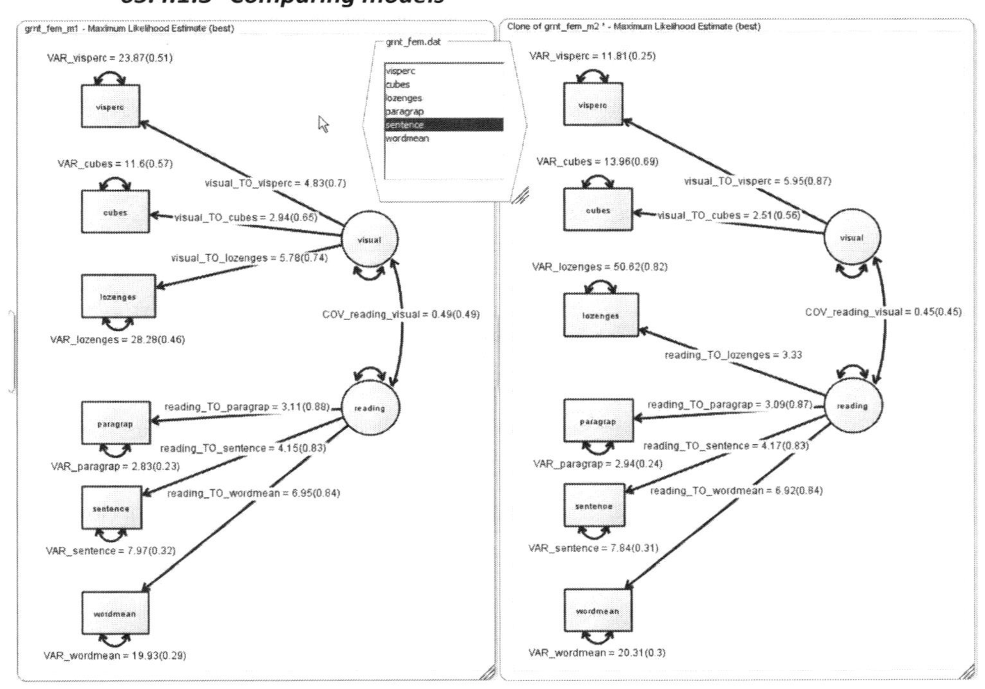

Ωnyx provides an interesting graphical approach to model comparison.

You start by defining the two, or more, models. An easy way to do this is to clone one from the other, achieved by selecting the **File->Clone Model** popup menu option which appears when you right mouse click within a model panel. In the example below I have cloned the model then changed the path from *visual_to_lozenges* to *reading_to_lozenges.*

To compare these models, where they both have the same number of parameters, we simply bring up the estimate summary box for both and compare the various fit indices. As shown below, the left hand side one has a smaller RMSEA value (☺) and smaller AIC and BIC values (☺) all indicating that it is a better fit.

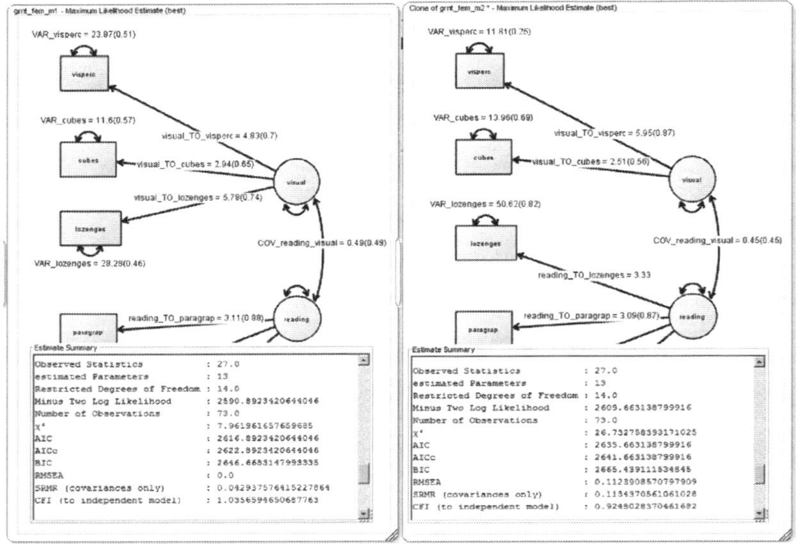

To compare models where in one of them one or more of the paths is constrained to a particular value (called a nested model = number of estimated parameters is different for each model) we use a likelihood ratio test, as discussed in the *Logistic Regression* chapter. To obtain this value in Ωnyx involves using a unique method, we simply right mouse click on one model and then drag to the comparator model. In the example below, I have constrained the covariance between the two latent variables to 0.1 and then compared the model to that where the covariance is estimated. Hovering over the black dot informs us that we have a -2log likelihood value of 8.002 with an associated p-value of 0.004; we can conclude from this that the two models are indeed different with the additional free parameter providing a better fit.

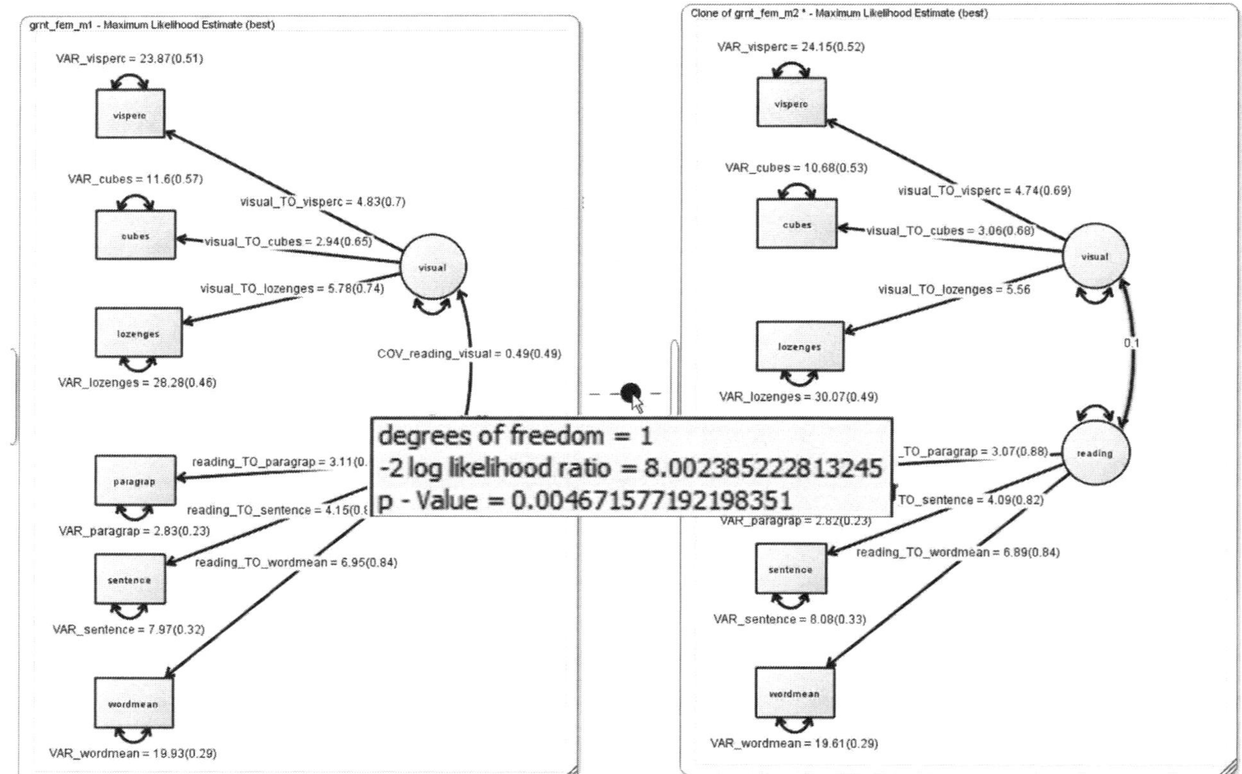

Ωnyx has many more features which I do not have room to discuss in this chapter, for further information visit the Ωnyx website.

65.4.2 OpenMx (formally Mx)

OpenMx is another SEM package. However this originally was a standalone application called Mx which has been redeveloped as an R library. The *OpenMx* code generated by Ωnyx for our CFA is shown below. I do not intend to discuss this, just noting that it is very different syntactically. The OpenMx website has extensive tutorials, and because people tend to learn either *lavaan* or *OpenMx* I will leave it at that.

```
# This model specification was automatically generated by Ωnyx
require("OpenMx");
modelData <- read.table(DATAFILENAME, header = TRUE)
manifests<-c("visperc","cubes","lozenges","paragrap","sentence","wordmean")
latents<-c("visual","reading")
model <- mxModel("grnt_fem_m1",
type="RAM",
manifestVars = manifests,
latentVars = latents,
mxPath(from="visual",to=c("visperc","cubes","lozenges"), free=c(TRUE,TRUE,TRUE), value=c(1.0,1.0,1.0) , arrows=1,
label=c("visual_TO_visperc","visual_TO_cubes","visual_TO_lozenges") ),
mxPath(from="reading",to=c("paragrap","sentence","wordmean"), free=c(TRUE,TRUE,TRUE), value=c(1.0,1.0,1.0) , arrows=1,
label=c("reading_TO_paragrap","reading_TO_sentence","reading_TO_wordmean") ),
mxPath(from="visperc",to=c("visperc"), free=c(TRUE), value=c(1.0) , arrows=2, label=c("VAR_visperc") ),
mxPath(from="cubes",to=c("cubes"), free=c(TRUE), value=c(1.0) , arrows=2, label=c("VAR_cubes") ),
mxPath(from="lozenges",to=c("lozenges"), free=c(TRUE), value=c(1.0) , arrows=2, label=c("VAR_lozenges") ),
mxPath(from="paragrap",to=c("paragrap"), free=c(TRUE), value=c(1.0) , arrows=2, label=c("VAR_paragrap") ),
mxPath(from="sentence",to=c("sentence"), free=c(TRUE), value=c(1.0) , arrows=2, label=c("VAR_sentence") ),
mxPath(from="wordmean",to=c("wordmean"), free=c(TRUE), value=c(1.0) , arrows=2, label=c("VAR_wordmean") ),
mxPath(from="visual",to=c("visual"), free=c(FALSE), value=c(1.0) , arrows=2, label=c("VAR_visual") ),
mxPath(from="reading",to=c("reading","visual"), free=c(FALSE,TRUE), value=c(1.0,1.0) , arrows=2, label=c("VAR_reading","COV_reading_visual") ),
mxPath(from="one",to=c("visperc","cubes","lozenges","paragrap","sentence","wordmean"), free=T, value=1 , arrows=1),
mxData(modelData, type = "raw")
);

result <- mxRun(model)
summary(result)
```

65.5 Structural Equation Modelling directly in R using the *sem* package

From the above lavaan example, I'm sure you realise that you could have also directly written the *lavaan* code in R and bypassed Ωnyx but at the same time I'm sure you can see the advantage of using Ωnyx. An R package originally developed long before *lavaan*, in 2001, called *sem* also allows the creation of SEM models in R directly. The *sem* package just uses two types of relation: **<->** to indicate variances/covariances and **->** to indicate regression paths. You then need to give each a name or specify it as being a fixed value using *NA* and also indicate the estimation starting/fixed value or allow the program to estimate it freely (*NA*). An example should help make this clearer. Because SEM model specification code is similar to the *lavaan* approach I have put them side by side below.

```
# lavaan code
onyx.model<-"
! regressions
  visual=~visperc
  visual=~cubes
  visual=~lozenges
  reading=~paragrap
  reading=~sentence
  reading=~wordmean
! residuals, variances and
covariances
  visperc ~~ visperc
  cubes ~~ cubes
  lozenges ~~ lozenges
  paragrap ~~ paragrap
  sentence ~~ sentence
  wordmean ~~ wordmean
  visual ~~ 1*visual
  reading ~~ 1*reading
  visual ~~ reading
"
onyx.model
result<-lavaan(onyx.model,
data=hozdata)
summary (result,
fit.measures= TRUE)
```

```
install.packages("sem") ; library(sem) # to install and load the sem package
hozdatamodel <- specifyModel()
visual -> visperc,          visual_TO_visperc,      NA #first factor
visual -> cubes,            visual_TO_cubes,        NA
visual -> lozenges,         visual_TO_lozenges,     NA
reading ->  paragrap,       reading_TO_paragrap,    NA # second factor
reading -> sentence,        reading_TO_sentence,    NA
reading -> wordmean,        reading_TO_wordmean,    NA
visperc <-> visperc,        err_visperc,            NA # errors
cubes <-> cubes,            err_cubes,              NA
lozenges <-> lozenges,      err_lozenges,           NA
paragrap <-> paragrap,      err_paragrap,           NA
sentence <-> sentence,      err_sentence,           NA
wordmean <-> wordmean,      err_wordmean,           NA
visual <-> reading,         COV_visual_reading,     NA # covariances
visual <-> visual,          NA,                      1  # fixed parameters
reading <-> reading,        NA,                      1

hozmodel1 <- SEM(hozdatamodel, data = hozdata)
summary(hozmodel1)
standardizedCoefficients(hozmodel1)
```

```
> summary(hozmodel1)

 Model Chisquare =  7.852894   Df =  8 Pr(>Chisq) = 0.4479704
 AIC =  33.85289
 BIC =  -26.47078

 Normalized Residuals
    Min.  1st Qu.   Median     Mean  3rd Qu.     Max.
-0.69510 -0.19380  0.00000 -0.02143  0.18000  0.68350

 R-square for Endogenous Variables
 visperc    cubes lozenges paragrap sentence wordmean
  0.4940   0.4275   0.5419   0.7736   0.6838   0.7081

 Parameter Estimates
                      Estimate  Std Error z value  Pr(>|z|)
visual_TO_visperc    4.8606192 0.8472456 5.736966 9.638767e-09 visperc <--- visual
visual_TO_cubes      2.9637940 0.5570414 5.320599 1.034261e-07 cubes <--- visual
visual_TO_lozenges   5.8237161 0.9681471 6.015321 1.795300e-09 lozenges <--- visual
reading_TO_paragrap  3.1331703 0.3492597 8.970890 2.941287e-19 paragrap <--- reading
reading_TO_sentence  4.1796433 0.5088761 8.213479 2.148706e-16 sentence <--- reading
reading_TO_wordmean  7.0004445 0.8316418 8.417620 3.842094e-17 wordmean <--- reading
err_visperc         24.2042982 6.0692766 3.988004 6.663159e-05 visperc <--> visperc
err_cubes           11.7627275 2.6196207 4.490241 7.114270e-06 cubes <--> cubes
err_lozenges        28.6680567 8.0011697 3.582983 3.396925e-04 lozenges <--> lozenges
err_paragrap         2.8731229 0.8805372 3.262920 1.102705e-03 paragrap <--> paragrap
err_sentence         8.0773857 1.8949870 4.262502 2.021508e-05 sentence <--> sentence
err_wordmean        20.2019195 5.0200027 4.024285 5.714877e-05 wordmean <--> wordmean
COV_visual_reading   0.4870169 0.1183007 4.116771 3.842174e-05 reading <--> visual

 Iterations =  83
```

There are many similarities between the two and I would recommend anyone who is thinking about using the *sem* package to begin by adapting the *lavaan* output from the Ωnyx application. The *SEM* package is very unhelpful when it comes to reporting errors providing, what I find to be, very cryptic messages. You can waste many hours only to discover the problem was a missing comma.

The results are provided opposite to demonstrate their equivalence to the other packages/ programs used.

There are also similar functions in the *SEM* package to produce the various residues and matrices:

```
> standardizedCoefficients(hozmodel1)
                        Std. Estimate
1    visual_TO_visperc     0.7028162    visperc <--- visual
2      visual_TO_cubes     0.6538466      cubes <--- visual
3   visual_TO_lozenges     0.7361554   lozenges <--- visual
4  reading_TO_paragrap     0.8795393   paragrap <--- reading
5  reading_TO_sentence     0.8269342   sentence <--- reading
6  reading_TO_wordmean     0.8414862   wordmean <--- reading
7          err_visperc     0.5060493    visperc <--> visperc
8            err_cubes     0.5724846      cubes <--> cubes
9         err_lozenges     0.4580753  lozenges <--> lozenges
10        err_paragrap     0.2264106  paragrap <--> paragrap
11        err_sentence     0.3161799  sentence <--> sentence
12        err_wordmean     0.2919009  wordmean <--> wordmean
13  COV_visual_reading     0.4870169    reading <--> visual
14                         1.0000000     visual <--> visual
15                         1.0000000    reading <--> reading
```

```
effects(hozmodel1)
# gives total/direct/indirect effects

hozmodel1$C # model-reproduced covariance matrix

hozmodel1$A # RAM A matrix

hozmodel1$P

residuals(hozmodel1)

standardizedResiduals(hozmodel1)

vcov(hozmodel1)
```

RAM P matrix
covariances amongst elements
same as S in Ωnyx

covariances amongst
all parameters

I have not shown all the results for the above options as they would produce results very similar to those shown previously.

Another useful command in the *sem()* function, is *objective =objectiveGLS* which instructs the program to produce generalized least squares estimates instead of the default maximum likelihood ones. This can be useful if there is a problem with the initial estimation process. Secondly you can get the *sem()* function to provide details for each iteration of the estimation process by adding the option *debug=TRUE*. The use of these options is shown below.

```
hozmodel2 <- sem(hozdatamodel, data = hozdata, objective=objectiveGLS, debug=TRUE)

hozmodel2
```

same as the mi measures in the lavaan package

```
> modIndices(hozmodel1)

 5 largest modification indices, A matrix:
wordmean<-visperc lozenges<-wordmean    cubes<-sentence
         2.485672          2.353646           2.141871
     visual<-cubes     reading<-cubes
         1.956953          1.956948

  5 largest modification indices, P matrix:
wordmean<->visperc wordmean<->lozenges  sentence<->visperc
          3.861991            2.506210            2.038260
     visual<->cubes      reading<->cubes
          1.956955            1.956949
```

The *modIndices()* function provides modification indices. By setting the *n.largest=n* you will obtain the nth largest values, the default is 5, as shown opposite.

The *sem* package has what are known as wrapper functions for the *sem()* function, this basically means that these other functions set a number of defaults so that you can specify the model by providing less information. One such wrapper function is *cfa()* standing for Confirmatory Factor Analysis. I have used this function below to create an equivalent cfa model to that produced above, called *cfa_hoz*, but with much less code. When you run this code the *sem* package informs us that additional error variances will automatically be added along with a covariance between the factors.

```
cfa_hoz <- cfa(reference.indicators=FALSE)

F1: visperc, cubes, lozenges

F2: paragrap, sentence, wordmean

hozmodel3 <- SEM(cfa_hoz, data = hozdata)

summary(hozmodel3)
```

I personally prefer to use the *sem()* function as you then know exactly what you are getting.

We have now considered both creating SEM models directly in Ωnyx, specifying them initially in Ωnyx, then repeating the analysis in R using the automatic code generated and finally specifying the model directly in R. I deliberately kept the example as simple as possible but now it is time to see a more typical example to demonstrate some of the representative characteristics of an SEM analysis.

The latest version of Ωnyx (v1.00) provides an export to the *SEM* package, producing the following code on the left, as with the previous export code you need to edit it by adding the dataframe you wish to use. Also the first call to the *sem()* function seems to create an error for this particular model.

Ωnyx export	Edited
require("SEM"); paths <- c("Education <-> Education", "Popden <-> Popden", "Nonwhite <-> Nonwhite", "Mortality <-> Mortality", "Education -> Mortality", "Nonwhite -> Mortality", "Nonwhite <-> Popden", "Education <-> Popden", "Education <-> Nonwhite", "Popden -> Mortality") parameter <- c("VAR_Education", "VAR_Popden", "VAR_Nonwhite", "VAR_Mortality", "Education_TO_Mortality", "Nonwhite_TO_Mortality", "COV_Nonwhite_Popden", "COV_Education_Popden", "COV_Education_Nonwhite", "Popden_TO_Mortality") values <- c("0.7026222048487266", "2108953.5033314778", "78.26043332764722", "1541.636507142136", "-25.50698060512576", "3.9999225532362965", "-163.73509363742735", "-286.0402052689569", "-1.5481333432299311", "0.007954486809572933") model <- array(c(paths, parameter, values), dim = c(10,3)) colnames(model) <- c("col1","col2","col3") dat <- read.table(DATAFILENAME, header = TRUE) result <- SEM(model = model, data = dat, fixed.x ="Intercept", raw = TRUE) result <- SEM(model = model, data = dat)	require("SEM"); paths <- c("Education <-> Education", "Popden <-> Popden", "Nonwhite <-> Nonwhite", "Mortality <-> Mortality", "Education -> Mortality", "Nonwhite -> Mortality", "Nonwhite <-> Popden", "Education <-> Popden", "Education <-> Nonwhite", "Popden -> Mortality") parameter <- c("VAR_Education", "VAR_Popden", "VAR_Nonwhite", "VAR_Mortality", "Education_TO_Mortality", "Nonwhite_TO_Mortality", "COV_Nonwhite_Popden", "COV_Education_Popden", "COV_Education_Nonwhite", "Popden_TO_Mortality") values <- c("0.7026222048487266", "2108953.5033314778", "78.26043332764722", "1541.636507142136", "-25.50698060512576", "3.9999225532362965", "-163.73509363742735", "-286.0402052689569", "-1.5481333432299311", "0.007954486809572933") model <- array(c(paths, parameter, values), dim = c(10,3)) colnames(model) <- c("col1","col2","col3") dat <- hozdata result <- SEM(model = model, data = dat) summary(result) standardizedCoefficients(result)

65.6 A complex example

Lynsky, Fergusson & Horwood, 1998, analysed the relationships between reports of tobacco, alcohol, and cannabis use in a birth cohort of New Zealand children studied until the age of 16 (n=1265 dropping to 913 for complete cases). Comparable information from their parents was also obtained. The researchers were interested in assessing the validity of a particular model known as Jessor and Jessor's Problem Behaviour Theory, which posits that proclivity to problem behaviours are due to three aspects; demographic social structure (peer afflictions, novelty seeking and parental illicit drug use); the perceived social environment and the personality system. This is in contrast to the stage or gateway theory where there is a progression from one substance to another. (Lynsky, Fergusson & Horwood, 1998 p.1004). They had therefore a general idea of something like that shown above.

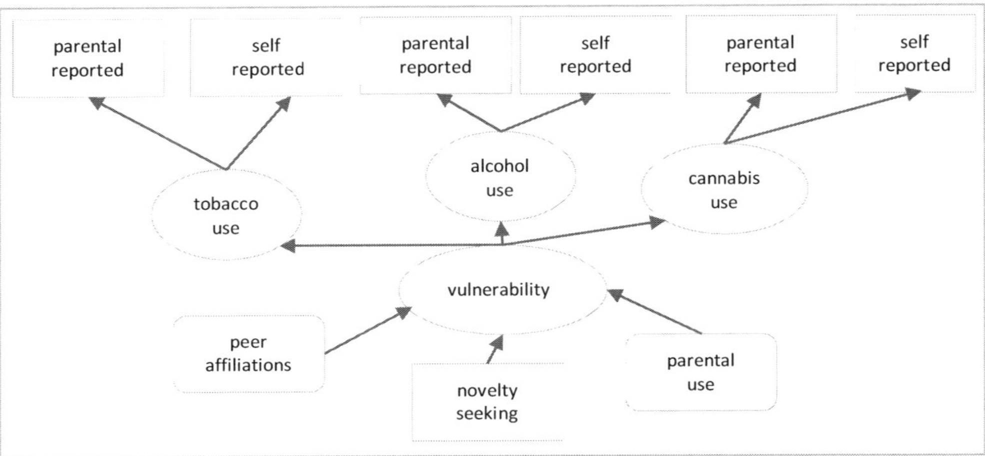

The final model produced from the study is shown opposite. It must be remembered that this is the final model of a long development process and I strongly advise you to read the original article. The parameter values are slightly different but comparable to the original as I have produced it using a subset of the original data (n=868) in the commercial SEM program EQS.

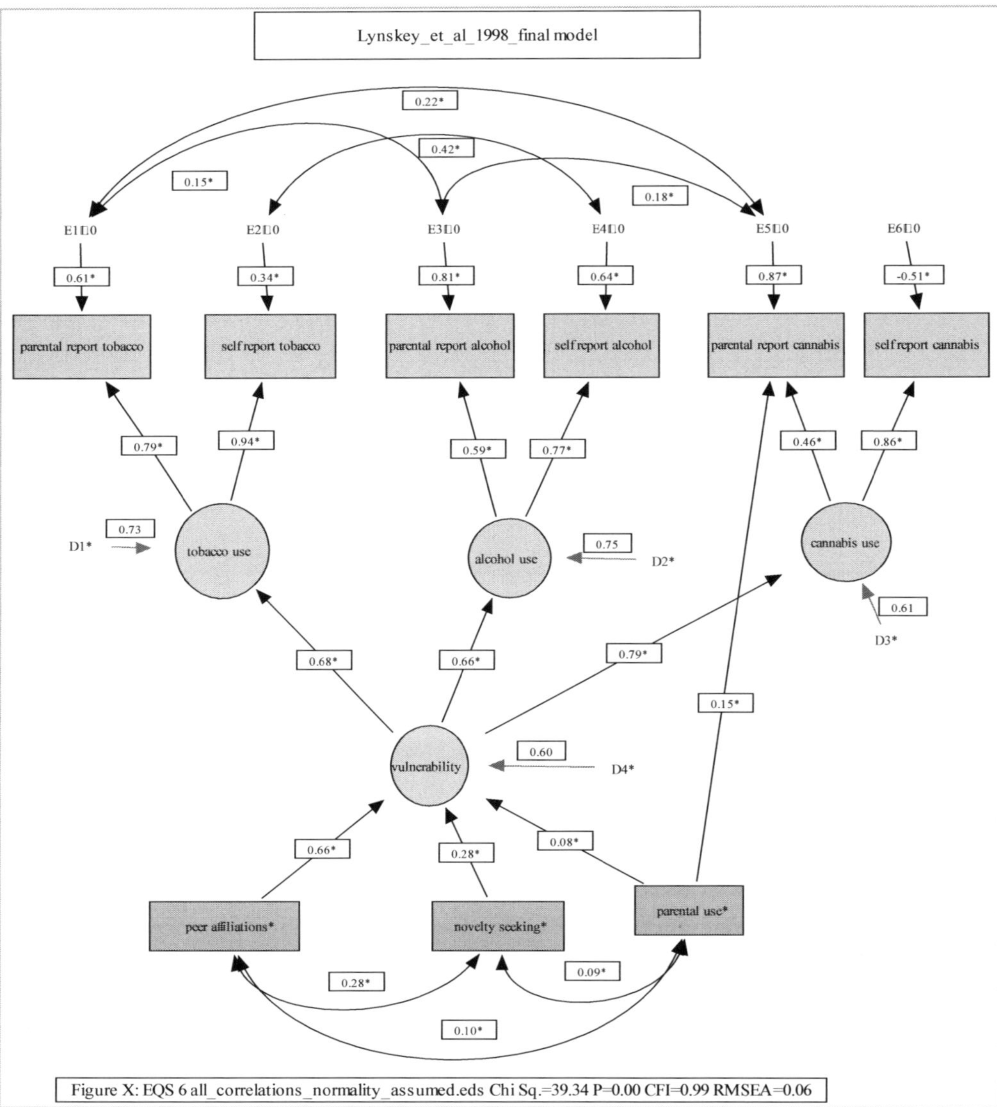

Figure X: EQS 6 all_correlations_normality_assumed.eds Chi Sq.=39.34 P=0.00 CFI=0.99 RMSEA=0.06

In EQS the error values are the squared roots of those produced in Ωnyx.
Therefore: 1 -(SEM error value)^2 = R^2.

There is a lot to take in concerning this complex model. Possible key points are:

1. Parental reports of tobacco, alcohol and cannabis use are correlated in contrast to self-reported cannabis use, which does not correlate with parental or other self reported measures.
2. Self-reported measures of tobacco, alcohol and cannabis use have higher standardised factor loadings (.77 to .94) compared to parental values (.46 to .79). Remember the higher the loading the greater the degree to which the measure is reflected in the factor.

3. Relationship between tobacco, alcohol and cannabis use (latent variable) and vulnerability - the standardised loading factor values here (.66 to .79) indicates a strong relationship between the three specific latent variables and the general vulnerability latent variable.

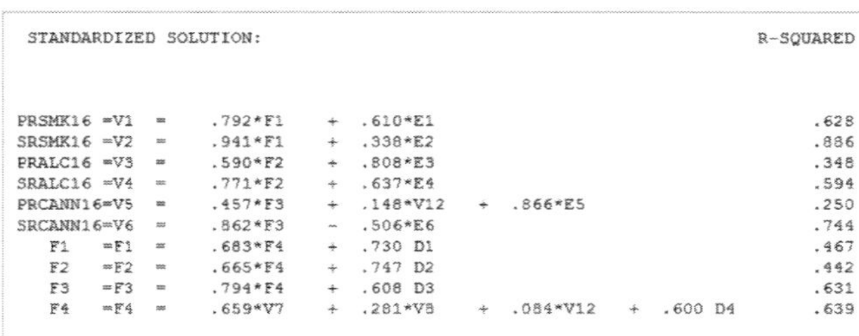

```
STANDARDIZED SOLUTION:                                                R-SQUARED

PRSMK16 =V1  =   .792*F1   + .610*E1                                    .628
SRSMK16 =V2  =   .941*F1   + .338*E2                                    .886
PRALC16 =V3  =   .590*F2   + .808*E3                                    .348
SRALC16 =V4  =   .771*F2   + .637*E4                                    .594
PRCANN16=V5  =   .457*F3   + .148*V12  + .866*E5                        .250
SRCANN16=V6  =   .862*F3   - .506*E6                                    .744
   F1   =F1  =   .683*F4   + .730 D1                                    .467
   F2   =F2  =   .665*F4   + .747 D2                                    .442
   F3   =F3  =   .794*F4   + .608 D3                                    .631
   F4   =F4  =   .659*V7   + .281*V8   + .084*V12  + .600 D4            .639
```

4. The error value associated with the vulnerability (.6) indicates how well the three predictors (peer affiliations, novelty seeking and parental drug) predict vulnerability. As $R^2 = 1-.6^2 = .64$ indicting that 64% of the variability in vulnerability can be accounted for by the predictors. For the full data set (Lynsky, Fergusson & Horwood, 1998) this value was reduced to 54%.

5. The asterisks (*) beside the parameter estimate indicates that the value is statistically significant.

6. The overall model fit measures indicate a well-fitting model with CFI =0.99 (good fit CFI≥0.95) and RMSEA=0.06 (RMSEA≤.05 very good fit).

65.7 Extending SEM

SEM techniques have been extended in many ways:

- Non-normal distributions
- Bayesian approaches
- Ordinal and binary data, mimicking factorial ANOVA and Logistic regression.
- Time series (also called latent growth models)
- Comparisons of multiple groups (means)
- moderation & mediation effects (see https://www.youtube.com/watch?v=mirI5ETQRTA)

A good introduction to the above techniques, except moderation and mediation, can be found in Schumacker & Lomax, 2010.

65.8 Reviewing SEM research

The following page provides a checklist, of aspects to consider when reviewing a SEM article. While all the points are valid, one or two of them, such as cross validation of the modal using a subset of the data is a rare occurrence.

Criteria for reviewing a article using SEM (adapted from Schumacker & Lomax 2006, p.251-5)

1. Data Preparation

1. Have you adequately described the population from which the random sample data was drawn?
2. Did you report the measurement level of your variables?
3. Did you report the descriptive statistics on your variables? [+ normality]
4. Did you test for multivariate normality?
5. Did you create a table with correlations, means, and standard deviations?
6. Did you have missing data if so why and what strategy did you use to overcome the problem.
7. Did you have outliers and did you resolve this by using robust statistics or deletion methods?
8. Did you resolve non-normality of a variable by some form of transformation?
9. Did you have problems with multi-collinearity among variables? If so how did you resolve it?
10. Did you specify the input data used (raw, correlations, covariances etc.)
11. Did you include the set of commands as an appendix so that others could carry out a similar analysis?

2. Model Specification

1. Did you provide a rationale for your study?
2. Did you explain why SEM rather than another approach was required?
3. Did you describe your latent variables
4. Did you provide a theoretical foundation for your model?
5. Did you theoretically justify alternative models?
6. Did you justify your sample size?
7. Did you clearly state your statistical hypotheses?
8. Did you discuss the expected magnitude and direction of expected parameter estimates?
9. Did you include a figure or diagram of your proposed model(s)?

3. Model Identification Issues

1. Did you calculate the number of distinct values in your sample matrix?
2. Did you calculate the number of parameters needing to be estimated (free parameters) [and constraints]
3. [Did you software provide you with appropriate messages indicating there was no identification problem.]

4. Model Estimation

1. How will you consider power and sample size?
2. What is the ratio of sample size to number of parameters?
3. What estimation technique is most suitable given your sample size and normality situation?
4. Did you encounter Heywood cases (negative variance) or other impossible values in the output?
5. Did the software use raw data or a matrix as input?
6. How did you scale the latent variable (reference variable or fix the variance of the latent variable).
7. Which SEM program and version did you use?
8. Did you encounter any convergence problems?
9. Did you report the R^2 values to indicate the total effect the independent variables had on each dependent variable.

5. Model Testing

1. Did you include a website providing more information?
2. Did you provide tables (American Psychological Association style, Schumacker p.241) providing details for each factor (i.e. Reliability rho) and for each indicator Loading and R^2
3. Did you use single items or composite scale scores?

Several other issues in the original list not considered here.

6. Model Modification

1. Did you compare alternative or equivalent models?
2. Did you clearly indicate how you modified the initial model?
3. Did you provide a theoretical justification for the respecified model?
4. Did you add or delete one path/ parameter at a time? [recommended]
5. Which statistics helped guide you through the process (Wald, Lagrange [good] or simple *t* tests [bad])
6. Did you provide parameter estimates and model fit indices for both the initial model and the respecified one(s)?
7. Did you report expected change statistics.
8. Your model is not the only model that fits the sample data, so did you check for equivalent models or theoretically justify your final model?
9. How did you evaluate and select the best model?

7. Model Validation

1. Did you replicate your SEM model analysis using another sample of data?
2. Did you cross-validate your SEM model by splitting your original sample of data?
3. Did you use bootstrapping [simulate the samples] to determine bias in your parameter estimates?
4. Did you report effect sizes and confidence intervals in addition to statistical significance testing?
5. Did you evaluate your results with regard to your original theoretical framework?

65.9 Tricks and Tips

This chapter has provided only an introduction but hopefully has helped you understand the basics of SEM and provided enough information to enable you to interpret findings from such an analysis. People who undertake SEM tend to learn a particular SEM program or R package and then stick to it. I hope by introducing a range of approaches in this chapter will have helped you decide which is the best one for you.

Understandably, much of the information available is aimed at using a certain package, particularly the many useful youtube videos. Notable exceptions are Loehin, 2004 and Kline, 2010. The companion website to Kline's book provides overheads for each chapter and an excellent set of web links to other SEM sites at:

http://www.guilford.com/cgi-bin/cartscript.cgi?page=etc/kline.html

There are many websites devoted to SEM techniques. Professor Jason Newsom, at Portland University, runs an excellent course on SEM with online resources and links to several SEM sites at: www.upa.pdx.edu/IOA/newsom . Another is David A. Kenny (Emeritus Professor, University of Connecticut, Department of Psychology) who keeps a useful, succinct webpage listing and explaining most of the model fit indices http://davidakenny.net/cm/fit.htm

There are the webpages associated with each of the R packages used in this chapter which provide further details and examples. The Ωnyx website provides some information which hopefully will be developed in future.

66 Summary

Model aspects

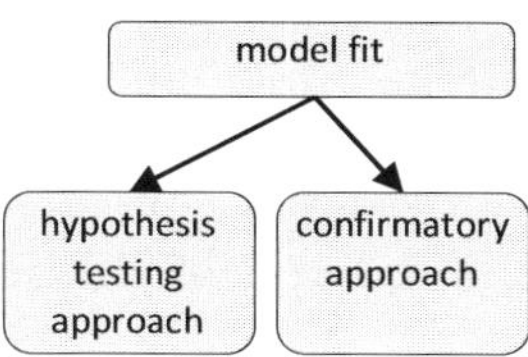

This book has presented a vast array of techniques and approaches to analysing data. Both the traditional hypothesis testing approach (p-values and critical values) along with the "*the new statistics*" Cumming, 2012, effect size/confidence interval style as used to interpret the results. The concept of a statistical model was stressed and how once again this could be interpreted from both these approaches. The various regression techniques used; ranging from the simple regression chapter to that introducing Structural Equation Modelling (SEM) highlighted this.

Typical stages of data analysis

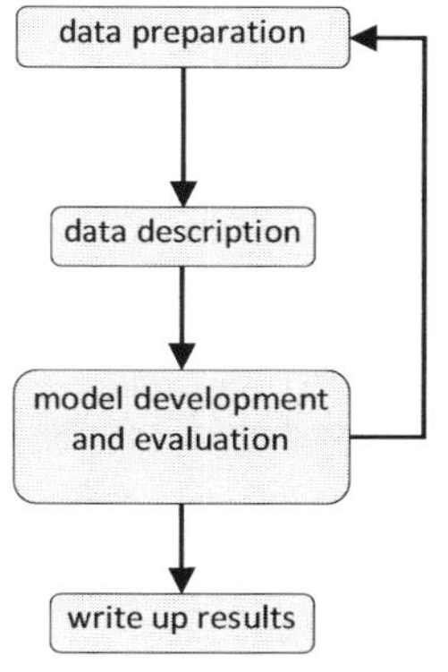

Those chapters that involved analysing a particular aspect of a dataset, followed the stages outlined in the diagram opposite.

Data preparation can take a significant proportion of the analyst's time as was demonstrated with a wide range of approaches. The simplest technique was importing a comma delimited file and possibly changing the format of particular variables, such as transforming numeric variables to R factors or the conversion to z scores to allow comparison across variables/samples. Another common technique demonstrated in the chapter *Comparing pre-post test means* involved the *reshape2* package to change the format from *wide* to *long*. This package was used again with a more complex example in the repeated measures chapter. The chapter also demonstrated the *subset()* function allowing the analysis process to use a subset of data for a single or range of, factor levels.

The data description stage focused on graphical rather than numeric (i.e. tabular) techniques. While R is espoused as being an excellent environment for developing graphical approaches, I've shown in this book that using raw R code does require some skill for the novice as demonstrated in the *repeated measures* chapter. Luckily R Commander offered some basic charts and Deducer, discussed in the publication quality graphic chapter, provided many additional options.

This book has developed and evaluated many different models and Implicitly in each chapter the variables selected for each model had the appropriate **level of measurement** and also a specific relationship with the other variables. One important facet of the relationship was that of dependency. In the earlier models, such as the one sample and two independent samples *t* test, values from different subjects were assumed to be *independent*. This contrasted to the repeated measures obtained from a single subject which were considered to be *dependent*. In the repeated measures chapter the idea of dependency was investigated further to model 'contexual' (nested) variables and the example demonstrated how they were analysed the same way as repeated measures. The chapter also demonstrated the dangers of assuming measures were independent when they were not.

By considering the **level of measurement** along with dependency we can create a diagram, similar to that frequently found in statistics books, showing the appropriate model for each combination.

66.1 Independent measures

The diagram below shows some of the models we have considered in this book where the output is considered to be independent. The only type of model we considered where there was the possibility of multiple outputs was in the *Structural equation modelling* chapter.

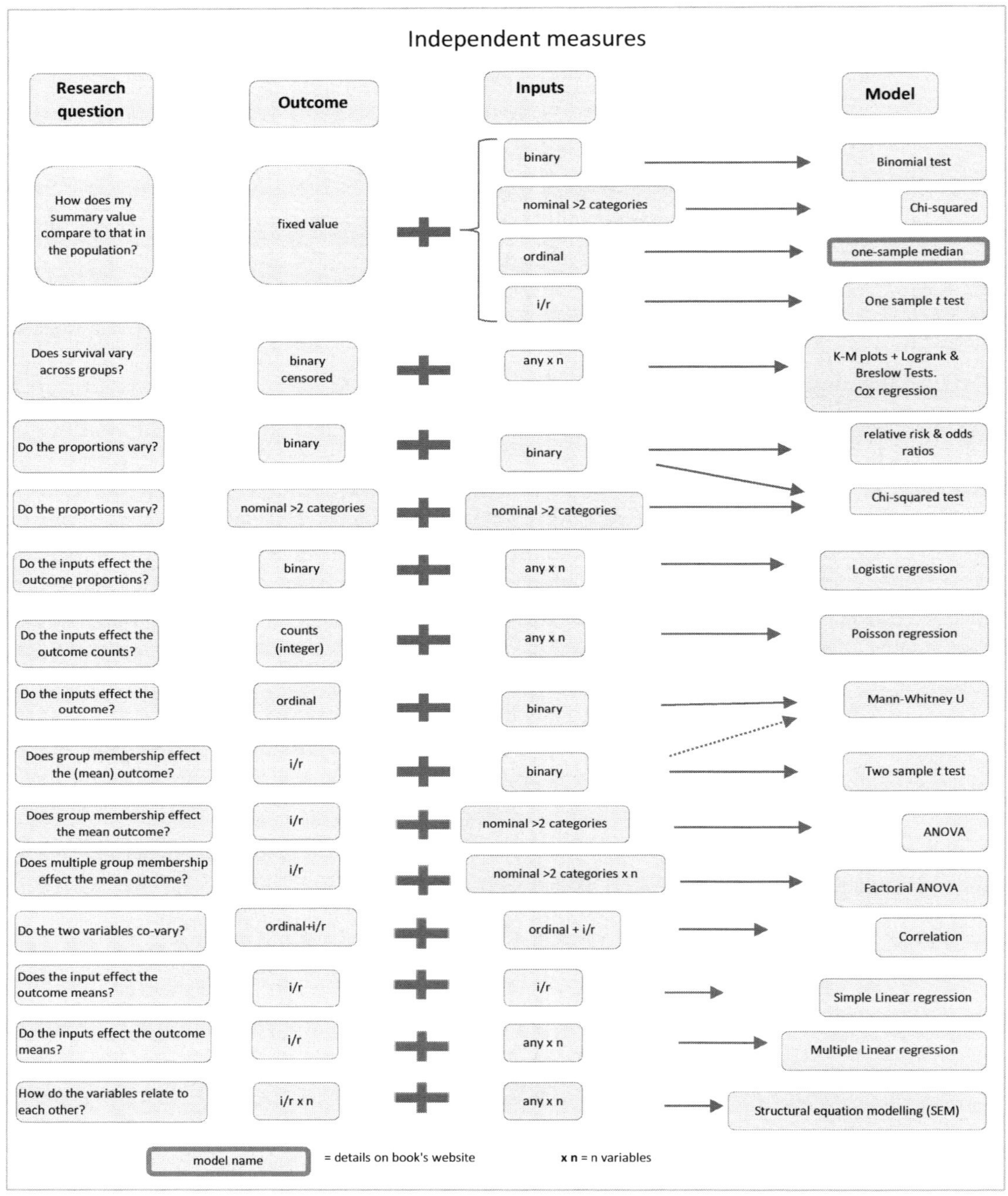

66.2 Dependent outputs (repeated/nested measures)

Traditionally repeated measures were less frequently met in the experimental setting because of the complexities involved in analysing such data. However, with the development of 'repeated' measures techniques, notably the multilevel modelling approach, developed primarily by the epidemiological community, analysis of such data has become more popular.

This book investigated a range of models involving repeated measures. Possibly the simplest was the Wilcoxon Matched Pairs statistic and the most complex the structural equation modelling approach (SEM). However, the SEM approach lacked many of the fine nuances that can be modelled, by way of design covariance matrices, with the multilevel approach.

While the typical stages of data analysis followed that described previously, with repeated measures there was an even greater focus in a range of graphical summaries before going on to produce a statistical model which often involved some tricky data manipulation.

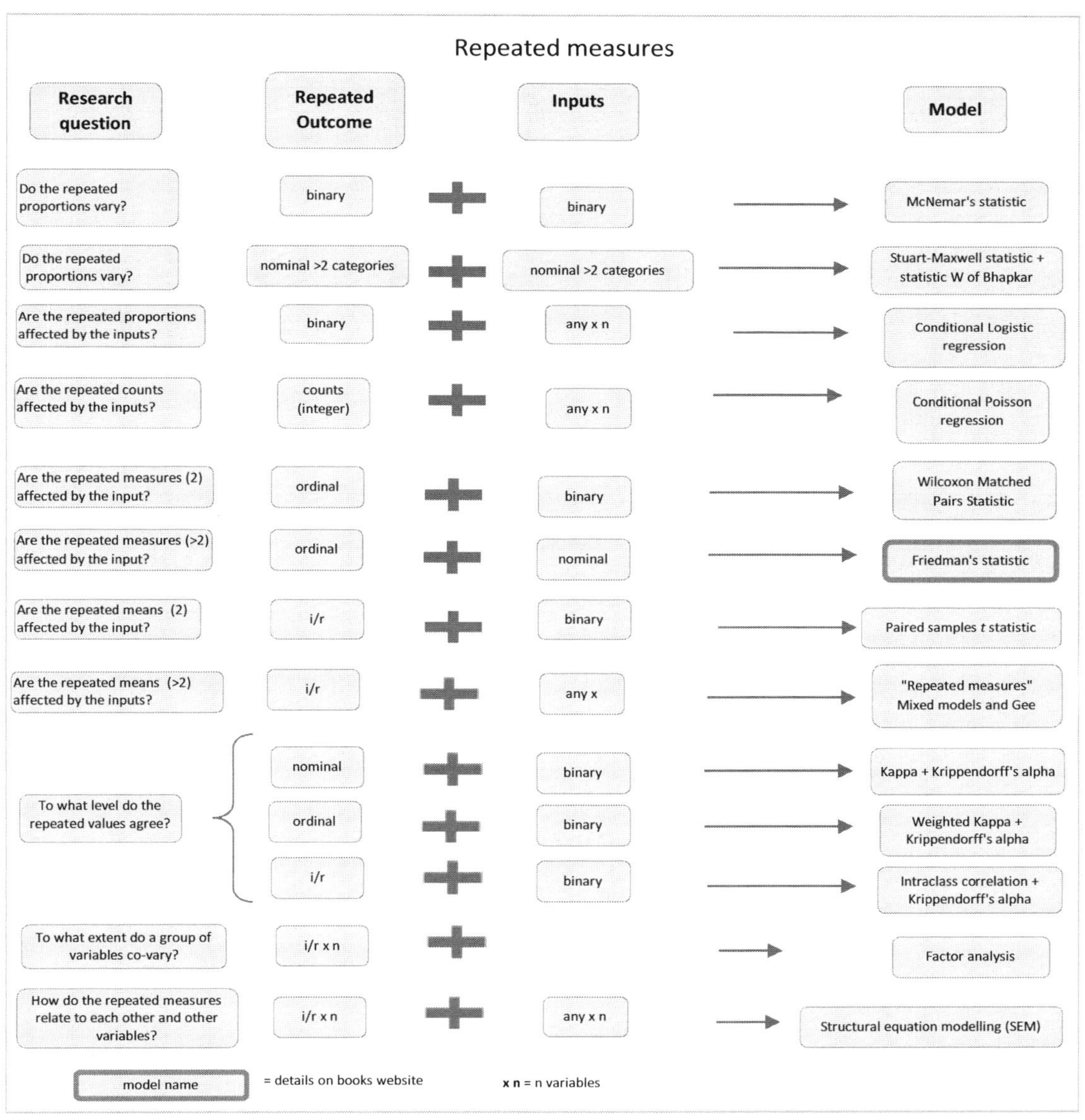

66.3 Moving on

Several of the techniques presented in this book, along with their implementation in R, are subject to ongoing development. I have attempted to select those that I feel will become more popular rather than those that will invariably wither on the vine.

Those wishing to learn more about a particular topic can start by looking at the accompanying *pdf* document to each R package found on the CRAN site. For traditional paper based introductions to R, Crawley, 2005, and Field, 2013, are good starting points. For a more interactive approach including a list of *websites/blogs and online* tutorials visit this books website www.robin-beaumont.co.uk/rbook and become part of the community.

67 Appendices

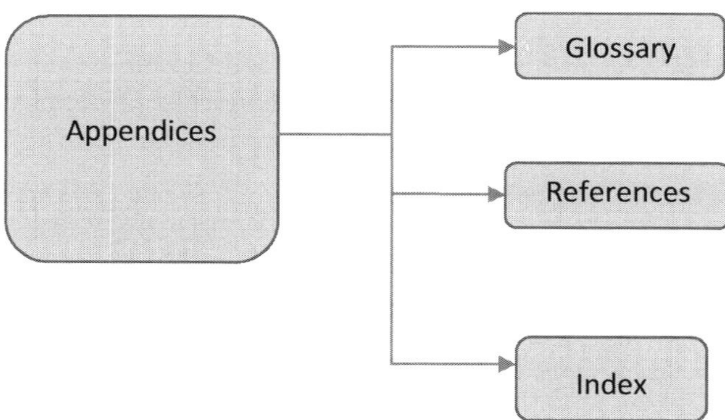

Additional chapters can be found on the book's website, www.robin-beaumont.co.uk/rbook.

Glossary

Attach(dataframe name)

Words in **bold** indicate a glossary entry

This R **function** makes it possible to refer to the individual column names directly rather than the dataframe name and then column name *(mydataframe$columnname)*. For example *weight* instead of *mydataframe$weight*. In R Commander the following menu option achieves this.

data->active data set->set active data set

The equivalent in R code is *attach(dataframe name)*. This makes the selected dataframe the active dataframe.

search() lists all the dataframes loaded. To un-attach a dataframe use the *detach(mydataframe)* function. Expert R programmers (not including myself!) avoid using *attach()* instead using one of the following strategies:

- Reference variables directly. For example *plot(dataframe$x ~ dataframe$y)*
- Specifying the dataframe for functions which support a *data =* option.
  For example *lm(y ~ x, data=mydataframe)*.
- Use the *with(dataframe, expression)* function, which returns the value of whatever expression is evaluated. For example *with (bb_long, mean(bdi_pre))*.

Binary variable

A variable that can only take one of two values, for example, status where the possible values are alive and dead. See **Level of measurement** and **Dichotomising variables**.

Boxplot

A boxplot, also called a box and whisker diagram, is a very useful graphical summary providing information about the:

- Range of values including extreme values marked with an '0'
- Median value (middle value when observations ranked)
- Middle 50% of values (interquartile range)

The presence/ absence of outliers along with the position of the median line within the box is useful for assessing the distribution of the data along with a histogram. For normally distributed data you would expect the median to appear roughly in the middle of the box and no/few outliers.

Comparing several boxplots side by side provides a very useful technique of assessing possible differences between the groups.

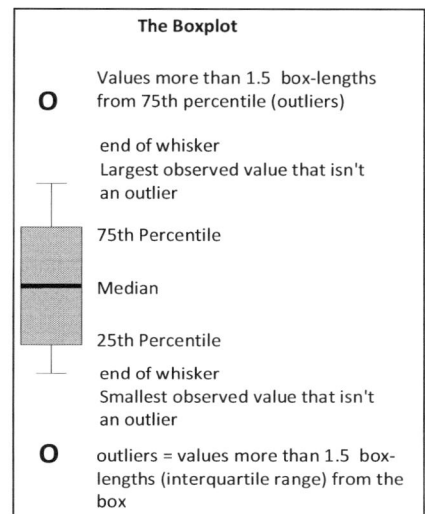

The Boxplot

O Values more than 1.5 box-lengths from 75th percentile (outliers)

end of whisker
Largest observed value that isn't an outlier

75th Percentile

Median

25th Percentile

end of whisker
Smallest observed value that isn't an outlier

O outliers = values more than 1.5 box-lengths (interquartile range) from the box

Confidence interval

Confidence intervals provide two values (an upper and lower value) constituting an interval that represents a certain degree of confidence of containing a particular summary statistic value (e.g. mean) over repeated trials. By using a confidence interval we are inferring (see **inference**) from our **sample** to a **population** value. We can create confidence intervals from any estimate we may create (variance, Standard Error of the mean, median etc). We can also control the level of 'confidence' we wish to set for the interval. We may only want to be 90% confident that a particular population value (i.e. parameter) is within the interval or we may want a higher level, say of 99% of certainty. The more certainty we require the wider the interval will be, although other important aspects also affect its width such as the size of the sample and reliability of the measure. The width of the interval is often called its precision, therefore wide intervals are often said to have low precision.

There are broadly two techniques for calculating confidence intervals, the traditional method uses various formulae (using the standard error, see **inference**) and assumes the estimated values are normally distributed. Alternatively, a computer technique called bootstrapping can be used (see the *re-sampling* chapter and Campbell, 2008 p.109 for details) which does not make this assumption. The conceptual formula for a confidence interval is:

estimated value -(confidence level) X (variability of estimated value) **to** estimated value +(confidence level) X (variability of estimated value)

Assume we have obtained a 95% confidence interval of the mean equal to 4 to 5.6 for a patient satisfaction score. This value can be interpreted thus; we are 95% certain that the true population mean would be within this interval, given that the sample was random.

Confidence intervals can also be used to guide decisions as part of the **inference** process and are used to evaluate **Clinical significance**.

Contrasts

Factor level coding schemes (http://www.ats.ucla.edu/stat/r/library/contrast_coding.htm) Several of the coding schemes require you to define them manually in R see the website for details		
Coding Scheme	**R code**	**Comparisons made between means**
Dummy Coding * (default)	contr.treatment()	Compares each level to the reference level, intercept being the cell mean of the reference group, when only input.
Simple Coding		Compares each level to the reference level, intercept being the grand mean, when only input.
Deviation Coding * (also called summation coding)	contr.sum()	Compares each level to the grand mean
Orthogonal Polynomial Coding *	contr.poly()	Orthogonal polynomial contrasts
Helmert Coding *	contr.helmert()	Compares levels of a variable with the mean of the subsequent levels of the variable
Reverse Helmert Coding		Compares levels of a variable with the mean of the previous levels of the variable
Forward Difference Coding		Compares adjacent levels of a variable (each level minus the next level)
Backward Difference Coding		Compares adjacent levels of a variable (each level minus the prior level)
User-Defined Coding *		User-defined contrast

Please read the glossary entry **Factor** for a general introduction. In regression analysis each level of a factor is classed as a separate group and the output provides estimates for each level, including a 'reference group'. R provides several methods of coding factors each of which provides a different way of interpreting the factor's levels. The default is dummy coding also called treatment coding.

The Institute of Digital Research and Education at UCLA (University of California, Los Angeles) provides an excellent tutorial concerning R and contrasts at http://www.ats.ucla.edu/stat/r/library/contrast_coding.htm the tutorial includes the following table, which I have adapted.

Several of the above contrast coding schemes are available from the R Commander menu option (asterisked above):

Data -> Manage variables in active dataset -> Define contrasts for a factor

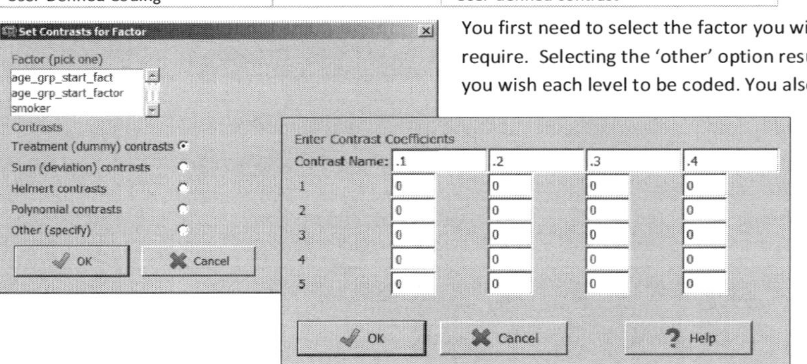

You first need to select the factor you wish to define a set of contrasts for, and then the type of contrasts you require. Selecting the 'other' option results in you being presented with a dialog box asking you to specify how you wish each level to be coded. You also have the option to add names for each level along the top, this can be very useful in aiding interpretation of results.

The *Multiple linear regression* chapter provides details of the R code equivalent to the above R Commander options.

Clinical significance

Clinical significance, also called clinical effect, is different to statistical significance. It is an effect size measure, sometimes interpreted with regard to a Minimal Clinically Important Difference (**MCID**). The MCID is the smallest treatment difference that would lead to a change in a patient's management. In chronic disease a similar concept is the Patient Acceptable Symptom State (PASS), see Kvien, Heiberg & Hagen, 2007 for details. Assessing if the MCID has been achieved is done by inspecting confidence interval plots. How the MCID is calculated is very much a matter of debate, stressing the importance of the clinician's expert knowledge.

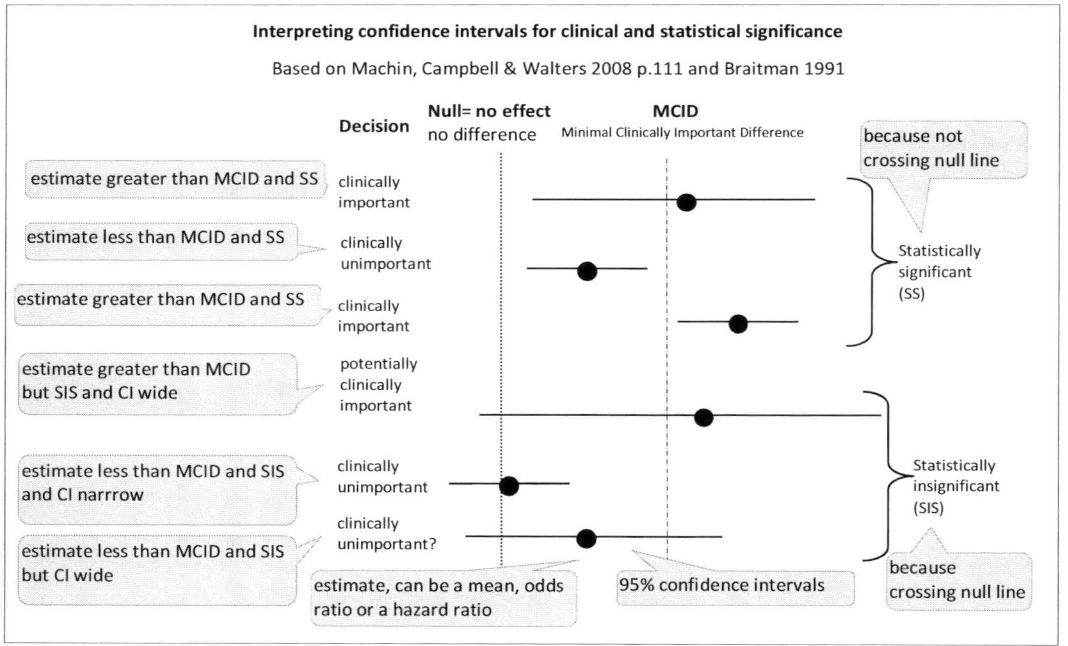

If the effect is less than the MCID then the effect is said to be clinically unimportant otherwise it is said to be either clinically important or clinically ambiguous depending upon the confidence interval result. The diagram below provides a visual explanation. For further details about clinical significance see Molnar, Man-Son-Hing & Fergusson, 2009.

Correlation ellipse

Whereas the confidence interval is for a single variable, for a set of points (i.e. a correlation) we can obtain a correlation ellipse. The area within the ellipse indicates a level of certainty of predicting a new observation in the population. Inspecting the shape is also useful, as a round ellipse indicates no correlation, the shape becoming more ellipsoid as the correlation increases.

Dataframe

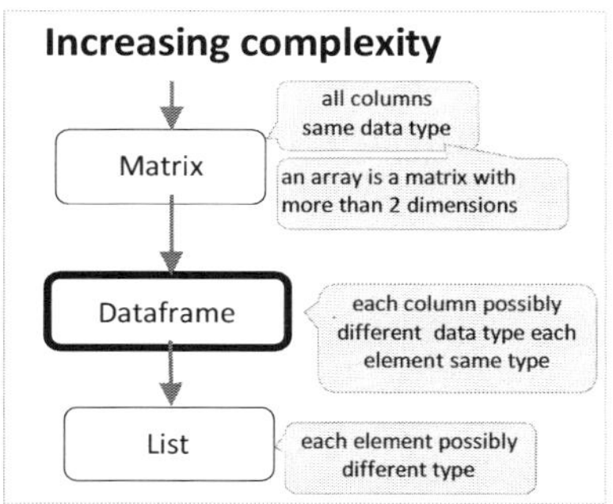

The dataframe is the most important data structure in R. Whatever format your original data is in it is usually converted to a dataframe to allow you to analyse it. Essentially a dataframe is a set of columns where each represents a variable (=vector) and each row a case. A situation where the data is said to be in *long* format. This is not always the situation, particularly with longitudinal data, where you have several repeated measurements for the same variable for the same case over time, and the data is said then to be in *wide* format. Further details of these two formats and how to convert between them can be found in the *Repeated measures* chapter.

In dataframes all the columns are the same length, missing values being indicated by the NA (i.e. not available) value. A related data structure in R is the list, which allows more flexibility. In a list each element can by a different type and length. A list might have a character as the first element and a vector as the second and a data frame as the third. At the opposite extreme is the matrix data structure where all the columns are of the same data type.

Density plot

A graph showing the distribution for one or more variables. The simplest way to think of it is as a smoothed histogram. This smoothing is achieved by assuming the dataset is a sample from a population and subject to the vagaries of random sampling. See **inference**.

Dichotomising variables

This is when a variable is recoded so that it only has two possible values. For example, recoding a weight variable which was originally recorded in kilograms into a **factor** with two levels, "acceptable" and "obese". Ideally when collecting data, the most complex **level of measurement** should be used for each variable and only afterwards should variables be dichotomised. Ideally the additional dichotomised variable should be kept alongside the original values.

Effect Size

Cohen's effect size measures				
situation	index	Effect size		
		small	medium	large
comparing means	d	.2	.5	.8
correlations	r	.1	.3	.5
F-test (ANOVA)	f	.1	.25	.4
F-test (MCR)	f^2	.02	.15	.35
Chi-square	w	.1	.3	.5

The effect size measure is an important statistic to report in any research and represents a measure of the difference between some hypothesised value and that obtained from an observed sample. The exact form the measure takes depends upon the circumstances and the **Level of measurement** of the variables under study. Jacob Cohen (1988) developed a set of effect size measures for most circumstances some of which have been adopted, others less so. Cohen also provided values for what he considered to be small, medium and large effects of each, of which the adoption and usefulness is of continuing debate.

A selection of other effect size measures, which also happen to be traditional statistics, is given in the table opposite.

Effect size measures are useful because they provide a method of quantifying the observable difference (in contrast to p-values, which do not do this, this being a widely held misconception) and form a major part of assessing **Clinical significance**.

Traditional effect size measures (adapted from Ellis 2010)	
situation	index
comparing means	*Glass's Δ*
	Hedges g
	eta squared (η^2)
correlations	partial correlation ($r_{xy,x}$)
(one variable binary)	point biserial correlation
multiple regression	R^2
	${}_{adj}R^2$
	change in $R^2 = \Delta R^2$
logistic regression	*logits*
	odds ratios
ANOVA/ANCOVA	eta squared (η^2)
MANOVA	partial eta squared (η^2)
Chi-square	*Phi coefficient* (φ)
	Pearson's C
	Cramer's V
	Goodman and Kruskal's lambda (λ)
	Kendall's tau (τ)

E Notation

The E is notation is a shorthand method of indicating a power of ten. Considering, 4.3E-04, this stands for $4.3 \times 10^{-4} = 0.00043$. The easy way of remembering what it means is to look at the last part, i.e. -04 then add a decimal point and the number, minus one of zeros. That is -4 indicates adding a decimal point and 3 zeros, -6 means a decimal point and 5 zeros. Such notation is used in the R statistics output, particularly for small p-values. Usually in reporting such numbers if the number is less than 0.0001 it is reported as simply >0.0001 .

As $10^0 = 1$ we can have $4.3 \times 10^0 = 4.3E0 = 4.3$.
Some typical large values are given opposite.

Power of 10	E notation	Number	term
10^{12}	1 E12	1,000,000,000,000	Trillion = thousand billion
10^9	1 E9	1,000,000,000	1 billion
10^6	1 E6	1,000,000	1 million
10^5	1 E5	100,000	1 hundred thousand
10^4	1 E4	10,000	ten thousand
10^3	1 E3	1,000	1 thousand
10^2	1 E2	100	1 hundred

Factor

A factor is one of the main data types in R, another being **numeric**. Examples of factors are the codes used to indicate a particular group an observation belongs to. The groups might be 'treatment' versus 'control' (often given the values '1' or '2') or, in another instance several different geographical locations (such as 'Leeds', 'London', 'Sheffield'). Each different value used to represent the different group is called a **level** so the number of levels a factor possesses indicates the possible number of groups for the factor. Such a variable is often called a grouping variable. Internally in R the values used to represent the level of the factor are numbers. These are associated with a text label for each level. Factors can be unordered reflecting a nominal level of measurement or ordered reflecting an ordinal level of measurement.

Within a particular dataset there may not be any observations for certain factor level and it is necessary to remove these factor levels from the dataset before carrying out certain statistical procedures. The chapter *A quick introduction to the R language* provides R specific information. Also, see glossary entries **Level of measurement** and **Contrasts**.

Function

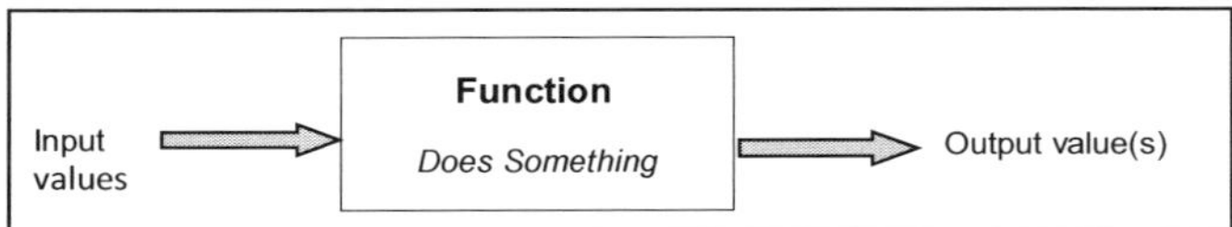

Functions are a core characteristic of R, common ones being the *mean()*, and *length()*. If you intend to write any R code beyond a few lines you probably need to learn how to write your own functions.

Functions receive one or more inputs (called arguments, *args* for short), do something and then return one or more outputs.

The structure of an R function

```
myfunction <- function(arg1,
arg2, ... )
{
statements
return(object)
}
```

The above diagram can be expressed in R, shown opposite. Notice the curly brackets indicating the beginning and end of the R code for the function. The three dots (. . .) represent a placeholder that is for other unspecified inputs.

The example below is for a function I have written called *samplemean()*, you send it an object which it gives the name *thedata* it then calculates the mean value and then returns this value:

```
samplemean <- function (thedata)
{
statistic <- mean(thedata)
print(statistic)
return (statistic)
}
```

this function I have called samplemean — takes one input — start the code for the function — calculate the mean and put it in an object called *statistic* — print, not usually necessary; helps you to see what is going on for debugging — return the statistic object — finish the code for the function

We would use the above function (technically we would say we *call the function*) by typing into the R Console window *samplemean(somedata)*. Typing the name of the function without the parentheses produces the code for the function in the window.

Inference, sampling distributions and standard errors (*se*)

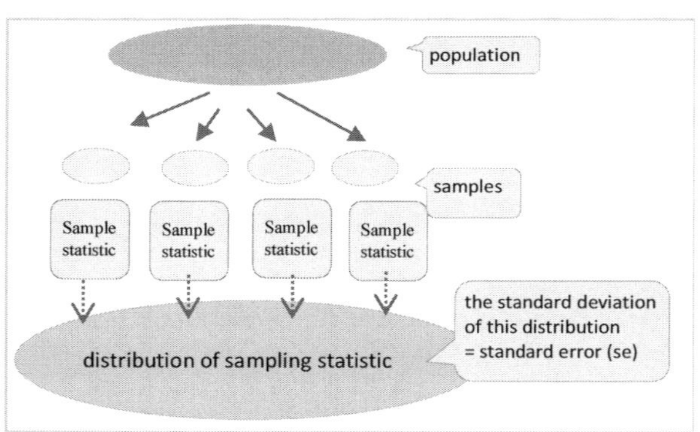

Inference is the process of making informed decisions about a population by observing one or more randomly selected samples from it. For technical reasons the summary values we obtain from our samples are called estimates or statistics and the corresponding inferred values in the population are usually called parameters.

The most common situation is to obtain some summary value (e.g. mean) from a single random sample which is known to be a good estimate of the equivalent population value. We them assume that we have obtained a very large number of these random samples of the same size from the population and have obtained a distribution of the statistic (known as a sampling distribution) from which we can calculate a Standard Deviation.

Unfortunately for technical reasons, in this situation the Standard Deviation is renamed the standard error (*se*) and, because here we are talking about the mean specifically, the Standard Error of the Mean (*sem*).

$se_{\bar{x}}$ = Standard Error of the Mean

$$se_{\bar{x}} = \frac{\sigma}{\sqrt{n}} = \frac{s}{\sqrt{n}} = \frac{\text{Standard Deviation}}{\text{square root of the sample size}}$$

This all sounds very complex but for some summary statistics when the sample is randomly selected from a population it is possible to infer its distribution, and therefore its *se* very accurately. For example, the *se* for the sampling distribution of the mean is given by the equation on the left. This is amazing, as it specifies we only have to take a single sample and we can accurately calculate the standard deviation of the distribution of means as if we had obtained a very large number of such samples. This measure of expected variability in the means forms a vital part to the *t* value which we use in many chapters of this book.

The *sem* value is dependent on the square root of the sample size, therefore as the sample size increases it gets smaller as demonstrated opposite for sample sizes of 1, 5, 25 and 50. The following R code produced the graph :

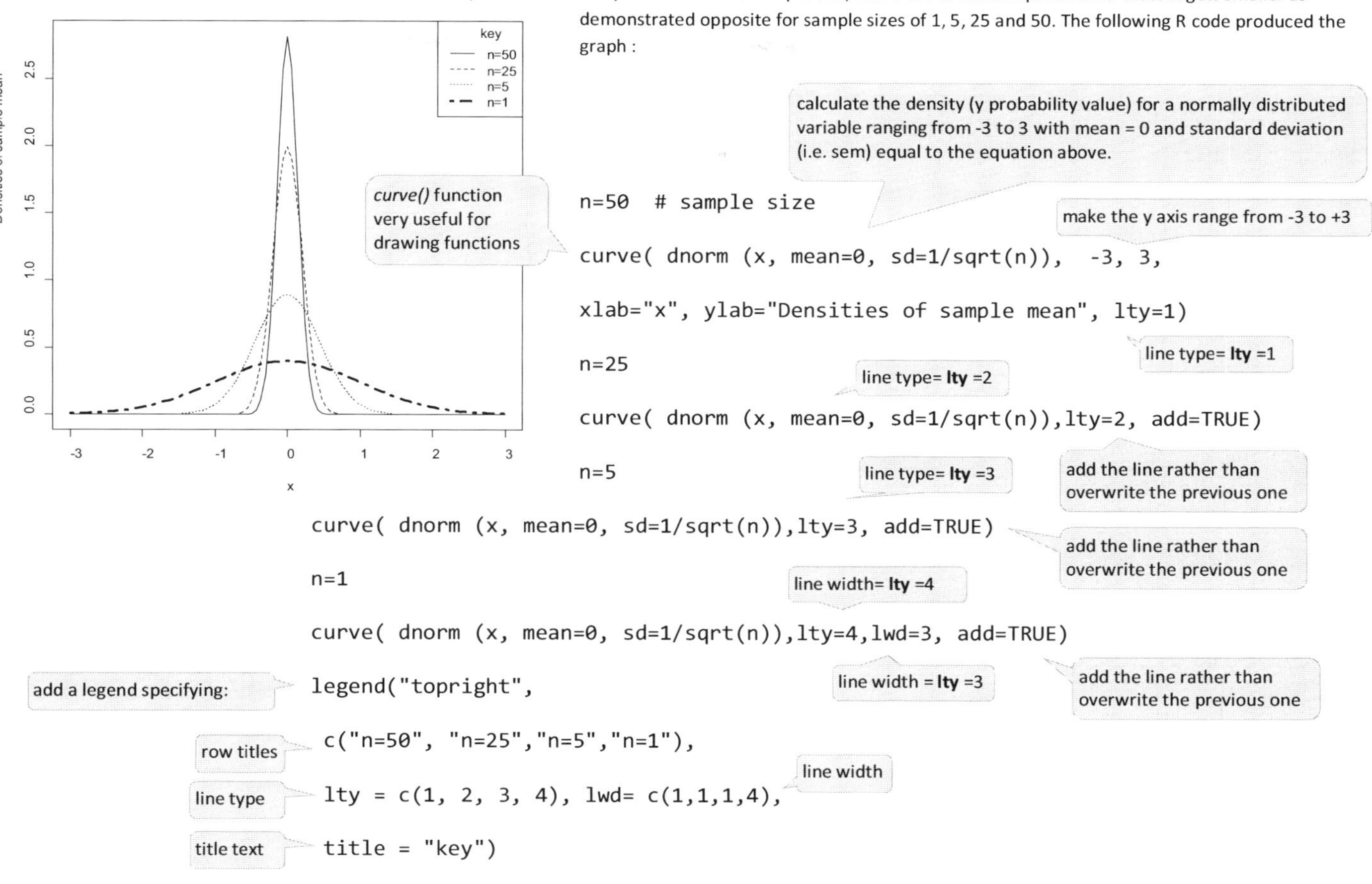
key
n=50
n=25
n=5
n=1
Densities of sample mean
x
curve() function very useful for drawing functions
calculate the density (y probability value) for a normally distributed variable ranging from -3 to 3 with mean = 0 and standard deviation (i.e. sem) equal to the equation above.
make the y axis range from -3 to +3
line type= lty =1
line type= lty =2
line type= lty =3
add the line rather than overwrite the previous one
add the line rather than overwrite the previous one
line width= lty =4
add the line rather than overwrite the previous one
line width = lty =3
add a legend specifying:
row titles
line type
line width
title text

```
n=50  # sample size
curve( dnorm (x, mean=0, sd=1/sqrt(n)),  -3, 3,
xlab="x", ylab="Densities of sample mean", lty=1)
n=25
curve( dnorm (x, mean=0, sd=1/sqrt(n)),lty=2, add=TRUE)
n=5
curve( dnorm (x, mean=0, sd=1/sqrt(n)),lty=3, add=TRUE)
n=1
curve( dnorm (x, mean=0, sd=1/sqrt(n)),lty=4,lwd=3, add=TRUE)
legend("topright",
 c("n=50", "n=25","n=5","n=1"),
 lty = c(1, 2, 3, 4), lwd= c(1,1,1,4),
 title = "key")
```

I could have also added some extra R code to the above to have first created the samples. For details of which see the chapters, *Creating datasets and distributions in R Commander and R,* and also, *Confidence intervals for effect sizes: noncentral distributions*.

Traditionally the estimated values for the population are considered to be fixed, that is given a single value. However, more recent statistical techniques (such as random effects models, hierarchal or multilevel models and random effects meta-analysis, to name but a few) allow the actual population parameter to follow a normal distribution. Such a parameter estimate is called a random effect, where random actually means usually a normal distribution here.

Likelihood and Maximum Likelihood

Let P_h be the probability that a certain coin lands heads up (H) when tossed (think of this as a parameter value).

So, the probability of getting two heads in two tosses (HH) is $P_h \times P_h$ If P_h =0.5 then the probability of seeing two heads is 0.25.

In symbols, we can say the above as:

$P(HH \mid P_h = 0.5) = 0.25$

Another way of saying this is to reverse it and say that

"the likelihood ($\mathcal{L}$) that P_h =0.5 , given the observation HH, is 0.25"; that is:

$\mathcal{L}(P_h = 0.5 \mid HH) = P(HH \mid P_h = 0.5) = 0.25$

While both likelihood and p-values are both probabilities they are not the same, in fact they are very different. A very simple but effective example demonstrates this (taken from Wikipedia).

With a **p-value** we have:

p-value = p(observed data| parameter) = given the parameter value then predict probability of observed data (+ more extreme ones)

Whereas with **likelihood**:

Likelihood = $\mathcal{L}$ (parameter | observed data) = given the data then predict the parameter value

The maximum likelihood estimate (MLE) attempts to find the parameter value which is the most likely given the observed data.

Notice that while both are conditional probabilities (that is the final value is worked out given that we already have a particular subset of results) the probability and the likelihood are calculated differently and can result in very different values.

Once we have a single likelihood value we can gather together a whole load of them by considering various parameter values for our dataset. Both the median and mean represented values where the deviations or squared deviations are at a minimum (see the online chapter *high school statistics revision* for details). That is if we had worked the deviation from any other value (think parameter estimate here) we would have ended up with a larger sum of deviations or squared deviations. In these instances it is possible to use a formula to find the value of the maximum likelihood. In the simple regression chapter we discussed how the value of the parameters *a* (intercept) and *b* (slope) were those parameter values which resulted from minimising the sum of squares (i.e. the squared residuals), incidentally using the mathematical technique of calculus. In other words our parameter values were those which were most likely to occur (i.e. resulted in the smallest sum of squares) given our data.

$\mathcal{L}(P_h=0.1 \mid HHT) = .027$
$\mathcal{L}(P_h=0.2 \mid HHT) = .096$
$\mathcal{L}(P_h=0.3 \mid HHT) = .189$
$\mathcal{L}(P_h=0.4 \mid HHT) = .288$
$\mathcal{L}(P_h=0.5 \mid HHT) = .375$
$\mathcal{L}(P_h=0.6 \mid HHT) = .432$
$\mathcal{L}(P_h=0.7 \mid HHT) = .441$
$\mathcal{L}(P_h=0.8 \mid HHT) = .384$
$\mathcal{L}(P_h=0.9 \mid HHT) = .243$
$\mathcal{L}(P_h=0.95 \mid HHT) = .135$

Taking another example consider the observation of two heads obtained from three tosses of a coin. We know we can get three different ordering THH, HTH, HHT, and by drawing a tree diagram we can see that for HHT we would have p x p x (1-p). Because the order does not matter we can simplify to 3x p x p x (1-p) to obtain all the required outcomes for two heads and a tail. To find the likelihood we simply replace p, the parameter value with a valid range of values - see opposite - we can then draw a graph of these results.

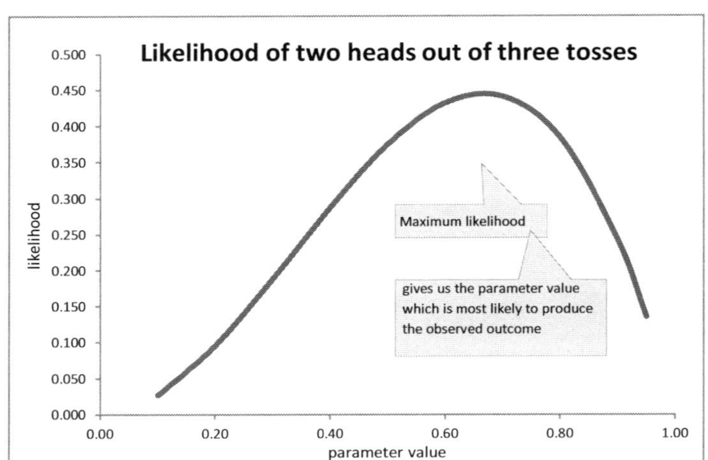

The graph shows that the likelihood reaches a maximum around the 0.7 value for the parameter. In other words to maximise the outcome of two heads and one tail for three tosses requires the coin to have a probability of falling heads up around 70% of the time.

In many situations in statistics, such as in Logistic regression, it is not possible to just apply a formula or use calculus to calculate the parameters (i.e. $b's$) instead the computer searches for the maximum likelihood value. This is achieved by the computer carrying out a search process.

For more information about likelihood and maximum likelihood see Shaun Purcell's introductory tutorial to likelihood at; http://statgen.iop.kcl.ac.uk/bgim/mle/sslike_1.html

Line type (lty R code)

Line **type** (lty) can either be specified as an integer or as one of several character strings. In R code you indicate which type of line to use by lty=6 or lty="twodash" etc.

Line types (lty)	
integer value: lty=0 etc	**character string: lty="blank" etc**
0	blank
1	solid (default)
2	dashed
3	dotted
4	dotdash
5	longdash
6	twodash

Line width (lwd R code)

Line **wid**th (lwd) is specified by using the R code lwd=positive number, the default value is 1.

LOESS line

In contrast to the standard least squares regression line which is straight, the LOESS line (locally weighted polynomial regression) is bendy and takes into account the local shape of the data. Many graphs in R produce these lines as well as the normal least squares (i.e. straight) regression lines.

Level of measurement

The level of measurement is a method of classifying observations (variables), of which there are two main systems, the statistical and the social science approach. To be able to classify a variable is a fundamental skill required before you can intelligently analyse or interpret any dataset.

Statisticians' classification

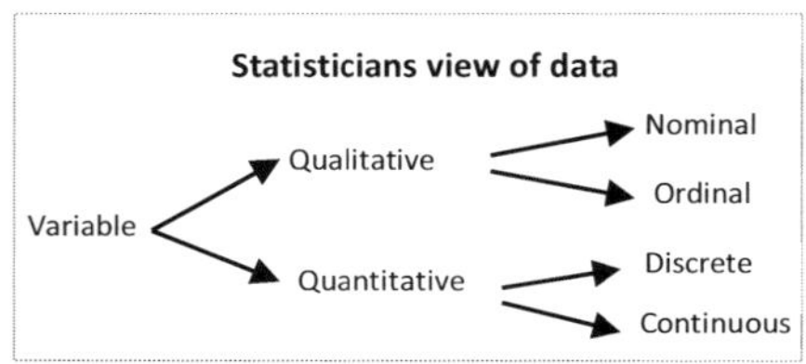

Statisticians classify data into two broad types - qualitative and quantitative, each of these is further sub-divided into two further types resulting in four basic types; Nominal, Ordinal, Discrete and Continuous. Eye colour is an example of qualitative data. This is because with qualitative data the 'values' are essentially words used to specify the categories. In contrast, with quantitative data, the values are numerical attributes which the data possesses itself. Qualitative data are often given numerical codes, but any arithmetic done with the codes will be meaningless, as will become obvious when one returns to the words behind the codes. For example, type of dwelling ('semi-detached', 'detached', 'terrace', etc.) is qualitative (nominal). The 'values' might be coded '1', '2', '3', etc., but although we can add '1' and '2', adding 'detached' and 'semi-detached' has no meaning, and certainly does not result in 'terrace'! This scale only allows us to state which particular category a data value belongs in (e.g. blue or green for eye colour) and count (enumerate) how many there may be in each category for a particular data set. Nominal data is therefore also often called *categorical* or *enumerate* data. The count for a particular category is called the *frequency*.

The term 'ordinal' is used when it is possible to order the various categories to create a *scale*. For example, the responses to a question asking 'How often do you have problems getting to sleep?' might be labelled 'every night' , 'most nights', 'some times', 'rarely' and 'never'. The correspondence between these words, although clearly graded in order, is questionable in terms of the relative distance between each. For example is 'most nights' a similar distance away from 'some times' as 'rarely' is to it in the opposite direction. Much effort is made to make such ordinal data possess a scale that approximates equal distances such as the common set of responses ('strongly disagree', 'disagree', 'neutral', 'agree', 'strongly agree') to attitudinal questions such as, 'I find statistics boring?'. It should be noted that even when efforts are made to make the scale have similar intervals, it does not make sense to perform mathematical operations on them such as 'disagree' + neutral = 'agree' etc.

The other major subdivision of data classified by statisticians is that dividing Quantitative data. Quantitative data, also called *numerical data* can be either discrete (for example, the number of children in a family) or continuous (for example, the height of an experimental subject in cm.).

Discrete data can only have values that are separated by impossible values, e.g. you cannot have half a child. Another example is shoe size, or for that matter most readymade clothes, sizes.

Continuous data can take any value within a range, e.g. While a particular height might be 217 cm. it could easily be 217.34 or even 2.17.345635 depending upon the accuracy of the measuring instrument.

While people are often told that discrete data consists of whole numbers (integers) this is not always the case. Take for example, the case of the number of questions answered correctly in a test on spelling. This will be represented by 'whole numbers', however you can represent the same data as a 'proportion of correct answers', for it is clear that 21 correct answers out of 30 is a discrete value that may nevertheless be represented as 0.7 when it is the proportion of correct answers.

While statisticians spend a great deal of time differentiating between discrete and continuous data, for our purposes continuous data is often treated as discrete data and it does not really cause too many problems.

In contrast to the above classification, social scientists classify data in a slightly different way which will be discussed next.

Social Scientists' classification

While social scientists use the Nominal and Ordinal classification they prefer to adopt another type of classification for the terms discrete and continuous. The third level of measurement they describe is one which possesses those characteristics described above for ordinal data, but in addition has equal sized intervals e.g: Fahrenheit, Celsius, available bank balance, are examples of interval level data.

The next most complex type of scale involves all of the above characteristics and in addition possesses an absolute zero point e.g. height, weight, distance, Kelvins. This, ratio measurement type, is the most complex of the measurement scale types.

Measurement level	**Examples**	**Name**	**Order**	**Equal Intervals**	**Absolute zero**
Nominal (simplest)	Star sign. Hair colour. Alive/dead. Male/female	X			
Ordinal	Fitness rating scale. How happy are you on a scale of 1 to 10?	X	X		
Interval	Fahrenheit	X	X	X	
Ratio (most complex)	Weight	X	X	X	X

Do not worry too much if you find it difficult to differentiate between interval and ratio data as this does not matter much. In contrast, the important thing is to be able to differentiate between Nominal, Ordinal, and (Interval/Ratio) level data. Since Stevens 1951, suggested this classification, several authors have criticised the degree of prominence it has achieved concerning which types of statistics are considered appropriate for each data type (Gaito, 1980).

The chart opposite provides the above information in summary form. An easy way to remember the levels of measurement is the word 'Noir'.

Normal distribution

See glossary entry **standard deviation**.

Numeric

The numeric data type in R is the most important type, the other being the **Factor**. The numeric data type in R allows the representation of Ordinal / Interval / Ratio Levels of measurement in R. While a Numeric variable can be "logical" (e.g. true/false=0/1), "integer" (e.g. 4) or "double" (e.g. 4.35) they are all stored the same way in R.

Odds

$$odds_{for\_event} = \frac{events}{none\_events}$$

$$odds_{against\_event} = \frac{none\_events}{events}$$

$$probability = \frac{events}{all\_outcomes}$$

An odd is rather a strange concept unless one is a gambler and even then, it is used in a slightly different way there compared to its use in statistics.

Odds relate to **binary variables**
= variables that can take only two values e.g. alive/dead etc.

Odds = proportion of events to none events for a binary variable

So an odd is just another proportion (i.e. ratio) but importantly it is not the same as a probability. The equations opposite illustrate this.

For the equations I have divided up the outcomes for the binary variable as none-events to events which I could easily have called event A and event B. An odds takes either the events or none events as the denominator depending on if you are talking about those *for* or *against* the event. This is in contrast to probabilities where the denominator is the total number of all outcomes.

Imagine you try out a new treatment on ten patients of which nine show no improvement and eventually die but one miraculously completely recovers. We say that the probability of death is therefore 9/10=.9 and probability of surviving is 1/10=.1 the important thing here is to note that the **denominator in both of these probabilities is the total number of outcomes.**

	probability	Odds for	Odds against
death (high)	9/10 = .9	9/1 = 9	1/9 =.1111
surviving (low)	1/10 = .1	1/9 = .1111	9/1 = 9

Low probability =
low odds in favour of = high odds against

High probability =
high odds in favour of = low odds against

In contrast the odds for (i.e. in favour of) death, is expressed as the deaths to non deaths ratio; deaths : none_deaths; 9:1 = 9/1 = 9. The odds in favour of death is 9 to one. The odds against death (i.e. odds in favour of surviving) is similarly, none_deaths to deaths 1:9 = 1/9 = 0.1111. **Notice here that the odds for and odds against are just the same proportion inverted neither having the total number of events in the denominator**. The table opposite summarises the discussion so far.

Notice that the probabilities add up to one, while the odds do not. However, the product does add up to one (i.e. .9 x .1111 = 1), also the odds for and against the event are related.

event A(odds for) = event B(odds against)
event B(odds for) = event A (odds against).

The way odds are used in gambling and horse riding creates confusion as those quote the odds **against** winning. A bet of twenty to one means the odds of losing is 20/1 = 20 and therefore the odds against losing (i.e. winning) is 1/20 = 0.05. In other words once in every twenty runs. In contrast, in medicine the odds for an event are usually quoted and the 'to one' is dropped (Campbell & Swinscow, 2009 p.25).

An example

A miraculous cure works for 1 in ten patients, therefore the odds for (i.e. in favour of) death is "9 to 1" = 9 : 1 = 9/1 = 9 statisticians would just say an odds of '9'.
Unfortunately just because it is not mentioned does not mean that it should be forgotten.

```
> probs <- seq(0,1, 0.05)
> odds <- round(probs/(1-probs) , 3)
> table <- data.frame(probs, odds)
> table
   probs   odds
1   0.00  0.000
2   0.05  0.053
3   0.10  0.111
4   0.15  0.176
5   0.20  0.250
6   0.25  0.333
7   0.30  0.429
8   0.35  0.538
9   0.40  0.667
10  0.45  0.818
11  0.50  1.000
12  0.55  1.222
13  0.60  1.500
14  0.65  1.857
15  0.70  2.333
16  0.75  3.000
17  0.80  4.000
18  0.85  5.667
19  0.90  9.000
20  0.95 19.000
21  1.00    Inf
```

Because in medical statistics odds are converted to the 'to one' standard we need to think about how we can consider the odds in favour of surviving which at present is *none_death* to *deaths* = 1 to nine = 1:9 = 1/9 = 0.1111. To convert this to the 'to one' standard we just divide each by the denominator (i.e. 9). Producing the ratio of *none_deaths* to *deaths* 1/9 :9/9 = a odd of one ninth to 1 which produces the same result as before = (1/9)/1 =0.1111.

While odds are not the same as probabilities there is a relationship between them given in the table and graphs below. The shaded row is the important one. The R code opposite produces a table for twenty values.

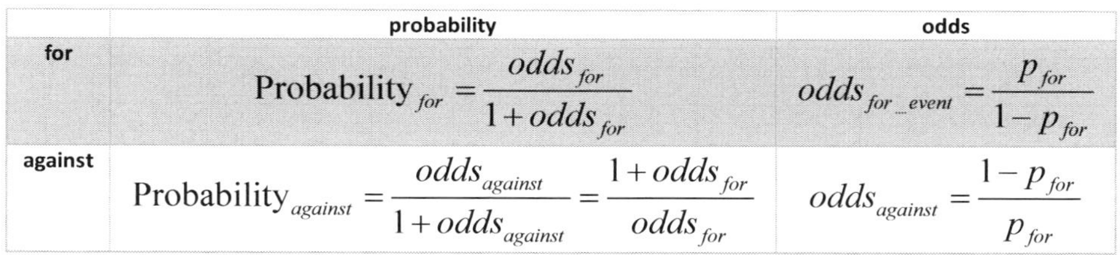

	probability	odds
for	$\text{Probability}_{for} = \dfrac{odds_{for}}{1+odds_{for}}$	$odds_{for\_event} = \dfrac{p_{for}}{1-p_{for}}$
against	$\text{Probability}_{against} = \dfrac{odds_{against}}{1+odds_{against}} = \dfrac{1+odds_{for}}{odds_{for}}$	$odds_{against} = \dfrac{1-p_{for}}{p_{for}}$

The equation of odds for an event is simply a swapped over version of the equation against the event, technically we say it is the **reciprocal**. Using the above equations it is possible to plot the relationship between odds (in favour of, or against an event) against probability.

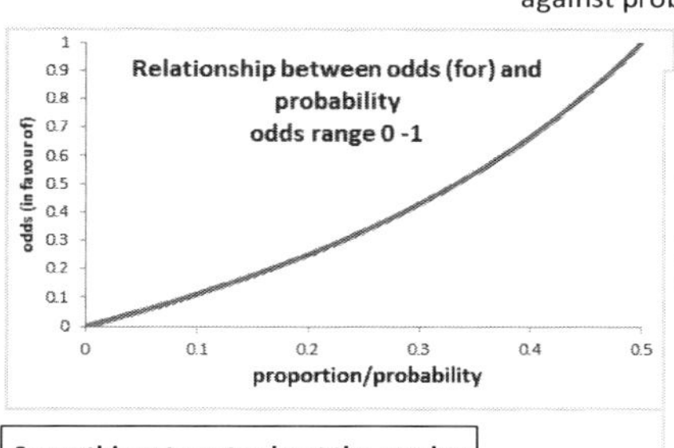

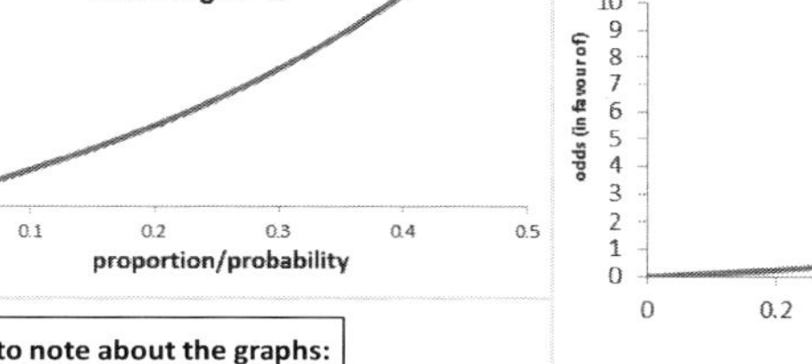

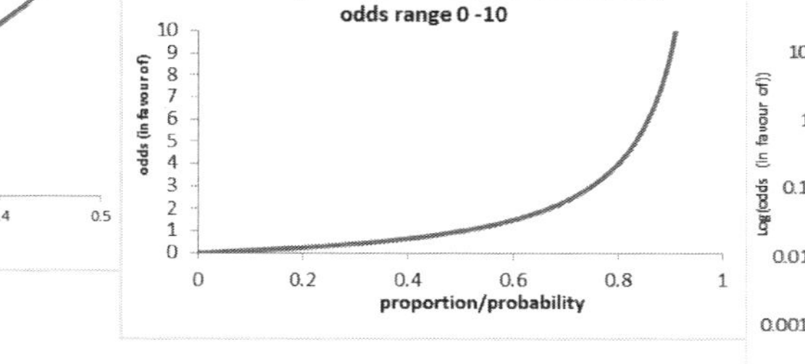

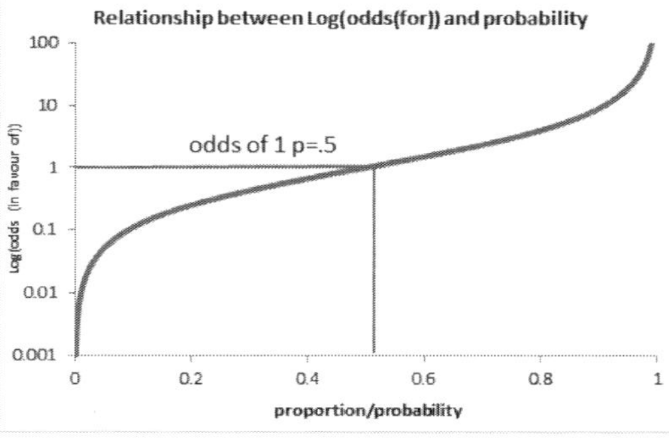

Some things to note about the graphs:

- Odds (in favour of) values range from zero to > 10 (in fact there is no upper limit)
- The probability varies between zero and 1
- As the probably gets smaller the odds (in favour of) also get smaller
- When the odds (in favour of) = 1 the probability = .5

Odds are very important in statistics for several reasons, two of which are; they do not have an upper value of 1 unlike probability. When you take the logarithm of the odd, a curve is produced which has a straight line for the middle probability values, a useful characteristic if you are trying to predict a binary outcome from one or more predictors, discussed in the logistic regression chapter.

Odds ratio

See also **Risk** and **Relative risk**. The odds ratio (**OR**) is the ratio of the odds (O) for the treatment group compared to that of the placebo group.

	Event	No event	
Placebo	a	b	a+b
Treatment	c	d	c+d

$OR=(P_{treatment}/(1-P_{treatment}))/(P_{placebo}/(1-P_{placebo}))$ = $O_{placebo}/O_{treatment}$ = (a/b) x (d/c) = (a/b) /(c/d) = ad/bc

Never smoked group	Heart attack	No heart attack	
Placebo	96	5392	5488
Aspirin	55	5376	5431

For the table opposite the odds are $O_{placebo}$ = 0.0178 $O_{treatment}$ = 0.0102 so the odds ratio is simply 0.0178/0.0102= 1.745098 (placebo to treatment) or 0.0102/0.0178= 0.5730337 (treatment to placebo) this latter value is very slightly less than the Relative Risk of 0.5789297. However, if the baseline (i.e. placebo) had not been so low there would have been a difference. Campbell & Swinscow, 2009 p.28, provide a nice table demonstrating how as the baseline risk increases the RR stays the same but the OR increases to reflect the change. See Sistrom & Garvan, 2004 for a more detailed discussion along with a chart indicating the inflation factor.

It is interesting to note that the shortcut formula ad/bc gives: (96 x 5376)/(5392 x 55) = 1.740275, is the placebo to treatment ratio whereas bc/ad gives .5730337, the treatment to placebo odds ratio.

There is a simple relationship between the two odds ratio values in the previous paragraph 1/1.745098= 0.5730337 this is not the case with relative risks. **The odds ratios have a reciprocal relationship, just like the odds for/against.**

The odds of having a heart attack in the placebo group is 1.7402 times it is in the aspirin group. Conversely the odds of having a heart attack in the aspirin group is .573 times that compared to the placebo group, that is the odds of having a heart attack in the aspirin group is around half that of the control group. Using software it is possible to obtain a confidence interval, details in the *Risk and Odds ratios* chapter.

The estimated parameter value for the null distribution of an odds ratio is 1 (both groups equivalent). As an approximate rule of thumb, if an alpha % confidence interval (CI) (100 - alpha) does not include 1 the result will also be statistically significant at the alpha level. For example if we have a 95% CI and it does not contain 1 within its range the p-value will be less than 0.05.

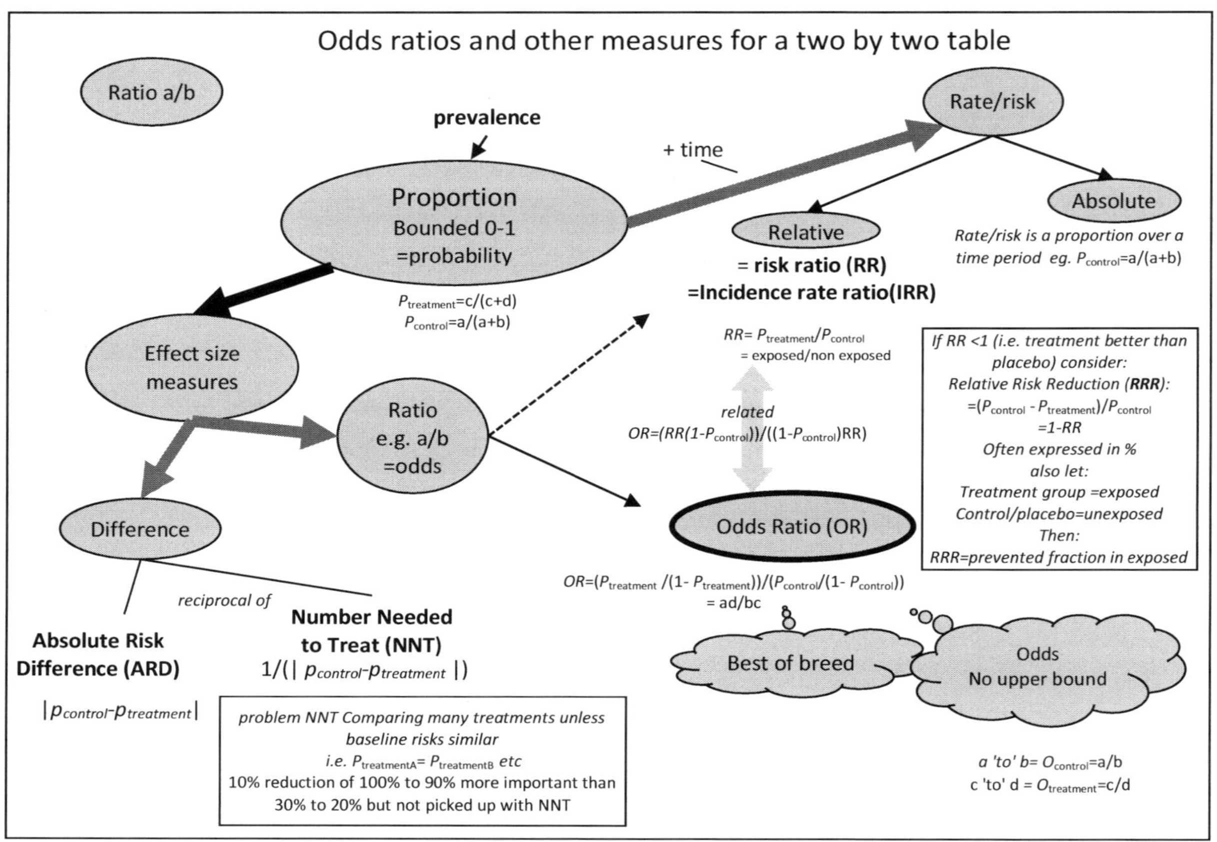

To answer the question whether you should report Relative risk or odds ratios see the section with that title in the *Risk and Odds ratios* chapter. The above mindmap illustrates the links between some of the two by two table measures.

P-value

1/p-value
=
statistical significance
=
effect size
x
sample size

Within this book there are many examples and explanations of p-values. Specifically each chapter within the samples & populations section provides an explanation of the p-value obtained for each model. The p-value can be thought of as a measure linking two aspects of the data (Rosnow & Rosenthal, 2003; Rosnow & Rosenthal 2009), effect size and sample size shown opposite. Therefore, with experiments with very large sample sizes even a miniscule (i.e. trivial) effect size will produce a statistically significant p-value. Thus, virtually anything can become statistically significant with a large enough sample size. Conversely, with a small sample size only an experiment with a very large effect size would produce a statistically significant result. This is a complex topic that requires understanding of probability, and specifically conditional probability. I would recommend that you start by watching the four youtube videos I have produced on the subject at: http://youtu.be/61UVCwe4z88

The *TeachingDemos* package contains a function that allows you to see the distribution of p-values when the effect size is zero (the 'null' hypothesis) or set to a specific value. An example is provided at the end of the *R and the Raspberry pi* online chapter.

Plot characters pch (R code)

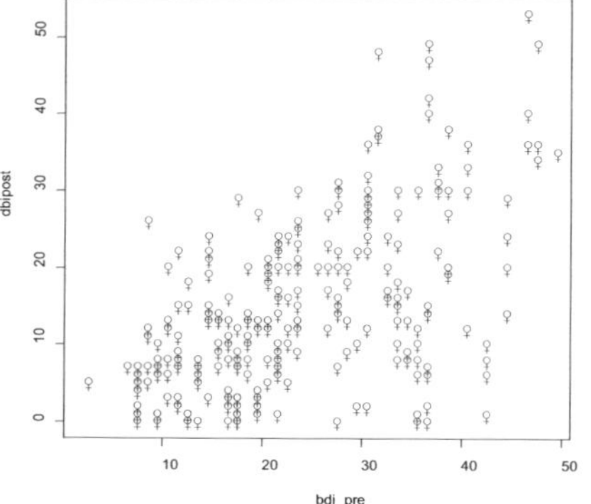

Plotting **ch**aracters (*pch*) provide a method of changing the marker on a graph for either the entire dataset or a subset. This is achieved by using the R code *pch*=number, for example *pch*=19 produces a red filled circle.

You can also request the Venus and Mars symbols to indicate males and females, as demonstrated in the plot on the left.

```
with (bb_long, plot(bdi_pre, dbipost, pch = -0x2642L)) # males

with (bb_long, plot(bdi_pre, dbipost, pch = -0x2640)) # females
```

Plot symbols 21 to 25 are the blue filled equivalents to symbols 15 to 19 (red filled).

Plot symbols in R;
col = " red3 ", bg= " slateblue3 "

0 1 2 3 4 5 6 7 8 9 10 11 12 13 14 15 16 17 18 19 20 21 22 23 24 25

Plot symbol colour col (R code)

Colour of the plot symbol (col) can be specified using *col=name* or number. *col="green" and col="red"* are typical examples.

Typing into the R Console window *colors()* produces a list of them. Alternatively, there is a nice online pdf listing the name and number for each colour at: http://research.stowers-institute.org/efg/R/Color/Chart/ColorChart.pdf

R commands

A group of R commands and functions form a block of text which R understands, called *R code*. R code is typed into either the R Console window followed by pressing the **Enter** key (↵) to run it, or alternatively it can be typed into the *Script* window of R Commander, which after highlighting it and clicking on the **Submit** button, runs it.

Reference group

A reference group is a subset of cases that provide a comparison with other groups. It is necessary when you calculate odds ratios (i.e. the reference group is given an odds ratio of 1) or in Cox Proportional Hazard (PH) regression when you create hazard ratios. In Cox PH you might have variables such as gender or treatment groups (i.e. here the reference group is given a hazard ratio of 1). Setting the reference group to the control or 'home' group makes interpreting results easier. See also the **factor** and **contrasts** entries and the *Multiple linear regression* chapter.

Relative Risk, IRR and RRR

See also **Risk**.

The relative risk is a ratio, where a ratio is one value divided by another. It is usually obtained by dividing the treatment risk by the control risk, which for the above table is $P_{treatment}/P_{control}$ 0.01012705/.01749271 = 0.5789297. The Relative Risk (**RR**) is also called the Incidence Rate Ratio (**IRR**). Most texts present the numerator as the treatment (i.e. exposed) group but one (Howells, 2007 p.154) uses it as the denominator.

Never smoked group	Heart attack	No heart attack	
Placebo	96	5392	5488
Aspirin	55	5376	5431

To calculate a valid Relative Risk you usually need some estimate of the population at risk, if your control group is either biased or too small the relative risk will be questionable.

If the relative risk is less than 1, that is when the treatment is better than the placebo it is sensible to consider another measure called the Relative Risk Reduction (**RRR**): **RRR** =$(P_{control} - P_{treatment})/P_{control}$ =1-RR

For the data in the above table the RRR is 1-0.5789297 = 0.4210703 and multiplying this value by 100 to express it as a percentage gives 42.107 telling us that there is 42% reduction in deaths in the treatment group for the period "from the late 1980's to the early 1990's". This is a reduction from just under 2% to 1% and one would need to consider the baseline risk when discussing this value. For example if it were some deadly disease such as heart attacks probably it is worthwhile but if it were a simple non life threatening condition (i.e. acne) would the intervention with this RRR be worthwhile? **Both the RR and RRR do not take into account the baseline risk** which in contrast the **odds ratio** does.

Risk, ARD, ARR and NNT

Never smoked group	Heart attack	No heart attack	
Placebo	96	5392	5488
Aspirin	55	5376	5431

	Event	No event	
control	a	b	a+b
Treatment	c	d	c+d

$AR_{treat} = p_{treat} = c/(c+d) = 55/5431 = 0.01012$

Shown opposite is the data from the Physicians' Health Study Data for the 10,919 physicians who had never smoked.

The proportion of those who received the placebo who had a heart attack is 96/5488 = 0.01749271, just under 2 percent in contrast to those who received aspirin of 55/5431=0.01012705 or 1 percent. We could express this algebraically; $P_{treatment}$=c/(c+d), $P_{control}$=a/(a+b). These proportions are also known as the **absolute risks (AR)**, strictly a risk is related to a *time period* and it is always a good idea to make sure you say what the period is. We know this data was collected "from the late 1980's to the early 1990's" possibly not the best time interval! It would be sensible to go back to the original paper and specify it at least in years. Note that the risk measure used both categories as the denominator so it is therefore a probability. This is in contrast to the **odds.**

The **odds** are worked out by just using the no event category, instead of the total, so we have for the placebo group $O_{placebo}$ = 96/5392= 0.0178. For the treatment group $O_{treatment}$ = 55/5376= 0.0102, algebraically this simply is a/b and c/d, so if b and d are large in comparison to the a and c we would expect the proportions and odds to be pretty similar as they are here. In other words, situations with rare diseases (= low prevalence rates) produce small absolute risks with similar values to the odds of the disease. (Agresti & Finlay, 2009).

The Absolute Risk Difference (**ARD**) is the risk difference between two groups, ignoring the possibility of a negative value (i.e. taking the absolute value),

$|p_1 - p_2| = |p_{placebo} - p_{treatment}|$, so for our control and treatment groups we have

| .01749271 - 0.01012705 | =.00736566 what does this tell us? It presents the reduction in risk in the treatment group. Considering this as a percentage, for every 100 patients treated .736566 of a patient did not suffer a heart attack compared to the placebo group. Here, where the treatment group has reduced the risk it is also called the **ARR Absolute Risk Reduction**.

Dividing 100 by this percentage value gives us the number needed to treat to prevent a single death, 100/.736566 =135.7652 so we would need to treat 136 patients to prevent a single death. This value is called the Number Needed to Treat (**NNT**).

Mathematically the NNT is: $1/(|p_1\text{-}p_2|)$ where p_1 = placebo group and p_2=treatment group, the same as the ARD but over 1 this is called the reciprocal of the ARD. Using this equation for our data gives, 1/.00736566 = 135.7652 in other words we need to treat 136 patients to save one life. The seemingly immediate attractiveness of the NNT means that it has gained rapid widespread popularity, however there are dangers associated with it. For specific detail of how to calculate the NTT see the *Number needed to treat/harm* chapter. For related concepts see **Relative Risk** and **Odds Ratio**.

Rprofile.site file

This text file is installed as part of R and provides information about default values to be used by R when it starts up. Editing it makes is possible to set a default paper size and working directory. To change the working directory permanently add the line, *setwd("C:/directory")* along with a carriage return at the end of the *Rprofile.site* file, substituting *C:/directory* for your desired directory.

Standard deviation

The standard deviation (SD) is a measure of spread that is valid for values that follow a normal distribution (the bell shape curve shown below). Approximately 68% of the scores are one standard deviation each side of the mean and 95% are within two standard deviations each side of it, which is four standard deviations in total. The larger the standard deviation the more spread out the scores are.

Mathematically the standard deviation, or more correctly the sample standard deviation, is the square root of the sample **variance**. A square root is the opposite ('inverse') to the square operator (e.g. the square root of 4 is 2 and the square of 2 is 4). Using it gets us back to the original units. So a variance of 21.43 years squared would produce a standard deviation of √21.4359 = 4.6299 years.

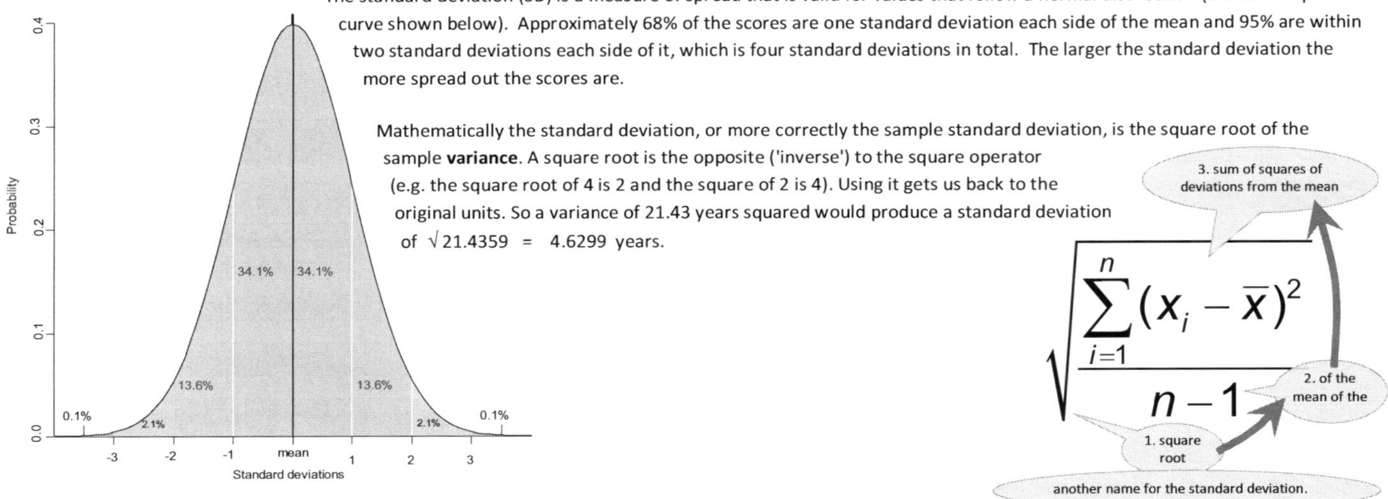

$$\sqrt{\frac{\sum_{i=1}^{n}(x_i-\bar{x})^2}{n-1}}$$

= Root Mean Square Deviation

Tab character

The tab character is often used to separate columns in data files. Such files are called *tab* delimited. You do not usually see anything when you insert a tab character, however, using a test editor such as the free *Notepad++* application displays them as orange arrows.

```
anxiety→finalexam
28→82
41→58
35→63
```

tab character

Variance

The variance is a measure of spread equal to the squared **standard deviation.** While it is an essential measure providing the basis of much statistical analysis, it is less easy to interpret as it represents squared values. Try explaining to someone what weight or height squared is?

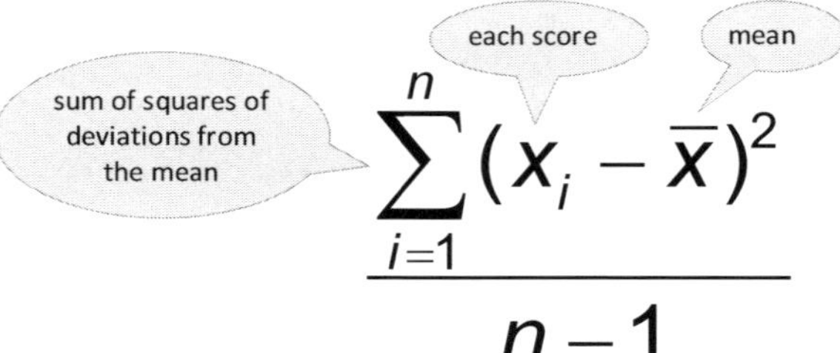

$$\frac{\sum_{i=1}^{n}(x_i-\bar{x})^2}{n-1}$$

Z value

The z-score of an observation is obtained by subtracting the sample mean value ($\bar{x}$) from it and then dividing by the standard deviation. Therefore a z-score of 1 indicates the observation is one standard deviation from the mean for that sample. Many tests produce z-scores allowing comparison with other peoples results, assuming the distribution is normal. It is them possible to calculate how typical or atypical a value is, see glossary entry **standard deviation**. Z-scores are also called standardised scores, z-values or normal scores.

References

Adelheid AMN Pexman PM 2010 (6th ed) Displaying Your Findings: A Practical Guide for Creating Figures, Posters, and Presentations. American Psychological Association (APA).

Adler J 2009 R in a Nutshell: A Desktop Quick Reference. O'Reilly Media.

Agresti A 2002 (2nd Ed.) Categorical Data Analysis. Wiley

Agresti A, Finlay B, 2008 (4th Ed.) Statistical Methods for the Social Sciences. Prentice Hall.

Altman DG 1998 Confidence intervals for the number needed to treat. BMJ, 317, 1309-1312

Altman DG, Andersen PK 1999 Calculating the number needed to treat for trials where the outcome is time to an event. BMJ, 319, 1492-1495

Amphanthong, 2012 The Comparison of Outlier Detection in Multiple Linear Regression Proceedings of the World Congress on Engineering. Vol.I WCE. July 4 - 6, 2012, London, U.K. Open access available from: http://www.iaeng.org/publication/WCE2012/WCE2012_pp336-341.pdf

Andrews J, Guyatt G, Oxman AD, Alderson P, Dahm P, Falck-Ytter Y, Nasser M, Meerpohl J, Post PN, Kunz R, Brozek J, Vist G, Rind D, Akl EA, Schünemann HJ. 2013 GRADE guidelines: 15. Going from evidence to recommendations: the significance and presentation of recommendations. J Clin Epidemiol. (Jan 9) 726 - 735 doi:10.1016/j.jclinepi.2012.03.013. Open access available from: http://www.gradeworkinggroup.org/publications/JCE_series.htm

Armitage P, Berry G, Matthews JNS 2002 (4th ed.) Statistical methods in Medical research. Blackwell.

Bagozzi RP, Yi Y 2012 Specification, evaluation, and interpretation of structural equation models. Journal of the Academy of Marketing Science [January 2012], 40(1), 8-34

Baguley T 2009 Standardized or simple effect size: What should be reported? British Journal of Psychology, 100, 603-617.

Bakeman, R 2005 Recommended effect size statistics for repeated measures designs. Behavior Research Methods, 37 (3), 379-384.

Bates R 2006 lme4: Mixed-effects modelling with R. Springer

Bates R Machler Bolker, date? Fitting linear mixed-effects models using lme4, Journal of Statistical Software.

Bentler PM, Chou CP 1987 Practical issues in structural equation modeling. Sociological Methods & Research, 16, 78–117.

Bickel R 2007 Multilevel Analysis for Applied Research: It's Just Regression! The Guilford Press.

Bland JM 1995 (2nd ed) An introduction to Medical Statistics. Oxford University Press.

Bland JM, Altman DG 1999 Measuring agreement in method comparison studies. Statistical Methods in Medical Research, 8, 136-160.

Bollen KA, Pearl J 2012 Eight myths about causality and structural equation models. Technical report r-393 April 2012 [also published in: In S.L. Morgan (ed.), Handbook of Causal Analysis for Social Research, Chapter 15, 301-328, Springer 2013] available from: ftp://ftp.cs.ucla.edu/pub/stat_ser/r393.pdf

Bollen KA, Curran PJ 2006 Latent Curve Models: A Structural Equation Perspective. Wiley.

Bollen KA 1989 Structural Equations with Latent Variables. Wiley.

Borenstein M, Hedges LV, Higgins JPT, Rothstein HR 2009 Introduction to Meta-analysis. John Wiley & Sons.

Bradford Hill A 1965 The Environment and Disease: Association or Causation? Proceedings of the Royal Society of Medicine, 58, 295-300. Available from http://www.ncbi.nlm.nih.gov/pmc/articles/PMC1898525/pdf/procrsmed00196-0010.pdf

Braitman H 1991 Confidence intervals: Assess both clinical significance and statistical significance. Annals of internal medicine 114 (6) 515 - 517

Breslow NE, Day NE 1980 Statistical Methods in Cancer Research. Volume I - The analysis of case-control studies. IARC press for WHO. http://www.iarc.fr/en/publications/pdfs-online/stat/index.php

Bretz F, Genz A, Hothorn LA 2001 On the numerical availability of multiple comparison procedures. Biometrical Journal, 43(5), 645--656.

Bretz F, Hothorn T, Westfall P 2010 Multiple Comparisons Using R, CRC Press, Boca Raton.

Brooks GP, Barcikowski RS 2012 PEAR Method for Sample Size Multiple Linear Regression. Viewpoints, regression, consider effect size to be an essential part of the calculation. Multiple Linear Regression Viewpoints, 38(2). http://mlrv.ua.edu/2012/vol38_2/Brooks-Barcikowski-38_2_proof_1-16.pdf

Brown H, Prescott R 2006 (2nd ed.) Applied Mixed models in Medicine. Wiley.

Cameron CA, Trivedi PK 2013 Regression Analysis of Count Data, 2nd edition, chapter 15 Essentials of Count Data Regression published online.

Campbell MJ, Machin D 1993 (2nd ed.) Medical statistics: A common sense approach. John Wiley & sons.

Campbell MJ, Swinscow TDV 2009 (11nd ed.) Statistics at Square One. BMJ books.

Campbell MJ 2006 (2nd ed.) Statistics at Square Two. BMJ books.

Canty AJ 2002 Resampling Methods in R: The boot Package. R News 2(3):2-7
URL http://CRAN.R-project.org/doc/Rnews/

Chan YH 2004a Biostatistics 202: Logistic regression analysis [using spss]. Basic statistics for Doctors. Singapore Med J 45(4) 149. Open access: http://www.sma.org.sg/smj/4504/4504bs1.pdf

Chan YH 2004b Biostatistics 201: Linear regression analysis [using spss]. Basic statistics for Doctors. Singapore Med J 45(2) 55. Open access: www.sma.org.sg/smj/4502/4502bs1.pdf

Chan YH 2005 Biostatistics 308: Structural equation modelling [using spss]. Basic statistics for Doctors. Singapore Med J 46(12) 675. Open access: www.sma.org.sg/smj/4612/4612bs1.pdf Lists are 18 of Chan's articles.

Chang W. 2013 R Graphics Cookbook. O'Reilly Media

Cohen J, Cohen P, West SG, Aiken LS 2003 (3rd ed.) Applied multiple regression/correlation analysis for the behavioral sciences. Hillsdale, NJ: Lawrence Erlbaum Associates.

Cohen J. 1988 (2nd ed.) Statistical power analysis for the behavioral sciences. Hillsdale, NJ: Lawrence Erlbaum Associates.

Collett D 1994 Modelling survival data in medical research. Chapman & Hall/CRC. Second edition, 2003 omits chapter on software packages.

Crawley M J 2005 Statistics: An introduction using R. Wiley

Crowder MJ, Hand DJ. 1990 Analysis of repeated measures. London: Chapman and Hall.

Cumming G. 2012 Understanding the New Statistics: Effect Sizes, Confidence Intervals, and Meta-Analysis. Routledge.

Cuthill I. 2013 R scripts for effect size calculation. Webpage available from http://www.bristol.ac.uk/biology/research/staff/cuthill.i/ See also Nakagawa & Cuthill 2007.

Daniel WW. 1991 Biostatistics: A foundation for analysis in the health sciences. Wiley.

Diggle P J Hegaerty P Liang K-Y, Zeger S. 2002 (2nd ed.) Analysis of Longitudinal Data. Oxford Science publications.

Doncaster CP, Davey AJH. 2007 Analysis of Variance and Covariance: How to Choose and Construct Models for the Life Sciences. Cambridge: Cambridge University Press. Companion website: http://www.southampton.ac.uk/~cpd/anovas/

Duncan TE, Duncan S C, Strycker L A. 2006 (2nd ed.) An Introduction to Latent Variable Growth Curve Modeling: Concepts, Issues, and Applications: Lawrence Erlbaum.

Eiseman B, Silen W, Bascom GS, Kauvar AJ. 1958 Fecal enema as an adjunct in the treatment of pseudomembranous enterocolitis. Surgery. (Nov) 44(5):854-9.

Erdfelder FF, Buchner E, Lang EA 2009 Statistical power analyses using GPower 3.1: Tests for correlation and regression analyses. Behavior Research Methods, 41, 1149-1160.

Everitt BS, Dunn G. 2001 (2nd ed) Applied multivariate data analysis. Arnold. London. Datasets available at: http://mnstats.morris.umn.edu/multivariatestatistics/data.html

Everitt BS, Hothorn T 2006 (1st ed) A handbook of statistical analyses using R. Chapman & Hall.

Everitt BS, Hothorn T 2010 (2st ed) A handbook of statistical analyses using R. Chapman & Hall.

Faul F, Erdfelder E, Lang AG, Buchner A 2007 G* Power 3: A flexible statistical power analysis program for the social, behavioral, and biomedical sciences. Behavior research methods, 39(2), 175-191

Field A 2009 Discovering Statistics Using SPSS. Sage

Field A 2013 Discovering Statistics Using R. Sage

Fisher RA 1915 Frequency Distribution of the Values of the Correlation Coefficient in Samples from an Indefinitely Large Population. Biometrika, Vol. 10, No. 4 (May, 1915), pp. 507-521

Fisher RA 1935 *The Design of Experiments*, New York: Hafner.

Fitzmaurice GM, Laird NM, Ware JM. 2004 Applied Longitudinal Analysis. Wiley

Fleiss JL, Levin B, Paik MC 2003 (3rd ed) Statistical Methods for Rates and Proportions. New York: Wiley.

Fox J, Weisberg S 2011 (2nd ed) An R Companion to Applied Regression. Sage.

Fox J 2006 Structural Equation Modeling With the sem Package in R." Structural Equation Modeling, 13:465-486

Friendly M, Kwan E. 2009 Where's Waldo: Visualizing Collinearity Diagnostics, The American Statistician, 63(1), 56-65.

Frobell RB, Roos HP, Roos EM, Roemer FW, Ranstam J, Lohmander LS 2013 Treatment for acute anterior cruciate ligament tear: five year outcome of randomised trial. BMJ 346:f232

Gaito J. 1980 Measurement Scales and Statistics: Resurgence of an Old Misconception. Psychological Bulletin. Vol 87(7), 564-567

Gardner M J, Altman DG 1989 Statistics with Confidence—confidence intervals and statistical guidelines. British Medical Journal.

Genizi A 1993 Decomposition of R^2 in multiple regression with correlated regressors. Statistica Sinica 3, 407–420. http://www3.stat.sinica.edu.tw/statistica/password.asp?vol=3&num=2&art=10

Glasziou P 2001 Which methods for bedside Bayes? Evid Based Med. 6:6 164-166. Open access: http://ebm.bmj.com/content/6/6/164.full.pdf+html

Goldstein H 2011 (4th ed) Multilevel statistical models. Wiley.

Gringras P, Gamble C, Jones AP, Wiggs L, Williamson PR, Sutcliffe A 2012 Melatonin for sleep problems in children with neurodevelopmental disorders: randomised double masked placebo controlled trial. BMJ 345:e6664

Grimm LG., Yarnold PR. 1995 Reading and understanding multivariate statistics. American Psychological Association.

Groemping U 2006 Relative Importance for Linear Regression in R: The Package relaimpo Journal of Statistical Software 17, Issue 1. http://www.jstatsoft.org/v17/i01

Guyatt GH, Oxman AD, Ali M, Willan A, McIlroy W, Patterson C 1992 Laboratory diagnosis of iron-deficiency anemia: an overview. Journal of General Internal Medicine. [Mar-Apr], 7(2) 145-53. Erratum in 1992 [Jul-Aug] 7(4) 423.

Halekoh U, Højsgaard S, Yan J. 2006 The package geepack for generalized estimating equations. Journal of Statistical Software. 15, 2

Hancock GR, Freeman MJ 2001 Power and sample size for the Root Mean square Error of Approximation of not close fit in structural equation modeling. Educational and Psychological Measurement, 61, 741–758.

Hand D, Crowder M 1996 Practical Longitudinal data analysis. Chapman & hall

Harkins S 2013 Five free Microsoft Project alternatives. Five Apps [blog], July 31, 1:29 PM PST.

Hart A 2001 Mann – Whitney test is not just a test for medians. BMJ [aug 18] 323 (7309) 391-393

Hayes AF, Krippendorff K 2007 Answering the call for a standard reliability measure for coding data. Communication Methods and Measures 1: 77-89. Available from: http://www.unc.edu/courses/2007fall/jomc/801/001/HayesAndKrippendorff.pdf

Hayes RJ, Moulton LH 2009 Cluster Randomised trials. Chapman Hall/CRC

Heck RH, Thomas SL & Tabata LN 2010 Multilevel and Longitudinal modelling with IBM SPSS. Routledge.

Hedeker D, Gibbons RD 2006 Longitudinal Data Analysis. Wiley. Website: tigger.uic.edu/~hedeker/ml.html

Henrich J, Heine S J, Norenzayan A 2010 The weirdest people in the world? Behav Brain Sci. (Jun) 33(2-3), 61-83; discussion 83-135. doi: 10.1017/S0140525X0999152X.

Higgins J, Thompson SG 2002 Quantifying heterogeneity in a meta-analysis. Statistics in medicine. 21 (11), 1539-1558.

Higgins J, Thompson SG, Deeks JJ, Altman DG 2003. Measuring inconsistency in meta-analyses. BMJ, *327* (7414), 557-560.

Hilbe JM 2011 (2nd ed.) Negative Binomial Regression. Cambridge University Press.

Hilbe JM 2014 Modeling count data. Cambridge University Press.

Hin, L.-Y. and Wang, Y-G. (2009). *Working-correlation-structure identification in generalized estimating equations*, Statistics in Medicine 28: 642-658.

Hindmarsh PC, Brook CDG 1987 Effect of growth Hormone on short normal children. British Medical Journal. 295 573 - 577

Hosmer DW, Hosmer T, le Cessie S, Lemeshow S 1997 A comparison of goodness-of-fit tests for the logistic regression model. Statistics in Medicine, 16, 965-980.

Hosmer DW, Lemeshow S 2000 (2nd ed.) Applied Logistic Regression. New York: Wiley.

Howell RD, Breivik E, Wilcox JB 2007 Is formative measurement really measurement? Reply to Bollen (2007) and Bagozzi (2007). Psychological methods, 12(2), 238-245.

Howell DC 2006 Statistical Methods for Psychology (Ise). Wadsworth.

Hox J 2010 (2nd ed) Multilevel analysis: Techniques and applications. Lawrence Erlbaum

Ip EHS 2001 Visualising Multiple regression. Journal of Statistics education. 9(1).

Jaeschke R, Guyatt G H, Dellinger P, Schünemann H, Levy M M, Kunz R, Norris S, Bion J 2008 GRADE working group. Use of GRADE grid to reach decisions on clinical practice guidelines when consensus is elusive. BMJ (Jul 31) 337:a744

Joreskog K G; Van Thillo, M. 1972 LISREL: A General Computer Program for Estimating a Linear Structural Equation System Involving Multiple Indicators of Unmeasured Variables. Educational Testing Service, Princeton, NJ. (RB-72-56). Available from: http://eric.ed.gov/?id=ED073122

Kabacoff R 2011 R in Action: Data Analysis and Graphics with R. Manning Publications Co. Shelter Island, NY 11964

Karp NA 2014 R Commander an introduction. Available from http://cran.r-project.org/doc/contrib/Karp-Rcommander-intro2.pdf

Kennedy PE 2002 More on Venn diagrams for regression. Journal of Statistics education. 10(1).

Kline R 2011 (3nd ed.) Principles and practice of Structural Equation Modelling. The Guilford Press. This book and is more discursive in style than Schumacker et al 2010.

Krippendorff K 2004 Reliability in Content Analysis: Some common Misconceptions and Recommendations. Human Communication Research 30, 3: 411-433. http://repository.upenn.edu/asc_papers/242

Kvien T K, Heiberg T, Hagen T K B. 2007 Minimal clinically important improvement/difference (MCII/MCID) and patient acceptable symptom state (PASS): what do these concepts mean? Ann Rheum Dis. 66(Suppl III) iii40–iii41

Landau S, Everitt B S. 2004 A handbook of statistical analyses using SPSS. Chapman & Hall.

Lee T, Cai L, MacCallum RC. 2012 Power analysis for tests of structural equation models. In In R. Hoyle (Ed.), Handbook of structural equation modeling (p.181–194). New York: Guilford Press.

Leeuw J D, Meijer E. 2008 (eds.) Handbook of Multilevel Analysis. Springer.

Lillie E O, Patay B, Diamant J, Issell B, Topol EJ, Schork NJ. 2011 The n-of-1 clinical trial: the ultimate strategy for individualizing medicine? Per Med. (2011 March) 8(2), 161–173. doi: 10.2217/pme.11.7

Loehlin J C. 2004 (4th ed.) Latent Variable Models: An Introduction to Factor, Path, and Structural Analysis. Psychology Press.

Lüdtke R, Kunz B, Seeber N, Ring J 2001 Test-retest-reliability and validity of the Kinesiology muscle test. Complementary Therapies in Medicine, 9 141-145.

MacCallum RC, Browne MW, Sugawara HM 1996 Power analysis and determination of sample size for covariance structure modeling. Psychological Methods, 1, 130-149.

Machin D, Campbell MJ. 2005 Design of studies for Medial Research. Wiley.

Machin D, Campbell MJ, Walters SJ. 2007 (4th ed.) Medical Statistics: A textbook for the health sciences. Wiley.

Mahon J, Laupacis A, Donner A, Wood T 1996 Randomised study of n of 1 trials versus standard practice. BMJ (April 27) 312(7038): 1069–1074. PMCID: PMC2350863

Man-Son-Hing M, Laupacis A, O'Rouke K, Molnar FJ, Mahon J, Chan K BY, Wells G 2002 Determination of the clinical importance of study results. Journal of general internal medicine. 17 (6) [june] 469-476

Marasco J, Doerfler R, Roschier L 2011 Doc, What are My Chances? The UMAP journal 32.4 http://www.myreckonings.com/modernnomograms/Doc_What_Are_My_Chances_UMAP_32-4-2011.pdf

Marschner I C 2011 glm2: Fitting generalized linear models with convergence problems. The R Journal, Vol. 3(2), 12-15.

Matthews J N S, Altman DG, Campbell M J, Royston JP 1990 Analysis of serial measurements in medical research. BMJ 300 230 – 235

McKelvey E M, Gottlieb JA 1976 Hydroxyldaunomycin (adriamycin) combination chemotherapy in malignant lymphoma. Cancer 38, 1484-1493

Michaëlsson K, Melhus H, Warensjö Lemming E, Wolk A, Byberg L 2013 Long term calcium intake and rates of all cause and cardiovascular mortality: community based prospective longitudinal cohort study. BMJ (13 February) 346 doi: http://dx.doi.org/10.1136/bmj.f228

Miles J 2003 A framework for power analysis using a structural equation modelling procedure. BMC Medical Research Methodology. ISSN 1471-2288 http://dx.doi.org/10.1186/1471-2288-3-27

Miles J, Shevlin M 2001 Applying regression & correlation: A guide for students and researchers. Sage publications. London.

Mirman, D 2014 Growth Curve Analysis and Visualization Using R. Chapman and Hall / CRC.

Molnar F J, Man-Son-Hing M, Fergusson D 2009 Systematic Review of Measures of Clinical Significance Employed in Randomized Controlled Trials of Drugs for Dementia. JAGS 57 (3) 536-546

Murrell P, Ihaka R 2000 An approach to providing mathematical annotation in plots. Journal of Computational and Graphical Statistics, 9, 582–599

Nakagawa S, Cuthill IC 2007 Effect size, confidence interval and statistical significance: a practical guide for biologists. Biological Reviews 82, 591–605. See also Cuthill 2013.

Norman GR, Streiner DL 2008 (3rd ed) Biostatistics: The bare Essentials. The peoples press.

Olejnik S, Algina, J 2003 Generalized Eta and Omega Squared Statistics: Measures of Effect Size for Some Common Research Designs Psychological Methods. 8(4), 434-447.

Oliver FR 1964 What do statistics show. Hodder and Stoughton. London.

Pagano RR 2010 (10th ed.) Understanding Statistics in the Behavioral Sciences.

Peat J, Barton B, Elliott E. 2009 Statistics workbook for Evidence-based Health care. John Wiley & Sons.

Pettersson E, Turkheimer E 2010 Item selection, evaluation, and simple structure in personality data. Journal of research in personality 44(4), 407-420.

Peto R, Pike MC, Armitage P, Breslow NE, Cox DR, Howard SV, Mantel N, McPherson K, J. Peto J, Smith PG 1977 Design and analysis of randomized clinical trials requiring prolonged observation of each patient. II. analysis and examples. British journal of cancer, 35(1), 1. Open access: http://www.ncbi.nlm.nih.gov/pmc/articles/PMC2025310/pdf/brjcancer00298-0013.pdf

Petrie A, Sabin C 2000 (1st ed.) Medical Statistics at a glance. Blackwell.

Phillips L J, Hammock R L, Blanton J M 2005 Predictors of Self-rated Health Status among Texas Residents. Preventing Chronic Disease 2(4) open access: http://www.ncbi.nlm.nih.gov/pmc/articles/PMC1435709/ or http://www.cdc.gov/pcd/issues/2005/oct/04_0147.htm

Pinheiro J C Bates D M 2000 Mixed-Effects Models in S and S-PLUS. Springer [excellent]

Pocock S, Clayton TC, Altman DG 2002 *Survival plots of time-to-event outcomes in clinical trials: good practice and pitfalls*. Lancet *359 1686 – 1689*

Rhemtulla M, Brosseau-Liard P, Savalei V 2012 When can categorical variables be treated as continuous? A comparison of robust continuous and categorical SEM estimation methods under suboptimal conditions. Psychological Methods. 17(3) 354-373.

Robinson C, Schumacker RE 2009 Interaction Effects: Centering, Variance. Inflation Factor, and Interpretation Issues. Multiple Linear Regression Viewpoints, Vol. 35(1).

Rosario RR 2010 Taking R to the Limit, Part II: Working with Large Datasets. Los Angeles R Users' Group. [August 17] presentation available online: http://www.bytemining.com/wp-content/uploads/2010/08/r_hpc_II.pdf

Rosnow RL, Rosenthal R 1995 Some things you learn aren't so: Cohen's paradox, Asch's paradigm, and the interpretation of interaction. Psychological science. 6(1) 3-9.

Rosnow RL, Rosenthal R 2003 Effect Sizes for experimenting psychologists. Canadian journal of experimental psychology (57) 221-237

Rosnow RL, Rosenthal R 2009 Effect sizes: Why, when, and how to use them. Zeitschrift für Psychologie/Journal of Psychology. 217(1), 6

Sacristan JA 2011 Exploratory trials, confirmatory observations: a new reasoning model in the era of patient-centered medicine. BMC Med Res Method. 11: 57

Sarkar D 2008 Lattice: Multivariate Data Visualization with R. Springer.

Scheaffer RL 2010 Categorical Data Analysis. Available from: http://courses.ncssm.edu/math/Stat_Inst/PDFS/Categorical%20Data%20Analysis.pdf

Scheffe H 1999 (reprint of the 1959 edition) The analysis of variance. Wiley-IEEE

Schoemann AM, Preacher KJ, Coffman DL 2010 Plotting power curves for RMSEA. Available from http://quantpsy.org/ . [Computer software].

Schulz KF, Altman DG, Moher D 2010 CONSORT 2010 Statement: updated guidelines for reporting parallel group randomised trials. BMJ. 2010 (Mar) 23(340). Available from: http://www.bmj.com/cgi/content/full/340/mar23_1/c332

Schumacker R, Lomax R 2010 (3nd ed) A beginners guide to Structural Equation Modelling. Lawrence Erlbaum Associates.

Schünemann HJ, Best D, Vist G, Oxman AD for the GRADE working group. 2003 Letters, numbers, symbols, and words: How best to communicate grades of evidence and recommendations? Canadian Medical Association Journal(CMAJ),169(7), 677-680

Sedgwick P 2011a Meta-analyses I BMJ, 342:d45 (Published 11 January 2011)

Sedgwick P 2011b Meta-analyses II BMJ, 341:d229 (Published 19 January 2011)

Sedgwick P 2011c Meta-analyses III BMJ, 342:d244 (Published 26 January 2011)

Sedgwick P 2011d Meta-analyses IV BMJ, 342:d540 (Published 02 February 2011)

Sedgwick P 2011e Meta-analyses V BMJ, 342:d686 (Published 09 February 2011)

Sedgwick P 2011f Meta-analyses VI BMJ, 342:d937 (Published 16 February 2011)

Sedgwick P 2011g Meta-analyses VII BMJ, 342:d1108 (Published 23 February 2011)

Sedgwick P 2011h Meta-analyses: funnel plots. BMJ, 343:d5372 (Published 31 August 2011)

Sedgwick P 2011i Meta-analyses: sources of bias. BMJ, 343:d5085 (Published 17 August 2011)

Sedgwick P 2012a How to read a forest plot. BMJ, 345:e8335 (Published 7 December 2012)

Sedgwick P 2012b Meta-analyses: tests of heterogeneity. BMJ, 344:e3971 (Published 13 June 2012)

Sedgwick P 2013 How to read a funnel plot. BMJ, 346 – (Published 1 March 2013)

Seen S. 1993 Suspended judgement: N of 1 trials. Controlled Clinical Trials 14:1-5

Sheather S J 2009 A Modern approach to regression with R. Springer. Website, includes R tutorials and datasets: http://www.stat.tamu.edu/~sheather/book/

Singer, JD, and JB Willett. 2003. Applied Longitudinal Data Analysis: Modeling Change and Event Occurrence. Oxford University Press Inc, USA. gseacademic.harvard.edu/alda/

Sistrom CL, Garvan CW 2004 Proportions, odds, and risk. Radiology. (Jan) 230(1), 12-9.

Smyth, GK 2011 Australasian Data and Story Library (OzDASL). http://www.statsci.org/data.

Spruance SL, Reid JE, Grace M, Samore M 2004 Hazard ratio in clinical trials. Antimicrobial agents and chemotherapy, 48(8), 2787-2792.

Staa TP, Goldacre B, Gulliford M, Cassell J, Pirmohamed M, Taweel A, Delaney B, Smeeth L. 2012 Pragmatic randomised trials using routine electronic health records: putting them to the test. BMJ. (Feb 7) 344:e55. doi: 10.1136/bmj.e55.

Steiger JH. 1990 Structural model evaluation and modification: An interval estimation approach. *Multivariate behavioral research*, *25*(2), 173-180.

Steiger JH. 1998 A note on multiple sample extensions of the RMSEA fit index.

Stevens SS. 1951 Mathematics, measurement, and psycho physics. In S S Stevens (Ed.) Handbook of experimental psychology (pp. 1 - 49) New York Wiley.

Tate RL, Perdices M, Rosenkoetter U, Wakim DU, Godbee K, Togher L, Skye McDonald S. 2013
Revision of a method quality rating scale for single-case experimental designs and n-of-1 trials: The 15-item Risk of Bias in N-of-1 Trials (RoBiNT) Scale. Neuropsychological Rehabilitation. 23(5) 619-638, doi: 10.1080/09602011.2013.824383

The GRADE* working group. 2004 Grading quality of evidence and strength of recommendations. BMJ 2004; 328:1490-1494 (printed, abridged version). For the non-abridged version of this article, please go to the BMJ web site http://bmj.bmjjournals.com/cgi/content/full/328/7454/1490 or download a full PDF version from this link: http://bmj.bmjjournals.com/cgi/reprint/328/7454/1490.pdf

Thompson, Laura A. 2006 S-PLUS (and R) Manual to Accompany Agresti's Categorical Data Analysis (2002) 2nd edition. Available at https://home.comcast.net/~lthompson221/

Thorpe KE, Zwarenstein M, Oxman AD, Treweek S, Furberg CD, Altman DG, Tunis S, Bergel E, Harvey I, Magid DJ, Chalkidou K 2009 A pragmatic-explanatory continuum indicator summary (PRECIS): a tool to help trial designers. CMAJ, 180:E47-57.

Twisk JWR 2003 Applied Longitudinal Analysis for Epidemiology: A practical Approach. Cambridge University Press.

Twisk JWR 2006 Applied multilevel analysis. Cambridge University Press.

University of Kent – Student Learning Advisory Service (UELT) ? Managing a Research Project. Accessed 12/11/2013. [Possibly written by A.Crump@kent.ac.uk] www.kent.ac.uk/learning/resources/studyguides/managingaresearchproject1.docx

Van Belle G 2008 (2nd ed.) Statistical Rules of Thumb. Wiley

Varmuza K, Filzmoser P 2009 Introduction to Multivariate Statistical Analysis in Chemometrics. CRC Press.

van Nood E, Vrieze A, Nieuwdorp M, Fuentes S, Zoetendal EG, de Vos WM, Visser CE, Kuijper EJ, Bartelsman JF, Tijssen JG, Speelman P, Dijkgraaf MG, Keller JJ 2013 Duodenal infusion of donor feces for recurrent Clostridium difficile. N Engl J Med. (Jan 31) 368(5):407-15.
doi: 10.1056/NEJMoa1205037.

Viechtbauer W 2010 Conducting Meta-Analyses in R with the metafor Package. Journal of statistical Software. Vol. 36, Issue 3, Aug. Available from: http://www.jstatsoft.org/v36/i03/

Viechtbauer W 2013 The metafor R Package website http://www.metafor-project.org/doku.php/metafor

Weisberg 2005 (3rd Ed.) Applied Linear Regression. J. Wiley & Sons, New York. Datasets available in the car R package.

Weiss R 2005 Modeling Longitudinal Data. Springer.

West BT, Welch KB, Galecki AT 2014 (2nd Ed.) Linear Mixed Models: A Practical Guide Using Statistical Software. Chapman & Hall. Book website: http://www.umich.edu/~bwest/almmussp.html. Also an R package containing the data sets for the book, WWGbook, has been posted on CRAN.

Wu L 2010 Mixed Effects Models for Complex Data. Chapman & Hall.

Wylie KR, Eardley I 2007 Penile size and the "small penis syndrome". BJU International. p. 1449–55. Available from: http://onlinelibrary.wiley.com/doi/10.1111/j.1464-410X.2007.06806.x/abstract

Zucker DR, Ruthazer R, Schmid CH 2010 Individual (N-of-1) trials can be combined to give population comparative treatment effect estimates: methodologic considerations. Journal of Clinical Epidemiology 63 1312e1323

Zuur AF, ·Ieno EN, · Walker NJ, Saveliev AA,·Smith GM 2009 Mixed Effects Models and Extensions in Ecology with R. Springer.

Index

R Reference Card

Based on those produced by Jonas Stein, Tom Short, Jim Robison-Cox, and Emmanuel Paradis
n= number (integer), *f1*=factor, *df* =dataframe,
v = vector, *s* = string, *f* = filename (string).*obj*=object.
Idem = same as previous entry.

Getting help

http://cran.r-project.org/ The Comprehensive R Archive Network
help(topic) documentation on topic
?.Machine help about maximum values on your machine
str(a) display the internal structure of *a*
summary(a) gives a "summary" of *a*, usually a statistical summary
ls() lists the objects in the search path; specify *pat="pat"* to search on a pattern
ls.str() str() structure for each variable in the search path
dir() show files in the current directory

Packages

install.packages("pkgs") download and install *pkgs*
update.packages checks for new versions and offers to install
library(pkg) loads *pkg*, if *pkg* is omitted it lists packages
detach("package:pkg") removes *pkg* from memory

Input and output

source("my.R") includes and executes my.R
data(f) loads specified dataset *f* (can be *.r*, *.rdata*, *.cvs*, *.txt*, or *tab*. To load package specific datasets, use *load(f, package =" package")*
data() list all available datasets
try(data(package = "pgt")) list the data sets in the *pgt* package
read.table(f) reads a file and creates a data frame from it; use *header=TRUE* to read the first line as a header of column names; use *as.is=TRUE* to prevent character strings from being converted to factors
read.csv(f, header=TRUE) Idem. but with defaults set for reading comma-delimited files
read.delim(f, header=TRUE) Idem. but with defaults set for reading tab-delimited files
save(f,...) saves the specified objects (...)
save.image(f) saves all objects
load(f) load the datasets written with *save()*, also use *load(file=file.choose())* for file pick box, to load a file from a website use: *load(url("http://www.......rdata"))*
cat(..., sep=" ") Concatenate and print ...; *sep* is the character between each
print(obj) prints *obj*, use *zero.print = "."* to print '.' instead of 0 in output, good for tables and matrices
format(x,...) format *x* for pretty printing, use *justify = "left"*. To specify minimum number of digits, *n* to the right of the decimal point use *nsmall = n*. To specify number of digits use *digits = n*
write.table(obj, file=f, row.names=TRUE, col.names=TRUE, sep=" ") prints *obj* after converting to a dataframe; if quote is *TRUE*, character or factor columns are surrounded by quotes ("); *sep* is the field separator; *eol* is the end-of-line separator; *na* is the string for missing values; use *col.names=NA* to add a blank column header to get the column headers aligned correctly for spreadsheet input
To exchange tables on windows via clipboard, use
df <- read.delim("clipboard")
write.table(df,"clipboard",sep="\t",col.names=NA)

Data creation

c(...) concatenate, creates a vector from individual values
from:to generates a sequence; ":" has the usual arithmetic precidence;
1:4 + 1 returns *[1] 2 3 4 5*
seq(from,to) generates a sequence *by=* specifies increment;
length= specifies desired length
seq(along=x) generates *1, 2, ...,* to *x*;
rep(x,times) replicate *x* times; use *each=* to repeat "each" element of *x* each times;
rep(c(1,2,3),2) is *1 2 3 1 2 3*;
rep(c(1,2,3),each=2) is *1 1 2 2 3 3*
data.frame(a=data1, b=datab) create a data frame consisting of two columns named *a* and *b*. Shorter vectors are recycled to the length of the longest
list(...) create a list *list(a=c(1,2),b="hi",c=3i)*
array(x, dim=) array with data *x*; specify dimensions like
dim=c(3,4,2); elements of *x* recycle if *x* is not long enough
matrix(x, nrow=, ncol=) matrix; elements of *x* recycle
factor(x, levels=) converts a vector *x* to a factor, use
exclude = NULL to include a *NA* level
*gl(n,k,length=n*k, labels=1:n)* generate levels (factors) by specifying the pattern of their levels; *k* is the number of levels, and *n* is the number of replications
expand.grid() a data frame from all combinations of the supplied vectors or factors
cbind(df1, df2), rbind(df1, df2) combine columns or rows for data frames

Data destruction

rm(obj) removes *myvar* object from memory
rm(list = ls(all = TRUE)) removes all objects from memory

Addressing vectors

v[n]	n^{th} element
v[-n]	all but the nth element
v[1:n]	first *n* elements
v[-(1:n)]	elements from *n+1* to the end
v[c(1,4,2)]	elements 1,4 2
v["name"]	element named "*name*"
v[x > 3]	all elements greater than 3
v[x > 3 & x < 5]	all elements between 3 and 5

v[x %in% c("a","and","the")] elements in the given set

Addressing lists

x[n]	list with *n* elements
x[[n]]	n^{th} element
x[["name"]]	element of the list named "*name*"
x$name	id.

Addressing matrices

x[i,j]	element at row *i*, column *j*
x[i,]	row *i*
x[,j]	column *j*
x[,c(1,3)]	columns *1* and *3*
x["name",]	row named "*name*"

Indexing data frames (matrix indexing plus the following):

df[["name"]]	column named "*name*"
df$name	id.

Variable conversion

as.array(x), as.data.frame(x), as.numeric(x), as.logical(x), as.complex(x), as.character(x), attemptS to convert to type; use *methods(as)* to display a complete list

Variable information

is.na(x), is.null(x), is.array(x), is.data.frame(x), is.numeric(x), is.complex(x), is.character(x), tests for type, use *methods(is)* to show a complete list
length(x) number of elements in *x*
dim(x) retrieve or set the dimension of an object; *dim(x) <- c(3,2)*
dimnames(x) retrieve or set the row/column (i.e. dimension) names
nrow(x) number of rows; *NROW(x)* is the same but treats a vector as a one-row matrix
ncol(x) and *NCOL(x) Idem* for columns
class(x) get or set the class *x*; *class(x) <- "myclass"*
unclass(x) remove the class attribute of *x*
attr(x,which) get or set the attribute of *x*
attributes(obj) get or set the list of attributes of *obj*

Data selection and manipulation

which.max(v), which.min(v) returns the index of the maximum or minimum element of *v*
rev(v) reverses the elements of *v*
sort(v) sorts the elements of *v* in increasing order; to sort in de- creasing order: *rev(sort(x))*
cut(x,breaks) divides *x* into factor levels; *breaks* is the number of levels or a vector of cut points
match(x, y) returns a vector of the same length as *x* with the elements of *x* which are in *y* (*NA* otherwise)
which(x == a) returns a vector of the indices of *x* if the comparison operation is true (*TRUE*), e.g. *x[i] == a* the values of *i* f which equal *a* (the argument of this function must be a logical variable)
na.omit(x) suppresses the observations with missing data (*NA*) (suppresses the whole line if *x* is a matrix or a data frame)
na.fail(x) returns an error message if *x* contains at least one NA *unique(x)* if *x* is a vector or a dataframe, returns a similar object but with the duplicate elements suppressed
complete.cases(x) returns only rows with no NA's
table(x) returns a table of counts of the differents values of *x* (typically for integers or factors)
split(x,f) divides vector *x* into groups based on *f* levels
subset(x, ...) returns a selection of x with respect to criteria
(..., typically comparisons: *x$V1 < 10*); if *x* is a dataframe, the option select gives the variables to be kept, or dropped if a minus sign is used
sample(x, size) resample randomly and without replacement size elements in the vector *x*, *replace = TRUE* allows resampling with replacement
prop.table(x,margin=) table entries as fraction of marginal table

Arithmetic

2^3 = 2 cubed
9^.5 = 3 square root same as: *sqrt(9)*
%/% Integer divide; 27 %/% 4 = 6

Statistical functions

sin(), cos(), tan(),asin(), acos(), atan(), atan2(), Angles are in radians, not degrees, (i.e. right angle is $\pi/2$)
exp()
log(v, base) computes the logarithm of *x* with base *base*
log10(v) base =10
range(x) same as *c(min(x), max(x))*
sum(x) sum of the elements of *x*
diff(x) lagged and iterated differences of vector *x*
prod(x) product of the elements of *x*
mean(x) mean of the elements of *x*
median(x) median of the elements of *x*
quantile(x,probs=) sample quantiles corresponding to the given probabilities (defaults: *0,.25,.5,.75,1*)
weighted.mean(x, w) mean of *x* with weights *w*
rank(x) ranks of the elements of *x*
var(x) or *cov(x)* variance of the elements of *x* (calculated on *n−1*); if *x* is a matrix or a dataframe, the variance-covariance matrix is calculated
sd(x) standard deviation of *x*
cor(x, y) correlation of vectors *x, y* To set case selection criteria set;
use = "complete" or *use = "pairwise"* To specify correlation type use; *method = "kendall"* or *"pearson"*, or *"spearman"*
cor(df1) or *cor(df1, df2)* correlation matrix within *df1* or between *df1* and *df2*
var(x, y) or *cov(x, y)* covariance between *x* and *y*, matrices or dataframes
choose(n, k) computes the combinations of *k* events among *n* repetitions = $n!/[(n-k)!k!]$
round(x, n) rounds the elements of *x* to *n* decimals
scale(x, center = TRUE, scale = TRUE) if *x* is a matrix, creates *z* scores, use *center=FALSE*, to centre only, use *scale=FALSE* to scale only
pmin(x, y,...) a vector which i^{th} element is the minimum of *x[i], y[i]..*,
pmax(x, y,...) Idem for the maximum
cumsum(v) a vector which i^{th} element is the sum from *x[1]* to *x[i]*
Many maths functions have a logical parameter: *na.rm=FALSE* to specify missing data (*NA*) removal

Matrices

t(x) transpose – Switch rows and columns
diag(x) diagonal
%*% matrix multiplication and scalar product
solve(a,b) solves *a* %*% *x* = *b* for *x*
solve(a) matrix inverse of *a*
rowsum(x) sum of rows for a matrix-like object; *rowSums(x)* is a faster version
colsum(x), colSums(x) Idem for columns
rowMeans(x) fast version of row means
colMeans(x) Idem for columns

Advanced data processing

apply(X ,INDEX, FUN=) a vector or array or list of values obtained by applying a function *FUN* to margins *(INDEX)* of *X*
lapply(X, FUN) apply *FUN* to each element of the list *X*
tapply(X, INDEX, FUN=) apply *FUN* to each cell of an array
given by *X* with indexes *INDEX*
by(data ,INDEX, FUN) apply *FUN* to dataframe data subsetted by *INDEX*
merge(a,b) merge two dataframes on common columns or row names
xtabs(a b,data=x) creates a table
aggregate(df, by, FUN) splits *df* into subsets, computes summary statistics for each, and returns the result in a convenient form; by is a list of grouping elements, each as long as the variables in *df*
stack(x, ...) transforms data from separate columns in a dataframe or list into a single column
unstack(x, ...) reverse of *stack()*
reshape(x, ...) reshapes a dataframe between 'wide' format with repeated measurements in separate columns of the same record and 'long' format with the repeated measurements in separate records; use *(direction="wide")* or *(direction="long")*

String manipulation

paste(s1,s2, sep=" ") concatenate vectors after converting to character; *collapse=* is an optional string to separate "*collapsed*" results, to paste without separator: *paste0(s1, s2)*
substr(s,start,stop) extract or replace substrings in a character vector, can also assign, using *substr(s, start, stop) <- value*
strsplit(s, split) split *s* according to the substring *split*
grep(pattern, s) search *s* for pattern; also see *?regex*
gsub(pattern, replacement, x) search *x*, find *(pattern)* and replace with *replacement*, *sub()* is the same but only replaces the first occurrence.
tolower(s), toupper(s) convert to lowercase or uppercase
match(x, table) returns a vector of the positions of first matches for the elements of *x* in *table*
x %in% table Idem but returns a logical vector
pmatch(x,table) partial matches for the elements of x among table
nchar(s) returns the number of characters in *s*

Statistical tests

aov(formula) analysis of variance model
anova(fit,...) analysis of variance (or deviance) tables for one or more fitted model objects allows comparison
density(x) kernel density estimates of *x*
,t.test(), pairwise.t.test(), power.t.test(), binom.test(), prop.test()
chisq.test()
wilcox.test(x, f, paired = FALSE) produces the Mann-Whitney U test, use *paired=TRUE* for the Wilcoxon signed rank test.
kruskal.test(x1, f) Kruskal-Wallis test for equal medians in *x* over groups *f*.
See: *help.search("test")*

Model formulas

Formulas use the form: output ~ inputA + inputB ...
Other formula operators are:
1 intercept, meaning output variable has its mean value when independent variables are zeros or have no influence
: interaction term * factor crossing, a*b is same as a+b+a:b ^ crossing to the specified degree, so (a+b+c)^2 is same as (a+b+c)*(a+b+c)
\- removes specified term, can be used to remove intercept *-1* as in *resp ~ a - 1*

Linear Models

lm(y1 ~x1, data = df)
If *x1* is quantitative (numeric), a regression of y1 on *x1*.
If *x1* is a factor, the analysis of variance.
Formula: the first argument of *lm()* can have the form
y ~x1 + x2 + x3 main effects for 3 predictors
y ~x1 + x2 + x1:x2 main effects and interactions
shorthand versions: y ~x1 *x2 or *y ~ (x1 + x2)^2*
To enforce arithmetic within a formula use I() as in
y ~x1 + I(x1^2) (quadratic in x1)
lm1 <- lm(formula1) a linear models object
summary(lm1) prints coefficient estimates and F test for $H_0 : \beta = 0$
update(lm1, formula2) shortcut to modify *lm1*
anova(lm1, lm2) gives Likelihood Ratio Test for nested models
predict(lm1, newdata = df2) prediction and confidence intervals for new *df2* values, omit *newdata* if you want predictions on supplied data
fitted(fit) returns the fitted values
df.residual(fit) returns the number of residual degrees of freedom
coef(fit) returns the estimated coefficients (sometimes with their standard-errors)
residuals(fit) returns the residuals
deviance(fit) returns the deviance
logLik(fit) computes the logarithm of the likelihood and the number of parameters
AIC(fit) computes the Akaike information criterion or AIC

Diagnostics:
par(mfrow=c(2,2)); plot(lm1) plots 4 plots; Residuals vs Fitted to look for curvature, Normal Q-Q plot to examine normality assumptions, Scale-Location plot to look for non-constant variance, Cook's distance plot to look for influential points
Many of the formula-based modelling functions have several common arguments: *data=* the dataframe for the formula variables, *subset=* a subset of variables used in the fit, *na.action=* action for missing values: "*na.fail*", "*na.omit*".

Mixed Models

Two packages *lme4* with *lmer()* function [easier to use best for crossed random effects], and *nlme* with *lme()* function [allows specification of correlated residuals, the R matrix]

lme4:
lmer(y ~x1 +x2 +x3 + (x3|subject), data=df, REML = FALSE, na.action= na.omit)
x1 to *x3* fixed effects. Values in brackets random effects.
Diagonal (uncorrelated, variances vary), use separate brackets or ||:
1|group + (0+x1|group) + (0+x2|group) = 1+(x1||x2)?
Unstructured (correlations and variances vary) all in one bracket: (1+x1 + x2|group)
Partially diagonal, specifying crossed random effects (g1 and g2) (1+x1|group) + (0+x2|group)
(x|f) is equivalent to (1+x|f).
To suppress the intercept use (0+x|f) or (x-1|f).

nlme:
lme(fixed=formula1, data=df1,random=formula2, corr = structure, weights= variance.structure)
random = ~ 1| g1 random intercept for each group
random = ~x1 | g1 random intercept & slope (over x1) for each group
corr= corCompSymm(form = ~ 1| g1) same correlation within group
corr = corAR1(form = ~ 1| Subj) AR1 correlations w/in *Subj*
weights= varIdent(form = ~ 1|Year) variance changes with year
weights= varPower(form = ~fitted() |g1) variance increases as power of E(Y), powers vary with group.

Generalized least squares:
gls(formula1,data=df1,corr=structure, weights=variance.structure).
Use corr and weights as with lme.
nlme() nonlinear mixed models

GEE:
library(geepack)
geeglm(y ~ x1 + x2, id=subject, family=gaussian, corstr = "independence")
correlation structures: *independence* (scaled identity) variance constant, correlation 0.
corstr=exchangeable (constant variance and correlations), *corstr=ar1* (constant variance, correlation drops), *corstr=unstructured* (different variance and correlation for each)

Distributions

First letter of functions signifies:
r(andom sample) ; p(robability density), c(umulative probability density),or q(uantile):

Gaussian (normal) rnorm(n, mean=0, sd=1)	**Exponential** rexp(n, rate=1)
Poisson rpois(n, lambda)	**Student (t)** rt(n, df)
Snedecor Fisher- (F) (χ^2) rf(n, df1, df2)	**Chi-squared** rchisq(n, df)
Binomial rbinom(n, size, prob)	**Logistic** rlogis(n, location=0, scale=1)
Lognormal rlnorm(n, meanlog=0, sdlog=1)	**Negative binomial** rnbinom(n, size, prob)
Uniform runif(n, min=0, max=1)	**Wilcoxon** rwilcox(nn, m, n), rsignrank(nn, n)

Programming

myfunction <- function(x1,v1) { . . .}
builds a function called *myfunction* with 2 args
e.g. sd <- function(x1){ sqrt(var(x1)) }
for (i1 in 1:n1) { stuff } repeat "stuff" n1 times
Logical Comparisons: ==, <=, >= Note 2 ='s. Usage:
if (condition1) {somestuff} else {otherstuff}
while (condition1) {stuff} repeat "stuff" until condition1 is false
break jumps out of a loop
switch avoids several if statements
next jumps to end of a loop
ifelse applies condition to every element of a vector
do.call(funname, args) executes a function call

Plotting

plot(y) plot of the values of y (on y-axis) ordered on the x-axis
plot(x=xv, y=yv) bivariate plot of *xv* (x-axis) and *yv* (y-axis)
hist(x) histogram of the frequencies of *x*
barplot(x) barplot of the values of *x*; use *horiz=FALSE* for hor- izontal bars
dotchart(x) if *x* is a data frame, plots a Cleveland dot plot (stacked plots line-by-line and column-by-column)
pie(x) circular pie-chart
boxplot(x~f1) boxplot *x* divided by *f1* (factor)
sunflowerplot(x, y) Idem for *plot()* but the points with similar coordinates are drawn as flowers which petal number represents the number of points
stripplot(x) plot of the values of x on a line (an alternative to *boxplot()* for small sample sizes)
coplot(x~y | z) bivariate plot of *x* and *y* for each value or interval of values of *z*
interaction.plot (f1, f2, y) if *f1* and *f2* are factors, plots the means of *y* (on the y-axis) with respect to the values of *f1* (on the x-axis) and of *f2* (different lines); the option *fun* allows the choice of summary statistic of *y* (default *fun=mean*)
matplot(x,y) bivariate plot of the first column of *x* vs. the first one of *y*, the second one of *x* vs. the second one of *y*, etc.
fourfoldplot(x) visualizes, with quarters of circles, the association between two dichotomous variables for different populations (*x* must be an array with *dim=c(2, 2, k)*, or a matrix with *dim=c(2, 2)* if *k = 1*)
assocplot(x) graph showing the chi-squared residuals for a two dimensional contingency table
mosaicplot(x) 'mosaic' graph of the residuals from a log-linear re- gression of a contingency table

pairs(x) if *x* is a matrix or a dataframe, draws all possible bivariate plots between the columns of *x*

The following parameters are common to many plotting functions:
add=FALSE if *TRUE* superposes the plot on the previous one
axes=TRUE if *FALSE* does not draw the axes and the box
type="p" specifies the type of plot,
"p": points,
"l": lines,
"b": points connected by lines,
"o" same as b but the lines are over the points,
"h": vertical lines,
"s": steps, the data are represented by the top of the vertical lines, "S": same as s but the data are represented by the bottom of the vertical lines
xlim=, *ylim=* specifies the lower and upper limits of the axes, e.g. *xlim=c(1, 10)* or *xlim=range(x)*
xlab=, *ylab=* annotates the axes, must be a character variable
main= main title, must be a character variable
sub= sub-title

Low-level plotting commands

points(x, y) adds points (the option *type=* can be used)
lines(x, y) adds lines
text(x, y, labels, ...) adds text given by labels at *coordinates (x,y)*; a typical use is: *plot(x, y, type="n"); text(x, y, names)*
mtext(text, side=3, line=0, ...) adds text given by *text* in the margin specified by *side* (see axis() below); *line* specifies the line from the plotting area
segments(x0, y0, x1, y1) draws lines from point (*x0,y0*) to point (*x1,y1*)
arrows(x0, y0, x1, y1, angle= 30, code=2) draws arrows at points (*x0,y0*) if *code=2*, at points (*x1,y1*) if *code=1*, or both if *code=3*; *angle* controls the angle from the shaft of the arrow to the edge of the arrow head
abline(a,b) draws a line of slope *b* and intercept *a*
abline(h=y) draws a horizontal line at ordinate *y*
abline(v=x) draws a vertical line at abcissa *x*
abline(lm.obj) draws the regression line given by *lm.obj*
rect(x1, y1, x2, y2) draws a rectangle which left, right, bottom, and top limits are *x1, x2, y1,* and *y2*, respectively
polygon(x, y) draws a polygon linking the points with coordinates given by *x* and *y*
legend(x, y, labels, lty=lty1, pch = pch1) adds a legend at coordinates *x, y*.
title() adds a title and optionally a sub-title
axis(side, vect) adds an *axis* at the bottom (*side=1*), on the left (*2*), at the top (*3*), or on the right (*4*); *vect* (optional) gives the abcissa (or ordinates) where tick-marks are drawn
rug(v) draws a rug - the data on the x-axis as small vertical lines
locator(n, type="n", ...) returns the coordinates (*x, y*) after the user has clicked *n* times on the plot with the mouse; also draws symbols (*type="p"*) or lines (*type="l"*) with respect to optional graphic parameters (...); by default nothing is drawn (*type="n"*)

Graphical parameters

These can be set globally with *par(...)*; many can be passed as parameters to plotting commands.
adj= controls text justification (*0* left-justified, *0.5* centred, *1* right- justified)
bg= specifies the colour of the background (e.g: *bg="red", bg="blue"*, . . ., colors())displays the list of the 657 available
bty= controls the type of box drawn around the plot, allowed values are: *"o", "l", "7", "c", "u" ou "]"*, box looks like the corresponding character; if *bty="n"* the box is not drawn
cex= a factor controlling the default size of texts and symbols; scale numbers on the axes, *cex.axis*, the axis labels, *cex.lab*, the title, *cex.main*, and the sub-title, *cex.sub*
col= controls the color of symbols and lines; use color names: *"red", "blue"* or as "*#RRGGBB*"; see *rgb(), hsv(), gray()*, and *rainbow()*; as for *cex* there are: *col.axis, col.lab, col.main, col.sub*
font= an integer which controls the style of text
1: normal,
2: italics,
3: bold,
4: bold italics
as for cex there are: *font.axis, font.lab, font.main, font.sub*
las= an integer which controls the orientation of the axis labels (*0*: parallel to the axes, *1*: horizontal, *2*: perpendicular to the axes, *3*: vertical)
lty= type of line, can be an integer or string
1: "solid", *2*: "dashed",
3: "dotted", *4*: "dotdash",
5: "longdash", *6*: "twodash", or
a string of up to eight characters (between "0" and "9") which specifies alternatively the length, in points or
pixels, of the drawn elements and the blanks, for example *lty="44"* will have the same effect as *lty=2*
lwd= width of line, a numeric default 1
mar= vector of 4 numeric values which control the space between the axes and the border of the graph: c(bottom, left, top, right), default values are c(5.1, 4.1, 4.1, 2.1)
mfcol= vector of the form c(nr,nc) which partitions the graphic window as a matrix of *nr* lines and *nc* columns, the plots are then drawn in columns
mfrow= Idem but the plots are drawn by row
pch= controls the type of symbol, either an integer between 1 and 25, or any single character within "". See the book glossary entry for a table.
ps= size in points of texts and symbols as integer
pty= a character which specifies the type of the plotting region, "*s*": square, "*m*": maximal
tck= specifies the length of tick-marks on the axes as a fraction of the smallest of the width or height of the plot; if tck=1 a grid is drawn
tcl= specifies the length of tick-marks on the axes as a fraction of the height of a line of text (by default *t=-0.5*)
xaxt= if *xaxt="n"* the x-axis is set but not drawn (useful in conjunction with axis(side=1, ...))
yaxt= if *yaxt="n"* the y-axis is set but not drawn (useful in conjunction with axis(side=2, ...))

Lattice (Trellis) graphics

xyplot(y ~ x|g1) scatterplot of *y* over *x* (optionally) separated by group *g1*
barchart(y~ g1) histogram of the values of *y* divided by *g1*
dotplot(y~x) Dot plot (stacked plots line-by-line and column-by-column)
densityplot(~x) density functions plot
histogram(~x) histogram of the frequencies of *x*
bwplot(y~x) "box-and-whiskers" plot
qqmath(~x) quantiles of *x* with respect to the values expected under a theoretical distribution
stripplot(y~x) single dimension plot, *x* must be numeric, *y* may be a factor
qq(y~x) quantiles to compare two distributions, *x* must be numeric, *y* may be numeric, character, or factor but must have two levels
splom(~x) matrix of bivariate plots
parallel(~x) parallel coordinates plot
*cloud(z~x*y|g1*g2)* 3d scatter plot

In the Lattice formula, *y x|g1*g2* has combinations of optional conditioning variables *g1* and *g2* plotted on separate panels. Lattice functions take many of the same arguments as base graphics plus also *data=* the dataframe for the formula variables and *subset=* for subsetting. Use *panel=* to define a custom panel function (see apropos("panel") and ?llines). Lattice functions return an object of class trellis and have to be printed to produce the graph. Use *print(xyplot(...))* inside functions where automatic printing doesn't work. Use *lattice.theme* or *trellis.par.get* and *trellis.par.set* to change. To set a white background use: *defaults.trellis.par.set(theme=col.whitebg())*

Dates and Times

The class *Date* has dates without times. POSIXct has dates and times, including time zones. Comparisons (e.g. >), *seq()*, and *difftime()* are useful. Date also allows + and −.
?DateTimeClasses gives more information. See also package *chron*.

as.Date(s) and *as.POSIXct(s)* to convert to the respective class; *format(dt)* converts to a string representation. The default string format is "2012-02-21". These accept a second argument to specify a format for conversion. Some common formats are:
%a, %A Abbreviated and full weekday name.
%b, %B Abbreviated and full month name.
%d Day of the month (01–31).
%H Hours (00–23).
%I Hours (01–12).
%j Day of year (001–366).
%m Month (01–12).
%M Minute (00–59).
%p AM/PM indicator.
%S Second as decimal number (00–61).
%U Week (00–53); the first Sunday as day 1 of week 1.
%w Weekday (0–6, Sunday is 0).
%W Week (00–53); the first Monday as day 1 of week 1.
%y Year without century (00–99). Avoid it.
%Y Year with century.
%z (read only) Offset from Greenwich; -0800 is 8 hours west of.
%Z (read only) Time zone as a character string (empty if not available).

Where leading zeros are shown they will be used on output but are optional on input. See *?strftime*.

Plotmath (for maths characters in text)

Used within the expression() function
[3] = makes 3 a subscript
2^3 = makes 3 a superscript
bar(xy) = xy with bar
chi = χ
hat(x) = adds a hat to x
infinity = infinity symbol
lambda = λ
nu = μ
over(x, y) = x over y
pi = π
x ~~ y = adds extra space
x ~y = keeps a space
For examples type into the R Console: *demo(plotmath)*
e.g.: text3 <- expression(survival~proportions:~hat(pi)[0.5]~versus~hat(pi)[0.7])
See: http://stat.ethz.ch/R-manual/R-devel/library/grDevices/html/plotmath.html

Microsoft office $\bar{x}$ and $\hat{x}$

You can type $\bar{x}$ by:
1. Pressing CTRL+F9 (shortcut to inserting a field into your Word document), braces appear
2. Copy and paste this expression between the braces. EQ \O(x,¯)
After doing this, right click on it and select "*Toggle field codes*" to view the result.
Above expression can be adapted.
If you have the "MS Reference 1" font installed it contains $\bar{x}$ which you can find by using the menu option insert->symbol